Y0-BWR-042

PHYSICAL PRINCIPLES OF OIL PRODUCTION

INTERNATIONAL SERIES IN PURE AND APPLIED PHYSICS

G. P. HARNWELL, *Consulting Editor*

Dr. Lee A. DuBridge was consulting editor of the series from 1939 to 1946.

PHYSICAL PRINCIPLES
OF OIL PRODUCTION

By MORRIS MUSKAT, Ph.D.

DIRECTOR OF PHYSICS DIVISION
GULF RESEARCH & DEVELOPMENT COMPANY

FIRST EDITION

NEW YORK   TORONTO   LONDON

McGRAW-HILL BOOK COMPANY, INC.

1949

PHYSICAL PRINCIPLES OF OIL PRODUCTION

DEDICATED TO

Fern

and

Phyllis, David, Bobby, and Rosalyn

PREFACE

This work is not a reorganized version of well-established principles and classical theories formulated to serve special pedagogical purposes. It is not a textbook at a particular grade or level and directed to a particular class of readers. It is an exposition of material that is in an active state of flux. Most of the content of this book appeared in the technical literature of the subject within the last fifteen years. Much of it is being currently extended, developed, and clarified. Some of it will no doubt need revision, correction, and replacement by better established concepts and interpretations.

It has been the immediate purpose of this work to formulate and correlate what appears to be known now about the physical principles and facts underlying the mechanics of oil production. It has been a more serious purpose to stimulate and encourage further research and study of the subject to fill in the many gaps in our present knowledge, to clarify the many aspects that are still subject to speculation and conjecture, to generalize the simplified and idealized treatments of special problems, and to improve the correlation between laboratory theories and field observations. Significant progress in these directions will evidently make much of this book out of date. However the author will consider the purposes of this work to have been achieved just to the degree to which it will accelerate its own obsolescence.

Handbook instructions for field operation are not given here. Nor will rule-of-thumb procedures for reservoir exploitation be listed. The subject is not susceptible to dogmatic generalization. If there is any general rule pertaining to it, it is that it is governed by no rule, beyond the basic laws of physics, which is universally valid for all oil-producing reservoirs.

The physics of oil production is unique among all applications of physics. It deals with objects—the oil reservoirs—that are variable continua, subject only to virtually infinitesimal sampling. Their physical histories are irreversible transients that can be observed only at isolated points "on the fly." They are controlled by a trinity of forces—hydrodynamic stress gradient, gravity, and interfacial—superposed on a trinity of phases—oil, gas, and water. Except for relatively microscopic laboratory investigations, any major and significant field experimentation creates an irreversible change in the state of the system and destroys the possibility of repeat or comparative experimentation on the same reservoir. And finally there are no identical specimens in the ensemble of the objects of study. Yet the goal toward which the study of this subject strives is the prediction and interpretation of the detailed behavior of *individual* reservoirs! Little wonder is it that the subject is hardly beyond the state of infancy.

The scope of this work comprises those physical principles controlling the behavior of oil fields and their implications when applied under conditions broadly simulating those which may obtain in oil-producing reservoirs occurring in nature. Where possible these implications are developed quantitatively, under stated simplifying assumptions, and expressed in analytical, graphical, or numerical form. And, where feasible, actual oil-field observations and data are presented as illustrative parallels of those anticipated from theoretical considerations. However, such parallelisms, suggested by the supplementary illustrations, should not be construed as establishing quantitative rules or generalizations applicable to broad classes or individual examples of fields that are superficially similar to the particular examples cited. For every such generalization that might be formulated, fully as many exceptions will be found as instances of agreement. Such exceptions will result from purely physical factors, which are at least quantitatively different from those tacitly or explicitly assumed in establishing the so-called "generalizations."

As may be inferred from the Contents, the treatment of the subject is developed as a sequence of four major parts. The first, comprised of Chaps. 1 to 3, presents general background material and summaries of available information on the fundamental physical properties of petroleum fluids and the oil-bearing rocks. With respect to the latter some of the experimental techniques are described, and the problems of interpreting the data obtained from the analysis of underground rock samples—cores—are discussed.

Chapters 4 to 6 give the formulation of the hydrodynamics of the flow of a single-fluid phase through porous media and present the solutions to a number of specific problems similar or related to those of interest in actual oil-producing systems. While extremely idealized in some respects, these solutions are nevertheless directly applicable to oil-reservoir operations under appropriate conditions.

Multiphase-fluid-flow principles and their applications to general oil-producing mechanisms are discussed in Chaps. 7 to 11. The experimental data establishing the physical concepts and characteristics of multiphase-fluid flow are reviewed and applied to the solution of several simple steady-state-flow systems. Capillary phenomena in porous media are discussed, and an analysis is given of the role they may play both in determining the initial-fluid distributions in oil reservoirs and in affecting their producing performance. A classification is made of the physical producing mechanisms of oil reservoirs. A detailed exposition is given of the theoretically expected performance of gas-drive reservoirs, including the effects of gas injection, gravity drainage, and partial water drives. This is supplemented by examples of actual performance records of fields produced by solution-gas-drive or modified-gas-drive mechanisms. Similar extended treatments are presented of complete-water-drive reservoirs, including illustrative observed histories of such fields.

The last three chapters treat secondary-recovery operations, the performance of condensate-producing reservoirs, the problem of well spacing, and the estimation of the recoverable reserves in oil fields. These, too, include the development of the theoretically expected behavior and comparative records of field observations.

While the great part of the material of this work is drawn from the published literature relating to the science of oil production, the personal association of the author with scientists and engineers in the oil-producing industry has, of course, contributed immeasurably to his views, concepts, and information. Especially instructive have been his many years of intimate collaboration with members of the Gulf Oil Corporation, and, in particular, those of the Gulf Research & Development Company. Dr. Paul D. Foote, Executive Vice-president of the Gulf Research & Development Company, and Dr. B. B. Wescott, Assistant to the Executive Vice-president, not only have given continued encouragement to the author's own work on the physics of oil production but have made constructive suggestions for improving the manuscript of this work. Dr. F. Morgan and D. W. Reed, as coworkers with the author, have cooperated in many of the developments of the subject that have been especially studied at the Gulf Research & Development Company. And particular acknowledgment is due to the assistance and efforts of Miss M. O. Taylor, who has checked most of the author's calculations quoted in the text and carried through many of the original numerical analyses. It is with regret that the author cannot explicitly acknowledge the contributions of the many others in the industry who have been most helpful on numerous occasions in discussions of the subject generally and in stimulating the continued researches for a number of years by the author and his collaborators in the science of oil production.

The happy circumstance by which the author has been privileged to work cooperatively with personnel of the Gulf Oil Corporation does not, of course, impose on them any responsibility of supporting this work. Nor is it thereby implied that the Gulf Oil Corporation indicates agreement with or approval in any form of any views, opinions, or propositions expressed in this work. It is a scientific treatise and not an exposition of policy. The responsibility for its contents, in general and in detail, is solely that of the author.

Morris Muskat

Pittsburgh, Pa.
July, 1949

CONTENTS

CHAPTER 1

INTRODUCTION

1.1. The Subject.—Up to Jan. 1, 1948, approximately 55,000,000,000 bbl of oil had been produced throughout the world. Sixty-four per cent, or 35,200,000,000 bbl, was produced in this country. The recoverable proved reserves have been estimated as equal to about 1.3 times that already produced, of which about one-third is in the United States. During 1947 the average daily rate of oil production in the United States was 5,196,000 bbl, or 62 per cent of the world's total.[1]

A very conservative estimate as to the average quantity of oil remaining underground in the oil fields which have already been economically depleted and abandoned is that this residual oil is as great in volume as that produced. If the future exploitation of the proved oil reserves should be associated with this unit ratio for the residual oil left in the producing horizons, there will still remain some 125 billion barrels of oil distributed throughout the developed oil fields after they have been condemned and abandoned. And this does not include the untold billions of barrels in strata that have "shows" of oil but that have been passed up as not being of commercial value.

Why? A quantitative answer to this question cannot be given. Nor is one justified in simply drawing the "obvious" inference that the process of oil production is inherently wasteful and inefficient. But an examination and study of the physical principles underlying the mechanics of oil production will provide an understanding and interpretation of these basic empirical facts of the oil-producing industry.

Will the ratio of unrecovered residual oil to that produced necessarily remain as high in the future as it has been in the past? Only a prophet would venture a dogmatic prediction. But an analysis of the physical laws and factors that affect this ratio will at least give an indication as to its probable future trend.

Must these residual billions of barrels of oil be considered as forever and irretrievably lost? This question, too, has no "yes" or "no" answer. On the other hand, here, too, an investigation into the physics of fluid flow through oil-bearing rocks yields quite definite information as to the degree to which one may successfully "recover the unrecovered oil" and the factors influencing such recovery.

[1] Cf. W. W. Burns, *Oil and Gas Jour.*, **46**, 196 (Jan. 29, 1948).

These are only a few of the questions upon which the material in the following chapters will have a bearing. However, in the discussion to follow, only the physical aspects of the general problems of oil and gas production will be considered. Moreover, these will be restricted to the oil- or gas-producing formations themselves. This limitation is not to be interpreted as implying an exaggerated importance of the fluid-mechanics phase of oil production in contrast to all other branches of the subject. On the contrary, the reader is urged to realize that what goes on in any particular oil reservoir will have no practical significance whatever unless the geologist and geophysicist locate the probable position of the oil field, unless the well driller and mud expert successfully sink the well bore to penetrate the oil-bearing formation, unless the petroleum-production engineer overcomes the multitude of problems involved in getting the oil to the surface at a reasonable cost, unless the refinery chemist succeeds in transforming the crude oil into marketable products, unless the transportation technicians provide effective distribution of the products, unless the marketing experts establish satisfactory sales facilities, and finally unless the advertising and sales engineers induce the consumer to purchase the crude oil in its many disguised forms. Indeed, it is because it is fully appreciated that it would require whole volumes to do justice to each of these other extremely important and fascinating aspects of the oil-producing industry that the reader is referred to the many excellent books on these individual topics and will not be misled here by sketchy and superficial résumés of matters deserving thorough and authoritative treatments.

1.2. Oil Reservoirs.—Oil[1] is produced from wells drilled into underground porous rock formations. The ensemble of wells draining a common oil accumulation or source or the surface area defined by the well distribution is termed an "oil field" or "oil pool." These are generally given names shortly after discovery and the beginning of development, for identification and distinction from other pools. The part of the rock that is oil productive is termed an "oil reservoir," although the names of the corresponding fields are often used in referring to the reservoir itself.

By virtue of the subsurface location of the reservoir rock, its entrained fluids are subject to elevated temperature and pressure—the reservoir temperature and reservoir pressure. The values of the latter at the time

[1] The term "oil" is used here to denote the general class of the heavier hydrocarbons that normally constitute a liquid phase—usually dark green or brown in color—at the surface. In practice, however, a distinction must be made between so-called "crude," or "black," oil, which is also liquid within the reservoir, and "condensate," which is in the vapor phase at the initial reservoir pressure and temperature but appears generally as a straw-colored or even colorless liquid at the surface. The source of the latter is usually referred to as a "condensate field" or "condensate reservoir" (cf. Sec. 2.5 and Chap. 13).

of discovery[1] are important physical parameters affecting the state and properties of the reservoir fluids. The reservoir temperature is that associated with the geothermal gradients in the geographical location of the field. A gross average of these is approximately a temperature increase of 1°F above mean ambient temperature per 60 ft of depth below the surface, though numerous exceptions have been found of abnormally low and high[2] temperatures, as compared with those predicted from the average gradient. Similarly, the initial reservoir pressures are commonly found to vary linearly with the reservoir depth, as if they were in equilibrium with a hydrostatic water column. Accordingly, these pressures vary with reservoir depth approximately by the hydrostatic gradient of 43 to 55 psi/100 ft, depending on the salinity and density of the equivalent water column. On the other hand, abnormally low or high[3] initial reservoir pressures, as compared with the expected range of hydrostatic pressures, are not rare exceptions. It is now well recognized that initial reservoir pressures and temperatures should be actually measured for each reservoir rather than estimated from empirical correlations. The very abnormalities that may be so disclosed may be of considerable significance in the subsequent study of the reservoir behavior.

While the term "oil reservoir" implies that the rock structure in question is definitely oil bearing and oil productive, the oil phase[4] itself in general will not comprise the exclusive fluid content of the void space. All samples

[1] While the reservoir temperature is generally assumed to remain constant throughout the producing life of a field, the reservoir pressure is a variable depending on the state of depletion of its initial fluid content, although the exact relationship will be unique to each reservoir and the nature of its producing mechanism.

[2] While excessively high temperatures are quite uncommon, the temperatures are generally abnormally low in West Texas, the Texas Panhandle, and in New Mexico (cf. C. E. Van Orstrand, "Problems of Petroleum Geology," p. 989, AAPG, 1934). Additional well temperature data are given by R. W. French, *API Drilling and Production Practice*, 1939, p. 653, who reports abnormally high temperatures at Lompoc, Calif.; by E. DeGolyer, *Econ. Geology*, **13**, 275 (1918), who discusses excessive temperatures in some wells in Mexico; and by E. A. Nichols, *AIME Trans.*, **170**, 44 (1947), who gives regional geothermal gradient contours in the Mid-Continent and Gulf Coast districts. Examples of very low reservoir temperatures, though not abnormal with respect to the geothermal gradients, are 61°F, at Norman Wells, Canada, and 11°F at Ukhta, in the Russian Arctic.

[3] Many abnormally high reservoir pressures have been observed along the Gulf Coast [cf. G. E. Cannon and R. C. Craze, *AIME Trans.*, **127**, 31 (1938), and G. E. Cannon and R. S. Sullins, *Oil and Gas Jour.*, **45**, 120 (May 25, 1946); cf. also E. V. Watts, *AIME Trans.*, **174**, 191 (1948), who discusses the general problem of excessive reservoir pressures, with special reference to the D-7 zone of the Ventura Ave. field, California, which had an initial pressure of 8300 psi at 9,200 ft]. Marked pressure deficiencies have been found in Kansas reservoirs and in some fields in West Texas.

[4] As noted in the footnote on p. 2, in condensate reservoirs the hydrocarbon content of the void space usually is originally in a gas phase, which contains both the gas and "oil" produced at the surface.

of oil-productive formations thus far brought to the surface and analyzed have been found to contain some water, presumably indigenous to the rock, immediately prior to exploitation. The amount of this water—usually[1] termed "connate" or "interstitial" water—has been found to range from about 2 per cent to more than 50 per cent of the total pore space. Connate water may well be considered as universally[2] associated with the oil itself. Moreover, all producing oil reservoirs contain gas in solution in the oil. And in many reservoirs the total gas content exceeds that which can be held in equilibrium in solution in the oil at the prevailing initial reservoir temperature and pressure, the excess generally[3] overlying the oil-saturated section as "gas caps." The gross content of an oil reservoir is thus initially a composite of at least two and often three fluid phases. They must all be recognized as integral parts of a common system. Indeed, it is their mutual interactions and reactions to the entry—artificial or natural—of similar extraneous fluids that give rise at the same time to the inherent complexity of oil-producing systems and the *raison d'être* of the science of reservoir engineering.

1.3. The Nature of Oil-bearing Rocks.[4]—We shall not enter here into the geochemical questions concerning the origin of oil. Nor shall we concern ourselves with the still controversial problem pertaining to the migration and accumulation of oil. For our purposes it will suffice to limit our interest to such rocks and underground strata—mainly sandstones, limestones, and dolomites—as now constitute oil-bearing reservoirs.

---

[1] While there appears to be some doubt regarding the literal etymological correctness of the term "connate," it has had such wide usage in the literature that there should be little danger of misinterpretation of its meaning and it will therefore also be used here interchangeably with "interstitial."

[2] Possible exceptions may occur in the case of oil masses trapped in actual fissures or cavernous voids, though it will obviously be very difficult to obtain conclusive data on this point.

[3] As a practical matter, it is virtually a universal assumption in all quantitative treatments of oil-reservoir performance, including those to be presented in this work, that except for a transition zone between the so-called "oil pay" and "gas cap," there is no free-gas phase initially disseminated in the main body of the oil-producing section. Under conditions of complete thermodynamic equilibrium the segregation and accumulation of the free gas into a continuous phase is to be expected. Although there appears to be no conclusive evidence invalidating this assumption, the existence of positive "proof" of its validity must be admitted to be still a moot question.

[4] For a comprehensive discussion of the subject matter of this and the following two sections, reference should be made to texts on petroleum geology. No attempt will be made here to provide anything more than the minimum required for establishing the terminology and the basic physical concepts related to the gross structure of oil-bearing reservoirs and the nature of the solid materials of which they are comprised. A detailed treatment of sedimentary materials is given by F. J. Pettijohn, "Sedimentary Rocks," Harper & Bros., 1949.

Such rocks are essentially sedimentary.[1] That is, they are either mechanical or chemical depositions of solid materials or simply the remains of animal or plant life. To the extent that they may play the role of oil-bearing reservoirs they must possess interstices between the solid particles or equivalent void regions in which oil could accumulate. The fraction of the bulk volume of the rock that is free for containing fluids is termed its "porosity."[2]

Porous sedimentary rocks represent an intermediate stage in the complete cyclical history of sediments, namely, deposition, lithification, metamorphism, and weathering, or disintegration. Except for the unconsolidated sands, which constitute some of the oil reservoirs in the Gulf Coast of the United States, California, and the Lake Maracaibo district of western Venezuela, all other reservoir rocks have been consolidated by lithification processes. Unless they are then subjected to premature weathering, they ultimately undergo complete metamorphism and are no longer considered as sedimentary, as they become crystalline throughout and lose their porosity. In particular, shales become slates, limestones are converted into marbles, pure sandstones become quartzites, and impure shales and sandstones become transformed to schists and gneisses.

Rocks resulting from mechanical deposition constitute the clastic sediments and include the gravels, sandstones, shales, clays, etc. They are granular accumulations composed of erosion fragments of older and larger rocks. The clays and shales, which are sedimentary deposits of the very fine grained materials, are not especially important as oil-producing reservoirs, even though they frequently are oil saturated and constitute about 80 per cent of all sedimentary rocks. For while fresh silt and clay deposits may have porosities as high as 85 per cent and surface clays often do have porosities of 40 to 45 per cent, they are very sensitive to the compacting effects of the overburden, *i.e.*, their overlying beds. As a result, these materials at appreciable depths will have lost the greatest part of their porosity and hence their capacity to hold petroleum fluids. In fact, the porosity of shales has been found[3] to decrease approximately exponentially with depth. Moreover, because of the very small size of the original grains that form the clays or shales, the interstitial pore openings left after compaction are so minute that whatever fluids do remain in them have

---

[1] It should be noted that, while virtually all oil-bearing rocks are sedimentary, not all sedimentary formations contain oil. On the other hand, only about 5 per cent of the lithosphere is comprised of sedimentary rocks [cf. F. W. Clarke, Data of Geochemistry, *USGS Bull.* 770, p. 34 (1924)].

[2] Methods of determining this basic property of porous media will be discussed in Sec. 3.1.

[3] L. F. Athy, *AAPG Bull.*, **14**, 1 (1930).

extremely small mobility and hence do not flow into the open well bores at rates corresponding to commercially profitable production.

In contrast to the clays and shales, the sands, sandstones, and sandy shales, deposited under water, are generally composed of appreciably larger fragments or grains. Moreover, they are but slightly compressed and compacted by their overburdens. Thus, typical oil-producing sands will pack under water to porosities of the order of 35 to 40 per cent. The application of a compacting pressure will reduce the porosity only negligibly—of the order of a few per cent[1]—until the crushing strength of the sand grains or strength of the cementing material is exceeded. Whatever difference in porosity does exist between the sandstone *in situ* at great depths and the random packing porosity of its constituent sand grains when brought to the surface is almost entirely due to the cementing materials, such as clays, gypsum, calcite, limonite, hematite, or quartz, deposited within the original pores by the circulating waters. The quantity of this cementing material and associated decrease in porosity will depend mainly upon the geological history of the deposit. Sandstones comprise about 15 per cent of the sedimentary component of the lithosphere.[2] Sandstones that constitute oil reservoirs of commercial value usually have porosities ranging from 10 to 35 per cent.

Practically all sandstones show bedding planes, which are due to the assortment and grading of the granular material during the course of transportation and deposition. The directional trend in the deposition may lead to an anisotropy in the fluid-transmitting capacity of the resulting rock mass. The successive layers of sediments are generally separated by bands of clay, shale, or micaceous material.

Aside from the cementing or binding material itself, sandstone rocks may differ in the amount and nature of solid matter present in the pores formed by the gross granular sand structure. Some of the major oil-bearing sandstone formations, such as the Oklahoma "Wilcox," and Woodbine in Texas, are comprised of "clean sands," in which the pore spaces are essentially free of solids. On the other hand, in some oil-producing districts, such as California and northwestern Pennsylvania, perhaps the majority of the oil-bearing rocks are "dirty" in some form and degree. The intergranular pore spaces in these sands are partly filled with argillaceous, silty, lignitic, or bentonitic material. This solid matter not only reduces the net porosity and petroleum-fluid-holding capacity of the rock but to an even more serious degree may affect the fluid-transmitting qualities of the porous medium (Secs. 3.7 and 8.6).

[1] H. G. Botset and D. W. Reed, *AAPG Bull.*, **19**, 1053 (1935); also C. B. Carpenter and G. B. Spencer, *U.S. Bur. Mines Rept. Inv.* 3540 (October, 1940).
[2] Clarke, *loc. cit.*

Although the variations in detailed characteristics of various sedimentary rocks as found in nature are virtually limitless in number, one may get a qualitative picture of the geometrical structure of a sand by consideration

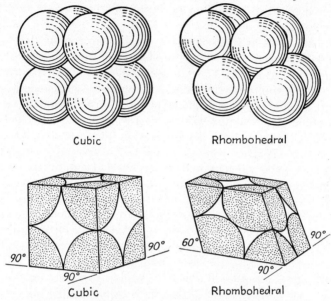

Fig. 1.1.   Sphere groups and unit cells of cubic and rhombohedral packings.   (*After Graton and Fraser, Jour. Geology, 1935.*)

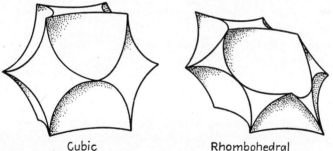

Fig. 1.2.   Unit voids of cubic and rhombohedral packings of spheres.   (*After Graton and Fraser, Jour. Geology, 1935.*)

of packings of spheres of uniform size.   These, too, are of infinite variety. However, it will suffice to note here two basic and extreme types, namely, the cubic and rhombohedral packing.   Unit cells of such packings are shown in Fig. 1.1, and the pores of these cells are drawn in Fig. 1.2.[1]   If

[1] The geometrical features of such idealized packings have been exhaustively studied by L. C. Graton and H. J. Fraser, *Jour. Geology*, **43**, 785 (1935).   Modifications of these two extreme types are also discussed in this paper.

R denotes the radius of the spheres, the volume of the unit cell in the cubic packing is $8.00R^3$, and $5.66R^3$ in the rhombohedral packing. Their porosities are 47.64 and 25.95 per cent, respectively, independent of the sphere size. Moreover, their minimum pore sections have areas $0.858R^2$ and $0.161R^2$, respectively.

Actual sandstones are, of course, of much more complicated structure than could be formed by packings of spheres of uniform size. For they are comprised of grains that are neither spherical nor of the same size. Moreover, these grains are generally held together by agglomerate masses of cementing material that is itself composed of very fine granular particles. While intergranular bridging will sometimes lead to locally high values for the resultant porosity, the deviations from the ideal spherical particles of uniform size will usually result in reduced porosities. Moreover, the rock mass as a whole will contain distributions and continuous graduations of pore sizes and shapes rather than a sharply defined geometry, such as is illustrated in Fig. 1.2. The magnitude of average grain diameters found in oil-producing sandstones generally lies between 0.005 and 0.05 cm, and the average pore diameters have been estimated to be of the order of one-fifth of these. Several photomicrographs[1] (15×) of some typical oil-productive sandstone samples are reproduced in Fig. 1.3.

Virtually all limestones are deposits precipitated from solutions. The latter are usually, though not always, sea waters. Limestones are often the remains of organic material, or they may be deposits of calcium carbonate enclosing marine organisms. Some forms are comprised of "oölitic" (egglike) masses of rounded grains, the layers of calcium carbonate sometimes covering shell fragments. Limestones represent about 5 per cent of the sedimentary rocks in the lithosphere.

The porosity of many limestones is developed[2] by a process of solution, the reverse of that leading to their formation. Such "secondary porosity" usually develops at erosion surfaces where the rock can be subjected to weathering and leaching by circulating waters.

A major source of "primary" porosity in limestones arises from the jointing, fracturing, or fissuring of the limestone mass. While such porosity is primarily formed as the result of stresses developed in geologic crustal movements, it is often further increased by solution processes.

When part of the calcium in limestones is replaced by magnesium, dolomites are formed. A crystalline shrinkage, up to 12 per cent, may

---

[1] These are taken from a paper by G. E. Archie, *AAPG Bull.*, **31**, 350 (1947). The numbers to the left of the porosity values are the permeabilities in millidarcys (cf. Sec. 3.5.).

[2] For a detailed discussion of porosity development in limestones, cf. W. V. Howard and M. W. David, *AAPG Bull.*, **20**, 1389 (1936).

result from this cation exchange, if occurring after lithification of the limestone, and give rise to joints and shrinkage cracks in the dolomite rock. Crustal movement may also lead to facturing in dolomites. Local dolo-

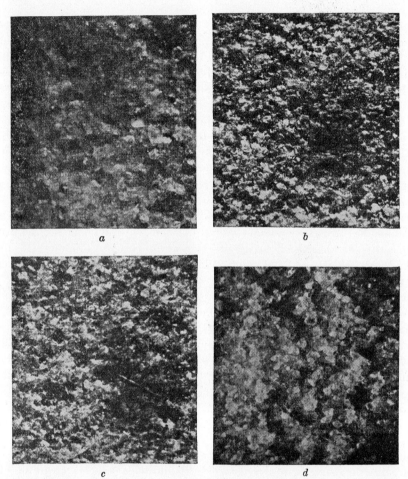

FIG. 1.3. Photomicrographs (15×) of samples of oil-productive sandstones. *a* (1240 md, 23.1% porosity), *b* (101 md, 20.9% porosity), and *c* (18 md, 12.4% porosity) are Wilcox sandstone; *d* (1048 md, 26.5% porosity) is Paluxy sandstone. (*After Archie, AAPG Bull., 1947.*)

mitization porosity apparently is the result of an excess of solution by ground waters over precipitation.[1]

Oölitic limestones often have a porous structure similar to sandstones. However, limestone porosity in the form of solution voids, fractures, or fissures is of a radically different type. By its nature it may be extremely

[1] Cf. K. M. Landes, *AAPG Bull.*, **30**, 305 (1946).

variable in its local characteristics. In some reservoirs the void space literally comprises solution caverns. In others it may be confined to the joints themselves or to the layers of rock immediately bounding the fractures. The main body of the limestone between the joints may have porosities as low as 3 to 5 per cent. In such "intermediate" limestones[1] there is superposed on the basic intergranular structure an independent system of voids—fractures, fissures, or solution cavities—that are sufficiently widely distributed as to dominate the gross properties of the rock, often both with respect to porosity and fluid-transmitting capacity. This fact must be kept in mind in the interpretation of the performance of limestone reservoirs. Photomicrographs[2] (15×) of a siliceous limestone and three oölitic (Smackover) limestone rock samples are reproduced in Fig. 1.4.

While most major oil reservoirs produce from sandstone or limestone and dolomite formations, other types of rocks occasionally have been found to be commercially productive. For example, the Lytton Springs, Tex., field produces from a porous and fractured serpentine. In the Texas Panhandle, oil has been found in granite wash, a basal conglomerate produced by weathering of the underlying granite core. At Florence, Colo., Salt Creek, Wyo., and Casmelia, Calif., some production is obtained from fractured shales. Basic igneous rocks form part of the reservoir at Furbero, Mexico. And some oil fields in Kansas and Oklahoma produce from porous residual chert breccia. In fact, in the general discussions of oil-reservoir performance to be presented in the following chapters no attempt will be made to identify explicitly the exact nature of the oil-bearing rock. The latter will usually be referred to synonymously as the "producing formation," "stratum," "pay," or "section," as is customary in the terminology of the oil-producing industry. And, for convenience, even the term "sand" will be occasionally used in a generic sense without necessarily implying that the rock is actually a sandstone. This, however, should not be construed as implying that the character of the reservoir rock is of minor importance. On the contrary, in applying the physical principles of reservoir behavior to specific fields, a knowledge of the nature of the producing formation is indispensable in planning their development program and interpreting their performance.

---

[1] Cf. A. C. Bulnes and R. U. Fitting, Jr., *AIME Trans.*, **160**, 179 (1945).

[2] Archie, *loc. cit.* Some very interesting photographs of plastic casts of the void space in typical limestones and dolomites are given in a paper by W. C. Imbt and S. P. Ellison, Jr., *API Drilling and Production Practice*, 1946, p. 364, in which is also included an extensive bibliography of publications on limestone and dolomite rock structure; cf. also W. F. Nuss and R. L. Whiting, *AAPG Bull.*, **31**, 2044 (1947).

1.4. The Confinement of Petroleum Reservoirs.—In the last section were discussed the types of detailed texture of rocks, comprising oil-

Fig. 1.4. Photomicrographs (15×) of samples of oil-productive limestones. *a* (1350 md, 18.5% porosity), *b* (103 md, 13.1% porosity), and *c* (0.9 md, 7.0% porosity) are Smackover limestone, *d* (16.4 md, 35.9% porosity) is Devonian Siliceous lime. (*After Archie, AAPG Bull., 1947.*)

producing formations, that gives them local fluid-holding capacity. It was tacitly assumed that these sedimentary structures possessed, in some degree at least, a fluid-transmitting capacity, *i.e.*, permeability. Rocks that may or do constitute oil-producing reservoirs evidently must possess both porosity and permeability. But an oil reservoir is far more than a rock having merely the capacity to hold and transmit oil.

An oil reservoir is a body of porous and permeable rock that *does* contain oil. To be of commercial interest, it must, of course, contain enough oil—*recoverable* oil—to justify the drilling and operation of at least one well to bring the oil to the surface. But the size is not a factor directly pertinent to its description as a physical system.

In addition to the axiomatic requirements that oil reservoirs must have at one time been accessible to the primary sources of the oil and must possess the capacity to hold and transmit fluids, they must constitute "traps" preventing the escape of the petroleum fluids once they have entered or developed within the reservoir rock. Otherwise, of course, they would not remain as oil reservoirs. The tendency for escape of the oil itself is usually an expression of a general buoyant force acting on the oil, which arises from its association with pressures of a hydrostatic type.[1] The same applies to any free-gas phase that may be trapped with the oil and becomes segregated to overlie the oil-saturated zone. In fact, it is just these same buoyant forces that lead to the gross separation of the intercommunicating masses of free gas, oil, and water according to their density within continuous reservoir strata. To prevent the upward seepage of the petroleum fluids the oil-bearing rock must have a blanket of effectively impermeable material forming its general upper boundary.

In principle, any of the rocks intimately associated with oil-bearing formations may serve as sealing blankets, provided that they are inherently substantially impermeable to fluid movement or are made so by excessive cementation and intergranular deposition. Thus highly cemented sandstones or their completely metamorphosed equivalents, quartzite, have served as seals in some fields. Likewise, pure limestones, chalks, and sandy limestones have been found as overlying covers for oil-bearing formations. Clays and shales and general argillaceous rocks, as sandy shales, shaly sandstones, and marls, constitute the most common oil-confining strata. Clays tend to be plastic and can follow crustal movements with a minimum of jointing and fracturing. And even jointing in shales is relatively uncommon, although in exceptional cases shales may be sufficiently fractured to serve as oil reservoirs.

It should be emphasized that the covering strata for oil reservoirs generally are not, nor need they be, completely impermeable barriers to fluid

[1] The natural tendency for the oil and free gas to expand beyond their confining volume against the average reservoir pressure that holds them in compression and of the dissolved gas to escape from solution is ever present in oil reservoirs. This, however, is directed uniformly outward and does not in itself give rise to a resultant force tending to induce an *upward* seepage.

flow. In the great majority of oil reservoirs the sealing blankets are simply water-filled strata of very small average grain and pore size. While the quantitative measures of their permeability[1] may be very low as compared with commercially productive oil-bearing strata, generally[2] they will not be strictly zero. The mechanism whereby they nevertheless prevent the upward seepage of the underlying oil and gas arises from the capillary interfacial flow resistance at the contact between the oil-bearing and sealing formations. This resistance is expressed by the "displacement pressure."[3] The latter is the pressure differential required to force the entry of a nonwetting phase into a porous medium saturated with a wetting phase. Virtually all rocks associated with oil reservoirs are preferentially wet by water.[4] In such rocks oil and gas are nonwetting phases. Aside from the effects of contact angle, the displacement pressure is proportional to the interfacial tension between the wetting and nonwetting phases and to the reciprocal of the maximum pore radius in the rock containing the wetting phase. It is because of the very low magnitudes of even the maximum pore radii of effective covering blankets that their displacement pressure may successfully resist the entry of oil or gas. As previously indicated, the force tending to induce such entry is largely the result of the buoyancy of the mass of oil and gas under hydrostatic pressures.[5] Hence it is of the order of magnitude of the thickness of the oil- and/or gas-saturated zone multiplied by the difference in density between the formation water and the petroleum fluids. Laboratory evidence indicates that these buoyant forces may well be exceeded by the displacement pressures in such fine-grained materials as constitute the common covering strata.

A fracture in the immediate rock blanket may have such a low displace-

---

[1] For a full discussion of this term and of means for expressing it quantitatively cf. Sec. 3.4.

[2] The reservoirs referred to here are those which are in substantial equilibrium with a hydrostatic column and have pressures equivalent to the hydrostatic-column head. Where the reservoir pressures are abnormally high or low to the extent of hundreds of psi, the oil-containing formation must be sealed everywhere by rocks of effectively zero permeability. On the other hand, even when upward oil or gas mass leakage prior to exploitation is prevented only by the displacement pressure, the apparent absence of substantial downward water leakage into the oil reservoir, after its pressure has been reduced by fluid withdrawals, implies a very low fluid-transmitting capacity for the covering strata.

[3] For further discussion of capillary phenomena cf. Sec. 7.8.

[4] The outstanding exception to this general rule appears to be the Wilcox Sand in the Oklahoma City field.

[5] While capillary interfacial forces between the water and oil (cf. Sec. 7.9) serve to counterbalance the buoyant forces within the oil-producing section itself, the buoyant forces will nevertheless be exerted against the overlying confining medium if the latter is completely water-saturated.

ment pressure as to permit direct entry of the oil and gas. However, the rock walls of the fracture may still prevent intrusion of the oil or gas into the main body of the sealing layer. And if the latter, in turn, is covered by another nonfractured tight formation, the leakage even through the fracture may be stopped, except for the volume required to permeate the fracture itself.

Even if physically sealed against a mass-leakage flow out of an oil reservoir, there is theoretically the possibility of loss of the petroleum fluids by diffusion processes. If the oil and/or gas phase is in direct contact with another fluid, a concentration gradient of the former will be set up in the latter, which will lead to a molecular transfer in the direction of low concentration. Although geologic time—millions of years—has been available for the action of such processes, it is extremely doubtful, because of the very low solubility of oil in water, that diffusion of the oil has taken place to an extent sufficient materially to deplete actual oil reservoirs. However, the diffusion of gas through water-saturated porous strata cannot be so easily ruled out. And, indeed, the explicit assumption that such diffusion does take place has been the basis of much recent work on the development of geochemical methods for prospecting for oil, in which attempts are made to determine the hydrocarbon seepage through the whole overburden to the surface. Without presuming to evaluate this type of evidence, it may be noted that in a number of oil fields the crude oil has been found to be greatly undersaturated with respect to gas content.[1] Such observations might be explained as indicating loss of the original gas content by diffusion. But they cannot represent definite evidence unless it is known that at the time of original accumulation the oil was fully saturated with gas or that subsequent to its accumulation the whole reservoir may not have undergone deeper burial. Moreover, in many reservoirs free-gas accumulations have been found overlying the oil zone at the time of discovery. Diffusion losses in these must have been at most of limited magnitude.

Until the situation is clarified by more complete data, it appears reasonable to consider the loss of reservoir gas by upward diffusion, to some degree at least, as a definite possibility in general and perhaps a probability in particular cases. However, from the point of view of the reservoir performance after development, it does not matter greatly if one knows how much gas, if any, may have been lost since the accumulation of the oil in the reservoir. For interpreting and predicting the producing be-

---

[1] In many of the fields in Kansas so little gas is produced with the oil that the oil is considered as entirely "dead." And at Smith Mills, Ky., the analysis of actual bottom-hole oil samples has shown the solution gas content to be only 2 ft^3/bbl oil, even though the sampling pressure was 850 psi.

havior of a reservoir it will suffice to know its gas content, free and dissolved, at the time of discovery and exploitation.

1.5. The Structural Classification of Oil Reservoirs.—The classification of petroleum reservoirs is necessarily arbitrary. No single system can encompass all aspects pertaining to their histories of development, their final physical state, and their performance during exploitation. A broad classification of the latter type will be given in Chap. 9 and will be used as the basis of the interpretation and analysis of oil-reservoir behavior. Indeed, it is the primary purpose of this work to provide the background, physical principles, and methods for the interpretation and prediction of reservoir performance. First, however, it is necessary to understand what "makes" oil reservoirs what they are. The manifold possibilities for creating an oil reservoir and structure provide the basis for the following[1] classification:

a. Reservoirs closed by local deformation of the strata
b. Reservoirs closed because of varying permeability of the rock
c. Reservoirs closed by a combination of folding and lack of adequate permeability
d. Reservoirs closed by a combination of faulting and lack of adequate permeability

The most common type of reservoir structure is a subclass of *a* above, namely, that in which the local deformation is a simple folding of the strata into closed anticlines or domes. An example of this type of reservoir is the Cushing field, Creek County, Okla., which is the longest anticline in Oklahoma. A structure map of the whole field and a section through one of its four local domes, Drumright, are shown in Fig. 1.5.[2] The Salt Creek field in Natrona County, Wyo. (Figs. 1.6a and b),[3] illustrates the rather frequently occurring reservoir trap in which the strata deformation is comprised of a combination of folding and faulting. The less common type of homoclinal structure offset by a fault provides the reservoir at Mt. Poso, Kern County, Calif. (cf. Fig. 1.7).[4] And if the trap is formed by

---

[1] This is the classification made by W. B. Wilson, "Problems of Petroleum Geology," pp. 433–445, AAPG, 1934 (cf. also K. C. Heald, "Elements of Petroleum Industry," pp. 26–62, AIME, 1940). Many others have been proposed, including those of F. G. Clapp, *Geol. Soc. America Bull.*, **28**, 557 (1917); V. Ziegler, "Popular Oil Geology," pp. 87–116, John Wiley & Sons, Inc., 1920; W. H. Emmons, "Geology of Petroleum," pp. 86–132, McGraw-Hill Book Company, Inc., 2d ed., 1931; F. M. Van Tuyl, "Elements of Petroleum Geology," p. 58, Denver Publishing Co., 1924; C. W. Sanders, *AAPG Bull.*, **27**, 539 (1943); S. Pirson, *Oil Weekly*, **118**, 54 (June 18, 1945); and O. Wilhelm, *AAPG Bull.*, **29**, 1537 (1945).

[2] These are taken from Emmons, *op. cit.*

[3] E. Beck, "Structure of Typical American Oil Fields," Vol. II, p. 589, AAPG, 1929.

[4] V. H. Wilhelm and L. W. Saunders, *California Oil Fields*, **12** (No. 7), 1 (1927).

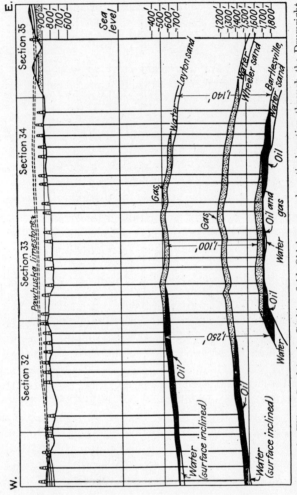

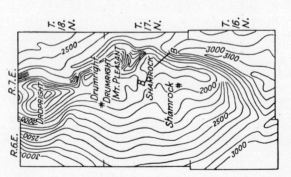

Fig. 1.5. Subsea depth structure contours on the Wilcox Sand in the Cushing field, Oklahoma, and vertical section through the Drumright dome. (*After Emmons, "Geology of Petroleum."*)

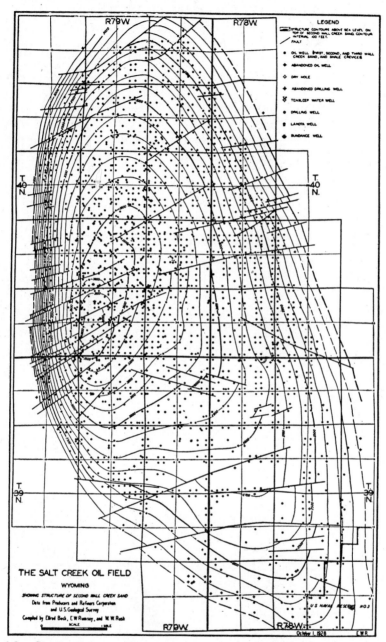

Fig. 1.6a. Structure contours, above sea level, on top of the Second Wall Creek Sand in the Salt Creek field, Wyoming. (*After Beck, "Structure of Typical American Oil Fields," Vol. II, 1929.*)

a seal against the flanks of a salt dome intrusion, the structural contours and section of the producing formation may be similar to that shown in Figs. 1.8[1] for the Fannett field, Jefferson County, Tex.

Reservoirs that are closed because of varying permeability of the rock occur in a great variety of forms. As a class they constitute the so-called "stratigraphic" traps of oil accumulation. While the field as a whole lies on a domal uplift, the producing section of Goose Creek, Harris County, Tex., is a striking example of reservoirs controlled by sand lensing (cf. Fig. 1.9).[2] A lenticular oil accumulation in calcareous rocks is illustrated by Fig. 1.10[3] showing a cross section through the East Dundas pool,

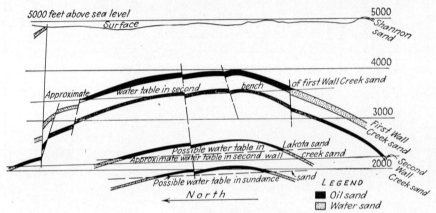

Fig. 1.6b. Vertical section in the Salt Creek field, Wyoming. (*After Beck, "Structure of Typical American Oil Fields," Vol. II, 1929.*)

Richland and Jasper Counties, Ill., producing from the McCloskey Limestone.

Perhaps the outstanding example of a reservoir comprised of a permeable section in an igneous intrusion is the Lytton Springs field, Caldwell County, Tex., as shown in Fig. 1.11.[4] Although only few reservoirs occur in igneous rocks, fissured and cavernous void space is occasionally found in shales and not infrequently in limestone reservoirs.

Stratigraphic trap reservoirs are often sealed by the overlap of relatively impermeable strata truncating the oil-bearing formation. The well-known and prolific East Texas field, in Gregg, Smith, Rusk, Cherokee, and Upshur Counties, Tex., in which the Woodbine Sand is sealed by the Austin Chalk

[1] These were prepared at the author's request by H. E. Minor.

[2] H. E. Minor, *AAPG Bull.*, **9**, 286 (1925).

[3] This section, prepared by P. Farmer, was made available through the courtesy of W. B. Wilson.

[4] D. M. Collingwood and R. E. Retzger, *AAPG Bull.*, **10**, 953 (1926).

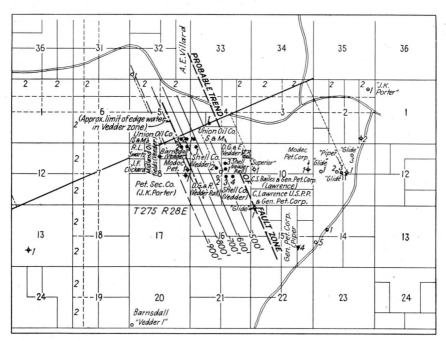

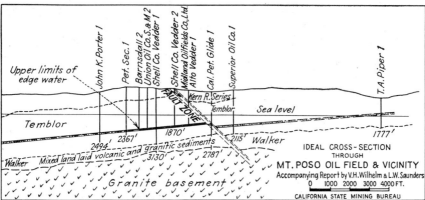

FIG. 1.7.   Vedder Zone structure contours and vertical section in the Mt. Poso field, California.   Contour numbers are subsea depths.   (*After Wilhelm and Saunders, "California Oil Fields," 1927.*)

(cf. Figs. 1.12)[1] illustrates this type of reservoir. Occasionally the oil-productive section is sealed and limited by a deposit of bitumen or equivalent viscous hydrocarbon within the oil-bearing stratum, as in the case

[1] H. E. Minor and M. H. Hanna, *AAPG Bull.*, **17**, 757 (1933); and C. E. Reistle, *API Drilling and Production Practice*, 1934, p. 96.

of the Lockwood-Talara area of Sunset-Midway, Kern County, Calif. (cf. Fig. 1.13).[1]

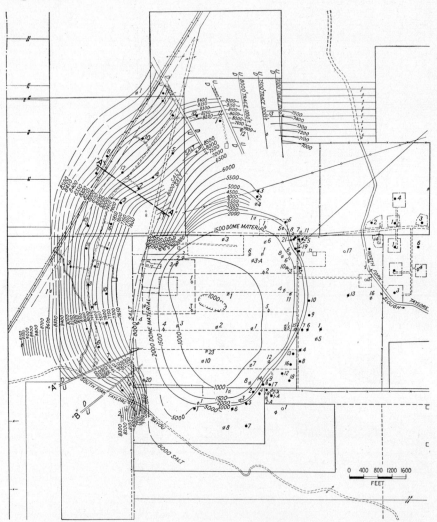

Fig. 1.8a. Structure contours (subsea depths) on top of the Frio Sand, in the Fannett field, Texas. (*After Minor.*)

Still other geometrical trap structures have been found to provide oil-producing reservoirs. And many oil fields are characterized by combinations of several types of closure. Those illustrated in Figs. 1.5 to 1.13 should serve as a basis for visualizing the great variety of physical and

[1] R. W. Pack, *USGS Prof. Paper* 116 (1920).

geometrical boundaries that may enclose and define the porous media and their entrained fluids comprising an oil reservoir. On the other hand, it

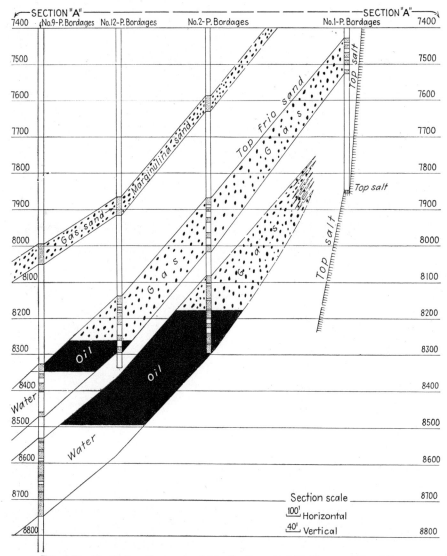

Fig. 1.8*b*.   Vertical section across *AA* in the Fannett field, Texas.   (*After Minor.*)

should be observed that, while an assembly of producing wells distributed over an "oil field" necessarily implies the existence of an underlying oil reservoir, the latter may be only one of a series of distinct reservoirs lo-

cated at various depths under the same or overlapping areal field "limits."
For example, in the Oficina field in eastern Venezuela, a surface area of
10,700 acres is underlain by some 85 individual productive oil and gas
reservoirs. Many of these lie in common stratigraphic levels between
4,000 and 6,100 ft containing one or more oil reservoirs. At near-by East
Guara there are some 40 distinct stratigraphic units and reservoirs between
4,225 and 7,000 ft underlying a surface area of only 1,400 acres. At
Seeligson, Jim Wells, and Kleberg Counties, Tex., at least 40 separate oil
and gas reservoirs are distributed in the Frio and Vicksburg formations

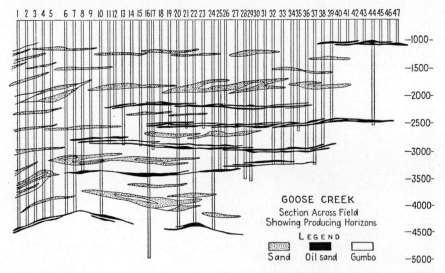

Fig. 1.9. A vertical section through the Goose Creek field, Texas. (*After Minor, AAPG
Bull., 1925.*)

between 4,300 and 6,800 ft, the productive surface area covering only
14,000 acres. And many of the California fields, which have total productive sections extending over more than 1,000 ft, are comprised of numerous
individual reservoirs even though they may be produced by a common
well system.

It must therefore be understood that, while the geographical location of
an oil reservoir may be properly associated with the name of the "oil
field" of which it is the sole member or one of the members, discussions of
production performance and history must refer to individual reservoirs.
The latter are the separate and noncommunicating oil-bearing units, although they may be brought into intercommunication by wells simultaneously exposed to several reservoirs and producing through the same conduit
to the surface. The separation may be due to shale strata or any non-

productive horizon that is substantially impermeable to gross cross-stratum movement. On the other hand, apparently distinct reservoirs are often

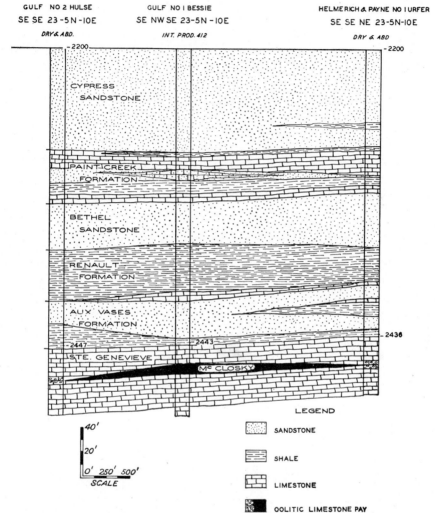

FIG. 1.10. Cross section through the East Dundas pool, Illinois. (*After Farmer.*)

bounded by common or merging water-bearing strata. In principle, these offer means for interaction, if not direct communication, between the oil reservoirs, as has actually been observed, for example, in the Arbuckle Limestone fields.[1] For practical purposes, however, such reservoirs should

[1] Cf. W. A. Bruce, *AIME Trans.*, **155**, 88 (1944); cf. also Sec. 11.10 (Fig. 11.29).

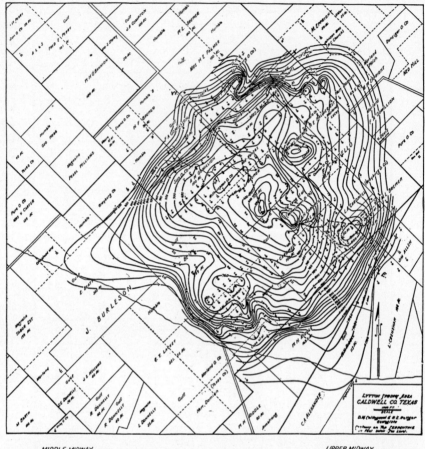

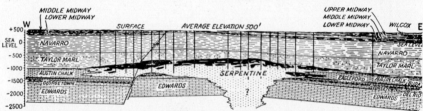

Fig. 1.11. (*Above*) structure contours (subsea depths) on top of the Serpentine formation and (*below*) vertical section in the Lytton Springs field, Texas. (*After Collingwood and Retzger, AAPG Bull., 1926.*)

still be considered in most cases as distinct, especially if their rock and fluid characteristics are different. Their mutual interactions via the common water reservoir may be treated either as a perturbation problem or in terms of a multiple outflow boundary representation of the water reservoir itself.

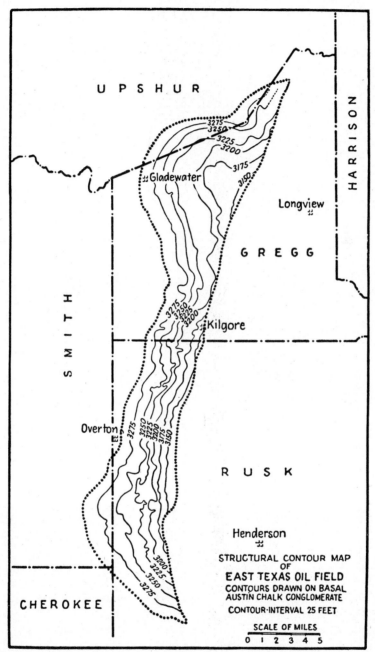

Fɪɢ. 1.12*a*.   Subsea depth structure contours, on the basal Austin Chalk Conglomerate, in the East Texas field, Texas.   (*After Minor and Hanna, AAPG Bull., 1933.*)

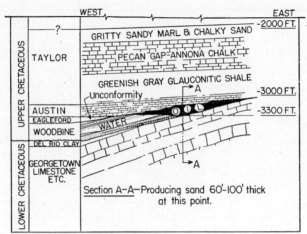

Fig. 1.12b. Vertical section in the East Texas field, Texas. (*After Reistle, API Drilling and Production Practice, 1934.*)

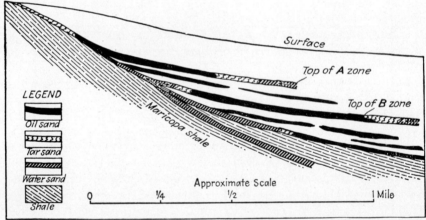

Fig. 1.13. Vertical section through the Lockwood-Talara area of the Midway-Sunset field, California. (*After Pack, USGS Professor Paper, 1920.*)

1.6. Reservoir Engineering.—The broad subject matter of this work is that currently referred to in the oil industry as "reservoir engineering," although the treatment here will be based on the physical rather than the engineering point of view. The ultimate goal underlying the development of the science of reservoir engineering is the attainment of a maximum efficiency in the exploitation of oil-bearing reservoirs. This implies the maximum recovery of oil at a minimum cost. While the consideration of economic factors may seem foreign to a discussion of "physical principles," it must be recognized that these same physical principles would be of very little interest—at least to the oil industry—unless they were applicable to actual oil reservoirs of commercial significance. Even the slightest pretense of realism would imply that the numerical factors used in the discussions lie in the range both of physical reality and of practical importance. It is true that in many applications absolute magnitudes need not be fixed, as ratios and dimensionless parameters can be used. Yet it would be sterile puritanism to insist that physical principles should not be applied to determine, for example, when a reservoir would have to be abandoned as unable to produce at profitable rates, just because such rates may depend on the market price of the oil. On the other hand, it should be emphasized that such economic considerations as occasionally may be referred to will be invoked only to supplement the technical factors. They are not to be construed as the primary framework of a structure to be clothed in scientific rationalization.

There are two major aspects of the science of reservoir engineering as related to the goal mentioned above. The one consists in those factors and characteristics of oil reservoirs and their performance which are inherent in the particular reservoir under consideration and in the basic physical mechanisms to which it may be subject. These are beyond the control of the "operator" of the field. But they must nevertheless be known and understood as thoroughly as possible to anticipate what the behavior of the reservoir may be and how to exercise such controls as may be feasible. They include the gross geometrical structure of the reservoir, its physical dimensions, its initial fluid content and distribution, the porosity of the rock, its permeability, its permeability-saturation relationship, the nature of the oil, the nature of the gas in solution, the reservoir temperature, the initial reservoir pressure, the gas saturation pressure, the character of contiguous water-bearing reservoirs, if any, and the uniformity or variability of the producing section within the reservoir. All these represent "initial" conditions and properties defining the particular reservoir of interest. All that can or should be done about them is to establish their magnitudes as accurately as possible. They may be interpreted as favorable or unfortunate but must be accepted simply as the contribution of nature.

While, obviously, the above-listed "fixed" properties of the reservoir will determine its inherent potentialities as an oil-producing system, there is still much left to the choice of the reservoir engineer with regard to the actual exploitation program to be undertaken. These include the location of the wells, their spacing, and the manner of completion and the withdrawal rates for the wells individually and the reservoir as a whole.[1] And once an initial development program has been established, changes as may be indicated by the field performance are subject to the control of the operator. Whether or not gas or water injection should be undertaken and how such "pressure-maintenance" operations should be conducted are matters to be decided by an evaluation of reservoir data and observations. The desirability of cycling a condensate field is subject to the decision of the operator. And while it now appears possible to plan and conduct the exploitation of most oil reservoirs in such a way as to make unnecessary "secondary-recovery" operations, there are many depleted reservoirs that were not so exploited and that can be profitably operated as secondary-recovery projects. The achievement of maximum efficiency in such operations also demands that they be conducted on the basis of a technically planned program.

Thus it will be seen that there is a wide scope for the application of reservoir-engineering principles, even though the basic reservoir characteristics are fixed in advance. On the other hand, just because these "initial conditions" will vary over wide ranges, the implications of the physical principles of reservoir behavior cannot be expressed as specific rules or procedures to be applied indiscriminately to particular reservoirs. It is indeed the primary function of the reservoir engineer to evaluate the many individual factors characterizing a particular reservoir and to determine their composite effect in modifying the performance that might be expected of the idealized prototypes illustrating the broad physical principles of oil production. Operating practices ideally suited to one field may lead to inexcusable inefficiency in another, in spite of their superficial similarity in some respects. Oil-producing reservoirs are not the product of a mass-production assembly line, "designed" and constructed to operate according to specifications. They are objects for individual study and analysis. And, in the light of such study, they should be developed and exploited to achieve the maximum returns inherent in their individual "personalities."

---

[1] Although in most states well and reservoir production rates are fixed by state regulatory bodies, these are generally based, at least in part, on the conclusions derived from an analysis of reservoir-engineering data regarding the most efficient development and operating methods.

CHAPTER 2

THE PHYSICAL PROPERTIES AND BEHAVIOR OF PETROLEUM FLUIDS

Although it will generally suffice in our considerations of the dynamics of oil production to treat the reservoir fluids as homogeneous simple gases or liquids, other phases of the subject require a more detailed knowledge of their thermodynamic properties. For example, the pressure-volume relationships of both the gas and liquid phases enter the problem of estimating the initial oil content and in predicting the production histories even of normal crude-oil reservoirs. And for the understanding of the performance of condensate or distillate fields, which are becoming more common as drilling depths are increased, a thorough knowledge of the physical equilibrium behavior of hydrocarbon systems is indispensable. We shall therefore discuss in this chapter the physical properties of petroleum fluids as static hydrocarbon systems in thermodynamic equilibrium, with no direct reference to the flow of these fluids through a porous reservoir and into producing wells. Unfortunately the treatment will be highly empirical. This, however, is necessary because of the great complexity of the problem and the lack of any unified theoretical correlation between the properties of various mixtures of hydrocarbons.[1]

While an extended review of empirical observations on hydrocarbon systems will be given in this chapter, it is not the purpose of the discussion to provide a handbook of numerical or graphical tabulations. In fact, much of the material to be presented in the following sections will be only of illustrative significance and will have no immediate or direct applicability to practical problems of oil production. On the other hand, since the whole subject matter of this work is concerned with mixtures of petroleum hydrocarbons, it seems appropriate to develop first a thorough understanding of their physiochemical properties, even though in the present state of the science of reservoir engineering it will usually suffice to characterize the fluid phases by gross parameters and empirically established functions.

---

[1] A systematic and thorough discussion of the basic thermodynamic characteristics of hydrocarbon systems has been given by B. H. Sage and W. N. Lacey in "Volumetric and Phase Behavior of Hydrocarbons," Stanford University Press, 1939. A number of the numerical and graphical illustrations used in this chapter have been taken from the above text or original papers by its authors.

2.1. One-component Systems.—Since we shall be primarily concerned with physical interactions and transformations between the gas and liquid phases of petroleum fluid systems, it will suffice to restrict the discussion essentially to the common hydrocarbon series. The main groups are the paraffins, or saturated chain hydrocarbons of composition C_nH_{2n+2}; the naphthenes, or saturated closed-ring compounds of composition C_nH_{2n}; the olefins, which are unsaturated chain hydrocarbons, also of composition C_nH_{2n}; and the benzene aromatics, or unsaturated closed-ring hydrocarbons, of composition C_nH_n. Still others are the polymethylenes, $(C_nH_{2n})_x$, the acetylenes, C_nH_{2n-2}, the terpenes, C_nH_{2n-4}, etc.[1] However, the natural gas phase associated with petroleum is largely composed of the first six members of the paraffin series, except for small amounts of sulfur-containing compounds—generally hydrogen sulfide—observed in some districts, water vapor, and rather anomalous instances where large concentrations of carbon dioxide or nitrogen have been observed. The lower volatile members of the other series are relatively unstable, while the higher members have such low vapor pressures that their detection in the gas phase is very difficult. The lower members of the paraffin series are methane, CH_4; ethane, C_2H_6; propane, C_3H_8; the butanes, C_4H_{10}; the pentanes, C_5H_{12}; the hexanes, C_6H_{14}; the heptanes, C_7H_{16}; the octanes, C_8H_{18}; etc. Because these are the predominant constituents of the gas phase, the great majority of thermodynamic studies, with respect to phase behavior, have been made on the paraffins. Accordingly, the illustrative examples to be used in the following discussion will also be confined to this series. The other series will be considered as being absorbed in the "heavy fractions" of the liquid phase[2] but will not be otherwise treated explicitly.

The basic empirical fact regarding the physical thermodynamic behavior of the petroleum hydrocarbons, and in fact of almost all pure substances in the appropriate ranges of temperature and pressure, is that they have a volume-pressure-temperature relationship of the type shown in Fig. 2.1 for ethane.[3] As will be obvious from this diagram, if the temperature is kept fixed, the volume first decreases rapidly with increasing pressure, then

---

[1] A detailed treatment of the chemical constitution of petroleum may be found in "The Chemical Technology of Petroleum" by W. A. Gruse and D. R. Stevens (2d ed., McGraw-Hill Book Company, Inc., 1942).

[2] As in the case of the gas phase the crude oil also generally contains, in addition to the pure hydrocarbons, small concentrations of oxygen-, nitrogen-, or sulfur-containing compounds, as well as inorganic salts, as impurities.

[3] Fig. 2.1, as well as Fig. 2.2, was plotted from data of B. H. Sage, D. C. Webster, and W. N. Lacey, *Ind. and Eng. Chemistry*, **29**, 658 (1937). Similar data of somewhat higher precision and covering greater pressure and temperature ranges have been since reported by H. H. Reamer, R. H. Olds, B. H. Sage, and W. N. Lacey, *Ind. and Eng. Chemistry*, **36**, 956 (1944).

drops sharply without any pressure increase—for temperatures below 90.1°F—and finally assumes but a very slow rate of volume diminution as the pressure increase is resumed. These three segments of the curves correspond, respectively, to the gas phase, the two-phase gas-liquid region, and the liquid phase. The boundary points separating these regions define the dotted curve. That part to the right of the maximum, separating the gas and two-phase regions, is termed the "dew-point" curve. When the pressure and volume of the fluid lie on this curve, it is said to be a "satu-

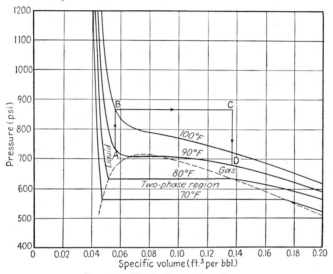

Fig. 2.1. The isotherms for ethane.

rated gas." For an attempt to increase the pressure of the saturated gas by decreasing the volume will only lead to a condensation of liquid. Moreover this liquefaction will proceed, on further decrease of volume, until all the gas phase has disappeared without the pressure rising at all. When the whole system has become liquid, it will be at its "bubble point." The curve through the various bubble points represents the states of the "saturated liquid." The sharp rise of the liquid segment of the isotherms evidently reflects the low compressibility of the liquid phase.

It will be noted that the straight two-phase segment decreases in length as the temperature rises. That is, as the temperature increases, the volume of the saturated gas becomes smaller while that of the saturated liquid becomes greater. Ultimately the length of the straight segment vanishes, and the isotherm merely has a horizontal tangent at the maximum of the dotted curve, where the dew point and bubble point merge. This point is the "critical point" of the system, and the isotherm temperature is the "critical temperature," which is the highest at which any two-phase region can exist. The corresponding pressure and volume are the "critical pressure" and "critical volume." From these definitions and from the nature of the corresponding isotherms, it is clear that for temperatures higher than the critical the fluid will persist in a single phase throughout the complete volume or pressure range. Whether this phase is considered gaseous

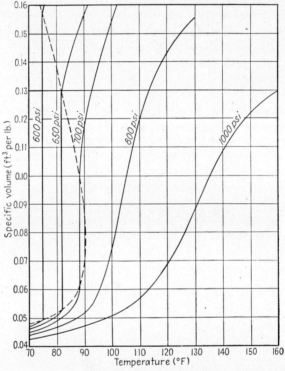

Fig. 2.2. The isobars for ethane.

or liquid is immaterial, although the convention is frequently made that it is to be termed gaseous if its volume exceeds the critical and liquid if the volume is less than the critical. Strictly speaking, however, at the critical point the properties of the gas and liquid phases become identical, and the interface between them disappears. In fact, by raising, at constant volume, the temperature of a liquid phase above the critical, then expanding, and subsequently dropping the temperature to its original value, as indicated by the path *ABCD* in Fig. 2.1, it is possible to end up with a gas without encountering any phase discontinuities.

The mutual relationships of the various isotherms may be visualized by plotting the same data as isobars, *i.e.*, constant-pressure curves. Such a plot for ethane is shown in Fig. 2.2. The physical interpretation of these curves will be evident on supposing the material to be contained in a vessel closed by a movable piston subjected to constant pressure. The curves of Fig. 2.2 then show what would happen, in the case of ethane, to the volume of the system—the position of the piston—as the temperature of the vessel is varied. Thus for pressures above the critical, as 800 psi, the volume increases continuously with increasing temperature. Although the small slope at the lower temperatures is suggestive of a liquid phase and the rapid rise of volume with temperature as the latter is increased corresponds to that of a gas phase, the ethane is, strictly speaking, a single-phase fluid throughout the whole temperature range. At pressures lower than the critical, however, the slopes of the curves are no longer continuous throughout. Here, at 650 psi, for example, the ethane volume rises slowly with increasing temperature when the latter is relatively low. This is typical of a liquid-phase behavior, and the ethane is indeed a liquid in this region. At 81.8°F, however, the volume may be expanded by more than a factor of 2, with no change in either temperature or pressure. This, of course, represents the vaporization of the liquid phase. After the vaporization has been completed, the raising of the temperature of the system can be resumed and it will be accompanied by a rapid rise in volume, characteristic of the gas phase into which the liquid has been vaporized. As in the case of Fig. 2.1 the curves through the points of slope discontinuity in Fig. 2.2 are the dew-point and bubble-point curves and refer to the saturated vapor and saturated liquid, respectively.

As the critical constants represent perhaps the most characteristic individual data determining the thermodynamic properties of one-component systems, the values for the paraffin hydrocarbons are presented in Table 1:

TABLE 1.—CRITICAL DATA FOR PARAFFIN HYDROCARBONS*

Compound	Mol. wt.	Crit. temp., °F	Crit. press., psia	Crit. vol., ft³/lb
Methane, CH_4	16.04	− 116.3	673.3	0.0989
Ethane, C_2H_6	30.07	90.1	708.5	0.0789
Propane, C_3H_8	44.09	206.3	617.5	0.0709
n-Butane, C_4H_{10}	58.12	307.6	529.2	0.0712
Isobutane, C_4H_{10}	58.12	273.2	542.4	0.0685
n-Pentane, C_5H_{12}	72.15	387.0	485.1	0.0690
Isopentane, C_5H_{12}	72.15	370.0	483.6	0.0685
n-Hexane, C_6H_{14}	86.17	454.6	433.6	0.0685
n-Heptane, C_7H_{16}	100.20	512.6	396.9	0.0665
n-Octane, C_8H_{18}	114.22	564.6	370.4	0.0690

* These have been recalculated from those listed in "Physical Constants of the Principal Hydrocarbons," 4th ed., 1943, by M. P. Doss. It may be noted, however, that in spite of the importance and common occurrence of these hydrocarbons there appears to be little agreement among the exact values for the critical data listed in the various published tabulations.

It will be noted that, whereas the critical temperature increases and the critical volume decreases (except for C_8H_{18}) with increasing molecular weight of the hydrocarbon, the critical pressure is a maximum for C_2H_6. For higher members of the series it is less and decreases to approximately half of the maximal value in the case of octane.

2.2. The Deviation Factors of Pure Hydrocarbon Gases.—From the quantitative point of view, the pressure-volume-temperature behavior of even the one-component system—the individual pure hydrocarbons—must be considered as a strictly empirical problem. No equations have been developed that quantitatively reproduce the empirically observed data throughout the complete ranges of the physical variables. And even for the gas or liquid phases separately the analytical equations referring to the various hydrocarbons have no simple physical interrelationship. In spite of this situation, however, it is useful to compare the actual behavior with that of so-called "ideal" systems. With respect to the gas phase, the ideal system is, of course, the "perfect" gas, which may be defined for our purposes as one with an equation of state

$$pv = \frac{RT}{M}, \tag{1}$$

where v is the volume per unit weight—the "specific volume"—, p the absolute pressure, T the absolute temperature, M the molecular weight, and R the gas constant per mole. Indeed, Eq. (1) is nothing more than the combination of Boyle's and Charles's laws, which were discovered empirically as describing the actual behavior of gases over moderate ranges of pressure and temperature.

Kinetic theory shows that Eq. (1) should describe the behavior of a gas which is composed of pointlike molecules with no mutual interactions except during collision. Clearly a real gas will approximate such an ideal system most accurately at low pressures and large molal volumes. This anticipation is confirmed by the fact that the hyperbolic variation of the isotherms, required by Eq. (1), is actually followed most closely at low pressures. The over-all deviation of the true behavior from that predicted by Eq. (1) may be conveniently represented by plotting the quantity, commonly termed the "compressibility factor"[1] or "gas deviation factor" namely,

$$Z = \frac{pvM}{RT}, \tag{2}$$

as a function of p and T. If Eq. (1) were obeyed, Z would obviously equal unity for all values of p and T.

A typical example of a set of Z vs. p isotherms is that for ethane shown

[1] Still another term for Z used in some quarters is "supercompressibility factor."

in Fig. 2.3.[1]   It will be seen that Z decreases from unity at low pressures to minimal values and then rises approximately linearly at the higher pressures.   The fact that the curves approach unity at vanishing pressures means that the gas becomes "perfect" at the low pressures, as previously suggested.   The deviation from ideal behavior ($Z = 1$) with increasing pressure develops more rapidly as the temperature is lowered.   For tem-

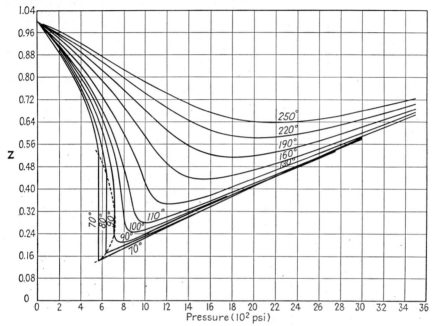

Fig. 2.3.   The deviation factors Z for ethane. Dashed curve represents states of saturated gas and saturated liquid.

peratures below the critical the curves break sharply at the dew-point curve and drop vertically through the region of liquid condensation until the bubble-point curve is reached.   Of course, in this region the nature of the system is so far removed from that of a perfect gas that the use of the latter as a reference system is indeed quite artificial.   However, because this discontinuous behavior passes smoothly into the continuous curves as the critical temperature is exceeded, it serves to complete the range of characteristics displayed by the hydrocarbon system under consideration. Likewise, the continuations of the two-phase vertical segments as the pressure becomes larger than the vapor pressure correspond to the liquid phase.   Nevertheless, they are so similar to the curves at high pressures at temperatures above the critical that they show quite instructively the

[1] Fig. 2.3 was calculated from the same data used in plotting Figs. 2.1 and 2.2.

breakdown in distinctions between liquid and gas phases above the critical pressure and temperature.

As already indicated, for temperatures higher than the critical the Z curves are continuous throughout. However, the initial rapid fall to a minimum and the subsequent slower and approximately linear rise persist until the temperatures become far removed from the critical. As the temperature continues to rise, the curves flatten out and indicate less deviation from the ideal-gas behavior. Ultimately the nature of the deviation becomes reversed, and the curves lie entirely above the unit line, although this behavior is not shown in Fig. 2.3.[1]

2.3. The Physical Origin of the Deviation Factors; van der Waals' Equation.—The fall of the Z curves below unity for moderate temperatures as the pressure is first increased has a very simple physical origin. This lies in the relatively long range attractive forces between the gas molecules —the so-called "van der Waals forces." These forces tend to contract the volume occupied by a molecular ensemble at a given pressure below that corresponding to a perfect gas in which the molecules exert no forces on each other. The v in Eq. 2.2(2) thus is less than the ideal value of Eq. 2.2(1), and Z is less than unity. In this sense Z is nothing more than the ratio of the actual volume of a gas to that of the same number of moles of a perfect gas at the same pressure and temperature. Or if the volume be considered as fixed, the van der Waals attractive forces may be visualized as detracting from the outward pressure exerted by the molecular ensemble constituting the gas. Z then represents the ratio of the pressure of the actual gas to that of an ideal gas of the same volume, temperature, and molecular weight. The values of Z less than 1 thus again reflect the effect of the intermolecular attractive forces. These forces will obviously become of less importance as the intermolecular distances increase. It is therefore to be expected that the deviation of Z from unity will diminish as the pressure is decreased or temperature increased, as is actually observed.

When the hydrocarbons become highly condensed so as to approximate incompressible fluids, the v in Eq. 2.2(2) may be taken as approximately constant and Z will increase linearly with p. This is the type of variation exhibited by the curves of Fig. 2.3 at high pressures. From the point of view of intermolecular forces, this behavior reflects a condition in which the intermolecular distances have become so small that the mutual repulsive forces greatly resist further decreases in volume.

An early attempt to take these intermolecular forces into account was

---

[1] The rise of the deviation factors above unity at high temperatures and the crossing of the isotherms at high pressures are shown in the generalized chart of Fig. 2.20.

that of van der Waals,[1] who proposed that the perfect-gas equation of state [Eq. 2.2(1)] be replaced by

$$\left(p + \frac{a}{v^2}\right)(v - b) = \frac{RT}{M} \tag{1}$$

Here a and b are constants characterizing the molecular properties of the individual gases. The symbol a is a measure of the intermolecular attractive forces and is of the order of 5 to 20 psi $[\text{ft}^3/\text{lb}]^2$. The symbol b represents the intermolecular repulsive forces in the sense that it is a measure of the actual molecular volume. From a detailed analysis of the implications of Eq. (1), it may be shown that b should be one-third of the critical volume and hence will be of the order of 0.02 ft^3/lb for the paraffin hydrocarbons.

Equation (1) leads to a number of interesting predictions. Among these are various interrelationships between the constants a and b and the critical constants. In particular, it implies that RT_c/Mp_cv_c, where the subscript c indicates the critical state, should have the value $\frac{8}{3}$. This requirement is very well satisfied for many gases if $3b$ be substituted for v_c and in fact to within 3 per cent in the case of the paraffin hydrocarbons.

Perhaps the most interesting feature of Eq. (1) is that it permits an interpretation suggestive of the liquid-condensation phenomenon and passage from the gas to the liquid phase as the gas is compressed. Thus at low pressures and large volumes, Eq. (1) reduces essentially to Eq. 2.2(1) and gives the ideal-gas hyperbolic p-v isotherm. At high pressures, however, where v becomes very small and approaches the value b, the term a/v^2 will ultimately again become small compared with p, so that the p-v isotherm will once more take a hyperbolic form but with a vertical asymptote $v = b$ instead of $v = 0$. This, of course, corresponds to the liquid state. Moreover, the transition between these limiting types of variation takes place continuously according to a cubic equation. While the actual maxima and minima in the isotherms below the critical point have no real physical counterpart, they do represent an approximation to the true behavior of considerable physical interest. In fact, if due care is exercised, it is actually possible to proceed past the normal dew points or bubble points for some distance into the normal two-phase region without the creation of the second phase, in accordance with the prediction of Eq. (1). On the other hand, even this unstable transition region disappears from the graphical representation of Eq. (1) at the critical temperature, which Eq. (1) predicts to have the value $T_c = \frac{8}{27}aM/Rb$. And for temperatures exceeding

---

[1] J. D. van der Waals, "Essay on the Continuity of the Liquid and Solid States," Leiden, 1873.

the critical, Eq. (1) leads to monotonic as well as continuous *p-v* isotherms, quite similar to those observed experimentally.

It should be emphasized that van der Waals' equation cannot be used to describe quantitatively the behavior of the pure paraffin hydrocarbons. Nevertheless, there is even now available no other equation comparable in scope with respect to its inclusion of the various characteristic properties of permanent gases or simplicity in physical interpretation. That this is so will be appreciated on noting that the equation often considered as providing the most precise description of the *p-v-T* behavior of pure gases, namely, the Beattie-Bridgeman[1] equation,

$$pv = \frac{RT(1 - C/vT^3)}{Mv}\left[v + B_o\left(1 - \frac{d}{v}\right)\right] - \frac{A_o}{v}\left(1 - \frac{a}{v}\right), \qquad (2)$$

has five empirical constants. Indeed, it is only very recently that any satisfactory kinetic-theory mechanism has been developed for explaining the details of the liquid-condensation process.[2]

The coefficients of thermal expansion or isothermal compressibility can be best obtained from the direct plots of the *p-v-T* data. It is interesting, however, to note that the *Z* curves give immediately the direction of the deviation of these coefficients from the perfect-gas values. Thus from the definition of *Z* by Eq. 2.2(2) we have for the thermal-expansion coefficient

$$\text{Coefficient of thermal expansion} = \frac{1}{v}\frac{\partial v}{\partial T} = \frac{1}{T} + \frac{1}{Z}\frac{\partial Z}{\partial T}. \qquad (3)$$

Since the perfect-gas value is $1/T$ and $Z > 0$, the sign of $\frac{\partial Z}{\partial T}$ determines whether the thermal-expansion coefficient is greater or less than that of a perfect gas. Using this criterion, it follows, by reference to Fig. 2.3, that in the true gas phase below the critical temperature the real gases have thermal-expansion coefficients exceeding that of an ideal gas. This also holds true at temperatures considerably higher than the critical and for pressures that are not too great. On the other hand, in the true liquid phase, at temperatures below the critical, the *Z* isotherms ultimately cross, and $\frac{\partial Z}{\partial T}$ becomes negative as the pressure is increased beyond the critical. The thermal-expansion coefficients then fall to the very low values generally associated with normal liquids. Similar behavior obtains at high pressures and temperatures above the critical.

[1] J. A. Beattie and O. C. Bridgeman, *Jour. Am. Chem. Soc.*, **49**, 1665 (1927), **50**, 3133, 3151 (1928).

[2] Cf. "Statistical Mechanics" by J. E. Mayer and M. G. Mayer (Chap. 14, John Wiley & Sons, Inc., 1940).

The compressibility coefficient is given by an equation similar to Eq. (3), namely,

$$\text{Coefficient of isothermal compressibility} = -\frac{1}{v}\frac{\partial v}{\partial p} = \frac{1}{p} - \frac{1}{Z}\frac{\partial Z}{\partial p}. \quad (4)$$

By reference to Fig. 2.3 it will be seen that as $\dfrac{\partial Z}{\partial p} < 0$ in the true gas phase the compressibility coefficients for the real gases will exceed those

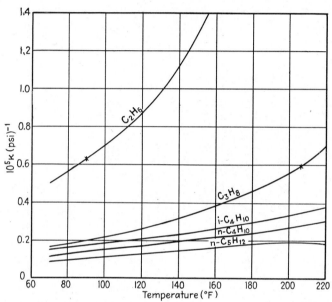

FIG. 2.4.   The variation with temperature of the compressibility κ of pure hydrocarbons at 2,500 psi.   Crosses indicate critical temperatures.

for the ideal gas. This characteristic will persist for temperatures exceeding the critical up to the Boyle point—the pressure at which Z is a minimum. For pressures exceeding the Boyle point, however, $\dfrac{\partial Z}{\partial p}$ will become positive, and the compressibility will fall below that of a perfect gas. Moreover, in the true liquid phase and even for temperatures appreciably exceeding the critical, the term $(1/Z)\dfrac{\partial Z}{\partial p}$ becomes almost as large as $1/p$, so as to leave a very small residual. In fact, as is to be expected, the resultant values of the compressibility are then of the same order of magnitude as that commonly associated with normal liquids. This is shown in Fig. 2.4, in which are plotted the compressibilities of ethane through n-pentane at a pressure of 2,500 psi.

The vapor pressures of the paraffin hydrocarbons, *i.e.*, the dew-point or bubble-point pressures, are plotted in Fig. 2.5 as a function of the temperature. The dashed boundary curve gives the critical points for these

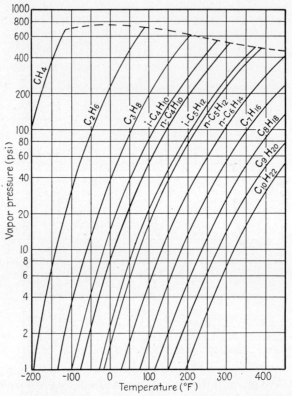

FIG. 2.5.   The vapor-pressure curves for the paraffin hydrocarbons.

hydrocarbons. Analytically, the data may be represented approximately by a Clapeyron equation, namely,

$$p = ae^{-b/T}. \tag{5}$$

They are frequently plotted with special scales (Cox charts[1]) adjusted to give converging straight lines for all the hydrocarbons.

2.4. Two-component Systems.—From a practical point of view two-component, or binary, hydrocarbon systems are of but little greater interest than the one-component systems. For in practice crude oils and the gases associated with them are universally found to be mixtures of many individual hydrocarbons. However, it will be instructive to consider in some detail the properties of binary systems, for even in such simple mixtures

[1] E. R. Cox, *Ind. and Eng Chemistry*, **15**, 592 (1923).

may be found examples of practically all the new features of the complex hydrocarbon fluids.

Before one can begin to discuss quantitatively the physical or thermo-dynamic properties of a binary system, it is necessary to specify the composition of the mixture. This refers not only to the identification of each of the two hydrocarbon components but also to the relative amounts of each in the composite system. The latter may be expressed in terms either of the fractional weight of the whole mass contributed by each or

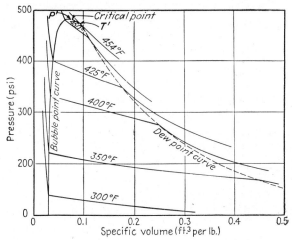

Fig. 2.6. The pressure-volume diagram for a mixture of *n*-pentane and *n*-heptane, containing 52.4 weight per cent *n*-heptane. (*After Sage and Lacey, "Volumetric and Phase Behavior of Hydrocarbons."*)

of the molar fractions of each component. The mole fraction of either component is simply the ratio of the number of moles of the particular component to the total number of moles comprising the whole system.[1] Of course, one of the important characteristics of the binary systems will be the variation of their thermodynamic properties with the composition. But in considering the effect of the basic variables of pressure and temperature the composition of the particular binary system in question must be specified and kept fixed.

The typical over-all behavior of actual binary mixtures is shown in Fig. 2.6, in which are plotted the isotherms for a *n*-pentane and *n*-heptane mixture containing 52.4 weight per cent of *n*-heptane.[2] In the gas phase, *i.e.*, to the right of the dew-point curve, the isotherms are rather flat and

---

[1] One may readily convert weight fractions w_i into mole fractions n_i in general multi-component systems by the relation $n_i = (w_i/M_i)/(\Sigma w_j/M_j)$, where M_i is the molecular weight of the *i*th component. Conversely, $w_i = n_i M_i / \Sigma n_j M_j$.

[2] Sage and Lacey, *op. cit.*, p. 78.

quite similar to those of the pure components. Likewise, the liquid-phase isotherms, to the left of the bubble-point curve, rise steeply with decreasing volume and are qualitatively similar to the pure-component liquid-phase isotherms. In the two-phase region, however, lying between the dew-point and bubble-point curves, the behavior deviates significantly from that of the individual pure components. Thus, whereas for the latter, as has been seen in Fig. 2.1, the two-phase states are characterized by horizontal segments (constant pressure), here the pressure is no longer constant. This means that after condensation first begins, on reaching the dew-point pressure, it is necessary to continue raising the pressure in order to obtain complete disappearance of the gas phase. In other words, here the bubble-point pressures are generally higher than the dew-point pressure, whereas they are equal for pure components. Since at the dew point the liquid phase present is only of infinitesimal volume, the composition of the gas at the dew point—and, of course, for the whole range of pressures below the dew point—is identical with that of the binary system as a whole. Likewise, the bubble-point liquid composition will also be the same as that of the composite binary system. In the two-phase region, however, the composition of the gas and liquid phases will in general be different from each other and from that of the system as a whole. And even the infinitesimal volumes of liquid at the dew point and gas at the bubble point will have different compositions from the dew-point gas or bubble-point liquid.

At the lower temperatures the dew-point and bubble-point curves are qualitatively quite similar to those for pure components. That is, the dew-point volumes decrease and bubble-point volumes increase as the temperature rises. However, when the critical temperatures are approached, very significant and important differences appear. In fact, the critical point itself assumes a different role. Whereas for pure substances it represents at the same time the state for which the gas and liquid phases have identical intensive properties as well as the highest temperature and pressure possible for the coexistence of two phases, it is only the former criterion that really defines the critical point in the case of binary and multicomponent systems. This new feature will become clear on following through in detail the p-v-T behavior near the critical region.

2.5. The Behavior of Binary Systems[1] in the Critical Region; Retrograde Phenomena.—An expanded scale diagram for a typical set of isotherms, in a pressure-volume phase plot, of multicomponent hydrocarbon systems in

---

[1] As noted before, the general qualitative features of binary systems, including the retrograde phenomena, have similar counterparts in the more complex multicomponent systems. In fact, the considerations of this section apply equally well to the latter, and no specific reference will be made to the binary character of the hydrocarbon system, except for Fig. 2.10.

the critical-point region is shown in Fig. 2.7.   The dew points lie along the dashed curve and bubble points along the solid curve, as indicated.   Now the basic criterion for a critical point, that the intensive properties of the gas and liquid phases are identical, is evidently equivalent to the definition that it is the junction point of the dew-point and bubble-point curves. This, therefore, fixes the critical point at C.   However, as the bubble-point curve has a maximum P', it follows at once that here the critical-

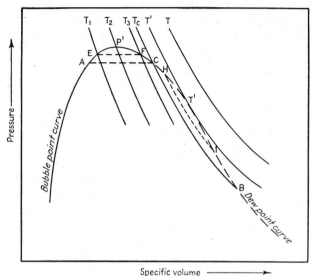

Fig. 2.7.   Typical isotherms of hydrocarbon mixtures in the neighborhood of the critical region.

point pressure $P(C)$ is no longer the highest possible for the coexistence of two phases.   For obviously by a slight reduction in pressure below P', at the temperature of P', T_2, a gas phase will appear and persist in equilibrium with the liquid at a pressure exceeding $P(C)$.   In fact, this situation obtains throughout the region bounded by $AP'CA$.

It will also be noted that the temperature of the isotherm through C, T_c, is lower than the maximum which permits a two-phase region, namely, that which is tangent to the dew-point curve, at T'.*   Thus we have here a region in which two phases can coexist even at temperatures exceeding that at C.   This region, bounded by $CT'BC$, also has no counterpart in the phase diagrams for one-component systems.

To see what occurs in these apparently anomalous regions one may focus attention upon an isobaric path, as EF.   Since E and F both lie on the bubble-point curve, the mixture at these points is entirely in the liquid

* This point T' is sometimes referred to as the "cricondentherm."

phase. In between, however, lies a two-phase region. Hence on increasing the temperature above T_1 a gas phase must develop. If this normal behavior were to continue as the temperature is raised to T_3, the amount of the gas phase would continually increase. But as F is also a bubble point, there can no longer be a gas phase when T_3 is reached. Thus it is clear that the growth of the gas phase on proceeding from E to F must have ceased at some intermediate temperature and that from there on to T_3 the gas-phase component decreased and finally disappeared altogether at T_3, that is, at F. As this latter behavior is the reverse of that encountered universally in one-component systems, and generally at lower temperatures even with multicomponent mixtures, it has been described as "retrograde." The whole process of passing from E to F is termed an "isobaric retrograde vaporization," although the actual retrograde phenomena take place only over part of the path.

An analogous "retrograde" phenomenon takes place within the region bounded by $CT'BC$. Here, for example, in traversing the isothermal path IH the rise in pressure above the dew point I first leads to a condensation of liquid. However, this liquid phase does not continue to grow throughout the path. For as H is also a dew point, the liquid formation must cease somewhere between I and H, and then on further rise of pressure the liquid phase must shrink and finally disappear as H is reached. This latter process is also "retrograde" when compared with the more familiar behavior in which the liquid-phase condensation continues as the pressure is increased. Again, although this retrograde behavior is limited to only part of the path IH, the whole process of passing from I to H is termed "isothermal retrograde condensation."[1]

While the p-v-T diagram of Fig. 2.7 is typical of many which show the retrograde phenomena, it does not represent the only conditions under which they may occur. Other such conditions may be conveniently illustrated by pressure-temperature diagrams, as shown in Fig. 2.8. For purposes of clarity only the two-phase boundary dew-point and bubble-point curves have been drawn. For pure components these two curves would, of course, coalesce into a single curve on the pressure-temperature plane and represent merely the vapor-pressure characteristic of the pure hydrocarbon (cf. Fig. 2.5).

Figure 2.8*a* gives the p-T diagram equivalent of that of Fig. 2.7. Here the critical point C lies between the point of maximum pressure P' and that of maximum temperature T'. In the region $BCT'B$, processes of

---

[1] The isobaric or isothermal paths are, of course, not the only ones exhibiting the retrograde behavior. Any path lying in the general retrograde-phase-diagram area with a monotonically changing basic variable will have a retrograde segment.

isothermal retrograde condensation occur, and those of isobaric retrograde vaporization occur in the region $ACP'A$.*

In Fig. 2.8b the critical point C lies at a pressure below both P' and T', and both P' and T' lie on the bubble-point curve. Hence for this case the region $CBT'C$ encloses the paths of isothermal retrograde vaporization, while $ACT'P'A$ borders the paths of isobaric retrograde vaporization.

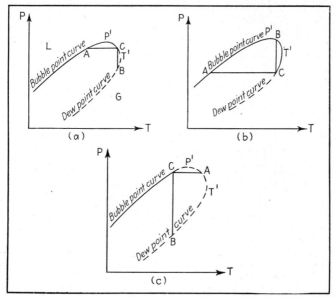

Fig. 2.8. Various types of pressure-temperature phase-diagram boundary curves giving rise to retrograde behavior.

When the maximal pressure P' lies on the dew-point curve and the critical pressure lies between P' and that for T', the retrograde regions are as indicated in Fig. 2.8c. Here isothermal retrograde condensation will occur along any complete vertical path bounded by $BCP'T'B$. In the region $CAP'C$, horizontal paths will lead to isobaric retrograde condensation. No significant change in the type of behavior will arise if the critical point in Fig. 2.8c is shifted to fall below the pressure for T'.

It is to be emphasized that the dew-point and bubble-point curves, joining at the critical point as shown in Fig. 2.8, enclose only the region of coexistence of two phases. Beyond and outside these boundary curves the

---

* For definiteness, the retrograde behavior in passing between points on a dew-point curve will be termed "condensation," whereas that occurring during passage between points on the bubble-point curve will be termed "vaporization." The exact terminology pertaining to retrograde phenomena is, however, not formally established as yet.

hydrocarbon mixture is in a single phase. From a strict thermodynamic standpoint, that is all that can be said about it. For continuous paths can be traversed from points such as G, in Fig. 2.8a, below the dew-point curve to L above the bubble-point curve, entirely outside of the border curves, without encountering any discontinuities in phase or the development of phase boundaries. Nevertheless, the region in the neighborhood of G is generally visualized as representing a gas phase, and that near L as a

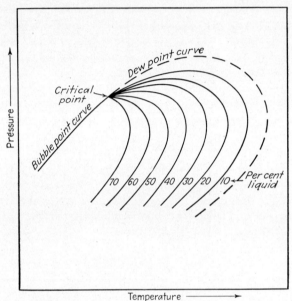

Fig. 2.9. Diagrammatic phase relations near the critical point showing the liquid-gas volume distribution.

liquid phase. For practical purposes such a distinction may be safely made at temperatures and pressures appreciably removed from those of the critical point. For until the critical conditions are approached the intensive properties of the single-phase fluids in the neighborhood of G and L are generally so different that the descriptive designations "gas" and "liquid" phases may be accepted as appropriate and should be free from misinterpretation. It is with this understanding that the terms gas and liquid are used throughout this work.

As will be clear from the previous discussion, the dew-point and bubble-point curves forming the two-phase boundary, as shown in Fig. 2.8, represent states in which the hydrocarbon system is either 100 per cent gaseous or 100 per cent liquid, respectively. In the region enclosed by such boundary curves the composite fluid system will be distributed between the gas and liquid phases in a manner illustrated qualitatively in Fig. 2.9, for a system corresponding to Fig. 2.8c. The course of the retrograde processes may be readily followed by noting the sequence of intersections of the constant-liquid-fraction curves as various paths are traversed with terminal points on the dew-point curve.

Although the retrograde phenomena, by the very nature of the term "retrograde," are frequently considered as anomalous and exceptional, they are, in fact, almost universal accompaniments of the phase behavior of multicomponent systems near their critical points. In fact, the retrograde behavior[1] can be entirely absent only if the critical point be at the same time the point of maximum pressure and maximum temperature for the coexistence of two phases, a situation that would involve an acute angled junction between the dew-point and bubble-point curves. Indeed, it is only because the operations commonly carried out with most fluid systems lie far removed from the critical regions that the retrograde processes do not occur as frequently as the familiar "normal" behavior. It should also be noted that an integral part of the appearance of the retrograde phenomena is the role played by the critical point as the junction of the dew-point and bubble-point curves (identity of gas and liquid phases), rather than as a limiting condition for the occurrence of two coexistent phases. In fact, as will be seen from Figs. 2.8 and 2.9, the isothermal retrograde processes can take place only at temperatures above the critical and below the maximum two-phase temperature—the cricondentherm. Likewise, the isobaric retrograde phenomena will be observed only between the critical pressure and the maximum two-phase pressure—the cricondenbar.

It will be recalled that at the dew point the gas-phase composition is that of the system as a whole. The same is true of the liquid phase at the bubble point. Within the two-phase region no such simple rules apply. Thus in passing along an isotherm from the dew point to the bubble point—hence, out of the retrograde region—the condensed-liquid phase will at first be relatively rich in the less volatile, or heavy, component of the mixture. As more liquid condenses, more of the volatile component goes into the liquid phase and the latter thus becomes less dense. At the same time, however, the heavy component in the gas continues to liquefy so as to make the gas phase lighter, except for the direct pressure effect in increasing the gas-phase density. This behavior is illustrated in Fig. 2.10, in which are plotted the calculated[2] composition, in mole fractions of the gas and liquid phases, of a mixture of propane and n-

---

[1] This refers to its limited sense in which either the temperature or the pressure is kept fixed while the other varies monotonically between the terminal points of the path.

[2] These calculations were made by an application of equilibrium ratios (cf. Sec. 2.9).

pentane of equal molar composite composition. The curves of Fig. 2.10 refer to a temperature of 250°F, for which the dew point is 275 psi and the bubble point is 440 psi. The total mole fraction in the liquid phase is also plotted in Fig. 2.10. It will be observed that, whereas at the dew point the gas phase has the composition of the mixture, the liquid phase which would be in equilibrium with the saturated gas will contain but 23.1 mole per cent of the propane and 76.9 mole per cent of *n*-pentane. Likewise, at the bubble point, where the liquid-phase composition is the same as that of the mixture, a gas phase in equilibrium with it will contain 72.8 mole per cent of propane and but 27.2 mole per cent of *n*-pentane.

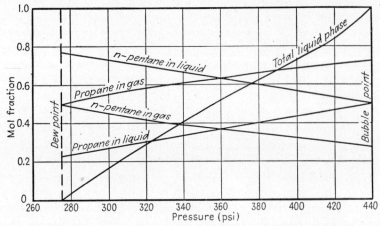

FIG. 2.10. The calculated variation of the gas and liquid compositions of a 50–50 mixture of propane and *n*-pentane, at 250°F, in passing from the dew point to the bubble point.

It may be noted here that the retrograde-condensation phenomenon provides the controlling mechanism for oil recovery from so-called "condensate fields." Whereas the fluid content of crude-oil reservoirs is generally comprised mainly of a liquid hydrocarbon phase at or above the bubble point, with or without an overlying "dry"[1] gas phase, condensate-producing reservoirs[2] contain a gas ("wet" gas) at or above the dew point. It is the formation of a liquid phase by retrograde condensation through paths in phase diagrams such as that of Fig. 2.9, as the reservoir fluid rises up the flow string and is reduced in temperature and to separator or stock-tank pressure at the surface, which is the source of the "condensate"

---

[1] The terms "dry" or "lean" and "wet" are used to indicate, respectively, that the gas has a low or high content of liquefiable hydrocarbons in the molecular-weight range of gasoline or heavier.

[2] In practice, condensate reservoirs usually constitute merely gas caps over crude-oil zones, although it is often convenient to treat them as distinct reservoirs.

production. As the terminal point (atmospheric conditions) falls in the two-phase region of the phase diagram, the retrograde-condensation process is not completely reversed and a residual liquid phase is recovered at the surface even without special processing of the produced fluid. While such phenomena are not to be considered as abnormal from a physical stand-

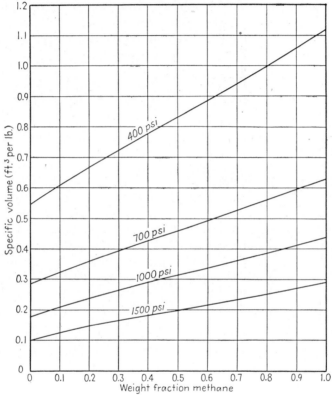

FIG. 2.11. The variation of the specific volume with methane concentration at 220°F in methane-ethane mixtures, at various pressures.

point, they actually are of little importance in the majority of oil-producing reservoirs. Accordingly, except when otherwise explicitly indicated, it will be assumed hereafter that the hydrocarbon systems refer to combinations of crude oil and "dry" gas, and the behavior of condensate-producing reservoirs will be discussed separately (cf. Chap. 13).

2.6. The Effect of Composition on the Phase Behavior of Binary Systems.—The considerations thus far have referred to the thermodynamic behavior of hydrocarbon binary systems of fixed over-all composition. The physical variables that suffice to fix a particular state of such mixtures are the pressure and temperature. It is the variations in these that give rise to the whole complex of volume and phase characteristics that together constitute the real thermodynamic description of the system. However, to complete the discussion it is necessary to take into account also the role played by the over-all composition which ultimately defines the actual hydrocarbon system under consideration. Here again it will not be possible to enter into the quantitative aspects of the subject, as these will vary with every binary system that could be formed from the simple hydrocarbons. Rather, we shall review only the qualitative effects on the thermodynamic properties of typical binary mixtures of changes in their composition.

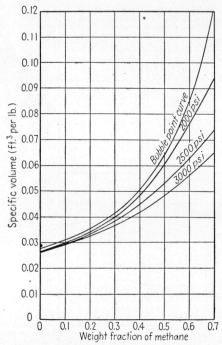

FIG. 2.12. The variation of the specific volume with methane concentration of methane-*n*-butane mixtures in the liquid phase at 70°F.

As might be anticipated on the basis of general principles, the volume of the gas phase of binary mixtures is essentially additive under moderate temperature and pressure conditions. That is, the resultant specific volumes of binary systems will approximate the arithmetic average of that of the two components, weighted according to the composition. This is illustrated for a mixture of methane and ethane in Fig. 2.11.[1]

For binary systems in the liquid phase the rule of additivity of volumes is usually well satisfied at temperatures below the criticals of both components. However, when the temperature is above the critical temperature of the more volatile component, the additivity rule may be but a poor approximation, as shown in Fig. 2.12[2]. for the case of methane and *n*-butane. Here the deviations from linearity—or the additive relationship—are quite marked even at pressures of 3,000 psi. Since even at 70°F the lighter component, methane, is far above its critical temperature, it tends to persist in its behavior as a single-phase gaseous fluid in spite of its

[1] Fig. 2.11 was calculated from Z data of B. H. Sage and W. N. Lacey, *Ind. and Eng. Chemistry*, **31**, 1497 (1939).

[2] This is plotted from data of B. H. Sage, R. A. Budenholzer, and W. N. Lacey, *Ind. and Eng. Chemistry*, **32**, 1262 (1940).

solution in the heavier component, which would normally be a liquid at
70°F and 2,000 to 3,000 psi.

The concept of a binary mixture as the equivalent of a pure component
with properties that are simple averages of those of its two constituents
becomes almost entirely meaningless—and at best highly artificial—in the
two-phase region and at conditions approaching the critical. For only by
hypothetical extrapolations into unstable regions could one attempt to

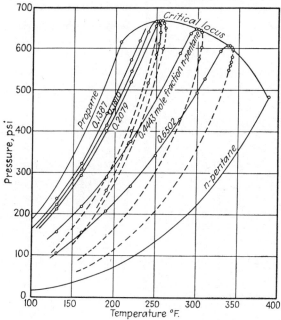

Fig. 2.13.   The pressure-temperature phase diagram for the propane-*n*-pentane system.
(*After Sage and Lacey, Ind. and Eng. Chemistry, 1940.*)

calculate an average property of a binary mixture under conditions where
the two separate components would be in different phases. In fact, the
only practical approach to the problem is then the empirical one.

An example of the direct experimental determination of the properties
of a set of binary systems of varying composition is shown in Fig. 2.13,[1]
which gives the boundary curves for various mixtures of propane and *n*-
pentane. These curves pass continuously from the vapor-pressure curve
of the one pure component to that of the other as the composition is varied.
However, no simple averaging procedure will give quantitatively the
properties of the mixtures as functions of their compositions. This will be
clear on reference to Fig. 2.14[2] giving the specific weights (weight per

[1] B. H. Sage and W. N. Lacey, *Ind. and Eng. Chemistry*, **32**, 992 (1940).
[2] *Ibid.*

unit volume) of the dew-point gas and bubble-point liquid, corresponding to the boundary curves of Fig. 2.13, as functions of the *n*-pentane content of the mixture.   Another example of this situation, when referring more directly to the critical properties, is provided by Fig. 2.15[1] for methane-*n*-butane mixtures.   Here the characteristic pressures obviously follow no linear variation with the methane concentration.   And even the ap-

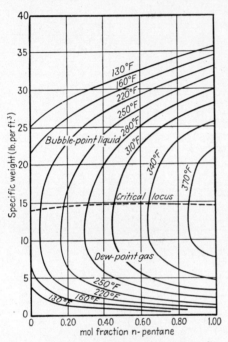

Fig. 2.14.   The specific weight-composition diagram for the dew-point gas and bubble-point liquid in propane-*n*-pentane mixtures.   (*After Sage and Lacey, Ind. and Eng. Chemistry, 1940.*)

proximately linear behavior of the critical temperature is of little significance, as the line will not pass through the limiting data for 100 per cent methane without the development of marked curvatures.

As indicated in Figs. 2.13 and 2.14, the critical parameters of binary mixtures of varying composition form continuous curves, termed the "critical locus."   In general, the critical temperature, as well as the cricondentherm, increases as the concentration of the less volatile component increases. The critical pressure in general rises to a maximum, higher than that of either pure component, as the concentration of either is increased from zero.

A summarized graphical representation of the critical pressures and temperatures for binary mixtures of the lower paraffin hydrocarbons is shown in Fig. 2.16.[2]   It

will be noted that the elevation of the maximal critical pressure over those of the pure components increases as the two components become increasingly dissimilar.   From the general nature of the critical locus it will be clear that, on changing composition so as to pass from one side of the maximum to the other, the character of the retrograde behavior in the critical region will change either from that corresponding to Fig. 2.8*a* to that of Fig. 2.8*c*, or vice versa.

[1] B. H. Sage, B. L. Hicks, and W. N. Lacey, *Ind. and Eng. Chemistry*, **32**, 1085 (1940).

[2] B. H. Sage and W. N. Lacey, "Volumetric and Phase Behavior of Hydrocarbons," p. 94.

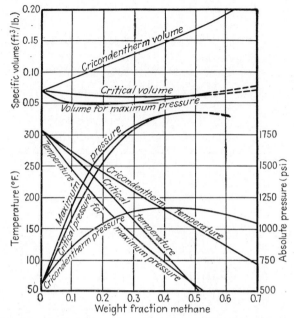

Fig. 2.15.   The effect of the composition of methane-*n*-butane systems on the pressure, temperature, and specific volume at the critical state, and points of maximum temperature and maximum pressure.   (*After Sage, Hicks, and Lacey, Ind. and Eng. Chemistry, 1940.*)

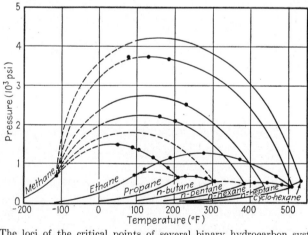

Fig. 2.16.   The loci of the critical points of several binary hydrocarbon systems.   (*After Sage and Lacey, "Volumetric and Phase Behavior of Hydrocarbons."*)

A convenient way of following the effects on the boundary properties of changing the composition of a binary system, as well as the effect on the individual phases when the pressure or temperature is varied in a binary mixture of fixed composition, is illustrated in Fig. 2.17. Here are plotted diagrammatically the bubble-point and dew-point pressures vs. composition for a fixed temperature. A little consideration will show that the terminal points of constant-pressure segments on these curves represent the compositions of coexisting gas and liquid phases.[1] Hence when a binary system

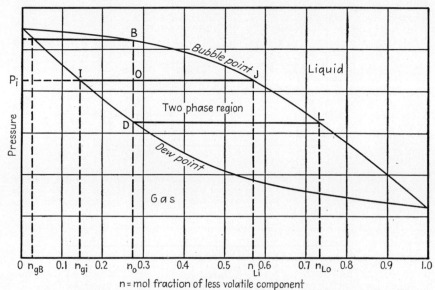

Fig. 2.17.   Illustrative pressure-composition phase boundary curves for a binary system.

defined by the composition n_o is at its dew point D, the liquid phase that would condense first on raising the pressure will have a composition n_{Lo}. At a pressure, as P_i, intermediate between the bubble point and dew point for the original composition, the gas-phase composition will be given by n_{gi} and that of the liquid phase by n_{Li}. Both of these will be lower than the corresponding values at the dew point. Finally when the bubble point B is reached, the liquid will have the over-all composition n_o and the last trace of gas will have taken on a composition n_{gB}. This sequence of composition changes as the pressure is raised means that both the gas and liquid phases become richer in the more volatile component in passing from the dew point to the bubble point.

[1] The same is true of the intersections of the dew-point and bubble-point curves in *p-T* diagrams, as in Fig. 2.13.

The effect of changing the over-all composition can be readily visualized by imagining the dew point or bubble point D or B to slide along its curve.   Thus as the concentration n of the less volatile component is increased in the system as a whole, so is it in the two phases, as is to be expected.   Moreover, as the concentration of either component becomes large as compared with the other, the spread in the composition of the two phases tends to diminish.   It may also be noted that the relative amounts of the gas and liquid phases may be easily determined from a diagram such as Fig. 2.17.   From elementary considerations it may be shown that the mole fraction of the whole system in the liquid phase for a pressure as P_i is

$$L = \frac{n_o - n_{gi}}{n_{Li} - n_{gi}} = \frac{OI}{IJ} \tag{1}$$

The mole fraction in the gas phase is, similarly,

$$G = \frac{n_{Li} - n_o}{n_{Li} - n_{gi}} = \frac{OJ}{IJ}, \tag{2}$$

so that

$$\frac{L}{G} = \frac{OI}{OJ}. \tag{3}$$

A similar graphical representation of the effect of over-all composition on the compositions of the separate gas and liquid phases is given by an isobaric temperature-composition diagram, as illustrated in Fig. 2.18. The composition variable also in Fig. 2.18 refers to the less volatile component.   The compositions of the coexistent gas and liquid phases are here, too, given by the terminal points on the dew-point and bubble-point curves of constant-temperature segments.   From this it follows that as the temperature is raised and the system vaporizes, both the gas and liquid phases become richer in the less volatile component.

The relative amounts of gas and liquid phases at any temperature between the dew point and bubble point are given here by a formula similar to that applying to Fig. 2.17.   Thus

$$L = \frac{OI}{IJ}; \qquad G = \frac{OJ}{IJ}, \tag{4}$$

and

$$\frac{L}{G} = \frac{OI}{OJ}. \tag{5}$$

It may also be noted from Fig. 2.18, as well as Fig. 2.17, that the compositions of the coexisting phases at any fixed pressure and temperature, as given by the points I and J, are independent of the relative amounts of the two components in the binary system as a whole as long as the pressure and temperature for the over-all composition do not lie outside of the boundary curves. The over-all composition, of course, will control the relative amounts of the gas and liquid phases. In fact, within the limits already indicated the composition of the mixture as a whole can be varied arbitrarily by changing the volumes of the separate phases while keeping their individual compositions fixed, corresponding to the assigned values of the pressure and temperature.

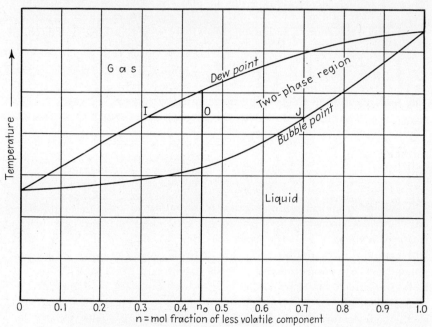

FIG. 2.18.   Illustrative temperature-composition phase boundary curves for a binary system.

From the fact that the diagrams of Figs. 2.17 and 2.18 covered the whole composition range it is clear that they referred to pressures and temperatures which were below the critical values for both components. When, however, this condition no longer obtains, the composition diagrams cannot extend over the whole composition range. Thus for temperatures lying between the critical temperatures of the two components the pressure-composition curve would be similar to that shown in Fig. 2.19. As the temperature for Fig. 2.19 is higher than the critical temperature for the light component, the border curves do not exist in the region of $n = 0$. In fact, there will be no two-phase region for the mixture in question and for the given temperature until the concentration of the heavier component exceeds n'.

For concentrations of the heavier component exceeding n' the interpretation of the diagram of Fig. 2.19 is similar to that of Fig. 2.17. In particular, for a system of over-all composition n_o the dew point will be at D and bubble point at B. The liquid-phase composition will vary from n_L to n_o, continually becoming leaner in the heavier component as the pressure is raised. In the case of the gas phase, however, the decrease in concentration of the heavy component proceeds only to $n'(P_i)$ and then increases to n_g at the bubble point.

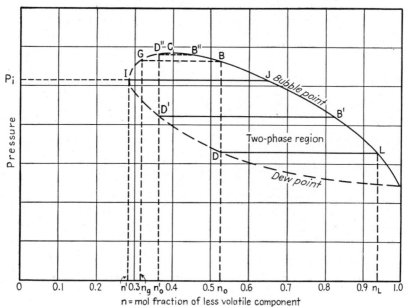

FIG. 2.19. Illustrative pressure-composition phase boundary curves for a binary system at a temperature between the critical temperatures of the components.

A radically different behavior is shown by a mixture whose over-all composition is n'_o, with original dew point at D'. Here, as the pressure is increased to P_i, the normal condensation takes place, with increasing molar fractions in the liquid phase. But as the pressure continues to rise, a point is reached where the amount of liquid phase begins to shrink and ultimately vanishes entirely as the second dew point D'' is reached. The last liquid phase remaining, at B'', is less rich in the heavy fraction than is that at B', although it is still richer than the system as a whole.

It will be recognized that the process of passing between D' and D'' is one of isothermal retrograde condensation, previously encountered in the consideration of the p-T diagrams. A similar qualitative analysis can be made of the temperature-composition diagram. Likewise, isothermal or

isobaric volume-composition diagrams may be used to represent the same
basic thermodynamic data. However, they are all essentially equivalent,
and the ones presented above should suffice to illustrate the significant
thermodynamic features of binary systems.

2.7. Multicomponent Systems; General Characteristics.—As previously
indicated, the new qualitative features introduced on passing from binary
mixtures to the complex multicomponent systems occurring in practical
crude-oil-production operations are rather minor.[1] However, the quantitative description of such multicomponent systems requires a considerable
extension of the simpler methods used for binary mixtures. The first
problem is that of specifying and identifying the particular hydrocarbon
system of interest. This arises from the practical limitations in analyzing
or separating complex hydrocarbon mixtures into all the individual components. Moreover, it is frequently not feasible to carry out such analyses
even to the limited extent that is physically possible. It is therefore
necessary to use empirical approximations and representations.

The compositions of the separate gas and liquid phases may be expressed
as either the weight or the mole fraction of the individual hydrocarbon
components. For practical purposes these are limited to pentanes, hexanes,
or heptanes, although for special purposes the fractionation may be carried
through the octanes or nonanes. In the case of liquids the residue beyond
the last hydrocarbon separated is generally characterized by a single
parameter, such as its average molecular weight. At moderate pressures
and temperatures the gas phase usually contains so little of the heavy
hydrocarbons that they may be conveniently lumped together with the
residue from the actual fractionation and denoted as "pentanes and
heavier," "heptanes and heavier," etc.

When the hydrocarbon system is entirely in the gas phase and at pressures and temperature not too near the critical values, its pressure-volume
behavior may be described by its deviation factor Z.* This is defined
in a manner quite similar to that previously introduced for the pure hydrocarbons, namely, by the equation

$$pv = \frac{ZRT}{M},\qquad(1)$$

---

[1] A radically new phenomenon sometimes encountered in multicomponent systems is
the presence of two liquid phases [cf. D. L. Katz, D. J. Vink, and R. A. David, *AIME
Trans.*, **136**, 106 (1940); D. F. Botkin, H. H. Reamer, B. H. Sage, and W. N. Lacey,
"Fundamental Research on Occurrence and Recovery of Petroleum," pp. 62 (1943)
and 42 (1944), API]. However, as such systems have not been fully explored as yet
and apparently occur only under rather specialized conditions, no attempt will be made
to discuss them here.

* As previously noted, Z is also often termed the "compressibility" or "supercompressibility factor."

where v refers to the specific volume of the gas (in cubic feet per pound) and $\overline{M}$ is its average molecular weight. This may be calculated from its mole-fraction composition as

$$\overline{M} = \Sigma n_i M_i, \tag{2}$$

where M_i is the molecular weight of the ith component, of mole fraction n_i. Or it may be computed directly from its specific gravity d, with respect to air at 60°F and 1 atm, by the formula

$$\overline{M} = 28.97d. \tag{3}$$

While Z, as defined by Eq. (1), can be readily determined experimentally, it is sufficient for most practical purposes to use the empirical correlations based on the so-called "pseudocritical" constants.[1] These are average critical constants of the system, weighted according to composition, *i.e.*,

$$p_c = \Sigma n_i p_{ci}; \qquad T_c = \Sigma n_i T_{ci}, \tag{4}$$

where p_{ci}, T_{ci} are the critical pressure and temperature of the pure ith component. By analogy with common thermodynamic usage the "pseudo-reduced" pressures[2] p_r and temperatures T_r are then defined by

$$p_r = \frac{p}{p_c}; \qquad T_r = \frac{T}{T_c}. \tag{5}$$

The quantities p_r and T_r having thus been determined, the value of Z is read from the chart, shown in Fig. 2.20,[3] which has been found to give predictions in close agreement with direct measurements. This chart is a correction of that obtained originally for methane.

If the detailed composition of the gas is not known, the pseudocritical constants may be estimated from an additional empirical correlation between these constants and the gravity of the gas, as shown in Fig. 2.21.[4] The values of p_r and T_r may then be again calculated by Eq. (5), and the value of Z read from Fig. 2.20.

---

[1] W. B. Kay, *Ind. and Eng. Chemistry*, **28**, 1014 (1936); G. G. Brown and D. E. Holcomb, *Petroleum Eng.*, **11**, 23 (February, 1940); M. B. Standing and D. L. Katz, *AIME Trans.*, **146**, 140 (1942).

[2] The use of "reduced" pressures and temperatures is an application of the "law of corresponding states."

[3] Cf. Standing and Katz, *loc. cit.* For an even more direct, though somewhat more approximate, method for determining Z an empirical chart and correction table devised by A. D. Brokaw [*AIME Tech. Pub.* 1375 (1941)] can be used. The empirical correlation charts for calculating deviation factors become somewhat inaccurate when the gas contains high concentrations of nitrogen or the intermediate hydrocarbons. Experimental data on the compressibility factors of nitrogen and natural gases containing appreciable concentrations of nitrogen have been recently reported by C. K. Eilerts, H. A. Carlson, and N. B. Mullens, *World Oil*, **128**, 129, 144 (June, July, 1948).

[4] Standing and Katz, *loc. cit.*

While in principle all the properties of a multicomponent liquid mixture are of necessity determined by a statement of the composition, there are for practical purposes no satisfactory simple means available for predicting its behavior from the composition. For many purposes, therefore, the gross properties themselves are used to characterize complex hydrocarbon

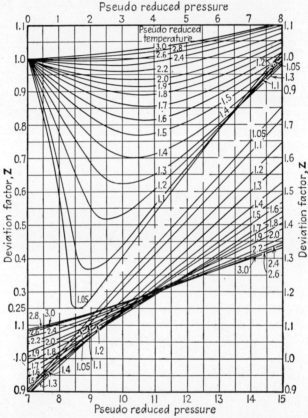

FIG. 2.20. A deviation-factor chart for hydrocarbon gases. (*After Standing and Katz, AIME Trans., 1942.*)

liquids. The most common of these are the gravity,[1] or density, and the average molecular weight. The latter is often determined by the freezing-point lowering of benzene. A number of gross properties of hydrocarbon liquids such as viscosity, thermal-expansion coefficient, and compressi-

---

[1] The "gravity" of an oil is generally expressed in "API degrees," computed from the specific gravity at 60°F, ρ, by the relation °API = $(141.5/\rho) - 131.5$. Conversely, $\rho = 141.5/(131.5 + °API)$. Thus API gravities ranging from 10 to 100° correspond to the range in specific gravity of 1.0000 to 0.6112.

bility show definite trends of variation with these parameters. Thus the viscosity generally rises, and compressibility, thermal-expansion coefficient, and gas solubility decrease as the density and average molecular weight increase. However, the quantitative features of such variations are not thus uniquely determined, and moreover they may be affected even more seriously by pressure and temperature than by the gravity alone.

In the consideration of the two-phase region of general multicomponent hydrocarbon systems a detailed graphical representation of the composition behavior would be entirely impractical. Fortunately, however, it is generally feasible with ordinary crude oils and gases to describe many of the physical characteristics and their variations in terms of the gross phases themselves. For these purposes the composite systems are defined by the relative amounts of gas phase and liquid phase, under normal pressure and temperature, of which the whole is composed. Such a phase composition may be expressed as mass percentages of the gas and oil (liquid) or more frequently in volumetric terms, commonly designated as the "gas-oil ratio," *i.e.*, the volume of gas (in cubic feet) associated with a unit volume (in barrels) of oil.

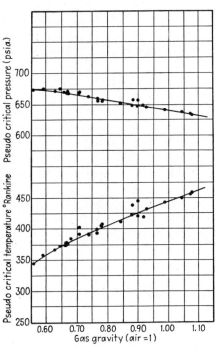

FIG. 2.21. The variation of the pseudocritical constants with the gas gravity. (*After Standing and Katz, AIME Trans., 1942.*)

Of course, the gas and oil are themselves further identified by a statement of their compositions, if known,[1] or frequently merely by their densities or gravities (with respect to air for the gas, and in API degrees for the oil). Moreover, these will change as the temperature and pressure are raised and gas dissolved in the oil changes. Not all the gas components will dissolve to an equal degree, and the liquid-phase components will enter the gas phase to varying degrees. Indeed, as is well known, in the case of ordinary crude-oil and gas hydrocarbon mixtures the gas will become leaner in the heavier components as the pressure is raised.

[1] In that case the gas-oil ratio also serves to define the composition of the composite system.

Nevertheless, these complications need not be taken into account explicitly except when the actual phase compositions are of primary interest.

The various significant features of the two-phase behavior of complex hydrocarbon mixtures can be best demonstrated by reference to specific empirical examples. Because of the difficulty and labor involved in getting the pertinent experimental data, the ranges of the variables covered are generally limited to those of major practical interest and do not cover the complete two-phase region. In fact, most of the data have usually been taken in the vicinity of the bubble-point curves. Because of the high boiling points of the heavier components of the crude oils, it is virtually impossible to obtain satisfactory dew-point data for such systems. Thus in Fig. 2.22[1] are shown typical volume-pressure isotherms for a mixture of gas and oil (39.9°API) from the Rio Bravo field, Kern County, Calif., near the bubble-point curve. The type of volume-temperature isobars generally observed are illustrated by Fig. 2.23,[2] which were obtained using a Dominguez field, Los Angeles County, Calif., crude oil (33.9°API) and 5.6 mass per cent of gas. The significant features of these sets of curves are so obvious as to need no discussion.

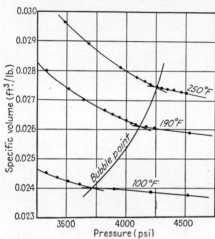

FIG. 2.22. The specific volumes near the bubble point of a mixture of oil and gas from the Rio Bravo field, containing 0.16 weight fraction of separator gas. (*After Sage and Reamer, AIME Trans., 1941.*)

While we shall not attempt to enter here into a detailed treatment of the changing composition of the gas and liquid phases as the pressure or temperature is varied, an indication of their over-all variation is provided by Fig. 2.24.[3] Here the gravities of the coexisting phases are plotted against the pressure, for a fixed temperature, for a mixture of a natural gas and crude oil. It will be observed that with increasing pressure the gas density rises and that of the liquid decreases. The former evidently is mainly a direct pressure effect, and the latter is due to the increased solution of gas in the liquid.

[1] B. H. Sage and H. H. Reamer, *AIME Trans.*, **142**, 179 (1941).

[2] B. H. Sage and W. N. Lacey, *Ind. and Eng. Chemistry*, **28**, 249 (1936).

[3] B. H. Sage and W. N. Lacey, "Volumetric and Phase Behavior of Hydrocarbons," p. 210.

Another interesting aspect of the behavior of actual gas-oil mixtures pertains to the effect of over-all composition, pressure, and temperature on

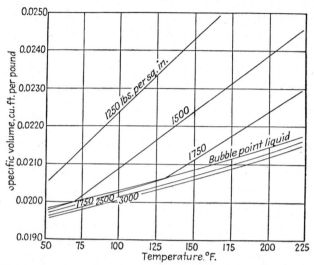

FIG. 2.23. The change of specific volume with temperature for a mixture of gas and oil from the Dominguez field, containing 5.61 weight per cent of gas. *(After Sage and Lacey, Ind. and Eng. Chemistry, 1936.)*

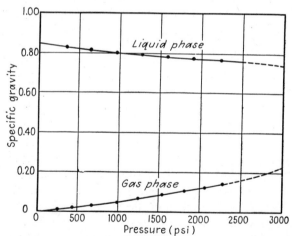

FIG. 2.24. The specific gravities of the coexisting phases in a mixture of natural gas and crude oil at 100°F as a function of pressure. *(After Sage and Lacey, "Volumetric and Phase Behavior of Hydrocarbons.")*

the "formation volume," which is the total volume occupied by a gas and oil mixture containing a unit volume of oil at standard conditions. For a fixed temperature and rather low gas-oil ratios the effect of gas-oil ratio is plotted for fixed pressures in Fig. 2.25[1] for the Dominguez oil previously, referred to (cf. Fig. 2.23). For a fixed pressure the effect of temperature is shown for the same system in Fig. 2.26.[2] The positive slope of the bubble-point curve indicates that the direct thermal expansion on temperature rise more than counterbalances the shrinkage due to the fall in gas-oil ratio and gas in solution.

The variation in gas in solution with pressure and temperature is illustrated by the curves of Fig. 2.27, obtained in recombination studies[3] on

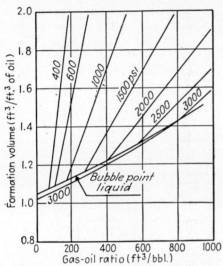

gas and oil (33.5°API) from the Oak Canyon field, Los Angeles County, Calif. It will be observed that the solubility decreases with increased temperature and that there is in this case an upward curvature at higher pressures. The initial steep rise in solubility is also to be noted.

These examples have been chosen to illustrate specific features of the general behavior of gas-oil mixtures. Data have been taken from different field studies to emphasize that the qualitative characteristics are much the same for all multicomponent hydrocarbon systems comprising gases and crude oils occurring in nature. Quantita-

FIG. 2.25. The formation-volume isobars of low gas-oil-ratio mixtures of oil and gas from the Dominguez field. (*After Sage and Lacey, API Drilling and Production Practice, 1935.*)

tively, however, the volumetric properties of gas-oil mixtures will vary with the nature of the crude oil and gas, as will be discussed in the next section.

2.8. The Prediction of the Volumetric Behavior of Gas and Crude-oil Systems.—The graphical data presented in the last section were obtained by experiments with actual gas and crude-oil systems. To determine such data quantitatively requires rather elaborate equipment and careful and tedious experimentation. We shall not enter here into a discussion of the experimental problems involved, except to point out that most of the published data on the general phase behavior of gas and oil mixtures have been obtained by recombining gas and oil samples taken from surface

[1] B. H. Sage and W. N. Lacey, *API Drilling and Production Practice*, 1935, p. 141.
[2] *Ibid.*
[3] H. G. Botset and M. Muskat, *Jour. Inst. Petroleum*, **30**, 351 (1944).

gas-oil separators and observing the volumetric and phase behavior as the pressure, temperature, or composition (gas-oil ratio) were varied. On the

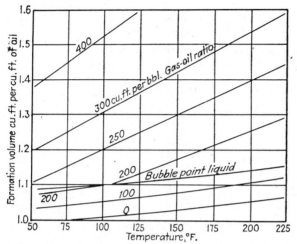

FIG. 2.26. The effect of temperature on the formation volume of low gas-oil-ratio mixtures of oil and gas from the Dominguez field, at a pressure of 1,000 psi. (*After Sage and Lacey, API Drilling and Production Practice, 1935.*)

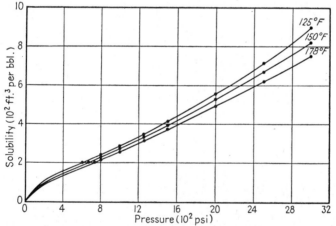

FIG. 2.27. The gas solubility vs. pressure curves for gas-oil mixtures from the Oak Canyon field, California. (*After Botset and Muskat, Jour. Inst. Petroleum, 1944.*)

other hand, a large part of the specific data available on the solubility of natural gases in their associated crude oils and on bubble-point liquid-formation volumes at reservoir temperatures have been obtained by the analysis of bottom-hole samples procured with suitable sampling equipment specially designed for this purpose. Although the necessary techniques are quite well established,[1] it is often necessary to make estimates of the fluid behavior when such analysis facilities are not available.

To predict the behavior of the dry gas phase alone it will in general suffice to apply the methods outlined previously in Sec. 2.7. In the case of the liquid phase, however, empirical correlations developed by Katz and Beal[2] on the basis of experimental data obtained on numerous actual

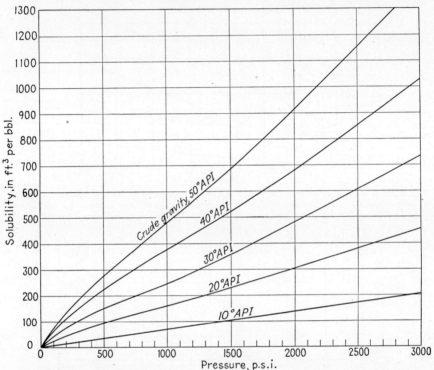

Fig. 2.28. The approximate variation of the solubility of natural gas in crude oils with pressure, at reservoir temperature. (*After Muskat and Taylor, Petroleum Eng., 1946.*)

[1] The latest form of the equipment used by Sage and Lacey and their collaborators in carrying through what are among the most accurate experiments performed of this type is described in *AIME Trans.*, **174**, 102 (1948). In addition to the extensive work of these authors, similar experiments have been reported by J. E. Gosline and C. R. Dodson, *API Drilling and Production Practice*, 1938, p. 423; Botset and Muskat, *loc. cit.;* and in a series of *U.S. Bur. Mines Rept. Inv.* by C. K. Eilerts and associates, *e.g.,* Nos. 3402, 3514, and 3642. The taking and analysis of bottom-hole samples are discussed by K. C. Sclater and E. R. Stephenson, *AIME Trans.*, 1929, p. 119; B. E. Lindsly, *AIME Trans.*, **92**, 252 (1931); P. G. Exline, *API Drilling and Production Practice*, 1936, p. 126; and D. L. Katz, *API Drilling and Production Practice*, 1938, p. 435.

[2] D. L. Katz, *API Drilling and Production Practice*, 1942, p. 137; C. Beal, *AIME Trans.*, **165,** 94 (1946).

gas–crude-oil systems should be quite useful. When only the saturation
pressure and crude-oil gravity are known, an estimate of the gas solubility

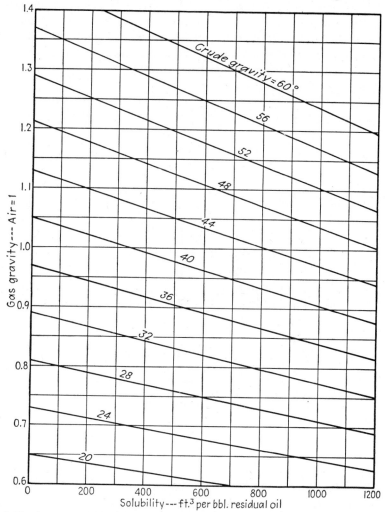

Fig. 2.29. A correlation chart between gas gravity, solubility, and crude-oil gravity. (*After
Katz, API Drilling and Production Practice, 1942.*)

may be made from Fig. 2.28, based on the curves of Beal but modified to
show the initial rise in solubility at low pressures.[1] These may involve

[1] M. Muskat and M. O. Taylor, *Petroleum Eng.*, **18**, 88 (December, 1946). While the
initial sharp rise in solubility is the usual occurrence, this is absent when the methane
concentration in the gas is exceedingly high (> 95 per cent), as has been observed in
several fields in Mississippi.

errors of about 25 per cent because of the entire neglect of gas gravity, reservoir temperature, and more detailed characterization of the crude oil. These curves do show, however, the correct orders of magnitude of the variation of solubility with pressure and with the gravity of the oil.

If the gas solubility is known or otherwise estimated, the gravity of the gas liberated from crudes of different gravity may be estimated from Fig. 2.29.[1] This, too, has been constructed from correlations of data from actual samples of natural gas and crude oil.

If the gas solubility is known, the oil shrinkage on gas evolution may be estimated from the curve in Fig. 2.30 with a probable error of 15 per cent.

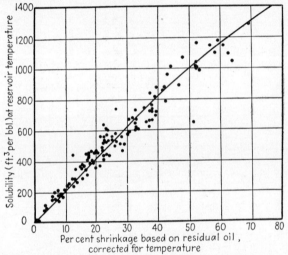

FIG. 2.30. Experimental data on the relation between the gas solubility and shrinkage of crude oils, corrected for the temperature of the residual oil. (*After Katz, API Drilling and Production Practice, 1942.*)

The shrinkage plotted in Fig. 2.30 is the percentage excess volume of the bubble-point liquid, at reservoir temperature and pressure, over its stock-tank (60°F) volume and hence equals 100 times the formation-volume factor for the bubble-point liquid minus 1.* The direct shrinkage effect of lowering the residual oil from reservoir temperature to 60°F has been taken into account in Fig. 2.30, according to the curves of Fig. 2.31. The latter also serve to show the increasing thermal expansion of the oil with increasing API gravity.

---

[1] Fig. 2.29, as well as Figs. 2.30–2.33 and 2.35, is taken from D. L. Katz, *API Drilling and Production Practice,* 1942, p. 137.

* A similar commonly used term is "shrinkage factor." This is the volume of stock-tank oil per unit volume of reservoir oil and hence equals the reciprocal of the formation-volume factor of the bubble-point oil.

When the gas gravity, crude gravity, gas solubility, and reservoir pressure and temperature are known, the oil shrinkage can be calculated with

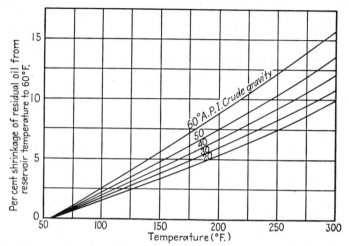

FIG. 2.31. The shrinkage in residual oils due to temperature changes. (*After Katz, API Drilling and Production Practice, 1942.*)

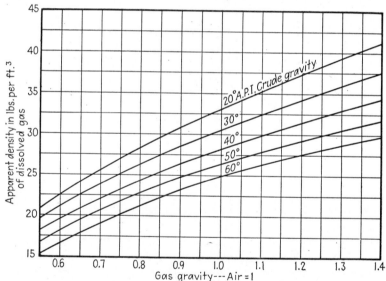

FIG. 2.32. The apparent density of natural gas dissolved in crude oils. (*After Katz, API Drilling and Production Practice, 1942.*)

a probable error of only 5 per cent, as outlined by Katz, as follows: The apparent density of the dissolved gas, as of 60°F and 1 atm, is first determined from the approximate empirical correlation curves of Fig. 2.32.

The total weight of dissolved gas in a barrel of crude divided by this apparent density gives the added volume, to the barrel of stock-tank oil, of the dissolved gas. The total weight of gas and oil divided by their combined volume gives the resultant density. The latter, which refers to 60°F and 1 atm, is corrected to reservoir pressure by use of Fig. 2.33

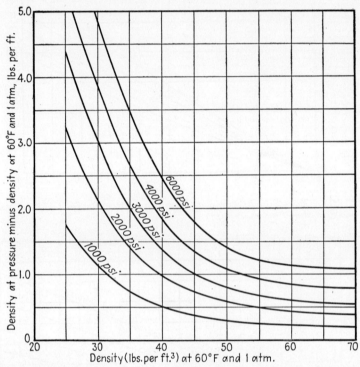

Fig. 2.33. The density change of crude oils with pressure. (*After Katz, API Drilling and Production Practice, 1942.*)

and to reservoir temperature with the aid of Fig. 2.34.[1] Dividing the resultant corrected density into the total weight of gas and oil will give the corrected bubble-point volume, and the percentage excess of the latter over the volume of 1 bbl will be the calculated shrinkage.

Finally, if the gas analysis is known in addition to the other physical data listed for the last procedure, the calculation of the oil shrinkage can be made even more precisely by adding separately the apparent liquid volumes for the individual gas components to obtain the composite liquid

[1] This is a revised plot of curves given by Katz, *API Drilling and Production Practice,* 1942, p. 137. In applying Figs. 2.32 to 2.34 account should be taken of the use of cubic feet as the volume unit in the expressions of density.

volume of the dissolved gas.   For this purpose the densities of all constituents heavier than ethane are taken as the normal densities[1] of the
pure components at 60°F and their vapor pressures, that is, 31.8, 36.1,
and 39.2 lb/ft³ for propane, the butanes, and the pentanes, and the appropriate density for the hexanes and heavier corresponding to the gravity of that

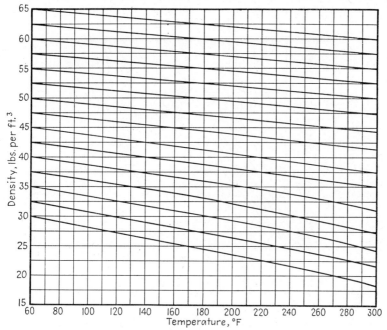

FIG. 2.34.   The density variation of crude oils with temperature.

residual component.   The effective densities of ethane and methane are
then determined as follows: From the total volume and weight of the
heavier than ethane components, an apparent density of the latter is
obtained.   The weight per cent of the ethane in the ethane and heavier
part of the liquid phase is next computed, and the apparent density of the
ethane is then determined by reference to the group of curves to the right of
Fig. 2.35.   In a similar manner one then computes the apparent density
of the ethane and heavier components, together with the weight per cent
of the methane in the whole system.   Reference to the group of curves

[1] These density values, which are also used below, are averages of those commonly
cited rather than precise, well-established constants.   The subsequent illustrative
calculations are themselves inherently of an approximate nature and are not to be
considered as accurate to the full ranges of the significant figures listed.

to the left of Fig. 2.35 gives the apparent methane density. With this value determined, the total volume occupied by and the density of the liquid phase, including the dissolved gas, may be computed. Upon correcting the latter for the reservoir pressure and temperature by Figs. 2.33 and 2.34, the final reservoir density is obtained. By converting this to the liquid-phase volume the formation-volume factor and shrinkage are immediately obtained.

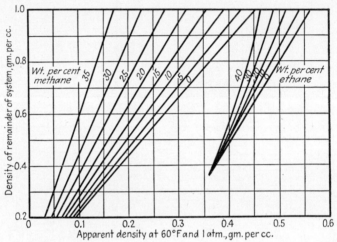

Fig. 2.35. The apparent density of ethane and methane in the liquid phase. (*After Katz, API Drilling and Production Practice, 1942.*)

To illustrate the rather involved procedure just outlined, the shrinkage will be calculated for the following hypothetical example:

Gas liberated.....................700 ft³/bbl
Crude gravity......................35°API
Reservoir temp.....................150°F
Reservoir (saturation) pressure.......2,500 psi

GAS ANALYSIS

	Mole %
Methane...........................	75
Ethane............................	10
Propane...........................	6
Butanes...........................	6
Pentanes..........................	2
Hexanes plus......................	1

The procedure for calculating shrinkage may then be tabulated as follows:

Component	Ft³/per 700 ft³	Mol. wt.	Wt., lb	Density in liquid phase		Vol. of liquid phase, ft³
				Lb/ft³	Gm/cc	
Methane.......	525	16	22.16	21.2	0.340	1.045
Ethane........	70	30	5.54	31.5	0.505	0.176
Propane........	42	44	4.88	31.8	.....	0.153
Butanes........	42	58	6.43	36.1	.....	0.178
Pentanes.......	14	72	2.66	39.2	.....	0.068
Hexanes plus....	7	90	1.66	42.1	.....	0.039
Crude oil.......	5.61	..	297.19	53.0	.....	5.610
Total........	......	..	340.52	46.85	.....	7.269
Propane plus	......	..	312.82	51.72	0.828	6.048
Ethane plus	......	..	318.36	51.15	0.819	6.224

As the weight per cent of the ethane in the ethane plus is 1.74 per cent, Fig. 2.35 gives for the apparent ethane density 0.505 gm/cc = 31.5 lb/ft³, which are the values listed in the table. It being noted now that the density of the ethane plus thus equals 0.819 gm/cc and the weight per cent of methane in the total equals 6.51 per cent, Fig. 2.35 shows the apparent methane density to be 0.340 gm/cc = 21.2 lb/ft³. On inserting this in the density column above, the methane volume becomes 1.045 ft³. The total volume is then 7.269 ft³, corresponding to a density at 60°F and 1 atm of 46.85 lb/ft³. Correcting for the reservoir pressure by Fig. 2.33, this becomes 47.65 lb/ft³; and then for the reservoir temperature by Fig. 2.34 the final density is found to be 45.4 lb/ft³. The total volume occupied by the 340.52 lb of reservoir fluid therefore is 7.500 ft³, or an excess of 1.89 ft³ over the barrel of stock-tank oil. The shrinkage is thus finally

$$(1.89 \times 100)/5.61 = 33.7\%.$$

As the excess volume of 1.659 ft³ at atmospheric pressure and temperature weighs 43.33 lb, the dissolved gas has an apparent composite density of 26.1 lb/ft³. This may be compared with the value 25.6 lb/ft³ indicated by Fig. 2.32 for a gas of gravity 0.810, which corresponds to the composition given above. It is to be noted that the latter value would have been used directly in the calculations if only the gas gravity, rather than its composition, had been given. The volume occupied by the dissolved gas would then have been computed as 1.693 ft³ so that the combined gas and oil volume would have been 7.303 ft³, at 60°F and atmospheric pressure. The equivalent over-all density would have been 46.63 lb/ft³. When corrected to reservoir pressure, by Fig. 2.33, the density would be 47.4 lb/ft³ and at reservoir temperature 45.3 lb/ft³. These latter are to be compared with 46.85, 47.65, and 45.4 lb/ft³ obtained previously by determining the apparent gas density from the detailed composition analysis. Finally, the corrected reservoir density of 45.3 lb/ft³ implies a volume of 7.517 ft³, or an excess of 1.907 ft³ over the barrel of stock-tank oil. The equivalent shrinkage is thus $(1.907 \times 100)/5.61 = 34.0\%$, as compared with 33.7 per cent obtained above.

As the choice of the original data used for these examples was purely arbitrary, they are not necessarily consistent with Figs. 2.28 and 2.29 pertaining to the relation of the solubility to the saturation pressure and of the former to the gas gravity. Nevertheless, it is of interest that the simple correlation curve of Fig. 2.30, of shrinkage vs. solubility, gives a value of 34.0 per cent for the shrinkage in almost exact agreement with that obtained by the detailed calculations.

Still more precise correlations of a purely empirical nature have been recently developed[1] from data on natural gases and crude oils from California fields. Using the notation S = solubility, γ_g = gas gravity, compared with air, γ_o = specific gravity of the stock-tank crude oil, $\bar{\gamma}$ = API gravity of crude, and T = temperature in degrees Fahrenheit, the bubble-point pressure is found to be a function only of the composite variable

$$(S/\gamma_g)^{0.83} \times 10^{0.00091T - 0.01257}.$$

If the latter be denoted by F, the bubble-point pressure above 400 psi will equal $18F$, to a close approximation. By inverting this relationship the solubility S can be readily calculated as a function of the pressure and the other variables. The formation-volume factor of the oil phase, β, has been similarly found to be a function of the quantity $1.25T + S\sqrt{\gamma_g/\gamma_o}$. On denoting this quantity by G the empirically established curve can be expressed by the equation

$$\log (\beta - 1) = 1.26 \log G - 9.5 \left(1 + \frac{10^3}{G^2}\right). \qquad (1)$$

And the combined formation volume of a unit of stock-tank oil plus its dissolved and liberated gas, at fixed pressures, is determined by the quantity $(S\sqrt{T}/\gamma_g^{0.3})\gamma_o^{2.9 \times 10^{-0.00027S}}$. While the complex structures of these composite variables evidently have no simple physical meaning and do not give a physical picture of the manner in which the expanded volumes of gas-saturated crudes are created, the above relationships do give predictions for the petroleum fluids that agree more closely with experimental data than those derived from the previously outlined correlation charts and synthesis procedures.

Although the density of residual or stock-tank oils at atmospheric pressure can be readily measured by standardized methods, it is often necessary

[1] M. B. Standing, *Oil and Gas Jour.*, **46**, 95 (May 17, 1947).

to estimate the density, especially in making calculations, from a knowledge of the composition. A curve giving the effective density of the paraffin hydrocarbons in the liquid phase as a function of the molecular weight is shown in Fig. 2.36. This curve may be used either by addition of the contributions of the individual components or by computing an average molecular weight from the composition and reading off the composite density directly. On the other hand, it should be recognized that the curve of Fig. 2.36 has no universal validity and should be considered only as semi-

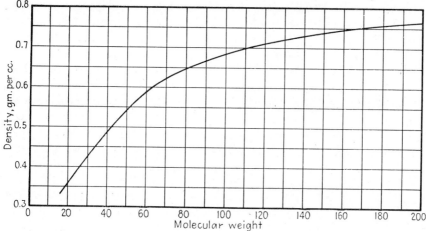

Fig. 2.36. The variation of the liquid-phase densities, at 60°F and vapor pressure, of paraffin hydrocarbon systems with the molecular weight.

quantitative. It does not take into account the nature of the hydrocarbon mixture in the region of high molecular weights and strictly applies only to the paraffin constituents of actual crudes. And for the low molecular weights the previous discussion and Fig. 2.35 show that the effective liquid-phase densities are quite sensitive to the composition of the remainder of the system.

The purely empirical approach presented above may be formulated in a more rigorous and formal manner by introducing the concept of the "partial volume." The latter is defined as the change in volume of a phase due to the addition of a unit weight of the component in question. It is presupposed that the original volume of the phase is so large that the addition of a unit weight of the particular component will not cause an appreciable change in the over-all composition. Upon denoting the partial volume of the ith component by $\overline{V}_i$, the weight fraction of this component in the phase by w_i the specific volume V (volume per unit weight) of the phase will accordingly be given by

$$V = \Sigma \overline{V}_i w_i. \tag{2}$$

A great many data have been gathered and correlated[1] for the partial volumes of the hydrocarbons in natural-gas and crude-oil systems. Unfortunately, however, aside from their dependence on the pressure and temperature of the phase they are also dependent on the composition of the phase and especially so in the case of the lighter components. It is for this reason that Fig. 2.36 cannot be used directly in determining the formation volumes of bubble-point gas-oil systems at high pressures containing appreciable concentrations of the lighter components. The apparent densities of the latter, as shown in Fig. 2.35, will depend on the nature of the heavier components; and if the methane and ethane are present in significant amounts, a neglect of the effect of the heavy components may lead to appreciable errors in computing the over-all density. On the other hand, at atmospheric pressures, where the concentrations of the lighter components are generally very small, even a rough estimate of their individual densities will suffice or one may simply compute an average molecular weight and use Fig. 2.36 directly, as previously indicated. In fact, the method described and illustrated above is essentially equivalent to Eq. (2) as applied to atmospheric conditions, and then translating to reservoir conditions by applying correction factors (Figs. 2.33 and 2.34) expressed in terms of the equivalent density. In any case, many more data will have to be obtained to make the direct method of partial volumes more practicable than that outlined above for predicting the volumes of gas-saturated crude oils.

Implicit in the above discussion has been the assumption that the solubility and shrinkage properties of crude-oil and natural-gas systems are independent of the thermodynamic paths followed between the terminal points, i.e., reservoir and atmospheric conditions. This, of course, will be true if the total hydrocarbon content be kept fixed and the gas and liquid phases are maintained in continuous contact. The gas evolved and shrinkage in oil volume as the pressure is reduced from the reservoir value to atmospheric would be determined only by these terminal states. Such a process (for fixed total composition) is termed "flash liberation" or "flash vaporization." An example of its occurrence in practice is the variation in pressure and temperature of the stream of oil and gas as it flows up the well bore to the surface under steady-state conditions.

If, however, during the pressure decline of the original saturated oil, part or all of the evolved gas is continually removed, as happens within oil-bearing reservoirs, the process is termed "differential liberation." Since the total composition is thus continually changing during the differential liberation, it is to be expected that when atmospheric conditions are

[1] B. H. Sage, B. L. Hicks, and W. N. Lacey, API Drilling and Production Practice, 1938, p. 402; also B. H. Sage and W. N. Lacey, API Drilling and Production Practice, 1939, p. 641.

reached the total gas evolved and residual volume of the oil will be different from what they would be if all the gas had been retained in contact with the liquid phase. And so it has been found by actual experimentation with both bottom-hole and recombined surface or separator samples of oil and gas, although in such experiments the continuous differential-liberation process is of necessity approximated by a stepwise procedure of lowering the pressure by finite increments, removing the evolved gas at fixed pressure, reducing the pressure again, etc.

In general it is found that the gas liberated by differential liberation is less than by flash liberation and the residual-oil volume greater. The latter implies that the formation-volume factor of the reservoir oil, or shrinkage as expressed in terms of the residual oil, derived by a differential-liberation process is lower than when the gas is released to the same terminal conditions by flash liberation. While these differences are of importance in the application of production data to the interpretation of reservoir performance, there are as yet available no general correlations relating the solubility and shrinkage data corresponding to the different gas-liberation processes. This matter will be discussed further in Sec. 9.5.

2.9. The Prediction of the Phase Behavior of Complex Hydrocarbon Systems; Equilibrium Ratios.—In the last section, methods were presented for computing, to various degrees of approximation, the volumetric behavior of the liquid phase of gas-oil systems, using information regarding the nature (gravity or composition) of the dissolved gas and the gross character (gravity) of the oil. There still remains the problem of predicting the compositions of the gas and liquid phases and the distribution between them of the various components of the composite hydrocarbon system. This problem is as yet far from solved, especially in the ranges of pressures and temperatures in the neighborhood of the critical regions of the composite systems. It is largely under the latter conditions that means for predicting phase distributions and compositions would be of the greatest value from the strictly production point of view, *i.e.*, in gas-condensate systems. Nevertheless it is of value to summarize the present status of the problem so as to permit immediate application to be made of additional data as they accumulate.

If hydrocarbon mixtures obeyed Raoult's solution law and the perfect-gas laws, the problem at hand could be solved readily as follows: According to these the partial pressure of each component, p_i, in the gas phase is related to the mole fraction of that component in the liquid phase, x_i, by the relation

$$p_i = P_i'' x_i, \qquad (1)^*$$

* It may be noted that for low molar concentrations of the components in the liquid phase, Henry's law, $x_i/(1 - x_i) = s_i p_i$, reduces to $x_i = s_i p_i$, s_i being the "solubility" of the ith component per unit partial pressure. It can therefore be used as an empirical equivalent of Raoult's law, leading to $K_i = 1/s_i P$ in place of Eq. (3).

and to its mole fraction in the gas phase, y_i, by

$$p_i = Py_i, \tag{2}$$

where P_i'' is the vapor pressure of the ith component at the temperature and total pressure P of the system. Upon introducing the notation

$$K_i = \frac{P_i''}{P} = \frac{y_i}{x_i}, \tag{3}$$

and denoting the mole fractions of the ith component in the composite system by n_i, the number of moles of all components in the gas phase by n_g, and that in the liquid phase by n_L, the sum being 1, it readily follows from Eqs. (1) to (3) that

$$n_i = \frac{n_L p_i}{P_i''} + \frac{n_g p_i}{P} = \frac{p_i}{P_i''}(n_L + n_g K_i),$$

so that

$$\left. \begin{aligned} x_i &= \frac{n_i}{1 + n_g(K_i - 1)}, \\ y_i &= \frac{n_i K_i}{1 + n_g(K_i - 1)}. \end{aligned} \right\} \tag{4}$$

For the given pressure and temperature of the system, the K_i's can be computed by their defining Eq. (3). For a fixed over-all composition, defined by the n_i's, the only unknown in the right-hand sides of Eqs. (4) will then be n_g. The latter is to be determined by trial and error so that

$$\Sigma x_i = \sum \frac{n_i}{1 + n_g(K_i - 1)} = 1, \tag{5}*$$

or

$$\Sigma y_i = \sum \frac{n_i K_i}{1 + n_g(K_i - 1)} = 1, \tag{6}$$

which are fully equivalent. The value of n_g that satisfies these equations then gives for the individual terms of the summation the mole-fraction composition of both the gas and liquid phases. By carrying through this procedure for other temperatures and pressures the complete phase and composition history of the system, defined by the n_i's, can be determined. Moreover, the dew-point curve may be calculated by observing that at the dew point $n_g = 1$, so that Eq. (5) becomes

$$\sum \frac{n_i}{K_i} = 1. \tag{7}$$

---

* An early presentation and discussion of the equivalents of these equations, together with an extensive bibliography of previous work, were given by D. L. Katz and G. G. Brown, *Ind. and Eng. Chemistry*, **25**, 1373 (1933).

Similarly at a bubble point, $n_g = 0$, and Eq. (6) requires that

$$\Sigma n_i K_i = 1. \tag{8}$$

By a trial-and-error variation of the pressure or temperature until Eq. (7) or (8) is satisfied, the boundary curves for the mixtures defined by the values of n_i may be established. It may be noted, incidentally, that if for the n_i's in Eqs. (7) and (8) are substituted the values of y_i and x_i, respectively, as given by Eqs. (4), the former will be automatically satisfied. This fact confirms the previously noted observation that coexisting gas and liquid phases must be at dew-point and bubble-point conditions for the compositions of the individual and respective phases.

In principle, the set of Eqs. (3) to (6) constitute the basis for the complete description of multicomponent hydrocarbon systems. The K_i's have been termed "equilibrium constants," with the implication that, as Eq. (3) indicates, each depends only on the pressure and temperature but is independent of the other components. This, however, unfortunately is a misnomer that was based on hope rather than experience. For the latter has shown that not only does the relation between the K_i's and the pressures P_i'' and P, indicated by Eq. (3), begin to break down at pressures appreciably exceeding atmospheric but, more seriously still, the true values of the ratios of the mole fractions, y_i/x_i, of the individual components depend on the composition of the remainder of the system.

It is possible to extend the range of theoretical prediction of the ratios K_i by introducing the concepts of fugacity and ideal solutions. The former may be defined by the equation

$$\log f = \frac{1}{RT} \int V \, dp, \tag{9}$$

the integral being evaluated with the aid of actual p-v-T data for the pure components in question, V representing the volume per mole. The significance of the fugacity is that it provides a correction to the partial pressures of the pure components when they do not obey the perfect-gas laws. The ideal solution is one obeying the rule of additivity of the volumes and heat contents of the individual components. For such systems it may be shown that

$$K_i = \frac{y_i}{x_i} = \frac{f_{ig}}{f_{il}}, \tag{10}$$

where the subscripts g, l refer to the gas and liquid phases. In fact, Eq. (10) may be taken as a definition of the ideal solution.

The use of these expressions has indeed permitted correct predictions of the values of K_i at pressures and temperatures far greater than by use of Eq. (3). However, their use involves graphical-extrapolation procedures for the liquid or gas fugacities when the total pressure in the system is either below or above the vapor pressure of any of the pure components. Moreover, in the vicinity of the critical state of the system, large deviations from ideal-solution behavior occur, and extrapolation methods are quite unreliable. Finally, the basic premise that the K_i's are independent of the over-all composition of the system vitiates any direct application of the ideal-solution and fugacity concepts, just as of Raoult's law, without the provision of correction factors and additional empirical information to take account of composition effects.

In view of this situation it seems best to place the whole problem on a purely empirical foundation and merely define the K_i's as

$$K_i = \frac{y_i}{x_i}, \tag{11}$$

the values of which are to be determined by direct experimentation. Of course, Eqs. (4) to (8) developed above will still remain valid. And the limiting behavior of the K_i as the pressure is reduced and the temperature becomes removed from the critical values will approach the predictions of Eqs. (10) and (3). Nevertheless, it seems appropriate to change the conventional designation of the K_i's as "equilibrium constants" to that of "equilibrium *ratios*," if only to emphasize that they are not constants but that rather they are dependent on the composition of the system as well as on pressure and temperature.

Acceptance of the empirical approach to the problem does not, of course, solve it. Indeed, an examination of the available data quickly reveals the pressing need for much more systematic experimentation. In many respects the data are still little more than fragmentary. Nevertheless, to provide at least some guide to the numerical values of the equilibrium ratios the published data will be briefly summarized, following Sage, Hicks, and Lacey.[1] Their data for methane are given in Table 2. The molecular weights heading the columns refer to that of the "less volatile constituent", *i.e.*, of the hexanes and heavier part of the system. The character of the latter is further limited by a value of 0.82 for its "viscosity-gravity" factor. This factor, which is frequently used to characterize oils, is defined by the equation

$$A = \frac{G - 0.10752 \log_{10}(S - 38)}{1 - 0.10 \log_{10}(S - 38)}, \tag{12}$$

where G is the specific gravity of the oil, referred to water at 60°F, and S is its viscosity at 100°F in Saybolt universal seconds. For less volatile

---

[1] B. H. Sage, B. L. Hicks, and W. N. Lacey, *API Drilling and Production Practice*, 1938, p. 386. No attempt has been made to bring these data up to date, and they are not to be considered as established constants.

TABLE 2.—EQUILIBRIUM RATIOS FOR METHANE

Abs. pressure, psia	Mol. wt.* = 60			Mol. wt. = 75			Mol. wt. = 100			Mol. wt. = 200			Mol. wt. = 300		
	100°F	160°F	220°F	100°F	160°F	220°F	100°F	160°F	220°F	100°F	160°F	220°F	100°F	160°F	220°F
14.7	⋯	⋯	⋯	224†	247	248	247	285	293	259	276	287	203	226	228
20	⋯	⋯	⋯	165	181	182	181	210	216	190	203	211	149	166	168
40	⋯	⋯	⋯	82.8	90.9	91.5	91.0	105	108	95.5	102	106	75.1	83.4	84.2
60	⋯	⋯	⋯	55.4	60.8	61.3	60.9	70.1	72.1	63.9	68.2	70.7	50.4	55.9	56.4
100	⋯	⋯	⋯	33.3	39.1	42.0	44.0	51.0	56.0	40.6	43.8	46.8	30.4	34.0	35.5
200	⋯	⋯	⋯	17.0	18.8	19.5	21.1	22.9	24.1	20.1	21.4	22.6	15.7	17.3	17.9
300	⋯	⋯	⋯	11.6	12.6	12.9	13.9	14.8	15.2	13.4	14.3	14.7	10.9	11.8	12.1
400	⋯	⋯	⋯	8.83	9.51	9.82	10.3	11.0	11.3	10.1	10.8	11.1	8.40	9.00	9.25
500	6.56	6.81	7.13	7.20	7.73	8.00	8.24	8.88	9.04	8.24	8.88	9.04	6.84	7.33	7.57
750	4.50	4.57	4.89	5.01	5.32	5.52	5.56	5.94	6.10	5.74	6.15	6.32	4.82	5.16	5.33
1,000	3.42	3.42	3.66	3.90	4.07	4.19	4.28	4.54	4.60	4.49	4.75	4.84	3.92	4.12	4.24
1,250	2.77	2.72	2.89	3.21	3.34	3.42	3.52	3.72	3.74	3.77	4.06	4.06	3.27	3.50	3.60
1,500	2.28	2.21	2.30	2.72	2.82	2.85	3.00	3.16	3.12	3.26	3.51	3.51	2.86	3.06	3.16
1,750	1.93	1.80	1.80	2.36	2.40	2.45	2.61	2.76	2.71	2.94	3.12	3.13	2.56	2.75	2.84
2,000	1.63	1.43	⋯	2.07	2.06	2.08	2.33	2.45	2.37	2.70	2.82	2.82	2.36	2.51	2.59
2,500	⋯	⋯	⋯	1.60	1.52	1.48	1.93	2.03	1.93	2.36	2.43	2.40	2.02	2.15	2.25
3,000	⋯	⋯	⋯	1.17	⋯	⋯	1.65	1.74	1.59	2.14	2.14	2.09	1.78	1.90	2.00

* These molecular weights refer to the "less volatile component".

† Values in the table apply directly only to systems whose less volatile constituent has a viscosity-gravity factor of 0.82.

constituents of general viscosity-gravity factor A the values of Table 2 are to be corrected by the formula

$$K_A = K_{0.82}[1 + 1.87(A - 0.82) + 13(A - 0.82)^2]. \qquad (13)$$

From Table 2 it will be seen that, for pressures up to about 100 psi, K indeed varies inversely as the pressure (to about 3 per cent) as implied by Eq. (3). Yet it also varies considerably with the molecular weight of the less volatile constituent. In fact, it first increases with the latter, reaches a maximum at molecular weights of about 150 to 200, and then decreases again. While it generally increases with increasing temperature, it may show maxima at the higher pressures, depending on the molecular weight of the less volatile constituent. Indeed, in the vicinity of the critical state both the absolute values of Table 3 and the correlation equation (13) lose their validity, and the molecular weight and character of the less volatile constituent become the controlling factors in fixing the values of K. This is a consequence of the fact that the critical state itself is greatly influenced by the nature of the less volatile constituent, and, by their definition, the equilibrium ratios must equal unity at the critical state.

The equivalent tabulation given by Sage, Hicks, and Lacey for ethane is reproduced in Table 3. The less extensive available experimental data

TABLE 3.—EQUILIBRIUM RATIOS FOR ETHANE

Pressure, psia	Mol. wt. = 100			Mol. wt. = 200			Mol. wt. = 300		
	100°F	160°F	220°F	100°F	160°F	220°F	100°F	160°F	220°F
14.7	37.1	56.5	75.6	38.7	57.7	77.2	41.2	56.9	73.7
20	27.3	41.5	55.6	28.5	42.5	56.9	30.3	41.9	54.2
40	13.8	20.9	27.8	14.4	21.4	28.6	15.3	21.1	27.3
60	9.24	14.0	18.6	9.66	14.3	19.2	10.3	14.2	18.4
100	5.65	8.50	11.2	5.88	8.71	11.7	6.26	8.66	11.2
150	3.84	5.73	7.56	3.99	5.90	7.88	4.26	5.90	7.67
200	2.93	4.35	5.72	3.06	4.49	6.04	3.28	4.52	5.90
300	2.05	2.96	3.89	2.13	3.09	4.13	2.27	3.15	4.13
400	1.61	2.30	2.96	1.66	2.41	3.22	1.79	2.46	3.23
500	1.35	1.90	2.42	1.38	2.00	2.68	1.50	2.07	2.71
750	1.04	1.35	1.70	1.04	1.46	1.93	1.16	1.56	2.02
1,000	.....	1.09	1.35	0.930	1.22	1.59	1.06	1.33	1.69
1,250	.....	.....	1.13	0.868	1.09	1.37	1.01	1.21	1.49
1,500	.....	.....	.....	0.811	1.01	1.24	0.974	1.15	1.38
1,750	.....	.....	.....	0.775	0.954	1.14	0.946	1.11	1.29
2,000	.....	.....	.....	0.755	0.913	1.08	0.932	1.07	1.23
2,500	.....	.....	.....	0.741	0.855	0.995	0.928	1.02	1.13
3,000	.....	.....	.....	0.740	0.830	0.945	0.930	0.970	1.06

do not yet permit a correlation with the chemical nature of the less volatile constituent, *i.e.*, its viscosity-gravity factor. Here again the inverse variation of K with pressure is followed quite well to pressures of about 100 psi. However, in contrast to the K's for methane, there is here a slower variation with the molecular weight of the less volatile constituent in the range listed. Moreover, whereas in the case of methane the value unity is approached only as the critical state is approached, here, as well

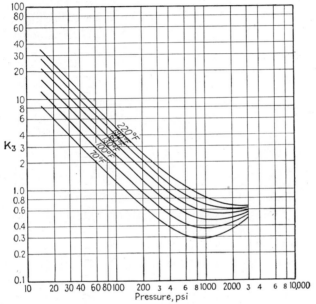

FIG. 2.37. Equilibrium-ratio isotherms for propane.

as for the components heavier than ethane, the equilibrium ratio first falls to unity at approximately the vapor pressure of the component, reaches a minimum, and then returns to unity at the critical state.[1] It is especially the behavior of the K vs. pressure curves in the region between the two values of unity and the location of that region that are greatly affected by the other components of the system. It should also be noted that the rise of equilibrium ratios after passing the minimum pressure

[1] Because of the limited pressure range covered, Table 3 does not reflect the development of the minimal values below unity and the subsequent tendency to return to unity in the case of ethane. However, this behavior is clearly shown in Figs. 2.37 to 2.40 for the heavier components, for which the minima occur at lower pressures. Moreover, whereas in the case of ethane the listed equilibrium ratios do not fall initially to unity until the critical pressure is exceeded, the value of unity is first reached for the heavier paraffins at pressures of the order of magnitude of their respective vapor pressures, when the temperatures are lower than the criticals.

implies a tendency for the component to begin reentering the gas phase in preference to the liquid. This phenomenon is that of retrograde vaporization (cf. Sec. 2.5). The sensitivity of these retrograde processes to the composition of the system is thus seen to be here reflected in the corresponding sensitivity of the equilibration ratios to composition at high pressures or in the vicinity of the critical state.

For the hydrocarbons heavier than ethane the experimental data are so meager that no correlation whatever with the character of the less volatile constituents has yet been developed. Accordingly they are presented only as functions of temperature and pressure. Sage and Lacey's

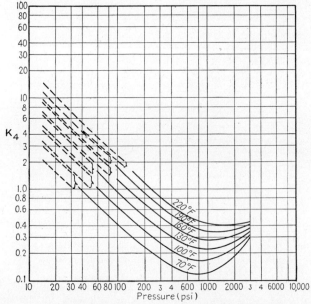

FIG. 2.38. Equilibrium-ratio isotherms for the butanes. Upper branches of the dashed segments apply to isobutane and the lower to *n*-butane.

tabulation for propane is shown graphically in Fig. 2.37 for several temperatures. Their data for the butanes are plotted in Fig. 2.38. Here, both *n*-butane and isobutane are grouped together for the higher pressures, and individual values are plotted, as dashed segments, only in the region where the ideal-solution behavior may be assumed. The values for the pentanes are presented in a similar manner in Fig. 2.39. The hexanes as a group are represented by the curves of Fig. 2.40.

In the case of the heptanes and heavier residuals only very few systematic investigations leading to quantitative data at high pressures have been published. Even these, however, instead of providing any definite graphical

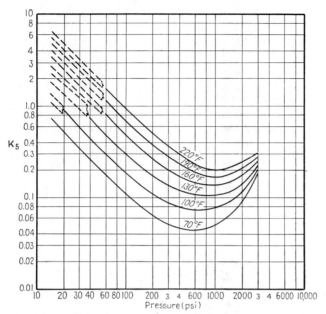

Fig. 2.39. Equilibrium-ratio isotherms for the pentanes. Upper branches of the dashed segments apply to isopentane and the lower to *n*-pentane.

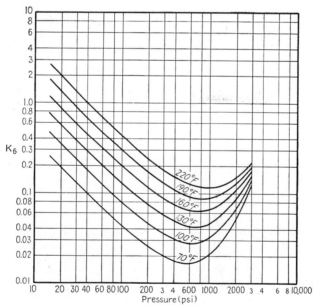

Fig. 2.40. Equilibrium-ratio isotherms for the hexanes.

or tabular set of values for the equilibrium ratios, serve mainly to emphasize their great variability and their extreme sensitivity to the nature of these residuals, especially at the higher pressures. This situation is illustrated by Fig. 2.41, in which are plotted the results at 120°F obtained from experiments with (1)[1] a Mid-Continent crude (38.4°API), (2)[2] a volatile distillate (57.5°API), (3)[3] a 49.9°API crude with a gas-oil ratio of 3,660 ft³/bbl, and (4) the same crude with a gas-oil ratio of 7,180 ft³/bbl. It is obvious from

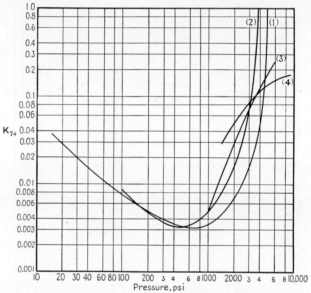

Fig. 2.41. The equilibrium-ratio curves for heptanes plus, at 120°F, for several hydrocarbon systems. (1) Mid-Continent crude (38.4°API). (2) Volatile distillate (57.5°API). (3) A 49.9°API crude with a gas-oil ratio of 3,660 ft³/bbl. (4) Same as (3) with a gas-oil ratio of 7,180 ft³/bbl.

these that the location of the critical point of the system (where $K_i = 1$) is the predominating factor in determining the shapes of the K_{7+} curves and their absolute values at the higher pressures. Of course, the requirement of convergence to the value unity at the critical point will also affect the equilibrium ratios for the lighter constituents. However, except for methane, they are usually present in rather low concentrations, so that the uncertainties in their equilibrium ratios will be of less importance in the solutions of the equilibrium equations, as Eqs. (5) to (8).

As an example of a set of equilibrium ratios for the C_7's, applicable at

[1] D. L. Katz and K. H. Hachmuth, *Ind. and Eng. Chemistry,* **29,** 1072 (1937).

[2] C. H. Roland, D. E. Smith, and H. H. Kaveler, *Oil and Gas Jour.,* **39,** 128 (Mar. 27, 1941).

[3] M. B. Standing and D. L. Katz, *AIME Trans.,* **155,** 232 (1944).

least to one type of gas-oil system, and as an illustration of the magnitudes involved, a chart for heptanes is plotted in Fig. 2.42. This is based on the data of Katz and Hachmuth[1] for a Mid-Continent crude and has been found to give approximate values for the "heptanes and heavier" in similar systems on multiplication by 0.15. While, as already indicated, they will not be applicable to other types of crude or to condensate oils at pressures exceeding the minimal point ($\sim$ 500 psi), they appear to give fair approx-

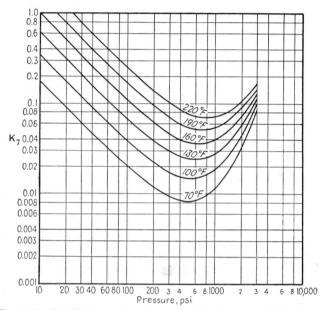

Fig. 2.42.   Equilibrium-ratio isotherms for the heptanes.   (*After Katz.*)

imations when used at lower pressures, such as in low-pressure stage separation calculations.

2.10. The Application of Equilibrium-ratio Data.—In principle it should be possible, by applying equilibrium-ratio data to Eqs. 2.9(5) to 2.9(8), completely to predict all significant features of the behavior of complex hydrocarbons systems. In practice, however, these theoretical possibilities are severely restricted. For example, the critical state of any system, defined by its composition, is immediately given by that pressure and temperature for which all the equilibrium ratios have the value unity. Such an interesting application of equilibrium-ratio data is, however, quite impractical. For to find the temperature and pressure at which the ratios converge to unity one must literally know in advance, from

[1] Katz and Hachmuth, *loc. cit.* The chart itself is due to Katz.

direct measurement or other correlations, the very critical data being sought. This, of course, is obvious on recalling the discussion in the preceding section of the variability in the equilibrium-ratio curves for the heavier components near the critical state and the use of knowledge of the latter as a control point for fixing the shapes of the curves. Indeed, to make any quantitative calculations in the vicinity of the critical state it is virtually necessary to have given the location of the critical point to permit an adjustment of the equilibrium-ratio curves for the heavier components to correspond to the type of system under consideration.

Nevertheless, many useful applications can be made in regions removed from the critical state. Thus, if the analysis of one phase in a reservoir be determined or known, the composition of the coexisting phase, if present, can be calculated. For example, if a sample of the reservoir bubble-point liquid be obtained and analyzed, the composition of the gas-cap gas, if present, can be immediately predicted by multiplying the mole fractions in the bubble-point liquid by the corresponding equilibrium ratios at the reservoir pressure and temperature. Or if the well has apparently been completed in a gas cap, its dew-point pressure can be calculated by applying Eq. 2.9(7) at the reservoir temperature. If this agrees with the reservoir pressure, the gas-cap fluid will be a saturated gas and the composition of a coexisting oil phase, if present, will be given by the individual terms in Eq. 2.9(7). The same procedure can be used to check if the crude oil from one well and free gas from another could have been in contact and hence coexistent in the same reservoir.

Another type of application is in the computation of the solubility-curve and phase relations between a natural gas and crude oil with the aid of equilibrium-ratio data. That is, for example, if one is given the gas-oil ratio for a producing system and the composition of the gas and crude oil, the bubble-point at reservoir temperature, the amount and composition of the free gas entering the well bore at the bottom-hole pressure, and the amount of liberated gas, its composition, and that of the oil at the separator pressure or atmospheric pressure can be calculated. The bubble-point pressure of the over-all system is computed using Eq. 2.9(8), after converting the original gas- and oil-composition data, together with the gas-oil-ratio observation, into an over-all mole-fraction-composition tabulation of the n_i's. By trial and error a value of the pressure can then be found such that the corresponding K_i's, for the reservoir temperature, will make $\Sigma n_i K_i$ equal unity. That pressure will be the bubble-point pressure.

To determine the phase distribution at the bottom-hole flowing pressure or in the separator, the previously determined values of the n_i's are now used in Eq. 2.9(5) or 2.9(6). With the aid of the K_i's corresponding to the pressure and temperature of interest these equations are solved by trial and error for n_g.* When so determined the n_g's will give the mole fraction of the composite system in the gas phase. The corresponding individual terms of the summation in Eq. 2.9(5) or 2.9(6) will at the same time represent the mole-fraction compositions in the liquid and gas phase, respectively. If these compositions are known, the volumetric characteristics and gas-deviation factors can be computed by the methods discussed in previous sections.

If the original data are given only in terms of the over-all composition, the ultimate atmospheric gas-oil ratio can also be computed, as just indicated. Moreover, if the gas and oil separation proceeds through one or more intermediate separator stages, a simple repetition of the above outlined basic calculation procedure will give all the information of interest. That is, the phase equilibrium is first computed, as above, for the highest pressure separator conditions. This will give the number of moles of gas, per unit mole of original bubble-point liquid or fluid mixture, which can be bled off that separator, together with the composition of the gas and the residual separator oil. In proceeding to the next separator stage or to the stock tank, the calculations are merely repeated, using for the over-all composition mole fractions n_i the values x_i previously found for the residual oil of the first separator and, of course, the new pressure and temperature of the second stage or the stock tank. The additional amount of gas liberated (the new value of n_g) is thus found, together with its composition and that of the new, or final, residual oil. From this latter the gravity and volume of the residual oil can be computed by the methods of Sec. 2.8. By calculating also the volume of the original bubble-point system the over-all shrinkage can be obtained. Moreover, from the amounts and compositions of the gas liberated at the various stages the gasoline content of the individual or combined streams of liberated gas is easily determined. These calculations can be repeated for different separator conditions to obtain the effect of the latter on the characteristics of the intermediate or end products of the pressure reduction.

A specific example of such applications is provided by the following tabulated computation of the results of the flash[1] vaporization of a bubble-point liquid of known composition to atmospheric pressure and 100°F.

The values of n_i are those defining the original total composition. The

---

* A rough guide for the choice of n_g in trial-and-error calculations at relatively low pressures (< 500 psi) is the inequality $n_1 < n_g < 1 - n_{7+}$. While not rigorous, this generally gives inclusive bounds for n_g.

[1] By following the above-outlined procedure, using small pressure increments, the corresponding results for differential gas liberation could be obtained.

TABLE 4.—COMPUTATION OF FLASH VAPORIZATION OF BUBBLE-POINT LIQUID

	n_i (mole fraction in bubble-point liquid)	K_i (14.7 psi and 100°F)	x_i (mole fraction in stock-tank oil)	y_i (mole fraction in liberated gas)
(1) Methane.....	0.4526	259	0.0031	0.8053
(2) Ethane......	0.0307	38.7	0.0014	0.0537
(3) Propane.....	0.0230	11.8	0.0033	0.0385
(4) Butanes.....	0.0250	3.81	0.0097	0.0370
(5) Pentanes.....	0.0204	1.06	0.0197	0.0209
(6) Hexanes.....	0.0175	0.47	0.0249	0.0117
(7) Heptanes plus	0.4308	0.035	0.9379	0.0329

listed K_i's were taken from tables and graphs of the preceding section, assuming the molecular weight of the less volatile component to be 200, and using the average values for normal and isobutane for K_4, and that for normal pentane for K_5. Upon using these data in Eq. 2.9(5) or 2.9(6) and solving for n_g it is found that $n_g = 0.5603$. For this value of n_g the individual terms in the summation of Eq. 2.9(5) are the x_i's listed in the fourth column above, and the corresponding terms in Eq. 2.9(6) are the y_i's listed in the last column.

To determine the gas-oil ratio equivalent of the mole fraction in the gas phase, namely, $n_g = 0.5603$, the following simplified procedure may be used: It is first noted that the above gas-liberation process leads to $0.5603/0.4397 = 1.2743$ moles of gas per mole of stock-tank oil, or 483.6 ft³ gas (at 60°F) per pound mole of stock-tank oil. From the composition of the stock-tank oil and the assumed molecular weight of 200 for the less volatile component, *i.e.*, hexanes and heavier, the average molecular weight is found to be 194.8. From Fig. 2.36 the atmospheric density, at 60°F, of a paraffin hydrocarbon of this molecular weight is 0.762 gm/cc or 47.57 lb/ft³ = 1.37 lb mole/bbl. The gas-oil ratio is, therefore, 483.6 × 1.37 = 663 ft³/bbl.

By carrying through the same procedure for a gas and oil system of similar composition, at a temperature of 70°F, but at various separator pressures, and then flashing the separator oil to atmospheric pressure, the results shown in Fig. 2.43 were obtained.[1] While the separator and stock-tank gas-oil ratios in terms of mole fractions were given directly by the solutions of Eq. 2.9(5), their conversion to cubic feet per barrel and the results on stock-tank gravity and shrinkage required additional volumetric calculations of the type discussed in Sec. 2.8. Curves such as those plotted

[1] S. E. Buckley, *AIME Trans.*, **127**, 178 (1938); cf. also D. L. Katz, *AIME Trans.*, **127**, 159 (1938).

in Fig. 2.43, whether determined by calculation or experimentally, can evidently be of great value in planning actual oil-field operation.

These examples[1] will illustrate the types of practical results that may be obtained by an application of equilibrium ratios. As the crucial data

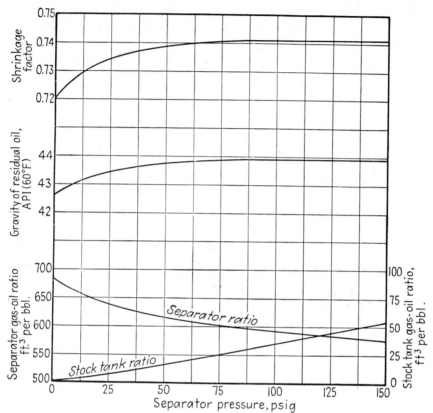

FIG. 2.43. The calculated effect of the separator pressure on the properties and relative amounts of oil and gas at 70°F separating from a fixed composition well fluid. (*After Buckley, AIME Trans., 1938.*)

involved are the values of the equilibrium ratios, care must be taken that those used are actually applicable to the type of gas-oil mixture under consideration.

2.11. The Viscosity of Petroleum Fluids.[2]—As will be seen in the fol-

[1] Still other applications to complete crude-oil and condensate processing systems are given in the University of Texas M.S. thesis of G. H. Fisher (1941).

[2] The detailed variation of the viscosity of water with temperature, at atmospheric pressure, can be found in standard handbooks. It varies from 1.13 cp at 60°F to 0.30 at 200°F. The effect of pressure alone is less than 1 per cent increase per 1,000 psi increase and can be neglected.

lowing chapters, a knowledge of the viscosity of the petroleum fluids is essential in evaluating quantitatively the dynamical behavior of these fluids in producing strata. We shall therefore summarize the available empirical information on the viscosity of hydrocarbon systems.

With respect to the gas phase of petroleum hydrocarbon mixtures, quite complete correlation charts have recently been developed.[1] Extrapolating observed data on the paraffin hydrocarbon gases at atmospheric pressure, the viscosity vs. molecular-weight isotherms are given by the curves of

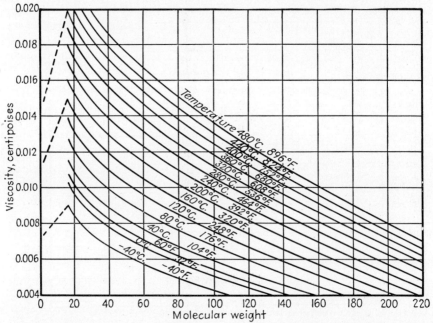

Fig. 2.44. The viscosity of paraffin hydrocarbon gases at atmospheric pressure. (*After Bicher and Katz, AIME Trans., 1944.*)

Fig. 2.44. The viscosities indicated by Fig. 2.44 are generally lower than those for most other common gases or vapors, except for the unique case of hydrogen. It will be noted, too, that the viscosities rise with increasing temperature and decrease with increasing molecular weight. Unless the natural gas contains more than 7 per cent nitrogen, the values read from Fig. 2.44 should be in error by no more than 3.5 per cent.

For pressures exceeding atmospheric a correlation and extrapolation of data obtained with binary mixtures of methane and propane give the

---

[1] L. B. Bicher and D. L. Katz, *AIME Trans.*, **155**, 246 (1944). Fig. 2.45 is also taken from this source.

results shown in Fig. 2.45. These curves predict the values as actually measured[1] for natural gases with an average deviation of 5.8 per cent. The two-phase-region boundaries indicated in Fig. 2.45 are those for the methane-propane system. Accordingly the curves should not be used near the critical or two-phase regions of the actual gas in question.[2]

In contrast to the curves of Fig. 2.44 for atmospheric pressure, Fig. 2.45 shows that at high pressures the viscosity increases with increasing molecular weight. Naturally, there is a transition region where it is essentially independent of molecular weight. These transition pressures increase with increasing temperature. The direct effect of temperature on viscosity is also reversed at high pressures from that at atmospheric. And again there is a transition region of pressures in which the viscosity varies but slowly with the temperature. These results are inherently empirical and cannot be predicted directly from the physical theory of fluid viscosity.[3] Yet it is of interest to note that the different variation of the viscosity of the gases at high pressure with temperature and molecular weight, as compared with atmospheric pressure, may be interpreted qualitatively as due to the fact that at the high pressures the gas begins to assume the properties of a liquid phase.

In the case of the liquid phase of petroleum hydrocarbon systems, the viscosity will depend not only on the temperature and absolute pressure but also on the amount of gas in solution. At atmospheric pressure an extrapolation of measured data on the paraffinic hydrocarbon liquids leads to the viscosity vs. molecular-weight isotherms shown in Fig. 2.46.[4] It will be noted that the viscosity increases with molecular weight and decreases with increasing temperature. Curves showing directly the variation with temperature for several typical crude oils at atmospheric pressure are plotted in Fig. 2.47. And a correlation chart[5] of the viscosity of gas-free crude oils at several temperatures as a function of the crude gravity, as obtained from measurements on hundreds of samples, is given in Fig. 2.48. The average deviation of the measured data from these curves was 24.2 per cent. The effect of pressure on the oil itself, without the gas that might be in solution, is illustrated by the curves of Fig. 2.49, obtained

---

[1] B. H. Sage and W. N. Lacey, *AIME Trans.*, **127**, 118 (1938).

[2] Their validity is also limited to gases containing small amounts of nitrogen. If the nitrogen content exceeds 5 per cent, a molal average of the nitrogen and hydrocarbon viscosities should be used, with Fig. 2.45 applied only to the hydrocarbon component.

[3] In principle the difficulty of theoretically predicting the viscosities of gases is only one of computation, provided that the intermolecular law of force is known. Ideal-gas kinetic theory leads to a viscosity independent of pressure.

[4] *Petroleum Eng.*, **15**, 159 (November, 1943).

[5] Beal, *loc. cit.*

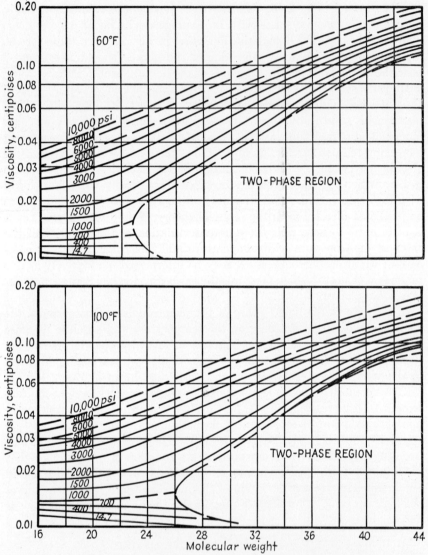

Fig. 2.45. The variation of the viscosity of natural gases with molecular weight, pressure,

by experiments[1] on a water-white refined fraction of a western crude oil. To show directly the relative effect of pressure for the different temperatures the viscosities are here plotted as ratios to the values at atmospheric pressure. It will be seen that the viscosity increases by 12 to 20 per cent

[1] B. H. Sage, J. E. Sherborne, and W. N. Lacey, *API Proc.*, **16**, Bull. 216, p. 40 (1935).

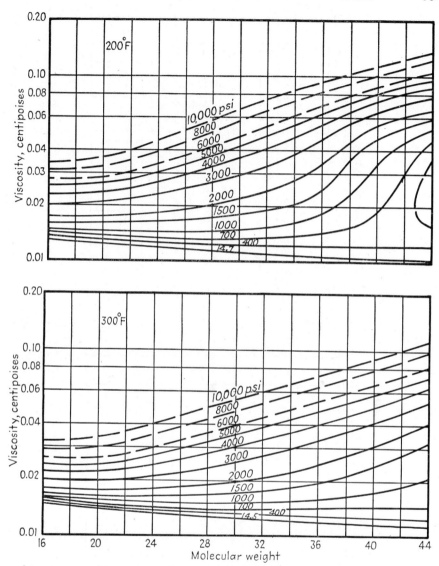

and temperature. (*After Bicher and Katz, AIME Trans., 1944.*)

per 1,000 psi increment, the exact value increasing somewhat with decreasing temperature and increasing mean pressure. The variation with pressure above the bubble point of gas-saturated crudes is of the same order of magnitude.[1]

All studies of the viscosities of gas-saturated or bubble-point crude oils

[1] Beal, *loc. cit.*

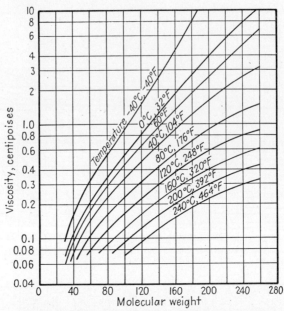

Fig. 2.46. The viscosity of paraffinic hydrocarbon liquids, at atmospheric pressure vs. the molecular weight. (*From Petroleum Eng., 1943.*)

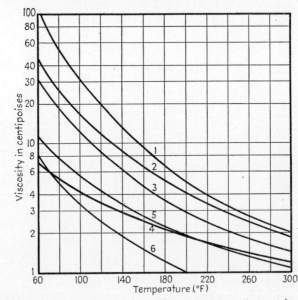

Fig. 2.47. The variation of the viscosity with temperature, at atmospheric pressure, of typical crude oils. (1) Fruitvale, Calif., 23.4°API. (2) Seal Beach, Calif., 26.8°API. (3) Ventura, Calif., 28.2°API. (4) Hobbs, N. Mex., 34.5°API. (5) Oklahoma City, Okla., 36.4°API. (6) East Texas, Tex., 40.2°API.

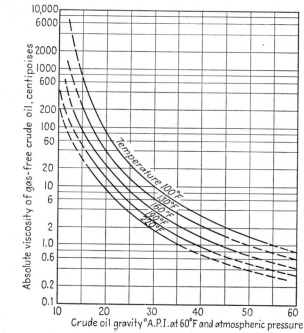

Fig. 2.48.   The variation of the viscosity of gas-free crude oils (atmospheric pressure) with the crude gravity at various temperatures.   (*After Beal, AIME Trans., 1946.*)

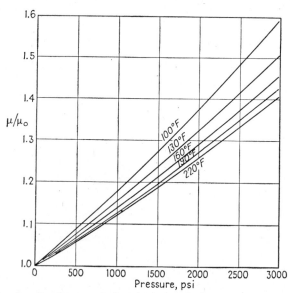

Fig. 2.49.   The viscosity of a crystal oil under pressure, μ, as compared with its viscosity at atmospheric pressure, μ_o.   (*After Sage, Sherborne, and Lacey, API Proc., 1935.*)

thus far made[1] show the effect of the dissolved gas at elevated pressures to more than counterbalance the direct increase in viscosity due to the pressure, illustrated in Fig. 2.49. Accordingly the viscosities decrease monotonically with the saturation pressure.

The results of a direct correlation of the viscosity vs. the gas in solution at reservoir pressure, for fixed gas-free viscosity, based on 351 measurements on 41 samples from 29 fields (20 in California), are plotted in Fig. 2.50.[2] As the average deviation of the measured data from the

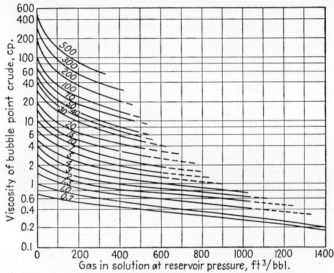

Fig. 2.50. The viscosities of gas-saturated crude oils at reservoir temperature and pressure. Numbers on curves are viscosities of gas-free crudes, at reservoir temperature, in centipoises. (*After Beal, AIME Trans., 1946.*)

correlation curves was only 13.4 per cent, they should provide a good approximation when direct measurements are not available.

It will be clear from the above discussion that the viscosities of hydrocarbon fluids are sensitive to the temperature and saturation-pressure conditions. Accordingly in dynamical problems of oil production in which the fluid viscosity is involved it will be necessary to establish the viscosity values under the conditions of interest in order to obtain results of quantitative significance.

[1] The earliest work on gas-saturated oils appears to be that of C. E. Beecher and I. P. Parkhurst, *AIME Trans.*, 1926, p. 51. But little additional work was done on the problem until some 10 years later, when Sage and Lacey undertook systematic investigations. Many of the published data now available were obtained by them.

[2] This is taken from Beal, *loc. cit.*

2.12. The Surface and Interfacial Tensions of the Fluids in Oil-bearing Rocks.—While there has been but little direct quantitative application of the numerical values of surface- and interfacial-tension data to production problems, except with relation to capillary phenomena (cf. Secs. 7.8 to 7.10), an appreciation of their magnitudes is helpful in an understanding of many of the details of production mechanisms. A brief survey will therefore be given of the available empirical data pertaining to the surface and interfacial tensions of the fluids of interest in oil production.

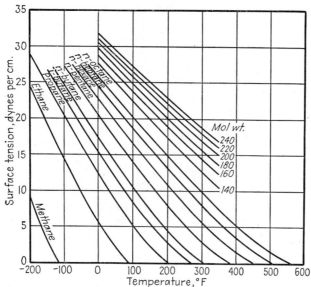

Fig. 2.51. The variation with temperature of the surface tension of paraffin hydrocarbons. (*After Katz, Monroe, and Trainer, Petroleum Technology, 1943.*)

For the oil phase in equilibrium with the atmosphere or its own vapor a correlation of experimentally determined surface-tension data on the lower paraffin hydrocarbons and extrapolation to higher molecular weights give the curves plotted in Fig. 2.51.[1] As is to be expected from general physical considerations, the surface tension is seen to decrease with decreasing molecular weight and increasing temperature. At 70°F the surface tensions of actual crude oils have been found to lie within the range of 24 to 38 dynes/cm.

The surface tension of pure water varies approximately linearly from 72.5 dynes/cm at 70°F to 60.1 dynes/cm at 200°F, with an average gradient of 0.095 (dyne/cm)/°F. In the case of oil-field brines, however, two

[1] D. L. Katz, R. R. Monroe, and R. P. Trainer, *Petroleum Technology,* **6,** 1 (Sept. 1, 1943).

opposing effects influence the surface tension. Inorganic mineral salts tend to increase it somewhat.[1] On the other hand, surface-active agents that may be dissolved in the brine, because of contact with oil, reduce the surface tension. These lead to an observed range at standard conditions of 59 to 76 dynes/cm.

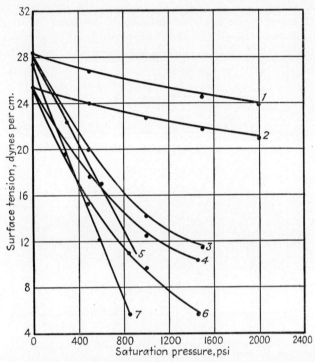

Fig. 2.52. The surface tension of crude oils and benzene as a function of the saturation pressure. (1) Sugarland, Tex., air. (2) Salt Creek, Wyo., air. (3) Sugarland, Tex., gas from Santa Fe Springs. (4) Salt Creek, Wyo., gas from Santa Fe Springs. (5) Sugarland, Tex., carbon dioxide. (6) Salt Creek, Wyo., gas from Bartlesville. (7) Benzene, carbon dioxide. (*After Swartz, Physics, 1931.*)

Data on the effect of dissolved gas on the surface tension of crude oils are rather meager. The original work of Beecher and Parkhurst[2] and the more recent studies of Jones[3] on an Iranian crude were carried out only to 500 to 600 psi. The results of experiments over a more extended pressure range,[4] at 88°F, are reproduced in Fig. 2.52. It will be seen that the

[1] Examples of the composition and concentration of the salt content of oil-field brines are given in Sec. 2.13.

[2] Beecher and Parkhurst, *loc. cit.*

[3] D. T. Jones, *AIME Trans.*, **118**, 81 (1936).

[4] C. A. Swartz, *Physics*, **1**, 245 (1931).

dissolved natural gas very greatly reduces the surface tension, so that at reservoir pressure the surface tension of the saturated crude may be expected to be much lower than at atmospheric pressure.   Comparison of curves I and II, obtained with air, and the others where natural gas or carbon dioxide was used shows that the amount and nature of the gas in solution are important factors in determining the magnitude of the surface-tension reduction.   On the other hand, the data of Jones at temperatures of 85, 95, and 105°F show that the direct effect of a temperature increase, as indicated in Fig. 2.51, more than counterbalances the effect of decreased gas solubility at the higher temperatures.   Hence at reservoir

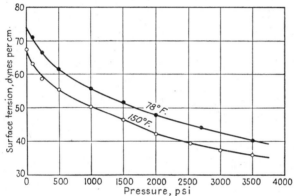

Fig. 2.53.   The surface tension of water against gas as a function of the saturation pressure. (*After Hocott, AIME Trans., 1939.*)

pressures and temperatures the actual surface tension of the crude may be even lower[1] than might be inferred from Fig. 2.52.   In fact, values of the order of 1 dyne/cm may well be expected at pressures and temperatures exceeding 3,000 psi and 150°F, respectively.   Of course, at the critical state of the hydrocarbon system the surface tension will vanish completely.

The effect of dissolved gas on the surface tension of a subsurface water has been measured[2] up to pressures of 3,500 psi, using a lean stripped natural gas.[3]   The results are reproduced in Fig. 2.53.   It will be seen

[1] Katz, Monroe, and Trainer, *loc. cit.*, have developed a procedure for calculating the surface tensions of hydrocarbons systems.   This, however, involves a knowledge of both the vapor- and liquid-phase compositions, their densities, and constants of the individual components, *i.e.*, the "parachors," related to the surface tensions of the separate components.

[2] C. R. Hocott, *AIME Trans.*, **132**, 184 (1939).

[3] *p-v-T* data on natural gas and water or brine systems have been reported by C. R. Dodson and M. B. Standing, *California Oil World*, **37**, 21 (Dec. 15, 1944).   These indicate solubilities in distilled water of the order of 15 ft³/bbl at 2,500 psi and 21 ft³/bbl

that the percentage reduction in surface tension is quite large, although less than in the case of crude oils.

While the interfacial interaction between the gas and oil in producing systems involves the surface tension of the oil in equilibrium with the gas, the interaction between the water and oil depends on the interfacial tension

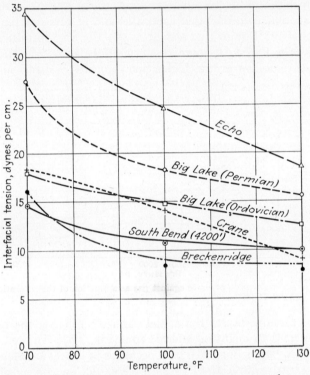

FIG. 2.54. Temperature–interfacial-tension plots for oil-water systems from several Northwest Texas fields. (*After Livingston, Petroleum Technology, 1938.*)

between these phases. An extended series of measurements has been reported[1] on the interfacial tensions of crude-oil–water systems from Texas

at 5,000 psi, with a small temperature variation. The effect of the brine salinity is to reduce the solubility by about 5 per cent per 1 per cent solid content of the brine. The dissolved gas was also found to increase the liquid compressibility by approximately 9 per cent per 10 ft³/bbl. The water content of the gas phase in equilibrium with the water-gas solution increases rapidly with increasing temperature, and with decreasing pressure below 1,500 psi, being about 1 bbl/10⁶ ft³ gas at 2,000 psi and 200°F. A solid brine content of 2 per cent reduces the water in the vapor phase by 5 per cent. Cf. also J. J. McKetta, Jr., and D. L. Katz, *AIME Trans.*, **170**, 34 (1947); and W. H. Ashby, Jr., and M. F. Hawkins, AIME meetings, October, 1948, Dallas, Tex., where somewhat different numerical data are reported.

[1] H. K. Livingston, *Petroleum Technology*, **1**, 1 (November, 1938).

oil fields. The data for several of the Northwest Texas fields are plotted in Fig. 2.54. The data for all the samples studied are listed in Table 5. In the latter are also included the surface tensions of the oil and water as well as their other physical constants. It will be observed that the variation of the interfacial tensions with temperature is generally somewhat more rapid than that of the surface tension of the crude oil, as would be

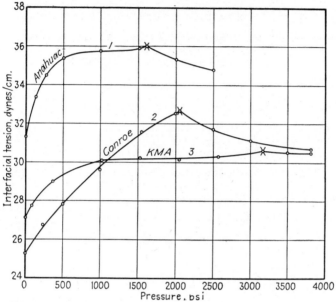

Fig. 2.55. The interfacial tension between water and oils from several fields as a function of the saturation pressure. Crosses indicate bubble points. (*After Hocott, AIME Trans., 1939.*)

anticipated from Fig. 2.51. While there appears to be no quantitative relation between the interfacial tension and the nature of the crude, the general trend of the data indicates lower interfacial tensions for the higher API gravity oils.

The effect of gas in solution and pressure on the interfacial tension of oil-water systems is illustrated by the data plotted[1] in Fig. 2.55. Curve 1 refers to a 33.5°API crude from the Anahuac field, liberating 640 ft³ gas per barrel when flashed[2] from 3,120 psi, 178°F, to 78°F and atmospheric pressure. A 36.9°API crude from Conroe gave the data for curve 2. This oil liberated 550 ft³ gas per barrel on flashing from 2,035 psi, 170°F, to 78°F and atmospheric. A 41.3°API crude from the KMA field which had a solubility of 650 ft³ gas per barrel at 1,570 psi, 130°F, gave the results

[1] Hocott, *loc. cit.*
[2] By flash liberation (cf. p. 76).

Table 5

Field	Formation	Depth, ft	Oil gravity, °API	Water sp. gravity, 60°F/60°F	μ_o 130°F	σ_o	σ_w	σ_{ow} 70°F	σ_{ow} 100°F	σ_{ow} 130°F
Breckenridge	Marble Falls	3,200	38.2	1.1299	3.51	28.8	67.6	19.0	10.9	
South Bend	Strawn	2,300	36.1	1.1037	4.73	29.9	61.5	29.1	21.4	9.6
Banyon	Austin	2,135	37.0	1.0270	3.97	29.3	72.5	24.4	17.4	
South Bend	Marble Falls	3,900	25.5	1.0106	24.37	31.8	71.4	24.5	16.9	12.9
Banyon	Austin	2,255	37.9	1.0280	3.89	28.9	72.1	16.9	13.6	16.7
Salt Flats	Edwards	2,700	34.9	1.0329	5.65	30.0	73.0	23.0	16.9	15.5
Driscoll	Catahoula	3,929	26.0	1.0520	4.86	32.4	61.4	20.4	16.0	
Wortham	Woodbine	2,800	38.3	1.0166	3.58	29.2	63.2	13.6	7.3	13.2
Wortham	Corsicana	2,200	22.4	1.0185	55.30	33.2	59.6	25.1	16.7	17.6
Mexia	Woodbine	3,000	36.4	1.0256	4.96	30.0	66.2	21.4	19.0	
Powell	Woodbine	3,000	22.9	1.0113	14.87	31.4	66.2	22.6	15.0	
Wortham	Woodbine	2,800	22.2	1.0218	0.73	33.3	66.0	25.8	15.6	
Mexia	Woodbine	3,000	36.6	1.0227	5.23	30.2	66.6	15.0	9.2	10.0
Breckenridge	Marble Falls	3,200	37.7	1.1362	3.34	28.9	70.1	16.2	8.5	
Breckenridge	Marble Falls	3,200	36.6	1.1249	3.97	29.4	74.1	15.5	11.3	10.1
South Bend	Marble Falls	4,200	38.6	1.0694	3.58	28.9	68.1	14.8	10.8	
Van	Woodbine	2,710	33.9	1.0483	10.09	29.0	61.7	18.1	16.2	14.3
Raccoon Bend	Cockfield	3,007	34.1	1.0268	3.82	31.6	69.8	24.7	14.6	
Tomball	Cockfield	5,541	41.6	1.0765	1.78	28.5	62.0	14.1	13.6	7.8
Van	Woodbine	2,710	35.0	1.0516	6.65	28.8	64.1	17.9	15.0	10.8
Saxet	Catahoula	4,308	26.2	1.0640	7.34	32.0	65.2	17.2	11.5	14.1
Catahoula	Catahoula	4,308	27.1	1.0602	6.35	32.3	66.5	20.9	16.5	8.7
Pierce Junction	Frio	4,325	29.4	1.1008	6.61	31.0	62.0	16.9	13.9	2.1
Pierce Junction	Frio	4,335	22.2	1.0716	17.80	32.6	64.1	20.7	12.9	9.6
East Texas	Woodbine	3,660	36.5	1.0444	4.54	28.2	68.6	19.7	10.9	13.9
East Texas	Woodbine	3,660	39.5	1.0433	102.20	27.5	70.2	31.4	17.9	
Goose Creek	Pliocene	1,470	14.2	1.0088	3.16	34.1	63.7	24.4	19.5	12.5
Goose Creek	Pliocene	2,040	21.1	1.0126	54.70	33.6	63.5	18.8	15.3	12.5
Goose Creek	Mio-Pliocene	2,560	21.2	1.0086	43.90	33.3	64.2	18.1	12.9	14.8
Talco	Glen Rose	5,000	23.0	1.0126	63.10	31.9	73.9	20.5	18.8	12.5
Big Lake	Ordovician	8,300	42.6	1.0671	3.20	28.5	63.3	18.1	14.8	15.7
Big Lake	Permian	3,000	38.0	1.0897	3.71	27.9	66.2	27.3	18.3	
Crane	Permian	3,500	31.1	1.0071	7.50	29.5	68.2	18.6	14.8	7.8
Echo	Frye (Pa.)	1,950	38.4	1.0773	2.91	27.8	49.5	34.3	24.6	18.6

μ_o = oil viscosity, cp. σ_o = surface tension of oil, dynes/cm.

σ_w = surface tension of water, dynes/cm.

σ_{ow} = interfacial tension of oil and water, dynes/cm.

plotted in curve 3.   In all cases the experiments were conducted at the corresponding reservoir temperatures, namely, 178, 170, and 130°F.   It will be noted that the interfacial tension of individual oil-water systems increases as the amount of gas dissolved increases but drops slowly as the pressure is raised above the bubble point.

2.13. Oil-field Waters.—As noted in Sec. 1.2, all oil- and gas-bearing rocks contain a water phase, generally termed "connate" or "interstitial" water.   Moreover, most oil-bearing reservoirs are bounded, in part at least, by contiguous strata containing water.   Such formation waters[1] have mineral-content compositions that are often characteristic of the geologic stratum and may be used to locate the source of waters being produced from a well.   They have also served to clarify the geological histories of oil accumulations.   While in some cases the oil-field waters appear to be similar to ocean waters, there are large variations among the compositions of brines from oil-bearing rocks.   The composition is usually expressed by the weight concentrations of the anions and cations in parts per million (ppm).   As the brine composition will enter explicitly in the treatment of specific reservoir-engineering problems to be presented in the following chapters only occasionally,[2] no further discussion of this subject will be presented here beyond the listing of typical analyses in Table 6.[3]

2.14. Summary.—The physical behavior of petroleum hydrocarbon fluids may be described in terms of the functional relationships between the pressure, volume, and temperature variables associated with these fluids.   The individual components, or pure hydrocarbons, are characterized by simple graphical representations similar to those obtaining for other pure compounds.   At constant temperature the volume of the gas phase decreases as the pressure is increased from initial low values until a

---

[1] The identity in composition of connate water and that in surrounding water strata is a moot question, though this is generally assumed to be the case.   Evidence supporting this assumption is cited by K. B. Barnes and R. W. Woods, *API Proc.*, **20**, *Bull.* 224, p. 140 (1939).   On the other hand, the salinity analyses of the interstitial water in cores taken with oil-base drilling fluid often show such variations within individual producing zones as to raise serious doubt that the connate water is in composition equilibrium either within the oil-producing formation or with bounding formation waters.

[2] Cf., for example, the interaction of water with clay-containing or "dirty" sands (Sec. 3.7).

[3] While these analyses are referred to as "typical," they should not be interpreted as averages characteristic of the formations from which the samples were taken.   Detailed discussions of water-analyses data for various producing districts in this country may be found in Part VI of "Problems of Petroleum Geology," AAPG, 1934, from which the analyses for samples 1, 2, 3, 11, 13, 14, 15, and 18 were taken.   Nos. 4, 5, 6, 8, 12, and 19 represent samples analyzed by L. C. Case.

"dew point" is reached at which liquid condensation begins. As the volume is further decreased, the pressure remains constant until all the

TABLE 6.—ANALYSES OF OIL-FIELD BRINES
(In ppm)

Field	Formation	Cl	SO₄	CO₃	HCO₃	Na	K*	Ca	Mg	Miscellaneous	Total
Sharon, Pa.	Berea Sand	6,740	0	0	250	3,440		700	160	....	11,290
Evans City, Pa.	3d Venango Sand,	88,820	180	0	40	38,660		13,790	2,220	....	143,710
Bell Run, W. Va.	Salt Sand	64,930	0	0	90	30,450		7,580	1,620	....	104,670
Kawkawlin, Mich.	Dundee Limestone	161,200	155	—	60	66,280	—	25,740	4,670	....	258,105
Reed City, Mich.	Marston Dolomite	156,235	265	—	20	59,080	—	28,440	5,155	....	249,185
Seminole, Okla.	Wilcox Sand	89,990	515	—	65	44,020	—	9,460	1,990	....	146,040
Glenn, Okla.	Arbuckle Limestone	101,715	120	—	60	50,345	—	10,160	2,120	....	164,520
Burbank, Okla.	Bartlesville Sand	107,895	—	—	35	50,000	—	14,340	1,875	....	174,145
Nikkel, Kans.	Hunton Limestone	76,797	207	—	61	40,284	—	5,440	1,790	....	124,579
East Texas, Tex.	Woodbine Sand	40,958	278	—	569	24,540		1,388	282	....	67,649
Yates, Tex.	San Andres Dolomite	2,518	2,135	251		1,624	—	587	288	Si: 18; Fe, Al: 24	7,445
Monument, N. Mex.	Grayburg Limestone	6,630	160	—	1,740	3,735	—	515	365	....	13,145
Shelby, Mont.	Madison Limestone	1,179	659	71	1,270	1,322	—	143	66	....	3,388
Frannie Dome, Wyo.	Tensleep Sand	27	2,303	0	691	51	—	760	240	....	4,022
Grass Creek, Wyo.	Frontier Sand	256	6	1,211	—	1,087	—	5	2	....	2,565
Fruitvale, Calif.	Fairhaven Sand	245	11	60	2,235	1,024		8	10	....	3,593
Edison, Calif.	Upper Duff Sand	79	4	29	648	299	—	17	1	....	962
Ventura Ave., Calif.	Pico-Repetto Sand	14,212	59	—	1,846	8,607	—	729	242	NH₃: 66; Si: 170; Fe, Al: 160	26,091
Bay City, Mich.	Salina Dolomite	403,207	0	—	1,208	21	21,362	206,300	7,300	Br: 3,500	642,798
Ocean waters (mean)	....	19,410	2,700	70	—	10,710	390	420	1,300	....	35,000

* In most brine analyses Na and K are determined together and reported as Na. For those cases, however, where the alkali content was listed as Na + K, the values are tabulated between the Na and K columns. A similar uncertainty obtains in the case of some of the reported values of CO₃ and HCO₃ concentrations.

gas disappears (the "bubble point"; cf. Fig. 2.1). Further decrease of volume requires a rapid increase of pressure, corresponding to the low compressibility of the liquid phase. The volume span of the constant-pressure segment, during which condensation or vaporization (on volume increase) takes place, decreases as the temperature is raised. Ultimately its length vanishes as the temperature is further increased, and the dew point and bubble point merge. The properties of the gas and liquid phases become identical, and the system is then in its "critical state." At still higher temperatures the hydrocarbon persists in a single phase, which is, strictly speaking, neither a gas nor a liquid.

The unique feature of the pressure, temperature, and volume behavior of the pure hydrocarbons is that the dew-point and bubble-point pressures are the same for fixed temperatures. The pressure-temperature diagram, therefore, consists of a unique and single-valued monotonic curve, namely, the vapor-pressure curve of the compound (cf. Fig. 2.5). This simplicity breaks down in the case of multicomponent and even binary hydrocarbon mixtures. During isothermal condensation of the gas phase the pressure increases, so that the bubble-point pressure exceeds the dew-point pressure (cf. Fig. 2.6). The curve in the pressure-volume plane representing states of the saturated vapor, *i.e.*, the dew-point curve, and that for the saturated liquid, *i.e.*, the bubble-point curve, will in general join at a pressure lower than the maximum of the combined boundary curve (cf. Fig. 2.7). This immediately implies that for binary and multicomponent systems the critical pressure and temperature, which lie at the junction of the dew-point and bubble-point curves, no longer represent the maximal values for which two phase conditions can exist. Rather, the critical state is defined by the identity of intensive properties of the coexisting gas and liquid phases. And in general the critical pressures and temperatures will be lower than the maximal pressure (cricondenbar) and temperature (cricondentherm) of the two-phase boundary curves.

If the pressure on a hydrocarbon mixture is maintained constant and at a value intermediate between the critical and maximal two-phase pressures, while the temperature or volume is varied monotonically so as to take the system from one bubble-point state to another or from one dew-point state to another, "retrograde" phenomena will be encountered. If both terminal states are bubble points, for example, the mixture will be wholly in the liquid phase at these states. However, during the passage between these bubble points the system will be in the two-phase region. Hence there will have occurred both the development of a gas phase and then its shrinkage and ultimate disappearance while the volume or temperature has continued to change monotonically. Now at lower pressures and temperatures the "normal" behavior of fluid systems is that an isobaric increase in temperature will lead to continued vaporization of the liquid phase. Such vaporization will also occur during the first stages of the temperature increase above that of the low-temperature bubble point in a retrograde-vaporization process. In addition, however, the later stages involve a condensation of the gas phase and its ultimate disappearance as the temperature is further increased to the high-temperature bubble point. It is this latter phenomenon that is expressed by the description "retrograde." On the other hand, from the nature of the boundary curves near the critical state it may be shown that such retrograde behavior will occur under appropriate conditions with virtually all multicomponent hydrocarbon mixtures. While these phenomena do not often arise in crude-oil systems, they are of fundamental importance in the production of condensate or distillate fields.

Both the dew-point gas and the bubble-point liquid have compositions identical with that of the system as a whole. But in the two-phase region the compositions of the coexisting phases differ from each other and from that of the composite system. The liquid first condensing from the dew-point gas will be relatively rich in the heavy components. As the pressure is raised and more liquid condenses, more of the volatile components enter the liquid phase, thus making it lighter. At the same time the removal of the condensable components lowers the molecular weight of the gas phase. At the bubble point where the liquid phase assumes the composition of the mixture the last trace of gas phase has a maximum concentration of the most volatile constituents (cf. Figs. 2.10 and 2.17).

For correlating and predicting the volumetric behavior of general multicomponent hydrocarbon systems, such as mixtures of natural gas and crude oil, it generally suffices to express the gross composition in terms of the relative amounts of gas and liquid phase at standard conditions. These are given by the "gas-oil ratio," the common units being cubic feet per barrel. The gas phase may be further identified by its composition, though often only its average molecular weight or specific gravity (with respect to air) will be available. Even the latter, however, suffices for a prediction of its behavior as a pure gas phase, *i.e.*, its equation of state. The latter may be expressed as that of an ideal gas with a correction factor, termed the "deviation factor," "compressibility factor," or "supercompressibility factor," multiplying the temperature term [cf. Eq. 2.7(1)]. Correlations have been developed for predicting these deviation factors from the "reduced" pressures and temperatures of the system (cf. Fig. 2.20). These latter are the actual pressures and temperatures divided by the "pseudocritical" pressures and temperatures, which are either composition weighted averages of those of the pure components or such as may be inferred from the gas gravity (cf. Fig. 2.21).

The liquid phase of complex hydrocarbon systems is even more difficult to describe completely by any simple representation. However, as a minimum, its gravity, average molecular weight, and some indication of its composition are essential for any significant characterization. On the other hand, even a complete statement of the composition through the heptanes cannot as yet provide quantitative predictions of its behavior near the critical region.

In the two-phase region of saturated-gas and crude-oil mixtures, direct experimentation on particular gas and oil samples suffices to show the qualitative features common to virtually all such systems.   Thus the specific volumes (reciprocals of the composite density) decrease with increasing pressure.   The rates of decrease, which are a measure of the compressibility, drop discontinuously at the bubble points at temperatures removed from the critical (cf. Fig. 2.22).   Similarly at fixed pressures the specific volume increases with the temperature, changes in slope again occurring at the bubble points, corresponding to the fall in thermal-expansion coefficient on entering the saturated-liquid region (cf. Fig. 2.23).   If the system is maintained at constant volume, the pressure increases as the temperature is raised, the slopes of the curves once more changing abruptly at the bubble points, as in the case of binary mixtures and even pure components.   And during an isothermal approach to the bubble point the density of each phase approaches the other in magnitude (cf. Fig. 2.24), though complete equality is not reached except at the critical point.

As is to be expected, the "formation volume" (the volume of a mixture containing a unit volume of oil at standard conditions) increases, at fixed pressure and temperature, with increasing gas-oil ratio.   The rate of increase is, of course, lower for the higher pressures (cf. Fig. 2.25).   The solubility, or volume of gas in solution in the bubble-point liquid, decreases as the temperature is raised but increases with increasing pressure.   The initial rise with pressure is generally quite steep and then tapers off to an approximately linear variation, which may, however, again become accelerated at pressures of several thousand psi (cf. Fig. 2.27).   Moreover, these solubilities increase with increasing crude API gravity.

For purposes of predicting the volumetric behavior of natural-gas and crude-oil systems under conditions removed from the critical state, approximate correlation charts and procedures are available.   Their accuracy depends, of course, on the completeness of the basic data known initially.   Thus, if only the crude gravity and saturation pressure are known, the gas solubility can be estimated to about 25 per cent (cf. Fig. 2.28) except at very low pressures.   If the crude gravity and total gas solubility are known, the gravity of the liberated gas can be predicted to an accuracy of 10 per cent (cf. Fig. 2.29).   From the solubility one can also estimate, to 15 per cent, the shrinkage of the bubble-point liquid on gas evolution (cf. Fig. 2.30).   Correlation graphs are also available for determining the direct thermal shrinkage of the residual oil on lowering its temperature from the reservoir temperature to standard (60°F) (cf. Fig. 2.31).

If the gas gravity, crude gravity, solubility, and reservoir pressure and temperature are specified, the oil shrinkage can be calculated with a probable error of only 5 per cent. This calculation is based on correlation curves giving the apparent liquid density, in crudes of various gravities, of the dissolved gas, as a function of the gas gravity (cf. Fig. 2.32). And if the gas composition is also known, the calculations can be further refined by adding up the volume contributions of the individual dissolved components to obtain the total volume of the bubble-point liquid. Moreover, the effective densities of the dissolved methane and ethane can be corrected for the effect of their concentrations in the system as a whole and the average density of the heavier constituents (cf. Fig. 2.35).

The prediction of the compositions of the gas and liquid phases of a composite hydrocarbon mixture, as its pressure and temperature are varied, requires a different type of approach. By a detailed composition analysis of the coexisting phases in actual mixtures at various pressures and temperatures, the ratios of the mole fractions in the gas phase to those in the liquid phase for the separate components have been computed and represented in tabular or graphical form as functions of pressure and temperature. If the hydrocarbon system obeyed ideal-solution and -gas laws, the values of these ratios could be directly calculated as functions only of the pressure and temperature. While the deviations of the observed ratios from the ideal behavior and calculated values were not surprising, it was hoped they would still remain independent of the over-all composition of the system and hence were termed "equilibrium *constants.*" As data accumulated, however, these ratios were found to be definitely dependent on the composition of the mixture as a whole, especially at higher pressures and, in the case of methane and ethane, even at low pressure (cf. Tables 2 and 3). To avoid the implication of the nonexistent constancy it seems more appropriate to refer to these quantities as "equilibrium *ratios.*"

Even if the equilibrium ratios be regarded as strictly empirical data, they can be used within the range of their validity to calculate the compositions of the coexisting phases of hydrocarbon mixtures and their dew points and bubble points [cf. Eqs. 2.9(4) to 2.9(8)]. In computing the phase compositions the over-all mole-fraction distribution between the two phases is determined first by trial and error. With this established, the mole-fraction compositions of the separate phases are then obtained by only slight additional calculation. Both for these and the boundary-point determinations it is, of course, necessary to preassign the over-all composition of the composite mixture.

It is possible by such procedures to calculate the results of the flash or differential liberation of the gas from a bubble point or two-phase system at high pressure and temperature, such as may obtain in the reservoir or at the bottom of the flow string, to lower pressures and temperatures, such as may correspond to stock-tank or separator conditions. Moreover the computations may be repeated in sequence, with appropriate modifications, so as to show the influence of various types and conditions of single or multiple separator stages on the amount and character of the gas liberated and oil recovered. Such applications (cf. Fig. 2.43) are especially amenable to treatment by these methods, as they generally involve pressure and temperature ranges for which the available equilibrium ratios are sufficiently complete to avoid uncertain extrapolations.

From the physical definition of the critical point it follows that the equilibrium ratios for all the individual components must converge to unity at the critical pressure and temperature. The fact that the location of the critical state itself is sensitive to the over-all composition of the hydrocarbon system makes it evident that the equilibrium ratios, as well, will be sensitive to the over-all composition in the region near the critical. Qualitatively, however, they have a characteristic pressure variation, at constant temperature, that is common to all systems thus far studied. At low pressures they all vary approximately inversely as the pressure, as would be predicted by ideal-solution and -gas-law theory. And within normal temperature ranges methane continues to decrease monotonically with increasing pressure toward the value unity. In the case of the other components, however, they first decline to a value unity at a pressure which is approximately equal to the vapor pressure of the pure component for propane through the heptanes, and then fall to minimal values in the pressure range of 500 to 1,000 psi, after which they increase and tend to approach unity once more as the pressure continues to rise. As is to be expected, they decrease with decreasing temperature and increasing molecular weight of the pure component. For practical purposes the individual components are generally distinguished only through the hexanes, the heavier residuals being lumped together as "heptanes plus." The ratios for this latter component are especially sensitive at high pressures to the nature of the oil that it represents.

In discussing the dynamical behavior of gas and oil during their flow through reservoir rocks, the fluid viscosity plays an important role. As in the case of the purely thermodynamic properties of hydrocarbon systems the viscosity is a function not only of the pressure and temperature but of the composition and phase as well and must, moreover, be treated from the empirical point of view. While for quantitative purposes direct measurements should be made on each particular system of immediate interest, there are sufficient data already available to provide at least a guide for estimating the viscosity when special measurements are not feasible.

The viscosities of hydrocarbon gases at atmospheric pressure are generally lower than those of other common gases or vapors, except for hydrogen. They increase with increasing temperature and decreasing molecular weight (cf. Fig. 2.44). They also increase as the pressure is raised above atmospheric. And at high pressures their atmospheric variations with temperature and molecular weight become reversed to simulate those characteristic of hydrocarbon as well as other liquids (cf. Fig. 2.45). Over the temperature range of 60 to 300°F the atmospheric-pressure viscosity of crude oils may vary by a factor of 50 (cf. Fig. 2.47), this factor generally decreasing as the 60°F viscosity decreases. The atmospheric-pressure (gas-free) viscosity at fixed temperature decreases monotonically with increasing crude gravity (cf. Fig. 2.48). Pressure alone will increase the viscosity of crudes by 10 to 20 per cent per 1,000 psi increment (cf. Fig. 2.49).

Gas dissolved in crude oils at elevated pressure will materially lower the viscosity. The viscosity of gas-containing oils may be correlated with the amount of gas in solution (cf. Fig. 2.50). The total decrease in viscosity caused by a given amount of solution gas increases with the gas-free viscosity.

The surface and interfacial tensions of the fluids in oil-bearing rocks are also of importance in understanding the detailed mechanisms of oil production, though their quantitative use is still quite limited. The observed surface tensions of crude oils at standard conditions range from 24 to 38 dynes/cm. They decrease with increasing temperature and with decreasing molecular weight (cf. Fig. 2.51).

Because of the opposing effects of inorganic mineral salts and dissolved surface-active agents, oil-field brines may have surface tensions either less or greater than pure water (72.6 dynes/cm at 70°F). Reported data range from 59 to 76 dynes/cm.

Dissolved gas lowers the surface tension of crude oils, the magnitude of the reduction depending on the amount and nature of the gas (cf. Fig. 2.52). At reservoir temperatures and pressures exceeding 150°F and 3,000 psi, gas-saturated oils may have surface tensions of the order of only 1 dyne/cm. While the solubility of natural gas in water is much less than in oil, that which is dissolved at elevated pressures in subsurface waters will also lead to very appreciable reductions in surface tension (cf. Fig. 2.53).

The interfacial tensions between oil and water also decrease with increasing temperature, though not so rapidly as in the case of the individual surface tensions (cf. Fig. 2.54). At standard conditions, measured values of the interfacial tension between crude oils and their brines generally range between 15 and 30 dynes/cm (cf. Table 5) and are usually lower for the higher API gravity crudes. The gas in solution increases the interfacial tension, but increasing the pressure above the bubble point will cause a decrease (cf. Fig. 2.55).

The inorganic content and composition of oil-field brines vary over extremely wide ranges—from concentrations so low as to make the water only brackish, to an extreme of 642,000 ppm and a specific gravity of 1.458 (cf. Table 6).   The brine composition is generally of but minor importance with respect to the thermodynamic phenomena associated with oil production.   However, such data are often of value in determining the source of water produced with the oil and in constructing the geological history of the oil-bearing reservoir.

CHAPTER 3

THE PROPERTIES OF OIL–BEARING FORMATIONS PERTAINING TO OIL PRODUCTION; CORE ANALYSIS

3.1. Porosity.—One of the most basic properties of practical interest of an oil-bearing rock is its volumetric capacity to hold petroleum fluids. The quantitative measure of this capacity is termed "porosity," which may be defined as 1 minus the fraction of the bulk volume of the rock comprised of solid matter. If the rock contains no fluid matter, the porosity is then that fraction of the bulk volume that is void space. These fractions are generally expressed as percentages.

The practical significance of the numerical value of the porosity will be readily appreciated on observing that one acre-foot[1] of rock of 10 per cent porosity *could* contain 775.8 bbl oil,[2] whereas the same volume of rock of 30 per cent porosity *could* hold 2,327.5 bbl oil. The porosities of oil-bearing and commercially productive sandstone formations generally lie between 10 and 35 per cent and thus have volumetric capacities ranging from 776 to 2,715.4 bbl/acre-ft.

The porosity of oil-productive limestones and dolomites covers a greater range. When cavernous or fractured they may have locally, in effect, 100 per cent porosity, although under such conditions the term loses its practical meaning. On the other hand, some oil-producing intergranular limestones and dolomites have porosities as low as 4 to 6 per cent.

From the inherent nature of consolidated rocks, it is apparent that, while their voids, or pores, will in general be interconnected, some may become sealed off during the course of cementation of the rock. The above definition, referring to the total void space, thus provides a measure of the "total porosity." It is clear, however, that only the "effective porosity," namely, that encompassing only the intercommunicating void space, is of practical interest with respect to oil production. It is common practice, therefore, to use the term porosity for that of "effective porosity,"

---

[1] An acre-foot is the common unit of volume for oil-reservoir rocks and simply equals the bulk volume of a slab 1 ft thick and 1 acre in area, and is equivalent to 43,560 ft³.

[2] This refers to the total physical volume. If this should be the content under reservoir conditions, its surface equivalent—stock-tank oil—would be less because of volume shrinkage as its solution gas is liberated and the temperature is reduced.

with the understanding that it is the latter which is implied. Moreover, as the sealed pore space seldom exceeds 10 per cent and is generally less than 5 per cent of the total void volume, simple routine determinations of total porosity are often considered and reported as being the equivalent of the effective porosity.

No attempt will be made here to describe in detail all the various methods that have been developed for determining porosity, although several will be briefly discussed to illustrate the types of measurement involved.[1] All, of course, are based on the equation underlying the definition of porosity,

$$f \text{ (in \%)} = \frac{100(V_b - V_g)}{V_b} = 100\left(1 - \frac{V_g}{V_b}\right),\tag{1}$$

where V_g is the grain volume of a sample of bulk volume V_b.

The bulk volume is usually determined by a displacement process.[2] If the pores are small and not subject to ready penetration by mercury, the latter may be used as the immersion liquid and the volume computed from the loss in weight on immersion or by direct observation of the volumetric displacement. If mercury penetration is feared, the sample may be covered with a thin coating of paraffin, collodion, or a similar material and an immersion liquid, as water, used for the loss in weight or displacement-volume measurement. Another alternative is that of first deliberately saturating the rock specimen with the liquid to be used for immersion, such as tetrachloroethane, and then directly observing the volumetric displacement when the sample is immersed in that liquid in a suitably designed volumeter, such as a Russell tube (cf. Fig. 3.1). To operate the latter, it is first inverted and the upper bulb filled with immersion liquid to the calibrated zero mark. The saturated sample is then placed in the stopper, the two bulbs joined, and the whole set upright, the

[1] Detailed techniques are given by G. H. Fancher, J. A. Lewis, and K. B. Barnes, *Pennsylvania State College Min. Ind. Exper. Sta. Bull.* 12 (1933) (this paper also contains an extensive historical review of the development of methods for determining both porosity and permeability); A. F. Melcher, *AIME Trans.*, **65**, 469 (1921), *AAPG Bull.*, **8**, 716 (1924); W. L. Russell, *AAPG Bull.*, **10**, 931 (1926); H. R. Brankstone, W. B. Gealy, and W. O. Smith, *AAPG Bull.*, **16**, 915 (1932); E. W. Washburn and E. N. Bunting, *Am. Ceramic Soc. Jour.*, **5**, 48, 112 (1922); C. J. Coberly and A. B. Stevens, *AIME Trans.*, **103**, 261 (1933); K. B. Barnes, *Pennsylvania State College Min. Ind. Exper. Sta. Bull.* 10 (1931), and *API Drilling and Production Practice*, 1936, p. 191. Some methods have even been patented, *e.g.*, E. O. Mattock, U.S. Patent No. 2,323,556 (July 6, 1943), and W. L. Horner, U.S. Patent Nos. 2,327,642 (Aug. 24, 1943) and 2,345,535 (Mar. 28, 1944).

[2] It is presupposed that the fluids in the rock sample have been extracted in an apparatus, such as a Soxhlet, by a suitable solvent, as carbon tetrachloride, butanol, pentane, or naphtha. If the sample is of simple and well-defined geometry, the bulk volume can, of course, be readily computed from its dimensions.

liquid rise in the graduated connecting tubes giving directly the bulk volume.

By crushing or grinding the dried bulk-volume sample the grain volume V_g can be determined similarly with the volumeter. An application of Eq. (1) will then give the porosity.

Another commonly used method is that of measuring the open-pore

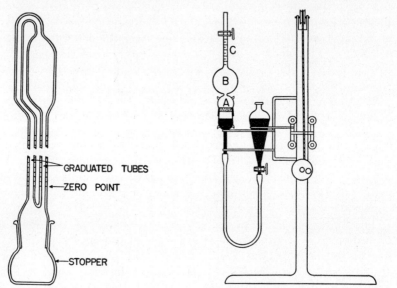

FIG. 3.1. The Russell apparatus for porosity determinations. (*After Russell, AAPG Bull., 1926.*)

FIG. 3.2. The Washburn-Bunting type of porosimeter.

volume $V_b - V_g$ directly and thus obtaining the value of the effective porosity. A convenient type of apparatus for this purpose is the Washburn-Bunting porosimeter illustrated in Fig. 3.2. Here the air in the pores of the rock specimen is first trapped at atmospheric pressure by letting the mercury rise past[1] the sample A (of 10 to 20 cc) through the expansion chamber B and the graduated capillary C to the stopcock, which is then closed. On lowering the mercury level below A the air in the pores expands into B. It is then displaced into the graduated tube and its volume measured at atmospheric pressure by raising the level bulb so that the mercury height in it and in the graduated tube are the same. The measured air is then expelled, and the process is repeated to determine the volume of the residual air trapped in the sample after the first expansion. The sum of the volumes obtained in each of such measurements until the

[1] This apparatus can be modified to eliminate the immersion of the sample in the mercury and thus avoid the possibility of mercury penetration.

residual air is negligible gives the total effective pore volume of the sample. On determining the bulk volume by a method such as has been previously described, the porosity is readily calculated.

The open-pore volume may also be determined by measuring the gain in weight resulting from a saturation of the dry sample with a suitable liquid of known density, such as tetrachloroethane. While the saturation process may be facilitated by preliminary evacuation of the core[1] before immersion in the saturation liquid, completeness of saturation may still require a long time, especially for tight rocks. The same difficulty and uncertainty as to the achievement of complete saturation arise in the method of determining the bulk volume by immersion of such saturated cores in a Russell volumeter.

Still another method for determining the effective porosity is based on a measurement of the aggregate composite grain volume. This may be carried out by placing the bulk sample in a chamber of accurately known volume, filled with gas or air at a known pressure, and measuring the final pressure when the chamber is placed in communication with another vessel that is of accurately known volume and that is initially evacuated or filled with the same gas at a known but different pressure. By a simple application of the gas laws the total grain volume V_g can be calculated by the formula

$$\frac{p_{i1}(V_1 - V_g)}{Z_1} + \frac{p_{i2}V_2}{Z_2} = \frac{p_f(V_1 + V_2 - V_g)}{Z_f}, \tag{2}$$

where p_{i1}, p_{i2} are the initial pressures in the two chambers, of volumes V_1, V_2, p_f is the final common pressure, and Z_1, Z_2, Z_f are the deviation factors[2] (from ideal-gas-law behavior) at pressures p_{i1}, p_{i2}, and p_f. The temperature is assumed the same and constant throughout. Or the gas in the original chamber may be bled off to the atmosphere through a gas meter or into a calibrated burette[3] and the grain volume calculated by the equation

$$\left(\frac{p_{i1}}{Z_1} - \frac{p_o}{Z_o}\right)(V_1 - V_g) = \frac{p_o V}{Z_o}, \tag{3}$$

---

[1] While the term "core" is used here to represent generally a rock sample, in the oil industry it usually refers to the cylindrical specimens of underground strata cut by special "core bits" and collected in "core barrels" in a drilling operation denoted as "coring." "Core analysis" comprises the processes of determining the properties and fluid contents of these cores, the actual measurements usually being made on broken fragments or specimens specially cut from the cores. Such shaped or prepared samples are also referred to as cores.

[2] Cf. Secs. 2.2 and 2.7.

[3] Cf. D. B. Taliaferro, T. W. Johnson, and E. J. Dewees, *U.S. Bur. Mines Rept. Inv.* 3352 (1937).

where V is the measured gas volume, p_o the atmospheric pressure, and Z_o the deviation factor at p_o. In using these methods the bulk volume must still be determined independently, by such procedures as have been previously outlined.

Finally, the porosity may be determined by measuring the bulk and grain densities γ_b and γ_g and computing f by replacing V_g/V_b in Eq. (1) by γ_b/γ_g. While this method, in principle, involves the additional step of weighing the sample used for getting the bulk and grain volumes, it permits a material simplification in routine analysis. This arises from the fact that the grain density often may be assumed constant,[1] as 2.65 to 2.70 gm/cc, for strata of similar geological character. All that needs to be measured for the individual samples is then the bulk density.

While determinations of the porosity as well as other physical characteristics of rock formations are generally made on cores deliberately cut for such purposes during drilling, a method has been developed for measuring the porosity of drill cuttings.[2] The new feature of this method is the use of a capillary diaphragm to remove the free surplus water from the surfaces of the water-saturated cuttings before their immersion in a calibrated vessel for determining their displacement and bulk volume. The grain volume is obtained during the saturation process in a calibrated graduate by measuring the weight (and volume) of water required to fill the graduate to a fixed mark. The need for the special technique in the bulk-volume determination arises from the relatively large surface area of the cuttings —those left on a No. 6 gauge screen—and potentially large error that might be caused if the free water adhering to the grains of rock were not removed.

3.2. The Fluid Content of Subsurface Rocks.[3]—Cores, or underground rock samples, when brought to the surface, are universally found to have entrained in their pores varying amounts of liquid. An important phase

[1] For example, consecutive cores whose grain densities were measured by C. R. Fettke [*AIME Petroleum Devel. and Technology*, 1926, p. 219] gave mean values for five cores from the Second Sand of the Venango group of 2.679, 2.666, 2.666, 2.661, and 2.658 gm/cc with average deviations of only ± 0.027, ± 0.011, ± 0.012, ± 0.010, and ± 0.009 for 20 to 26 samples in each group. Moreover, some measurements on large intergranular limestone cores reported by F. B. Plummer and P. F. Tapp [*AAPG Bull.*, **27**, 64 (1943)] indicate that even these will show average variations of only about 1 per cent among samples from the same stratum.

[2] M. A. Westbrook and J. F. Redmond, *AIME Trans.*, **165**, 219 (1946).

[3] Detailed descriptions of techniques for this phase of core analysis will be found in the papers by E. S. Hill, *Pennsylvania State College Bull.*, **19**, 41 (1935); W. L. Horner, *Petroleum Eng.*, **6**, 33 (April, 1935); Barnes, *loc. cit.*; H. C. Pyle and P. H. Jones, *API Drilling and Production Practice*, 1936, p. 171; H. G. Botset, *Petroleum Technology*, **1**, 1 (August, 1938); M. D. Taylor, *Oil and Gas Jour.*, **37**, 59 (June 16, 1938); D. B. Taliaferro and G. B. Spencer, *U.S. Bur. Mines Rept. Inv.* 3535 (1940); S. T. Yuster, *Oil Weekly*, **113**, 20 (Mar. 20. 1944); cf. also R. J. Schilthuis. *AIME Trans.*, **127**, 199

of core analysis is the determination of the nature and amount of this liquid content. The method used for this purpose is generally that either of retorting or of solvent extraction and distillation.

For retort determinations, 100 to 200 gm of rock sample, broken into small pieces, is placed in a metal (usually copper or cast iron) retort, and heat is applied to distill off the liquids contained in the rock. A typical

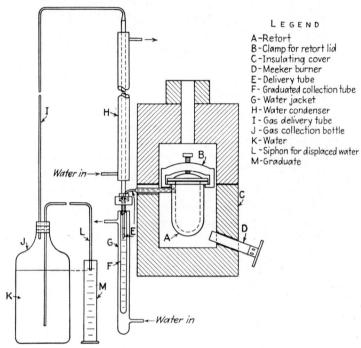

L E G E N D

A –Retort
B –Clamp for retort lid
C –Insulating cover
D –Meeker burner
E – Delivery tube
F – Graduated collection tube
G – Water jacket
H – Water condenser
I – Gas delivery tube
J – Gas collection bottle
K – Water
L – Siphon for displaced water
M – Graduate

Fig. 3.3.   A diagram of a retort apparatus for fluid-content determination.   (*After Yuster and Levine, Oil Weekly, 1938.*)

construction and arrangement of connections are shown in Fig. 3.3.[1] The heating is carried out in two stages. The first, at a temperature of 350 to

(1938), where a method of water determination is described in which the water vaporized from a crushed sample in a combustion boat is passed through and absorbed by a dehydrite tube. From the gain in weight of the latter and total loss in weight of the combustion boat on additional heating to expel the oil, both the water and oil contents are readily determined. This method has been patented by M. Williams, U.S. Patent No. 2,269,569 (Jan. 13, 1942). Other patents relating to methods for determining fluid contents of rocks are K. H. Clough, U.S. Patent No. 2,095,056 (Oct. 5, 1937); W. L. Horner, U.S. Patent Nos. 2,282,654 (May 12, 1942) and 2,345,535 (Mar. 28, 1944); M. C. Leverett, U.S. Patent No. 2,330,721 (Sept. 28, 1943); F. C. Kelton, U.S. Patent No. 2,352,638 (July 4, 1944).

[1] S. T. Yuster and J. S. Levine, *Oil Weekly,* **89,** 22 (May 23, 1938). The flame heating shown in this figure can, of course, be replaced by electric heating coils.

400°F, serves to vaporize the water[1] and all but the heaviest fractions of the oil that may be contained in the pores of the rock. This requires a period of 40 min to 1 hr. The heavy oil residuals are then distilled off by raising the temperature to about 1100°F and maintaining it for ½ hr or more. The material vaporized in these processes is collected and measured in a graduated receiver tube at the bottom of a condenser column into which the vapors pass from the retort. To facilitate separation between the oil and water the contents of the receiver tube may be centrifuged.[2] These volumes are usually expressed as percentage saturations of the pore space. The latter may be computed from the bulk volumes of the retort samples—directly by a displacement method, or by calculation from the weight and bulk density—combined with the value of the porosity as measured on a neighboring sample.

For accurate work a number of corrections have to be applied in using this method. These include the excess collected water due to water of crystallization of the rock that is driven off in heating, the loss in oil recovery due to cracking and coking, and the correction for the change in gravity of the oil after distillation as compared with that in the pores of the rock before distillation. To obtain the proper correction data, blank and calibration tests must be run on synthetic mixtures or aggregates of the actual rock, oil, and water.

The procedure for determining fluid saturations by extraction consists essentially in a combination of the ASTM distillation[3] and Soxhlet extraction methods. A convenient form of the apparatus in which these two steps are carried out simultaneously is that illustrated in Fig. 3.4.[4]

---

[1] It may be noted that the presence of water in oil-field cores was not always a fact accepted axiomatically. Among the earliest determinations of core water are those of C. R. Fettke, *loc. cit.* and *AIME Trans.*, **82**, 221 (1928). While water saturations of 26 to 48 per cent were found in cores from the Second Sand of the Venango group, near Oil City, Pa., and as much as 57 per cent in a core of a Bradford Sand, the low salt content of the cores and the fact that the fields had produced little or no water, except around the edges, cast doubt on the hypothesis that any significant part of the core water was of connate origin. N. T. Lintrop and V. M. Nickolaef (*AAPG Bull.*, **13**, 811, 1929) appear to have been the first who explicitly concluded from their observations that virgin oil-bearing rocks contain an appreciable content (20 to 30 per cent) of connate water and performed experiments with sands indicating that such residual-water contents would be expected after the entry of oil into a water-saturated porous medium.

[2] The above-described details should not be considered as rigid instructions. Different laboratories vary in the exact procedures followed; *e.g.*, the collection bottle for noncondensable vapors, shown in Fig. 3.3, is generally omitted in apparatus set up for routine work.

[3] Cf. Standard Method of Test for Water in Petroleum Products and Other Bituminous Materials, ASTM Designation D95–30, III, p. 230, 1939.

[4] This construction is similar to that described by Yuster, *loc. cit.*

In a porous weighed thimble 50 to 100 gm of sample, broken into small pieces, is placed, and the combined weight is accurately measured to determine the original weight of the sample. The thimble is then suspended in a flask containing a suitable solvent, such as light naphtha, which is immiscible with water but miscible with oil and which has a boiling point near that of water.[1] The flask is connected to a reflux condenser fitted with a water trap. Heat is applied to the flask until both the collection of water in the trap and extraction of the oil in the samples by the solvent are completed. The latter stage may require 4 or 5 times as long a period as for the water distillation, the total time usually extending from 2 to 5 hr,[2] but often requiring even 12 to 24 hr for complete extraction. On completion of extraction the heat is removed, and after slight cooling the thimble is removed and the solvent driven off in a drying oven at about 105°C. The thimble is then reweighed, and the total loss in weight is thus determined. That part of the total due to the water distillation is given by the volume or weight of the water collected in the trap.

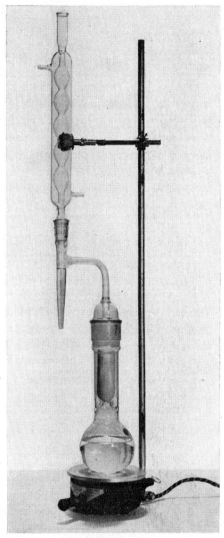

FIG. 3.4.   An extraction apparatus for fluid-content determination.

[1] Other solvents that may be used for the distillation of the water are toluene, tetrachloroethane, and xylene.

[2] Such details of technique as extraction or heating times, size of samples, and temperatures should not be considered as rigidly fixed but must be adjusted to suit the particular character of the rock sample being analyzed.

The difference represents the weight of the extracted oil. From a knowledge of the gravity of the oil involved its volume may be computed.[1] These volumes of water and oil contents may again be converted into percentages of pore saturation by combining with data on the bulk density and porosity of neighboring rock samples, together with the measured weight of the dried sample used for the extraction.

Here, too, additional refinements may be made by applying corrections to the collected distilled water volume due to the presence of salts in the water before extraction and by more direct determinations of the oil density[2] after distilling off the solvent from the residual solvent-oil extract.

When the rock samples are very small or have abnormally low oil saturations, the accuracy of the oil-content determination can be increased by using a vacuum distillation procedure.[3] This makes possible the use of lower distillation temperatures, all-glass apparatus, and gravimetric methods for measuring the fluid loss. The evacuated core is left clean by the distillation process so that it can be used also for porosity and air-permeability measurements, although the loss of water of crystallization during distillation may result in spurious values for the permeability to water. This method should be of special value in the analysis of samples from condensate producing formations.

3.3. Salinity.—Measurements of the salinity of core waters are based on the assumption that all the soluble chlorides of the rock sample are present in the interstitial waters. Accordingly the procedure consists merely in leaching a weighed (5 to 20 gm) and crushed sample with 100 to 150 cc distilled water, until it can be assumed that all the chlorides have been dissolved. The water and sand are then separated by filtration, and a suitable portion of the filtrate is titrated.[4] The resulting chloride content is expressed as grains of sodium chloride per gallon of water or as ppm (1 gr/gal = 17.1 ppm) after having previously determined the quantity of core water per unit weight of rock sample.

---

[1] If a volatile solvent, as pentane, is used for the oil extraction, the oil volume may also be determined by carefully distilling the solvent from the oil-solvent solution. This implies, of course, that the water has been previously removed by a higher boiling point fluid, as in the conventional ASTM procedure.

[2] The determination of the oil gravity during the course of the general core analysis, though of somewhat limited accuracy, is itself of considerable practical importance in the over-all evaluation of the section cored. If the amount of oil recovered from the cores is too small for conventional gravimetric density determination, the index of refraction may be used to estimate the API gravity, subject to calibration correlations [cf. H. D. Hedberg, *AAPG Bull.*, **21**, 1464 (1937)].

[3] C. M. Beeson and N. Johnston, *AIME Trans.*, **165**, 116 (1946).

[4] The salt content can also be determined by a conductivity measurement [cf. A. P. Clark, Jr., *Producers Monthly*, **11**, 11 (July, 1947)].

3.4. Permeability—Darcy's Law.—It will be obvious even on the most cursory consideration that, in addition to measures of the volumetric capacity of petroleum-bearing rocks and of their fluid contents, the ability of the rock to transmit fluids—ultimately to the producing wells—is an equally significant characteristic. Indeed, most of the procedures already outlined for determining the porosity and fluid saturations have been based upon the a priori assumption that the fluids entrained in the rocks can be extracted from them. In fact, it is just to the extent that this assumption is inherently satisfied that measurements of the static volumetric properties of a rock are warranted and have basic meaning from a practical point of view.

In the oil industry the term universally used to describe this capacity of rocks to transmit fluids is "permeability." To go beyond this qualitative definition it is necessary to provide an empirical basis for its quantitative determination. Darcy's law constitutes such a basis.

Darcy's law, first formulated by H. Darcy[1] in 1856, states that the rate of flow of a homogeneous fluid through a porous medium is proportional to the pressure or hydraulic gradient and to the cross-sectional area normal to the direction of flow and inversely proportional to the viscosity of the fluid. The law as stated in this form involves both limitations as well as generalizations with respect to the statement originally propounded by Darcy. Among the former is the reference to *homogeneous*, or single-phase, fluids, with the implication that for heterogeneous, or multiphase-fluid, systems the law may not be valid.[2] While Darcy, too, was concerned only with a single-phase fluid, namely, water, the question of the validity of his empirically established relationship in case the porous medium may also contain a second phase, as air evolved from the water, was not contemplated by him—and, of course, for good and obvious reasons. Moreover, Darcy, working with unconsolidated sands, did not envisage the possibility of the fluid interacting with and perhaps hydrating the porous medium. Yet the absence of such effects must be explicitly assumed in the above-stated formulation of the law.

Of greater significance, however, are three generalizations of Darcy's original formulation implied by the above statement. The first is the generalized use of the term "fluid." This encompasses any liquid—not only water—and all gases not interacting with the porous medium. The distinction between them is expressed only by the differences in their viscosity, which is explicitly introduced in the relationship between rate of flow and driving force. The residual coefficient of proportionality, the

---

[1] H. Darcy, "Les Fontaines publiques de la ville de Dijon," 1856.

[2] The generalizations required in the treatment of multiphase-fluid systems will be discussed in Chap. 7.

permeability, is thus independent of the nature of the fluid and is determined solely by the structure of the porous medium. Conversely, this permeability coefficient constitutes the complete dynamical characterization of the porous material as a fluid-transmitting medium.

The second generalization pertains to the inclusion as measures of the driving force both pressure and hydraulic gradients. Naturally only the latter were considered by Darcy, who experimented exclusively with water. While, strictly speaking, in the oil industry gravitational or fluid-head forces are almost universally present as driving agents affecting the movement of oil, gas, and water, in the great majority of practical oil-production systems the predominating driving forces are those due to fluid pressure gradients. Moreover, because the flow resistance of petroleum-bearing rocks is generally much greater than that of the filter beds studied by Darcy, the laboratory study of Darcy's law and determination of the permeability of the former often demand that, to obtain flow rates of measurable magnitude, pressure gradients be used which are correspondingly much greater than could be provided by fluid heads of reasonable height. Accordingly the above statement of Darcy's law implies that pressure gradients and hydraulic heads are equivalent, interchangeable, and superposable.

Finally, whereas Darcy's experiments and empirical representation pertained to essentially linear-flow systems, the above generalized formulation is to be construed as applying also to multidimensional-flow systems, the rate of flow and driving forces being considered as vectors. Admittedly, this constitutes largely a heuristic generalization, which is supported mainly by the lack of evidence to the contrary. Indeed, practically all laboratory flow investigations have been conducted with linear systems, and more general types of flow for which the law has been verified have been limited almost exclusively to axially symmetrical radial systems. Yet the intuitive argument for the validity of the generalized proposition seems so compelling that all doubt has been cast aside, subject only to the discovery of hitherto unsuspected empirical facts.

The above considerations make it clear that the generalized Darcy's law for homogeneous fluids may be stated analytically as

$$v_s = -\frac{k}{\mu}\left(\frac{\partial p}{\partial s} - \gamma g \cos \theta\right), \tag{1}$$

where v_s is the volume flux (per unit area) in the direction s, making an angle θ with the vertical, of a fluid of viscosity μ and density γ, p is the fluid pressure, and g is the acceleration of gravity. Equation (1) is a differential relationship pertaining to any point within the porous medium, so that v_s is to be considered as a function of that point, as must also p,

$\frac{\partial p}{\partial s}$, θ, and even γ, if the fluid is compressible. μ is also to be considered as variable and ultimately dependent on s, if conditions dictate that such variations should be taken into account.

While the permeability coefficient k is the unique factor in Eq. (1) representing the porous medium, it may, of course, vary from point to point in the medium, if the latter is not uniform throughout.   In such circumstances k must be considered as a variable in integrating Eq. (1). Moreover, it is to be recognized that the medium may be anisotropic, in which case the value of k to be used in Eq. (1) is that corresponding to the direction s.   As it stands, Eq. (1) refers to an *isotropic* medium.

Written out explicitly for a three-dimensional cartesian coordinate system, Eq. (1) takes the form

$$
\left.
\begin{aligned}
v_x &= -\frac{k_x}{\mu}\frac{\partial p}{\partial x}, \\
v_y &= -\frac{k_y}{\mu}\frac{\partial p}{\partial y}, \\
v_z &= -\frac{k_z}{\mu}\left(\frac{\partial p}{\partial z} - \gamma g\right),
\end{aligned}
\right\}
\tag{2}
$$

where the z axis is directed downward and the possibly different permeabilities in the different directions are so designated explicitly.

Although the above formulation of Darcy's law has been accepted throughout the oil industry as generally valid and will be so considered here, it is desirable to define more closely the conditions under which its validity may be assumed.   The first, of course, is that the fluid system under consideration be single-phase, or homogeneous.   Otherwise, as will be seen later (Chap. 7), the permeability k will depend on the fluid distribution within, as well as the structure of, the porous medium, although the formal relationship between rate of flow and pressure gradient will remain the same.

Second, Eq. (1) or Eqs. (2) will ultimately break down if the rate of flow or pressure gradient is increased indefinitely.   The deviation from these equations that then develops corresponds to an apparent decrease in the permeability or to a less than linear increase of rate of flow with pressure gradient.   By analogy with experience in hydraulics the range of fluid velocities over which Darcy's law is obeyed is termed "viscous," and that of higher velocities over which it breaks down is referred to as "turbulent." This analogy is quite logical, since, in a manner quite similar to that well established in hydraulics, it has been found that the transition region between the two conditions of flow is rather well fixed by the Reynolds number.   The latter, a dimensionless quantity, is simply the quantity

$dv\gamma/\mu$, where v is the mean velocity, γ the density, and μ the viscosity of the fluid and d is a linear dimension measuring the size of the passageway. For hydraulic systems and, in particular, flow through pipes, d is taken as the pipe diameter, and the transition between viscous and turbulent flow takes place at a Reynolds number of the order of 2,000. For lower values the velocity is directly proportional to the pressure or hydraulic gradient, whereas for higher Reynolds numbers the velocity rises approximately as the square root of the pressure gradient. Moreover this transition takes place abruptly as the critical Reynolds number is exceeded.

In the case of flow through porous media the transition is rather gradual. This is undoubtedly due to the spread in sizes of the pores and passageways, so that the turbulence disseminates gradually from the larger to the smaller pores as the velocity is increased. Indeed, this very distribution of pore sizes makes the definition of the dimension d rather arbitrary. Moreover, the pore diameters themselves are very difficult to measure. As a practical necessity, therefore, d is generally taken as an average[1] grain (rather than pore) diameter. With such values of d, it has been found that the transition region lies at values of Reynolds number of the order of 1 to 10. A "friction-factor" chart demonstrating this observation is shown in Fig. 3.5.[2] Here the ordinates are the friction factors $\zeta = d\Delta p/2L\gamma v^2$, commonly used in pipe-flow analysis, $\Delta p/L$ representing the pressure gradient. It will be clear that the linear segments with 45° slopes in Fig. 3.5 imply a proportionality between v and $\Delta p/L$, as required by Darcy's law. It will be noted that Fig. 3.5 includes data for both consolidated and unconsolidated sands and also for both gases and liquids.

For a complex system as the multiply connected flow channels of a porous medium it is not to be expected that the simple Reynolds number criterion, even if it could be exactly defined, could alone uniquely describe the character of the flow. Evidently the distribution of pore and grain sizes, the shapes of the grains and geometrical characteristics of the pores, and the effect on the latter of cementing material are all factors that will influence the nature of the fluid flow. Yet for practical purposes it is not necessary to attempt to take these factors into account provided that a conservative limit is chosen for the Reynolds number below which Darcy's law will be assumed to hold. Such a value appears to be 1.* Accordingly,

---

[1] The average commonly used is the root-mean-cube over a distribution determined by a Tyler or U.S. Standard sieve analysis.

[2] Fancher, Lewis, and Barnes, *loc. cit.*

* In Sec. 2.2 of "The Flow of Homogeneous Fluids through Porous Media" by M. Muskat (McGraw-Hill Book Company, Inc., 1937) is cited additional evidence for the validity of Darcy's law below Reynolds numbers of the order of 1. A still more recent confirmation is provided by the data of H. E. Rose [*Proc. Inst. Mech. Eng.*, **153**, 141 (1945)] on relatively coarse unconsolidated porous materials.

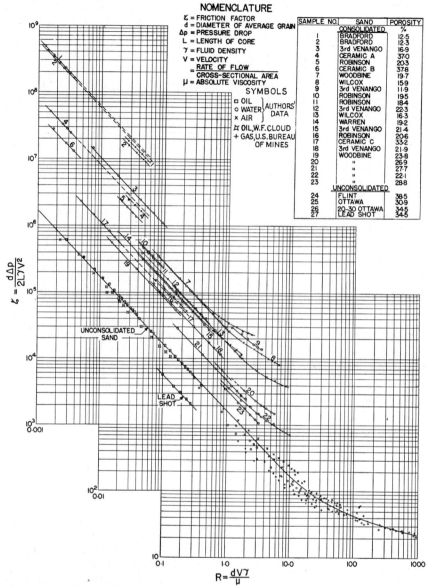

NOMENCLATURE

ζ = FRICTION FACTOR
d = DIAMETER OF AVERAGE GRAIN
Δp = PRESSURE DROP
L = LENGTH OF CORE
γ = FLUID DENSITY
V = VELOCITY
= $\dfrac{\text{RATE OF FLOW}}{\text{CROSS-SECTIONAL AREA}}$
μ = ABSOLUTE VISCOSITY

SYMBOLS

□ OIL   }
○ WATER } AUTHORS'
× AIR   } DATA
⋈ OIL, W.F. CLOUD
+ GAS, U.S. BUREAU
      OF MINES

SAMPLE NO.	SAND	POROSITY %
	CONSOLIDATED	
1	BRADFORD	12·5
2	BRADFORD	12·3
3	3rd VENANGO	16·9
4	CERAMIC A	37·0
5	ROBINSON	20·3
6	CERAMIC B	37·8
7	WOODBINE	19·7
8	WILCOX	15·9
9	3rd VENANGO	11·9
10	ROBINSON	19·5
11	ROBINSON	18·4
12	3rd VENANGO	22·3
13	WILCOX	16·3
14	WARREN	19·2
15	3rd VENANGO	21·4
16	ROBINSON	20·6
17	CERAMIC C	33·2
18	3rd VENANGO	21·9
19	WOODBINE	23·8
20	"	26·9
21	"	27·7
22	"	22·1
23	"	28·8
	UNCONSOLIDATED	
24	FLINT	38·5
25	OTTAWA	30·9
26	20-30 OTTAWA	34·5
27	LEAD SHOT	34·5

$\zeta = \dfrac{d\Delta p}{2L\gamma V^2}$

UNCONSOLIDATED SAND

LEAD SHOT

$R = \dfrac{dV\gamma}{\mu}$

FIG. 3.5. Friction-factor vs. Reynolds number data for the flow of homogeneous fluids through porous media. (*After Fancher, Lewis, and Barnes, Pennsylvania State College Min. Ind. Exper. Sta. Bull., 1933.*)

and by analogy with hydraulic terminology, it will hereafter be assumed that the flow will be "viscous" and Darcy's law will be obeyed provided that the Reynolds number $dv\gamma/\mu$ does not exceed unity. Conversely, the flow will be considered to be viscous when it is described by Darcy's law.

Fortunately, the limit of unity in the Reynolds number for the range of validity of Darcy's law encompasses the great majority of homogeneous-flow systems, producing gas or oil, of practical interest. Thus in a 6-in.-diameter well producing 1,000 bbl/day of an oil of 1 cp viscosity and a gravity of 30°API (0.876 gm/cc) from a 10-ft sand the flux velocity at the exposed sand face will be 0.126 cm/sec and the Reynolds number only 0.55, even if the effective d of the sand grains and pores is 0.5 mm. And within the sand body the Reynolds number will decrease inversely as the radial distance from the well bore, so that at 10 ft it will have fallen to only 0.014. Likewise, for a 10-ft pay producing 500,000 ft^3/day of a gas of gravity 0.70 as compared with air, and viscosity 170 micropoises, at the formation temperature of 150°F and bottom-hole pressure, the Reynolds number at the well bore will be 2.69 and 10 ft within the sand only 0.067.

If under special circumstances the flow should become turbulent, the apparent sand resistance will be higher than computed by using Darcy's law. One may then attempt to correct the latter by generalizing the linear relationship between rate of flow and pressure gradient by formulas such as

$$v_s = c\left(\frac{\partial p}{\partial s}\right)^n, \quad n < 1, \quad \text{or} \quad \frac{\partial p}{\partial s} = av_s + bv_s^2. \tag{3}$$[1]

These, however, require special determinations for the constants c and n, or a and b, for each individual sand. Moreover, even when the constants are determined, the application of these equations to the actual calculation of flow characteristics becomes so difficult that only systems of the simplest geometry can be treated effectively. As a practical procedure, therefore, it seems best to revert even in such situations to the use of Darcy's law, with the understanding that the flow resistances thus computed will be too low in proportion to the extent to which the turbulent flow permeates the system and predominates over the viscous flow.

3.5. The Permeability Coefficient—The Darcy.—Accepting Darcy's law [Eqs. 3.4(2)][2] as the basic foundation for the treatment of homogeneous-flow systems, attention must next be focused on the permeability coefficient, *i.e.*, the constant of proportionality in Eqs. 3.4(2). The first point to be

[1] An early study of transients in nonviscous radial gas-flow systems, governed by the first of Eqs. (3), was reported by M. Muskat and H. G. Botset, *Physics*, **1**, 27 (1931). Detailed numerical calculations on the steady-state production capacities of radial gas-flow systems under turbulent-flow conditions have been made by J. R. Elenbaas and D. L. Katz, *AIME Trans.*, **174**, 25 (1948).

[2] The reference to Eqs. 3.4(2) in this section relates to the formal structure of the equations rather than the subdivision into the components appropriate to anisotropic media.

ascertained is the dimensional structure of this factor. Perhaps the simplest procedure would be by referring to Eqs. 3.4(2) and imposing the requirement of dimensional homogeneity on both sides of the equations. It is more instructive, however, to appeal to more general dimensional considerations. These, based on well-established principles of the theory of dimensional analysis,[1] immediately lead to the result that in any flow system the pressure gradient $\Delta p/\Delta s$ must be related to the fluid velocity v, the density γ, viscosity μ, and a linear dimension d according to the equation

$$\frac{\Delta p}{\Delta s} = \text{const } \frac{\mu^2}{\gamma d^3} F\left(\frac{dv\gamma}{\mu}\right),$$ (1)*

in which only the value of the dimensionless constant and the form of the function F may depend on the particular system in question. The argument of F will be recognized as the Reynolds number. Hence when the latter is of the order of or less than 1, the linear relationship between v and $\Delta p/\Delta s$, as expressed by Darcy's law, requires that F be linear in its argument, thus reducing Eq. (1) to

$$\frac{\Delta p}{\Delta s} = \text{const } \frac{\mu v}{d^2}, \quad \text{or} \quad : \quad v = \text{const } \frac{d^2}{\mu} \frac{\Delta p}{\Delta s}.$$ (2)

From Eq. (2) it is seen that in any "viscous-flow" system the constant of proportionality, except for the viscosity, between the velocity and pressure gradient must have the dimensions of length squared. As an example, it may be noted that for viscous flow through an open tube of diameter d the second of Eqs. (2) becomes equivalent to Poiseuille's law, the constant having the value $\frac{1}{32}$ and v denoting the average fluid velocity over the section of the tube. For flow through porous media a comparison of Eq. (2) with Darcy's law [Eqs. 3.4(2)] shows that

$$k = \text{const } d^2 = [L]^2.$$ (3)

It follows that the permeability has the dimensions L^2. As the only length of physical significance associated with the structure of a porous

---

[1] Cf., for example, P. W. Bridgman, "Dimensional Analysis," Chaps. VI and VII, Yale University Press, 1931.

* Eq. (1) only serves to group the independent dynamical variables according to dimensional requirements. Terms that are inherently dimensionless, such as the porosity, could be included in Eq. (1) either as factors or as parts of the argument of F. In fact the inclusion of a power of the porosity as a factor of $dv\gamma/\mu$ has been rather common practice in hydraulic studies of porous materials. While this appears to facilitate somewhat the correlation of data on different porous media, its physical significance cannot be determined by dimensional analysis alone. Moreover it involves the introduction of additional empirical constants. In any case the ultimate criterion for ensuring viscous-flow conditions as derived from the generalized forms of the Reynolds number is essentially equivalent to that indicated in the last section.

medium is its grain, or pore, diameter, Eq. (3) thus also implies that the permeability is proportional to the square of the grain diameter. The remaining dimensionless constant obviously depends on additional detailed features, as the porosity, grain-size distribution, shapes of the grain, and cementing material, if the medium is consolidated. While a great deal of work has been done on attempts to develop correlations between these factors and the resultant permeability,[1] they all appear to be limited in validity to the special types of grain assemblies used in the investigation. For practical purposes, therefore, it is far safer to determine directly the value of the permeability for each sand or porous medium empirically than to calculate it by a formula of uncertain accuracy and validity. Moreover, as will be shown below, the direct determination is so simple that it will in general take much less time than the measurement of such auxiliary factors as porosity, grain-size distribution, and grain-shape factors.

To make specific use of the permeability concept and of Darcy's law it is necessary to define the unit by which the magnitude of the permeability may be expressed. A simple guide to such a definition is Darcy's law itself, *i.e.*, Eqs. 3.4(2), from which it would appear that a porous medium would have a unit permeability when a unit pressure gradient will induce a unit rate of flow, as volume per unit area, of a fluid of unit viscosity. For a consistent set of units such a definition would describe an absolute unit. It turns out, however, that in the cgs system such a definition would lead to inordinately small numerical values of the permeability for porous media constituting oil- and gas-bearing rocks. Accordingly, as a practical measure a compromise semiabsolute unit has been adopted[2] in which the atmosphere has been taken as the unit of pressure and the centipoise as the unit of viscosity. By general acceptance it is now[3] known as the "darcy." Its explicit definition may be stated as follows: The permeability of a porous medium is one darcy if through it will flow 1 cc[4] per second, per square centimeter of cross section, of a fluid of 1 cp viscosity under the action of

---

[1] Cf., for example, the statistical analysis of the effect of size parameters on the permeability of unconsolidated sands reported by W. C. Krumbein and G. D. Monk, *AIME Trans.*, **151**, 103 (1943).

[2] Cf., for example, "Standard Procedure for Determining Permeability of Porous Media," API, April, 1942.

[3] The name and units were first proposed by R. D. Wyckoff, H. G. Botset, M. Muskat, and D. W. Reed [*Rev. Sci. Instr.*, **1**, 394 (1933) and *AAPG Bull.*, **18**, 161 (1934)] after it became established that the permeability could be considered as a well-defined empirical constant.

[4] For all practical purposes, and throughout this work, the cubic centimeter may be considered as exactly equivalent to the milliliter, which is often used in defining absolute units.

1 atm/cm pressure gradient. In further recognition of the practical range of actually measured permeabilities the subunit millidarcy (1 md = 0.001 darcy) is generally used in expressing the numerical values of the permeability. Commercial oil- or gas-bearing sandstones generally have permeabilities in the range of 5 to 5,000 md. The permeabilities of oil-producing intergranular limestones sometimes average as low as 1 md, whereas the apparent permeability of fractured limestone samples may be found to be many thousands of millidarcys.

Although the units darcy and millidarcy as above defined have had a virtually universal acceptance by the oil industry, they have, of course, no inherent uniqueness. Other definitions have been suggested, and different ones are actually used in other technical fields, as ground-water hydrology, textiles, and ceramics, where permeabilities of porous media are also of practical importance. Accordingly, to provide at least some bridge between the numerical values to be used here and those of interest in allied sciences, conversion factors between various possible definitions are listed in Table 1.

3.6. The Measurement of Permeability.[1]—The previous discussion has provided the basic principles that must underlie a quantitative determination of the permeability of porous media. Indeed, all that is necessary is a measurement of the rate of flow of a fluid of known viscosity through a sample of the medium, of well-defined geometry, under a measured pressure differential.[2] A convenient form of core holder and arrangement of the apparatus for gas-flow measurements, including a pressure gauge, a dial water manometer, and flowmeters (of the rotameter type) is shown in Fig. 3.6.[3] The rock may be cut by a diamond drill into a cylindrical form,

---

[1] It is assumed in the following discussion that the core samples are consolidated. Much work has been done in attempting to recompact loose sand to simulate the consolidated sand structure as it may have existed in the reservoir (cf. N. Johnston, *API Drilling and Production Practice*, 1941, p. 180, and S. H. Rockwood, API meetings, San Antonio, Tex., April, 1948). While some success has been achieved in obtaining thereby good estimates of porosity, by the use of correction factors, techniques required for duplicating original permeability values are not yet satisfactorily established.

[2] While the term "differential" connotes a limiting infinitesimal value, it will be used in this work to represent the actual difference of pressure over any linear distance of interest, as is common practice in the oil industry.

[3] The details of technique and apparatus generally vary with the personal preferences of the experimenter. No attempt will be made to discuss here the special features of the various "permeameters" used in different laboratories or sold commerically. A safe guide for the construction of the apparatus and its use will be found in the Permeability Code 27, API, April, 1942, prepared by a special committee of the API. As in the case of other phases of core analysis, some of the detailed procedures for permeability have been patented. These include G. S. Bays, U.S. Patent No. 2,293,488 (Aug. 18, 1942); J. A. Lewis, U.S. Patent No. 2,348,985 (May 16, 1944); and K. L. Hertel, U.S. Patent No. 2,352,836 (July 4, 1944).

TABLE 1.—CONVERSION TABLE FOR PERMEABILITY UNITS

Units	darcy $\dfrac{1\ cc}{sec.\ cm^2\ (atm/cm)}$	$\dfrac{1\ cc}{sec.cm^2\ (dyne/cm^2)/cm}$	$\dfrac{1\ ft^3}{sec.ft^2\ (atm/ft)}$	$\dfrac{1\ ft^3}{sec.ft^2\ (psi/ft)}$	$\dfrac{1\ cc\ H_2O\ (20°C)}{sec.cm^2\ (1\ cm\ H_2O)/cm}$	$\dfrac{1\ gal\ H_2O\ (20°C)}{min.ft^2\ (1\ ft\ H_2O)/ft}$	$\dfrac{1\ bbl\ H_2O\ (20°C)}{day.\ ft^2\ (psi/ft)}$
$\dfrac{darcy}{\dfrac{1\ cc}{sec.\ cm^2\ (atm/cm)}} =$	1	9.8692×10^{-7}	1.0764×10^{-3}	7.3243×10^{-5}	9.6301×10^{-4}	1.4181×10^{2}	1.1215
$\dfrac{1\ cc}{sec.\ cm^2\ (dyne/cm^2)/cm} =$	1.0132×10^{6}	1	1.0906×10^{3}	74.211	9.7576×10^{2}	1.4369×10^{4}	1.1364×10^{6}
$\dfrac{1\ ft^3}{sec.\ ft^2\ (atm/ft)} =$	9.2903×10^{2}	9.1688×10^{-4}	1	6.8046×10^{-2}	0.89467	13.174	1.0419×10^{3}
$\dfrac{1\ ft^3}{sec.\ ft^2\ (psi/ft)} =$	1.3653×10^{4}	1.3474×10^{-2}	14.696	1	13.148	1.9360×10^{2}	1.5312×10^{4}
$\dfrac{1\ cc\ H_2O\ (20°C)}{sec.\ cm^2\ (1\ cm\ H_2O/cm)} =$	1.0384×10^{3}	1.0248×10^{-3}	1.1178	7.6063×10^{-2}	1	14.726	1.1646×10^{3}
$\dfrac{1\ gal\ H_2O\ (20°C)}{min.\ ft^2\ (1\ ft\ H_2O/ft)} =$	70.519	6.9596×10^{-5}	7.5906×10^{-2}	5.1651×10^{-3}	6.7910×10^{-2}	1	79.087
$\dfrac{1\ bbl\ H_2O\ (20°C)}{day.\ ft^2\ (psi/ft)} =$	0.89165	8.7999×10^{-7}	9.5977×10^{-4}	6.5308×10^{-5}	8.5865×10^{-4}	1.2644×10^{-2}	1

First four units refer to a fluid of 1 cp viscosity; viscosity of water at 20°C is taken as 1.005 cp.

of the order of 2 cm in diameter and 1 to 2 cm in length, either parallel or perpendicular to the bedding plane, depending on the direction of flow of interest. The end faces may be prepared by cutting with an abrasive wheel or fracturing. In the former case plugging of the faces may be pre-

FIG. 3.6. A photograph of apparatus for measuring permeability.

vented by presoaking the core in water or other liquid that does not react with the cementing material and also feeding this liquid to the cutting wheel. To ensure that during measurement the fluid will be forced to flow only through the prepared core, the latter may be inserted into a close-fitting tapered thiokol or rubber stopper, which is squeezed against the sides of the core when placed in the core holder and the assembly is clamped or screwed tight.[1]

[1] For special flow experimentation with consolidated rock samples the sealing of the exterior surfaces may be accomplished by casting the specimen, cylindrical or rectangular, in a plastic, as lucite or bakelite [cf. R. G. Russell, F. Morgan, and M. Muskat,

In some laboratories rectangular-shaped permeability samples are used and sealed by suitably shaped rubber stoppers. The basic technique of the measurements and formula for calculating the permeability are essentially the same as in using cylindrical plugs. However, if the permeability is determined by radial-flow tests on cylindrical samples with central axial holes, such as are often used for water-flooding experiments, the radial-flow formulas to be given below must be used to calculate the permeability.

The pressure differential impressed over the core length may be determined with gauges or manometers, and the rate of flow by calibrated flowmeters or by collecting the effluent from the core in suitable graduated vessels. The fluid temperature should be observed so as to be able to fix its viscosity.

In principle the measurements may be made either with liquids or gases. If a liquid is used, one must ascertain first that it is inert to the core and does not affect the cementing material.[1] To make the results comparable with those which would be obtained using a gas, trapping of the air in the core must be eliminated. This may be done by evacuating the core prior to the first entry of liquid in it, and also deaerating the test liquid. To avoid interference by capillary forces at the core faces it is desirable to mount the core vertically with the input end at the bottom, thus ensuring that the top face is always covered with liquid. Care should also be taken to prevent plugging of the core by dissolved or suspended matter in the test liquid, which might be filtered out and retained at the surface or in the interior of the core.

When a gas is used, it should be filtered and dried. The absolute pressure as well as the differential pressure must be recorded. Of course, in all cases the original fluids entrained in the core must be thoroughly extracted before attempting a homogeneous-fluid permeability measurement.

The computation of the permeability values from the observed data[2]

AIME Trans., **170**, 51 (1947)]. In still another method the cores are inserted in thin rubber sleeves, which are pressed tight against the outside of the core by application of external pressure [cf. G. L. Hassler, R. R. Rice, and E. H. Leeman, *AIME Trans.*, **118**, 116 (1936)].

[1] Cf. Sec. 3.7 for a discussion of permeability measurements when the fluid interacts with the rock specimen.

[2] The formulas given here are based on the assumption that the permeability is measured under steady-state flow conditions, as is current common practice throughout the oil industry. It is possible, however, also to make permeability determinations from measurements of the decline in head of a column of liquid draining through the rock sample, or the fall in pressure in a gas chamber bleeding through the specimen, or the rate of pressure equalization between closed gas-collecting and -supply vessels connected through the sample [cf. J. C. Calhoun, Jr., *Petroleum Eng.*, **18**, 103 (February, 1947)].

may be made by using formulas for liquid- and gas-flow measurements, which may be developed as follows:

Liquids: Linear Flow.—For a uniform sample of constant cross section the average flux v must be constant, and hence by Darcy's law [Eqs. 3.4(2)] so must be the pressure gradient. The first of Eqs. 3.4(2) may then be written

$$k = \frac{\mu v}{dp/dx} = \frac{\mu v}{(P_1 - P_2)/L},\qquad(1)$$

where P_1, P_2 are the terminal pressures across the core, L is its length, and the sign has been adjusted to give positive numerical values. Upon denoting by Q the total flux as volume per unit time and by A the cross section of the core, so that $v = Q/A$, Eq. (1) becomes

$$k = \frac{\mu Q L}{A(P_1 - P_2)}.\qquad(2)$$

Equation (2) will give the permeability in darcys if μ is expressed in centipoises, Q in cubic centimeters per second, L in centimeters, A in square centimeters, and $P_1 - P_2$ in atmospheres.

Gases: Linear Flow.—Here, because of the expansion of the gas as it falls in pressure in flowing through the core, the volume flux v will no longer be constant. What will be constant is the mass flux of gas γv. For isothermal expansion of the gas, which should always obtain except at very high velocities, the density γ will be proportional to the pressure,[1] thus making pv or $p\,dp/dx$ constant, by Darcy's law. Accordingly,

$$2p\frac{dp}{dx} = \frac{dp^2}{dx} = \text{const} = \frac{P_1^2 - P_2^2}{L} = \frac{2\overline{P}(P_1 - P_2)}{L},$$

where $\overline{P}$ is the mean pressure $(P_1 + P_2)/2$.

Upon applying again the first of Eqs. 3.4(2), it follows that

$$k = \frac{\mu v}{dp/dx} = \frac{2\mu p v L}{P_1^2 - P_2^2} = \frac{\mu p v L}{(P_1 - P_2)\overline{P}},\qquad(3)$$

where v refers to the velocity at the pressure p. Now, as previously indicated, pv is proportional to the mass velocity of the gas. If the total mass rate of flow be denoted by Q_m, so that $Q_m = \gamma v A = \gamma_o p v A$, Eq. (3) can be rewritten as

$$k = \frac{2\mu Q_m L}{\gamma_o A(P_1^2 - P_2^2)} = \frac{2\mu Q_a L}{A(P_1^2 - P_2^2)},\qquad(4)$$

γ_o being the density at atmospheric (unit) pressure, and Q_a the *volume* rate of flow at atmospheric pressure. Noting also that $pv/\overline{P}$ is the velocity

---

[1] It is assumed here that over the pressure range used the "deviation factor" of the gas (cf. Secs. 2.2 and 2.7) is constant.

at the mean pressure $\bar{P}$, Eqs. (3) and (4) can be given the alternative form

$$k = \frac{\mu \bar{Q} L}{A(P_1 - P_2)}, \tag{5}$$

where $\bar{Q}$ is the volumetric flow rate as measured at $\bar{P}$. This latter form is evidently identical with Eq. (2) for liquid measurements, except for the special interpretation of $\bar{Q}$. In the above equations the value of k will be obtained in darcys if the units for the terms on the right-hand side are the same as those indicated in the case of Eq. (2).

Liquids: Radial Flow.—Since a general discussion of the radial flow of homogeneous fluids will be presented in Sec. 5.1, only the final formulas for permeability calculations from radial-flow experiments will be given here. Assuming that the cylindrical annular specimen has an external radius r_e and an internal radius r_i and is of length L, k can be calculated from a liquid-flow test by the formula

$$k = \frac{\mu Q \log r_e/r_i}{2\pi L(P_e - P_i)}, \tag{6}$$

where P_i, P_e are the pressures at r_i, r_e, Q is the rate of flow, and μ the viscosity. If the flow is actually homogeneous and viscous, the direction of flow will not affect the calculated value of k. The units for μ, Q, L, $P_e - P_i$, on the right-hand side of Eq. (6) should be the same as in Eq. (2) to give k in darcys, except that r_e, r_i should be simply in the same units and the natural logarithm should be used.

Gases: Radial Flow.—For gas-flow radial-permeability measurements the appropriate formula for computing k may be readily shown to be

$$k = \frac{\mu Q_m \log r_e/r_i}{\pi L \gamma_o(P_e^2 - P_i^2)} = \frac{\mu \bar{Q} \log r_e/r_i}{2\pi L(P_e - P_i)} = \frac{\mu Q_a \log r_e/r_i}{\pi L(P_e^2 - P_i^2)}, \tag{7}$$

where Q_m, $\bar{Q}$, Q_a represent, respectively, the mass flux (in grams per second), the volume flux at the mean pressure, and the volume flux at atmospheric pressure. If the latter two are expressed in cubic centimeters per second, μ in centipoises, L in centimeters, and P_e, P_i in atmospheres, the computed value of k will be in darcys.

By the use of the above equations the permeability can be calculated from a single set of observations of pressure differential and associated flow rate. It is desirable, however, always to make several independent determinations for different pressure differentials. In using liquids, such check measurements will not only give an indication of the inherent accuracy of the experiments but will also provide a confirmation of the assumption that the conditions of flow were viscous, which underlies both Darcy's law and the above equations. The constancy of the value of k, calculated by Eq. (2), or a linearity in the plot of Q vs. $P_1 - P_2$ is a necessary and sufficient condition for the validity of the assumption.

If a gas is used for the measurements, there is still another reason for making repeat measurements at various pressure differentials or, more specifically, at different mean pressures, especially for low-permeability samples. For careful experiments[1] have shown that even under viscous-flow conditions the permeability for gas flow as calculated by Eq. (4) or (5) is not constant but depends on the mean pressure. Moreover, it is greater than the value found for the same sample using inert liquids. In particular, it increases with decreasing mean pressure, apparently owing to molecular slippage of the gas at the solid grain surfaces, according to an equation of the form[2]

$$k_a = k_o\left(1 + \frac{b}{\bar{P}}\right), \tag{8}$$

where k_a is the value calculated by Eqs. (4) or (5) and k_o and b are constants. This is the type of relation to be expected from molecular considerations involving the well-known slippage phenomenon, long before[3] observed in capillary-tube experiments.

Of particular significance is the fact that the constant k_o, which is the limiting value of k_a at infinite mean pressures, is found to be equal, within experimental error, to the permeability as measured with liquids. Accordingly, it appears that the basic postulate of independence of k of the nature of the fluid, even under homogeneous fluid conditions, is not strictly accurate, except under the limiting condition of infinite mean pressure in the case of gas flow. Hence to determine with gases a value for k completely equivalent to that which would be obtained with inert liquids, it is necessary to make measurements at several values of $\bar{P}$, plot the results calculated by Eq. (4) or (5) [or Eq. (7)] vs. $1/\bar{P}$, and locate the vertical intercept by extrapolation. It is this intercept that will represent the true permeability of the porous medium and be independent of the nature of the fluid.

From a practical point of view it is pertinent to note that the variation of the apparent gas permeability with the mean pressure is large only for tight samples. As would also be expected from consideration of the slip-

---

[1] L. J. Klinkenberg, *API Drilling and Production Practice*, 1941, p. 200; H. Krutter and R. J. Day, *Oil Weekly*, **104**, 24 (Dec. 29, 1941). Even earlier studies on the variation of the permeability to gas with the mean pressure, and its interpretation in terms of slip effects, were reported by H. Adzumi, *Bull. Chem. Soc. Japan*, **12**, 304 (1937); cf. also W. D. Rose, API meetings, Chicago, Ill., November, 1948, for a general review of both early and recent work on gas slippage phenomena in porous media.

[2] While Eq. (8) is usually obeyed exactly, within experimental errors, as yet unexplained nonlinearities in k_a vs. $1/\bar{P}$ plots are sometimes observed [cf. J. C. Calhoun, Jr., and S. T. Yuster, *Producers Monthly*, **11**, 32 (August, 1947)].

[3] A. Kundt and E. Warburg, *Poggendorgs Ann. Physik*, **150**, 337, 525 (1875).

page phenomenon, experiment[1] has shown that the constant b in Eq. (8) generally decreases with increasing permeability k_o. Hence for samples of moderate or high permeability the variation of the apparent permeability with mean pressure can often be ignored, unless there are other reasons demanding a high precision in the permeability value. Because of the greater difficulties in making liquid permeability measurements, due to

TABLE 2

Core sample	Gas	Permeability for gas at		Permeability for isooctane, md	b at $\overline{P} = 1$
		Atm pressure, md	Infinite pressure, md		
A	Air	28.2	23.6	23.66	0.195
F	Air	195	170	170	0.147
H	Air	1,406	1,347	1,353	0.044
C	Hydrogen	50	32	32.1	0.563
C	Air	45	32	32.1	0.406
C	Carbon dioxide	42	32	32.1	0.313
L	Hydrogen	5.64	2.75	2.55	1.051
L	Nitrogen	4.41	2.75	2.55	0.604
L	Carbon dioxide	3.84	2.75	2.55	0.396
M	Hydrogen	15.92	11.10	10.45	0.434
M	Nitrogen	13.65	11.10	10.45	0.230
M	Carbon dioxide	12.83	11.10	10.45	0.156
N	Hydrogen	20.8	14.76	14.68	0.409
N	Nitrogen	18.75	14.76	14.68	0.270
O	Hydrogen	44.70	35.50	36.20	0.259
O	Nitrogen	41.65	35.50	36.20	0.173
P	Nitrogen	68.9	60.2	61.2	0.145
R	Nitrogen	182.3	166.6	166.1	0.094
S	Nitrogen	223.0	204.3	190.7	0.092

problems of eliminating air trapping and core plugging, routine permeability analyses are generally made with air, in spite of the possible complications due to the slippage effect.

Some of the data of Klinkenberg showing the difference between the permeability at atmospheric pressure and the limiting value at infinite mean pressure and the equivalence of the latter to those determined with

[1] While the data listed in Table 2 show only a general trend of decreasing b with increasing k_o, the experiments of Krutter and Day (loc. cit.) on cores from a single sand (Second Venango) gave a well-defined inverse-power relation between b and k_o. Theoretically b should vary as $1/\sqrt{k_o}$, since b is essentially proportional to the reciprocal of the mean pore radius, but data on different sands give different slopes for the log-log plots of b vs. k_o. A simple interpretation of the numerical value of b is that it is equal to the mean pressure at which the apparent gas permeability k_a is just twice the extrapolated or inert liquid permeability.

isooctane are listed in Table 2. Also included are the values of b corresponding to the listed permeability values.[1]

3.7. The Permeability of "Dirty" Sands.—It was explicitly indicated in the discussions thus far relating to the definition and measurement of permeability that the considerations were based on the assumptions that the fluid was homogeneous, that the flow was viscous, and that the fluid did not interact with the porous medium. The severe limitation, from the point of view of actual oil-producing systems, implied by the first assumption is fully recognized and will be considered in detail in Chap. 7 and the following chapters. On the other hand, while strictly homogeneous-fluid systems perhaps never occur in actual oil-producing reservoirs, the measurement of the homogeneous-fluid permeability provides a convenient and useful reference base for the consideration of permeabilities for multi-phase-flow systems.

The assumption of viscous flow does not demand a serious restriction on the validity of the flow analyses of practical oil-producing systems based on it. For, as noted in Sec. 3.4, there is good reason to believe[2] that except possibly in the immediate vicinity of well bores producing at very high rates the flow should lie in the viscous region.

On the other hand, the assumption that the fluid does not interact with the porous medium, has, in recent years, been found to be completely invalid in the case of many oil-producing horizons. This situation has been especially emphasized in the case of California reservoir rocks,[3] where it was found that the permeability of extracted core samples to water was generally much lower than to air. While no simple correlations were found among the many measurements, the permeability to salt water, when lower than to air, was usually higher than to fresh water. In fact, in many cases it was observed that the permeability to fresh water was virtually zero. Some typical published data[4] are given in Table 3.

---

[1] The variations of the b in Table 2 for the same samples with the nature of the gas are also in accord with the slippage-effect interpretation (cf. Calhoun and Yuster, and Rose, *loc. cit.*). It is to be noted, however, that the limiting k_o values are independent of the gas used.

[2] While recent work of L. Grunberg and A. H. Nissan [*Jour. Inst. Petroleum*, **29**, 193 (1943)] appears to cast serious doubt on the inherent applicability of Darcy's law in the region of very low fluid velocities, their results have not been confirmed by others and their experimental data seem open to question [cf. J. C. Calhoun, Jr., and S. T. Yuster, *Producers Monthly*, **11**, 22 (September, 1947) and Rose, *loc. cit.*].

[3] N. Johnston and C. M. Beeson, *AIME Trans.*, **160**, 41 (1945). The earliest report of differences between permeabilities to air and water in consolidated sands, after correcting for the viscosities, and its interpretation as the result of the hydration of the intergranular clays appears to be that of Fancher, Lewis, and Barnes, *op. cit.* Very much earlier, however, this effect had been observed and reported with respect to soils [cf. W. H. Green and G. A. Ampt, *Jour. Agr. Sci.*, **4**, 1 (1911)].

[4] Johnston and Beeson, *loc. cit.*

TABLE 3.—PERMEABILITIES OF SANDSTONE CORES TO AIR, SALT WATER, AND FRESH WATER

(In md)

Air	Salt water	Fresh water	Air	Salt water	Fresh water
18,800	15,800	15,100	173	74	0.8
1,690	1,690	1,670	112	0.8	0.5
3,540	2,093	2.4	105	0.9	0.9
34,800	23,600	9.9	81	76	66
2,560	216	0.0	9.5	6.3	5.7
1,020	114	20	92	89	12
1,490	0.45	0.0	31	31	12
645	573	568	28	3.3	0.0
565	505	210	6.9	6.1	0.06
438	360	4.9	5.5	0.07	0.07
175	153	3.9	5.8	0.2	0.0
705	147	0.0			

A series of measurements in 15 samples along 23 ft in a well bore, tabulated in Table 4, shows strikingly the contact—between samples 10 and 11—between a massive clean sand and a laminar sediment containing hydratable material.

TABLE 4.—AIR AND WATER PERMEABILITY SHOWING CONTACT BETWEEN CLEAN AND DIRTY SAND

(In md)

Sample no.	Permeability to		
	Air	Salt water	Fresh water
1	3,970	635	217
2	4,790	2,020	1,650
3	4,490	965	313
4	2,640	734	520
5	4,810	2,670	2,340
6	5,280	2,400	1,980
7	5,730	2,270	1,400
8	10,900	4,660	1,740
9	6,600	3,250	2,680
10	5,270	2,995	2,830
11	5,650	2,210	7.4
12	7,760	2,640	10.0
13	2,430	1,190	2.6
14	4,070	1,290	7.5
15	2,260	1,070	90

The cause of this phenomenon is generally attributed to the sands being "dirty" and to the fact that the gross pores between the cemented sand grains contain material which reacts with water. This intergranular material appears to be comprised largely of clays and has been so identified in many instances by mineralogical examination and X-ray diffraction. Argillaceous or bentonitic materials are known to hydrate and swell on exposure to water. Such swelling could obviously reduce the permeability of the rock by enormous factors. Moreover, studies of producing sands from Pennsylvania[1] have shown that the permeability decrease may be correlated with the pH of the water, being less for low pH waters. The processes of flocculation and deflocculation of the clays, which are, of course, intimately related to hydration effects, also appear to play an important role, the composite clay-water interactions being sensitive both to the nature of the intergranular fines and the ionic composition of the water.[2]

Regardless of the detailed mechanism involved, the existence of this phenomenon must be recognized in the evaluation and application of permeability measurements. In California it appears to provide at least a partial explanation for the gross discrepancies, by factors of 10 to 50, found between observed well productivities and those anticipated from air-permeability measurements on cores from the producing formations.[3] While such extreme effects have not been reported in other producing districts, there is evidence that silty and dirty sands are not uncommon among the Mid-Continent oil fields, some of the consolidated sand reservoirs in the Gulf Coast, and at least one major field in eastern Venezuela. In most cases these sands have rather low air permeabilities. But as Table 3 shows, even very permeable sands may also be greatly affected by water.

It is of practical interest to note that the reaction between water and dirty sands is usually a reversible phenomenon. Except as the water may cause a complete disintegration of the rock or a gross transport of the clay content, the continued flow of any particular fluid will restore the permeability to that fluid—in order of magnitude at least—regardless of its previous history. In particular, drying or extracting a core carrying water at a very low permeability will generally restore the air permeability to a value substantially equal to or perhaps even greater than its initial air permeability. In Fig. 3.7 are plotted the results of a series of measure-

---

[1] J. N. Breston and W. E. Johnson, *Producers Monthly,* **9,** 19 (October, 1945).

[2] In the voluminous general literature on clays and their interactions with water, recent discussions of the subject from the point of view of oil-production problems may be found in the papers of J. C. Griffiths, *Jour. Inst. Petroleum,* **32,** 18 (1946); T. F. Bates, R. M. Gruver, and S. T. Yuster, *Producers Monthly,* **10,** 16 (August, 1946); L. E. Miller, *Producers Monthly,* **11,** 35 (November, 1946); and R. V. Hughes, *Producers Monthly,* **11,** 13 (February, 1947).

[3] Cf. Sec. 8.6 for a detailed discussion of this matter.

ments of permeability to oil, salt water, and fresh water on two cores drilled with an oil-base fluid from the Stevens Sand in the Paloma field, Calif.[1] The rather striking parallelism in the behavior of the two cores, 2 ft apart, and the gross reversibility of the permeability are to be noted. This reversibility provides the possibility of restoring the flow capacity of a well suffering from a "water block" by drying out the region about the well bore by dry-gas injection.

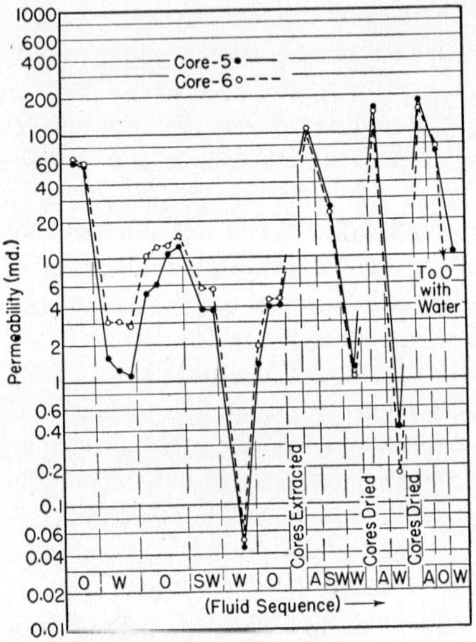

Fig. 3.7. The permeability history of two Stevens Sand cores (dots and circles) when subjected to and measured with various fluids. Code: *A*—air; *O*—oil; *W*—fresh water; *SW*—salt water. (*After Miller, Morgan, and Muskat, Producers Monthly, 1946.*)

It will be clear from this discussion that the permeability to air cannot be assumed as a complete dynamical index of the flow capacity of an actual oil-productive rock unless it is known to be "clean." When this has not been established, tests on the permeability to water should be made. If a reaction of the core with water is found, this must be taken into account in all practical applications,[2] in view of the fact that all reservoir rocks initially contain some water even when oil productive. On the other hand, since they are so simple to measure, it will still be useful to determine the air permeabilities even for dirty sands to provide a reference base for comparison with other measurements.

3.8. Some Practical Aspects of Core Analysis.—While no attempt has been made in the above discussion to present detailed instructions for the various core-analysis measurements, reference to the cited literature will show that they can be carried out as accurate experimental techniques. It is important to have an appreciation of the value of precision in the determinations, as well as the conditions that may affect the significance

[1] Fig. 3.7 is taken from K. T. Miller, F. Morgan, and M. Muskat, *Producers Monthly*, **11**, 31 (November, 1946).

[2] A "wet" permeability, *i.e.*, the permeability to air with the core containing its connate-water saturation, will then give a more pertinent measure of the fluid-transmitting capacity of the rock under actual reservoir conditions, although the quantitative significance of such data must also be evaluated with care.

of the data. In this connection it must be realized first of all that, no matter how complete the coring and precise the data, one is still limited to an examination and study of rock samples which can constitute at the most a fraction of the total reservoir volume only of the order of 0.0001 per cent.

The tremendous statistical problem arising from this circumstance would in itself appear to make the determination of the true character of a reservoir formation an utterly hopeless task. And this would indeed be the case if a reservoir rock were inherently an ensemble of totally independent components whose inclusion in the whole were entirely a matter of chance. Actually, however, geologic strata are the results of dynamical processes subjecting the individual components to substantially similar histories. Moreover, these same processes of transportation, deposition, compaction, cementation, etc., automatically tend to favor a general classification among the primary constituents so as to induce a large-scale homogeneity in the resultant porous medium. Individual underground strata are thus comprised of ensembles of elements that have already been exposed to selective groupings and common environmental factors. It is because of this a priori common denominator in the basic components that the almost infinitesimal sampling provided by the coring of wells can give any significant numerical measure of the rock properties. It is only in the light of such considerations that one can accept comparisons of actual core-analysis data in neighboring wells, which often show the lateral variations in porosity and permeability along continuous lithologic strata to be rather gradual and of limited magnitude, as reflecting the actual nature of the rock. The observation that fluid saturations usually vary but slowly over the areas of continuous geologic horizons, except when obvious gross changes in fluid content (as entry into a gas cap or water zone) are encountered, has a similar basis. From a practical standpoint, therefore, the small areal sampling of a reservoir as obtained by coring may still suffice for a description of the gross average properties of the producing formation. On the other hand, the ultimate limitation imposed thereby on the quantitative applicability of core-analysis data cannot be totally ignored.

With respect to the matter of vertical sampling for core analysis a balance must be made between the cost of coring and analysis and the value of the information obtained. This will depend on the type of formations involved and the amount of core-analysis data already available from other wells penetrating the same strata. Since even when the operator drilling the well has no established core-analysis laboratory of his own the cost of the coring and delay in drilling or well completion will still generally far exceed the cost of the analysis, such cores as are taken should be given a complete[1] analysis, unless it is evident from inspection of the samples that they do not represent commercially productive strata. The density of sample analysis should be of the order of 1 sample per foot[2], as is common practice, unless it is known that the formation is quite uniform over appreciable depth intervals or if, conversely, under special circumstances very localized changes in rock structure must be detected.

Regarding the question as to how much coring should be done, a sound policy would demand that all wildcats and outpost wells should be cored throughout the prospective producing section. In "proven" territory or within the established limits of a reservoir the coring program should be coordinated with current information on the nature of the reservoir rock and other means of logging or evaluating the producing section.[3] If electrical logs, for example, show good well-to-well correlation and the core analyses of the wells cored can be correlated with the electric logs of the same wells, such correlations should be used as a key for interpretation of the electric logs taken on uncored wells. On the other hand, if the reservoir rock is lenticular, the coring density should be increased to obtain representative analyses in each reservoir unit. While recent developments of side-wall coring have expanded the possibility of analyzing the nature of reservoir rocks beyond the time of actual drilling, it is during drilling that the most satisfactory cores can be obtained. In case of doubt it is advisable to "play safe" and favor the maximum economic degree of coring rather than to follow a minimum coring policy. On the other hand, even when samples are not susceptible to quantitative core analyses, as in the case of highly fractured or cavernous limestones, coring may be

---

[1] It may be noted here that "complete" analyses should include at least some sample determinations of the vertical permeability or that normal to the bedding plane when practicable. The vertical permeability has a direct bearing on problems of cross-bedding fluid flow, partially penetrating well production, bottom-water drives, and other related matters pertaining to oil production. From measurements made thus far—all too few—it appears that generally the permeability normal to the bedding plane is somewhat less, though of the same order of magnitude, than the horizontal permeability. However, observations of marked anisotropy are not at all infrequent.

[2] In general, the much more limited variations in porosity than permeability (cf. Figs. 3.9 and 3.10) would appear to justify a lower sampling density for the former. However, the importance of the porosity as a factor in determining the oil reserves, as the unit of volume in describing reservoir performance, and in computing the relative equities of competitive interests in arranging for unitized operation programs, makes the accuracy in the average porosity of more practical value than that of the permeability. Moreover, in formations, as fractured limestones, where permeability measurements may be inherently of questionable validity, the porosity often will still represent a physically significant and important datum.

[3] Of course, during the initial development of a field the coring density should be high so as to facilitate the development of correlations with other logging means.

justified by the information that can be derived from a detailed geological study of the samples. Of course, a geological inspection and evaluation of the cores should accompany all coring operations.

When practicable, analyses of the cores at the well site are to be preferred to shipment for analysis in central laboratories. Not only are the questions of the effect of shipment on the condition of the cores thus eliminated, but the results can be more quickly applied in guiding further coring, drilling, or completion procedures. Immediate consultation with technical and skilled personnel at the well on questions raised by the core-analysis data are also facilitated. When shipment of the cores to a central laboratory for analysis is necessary, the cores should be placed in sealed containers immediately on removal from the core barrel, to minimize weathering and evaporation of the fluid contents.[1] After scraping off the surface mud the core should be wrapped in foil or coated with wax and packed firmly in the container. Weathered cores are literally worthless for fluid-content analysis.

The question of the accuracy with which the analyses should be made merits consideration. While no fixed rule can be given, it must be recognized that, from a practical point of view, time and effort for extreme precision are not warranted. Of course, the analyses should not be undertaken at all if the methods used give data subject to order-of-magnitude uncertainties. Moreover, it is essential that systematic errors of appreciable magnitude be eliminated. Nevertheless, the basic fact is that all the features of the rock which are measured in core analyses are often so variable in passing from sample to sample along the well bore that the exact numerical data for a single sample are of little importance. What are significant are the average values for a set of neighboring similar samples or the large differences between adjacent groups of samples, which may indicate definite changes in type of strata or transition zones with respect to fluid content. Moreover, as will be seen in Sec. 3.10, the physical interpretation of the fluid-content determinations often is so beset with uncertainties that precise and laborious measurements may become but little more than futile gestures. Indeed, from a realistic point of view it seems that, while for research or standardization purposes higher accuracy may be desirable, two significant figures in the data are all that can be basically justified for routine analysis. The implication of this conclusion is not a condemnation of core analyses or an encouragement of careless and crude techniques, but rather an invitation to serious effort directed toward making core-analysis interpretation and use a more exact

[1] To further ensure against the loss of core fluids during shipment a method of quick freezing of the cores at the well site has recently been developed [cf. J. D. Wisenbaker, *Oil Weekly*, **124**, 42 (Jan. 20, 1947)].

science than it is at present. Moreover, it should serve to emphasize the need for a more quantitative evaluation of the other production data that are involved in reservoir analysis.

Finally, it may be observed that in addition to a tabular listing of the core-analysis data, a graphical or logging representation should be made to facilitate interpretation and correlation. On such representations may be included the geological and such other logs as have been made for the section cored. An example of such a composite plot is shown in Fig. 3.8.

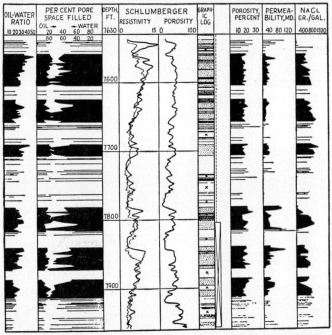

FIG. 3.8. A typical log of core analysis and related data from a well in California. (*After Pyle and Sherborne, AIME Trans., 1939.*)

Additional typical permeability, porosity, and fluid-content data are plotted in Fig. 3.9.[1] Available related data pertaining to the drilling, mode of completion, and subsequent testing or production information should be recorded to provide as complete a picture as possible of the condition of the cored section and completed well.

[1] Figure 3.8, for a well in California, is taken from H. C. Pyle and J. E. Sherborne, *AIME Trans.*, **132**, 33 (1939). The "Schlumberger" traces give the resistivity and self-potential logs (cf. Sec. 3.11). Figure 3.9 gives sample data from a well in Mississippi and includes probable water, condensate, and oil-productive sand sections, as indicated to the right of the fluid-saturation plots. As may be seen from Fig. 3.9 a linear-permeability scale may be inconvenient when the total permeability range is large. The use of a log scale may then be advantageous.

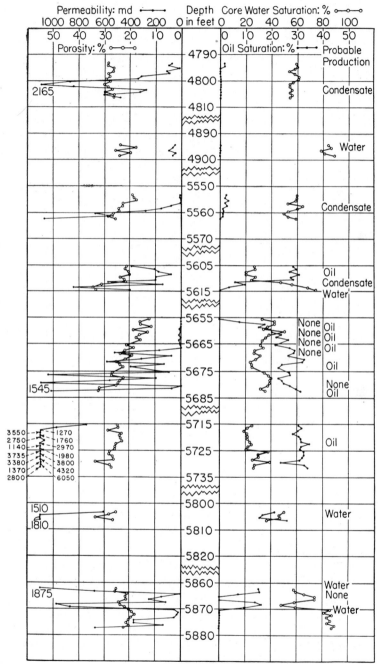

Fig. 3.9. Typical segments of core-analysis data from a well in Mississippi, cored with a water-base mud.

3.9. Practical Applications and Interpretation of Porosity and Permeability Data.—In the preceding sections have been presented the principles underlying the determination of the various types of data embodied in the general practice of core analysis. As both the process of coring and the analyses of the cores are obviously expensive, it is pertinent to inquire as to the value that the data so obtained may have. Unfortunately, at this stage of our presentation of the physical principles of oil production it is impossible to give a quantitative and explicit list of applications of core-analysis data.[1] For it is in the following chapters that specific reservoir-engineering problems will be treated. And it is in the discussion of these problems that the results of core analyses will be assumed as the governing factors determining the quantitative features of the solutions and conclusions to be derived. Indeed, the numerical values of porosity, permeability, and fluid saturations as determined by core analysis, together, of course, with the properties of the petroleum fluids, will constitute the framework of special conditions characterizing specific producing systems and resulting in particular production histories. Here we shall therefore survey only briefly in a qualitative manner the types of problems to which core-analysis data have immediate application.

In considering these questions it should be observed that, while all the numerical results obtained may be inherently determined with considerable precision, they encompass two distinct types of information with respect to physical significance. The first, including porosity and permeability (and grain-size distribution, if this determination be made), represent basic intrinsic characteristics of the rock. The second class, composed of the fluid saturations and the salinity, pertain directly only to the core as brought to the laboratory and must be supplemented with interpretative considerations to reconstruct the fluid conditions in the rock as they may have actually occurred underground.

As pointed out in Sec. 3.1, the porosity of a rock as determined by core analysis immediately provides a quantitative measure of the total volumetric capacity of the rock. While, from a practical point of view, experience has already shown that virtually all sandstones have high enough porosities to be of commercial value, if they are oil saturated and are sufficiently permeable, the quantitative differences cannot be ignored if economically sound evaluations of oil-bearing formations are to be made. Although producing and recovery efficiencies are essentially independent of the porosity (cf. Chap. 14), the absolute volumetric recoveries are, for practical purposes, directly proportional to the porosity. The range of

---

[1] For specific examples of direct applications cf. Pyle and Sherborne, *loc. cit.*, where a detailed discussion is given of almost all phases of core analysis, including the techniques of measurement used by these authors.

porosities of 10 to 30 per cent generally encountered may thus involve a factor of 3 in the actual recovery, or a variation from 150 to 450 bbl/acre-ft in typical gas-drive fields.

Permeability values also reflect directly a characteristic of the rock as it exists underground. However, when measured with air and expressed as the permeability for a homogeneous fluid, it is in itself largely an abstract datum. For the oil flowing into a well will practically always appear to be entering the well bore from a formation of lower permeability than the homogeneous-fluid value measured in the laboratory. This is a consequence of the virtually universal occurrence of interstitial water in oil-bearing formations, which reduces the permeability to the oil. In addition, the presence of free gas in the pores, which may be evolved from the oil on pressure release, will also tend to restrict the rate of oil flow. These effects on the "productivity index" of actual producing formations limit the quantitative significance of the simple homogeneous-fluid values of the permeability as determined by conventional core analysis.[1] Nevertheless, the orders of magnitude of such permeability data do serve to distinguish between commercially valuable oil-bearing strata and those which would not warrant exploitation. Thus, for example, whereas the higher priced oil of the Pennsylvania fields makes profitable the exploitation even of oil-bearing sandstones with permeabilities of only 10 md, such tight formations would generally be passed over in other districts producing lower priced oil unless the strata were of great thickness and under high pressure to give flowing production. Moreover, the homogeneous-fluid permeabilities provide a basis for correction for the effects of interstitial water and free gas, as the latter can often be approximated by magnitudes essentially independent of the absolute value of the permeability for pays of similar geological definition. On the other hand, when the rock is known to be dirty, permeabilities to be used in practical applications must be determined directly under conditions simulating those in the reservoir, *i.e.*, with the samples containing their normal saturation of connate water.

Because of its rapid and wide variations as the subsurface strata are penetrated, the permeability can often be used as a correlation index of the strata penetrated by different wells in a single field. The permeability log may also serve to locate tight streaks or shale breaks that may have been missed by other well-logging means. Such information is obviously of great value in choosing proper completion methods. Finally the permeability profiles will reveal the broad stratification in productivity and make possible general predictions regarding the sequence of depletion among the various strata or the most likely zones for early water intrusion if all the pays be produced simultaneously.

[1] This will be further discussed in Chap. 8.

3.10. The Interpretation of Fluid-content Data.—While the laboratory techniques for fluid-content measurements, as described in Sec. 3.2, can be carried out with satisfactory precision, it must be recognized that in the case of high-pressure or undepleted reservoirs the fluid saturations so determined represent the final conditions assumed by the fluids involved following two very severe ordeals. The first is an exposure of the virgin rock and its entrained fluids to the drilling mud ahead of the coring bit, as well as to peripheral contamination by the drilling fluid during the cutting of the core. As the mud pressure is generally greater than the formation pressure, the drilling fluid will naturally tend to flow into and pass through the rock before the latter enters the core barrel. A displacement of some of the original rock fluids is usually unavoidable. Thus, by the time the core is in the core barrel, it no longer constitutes a true sample of the virgin rock.

During the passage of the core to the surface an additional change in fluid content takes place. Commercially used core barrels are not pressure-tight. When they are brought to the surface, the pressure on the fluids entrained in the cores will be released. Gas dissolved in the oil remaining in the pores of the rock (or in the water) will be evolved and will expel some of the liquid on the way up the hole. On arrival at the surface the core in the core barrel will be depleted of its pressure and possibly of most of its original fluids. The laboratory fluid-content analysis thus constitutes only an autopsy on the remains resulting from processes of unknown magnitude.

It is nevertheless instructive to review the physical pictures that have been developed regarding these processes and the empirical correlations that have grown out of studies of the fluid-content data as actually determined for the depleted cores. Because there is very little control on the details of the fluid-interaction processes occurring during coring, the resultant states of the cores when arriving at the surface will be subject to rather wide ranges of variation. The numerical values of the fluid saturations therefore will usually have only statistical significance. Yet there can be little doubt that to an experienced core analyst the fluid-content data can serve as valuable guides for the prediction of the productive potentialities of the formations being cored.

Assuming that a water-base mud has been used during the coring, it is clear that the residual-oil saturation found on core analysis, when corrected for the shrinkage due to gas evolution, represents only a minimum for the original oil content. And the latter may well be two to three times the former. Yet experience in correlating such residual data with subsequent well behavior shows that their magnitudes do give reliable indications of the gross saturation characteristics of the formations. Thus values of 1 to 4 per cent are frequently found in rotary cores obtained with water-base muds from condensate-productive pays. Residual-oil saturations of 5 to 10 per cent usually result from the analysis of similarly taken cores in gas caps near the gas-oil contact. Cores from strata in the water-oil transition zone, or water-producing but oil-bearing formations, often show residual saturations of 8 to 15 per cent. And so-called "oil-saturated" oil-producing zones generally give values of 10 to 40 per cent for the residual-oil saturations. These, of course, represent only average or typical results obtained from core analyses in particular oil-producing districts. They are not to be interpreted as universal rules, and quite different ranges may be found in the exploration of new or different territories or formations. The important point is that useful correlations of this type can generally be established by a close study of the fluid-content-analysis results obtained in limited regions or groups of formations of similar geological histories.

When cable-tool cores[1] are analyzed, the oil saturations will often be higher, because of lesser flushing,[2] than if the same formation had been cored with rotary tools. Yet unless the hole is completely free of water, there may still be some water invasion of the cores.[3] A direct comparison of the oil saturation of cores from the same formation taken in one case with a mud and in the other dry (no water or mud in the hole) is shown in Fig. 3.10.[4] For this case, at least, the residual oil when the water-base mud was used was only about 40 per cent of that obtained when the cores were cut dry.

From the many core analyses that have been made on samples taken with conventional core barrels using water-base muds a number of general correlations have been established with respect to the degree of contamination and invasion by the drilling fluid. It has been found that it will in general be less in depleted pays than in virgin formations. It is greater in high- than in low-permeability strata. It is more serious in the outer parts than in the center of the core. Large-diameter cores are less susceptible to water flushing than small-diameter cores. And cores taken immediately below shale breaks may be virtually free of contamination.

---

[1] For a discussion of the mechanical aspects of coring and of the types of tools and core barrels commonly used, see L. C. Uren, "Petroleum Production Engineering—Oil Field Development," McGraw-Hill Book Company, Inc., 3d ed., 1946.

[2] While this has been a rather common assumption for some time, more recent data cast serious doubt regarding its general validity.

[3] Cf. C. R. Fettke, *AIME Trans.*, **82**, 235 (1930). While water flushing is generally avoided in "chip coring," the rock samples that are obtained are considerably smaller than in conventional coring [cf. H. M. Ryder and D. T. May, *Producers Monthly*, **2**, 16 (1938)].

[4] J. A. Lewis, W. L. Horner, and M. H. Stekoll, *Petroleum Eng.*, **12** (No. 10), 165 (1941).

If the cores are cut with oil or oil-base mud, oil invasion and subsequent depletion will also change the original oil content of the rock. There is no evidence, however, that the connate water is disturbed and ultimately expelled on pressure release if the core is taken above the water-oil transition zone. Moreover, this is not to be expected, for above such transition

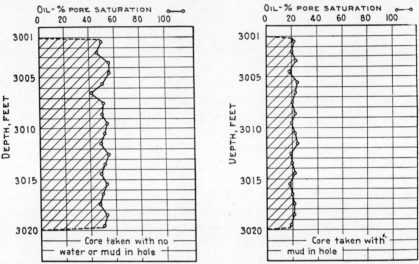

FIG. 3.10. A comparison of the residual-oil saturation found in cores, with and without mud in the hole during coring. (*After Lewis, Horner, and Stekoll, Petroleum Eng., 1941.*)

zones the connate water is presumably in a state of "irreducible water saturation" and should have negligible or strictly zero permeability. It is therefore generally assumed that the interstitial water is preserved intact throughout the coring and pressure release, if the coring fluid is not aqueous. Accordingly the interstitial-water saturation as determined from such cores is considered as truly representative of the undisturbed reservoir water content. The oil saturation corresponding to this undisturbed state is then simply taken as 1 minus the water saturation.

The resultant situation regarding the residual-fluid contents of core samples is thus essentially as follows. Due to drilling-water invasion, cores taken with conventional core barrels and water-base muds give minimal values of the original-oil saturation and core-water contents that are maximal compared with the true connate water. Because the drilling fluid may also displace some of the connate water (cf. Sec. 3.11), the residual connate water in such cores may be appreciably lower than their original-water saturations. Oil-base mud cores in undepleted formations also show minimal values of oil content but should give substantially correct interstitial-water saturations, above the water-oil transition zones.

These observations may be further crystallized by a tabulation of *possible* fluid-saturation states at various stages of the coring process and under different coring conditions in an undepleted reservoir, as follows:

TABLE 5.—POSSIBLE STATES OF FLUID SATURATION AT VARIOUS STAGES IN THE HISTORY OF CORES FROM VIRGIN OIL RESERVOIRS

(In %)

	Water-base Mud					
	Full flushing			No flushing		
	Oil	Core water	Connate water	Oil	Core water	Connate water
Sample prior to coring	65	35	35	65	35	35
Sample in core barrel at bottom of hole..	25–30	70–75	25–30	65	35	35
Sample at surface....	15–25	45–50	20–30	30–35	35	35

	Oil-base Mud					
	Full flushing			No flushing		
	Core oil	"Connate" oil	Core water	Core oil	"Connate" oil	Core water
Sample prior to coring	65	65	35	65	65	35
Sample in core barrel at bottom of hole..	65	10–40	35	65	65	35
Sample at surface....	35–45	8–25	35	30–35	30–35	35

One of the primary purposes of core analysis is evidently the reconstruction of the actual fluid distributions in the rock before its penetration by the drill. The value of such information is twofold. First it provides an immediate evaluation of the total reserves in the formation, taking into account, of course, the porosity of the pay and the shrinkage of the oil on being brought to the surface (cf. Sec. 14.15). Thus, for example, on subtracting from unity the true value of the connate-water saturation, one would obtain directly the original-oil saturation and the oil content of the stratum being cored. Second it often makes possible a prediction of the character of the production from the formation. For as will be seen in Chap. 7, porous media containing more than one fluid phase will permit ready flow of the individual phases only if their saturations exceed rather fixed minimal values. Indeed, the observation that a sand containing 30 per cent connate water produces clean oil or that one with an oil saturation of 15 per cent produces only brine is nothing more than an expression of that basic physical fact. Hence, by reconstructing the virgin-fluid-distribution conditions and comparing them with independently determined criteria for the flow of the individual phases, it may be possible to predict, without the aid of formation testing, whether the strata will produce free gas, oil, or water or possibly a combination of these. As previously suggested and as will be discussed more fully in Chap. 8, the productivity index for the formation can also be estimated by combining heterogeneous-fluid-permeability data with the fluid-saturation determinations (cf. Sec. 8.5).

In lieu of trustworthy methods for determining the true connate-water saturation, which have been developed only recently (cf. Sec. 3.11), attempts have been made to correlate empirically the fluid saturations as measured at the surface and the results of actual production tests. It has been found that such correlations can often be established with considerable reliability for groups of strata of similar physical structure. These may consist merely of the upper limits in total water saturation for the formation to produce oil, such as are indicated by the following table.[1]

TABLE 6.—SUGGESTED LIMITS OF TOTAL CORE WATER IN OIL-PRODUCTIVE SANDS

Permeability, Md	Total Max. Water Saturation, %
10	65
50	63
100	60
500	48
1,000	41

Similar empirical criteria have been suggested in terms of the residual-oil saturations. For example, from studies[2] of the relationship between the directly measured core fluids and connate-water saturations, as determined by the electrical-resistance method (cf. Sec. 3.11), the following gross correlations[3] have been found for samples that have been flushed by drilling water:

TABLE 7.—FLUID CONTENTS OF FLUSHED CORES IN OIL AND GAS RESERVOIRS

Connate water, %	Oil reservoir		Gas reservoir	
	% oil	% water	% oil	% water
60	4	85	—	85
40	10	65	—	65
25	15	50	0–2	50
10	20	30	0–2	30

[1] These numbers, which are to be considered as illustrative only, were read from a curve by Pyle and Sherborne (loc. cit.) relating the permeability to average water saturations as measured on cores from wells in California.

[2] G. E. Archie, AAPG Bull., 31, 350 (1947).

[3] Similar correlations, in graphical form, have been given by R. C. Earlougher, API Proc., 24, Production Bull. 230, p. 323 (1943).

These values, of course, have no universal significance and may well be inconsistent with other related data used for illustrative purposes in the above discussion. They are suggestive, however, of the possibility of distinguishing between gas- and oil-producing reservoirs on the basis of the core-fluid determinations.

The oil-water ratio for the fluids found in core samples has also been used as an index of the nature of the fluid to be produced from the rock.[1] Correlations of this type evidently must take into account and will be affected by the type of coring fluid, character of the oil, rock permeability, etc. For example, in California the range of oil-water ratios in cores taken with water-base muds above which the formation will produce oil has been found to be 0.05 to 0.35,[*] whereas in the Mid-Continent[†] district the limiting ratio appears to lie in the range 0.35 to 1.0[‡]. No single formula or rule governing all conditions is to be expected.

In addition to the implications of the reconstructed virgin-fluid distributions with respect to the oil reserves in the rock and the nature of the fluid to be produced, valuable inferences pertaining to oil recovery can often be drawn from the values of the residual core saturations as measured in the laboratory. For the processes of mud invasion during coring and pressure release on rising to the surface are evidently quite similar to those of water flooding and gas-drive production such as might occur in the reservoir as a whole. The fluid-content determinations may therefore be indicative of the magnitudes of the recoveries from the corresponding types of production mechanism. In particular, if the core has been subject to complete drilling-water flushing, the residual oil will be that remaining after it has been subjected to both the water-flooding and gas-depletion processes and hence should correspond to the oil which is physically nonrecoverable except by mining. The oil that is recoverable by water flooding and gas depletion would thus correspond to the difference between the original- and residual-oil saturations, corrected, of course, for the shrinkage. Upon denoting these by ρ_{oi} and ρ_{or}, that of the connate water by ρ_w, and the formation-volume factor of the saturated oil by β, these considerations will give

$$\text{Physically recoverable oil} = \frac{\rho_{oi}}{\beta} - \rho_{or} = \frac{1 - \rho_w}{\beta} - \rho_{or}, \qquad (1)$$

[1] Methods of making such predictions have even been patented: *e.g.*, J. A. Lewis and W. L. Horner, U.S. Patent No. 2,225,248 (Dec. 17, 1940); W. L. Horner, U.S. Patent No. 2,296,852 (Sept. 29, 1942).

[*] Johnston, *loc. cit.*

[†] K. B. Barnes and R. W. Woods, *Proc. API*, **20**, *Production Bull.* 224, p. 140 (1939).

[‡] Correlations of this type, as well as those represented by Tables 6 and 7, are also of value in determining water-oil and gas-oil contacts.

expressed as a fraction of the pore space. This equation may also be used
to estimate the recovery by water drives or water flooding alone, on the
assumption that the additional process of gas evolution and pressure release while the core is being brought to the surface expels only water but
not the oil left after the drilling-fluid flushing. From data using the
pressure core barrel[1] this assumption appears to be reasonably safe for low-
permeability cores but may be considerably in error for cores of high
permeability. Moreover, unless the core has actually been flooded Eq. (1)
will give only lower limits for the physically recoverable oil.

The recovery due to pressure depletion alone, *i.e.*, for an internal gas-
drive mechanism,[2] may be estimated from the total free-gas space in the
core. This involves the assumption that the gas dissolved in the residual
oil, at the bottom of the hole, is able to expel as much liquid as would the
gas in the original oil-saturated rock. However, this assumption should
not lead to serious errors, since both laboratory evidence and theoretical
calculations indicate that solution gas-drive recoveries are not very sensitive to the total dissolved-gas content of the oil, in the range normally
encountered in saturated[3] crudes. Accordingly the order of magnitude of
the solution or internal gas-drive recovery may be approximated by

$$\text{Solution gas-drive recovery} = \frac{1 - \rho_w}{\beta} - (1 - \rho_w - \rho_g), \qquad (2)$$

where ρ_g is the free-gas saturation.

Of course, estimates made by Eq. (1) or (2) are not to be considered as
quantitative predictions. For while the processes of water flushing and
pressure release in cores are indeed similar to those occurring in large-scale
reservoirs, the quantitative features are quite different in the two cases.
Thus the pressure gradients during flushing are undoubtedly much greater
in the core than in the reservoir. Likewise the pressure gradients and time
rates for the pressure release of the core are certainly higher by orders of
magnitude than the corresponding processes in the reservoir as a whole.[4]
Moreover, the lack of uniformity of large-scale formations and economic

[1] J. J. Mullane, *API Drilling and Production Practice*, 1941, p. 163. On the other hand,
some of the extremely low oil saturations reported for cores from California productive
formations suggest that the rapid evolution of gas from the oil trapped by drilling-fluid
invasion may lead to additional oil expulsion even in tight rocks.

[2] The various oil-producing mechanisms will be defined and discussed in detail in
Chapters 9 to 11.

[3] Exceptional cases, such as are found in some Kansas fields, where the solution-gas
content is only of the order of 2 to 25 ft³/bbl, are, of course, to be excluded from this
generalization. However, in such reservoirs, gas-drive recoveries would be anyway
too low to be of practical interest.

[4] Laboratory studies [H. G. Botset and M. Muskat, *AIME Trans.*, **132**, 172 (1939)]
simulating the process of pressure release in an oil-saturated core when being brought
up a well bore showed that the residual oil in the tested cores decreased with increasing
rate of pressure decline, except when the latter was of the order of 1 psi/min or less.

factors involved in actual production operations will tend to lead to lower yields from the reservoir as a whole than indicated by individual cores, with respect to both water- and gas-drive recoveries. Certainly the drawing of "recovery logs," calculated by Eqs. (1) and (2) and paralleling the fluid-saturation data, such as are shown in Figs. 3.8 and 3.9, would be totally unjustified. Nevertheless, experience has shown that if these limitations are clearly understood, interpretations of core-saturation data such as are outlined above can provide very valuable guides in estimating reservoir recoveries.

It should be noted that underlying most of the previous discussion on methods of studying rock formations has been the tacit assumption that the rocks are sandstones or possibly shales. Admittedly this is a very severe limitation, for many large oil fields, distributed through all the major producing states except California and Pennsylvania, drain limestone reservoirs. Much intensive study has recently been devoted to limestone reservoirs and cores. Nevertheless, it must be emphasized that unless the limestone is of the oölitic type or otherwise shows a microscopic structure and degree of uniformity comparable with sandstone rocks, the core analysis will be of questionable value, especially with respect to permeability data. The problem is essentially one of sampling. If the limestone is fractured or creviced and the production is obtained largely by virtue of these fractures, then a few cubic centimeters[1] of the very tight and unfractured rock could hardly give a correct picture of the actual producing capacity of the rock. On the other hand, a core that does cut across a fracture will give permeabilities that are so sensitive to the detailed structure and location of the fracture as to be likewise of little significance. Accordingly, while the core-analysis procedures discussed in this chapter will still be useful, when applied to limestones, for correlation purposes, and in providing a qualitative picture of the nature of the rock and its fluid contents, great caution must be used in making quantitative applications.[2]

Finally, it should be observed that the basic problem of the interpretation of the fluid saturations of cores taken in planning secondary-recovery programs is not automatically solved even if the true value of the connate-water saturation is established. For evaluating proposed water-flooding or gas-repressuring projects it is important to know the oil content of the rock in the state of reservoir depletion when the secondary-recovery opera-

---

[1] An attempt to minimize this sampling problem has been made by Plummer and Tapp, *loc. cit.*, by using cores of 200 cc or more in volume. The techniques are essentially the same as for conventional core analyses, except for the larger sizes of the main equipment items.

[2] For a detailed discussion of limestone-core-analysis data, cf. A. C. Bulnes and R. U. Fitting, Jr., *AIME Trans.*, **160**, 179 [1945; cf. also B. Atkinson and D. Johnston, *Journal of Petroleum Technology*, **11**, 1 (September 1948)].

tions are being undertaken.[1] While this could be determined, in principle, by subtracting the cumulative production from the initial-oil content, this is not always feasible, as the required data are often not available in the older depleted fields. Moreover, such computations would give only average values over the whole producing section.

Without repeating the detailed considerations already presented above, it appears that the safest coring procedure for determining oil contents in depleted fields is that of cable-tool coring, with a minimum of fluid in the hole, or chip coring. While rotary cores have been reported as satisfactory in some instances, drilling-water flushing is usually considered to be more probable with rotary drilling than when cable tools are used.[2] Of course, coring with oil will be of no particular value even if used with a tracer, except to check on the connate-water content.

3.11. Connate-water Saturations.—It has been seen in the last section that although the liquid contents of cores cut with water-base muds when brought to the surface are usually of little value in determining the original interstitial-water saturation, empirical correlations of the measured residual core liquids often serve to give very useful information regarding the general reservoir conditions and its probable productive potentialities. These alone fully justify the determination of the fluid saturations in routine core analysis. On the other hand, for the purpose of establishing the actual connate-water saturation in the undisturbed reservoir rock it is necessary to apply independent procedures.

Before, however, discussing these it will be instructive to review two methods, based on the tracer principle, that have been used for connate-water determinations, although they are now recognized as being of questionable significance at best. In the first, the chloride ion or mineral content of the original interstitial water, as indicated by an analysis of the brines from the same formation,[3] is used as the tracer. By measuring the chloride content of the core, as described in Sec. 3.3, and translating the latter into the salinity of the residual water, the interstitial-water saturation ρ_w left in the core may then be calculated from the measured total core water saturation ρ_c by the formula

$$\rho_w = \frac{(c_c - c_m)\rho_c}{c_w - c_m},$$ (1)

[1] Cf. Sec. 12.16 for a fuller discussion of this subject.

[2] Cf. R. C. Earlougher, *API Drilling and Production Practice*, 1944, p. 72. On the other hand, some recent studies indicate that flushing in cable-tool cores may be just as serious as in rotary cores [H. M. McClain, *Oil and Gas Jour.*, **45**, 152 (Apr. 26, 1947)].

[3] The identity of the salt content or composition of the interstitial and bounding-formation waters is an assumption underlying this method. There is only meager evidence for its validity, and some recent data on cores obtained with oil-base muds at least cast doubt on its universality.

where c_c, c_m, and c_w denote the chloride ion concentrations or salinities of the residual water, the coring mud, and the true interstitial water.

While this type of determination is frequently made in routine core analysis, it has often been found to give what are obviously too low values for the original interstitial water, when the cores were taken with a water-base mud.[1] The reason appears to lie in the fundamental assumption that the connate waters and the total original mineral content remain in the core throughout its history of excision by the drill and pressure release on its way to the surface. However, analyses of cores taken with a pressure core barrel,[2] in which the cores are brought to the surface under pressure, raise serious doubt about these assumptions. Indeed, these studies show that on pressure release an appreciable amount of the chloride content of the core appears in the mud in the core barrel. In some cases the corresponding amount of interstitial water that had apparently been expelled on releasing the pressure amounted to one-fourth to one-third the total original content.[3] Under such conditions, even assuming that the salt concentration of the native brine is accurately known, the residual chloride contents of the cores will usually give only minimal values for the original-water saturations.

A direct measure of the degree of drilling-water invasion can be obtained by adding a water-soluble tracer, such as dextrose,[4] to the drilling mud. On analysis of the tracer concentration in the residual core water an application of Eq. (1), with c_w set equal to 0, will give the residual connate water, and by subtraction from the total core water that due to mud invasion. By such tests on a well in California it was found that an average of 29.1 per cent of the total core water was drilling fluid (15.2 per cent of the pore space) and that the average residual-connate-water saturation was 37.1 per cent. The average residual-oil saturation was 30.0 per cent.

Unfortunately, however, the use of aqueous mud tracers also may give only minimum values for the original interstitial water, owing to possible expulsion of the connate water on pressure release. And more serious still are the implications of recent experiments,[5] using radioactive tracers, on

---

[1] Mullane, *loc. cit.*

[2] *Ibid.*

[3] The fact that pressure release in the core barrel does expel the interstitial water, whereas the same formation may yield clean oil on production, is apparently due to a redistribution of the connate water in the core during drilling and mud invasion so as to leave it in a mobile state. This explanation is also suggested by observations on the mobility of connate water when subjected to water-flooding action (cf. Russell, Morgan, and Muskat, *loc. cit.*).

[4] Pyle and Jones, *loc. cit.;* cf. also W. L. Horner, *Oil Weekly*, **78**, 29, 71 (July, 1935); Mullane, *loc. cit.;* and Clark, *loc. cit.*

[5] Russell, Morgan, and Muskat, *loc. cit.;* cf. also W. S. Walls, *API Drilling and Production Practice*, 1941, p. 178.

the displacement of connate water in oil-containing cores on entry of extraneous flooding waters. These showed that, after a water-throughflow volume equal to the pore volume of the sample, 70 to 80 per cent of the original connate-water content will generally be displaced. It would thus appear that a core which has been completely flushed by the drilling fluid should contain but a small fraction of its initial interstitial water. Fortunately, the drilling-water invasion generally seems to be insufficient to flush out the bulk of the connate water, which can occur only after the major part of the floodable oil has been displaced. On the other hand, it will be unsafe to assume that this delicate balance always obtains. And unless there is direct evidence of the absence of significant oil flushing and water invasion, connate-water determinations on cores drilled with water-base mud must be considered as representing minimal values for the original interstitial-water content.

Proceeding now to methods that are not complicated by water-displacement uncertainties, the simplest is evidently that of analyzing cores cut with oil or oil-base mud as the drilling fluid by the same techniques used for determining the saturation of cores cut with water-base muds. As noted in the last section, such measurements should give correct values of the interstitial-water saturation, excepting only when the core is obtained from the water-oil transition zone. In fact, at the present time oil-base cores provide the basic reference for testing the value and reliability of other and less direct methods.

An entirely different principle of connate-water determination is that based on a quantitative interpretation[1] of the electrical resistivity of the producing formation. The resistivity measurements are generally made by electrode systems lowered in the open (uncased) well bore as part of standard "electrical-logging" procedures.[2] If the specific resistivity of the mud fluid used during the logging is known, the measured resistance, recorded at the surface, between the electrodes in the well bore can be translated into an equivalent specific resistivity of the formation lying between the electrodes. If the latter be denoted by r_a, the connate-water saturation ρ_w can be computed from the relation

$$\frac{r_a}{r_o} = F(\rho_w), \tag{2}$$

[1] G. E. Archie, *AIME Trans.*, **146**, 54 (1942); R. H. Zinser, *AIME Trans.*, **151**, 164 (1943).

[2] Such logs, including those of the "self-potential," are valuable aids in identifying formations with respect to whether they are sands, shales, or limestones and in predicting whether they will be oil and gas productive or "wet." The great majority of the wells drilled today are logged electrically before final completion; cf. C. Schlumberger, M. Schlumberger, and E. G. Leonardon, *AIME Tech. Pub.* 503 (1934); H. C. Doll, J. G. Legrand, and E. F. Stratton, *Oil and Gas Jour.*, **46**, 297 (Sept. 20, 1947).

where r_o is the specific resistivity of the formation if fully saturated with the formation water and $F(\rho_w)$ is an empirically established function that appears to be substantially the same for all water-wet porous media thus far studied.[1] The value of r_o is evidently proportional to the specific resistivity of the formation water, which can be determined by direct measurement or from the composition of its salt content,[2] and is otherwise a function of the type of rock and its porosity and permeability. These latter relationships have also been developed empirically.[3]

While the application of this method requires careful consideration of complicating effects of mud filtration into the formation logged, the finite thickness of the strata in question, etc., it has reportedly given satisfactory results in sandstones[4] when used under favorable conditions (cf. Table 8). It offers the advantages with respect to connate-water determinations both of speed and of an automatic type of averaging both areally and vertically over the separation of the resistance-measuring electrodes. And except for the effect of the mud invasion it represents a measurement of the state of the reservoir undisturbed by the processes of drilling and pressure depletion.

Even when the absolute evaluations of the apparent and specific formation resistivities are subject to question, it may be possible to establish direct correlations between independently determined connate-water saturations and the apparent resistivities. The electrical logs can then be used for obtaining connate-water saturations from well bores that have not been cored. Electrical logs are generally run anyway for identification and lithologic-correlation purposes. Such additional information as may be derived from them regarding the interstitial-water saturation will require mainly interpretive analysis of data that will usually be available as part of the logging record.

A promising method that has recently been developed for connate-

---

[1] R. D. Wyckoff and H. G. Botset, *Physics*, **7**, 325 (1936); M. Martin, G. H. Murray, and W. J. Gillingham, *Geophysics*, **3**, 258 (1938); J. J. Jakosky and R. H. Hopper, *Geophysics*, **2**, 33 (1937); M. C. Leverett, *AIME Trans.*, **132**, 149 (1938). An approximate analytic form for $F(\rho_w)$ is $1/\rho_w^2$.

[2] This in itself is a source of some uncertainty in the use of the electrical-resistivity method, and the determination of r_o is often of questionable accuracy. However, recent studies of the self-potential logs indicate that the specific resistivity of the formation water can also be computed from the self-potential data and the mud-filtrate resistivity [cf. M. J. R. Wyllie, *Journal of Petroleum Technology*, **1**, 17 (1949)].

[3] G. E. Archie, *loc. cit.*, and *AAPG Bull.*, **31**, 350 (1947).

[4] Some applications have also been made to limestones, although electrical logs often are difficult to interpret and show little "character" in limestones and dolomites. On the other hand, even in sandstones a quantitative translation of the measured apparent resistivity into the actual formation specific resistivity, r_a, may be an extremely complex and uncertain procedure in highly stratified zones.

water determination is based on the measurement of the "irreducible water saturation" that a core can hold. The latter is its residual-water content when subjected to a capillary pressure[1] equal to or greater than the differential head between a water and reservoir oil column extending from the reservoir water table to the location of the core. The capillary pressure

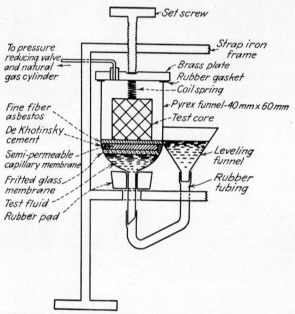

FIG. 3.11. A capillary-pressure apparatus for determining connate-water saturations. (*After McCullough, Albaugh, and Jones, API Drilling and Production Practice, 1944.*)

represents the magnitude of the pressure discontinuity across the interface between the water phase and whatever other fluid is in contact with it.

The measurements may be made by using an apparatus such as is shown diagrammatically in Fig. 3.11.[2] The test core is first extracted, dried, and weighed. It is then saturated with water or brine, weighed again, and placed on the asbestos pad, which facilitates capillary contact with the

[1] The general subject of capillary phenomena and the physical considerations underlying this method will be discussed in Secs. 7.8 to 7.10. Only the technique and results of connate-water determinations by the capillary-pressure method will be reviewed here.

[2] This is taken from J. J. McCullough, F. W. Albaugh, and P. H. Jones, *API Drilling and Production Practice*, 1944, p. 180. Similar apparatus and techniques have been described by G. L. Hassler and E. Brunner, *AIME Trans.*, **160**, 114 (1945) and O. F. Thornton and D. L. Marshall, *AIME Trans.*, **170**, 69 (1947) and have been patented by M. C. Leverett (U.S. Patent No. 2,330,721, Sept. 28, 1943). While these applications to the determination of connate-water saturations in oil-producing formations were not made until rather recently, the same basic principles and techniques had long before been established and widely used in the study of soil moisture.

porous porcelain capillary membrane. The latter, also saturated with water, is in contact with a free water table through the supporting glass membrane and connecting tube to the leveling funnel. After tightening the cover plate, air or gas (or oil) is applied to the core chamber. The pressure applied is the estimated capillary pressure at the location of the core. The core is removed periodically and weighed to check whether or not equilibrium has been established. The residual-water content when equilibrium is attained[1] is considered as equivalent to the connate-water content of the core in its original location in the reservoir. From the difference in the dry and fully saturated core weights or from an independent porosity measurement the residual-water content can be expressed as a fractional saturation.

As an alternative to removing and weighing the core to check on its approach to and attainment of equilibrium the leveling funnel may be replaced by a suitable graduated tube or capillary and observations made on the change in the water level.[2] The latter will give a measure of the displacement of water from the core. Moreover, when the maximum required capillary pressure is less than 1 atm, the application of a positive pressure to the core can be replaced by an equivalent suction imposed on the free-water volume in contact with the bottom side of the capillary diaphragm.

A disadvantage of the capillary-pressure technique, as described above, for routine measurements is the long time required for reaching the final equilibrium saturation because of the continuously decreasing permeability to the water as its saturation is decreased. This may be especially serious in tight cores. This situation may be alleviated to some extent by centrifuging the core while in contact with a free-water volume through a capillary membrane. The centrifugal acceleration acting on the water in the core is equivalent to an increased gravity or pressure head for inducing the removal of the water content against the retaining action of the capillary forces. The change in water content can be followed by observations on the electrical conductivity of the core,[3] after adding a salt to the

---

[1] If there is question about the water content representing the "irreducible water saturation," the tests may be repeated at a higher pressure and the equilibrium residual-water saturation redetermined. On the other hand, from a strict point of view the equilibrium residual-water content at the correct capillary-pressure level should represent the true connate-water saturation even if it is not quite so low as the irreducible water saturation. Such intermediate water contents represent states in the transition zone between the region of complete water saturation and that of maximum and constant oil saturation, for a uniform formation.

[2] Care, however, must be taken that the displacing fluid does not break through the capillary diaphragm and affect the volumetric-displacement readings.

[3] McCullough, Albaugh, and Jones, *loc. cit.*

water, via slip-ring contact with an appropriate measuring circuit or stroboscopic illumination[1] of glass pipettes attached to the core for collecting the drained water.

Examples of comparative measurements of connate-water saturations by the capillary-pressure technique and other methods are listed in Table 8.

The first 4 samples were oil-base cores from the Dominguez field, California.[2] The group of 15 from which these were chosen gave an average connate-water saturation of 27.8 per cent by the capillary-pressure method and 27.7 per cent by distillation. Samples 5 to 8 also were taken in the Dominguez field and are listed to show especially the comparison between the two types of capillary-pressure determinations, although extraneous factors appear to have influenced the results. Oil-base cores 9 to 12, from the Ventura field, California, give additional comparative results for the centrifugal capillary-pressure method with the core water measurements. Simultaneous tests of three methods are illustrated by samples 13 to 18. These were chosen[3] from a group of 20, from a Gulf Coast reservoir, which gave average values of 35, 43, and 36 per cent for the capillary-pressure, distillation, and salinity methods. The higher saturations found by distillation were apparently due to the inclusion of water of crystallization, released by the retorting. On the other hand, the salinity measurements agreed almost exactly with the capillary-pressure data, which represents perhaps indirect confirmation of the absence of connate-water displacement in oil coring.

The comparative tests of electrical-log and capillary-pressure determinations (items 19 to 21) each represents averages of 4 to 7 samples, also from a well in the Gulf Coast. These indicate that the two can give equivalent results under favorable conditions.

---

[1] Hassler and Brunner, *loc. cit.* These authors also give the theory and equations for evaluating the centrifuge data and show that the variation of the water saturation with the equivalent capillary pressure (the capillary-pressure curve) so obtained agreed with that determined by an application of suction pressure to the capillary diaphragm. It may be noted that the centrifuge method had also been used in early soil-moisture investigations. Cf., for example, L. J. Briggs and J. W. McClane, *U.S. Dept. Agr. Bur. Soils Bull.* 45 (1907); M. B. Russell and L. A. Richards, *Soil Science Soc. Proc.*, **3**, 65 (1938). Another form of the capillary-pressure method which recently has been developed for obtaining connate-water saturations rather rapidly involves the injection of mercury into the core and determining the pore space remaining unfilled at the desired equivalent capillary pressure of injection [cf. W. R. Purcell, *Journal of Petroleum Technology*, **1**, 39 (1949)]. While the capillary phenomena underlying this procedure are not identical to those controlling capillary desaturation, the reported experiments indicate that for most of the cores studied both methods will give substantially the same results at equivalent capillary pressures of about 1 atm.

[2] McCullough, Albaugh, and Jones, *loc. cit.* Samples 1 to 12 are all taken from this reference.

[3] Samples 13 to 21 are taken from Thornton and Marshall, *loc. cit.*

Samples 22 to 30, all of which refer to cores from the same well in the Hawkins field,[1] Texas, are of interest in illustrating the effect of the dis-

TABLE 8.—COMPARATIVE MEASUREMENTS BY VARIOUS METHODS OF CONNATE-WATER SATURATIONS

Sample No.	Capillary pressure			Distillation	Electrical resistivity	Salinity
	Air displacement	Oil displacement	Centrifugal			
1	31.5	. . . .	. . . .	26.0		
2	24.5	. . . .	. . . .	26.5		
3	27.0	. . . .	. . . .	24.0		
4	22.0	. . . .	. . . .	25.5		
5	27	. . . .	23	24.0		
6	31	. . . .	26	35.0		
7	27	. . . .	22	28.5		
8	28	. . . .	23	26.5		
9	. . . .	. . . .	53	53		
10	. . . .	. . . .	51	48		
11	. . . .	. . . .	52	38		
12	. . . .	. . . .	34	35		
13	74	. . . .	. . . .	89	. . . .	71
14	23	. . . .	. . . .	28	. . . .	23
15	37	. . . .	. . . .	52	. . . .	37
16	33	. . . .	. . . .	29	. . . .	34
17	38	. . . .	. . . .	53	. . . .	36
18	19	. . . .	. . . .	19	. . . .	20
19	19	. . . .	. . . .	. . . .	19	
20	11	. . . .	. . . .	. . . .	11	
21	8	. . . .	. . . .	. . . .	10	
22	. . . .	25.5	. . . .	20.9		
23	. . . .	7.4	. . . .	5.8		
24	. . . .	17.5	. . . .	13.8		
25	. . . .	19.3	. . . .	19.2		
26	. . . .	13.8	. . . .	10.5		
27	19.0	. . . .	. . . .	17.3		
28	12.0	. . . .	. . . .	12.4		
29	14.4	. . . .	. . . .	13.8		
30	20.0	. . . .	. . . .	20.9		

placement fluid on capillary-pressure determinations. While air-displacement capillary-pressure experiments gave saturations agreeing quite well with the distillation data on the oil-base cores, the saturations were generally higher when oil was used as the water-displacing medium. On the

[1] These are taken from W. A. Bruce and H. J. Welge, *Oil and Gas Jour.*, **46**, 223 (July 26, 1947).

other hand, measurements on oölitic limestone samples from the Magnolia field, Arkansas, gave lower connate-water saturations when oil was used than when air was the displacing fluid.

While from a practical point of view the implication of these observations is simply that the fluids actually of interest should be used in the capillary-pressure measurements,[1] they do create some uncertainty as to the absolute significance of the state of "irreducible water saturation" which is presumably created in the capillary-pressure experimentation. Moreover, comparative studies[2] in the North Belridge Wagonwheel field and other reports indicate that for low-permeability cores the capillary-pressure data may give appreciably higher saturations than oil-base cores, whereas the reverse may occur in cores of high permeability. Undoubtedly, much study will be needed to fully clarify all aspects of the capillary-pressure method of connate-water determination. Nevertheless, the investigations already made, as illustrated[3] by Table 8, do seem to indicate that this method, even as developed thus far, will usually give at least a fair approximation to the true interstitial-water content of oil-bearing formations.

The absolute magnitude of the connate water in oil-producing strata varies widely among different formations even when the gross characteristics, as porosity and permeability, are similar. It is probably determined by the history of the oil accumulation in the original water-saturated rocks, the viscosity of the oil,[4] the interfacial tension between the oil and water, grain-size distribution of the rock, proximity to the water-oil interface of the stratum cored, clay content of the rock, and the detailed geometry of the pore space. Nevertheless, as is to be expected, in individual geological horizons it generally tends to increase with decreasing permeability, as shown by the curves of Fig. 3.12.[5] Gas-bearing zones overlying oil-saturated pays usually have water saturations comparable with those in the oil-bearing formations.

[1] It is possible, too, that the different interfacial-tension values obtaining under actual reservoir conditions may influence the capillary-pressure equilibrium in the undisturbed reservoir rocks.

[2] C. Beal, *API Drilling and Production Practice*, 1944, p. 187; cf. also S. T. Yuster and C. D. Stahl, *Producers Monthly*, **8**, 24 (December 1948).

[3] It should be emphasized, however, that Table 8 gives only examples of reported data and does not necessarily represent typical comparative results of statistical significance with respect to any of the methods listed.

[4] These factors should be of significance, mainly with respect to the possibility that complete equilibrium may not have developed by the time the reservoir is discovered and opened to production.

[5] Schilthuis, *loc. cit.*, is the source for the data on East Texas, Anahuac, and Tomball. The curve for the Wasson Dolomite was taken from Bulnes and Fitting, *loc. cit.* Those for Magnolia and Elk Basin are reproduced from Bruce and Welge, *loc. cit.* The curve for the Dominguez field is replotted from Johnston, *loc. cit.*

While the differences shown in Fig. 3.12 between the curves for different fields should serve to emphasize that there is no single and universal relationship between connate-water saturations and permeability, the gross trends can be at least rationalized. The general decline of water content with increasing permeability is evidently an expression of the fact that the capillary forces maintaining the water saturation against the fluid head above the water table increase with decreasing average pore radius. This is to be expected from basic physical principles.[1] The exceptionally high

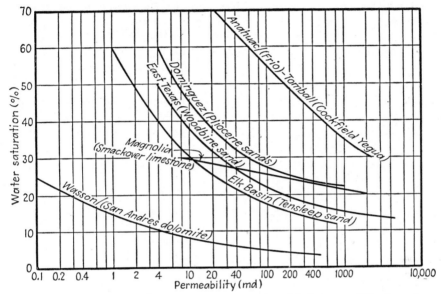

Fig. 3.12.   The variation of the connate-water saturation with the permeability in various fields.

saturations at Anahuac and Tomball are probably due to the presence of intergranular clays in the producing formations. Correlation between the clay content and the connate-water saturation has been reported for a major field in eastern Venezuela. The very low values listed for the Wasson Dolomite, which are typical of those found in many West Texas limestone and dolomite fields, suggest that the cementing material in these rocks probably fills out the angular regions between the basic granular particles so as to minimize the resultant average curvatures in the residual pore voids. The wettability of the solid surfaces may also be an important factor.

Since the permeability alone will not uniquely determine the connate-water saturation, the experimental data on which the curves of Fig. 3.12

[1] Cf. Sec. 3.7.

are based show considerable scattering. Nevertheless, for a fixed type of lithologic formation such relationships as plotted in Fig. 3.12 undoubtedly do have statistical significance. Hence, when such a curve is established for a particular geologic zone of interest, it may be used in deriving the gross average connate-water content of the formation when the statistics of the permeability distribution have been determined. Moreover, it can be used to estimate the local volumes of the interstitial-water saturation from the permeabilities of the individual samples and thus make it unnecessary to measure the former directly for every sample by such time-consuming procedures as the capillary-pressure method.

3.12. Core-analysis Results on Porosity and Permeability.—An appreciation of the concepts of porosity, permeability, fluid saturations, etc., which have been discussed in this chapter, may be crystallized in terms of the numerical values associated with these terms. Unfortunately, however, no well-defined set of "typical" magnitudes can be given for these quantities. For they not only vary from formation to formation, but also from field to field draining the same geologic stratum. And even in a single well, while penetrating a particular zone, the variations in the actual core-analysis data from sample to sample may be so large, especially with respect to permeability, that simple averaging over the whole section may be unjustified and the supposedly single stratum must be considered as a composite of several distinct rock layers. Indeed, it is much easier to exhibit the variability in core-analysis data than to provide average values of any significance. Nevertheless, a few numerical examples of the results of porosity and permeability measurements will be given so as at least to indicate the orders of magnitude involved.

In order to have some basis for comparison with actual data as obtained on consolidated subsurface cores, it is useful to note the magnitudes of porosity and permeability that might be found for unconsolidated sands. It may be proved that if such unconsolidated sands were composed of spherical grains of uniform size, the porosity would range from 26.0 to 47.6 per cent, depending on the type of packing of the spheres. For non-spherical grains of varying sizes the porosity will generally be lower than the upper limit for uniform spheres and may even be reduced to practically zero by the addition of silts and fines. On the other hand, if there is extensive bridging between the grains, as may occur in the case of powders, abnormally high porosities—up to as much as 70 per cent—may be obtained.

Except for fine powders the permeabilities of unconsolidated sand assemblies will usually be considerably higher than those found for consolidated sand cores. Such values are illustrated by the upper eight rows of data recorded in Table 9, which were obtained for a series of sand samples of limited grain size and packed to the same porosity. The permeabilities have been listed in darcys to emphasize the high values as compared with those characteristic of actual rock samples.

The various tables and figures given in previous sections already include a number of examples of typical core-analysis data as obtained from oil-productive sands. Some additional porosity and permeability data, re-

TABLE 9.—POROSITIES AND PERMEABILITIES OF UNCONSOLIDATED SANDS

Sand mesh	Porosity, %	Permeability, darcys
30–40.................	40	340
40–50.................	40	66
50–60.................	40	43
60–70.................	40	31
70–80.................	40	26
80–100................	40	10
100–120...............	40	9.9
120–240..............	40	9.3
Fine heterogeneous......	30–35	1–10
Silts*..................	36–41	5–180
Fine powders†..........	37–70	0.01–0.1

* V. I. Vaidhianathan, and H. R. Luthra, *Punjab Irr. Research Inst. Research Pub.*, **2**, No. 2 (1934).
† R. N. Traxler, and L. A. H. Baum, *Physics*, **7**, 9 (1936).

ferring mainly to Pennsylvania and Mid-Continent fields and chosen from the extensive compilation of Fancher, Lewis, and Barnes,[1] are listed in Table 10.

An over-all graphical picture of the range of measured porosities and permeabilities is provided by Fig. 3.13,[2] in which are plotted the data for about 2,200 sandstone samples, including some 1,300 from Gulf Coast fields. A similar plot of data from limestone and dolomite reservoirs in West Texas and New Mexico is shown in Fig. 3.14.

3.13. Summary.—The properties of oil-bearing formations pertinent to oil production are those usually determined in the procedures termed core analysis. These include the porosity, permeability, fluid content of the rock, and salinity of the formation waters.

The porosity, which represents the volumetric fluid-holding capacity of the rock and is expressed by the fraction of the bulk volume available for

[1] Fancher, Lewis, and Barnes, *loc. cit.* The permeabilities of most of the samples listed were measured with water, and parallel to the bedding planes in all cases. The original compilation contains many comparative tests with air and also permeabilities normal to the bedding planes.

[2] Both Figs. 3.13 and 3.14 are taken from Bulnes and Fitting, *loc. cit.* The boundary lines on these figures are drawn to limit the regions where the rocks may be either intermediate or intergranular or definitely intermediate (cf. Sec. 1.3).

Table 10.—Porosity and Permeability Data from Oil-productive Formations

Sample No.	Sand	State	Field or locality	Porosity, %	Permeability, md
1	Bradford	Pennsylvania	Bradford	12.5	3.13
2	Bradford	Pennsylvania	Bradford	12.3	3.48
3	Bradford	Pennsylvania	Bradford	9.6	0.51
4	Bradford	Pennsylvania	Bradford	16.4	2.81
5	Bradford	Pennsylvania	Bradford	22.7	131
6	Bradford	Pennsylvania	Kane	8.4	*
7	Bradford	Pennsylvania	Kane	14.3	5.02
8	Bradford	Pennsylvania	Kane	12.1	0.09
9	Bradford	Pennsylvania	Kane	11.0	12.4
10	Bradford	Pennsylvania	Kane	11.3	1.13
11	Second Venango	Pennsylvania	Oil City	20.1	30.3
12	Second Venango	Pennsylvania	Oil City	13.6	23.3
13	Second Venango	Pennsylvania	Oil City	10.6	0.84
14	Second Venango	Pennsylvania	Oil City	8.2	12.7
15	Second Venango	Pennsylvania	Oil City	23.6	136
16	Third Venango	Pennsylvania	Oil City	16.9	65.9
17	Third Venango	Pennsylvania	Oil City	11.9	1,130
18	Third Venango	Pennsylvania	Oil City	21.4	315
19	Third Venango	Pennsylvania	Oil City	11.0	541
20	Third Venango	Pennsylvania	Oil City	7.7	40.8
21	Third Venango	Pennsylvania	Grand Valley	12.0	0.99
22	Third Venango	Pennsylvania	Grand Valley	23.6	100
23	Third Venango	Pennsylvania	Grand Valley	22.5	70.7
24	Third Venango	Pennsylvania	Grand Valley	22.1	235
25	Third Venango	Pennsylvania	Grand Valley	15.4	2.48
26	Kane	Pennsylvania	Kane	15.1	23.2
27	Kane	Pennsylvania	Kane	14.2	1.37
28	Kane	Pennsylvania	Kane	22.1	216
29	Kane	Pennsylvania	Kane	18.1	87.8
30	Kane	Pennsylvania	Kane	18.7	28.3
31	Clarendon	Pennsylvania	Warren	10.1	*
32	Clarendon	Pennsylvania	Warren	10.0	0.60
33	Clarendon	Pennsylvania	Warren	18.5	3.38
34	Clarendon	Pennsylvania	Warren	8.8	*
35	Clarendon	Pennsylvania	Warren	13.5	0.58
36	Speechley	Pennsylvania	Oil City	10.3	2.45
37	Speechley	Pennsylvania	Oil City	13.6	36.6
38	Speechley	Pennsylvania	Oil City	7.6	0.16
39	Speechley	Pennsylvania	Oil City	3.7	0.02
40	Speechley	Pennsylvania	Oil City	11.0	51.0

Sample No.	Sand	State	Field or locality	Porosity, %	Permeability, md
41	Woodbine	Texas	East Texas	23.8	111
42	Woodbine	Texas	East Texas	26.9	2,500
43	Woodbine	Texas	East Texas	22.1	2,390
44	Woodbine	Texas	East Texas	23.4	387
45	Woodbine	Texas	East Texas	8.1	*
46	Johnson	Oklahoma	Oklahoma City	11.7	24.7
47	Johnson	Oklahoma	Oklahoma City	21.3	692
48	Johnson	Oklahoma	Oklahoma City	15.7	258
49	Johnson	Oklahoma	Oklahoma City	15.3	88.8
50	Johnson	Oklahoma	Oklahoma City	15.5	464
51	Cromwell	Oklahoma	Little River	16.9	53.8
52	Cromwell	Oklahoma	Little River	19.9	96.8
53	Cromwell	Oklahoma	Little River	21.3	359
54	Cromwell	Oklahoma	Little River	20.9	159
55	Cromwell	Oklahoma	Little River	23.2	314
56	Wanette	Oklahoma	Wanette	22.7	482
57	Wanette	Oklahoma	Wanette	17.4	199
58	Wanette	Oklahoma	Wanette	14.2	18.3
59	Wanette	Oklahoma	Wanette	14.8	11.4
60	Wanette	Oklahoma	Wanette	18.1	52.7
61	Glen Rose	Louisiana	Caddo	8.9	—
62	Glen Rose	Louisiana	Caddo	24.8	1,460
63	Glen Rose	Louisiana	Caddo	24.3	341
64	Glen Rose	Louisiana	Caddo	7.2	*
65	Glen Rose	Louisiana	Caddo	16.2	3.89

* Sample was impermeable under conditions of test.

holding fluids, involves measurements of both the bulk and pore volumes. The former is usually measured by displacement methods, by immersion in mercury of the coated or uncoated sample or in another liquid after saturation with that liquid. The pore volume may be determined directly by measuring the gas or liquid volumes the sample can hold or indirectly by establishing first the volume occupied by the solid substance of the rock. Most commercially productive sandstones have porosities in the range of 10 to 35 per cent. The lower range of porosities of oil-producing limestone reservoirs extends to 4 to 6 per cent. Cavernous limestones may locally have porosities that in effect are 100 per cent.

The fluid contents of subsurface rocks are generally determined by simply retorting (cf. Fig. 3.3) or extracting (cf. Fig. 3.4) samples brought to the surface. These, however, do not represent the fluid composition of the reservoir rock in its undisturbed virgin state. For the use of drilling

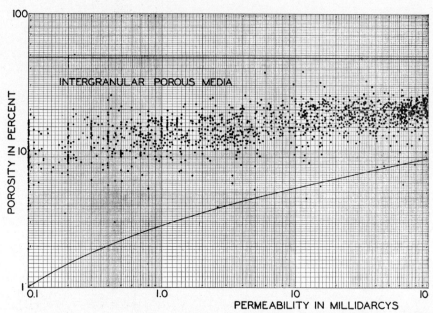

FIG. 3.13. Porosities and permeabilities of about 2,200 sand and sand-

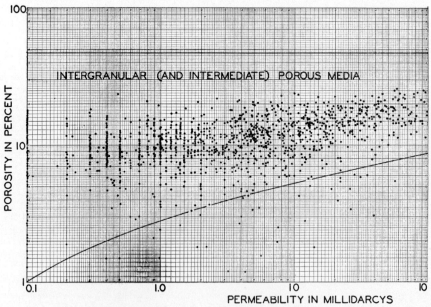

FIG. 3.14. Porosities and permeabilities of about 1,200 dolomitic lime-

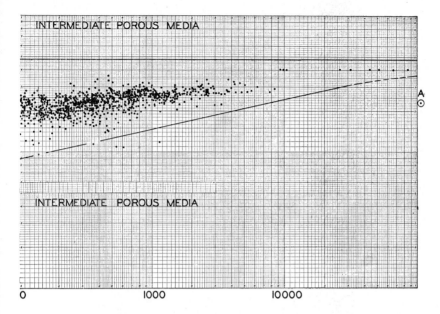

stone specimens. (*After Bulnes and Fitting, Jr., AIME Trans., 1945.*)

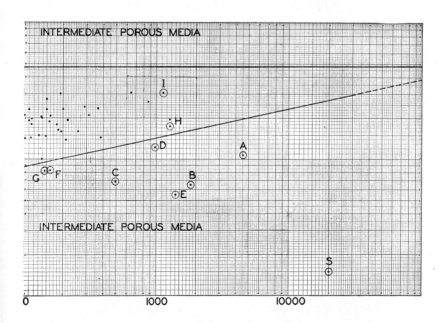

stone specimens. (*After Bulnes and Fitting, Jr., AIME Trans., 1945.*)

mud during the process of cutting the samples usually leads to a penetration of the cut rock (cores) by the drilling fluid and a corresponding displacement of its initial liquid content. Moreover, the release of formation pressure and gas evolution as the sample is brought to the surface also result in an expulsion of some of the liquid the sample contained at the bottom of the hole.

The measure of the fluid-transmitting capacity of reservoir rocks is based on Darcy's law, which states that the rate of flow of a homogeneous fluid is proportional to the pressure and/or hydraulic gradient and varies inversely as the fluid viscosity. By definition, this law is valid only in the region of viscous flow. The latter obtains under conditions where the Reynolds number associated with the flow is of the order of or less than 1. This limit covers virtually all flow systems of practical interest with respect to oil production, except for regions in the immediate vicinity of wells producing at very high rates, where the flow may be turbulent. The constant of proportionality in Darcy's law between the rate of flow and pressure gradient is termed the permeability of the porous medium. It has the dimensions of length squared. Its numerical value will be in darcys if the rate of flow be expressed as cubic centimeters per second per square centimeter, the viscosity in centipoises, and pressure gradient as atmospheres per centimeter. Commercially productive sandstones usually are found to have permeabilities in the range of 5 to 5,000 millidarcys (0.005 to 5 darcys). The permeabilities of some producing limestone reservoir rocks are even lower than 1 millidarcy.

The permeability is determined on extracted samples of rock, of simple geometry and known dimensions, by merely observing the rate of throughflow of a fluid at a measured pressure differential. It is generally simpler and more convenient to use air as the test fluid. This will give values equivalent to those which would be obtained by using liquids which do not interact with the rock, although for tight samples a slippage effect of the gas with respect to the grain boundaries has to be taken into account for accurate determinations. On the other hand, in the case of dirty, or clay-containing, sands, such as are commonly found in California and in some of the Pennsylvania fields, the measured permeabilities to water may be much lower than to air. For practical applications to such reservoir rocks the measurements should be made with water of composition similar to the formation water, although the air permeabilities of dried samples will still be useful for comparative purposes.

Because of the great variability among core-analysis data on individual samples, great precision is not warranted for routine analytical purposes. All wildcats or exploratory wells should be cored throughout the productive section. All the cores from the oil-productive strata should be subjected to complete analysis. Coring may be minimized in developing proven regions of a reservoir. But attempts should be made to correlate the core-analysis data from key wells with other means of identifying and evaluating the reservoir rock, such as electrical-log records. All available information regarding the reservoir rock and its fluid contents should be studied together to derive a composite representation of its physical properties and structure.

Core-analysis data provide the basic numerical framework for defining the unique properties of a particular reservoir and predicting and interpreting its performance. Knowledge of its porosity and a reconstruction of its initial-fluid composition permit an immediate computation of its original-oil content per unit of rock volume. The permeability data are essential to making estimates of the productive capacities of the wells. They are especially significant in revealing the stratification of the composite productive section, which will determine differential rates of depletion and susceptibilities to extraneous fluid encroachment or channelling.

Because the common procedure of cutting cores with rotary tools and water-base muds makes it impossible to ensure the complete absence of drilling-fluid invasion and the subsequent expulsion of the core fluids because of pressure release, it is difficult accurately to reconstruct the undisturbed reservoir-rock fluid composition from measurements of the fluid content of the surface cores. The residual-oil saturations and connate-water contents of such cores will be only minimal values as compared with the original saturations of oil and connate water. And the total water contents of the cores will be greater than the initial connate-water saturations. Useful empirical correlations have been nevertheless developed using the data as obtained from mud-invaded and depleted cores. Means for distinguishing between gas- and oil-productive strata have been reported on the basis of the residual-oil saturations. In some districts the ratio of the residual-water to -oil saturations has been found to provide an approximate index for predicting whether the formation will produce oil or water. If the core is known to have been subjected to complete drilling-water flushing, the residual oil found in the core may be considered as representing the ultimate physically unrecoverable oil saturation even under perfect water-drive operation of the reservoir. And the free-gas saturation of the pressure-depleted core should indicate the approximate free-gas saturation that may be developed by solution-gas pressure-depletion[1] operation of the reservoir. Knowledge of the ultimate residual-oil or free-gas saturation permits ready calculation of the corresponding oil recoveries. On the other hand, such calculations should be made with

[1] The various reservoir production mechanisms will be discussed generally in Chap. 9 and in more detail in the later chapters.

caution, as the detailed conditions of fluid displacement and expulsion in cores are quantitatively quite different from those obtaining during gross reservoir depletion processes.

To determine the actual connate-water saturation of the undisturbed reservoir rock, neither the use of tracers in the mud nor measurements of the salt content of the core water will give trustworthy results because of the possibility of displacing at least part of the connate water by the invading drilling fluid. The only direct method that appears to be free of such complications requires the use of oil or oil-base mud as the coring fluid, which will leave the connate-water content undisturbed, provided that the core is taken above the water-oil transition zone. An indirect procedure, however, in which the electrical resistance of the formation as measured by electrical-logging methods is interpreted in terms of the connate-water saturation, also has given satisfactory results under favorable conditions.   Another method that offers promise of giving satisfactory connate-water saturations regardless of the coring fluid or tools is based on the capillary-pressure phenomenon. It is postulated that the connate-water saturation at any point in an oil reservoir is simply that which the rock can hold by capillary forces in equilibrium against the differential gravity head, between water and oil, corresponding to the height of the rock sample above the water-saturated section. The connate-water saturation is then determined by extracting the core, resaturating it with water, and measuring the residual-water content after it has been subjected to a capillary displacement pressure equal to the differential gravity head. While some detailed features of this method still require clarification, a number of comparisons with directly determined saturations from oil-base cores show it to be in most cases of sufficient accuracy from a practical standpoint (cf. Table 8).

As would be anticipated from the physical basis of the capillary-pressure method for determining connate-water saturations, the latter have been found to decrease with increasing permeability (cf. Fig. 3.12). Such relationships have been established for different formations and appear to have statistical significance. The absolute positions of the curves of connate-water saturation vs. permeability reflect largely the microscopic pore geometry of the rock and the presence of intergranular clays. Limestones generally have lower water contents than sandstones of comparable permeability.

CHAPTER 4

BASIC DYNAMICAL EQUATIONS FOR HOMOGENEOUS-FLUID-FLOW SYSTEMS

4.1. Introduction.—Now that the physical characteristics of the porous media and their entrained fluids, of interest in oil-producing systems, have been discussed, we are ready to proceed with a quantitative formulation of their flow behavior. In doing so it will be convenient to consider first homogeneous-fluid systems, that is, those in which there is present only a single mobile fluid phase. In view of the almost universal occurrence of connate waters within the pores of oil- or gas-producing formations and the fact that the oil itself always contains some dissolved gas, the treatment of homogeneous-fluid systems may appear to be of only academic interest. Nevertheless, their detailed study is of value for a number of reasons.

Perhaps the most realistic excuse for an extended consideration of single-phase-flow systems is simply that the more general and practical heterogeneous-flow conditions are too difficult for rigorous mathematical treatment. In fact, as will be seen in Chap. 8, only very few such problems have as yet been analyzed, and virtually all these have referred to "steady states." As a result one is practically forced to resort in many cases to a simplification of the practical fluid-mixture problems to their counterparts under homogeneous-flow conditions, in order to get at least a guide to an understanding of the behavior and performance of the actual systems. Of course, a clear understanding must be had of the significance of the simplification and of at least the qualitative effect it will have on the results. The latter is generally rather easy to predict, so that the conclusions drawn for the simplified problem can usually be considered as limiting results with respect to those applying directly to the practical and more complex situation.

In addition to having the pragmatic virtue of simplicity, homogeneous-fluid systems actually do constitute close approximations to the real situations in a number of cases of practical interest. For the real dynamical criterion defining the homogeneous-fluid system is that it involves only a single *mobile* phase. Hence, for example, a formation producing only free gas will constitute a homogeneous-flow system even though it may

177

contain 50 per cent connate water. In fact, as long as the latter is not produced[1] it may be considered as literally frozen and as part of the pay, leaving only the gas as the fluid of interest. Of course, the homogeneous-fluid permeability for the gas and the effective porosity of the porous medium must be corrected for the presence of the connate water. But the dynamical features of the gas production and flow should be fully described by homogeneous-fluid considerations. Similarly, a formation producing an undersaturated oil or one in which the pressures are maintained to prevent gas evolution can be treated as a homogeneous-fluid system regardless of the connate water in the pay, as long as the water is not actually moving through the formation. Finally, the study of the fluid movement within water reservoirs bounding and feeding into oil-producing strata falls within the scope of the homogeneous-fluid representation. Indeed, the study of such water reservoirs is becoming of increasing importance as the control and possibilities of utilization of natural water drives are being more widely applied.

No attempt will be made to distinguish between those phases of the treatment which have immediate application per se to practical producing conditions and those which provide mainly a simplified prototype or limiting case for the corresponding more complex heterogeneous-flow system. In fact, it would not always be possible to do so. Many of the problems will fall in either category depending on the specific application in question. A clear understanding of the significant physical elements of the problems will suffice to determine the degree of applicability of the homogeneous-fluid treatment to the corresponding practical situations of interest.

4.2. The Structure of Hydrodynamic Systems.—Fluid dynamics is comprised of three basic types of physical laws. These are (1) the conservation of matter, (2) the thermodynamic equation of state of the fluid in question and of its condition of flow, and (3) the law of force to which individual fluid elements are subject. The composite resultant of these laws defines the structure of the hydrodynamic system.

The law of conservation of matter is, of course, a general and universal pillar of *macroscopic* physics. Its statement, that the total mass of any closed system must remain constant, may be taken as virtually axiomatic regardless of the particular nature of the physical problem. However, for application to hydrodynamic systems it is convenient to express it as follows: The net mass flux, per unit time, through any infinitesimal volume element of the fluid system is equal to the free volume of the element multiplied by the rate of change of the fluid density within it. Expressed

---

[1] It is also presupposed that the solubility of the gas in the connate water and its evolution on pressure decline may be neglected.

analytically with reference to porous media, this statement becomes the "equation of continuity," *i.e.*,

$$\text{div}\ (\gamma \bar{v}) \equiv \frac{\partial}{\partial x}\ (\gamma v_x) + \frac{\partial}{\partial y}(\gamma v_y) + \frac{\partial}{\partial z}\ (\gamma v_z) = -f\frac{\partial \gamma}{\partial t}, \tag{1}*$$

where v_x, v_y, v_z are fluid-velocity components in the cartesian coordinate system (x,y,z), γ is the fluid density, f is the porosity of the medium at (x,y,z), t is the time variable, and $\bar{v}$ is the resultant vector fluid velocity.

The right-hand side of Eq. (1) provides an accounting for time variations within the system although it implies the tacit assumption that the porosity f is a time-independent property of the medium and is not affected by compaction, the fluid, or any of the dynamical variables. Hence if conditions are known not to vary with time, *i.e.*, if the state of flow is "steady," the right-hand side of Eq. (1) may be set equal to zero, so that

$$\frac{\partial}{\partial x}\ (\gamma v_x) + \frac{\partial}{\partial y}\ (\gamma v_y) + \frac{\partial}{\partial z}\ (\gamma v_z) = 0. \tag{2}$$

The thermodynamic equation of state of the fluid is also a component common to all hydrodynamic systems, whether or not they involve porous media. For this equation merely defines the physical nature of the fluid of interest and the thermodynamic conditions under which it is flowing. Each phase of the definition will be expressed by a relationship between the fluid density γ, its pressure p, and temperature T, such as

$$\Phi(\gamma,p,T) = 0. \tag{3}$$

Thus the physical definition of incompressibility of the fluid will imply for Eq. (3) the simple form

$$\gamma = \text{const.} \tag{4}$$

Or if the fluid is an ideal gas, its "equation of state" will be

$$\gamma = \frac{wp}{RT}, \tag{5}$$

where w is its molecular weight and R is the gas constant per mole.

Similarly, in defining the thermodynamic conditions of flow, Eq. (3) may be expressed, for example, by

$$T = \text{const}, \tag{6}$$

if the flow is isothermal. Or if the density and pressure variations be adiabatic, and the fluid is an ideal gas, Eq. (3) must be given the form

$$T = \text{const}\ \gamma^{m-1}, \tag{7}$$

---

* The symbol div (or $\nabla \cdot$) represents the differential operator defined by the intermediate expression of Eq. (1).

where m is the ratio of the specific heat at constant pressure to that at constant volume.

On eliminating one of the variables γ, p, T by combining the two equations of the type of Eq. (3), that is, Eqs. (4) or (5) and (6) or (7), a single equation in two variables is obtained as the expression of both the equation of state and thermodynamic condition of flow. As a practical matter the temperature is the variable generally eliminated, thus reducing Eq. (3) to the form

$$\Phi(\gamma, p) = 0. \tag{8}$$

As an illustration of the immediate applicability of this type of consideration it may be noted that, if the fluid be taken as incompressible, the corresponding special form of Eq. (8), namely, Eq. (4), can be inserted directly in Eq. (1) to give

$$\text{div } \bar{v} = \frac{\partial v_x}{\partial x} + \frac{\partial v_y}{\partial y} + \frac{\partial v_z}{\partial z} = 0. \tag{9}$$

Or if the fluid system is comprised of an ideal gas flowing adiabatically, the combination of Eqs. (5) and (7) gives as the equivalent of Eq. (8)

$$\gamma = \text{const } p^{1/m}, \tag{10}$$

and Eq. (2) becomes

$$\text{div } (\bar{v} p^{1/m}) = \frac{\partial}{\partial x} (v_x p^{1/m}) + \frac{\partial}{\partial y} (v_y p^{1/m}) + \frac{\partial}{\partial z} (v_z p^{1/m}) = 0. \tag{11}$$

This equation will also apply for the special case of isothermal gas flow ($m = 1$).

While in general two- and three-dimensional-flow systems Eq. (1) or even Eqs. (9) and (11) in themselves give no direct physical description of the nature of the flow behavior, they do provide immediately physical results of interest in the case of one-dimensional systems. Thus if the flow is confined to the x direction and the fluid is incompressible, Eq. (9) gives at once

$$\frac{\partial v_x}{\partial x} = 0; \qquad v_x = \text{const.} \tag{12}$$

Likewise, for steady-state adiabatic unidirectional ideal-gas flow along x, Eq. (11) implies

$$\frac{\partial}{\partial x} (v_x p^{1/m}) = 0; \qquad v_x = \frac{\text{const}}{p^{1/m}}. \tag{13}$$

Eq. (13), with $m = 1$, was used in Sec. 3.6 for developing permeability-calculation formulas for gas-flow measurements.

Similarly, for pure radial flow, *i.e.*, where the flow is restricted to motion along the radius vectors issuing from a central axial origin, as in the case of an idealized flow into a well bore, the analogue of Eq. (12) for an incompressible fluid is

$$\frac{\partial}{\partial r}(rv_r) = 0; \qquad v_r = \frac{\text{const}}{r}, \tag{14}$$

where r is the radial distance from the axis. And for the corresponding case of adiabatic ideal-gas flow,

$$\frac{\partial}{\partial r}(rv_r p^{1/m}) = 0; \qquad v_r = \frac{\text{const}}{rp^{1/m}}. \tag{15}$$

Of course, Eqs. (12) to (15) are nothing more than specialized expressions of the law of conservation of matter and could have been written down at once on the basis of direct physical considerations. In more general situations, however, the procedure of deriving such equations by the analytical processes illustrated here is a virtual necessity because of the complexity of following through the fluid motion in purely physical language.

4.3. The Force Reactions of Fluid Systems.—While the combination of Eqs. 4.2(1) and 4.2(8), as illustrated by the derivations of Eqs. 4.2(12) to 4.2(15), does represent a major step in the analytical description of hydrodynamic systems, it is still in no sense complete. For as Eqs. 4.2(13) and 4.2(15) show explicitly, this procedure leads only to a relation between the velocity components and the pressure. They do not provide explicit expressions for the variation of either through the flow system. And even in the case of Eqs. 4.2(12) and 4.2(14), no indication is given of the pressure distribution. To proceed further and achieve a complete formulation of the hydrodynamic structure by which the velocity and pressure distribution may be individually derived, the above considerations must evidently be supplemented by a statement of the "law of force" that the fluid system must obey.

From a microscopic point of view the flow of a homogeneous fluid through a porous medium is merely a special case of the classical hydrodynamics of viscous fluids. The "law of force" of the latter should accordingly apply also to problems of fluid flow through a porous medium. This law, which is nothing more than a hydrodynamic expression—in differential form—of the basic Newtonian theorem that the force is equal to mass times acceleration, states that

$$\left. \begin{aligned} \gamma \frac{Dv_x}{Dt} &\equiv \gamma \left(\frac{\partial v_x}{\partial t} + v_x \frac{\partial v_x}{\partial x} + v_y \frac{\partial v_x}{\partial y} + v_z \frac{\partial v_x}{\partial z} \right) = \mu \, \nabla^2 v_x + \frac{1}{3} \mu \frac{\partial \theta}{\partial x} - \frac{\partial p}{\partial x} + F_x, \\ \gamma \frac{Dv_y}{Dt} &\equiv \gamma \left(\frac{\partial v_y}{\partial t} + v_x \frac{\partial v_y}{\partial x} + v_y \frac{\partial v_y}{\partial y} + v_z \frac{\partial v_y}{\partial z} \right) = \mu \, \nabla^2 v_y + \frac{1}{3} \mu \frac{\partial \theta}{\partial y} - \frac{\partial p}{\partial y} + F_y, \\ \gamma \frac{Dv_y}{Dt} &\equiv \gamma \left(\frac{\partial v_z}{\partial t} + v_x \frac{\partial v_z}{\partial x} + v_y \frac{\partial v_z}{\partial y} + v_z \frac{\partial v_z}{\partial z} \right) = \mu \, \nabla^2 v_z + \frac{1}{3} \mu \frac{\partial \theta}{\partial z} - \frac{\partial p}{\partial z} + F_z, \end{aligned} \right\} \tag{1}$$

where the differential operator ∇^2 is defined by

$$\nabla^2 \equiv \frac{\partial^2}{\partial x^2} + \frac{\partial^2}{\partial y^2} + \frac{\partial^2}{\partial z^2}, \tag{1a}$$

μ is the fluid viscosity, assumed to be constant, θ is the rate of volume dilatation, defined by

$$\theta = \mathrm{div}\ \bar{v} = \frac{\partial v_x}{\partial x} + \frac{\partial v_y}{\partial y} + \frac{\partial v_z}{\partial z}, \tag{1b}$$

and F_x, F_y, F_z are the components of any body forces, such as gravity, that may be acting on the fluid.

The terms D/Dt of the left-hand side of Eqs. (1) represent the acceleration components and include the local accelerations, as $\dfrac{\partial v_x}{\partial t}$, plus the contributions in fluid-velocity changes due to the motion of the fluid elements. The first two terms on the right-hand side comprise the viscous-force reactions opposing the acceleration and are derived in standard hydrodynamics texts.[1] The pressure-gradient terms give the effect on the motion of pressure variations in the system, and those due to body forces are given by the terms F_x, F_y, F_z. On combining Eqs. (1) with Eqs. 4.2(1) and 4.2(8), one obtains the analytical representation of the classical Stokes-Navier[2] hydrodynamics.

In principle this system of equations should also tell the whole story of homogeneous-fluid flow through porous media. It turns out, however, that this observation is of only academic interest and of virtually no practical significance. Indeed, even for systems of fluid flow containing no porous media the formal structure of the set of Eqs. (1), 4.2(1), and 4.2(8) is practically useless in its general form. Because of the nonlinearity of the left-hand side of Eqs. (1) it is literally impossible to solve them analytically except for a very few cases of special geometric simplicity. One is forced to resort to such approximations as may be valid under particular simplifying conditions. For example, a common procedure used in treating liquid-flow systems is to drop entirely the quadratic nonlinear velocity terms on the left-hand side of Eqs. (1). Even so the analytical treatment remains quite formidable unless the macroscopic geometry of the system is rather simple and possesses some degree of symmetry.

When an attempt is made to apply this classical hydrodynamic structure, even with all its usual approximations, to a porous medium, the problem

---

[1] Cf., for example, H. Lamb, "Hydrodynamics," 6th ed., p. 577, Cambridge University Press, 1932.

[2] C. L. M. H. Navier, *Ann. chim. phys.*, **19**, 234 (1881); G. G. Stokes, *Trans. Cambridge Phil. Soc.*, **8**, 287 (1845).

assumes an entirely different order of complexity. It is important to understand the reason for this. For it represents the crucial and unique major significant feature by which the flow through a porous medium is to be differentiated from that described by the classical hydrodynamics.

It will be observed that even if Eqs. (1), together with Eqs. 4.2(1) and 4.2(8), could be given formal solutions, they would be barren of physical meaning without further specifications. Such solutions would be merely analytical expressions containing terms that *all* porous-media-flow systems must formally satisfy. As is known from the theory of differential equations, they would contain in addition unspecified or arbitrary functions and constants. Indeed, there would be a multiple infinity of such solutions, each expressing *possible* pressure and velocity distributions. Which distributions will pertain to a specified practical and physical flow system? The answer will obviously be determined by those features of that particular flow system which are unique and distinguish it from other possible flow systems. These are evidently the state of the fluid at the initial instant of observation, the over-all boundaries delimiting the fluid field, and the conditions imposed at these boundaries. The first of these is expressed by the "initial conditions" and the last by the "boundary conditions." As the former enter a problem only when it is subject to time variations or transients and play a similar role in both classical and porous-flow hydrodynamic systems, they will not be considered further at this point.

Even for steady-state conditions of flow, it is clear that to specify a particular problem to be solved the geometry of the boundary surfaces confining the fluid or that part of it of primary interest must be defined. This would consist, for example, in stating that the fluid is moving through a capillary or pipe of given diameter, or perhaps through the annular channel between two cylindrical surfaces, if this be the actual fluid system in question.

The geometrical definition of the boundaries having been fixed, the conditions imposed on or obtaining over these surfaces must be stated explicitly. For example, for flow through a cylindrical capillary or pipe a condition to be imposed on the solutions for the velocity distribution is that the velocity at the pipe surface be parallel to it. Moreover, the value of the velocity at the surface must agree with preassigned conditions, *i.e.*, it must vanish if the fluid is known to adhere to the surface, or it must satisfy fixed slippage relationships. Finally the pressure distribution representing the driving force acting on the fluid must be specified over surfaces cutting across the direction of flow and enclosing the length of flow channel of interest. Obviously the detailed features of the particular flow system under consideration will depend on the exact numerical specifications of the above-mentioned conditions, and only by imposing them on the general solution of the hydrodynamic equations (assuming this to be available) can unique pressure and velocity distributions be determined.

On carrying over these considerations to porous-media problems the reason why they cannot be treated by the classical hydrodynamic equations (1) becomes evident. For all that can be definitely specified about the geometry of a porous medium are its gross dimensions and shape. The myriads of interstitial and multiply connected grain surfaces are obviously beyond any conceivable power of quantitative description. And yet in comparison with these internal surfaces the gross surface area is virtually neglible. Thus 1-ft-radius sphere of rock composed of spherical grains of 0.01 cm radius would have a gross external surface area that is less than 0.05 per cent of the total interior grain surface. And even if by some miraculous insight the detailed geometry of the almost infinite variety of these intertwining labyrinths could be crystallized into a quantitative description, it would be utterly futile to attempt finding solutions of Eqs. (1) that would be amenable to adjustment for satisfying the boundary conditions at these surfaces.

It will thus be clear that an approach to the treatment of fluid flow through porous media by the classical hydrodynamics must be condemned from the start as being totally devoid of any hope of success. On the other hand, the same considerations leading to this inevitable conclusion provide the key to an effective and practical attack on the problem. This lies in the observation that porous media contain an enormous internal surface area as compared with that represented by their macroscopic dimensions. As a result the viscous-resistance reactions will be far greater than the inertia and acceleration forces, so that the left-hand side of Eqs. (1) can be simply dropped. The "law of force" for porous media will therefore involve only the dynamical equilibrium between viscous shearing resistances and driving forces, as pressure gradients. Moreover, the futility of describing and mathematically adjusting solutions to fit boundary conditions at the whole multitude of individual internal surface elements shows that the resistance terms in Eqs. (1) must be replaced by their statistical equivalent, averaged over large numbers of grains or pores such as will extend over normal macroscopic dimensions. The complex ensemble of multiply connected solid and free-space microscopic volume elements must be considered as a locally homogeneous and uniform continuum, showing such resistance to fluid flow as is statistically and macroscopically equivalent to the actual resultant of the individual microscopic components of the whole.

The establishment of such macroscopic equivalents is evidently an empirical problem. And this was the basic purpose behind Darcy's original experimentation. The beautiful simplicity of his findings on water flow through sand filters and their extension to other porous media and fluids, namely,

$$\bar{v} = -\frac{k}{\mu}\,\nabla\,p, \tag{2}$$

obviously justifies fully the reasoning underlying its development. It is, then, the "law of force" for porous media. It is not a fortuitously simple solution of Eqs. (1). Rather, it constitutes only a statistical and macroscopic resultant of the infinitely complex solutions of Eqs. (1), whatever they may be. And the numerical expression for the resultant effect of the particular assembly of grains and cementing material comprising any specific rock element is crystallized in the magnitude of the permeability k.

As discussed in Sec. 3.4, what is now known as "Darcy's law" represents a significant generalization of Darcy's original empirical correlation between the rate of flow of water through a filter bed and the head of water. When including the effect of "body" forces, such as gravity, represented by the vector $\bar{F}$, it may be written as

$$\bar{v} = -\frac{k}{\mu}\,(\nabla\,p - \bar{F}), \tag{3}$$

where the medium is assumed to be isotropic. If it is anisotropic,[1] the equivalent of Eq. (3) is obtained by considering k as having its individual values along the coordinate axes when Eq. (3) is written out in terms of its components. If the medium is both isotropic and of uniform permeability and if the force $\bar{F}$ has a potential V, so that $\bar{F} = -\nabla V$, one may introduce the function Φ, defined by

$$\Phi = \frac{k}{\mu}(p + V), \tag{4}$$

so that Eq. (3)[2] becomes

$$\bar{v} = -\nabla\,\Phi. \tag{5}$$

It thus appears that Φ is a "velocity potential"—a function whose negative gradient gives the velocity vector.

It is to be observed that these macroscopic equivalents [Eqs. (3) and (5)] of the Stokes-Navier classical-hydrodynamics law of force provide for an explicit separation between the permeability k and viscosity μ. This, however, is to be expected, as the averaging process leading to Darcy's law

---

[1] As the great majority of the problems to be treated in the following chapters will not involve the question of isotropy, the assumption of isotropy will be hereafter presupposed wherever k appears, except as it is explicitly stated otherwise.

[2] The equivalence of Eqs. (3) and (5) implies that μ as well as k is constant. It is possible so to generalize the definition of the potential [Eq. (4)] as to include the case of variable k/μ. However, this would be of but little practical value, since the situations where continuous areal variations of k/μ are important generally involve heterogeneous-fluid flow.

evidently pertains only to the structure of the porous medium and not to the fluid it is carrying.[1] The latter, therefore, will retain its characterization by its viscosity μ as if it were part of a classical-hydrodynamics system.

Thus we have substantially completed the development of the structure of the hydrodynamics of homogeneous-fluid flow through porous media. The basic dynamical element, namely, Darcy's law, is its unique feature. By its nature it masks the detailed microscopic phenomena relating to individual pores and provides only an equivalent statistical representation of the flow behavior as if the porous medium were a locally homogeneous continuum. As will be seen in later discussions of multiphase-fluid systems, at least a qualitative understanding of the processes taking place in individual pores is indeed vitally essential. Nevertheless even there, as well as here, the locally macroscopic representation of the flow phenomena suffices fully to describe most practical aspects of oil and gas production.

4.4. The Equations of Motion.—By combining now the basic elements of the hydrodynamic system developed above, the equations of motion for various types of fluid system are readily derived. Thus upon applying first the dynamical equation [Eq. 4.3(3)] to the equation of continuity [Eq. 4.2(1)], it follows that

$$\nabla \cdot \left[\gamma \frac{k}{\mu} \nabla (p + V) \right] = f \frac{\partial \gamma}{\partial t}, \tag{1}$$

where V is the potential of the body force vector. Under most conditions of practical interest in homogeneous-fluid flow, μ may be taken as independent of the pressure and may therefore be taken out of the brackets. For homogeneous media, k may also be taken out of the brackets. As we shall confine ourselves to uniform media in considering homogeneous-fluid systems, except when dealing with certain special problems (Chap. 6), k will hereafter be taken as constant unless otherwise specified. Equation (1) then takes the form

$$\nabla \cdot [\gamma \, \nabla (p + V)] = \frac{f\mu}{k} \frac{\partial \gamma}{\partial t}. \tag{2}$$

To obtain a differential equation in the single variable γ or p the equation of state of the fluid must be introduced. As a sufficiently general equation including all homogeneous fluids of practical interest and all thermodynamic types of viscous flow we may take[2]

$$\gamma = \gamma_o p^m e^{\kappa p}. \tag{3}$$

---

[1] The assumption is here made, as in classical hydrodynamics, that the fluid does not affect the medium in which it is carried, although this does not always obtain in practice (cf. Sec. 3.7).

[2] M. Muskat, *Physics*, **5**, 71 (1934).

The particular fluids of physical significance may then be classified as follows:

Liquids: $m = 0$;
   Incompressible liquids: $\kappa = 0$,
   Compressible liquids: $\kappa \neq 0$;
Gases: $\kappa = 0$;
   Isothermal expansion: $m = 1$,

   Adiabatic expansion: $m = \dfrac{\text{specific heat at const. vol.}}{\text{specific heat at const. pressure}}$

Applying these to Eq. (2) with the assumption that gravity is the only body force acting on the fluid, so that $V = \gamma gz$* (if z is directed upward), it follows that for
Incompressible liquids:

$$\nabla^2 \, p = \frac{\partial^2 p}{\partial x^2} + \frac{\partial^2 p}{\partial y^2} + \frac{\partial^2 p}{\partial z^2} = 0 = \nabla^2 \, \Phi, \tag{4}\dagger$$

where Φ is defined by Eq. 4.3(4);
Compressible liquids:

$$\nabla \cdot \left(\frac{1}{\kappa} \nabla \gamma + \gamma^2 g \, \nabla \, z \right) = \frac{f\mu}{k} \frac{\partial \gamma}{\partial t}. \tag{5}$$

It may be noted that the terms $\gamma^2 g \, \nabla \, z$ and $(1/\kappa)\nabla \, \gamma$, in Eq. (5), are in the ratio of the vertical body force due to gravity to that due to the fluid pressure gradients. When these forces are of comparable magnitude and the compressibility of the liquid is of physical significance, an accurate treatment would therefore require the solution of Eq. (5) as given. However, such a treatment will in general be very difficult because of the non-linearity of the equation. For practical purposes, therefore, it is necessary to discuss the phases of the compressibility and the gravity component of the flow separately, *i.e.*, the former by Eq. (5), with the neglect of the term involving g, and the latter by the solution of Eq. (4). The errors in this approximation will not be serious from a practical point of view. In those cases where the pressure gradients will be of sufficiently large magnitude to lead to appreciable variations of the liquid density over small distances, these same gradients will also be large as compared with the gravity-head gradient γg.‡ And if the density variations in homogeneous-

---

* For compressible liquids, V should be set equal to $g \int \gamma \, dz$.

† The derivation of this equation, as governing the flow of incompressible liquids through porous media, was first given by C. S. Slichter, *USGS 19th Ann. Rept.*, 1897–1898, p. 330, where a number of solutions for specific problems are also presented.

‡ Even if $(1/\gamma)\dfrac{\partial \gamma}{\partial x} = 10^{-5}$, the associated pressure gradients will be larger than the gravity gradient γg by a factor of the order of 100.

liquid flow are of appreciable magnitude, essentially because of the extended dimensions of the flow system (cf. Sec. 11.3), the major component of the flow will be horizontal and the gravity component again will not be of primary importance. Hence in homogeneous-fluid problems where gravity will play a significant role the effects of the compressibility will be of lesser importance, and the analysis can be safely based upon Eq. (4), whereas in those where the effects of the liquid compressibility are the predominant features the force of gravity usually can be left out of consideration. Where the liquid flow will actually be considered as that of a compressible liquid, we shall therefore drop the term with g in Eq. (5) and take the system to be governed by the equation

$$\nabla^2 \gamma = \frac{\partial^2 \gamma}{\partial x^2} + \frac{\partial^2 \gamma}{\partial y^2} + \frac{\partial^2 \gamma}{\partial z^2} = \frac{f \kappa \mu}{k} \frac{\partial \gamma}{\partial t}. \tag{6}$$

It should be noted that the linearity of Eq. (6) is the result not only of the neglect of gravity effects but also of the particular form of Eq. (3), with $m = 0$, used as the relation between the fluid density and pressure. The latter implies that the compressibility is the constant κ, independent of the pressure. Such a relationship, in a strict sense and over arbitrary pressure ranges, is not to be expected physically, nor is it observed empirically. Measurements on oils above their bubble points give compressibilities decreasing with increasing pressure. Published data on water show similar trends. While the latter may be roughly expressed as a linear decline of the compressibility with increasing pressure, this would also be an approximation valid only over a limited range. The same applies to an expression of the density as a linear function of the pressure. Moreover, all such approximate representations would lead to nonlinear equations in both the density and pressure when introduced into Eq. (2). The linearization of these equations would then require further approximations in the dropping of the nonlinear terms. On the other hand, since over the limited range of pressure of several hundred atmospheres, to which Eq. (2) will find application in actual oil-producing reservoirs, the compressibilities do not change greatly, the density function of Eq. (3), with $m = 0$, in which κ is considered as the average compressibility over the pressure interval of interest, will provide an approximation but little different from any other. As this representation also directly leads to the linear classical equation in γ [Eq. (6)], when gravity is neglected, it will be used throughout this work as the basis for the analytical treatment of problems in which liquid-compressibility effects must be taken into account (cf. Chap. 11), although, again for reasons of convenience, the linear density vs. pressure approximation will be assumed in the discussion and construction of the electrical-analyzer method for treating compressible-liquid systems (cf. Sec. 11.8).

Gases:

$$\nabla^2 \gamma^{(1+m)/m} = \frac{\partial^2 \gamma^{(1+m)/m}}{\partial x^2} + \frac{\partial^2 \gamma^{(1+m)/m}}{\partial y^2} + \frac{\partial^2 \gamma^{(1+m)/m}}{\partial z^2} = \frac{(1+m)f\mu\gamma_o{}^{1/m}}{k}\frac{\partial \gamma}{\partial t}. \quad (7)$$

These are the fundamental differential equations that we shall take as the basis of the treatment of the various problems of homogeneous-fluid flow through porous media of practical significance. It will be seen at once that for incompressible liquids the time variations disappear, so that there can be no time transients or nonsteady states within the system unless the conditions at the boundaries vary with the time. Both the pressure and velocity potential Φ obey "Laplace's equation."

In the case of compressible liquids the fundamental equation [Eq. (6)] does involve the time and is, in fact, identical in form with the "Fourier heat-conduction equation."[1] Its steady-state[2] form is, however, identical with that for incompressible liquids, with the density playing the role of the pressure or velocity potential.

Equation (7) for gases also contains the time and therefore permits non-steady states. However, it is nonlinear in that it involves the dependent variable γ to a power other than the first and cannot be solved rigorously in closed form. The treatment of transient-gas-flow systems must therefore be approximate, although here, too, the steady state is governed by Laplace's equation in the dependent variable $\gamma^{(1+m)/m}$. On the other hand, since the dynamics of gas reservoirs as such is only of minor interest in the study of oil-production problems, no special discussion of them will be presented in this work.[3]

Laplace's equation (4) will be taken as the basis of all the analysis of the next two chapters, in which will be considered problems of the steady-state flow of liquids. While this equation resulted from the assumption that the liquid is strictly incompressible, it will in general represent a good approximation for actual liquids except when they have an abnormally high compressibility or when the dimensions of the flow system are very large (cf. Sec. 11.3). Usually, however, it will suffice, in view of the very low compressibility of normal liquids, to treat homogeneous-liquid-flow systems

[1] Cf. H. S. Carslaw, and J. C. Jaeger, "Conduction of Heat in Solids," Oxford University Press, 1947.

[2] The term "steady state" is used to denote the condition of flow in which the significant dynamical variables (the pressure or density and velocity) do not vary with the time, so that all terms in the fundamental differential equations involving $\frac{\partial}{\partial t}$ may be set equal to zero.

[3] Several approximate treatments of gas-flow transient problems are given in M. Muskat, "The Flow of Homogeneous Fluids through Porous Media," Chap. XI, McGraw-Hill Book Company, Inc., 1937, where a number of steady-state systems are also briefly reviewed.

in the steady state explicitly as problems in the flow of incompressible liquids and hence as governed by Eq. (4).

Except when the compressibility inherently enters as a significant feature of the problem, the conditions for which are discussed in Sec. 11.3, the steady-state solutions of Eq. (4) may be considered as subject to synthesis in continuous sequences to correspond to variations with time in the boundary conditions, the time entering in all the expressions as a parameter rather than as an independent variable. Each instantaneous pressure distribution and associated flux will correspond to the boundary conditions at the same instant, as if these conditions had been maintained previously for an indefinitely long time. Although this treatment of time variations will be rigorous only if the liquid is strictly incompressible, it will also suffice, from a practical point of view, in the discussion of problems where the compressibility could be neglected if the system were actually in a steady state.

As in the case of the classical hydrodynamics, the application of the "equations of motion" to a specific problem presupposes that the latter has been explicitly defined with respect to its unique geometrical and physical characteristics. In particular, these comprise (1) a geometrical definition of the boundaries of the region in space for which a solution is desired; (2) a specification of the "boundary conditions" imposed on and to be satisfied at the boundary; and (3) an assignment of the "initial conditions," *i.e.*, the density or pressure distribution at the initial instant of the history, if the fluid is compressible and nonsteady states are involved. For the boundary conditions the values of the pressure, velocity potential, or density, or their normal gradients, or a linear function of both must be assigned at all points of the boundary surface. That solution of the appropriate equation of motion, among Eqs. (4) to (7), that satisfies these conditions will then give the complete physical description of the corresponding particular flow system and problem of interest.

4.5. Other Coordinate Systems.—As all the specific problems for homogeneous-fluid systems to be discussed in the next two chapters will refer to steady-state conditions, it will be convenient to list here the most common transformations of the basic equations that will be used in treating problems involving special types of geometrical symmetry. These will include the cylindrical- and spherical-coordinate systems. Since under steady-state conditions the equations of motion in isotropic homogeneous media formally reduce to Laplace's equation for both liquids and gases, the transformations will be exhibited only for the particular equation

$$\nabla^2 \Phi = \frac{\partial^2 \Phi}{\partial x^2} + \frac{\partial^2 \Phi}{\partial y^2} + \frac{\partial^2 \Phi}{\partial z^2} = 0, \tag{1}$$

where Φ may be interpreted as any of the appropriate dependent variables applying to the various fluid systems discussed in Sec. 4.4. In addition, on the assumption that Φ represents a velocity potential, expressions will be given for the velocity components in the new coordinates. These transformations are the following:

Cylindrical coordinates (r,θ,z) (Fig. 4.1):

$$r = \sqrt{x^2 + y^2}; \qquad \theta = \tan^{-1}\frac{y}{x}; \qquad z = z,$$
$$x = r\cos\theta; \qquad y = r\sin\theta; \qquad z = z. \tag{2}$$

$$v_r = -\frac{\partial\Phi}{\partial r}; \qquad v_\theta = -\frac{1}{r}\frac{\partial\Phi}{\partial\theta}; \qquad v_z = -\frac{\partial\Phi}{\partial z}, \tag{3}$$

$$\nabla^2\Phi = \frac{1}{r}\frac{\partial}{\partial r}\left(r\frac{\partial\Phi}{\partial r}\right) + \frac{1}{r^2}\frac{\partial^2\Phi}{\partial\theta^2} + \frac{\partial^2\Phi}{\partial z^2} = 0. \tag{4}$$

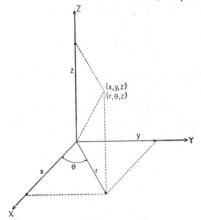

Fig. 4.1. The polar coordinate system.

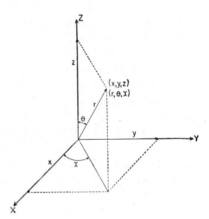

Fig. 4.2. The spherical coordinate system.

Spherical coordinates (r,θ,χ) (Fig. 4.2):

$$r = \sqrt{x^2 + y^2 + z^2}; \qquad \theta = \tan^{-1}\frac{\sqrt{x^2 + y^2}}{z}; \qquad \chi = \tan^{-1}\frac{y}{x},$$
$$x = r\sin\theta\cos\chi; \qquad y = r\sin\theta\sin\chi; \qquad z = r\cos\theta. \tag{5}$$

$$v_r = -\frac{\partial\Phi}{\partial r}; \qquad v_\theta = -\frac{1}{r}\frac{\partial\Phi}{\partial\theta}; \qquad v_\chi = -\frac{1}{r\sin\theta}\frac{\partial\Phi}{\partial\chi}. \tag{6}$$

$$\nabla^2\Phi = \frac{1}{r^2}\frac{\partial}{\partial r}\left(r^2\frac{\partial\Phi}{\partial r}\right) + \frac{1}{r^2\sin\theta}\frac{\partial}{\partial\theta}\left(\sin\theta\frac{\partial\Phi}{\partial\theta}\right) + \frac{1}{r^2\sin^2\theta}\frac{\partial^2\Phi}{\partial\chi^2} = 0. \tag{7}$$

These equations show that, if the fluid system is symmetrical about an axis, the symmetry is readily expressed in the system of cylindrical coordinates by taking z as the axis of symmetry and setting $\frac{\partial\Phi}{\partial\theta} = 0$. On the other hand, if it possesses spherical symmetry, this is naturally expressed by taking the origin of a system of spherical coordinates at the center of symmetry and dropping the terms in $\dfrac{\partial \Phi}{\partial \theta}$ and $\dfrac{\partial \Phi}{\partial \chi}$ in Eq. (7).[1]

4.6. Summary.—As in the case of all fluid-flow systems, the problem of homogeneous-fluid flow through porous media may be described in terms of a hydrodynamic structure composed of (1) the law of conservation of matter, (2) the thermodynamic equation of state of the fluid and of its condition of flow, and (3) the law of force to which the fluid elements are subject. The first may be expressed by the "equation of continuity" [Eq. 4.2(1)] and simply requires that in closed systems there is no net creation or destruction of fluid mass. The second serves to define the nature of the fluid, whether it be a gas or liquid, and its physical parameters [cf. Eq. 4.4(3)] and the thermodynamic character of the flow, *i.e.*, isothermal, adiabatic, etc. [Eq. 4.2(3)].

While the first two are common to all hydrodynamic systems, such as the classical Stokes-Navier hydrodynamics, the law of force constitutes the unique feature for flow through porous media. This is so, however, only because of the high order of complexity of the basic physical problem. From a microscopic point of view the fluid flow in the individual pores of a porous medium must obey[2] the classical hydrodynamic equations [Eqs. 4.3(1)]. But it is impossible to apply these equations effectively, for two basic reasons. In the first place, from a practical point of view it is not feasible to give a quantitative description of the microscopic geometry of the flow system, as defined by the myriads of complicated passageways bounded by the ensemble of grains and their cementing material. Such definitions of the detailed boundaries of the flow system are, of course, required for fixing the solutions of the governing differential equations so as to correspond to the particular problem of interest. Second, previous studies of the classical hydrodynamics equations show that they are far too complex to permit integration for any such systems of irregular and multiply connected flow channels, comprising the pores of oil-bearing rocks, even if they could be quantitatively defined.

Accordingly it has been necessary to visualize the problem of fluid flow through porous media from a statistical point of view. The medium is considered as a continuum, offering such macroscopic resistance to the fluid flow as is statistically equivalent to the resultant of the individual resistance contributions of the various flow channels in the actual rock. This statistical representation has been established empirically, and is expressed by Darcy's law [Eq. 4.3(2)] to the effect that the macroscopic fluid velocity

---

[1] The transformation to general curvilinear coordinates is given in Appendix II.
[2] If the flow is viscous.

(volume flux rate per unit area) is directly proportional to the pressure gradient acting on the fluid. On generalizing the original experimental results to include the effects of both pressure gradients and those due to body forces, as gravity, and so as to apply to general three-dimensional-flow systems, the "law of force" for flow through porous media is readily formulated [Eq. 4.3(3)].

The structure of the porous medium is represented by the constant of proportionality in Darcy's law and is termed its permeability. It must be determined empirically.[1] In fact, such determinations themselves are essentially equivalent to specific verifications of Darcy's law for the particular rock sample involved. The nature of the fluid enters in Darcy's law only through its viscosity. Because of the tremendous internal surface area of porous media the inertia forces, involving the fluid density, will be negligible as compared with the viscous shearing resistance. It is for this reason that the fluid density does not enter in the "law of force" equation.

The law of force having been established, the fundamental equations of motion are readily derived on combining with it the equation of continuity and the equations of state. It is found that for incompressible liquids the basic equation is Laplace's equation in the fluid pressure or velocity potential [cf. Eqs. 4.4(4) and 4.3(4)]. For compressible-liquid systems that are not dominated by gravity effects the fluid density will obey the heat-conduction equation [Eq. 4.4(6)] if the compressibility be taken as constant. And for gas flow the equation for transient systems is nonlinear but reduces to Laplace's equation in $\gamma^{(1+m)/m}$ for steady states [cf. Eq. 4.4(7)], where γ is the gas density and m is the exponent defining the type of gas expansion.[2] On applying the appropriate boundary and initial conditions for the porous medium, when considered as a macroscopic continuum, the solutions of these equations will give complete descriptions of all the pertinent physical characteristics of the corresponding homogeneous-fluid-flow systems.

---

[1] For a discussion of the technique involved in these measurements, cf. Sec. 3.6.

[2] In all these cases it is assumed that the fluid viscosity is constant and the medium is isotropic and homogeneous.

CHAPTER 5

INDIVIDUAL WELL PROBLEMS IN UNIFORM STRATA WITH HOMOGENEOUS-FLUID FLOW

The basic producing element of an actual oil field is evidently the well bore penetrating the oil-bearing formation. It is therefore pertinent to apply first the equations of motion developed in the last chapter to determine the flow characteristics of such production units. Because of the basically different type of analysis and physical implications involved in the consideration of incompressible- and compressible-fluid systems, only the former will be treated here,[1] and the discussion of the latter will be deferred to a later chapter (Chap. 11). Moreover, in the present chapter the discussion will be restricted to wells penetrating uniform and isotropic[2] strata producing a homogeneous liquid, as an undersaturated oil.

5.1. Radial Flow.—The simplest problem of the type referred to above is that of perfect radial flow of a homogeneous fluid into a well. Such flow will obtain if the well completely penetrates the rock stratum and the distant fluid source acts uniformly in all directions radiating from the axis of the well bore. As gravity effects will not enter in this problem and the system has axial symmetry, it will be described by Laplace's equation in the fluid pressure in cylindrical coordinates [cf. Eq. 4.5(4)], namely,

$$\frac{1}{r}\frac{\partial}{\partial r}\left(r\frac{\partial p}{\partial r}\right) = 0. \tag{1}$$

The general solution of Eq. (1) is

$$p = c_1 \log r + c_2, \tag{2}$$

where c_1 and c_2 are constants of integration. The latter are determined by applying the boundary conditions of the problem, namely, the requirement that the pressure assume the preassigned constant values at the physical boundaries of the system. One of these is evidently the well bore, defined by its radius r_w, and where the pressure may be supposed to have the value p_w. While the other is in principle arbitrary, it may be taken as any distant

---

[1] As the fundamental differential equation is also Laplace's equation for the steady-state flow of compressible liquids and gases (cf. Sec. 4.4), the latter will be formally subject to the solutions to be developed here for incompressible liquids.

[2] An exception will be made in Sec. 5.6 to provide a more complete treatment of wells completed with perforated casings.

cylindrical surface, of radius r_e, over which the pressure may be assumed to have a substantially uniform value p_e, as indicated in Fig. 5.1. Expressed analytically, these physical requirements are

$$p = p_w \quad : \quad r = r_w, \\ p = p_e \quad : \quad r = r_e. \tag{3}$$

It is readily verified that if c_1 and c_2 in Eq. (1) are so chosen that p takes the form

$$p = p_w + \frac{p_e - p_w}{\log r_e/r_w} \log \frac{r}{r_w}, \tag{4}$$

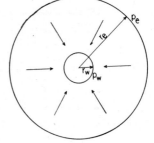

FIG. 5.1.

Eqs. (3) will be satisfied. Accordingly Eq. (4) represents the unique and complete pressure distribution for the radial-flow system under consideration. It will be observed that the pressure varies logarithmically between the bounding radii.

The velocity distribution may be obtained by applying Eq. 4.5(3), on noting that the velocity potential here is $(k/\mu)p$. Accordingly the radial velocity is given by

$$v_r = -\frac{k}{\mu} \frac{\partial p}{\partial r} = -\frac{k}{\mu r} \frac{(p_e - p_w)}{\log r_e/r_w}, \tag{5}$$

and hence varies inversely as the radius, with an absolute value proportional to the pressure differential over the system. From Eq. (5) the total fluid flow through the formation and into the well bore is readily calculated[1] to be

$$Q = -h \int_0^{2\pi} r v_r \, d\theta = \frac{2\pi k h (p_e - p_w)}{\mu \log r_e/r_w}, \tag{6}$$

where h is the thickness of the stratum.

As is to be expected, Q is directly proportional to the permeability k, to the thickness, and to the pressure differential and inversely proportional to the fluid viscosity μ. Its unique feature is the logarithmic variation with the boundary radii, which implies that moderate changes in these will affect the flow rate but slightly. Thus to increase Q by a factor of 2 by changes in the boundary radii the ratio of the latter must be reduced to the value

$$\frac{r_e'}{r_w'} = \sqrt{\frac{r_e}{r_w}}. \tag{7}$$

---

[1] While the same convention is used here in fixing the sign in Darcy's law as indicated in the footnote on p. 197, an additional adjustment of sign will be made in evaluating the flux so that it will have a positive value.

For example, to double the flow capacity of a radial system with a well radius $r_w = \frac{1}{4}$ ft and external radius $r_e = 660$ ft the ratio of the new radii must be made equal to 51.38. If the external radius be kept fixed, the well-bore radius must be increased to 12.84 ft. And if only the former is varied, it must be shrunk to the same value, 12.84 ft. It is thus also evident that the exact dimensions of the well bore or external boundary will have but little effect on the production capacity of the well.

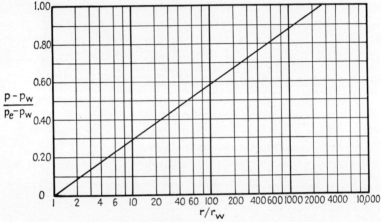

Fig. 5.2. The calculated pressure distribution in steady-state homogeneous-fluid radial-flow systems. p = pressure at radius r; p_w = pressure at internal boundary, of radius r_w; d_e = pressure at external boundary, of radius $r_e = 2{,}640 r_w$.

In terms of Q the pressure and velocity distributions [Eqs. (4) and (5)] may be expressed as

$$p = p_w + \frac{Q\mu}{2\pi kh} \log \frac{r}{r_w}, \tag{8}$$

$$v_r = -\frac{Q}{2\pi hr}. \tag{9}$$

To illustrate the nature of the pressure distribution, Eq. (4) is plotted in relative terms in Fig. 5.2. The rise in pressure above the well pressure, $p - p_w$, at a radius r, and expressed as a fraction of the total pressure drop, $p_e - p_w$, is plotted as the ordinate. The abscissas are the radii divided by the well radius. The over-all radii ratio r_e/r_w has been taken as 2,640. A semilogarithmic plot has been used to emphasize the logarithmic variation of the pressure distribution. It will be noted that 29.2 per cent of the total pressure drop occurs between the well and a radius = $10 r_w$, or $2\frac{1}{2}$ ft for a $\frac{1}{4}$-ft well radius. And 58.5 per cent of the drop takes place between the well and $100 r_w$, or 25 ft for a $\frac{1}{4}$-ft well radius. It will be thus clear that in radial-flow systems the pressure gradients and flow resistance are highly concentrated about the well bore.

Equation (6) will give the rate of flow Q, in cubic centimeters per second, provided that k is expressed in darcys, h in centimeters, μ in centipoises, and p_e, p_w in atmospheres. While the final value of Q so computed can be converted to barrels per day by multiplying by the factor 0.5434, it is more convenient for practical purposes to use the following units: k in millidarcys, h (and r_e, r_w) in feet, p_e, p_w in psi, and μ in centipoises. With these units the value of Q in surface barrels per day will be given by

$$Q = 0.007082 \frac{kh(p_e - p_w)}{\mu\beta \log_e r_e/r_w} \text{ bbl/day} = 0.003076 \frac{kh(p_e - p_w)}{\mu\beta \log_{10} r_e/r_w} \text{ bbl/day}, \quad (10)$$

where the formation-volume factor β has been introduced to correct for the possible shrinkage of the volume of the fluid entering the well bore to its measured volume at the surface.

Similarly, if Q is expressed in barrels per day and h and r in feet, the velocity v_r in the formation in feet per second* will be given by

$$v_r = -\frac{1.034 \times 10^{-5} Q\beta}{hr} \text{ ft/sec.} \quad (11)\dagger$$

From Eq. (10) it is readily found that a $\frac{1}{4}$-ft-radius well in a 10-ft stratum of 500 md permeability will produce oil of 2 cp viscosity and shrinkage of 20 per cent at a surface rate of 374.5 bbl/day if the pressure drop is 200 psi between the well and an external radius of 660 ft. From Eq. (11) it then follows that the velocity at a radius r will be $4.65 \times 10^{-4}/r$ ft/sec. This implies that at the well bore the velocity will be 1.86×10^{-3} ft/sec, 4.65×10^{-4} ft/sec at a radius of 1 ft, and 7.04×10^{-7} ft/sec at the external (inflow) boundary.

Although the above derivations were based upon the direct analytical treatment of the problem of radial flow, it is of interest to outline a somewhat more physical method, which is perhaps more illuminating than that of manipulating the differential equation (1). Here one begins with the integrated equation of continuity, which states that the total flow through any cylindrical surface coaxial with the well bore is a constant, so that

$$2\pi r v_r h = \text{const} = -Q.$$

Hence it follows at once that

$$v_r = -\frac{Q}{2\pi r h}.$$

* This velocity is, of course, of the same nature (flux per unit area, *i.e.*, cubic feet per square foot) as that used in Darcy's law. To convert it into an actual velocity of advance of a fluid front it must be divided by the porosity.

† The negative sign here and in Eqs. (5) and (9), when $p_e > p_w$, merely indicates that the fluid is flowing toward decreasing values of r.

Applying now Eq. 4.5(3),

$$v_r = -\frac{k}{\mu}\frac{\partial p}{\partial r} = -\frac{Q}{2\pi rh},$$

so that by integration, again

$$p = \frac{Q\mu}{2\pi kh}\log\frac{r}{r_w} + p_w. \tag{8}$$

These results are, of course, identical with those obtained before. The derivation, however, shows clearly that the first integral of the differential equation for p simply takes account of the equation of continuity, whereas the second integration corresponds to the application of Darcy's law to the velocity distribution and the derivation therefrom of the pressure distribution.

5.2. Unsymmetrical Flow into[1] a Well Bore.—From a practical point of view the case of strictly radial flow, which implies an exactly uniform pressure imposed on a circular boundary concentric with the well surface, is evidently too idealized to correspond to such situations as are likely to occur in practice. Rather it is to be anticipated that even such flow systems as contain but a single well will in general have nonuniform pressure distributions over their external boundaries, that the wells will not in general lie at the centers of their external boundaries, and that finally the boundaries themselves over which the pressure distributions are preassigned and known will be other than circular in shape. In all these cases the flow into the wells will be unsymmetrical, and the pressure distributions will depend upon both the azimuthal and radial coordinates. In this section we shall therefore consider the effect of such asymmetries on the production capacities of individual wells.

If both the well bore and external boundary are exactly cylindrical and concentric but are subjected to nonuniform pressure distributions, the purely radial pressure distribution [Eq. 5.1(2)] must be generalized to include the azimuthal coordinate θ. By Eq. 4.5(4) we shall then have

$$\frac{1}{r}\frac{\partial}{\partial r}\left(r\frac{\partial p}{\partial r}\right) + \frac{1}{r^2}\frac{\partial^2 p}{\partial\theta^2} = 0. \tag{1}$$

As the general solution of this equation will have the form

$$p = c_0 \log r + \sum_{-\infty}^{+\infty} r^n(a_n \sin n\theta + b_n \cos n\theta), \tag{2}$$

---

[1] It is to be noted that both here and in the case [treated in ;the preceding section the same formal results will apply when the liquid is injected into the well and flows into the surrounding strata.

it is convenient to express the boundary conditions in Fourier series[1] form,
as

$$r = r_w \quad : \quad p = p_w(\theta) = \Sigma(w_n \sin n\theta + x_n \cos n\theta)$$
$$r = r_e \quad : \quad p = p_e(\theta) = \Sigma(e_n \sin n\theta + f_n \cos n\theta). \quad \left. \begin{matrix} \\ \\ \end{matrix} \right\} \quad -\pi \leqslant \theta \leqslant \pi. \quad (3)$$

where the geometry of the system is defined by Fig. 5.3.

To determine the rate of flow it is not necessary to evaluate explicitly all
the coefficients a_n, b_n. It will suffice to
observe that since the radial velocity
component is still given by

$$v_r = -\frac{k}{\mu} \frac{\partial p}{\partial r}, \quad (4)$$

the production (or injection) rate Q
will have the value

$$Q = -h \int_{-\pi}^{+\pi} r v_r \, d\theta = \frac{2\pi k h c_0}{\mu}, \quad (5)$$

where h is again the thickness of the
producing stratum. On adjusting the
constant c_0 in Eq. (2) so that the latter

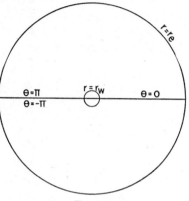

Fig. 5.3.

will satisfy the boundary conditions [Eqs. (3)], it is readily found that

$$c_0 = \frac{f_0 - x_0}{\log r_e/r_w}. \quad (6)$$

Moreover, it is clear from inspection of Eqs. (3) that the constants x_0
and f_0 are equal to the averages $\bar{p}_w$, $\bar{p}_e$ of the variable boundary pressures
$p_w(\theta)$ and $p_e(\theta)$, respectively. It follows, then, that

$$Q = \frac{2\pi k h (f_0 - x_0)}{\mu \log r_e/r_w} = \frac{2\pi k h (\bar{p}_e - \bar{p}_w)}{\mu \log r_e/r_w}. \quad (7)^*$$

It is thus seen that even though the flow is unsymmetrical between the
two coaxial cylindrical boundaries, its magnitude will be exactly the same
as if there were complete symmetry and the averages of the variable
boundary pressures were applied uniformly over their respective boundaries. By inverting Eq. (7) as

$$\bar{p}_e = \bar{p}_w + \frac{\mu Q \log r_e/r_w}{2\pi k h}, \quad (8)$$

[1] Cf., for example, H. S. Carslaw, "Fourier Series and Integrals," 3d ed., The Mac-
millan Company, 1930, or M. Muskat, "The Flow of Homogeneous Fluids through
Porous Media," Sec. 4.3, McGraw-Hill Book Company, Inc., 1937.

* Here again, in using Eq. (7) numerically to obtain the rate of production as measured at the surface, the right-hand side must be divided by the formation-volume
factor β.

it may be used to estimate the average pressure at a radius r_e about a well producing a liquid of viscosity μ, at a reservoir-volume rate Q, from a formation of thickness h and permeability k.

If the cause of the asymmetry in flow and pressure distributions is that the well and external boundary are not concentric, the rate of flow through the system will still be given by a radial-flow type of formula. In particular, if the center of the well bore is displaced by a distance d from the center of the external boundary, it may be shown, by using an analysis[1] based on the theory of "Green's function" or on the "method of images" that Q will be given by

$$Q = \frac{2\pi kh(p_e - p_w)}{\mu \log (r_e/r_w)[1 - (d^2/r_e^2)]}, \tag{9}$$

where the boundary pressures p_e, p_w have been assumed constant. It will be noted that the eccentricity between the well bore and external boundary is simply equivalent to decreasing the radius of the latter by the factor $1 - (d^2/r_e^2)$ or increasing the well-bore radius to $r_w/[1 - (d^2/r_e^2)]$. However, the resultant increase in flow capacity is quite small, and indeed negligible, unless the well is so close to the external boundary as to make the whole concept of radial flow entirely inapplicable. Thus for $r_e/r_w = 2,640$, if $d/r_e = 0.1$ the increase in flow capacity is only 0.13 per cent; and even if the well is halfway between the external boundary and the center of the latter, the increase is still less than 4 per cent. Accordingly this type of perturbation from exact radial-flow conditions may be entirely ignored under practical circumstances.

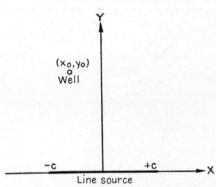

Fig. 5.4. Diagrammatic representation of a finite line source and single well system.

Finally it may be observed that a radial-flow type of formula will give the flow capacity of a well even if the external boundary is no longer circular but has a radically different geometry. For example, in the extreme case where the external boundary is merely a finite linear-fluid source, as indicated in Fig. 5.4, the solution of the basic equation 4.5(1) by an application of the theory of "conjugate functions" leads[2] to a value of Q given by

$$Q = \frac{2\pi kh(p_e - p_w)}{\mu \log \dfrac{4y_0}{r_w} \left[\dfrac{\sqrt{(c^2 - r_0^2)^2 + 4y_0^2 c^2}}{c^2 - r_0^2 + \sqrt{(c^2 - r_0^2)^2 + 4y_0^2 c^2}} \right]}, \tag{10}$$

[1] Cf. Muskat, *op. cit.*, Secs. 4.6 and 4.7.
[2] Cf. *ibid.*, Sec. 4.9.

where y_0 is the normal distance of the well from the line source, r_o is its distance to the center of the line source, and c is the half length of the latter. When the line source is of infinite length ($c = \infty$), Eq. (10) reduces to

$$Q = \frac{2\pi kh(p_e - p_w)}{\mu \log 2y_0/r_w},$$ (11)

thus showing the infinite line source to be equivalent to a cylindrical external source boundary of radius $2y_0$. Other special cases of Eq. (10) show similar types of equivalence.

The general implication of these considerations and examples is that regardless of details of boundary geometry and pressure distribution the homogeneous-fluid (liquid) production capacity of a well (measured at the surface) completely penetrating a uniform porous stratum will be given by the formula

$$Q = \frac{2\pi kh(\overline{p_e} - \overline{p_w})}{\mu\beta \log \overline{r_e}/r_w},$$ (12)

where $\overline{r_e}$ is any reasonable average of the distance of the external boundary from the well and $\overline{p_e}$, $\overline{p_w}$ are the averages of the actual pressure distributions over the external and well boundaries. The reason why $\overline{r_e}$ need not be known or specified accurately is that it enters in Eq. (12) logarithmically. Hence, as noted in Sec. 5.1, moderate uncertainties or errors in its value will have but a minor effect on the calculated production capacity.

5.3. Wells Just Penetrating Producing Strata; Spherical Flow.—The assumption of complete penetration of the well through the producing section, which underlay the above discussion, will evidently not always obtain in practice. It is often desirable and necessary to bottom a well without completely penetrating the formation. And at times the penetration may be so small, as compared with the pay thickness, that the flow into the well may be considered as essentially spherical, $i.e.$, as convergent toward a point, the well bottom.

A spherical-flow system must evidently have a vertical component. It is therefore necessary, in principle, to take into account the effect of gravity. Accordingly, the fluid motion will be described by the velocity potential

$$\Phi = \frac{k}{\mu}(p - \gamma g z),$$ (1)

where the vertical coordinate z is directed downward and the porous medium is taken as homogeneous and isotropic. Under conditions of complete spherical symmetry, Laplace's equation in spherical coordinates [Eq. 4.5(7)] will reduce to

$$\frac{1}{r^2}\frac{\partial}{\partial r}\left(r^2\frac{\partial\Phi}{\partial r}\right) = 0.$$ (2)

Its solution is easily verified to be

$$\Phi = \frac{c_1}{r} + c_2,$$ (3)

where c_1 and c_2 are constants of integration.

This is the general distribution function for the potential Φ in a spherical-flow system. Its characteristic feature is that Φ varies inversely with the radius r. This is evidently a more rapid variation than the logarithmic dependence on r characteristic of the two-dimensional radial-flow system [cf. Eq. 5.1(2)]. To determine the two constants c_1, c_2, Eq. (3) may be forced to conform to a specific case defined by the boundary conditions (cf. Fig. 5.5):

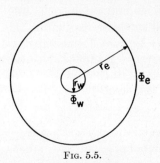

FIG. 5.5.

$$\Phi = \Phi_w \quad : \quad r = r_w, \\ \Phi = \Phi_e \quad : \quad r = r_e.$$ (4)

To satisfy these, Φ must have the explicit form

$$\Phi = \frac{\Phi_e - \Phi_w}{(1/r_e) - (1/r_w)}\left(\frac{1}{r} - \frac{1}{r_w}\right) + \Phi_w.$$ (5)

It is thus seen that Φ is proportional to the potential difference $\Phi_e - \Phi_w$ between the spherical boundaries at r_e, r_w.

The velocity in the system is, as usual obtained, by differentiation,

$$v_r = -\frac{\partial \Phi}{\partial r} = \frac{\Phi_e - \Phi_w}{(1/r_e) - (1/r_w)}\frac{1}{r^2},$$ (6)

and the total flow through the system is given by

$$Q = -\int_0^{2\pi} d\chi \int_0^{\pi} r^2 \sin \theta v_r \, d\theta = \frac{4\pi(\Phi_e - \Phi_w)}{(1/r_w) - (1/r_e)}.$$ (7)

Hence Φ and v_r can be rewritten as

$$\Phi = \frac{Q}{4\pi}\left(\frac{1}{r_w} - \frac{1}{r}\right) + \Phi_w,$$ (8)

$$v_r = -\frac{Q}{4\pi r^2}.$$ (9)*

It is seen from these that both Φ and v_r vary directly as the flux Q, as they do in the two-dimensional case [cf. Eqs. 5.1(8) and 5.1(9)], although their

---

* Just as in the case of radial flow, Eq. (9) could have been written down at once from the integrated form of the equation of continuity, namely, that $Q = -4\pi r^2 v_r = $ const. A single integration, after replacing v_r by $-\dfrac{\partial \Phi}{\partial r}$, will then give Eq. (8) at once.

variation with r is much more rapid here. This may be seen more clearly from the curves for Φ and v_r in Fig. 5.6 for the case where

$$p_w = 1, \quad p_e = 11 \text{ atm}, \quad \text{at } z = 0; \quad \text{and} \quad \frac{k}{\mu} = 1,$$

so that:*

$$\Phi_w = 1 \quad : \quad r_w = \frac{1}{4} \text{ ft}; \qquad \Phi_e = 11 \quad : \quad r_e = 660 \text{ ft}.$$

Hence

$$\Phi = 11.004 - \frac{2.501}{r}(r \text{ in ft}).$$

$$v_r\dagger = - \frac{0.002689}{r^2} \text{ ft/sec}; \qquad Q = 520.6 \text{ bbl/day}.$$

A comparison with the dashed curves, which refer to the radial-flow case for the same boundary conditions, will show at once that the pressure gradients and velocities in the case of spherical flow are localized near the small-radius boundary with a much greater concentration than the already highly concentrated pressure gradients and velocities in the radial-flow system. Another important difference lies in the value of Q. Thus for practical cases where $r_w \ll r_e$, Eq. (7) gives

$$Q \simeq 4\pi(\Phi_e - \Phi_w)r_w, \qquad (10)$$

with a rate of flow varying as the radius r_w, whereas in the case of radial flow Q varies in the much slower logarithmic manner [cf. Eq. 5.1(6)]. Further, Eq. (10) shows that the flux in the case of spherical flow is independent of the external-boundary radius as long as it is large compared with r_w. In the radial-flow case, it will be recalled, Q varies logarithmically with both the external and the well radius.

As previously indicated, the practical significance of the problem of spherical flow is that it corresponds to a well (of small radius) just tapping a relatively thick sand. This correspondence will appear as a special case of the analysis of the next section. Equation (10) itself shows Q to be

* It should be observed that the apparently indirect specification of the boundary conditions for Φ by a preliminary assignment of the values of the pressures is not a complication inherent in the use of the potential function Φ but arises rather from the superficial concept of the pressure as the quantity of fundamental physical significance even in three-dimensional systems. In reality, however, the potential function is of primary significance, when gravity is taken into account, although both the pressure and Φ satisfy Laplace's equation. For if in the above system the pressure had been specified as constant over $r = r_w$, r_e, its distribution would be of the same form as Eq. (5) and hence spherically symmetrical. The velocity distribution, however, would no longer be radial, and hence the system as a whole would really not be spherically symmetrical.

† Here, as in the case of Eq. 5.1(11), the units for v_r are actually (ft.^{3}/ft.^{2})/sec., r being expressed in feet.

independent of the radius of the external boundary and hence of its exact shape, provided that the external-boundary radius is large as compared with the well radius.[1] On the other hand, the value of Q is sensitive not

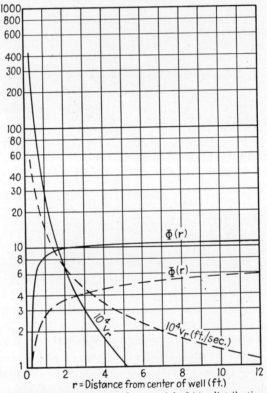

Fig. 5.6. The calculated velocity v_r, and potential, $\Phi(r)$, distributions in steady-state homogeneous-fluid spherical- (solid curves) and radial-flow (dashed curves) systems, for boundary values $\Phi(\frac{1}{4}$ ft$) = 1$; $\Phi(660$ ft$) = 11$.

only to the dimensions of the interior surface but also to its shape. Thus it may be shown that if the bottom of the well be considered as a flat disk, of radius r_w, rather than a hemisphere as assumed above, the flow capacity of the system will be cut exactly in half, i.e., to $\pi(\Phi_e - \Phi_w)r_w$.* Because

---

[1] The assumption of a uniform external-boundary potential Φ_e can also be dropped if Φ_e in Eqs. (7) and (10) be replaced by the average of the actual potential over the external boundary. This result can be derived in a manner entirely analogous to that developed in Sec. 5.2 for two-dimensional-flow systems, the Fourier series there being replaced here by the corresponding functions—the surface spherical harmonics—of the polar and azimuthal angles θ and χ. For details see W. E. Byerly, "Fourier's Series and Spherical Harmonics," Chap. VI., Ginn & Company, 1893.

* In the practical case of a hemispherical well bottom draining only the strata underlying its diametral plane the rate of flow will evidently be half of that given by Eq. (10)

of the highly convergent character of the flow near the well surface, merely the addition of the hemispherical cap of porous material, of radius r_w, in the fluid path will double the over-all flow resistance of the system.

Finally it is of interest to compare the production capacity of a well producing by spherical flow with one producing by radial flow, with the same potential drop $\Delta\Phi$. Upon denoting the production rate of the former by Q_s and that of the latter by Q_r, the above results show that

$$\frac{Q_s}{Q_r} = \frac{r_w}{h} \log \frac{r_e}{r_w}, \tag{11}$$

where h is the formation thickness in the case of radial flow. For the numerical example considered above, it follows that

$$\frac{Q_s}{Q_r} = \frac{1.97}{h}(h \text{ in ft}), \tag{12}$$

so that for a sand thickness of 50 ft the radial-flow system will produce at a rate 25 times as high as the spherical-flow system, when flowing under the same potential drop. In view of these radically different production capacities of radial-flow (completely penetrating wells) and spherical-flow systems ("nonpenetrating" wells) it is clear that the only conditions under which one should deliberately plan to produce from the latter type of flow system would be those where bottom waters underlie the oil zone and the difficulties of premature water break-through would ensue from high well penetrations (cf. Secs. 5.9 and 11.5).

5.4. Partially Penetrating Wells in Isotropic Formations; Potential Distributions.—To proceed now to the more practical and more frequently occurring condition of partial well penetration, the analytical problem becomes considerably more complex. However, it will be instructive to outline the analysis of this problem, although many of the details will be omitted for the sake of brevity.[1]

It will be supposed, as usual, that the well is symmetrically placed with respect to the surrounding pay, at the boundary of which the potential is maintained at a uniform value. The system will then be radially symmetrical, and the natural coordinates will be those of the cylindrical system. Specifically it will be supposed that a well of radius r_w penetrates an isotropic stratum of thickness h to a depth b. The formation is bounded externally[2] by the circle $r = r_e$ concentric with the well and is bounded at the top and bottom by impermeable strata (cf. Fig. 5.7).

---

[1] The treatment presented here and in the following section follows that given by M. Muskat, *Physics*, **2**, 329 (1932), for the analogous electrical problem of an electrode partially penetrating a large cylindrical conducting disk.

[2] Here too the "external" circular boundary is to be interpreted not as a physical limit to the extension of the stratum but rather as a geometrical surface over which the value of the pressure or potential may be considered as known.

Analytically this problem may be restated as that of finding a solution Φ to the system of equations [cf. Eq. 4.5(4)]

$$\frac{1}{r}\frac{\partial}{\partial r}\left(r\frac{\partial\Phi}{\partial r}\right)+\frac{\partial^2\Phi}{\partial z^2}=0, \tag{1}$$

$$\left.\begin{aligned}v_z=-\frac{\partial\Phi}{\partial z}=0 &\quad:\quad z=0,h,\\ \Phi=\text{const}=\Phi_w &\quad:\quad r=r_w,\quad z\leqslant b,\\ \Phi=\text{const}=\Phi_e &\quad:\quad r=r_e.\end{aligned}\right\} \tag{2}$$

The first of Eqs. (2) is equivalent to the requirement that no fluid pass across the upper and lower faces of the stratum, since it is bounded by

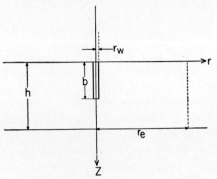

FIG. 5.7. A diagrammatic representation of a partially penetrating well.

impermeable formations. The second boundary condition means that the well surface is kept at a uniform potential, as it will be if the well bore is full of oil, at least to the top of the pay. The third condition repeats the assumption that the external boundary of the formation is at a uniform potential.

In carrying through the analysis it is convenient to introduce as a unit of length one that is equal to twice the formation thickness h. The variables may then be redefined in dimensionless terms as

$$\rho=\frac{r}{2h};\qquad w=\frac{z}{2h};\qquad x=\frac{b}{2h};\qquad \rho_w=\frac{r_w}{2h};\qquad \rho_e=\frac{r_e}{2h}. \tag{3}$$

Equation (1) is evidently unchanged by this transformation.

To find a solution of Eq. (1) satisfying the boundary conditions of Eqs. (2) one may visualize the well to be comprised of a continuous distribution of flux elements, each of which is provided with a suitable series of images (reflections) in the bounding planes such that the net flow across the latter cancels to zero. Now such a fluid-source element of strength $q\,d\alpha$, placed along the w (well) axis at the distance α from the top of the stratum, will be represented by a contribution to the resultant potential distribution given by

$$d\Phi = \frac{q\,d\alpha}{\sqrt{\rho^2 + (w - \alpha)^2}}, \tag{4}$$

which is readily verified to be a solution of Eq. (1).

A single term such as this, however, will not satisfy the condition of zero flow across the bounding planes $w = 0, \frac{1}{2}(z = 0,h)$. To cancel the flow implied by Eq. (4) across these boundaries, it is necessary to add a series of images of the real flux element on the well axis composed of an infinite array of equal strength but "virtual" flux elements placed on the well axis at distances $\pm n \pm \alpha$, where n goes from 0 to ∞ (cf. Fig. 5.8). The resultant potential due to this array is simply the sum of the potentials of the type of Eq. (4) due to the individual elements. Upon constructing this summation and expanding as a power series in ρ, for use in the immediate vicinity

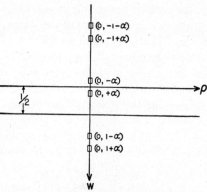

FIG. 5.8. A part of the image system for a flux element along the axis of a partially penetrating well, in a dimensionless polar coordinate system.

of the well, it is found that the direct effect of the flux element $q\,d\alpha$ at $(0,\alpha)$ and its images may be expressed as

$$d\Phi = q\,d\alpha \left\{ \frac{1}{[\rho^2 + (w - \alpha)^2]^{\frac{1}{2}}} + \frac{1}{[\rho^2 + (w + \alpha)^2]^{\frac{1}{2}}} - \Psi(1 - w - \alpha) \right.$$

$$- \Psi(1 - w + \alpha) - \Psi(1 + w + \alpha) - \Psi(1 + w - \alpha)$$

$$- \tfrac{1}{2}\rho^2[\zeta(3,1 - w - \alpha) + \zeta(3,1 + w + \alpha) + \zeta(3,1 - w + \alpha)$$

$$\left. + \zeta(3,1 + w - \alpha)] + 0(\rho^4) \right\} \tag{5}$$

where Ψ is a function defined in terms of the Γ function[1] by

$$\Psi(y) = \frac{\Gamma'(y)}{\Gamma(y)} = -0.5772 - \frac{1}{y} + \lim_{n\to\infty} \sum_{1}^{n} \left(\frac{1}{m} - \frac{1}{y + m} \right),$$

$$\zeta(s,y) = \sum_{0}^{\infty} \frac{1}{(n + y)^s}. \tag{6}$$

[1] Cf. E. T. Whittaker and G. N. Watson, "Modern Analysis," 4th ed., Chap. 12, Cambridge University Press, 1927.

For large values of ρ, that is, values of the order of 1, Eq. (5) converges very slowly. However, a direct solution of Laplace's equation for a system composed of the above set of images, namely,

$$d\Phi = 4q\,d\alpha\left[2\sum_1^\infty K_0(2n\pi\rho)\cos 2n\pi w\cos 2n\pi\alpha + \log\frac{2}{\rho}\right], \qquad (7)$$

where K_0 is the Hankel function[1] of order zero, is especially convenient numerically. Since this function decreases exponentially for large arguments, one or two terms of the series in Eq. (7) suffice for practical purposes and even for ρ as small as 0.5.

Now the series of Eqs. (5) and (7) are potential functions representing the effect of a single flux element $q\,d\alpha$ at $(0,\alpha)$ and its images. To get the potential due to a well of depth b, one must distribute such real flux elements over the whole length of the well and likewise their image systems along the well axis. By assuming that the real source strength q representing the well is uniform along its length and integrating Eqs. (5) and (7) between 0 and $\alpha = x = b/2h$, the results may be expressed as follows:
For small values of ρ,

$$\Phi = q\left\{-\log\frac{\Gamma(1+w+x)\,\Gamma(1-w+x)}{\Gamma(1-w-x)\,\Gamma(1+w-x)} + \log\frac{w+x+[\rho^2+(w+x)^2]^{\frac12}}{w-x+[\rho^2+(w-x)^2]^{\frac12}}\right.$$
$$-\tfrac14\rho^2[\zeta(2,1-w-x)-\zeta(2,1-w+x)+\zeta(2,1+w-x)$$
$$\left.-\zeta(2,1+w+x)]+0(\rho^4);\right\} \qquad (8)$$

For large values of ρ,

$$\Phi = 4q\left[\frac{1}{\pi}\sum_1^\infty\frac1n K_0(2n\pi\rho)\cos 2n\pi w\sin 2n\pi x + x\log\frac{2}{\rho}\right]. \qquad (9)$$

These solutions have been developed so as to give no flux across the formation faces. To see whether or not they are the final answer to the problem of Eqs. (1) and (2), one must test them to determine whether or not they satisfy the last boundary conditions of Eqs. (2). Considering, first, the requirement that Φ be constant at $r = r_e$, it is seen from Eq. (9) that though this is not satisfied exactly[2], the variation of Φ with w or z for $\rho > 1$ may be safely neglected for all practical purposes. This is due

[1] *Ibid.*, p. 373.

[2] The strict constancy of Φ at the external radius $\rho = \rho_e$ could be obtained by simply replacing the terms $K_0(2n\pi\rho)$ in Eqs. (7) and (9) by $K_0(2n\pi\rho) - K_0(2n\pi\rho_e)I_0(2n\pi\rho)/I_0(2n\pi\rho_e)$, where I_0 is the zero-order Bessel function of the third kind, which, in contrast to K_0, becomes exponentially large for large arguments and equals unity for zero argument (Whittaker and Watson, *loc. cit.*). However, the added term will be of insignificant numerical magnitude for cases of practical interest.

to the fact that, for $\rho > 1$, $K_0(2n\pi\rho)$ is so much smaller than $x \log 2/\rho$ (except for the special values $\rho \sim 2$) that the whole trigonometric series may be dropped.

On checking, however, whether or not Eqs. (8) and (9) satisfy the final boundary condition that Φ be uniform over the well surface, *i.e.*, over $\rho = \rho_w$, $w \leqslant x$, it is found that this condition is not satisfied. An analysis of the difficulty shows that the reason lies in the assumed uniformity of the flux density $q(\alpha)$ along the well axis. While it is quite difficult to de-

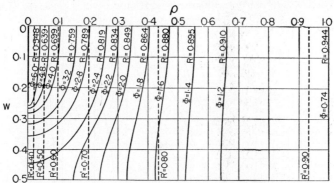

FIG. 5.9. The calculated steady-state homogeneous-fluid potential distribution about a well of radius = 1/500 of the formation thickness, and 50 per cent penetration. R = fraction of total potential drop (= 13.28). External = boundary radius = 4 × formation thickness. Unit of distance = 2 × formation thickness. Dashed curves and R' correspond to a strictly radial-flow (complete well penetration) system. (*From Physics, 1932.*)

rive directly a continuous flux distribution such that the well potential will be exactly constant,[1] a high degree of uniformity can be achieved by a suitable superposition of discrete flux segments of different strengths, increasing as the bottom of the well is approached. The increase evidently represents the additional flow entering the lower parts of the well bore coming from the sand not penetrated by the well.

An example of the resultant potential distribution so found is shown in Fig. 5.9 for the case of a 50 per cent penetrating well, of radius equal to 1/500 of the formation thickness, and the latter equal to $\frac{1}{4}$ the external boundary radius. The concentrated character of the potential drop about the well is indicated by the values of R, which give the fraction of the total drop across the pay. By comparison with the corresponding values for strict radial flow, indicated by the dotted equipotentials and fractions R', the markedly higher concentrations for the partially penetrating well distributions become evident. It is to be noted, however, that the equipotentials for the partially penetrating case rapidly change to a radial type,

[1] A formal exact solution can be expressed by means of an integral equation. The synthesis analysis used here is essentially equivalent to an approximation solution of the integral-equation formulation of the problem.

and can hardly be distinguished from those for a radial system, at a distance from the well equal to only twice the sand thickness. Obviously, for increasing penetrations this change to a radial character will take place even more rapidly.

For the other extreme, when the well just enters the top of the permeable stratum, the corresponding system of equipotentials is given in Fig. 5.10. It will be noted that in this case the equipotentials near the well are closely spherical, as one should expect. But here, again, on receding from the well

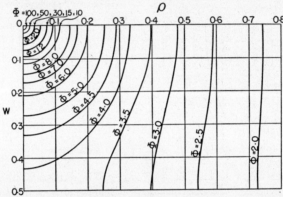

FIG. 5.10. The calculated steady-state homogeneous-fluid potential distribution about a "nonpenetrating" well. Unit of distance = 2 × formation thickness. (*From Physics, 1932.*)

the equipotentials flatten and assume a radial character, although, of course, not so rapidly as when the wells actually penetrate the pay for a considerable fraction of its thickness.

5.5. Partially Penetrating Wells in Isotropic Formations; Production Capacities.—Now that the nature of the potential distributions existing about the partially penetrating wells and the corresponding solutions of Eqs. 5.4(1) and 5.4(2) are established, the computation of the numerical values of the production rates to be expected from such systems can be readily made. For this purpose it is useful to derive a more exact interpretation of the flux density q appearing in the previous equations. This is found by simply computing the actual flux in a partly penetrating well with a uniform value of q. Thus, by using Eq. 5.4(9) and recalling the definitions of ρ and w of Eq. 5.4(3), it follows that

$$Q = -4\pi h \int_0^{1/2} \rho \frac{\partial \Phi}{\partial \rho} \, dw = 8\pi h q x = 4\pi q b, \qquad (1)$$

since the series terms in Eq. 5.4(9) vanish on integration. q is therefore seen to be $1/4\pi$ of the actual flow into[1] the well per unit of its length exposed

---

[1] The sign of q in Eq. 5.4(9) has been changed so as to make the well an outflow surface for the pay.

to the producing section. Hence in the type of solution for the potential distribution, previously indicated, in which the flux density is approximated by a superposition of separate flux elements of strengths q_m and extending to depths x_m, Q will be given by

$$Q = 8\pi h \Sigma q_m x_m, \tag{2}$$

since the potential will be given by a sum of terms as in Eq. 5.4(9).

Now at the external boundary, one may, to a very close approximation, drop the series in Eq. 5.4(9) and express the potential, with q taken as a positive quantity, as

$$\Phi_e = - 4qx \log \frac{2}{\rho_e}.$$

Hence, in the case of the superposed system of flux elements, Φ_e will become

$$\Phi_e = - 4\Sigma q_m x_m \log \frac{2}{\rho_e} = - \frac{Q}{2\pi h} \log \frac{2}{\rho_e}. \tag{3}$$

Upon denoting explicitly the potential over the well surface by Φ_w, the difference in Φ between the well and ρ_e will be

$$\Delta\Phi = \frac{Q}{2\pi h}\left(- \frac{\Phi_w}{4\Sigma q_m x_m} - \log \frac{2}{\rho_e}\right). \tag{4}$$

Finally, upon reverting to the original units of length and denoting the formation thickness by h, the value of the production rate in reservoir measure takes the form

$$Q = \frac{- 2\pi h \ \Delta\Phi}{[\Phi_w(r_w)/4\Sigma q_m x_m] + \log 4h/r_e}. \tag{5}$$

To use this formula one must know, in addition to the preassigned physical constants r_w, r_e, and h, the values of Φ_w and $\Sigma q_m x_m$. While direct evaluations for the q_m's have been made, it has been found that the following approximation will give values of Q accurate to $\frac{1}{2}$ per cent: One may assume the flux density over the well to be uniform and then take as the potential Φ_w an "effective average," which turns out to be that at three-fourths the distance from the top of the stratum to the bottom of the well.

To get the approximate formula, one may thus replace in Eq. (5) $\Sigma q_m x_m$ by qx, and Φ_w by the value of Φ in Eq. 5.4(8) at $w = \frac{3}{4}x$. On dropping terms of the order of ρ^2, it is found in this way that

$$Q = \frac{2\pi k h \ \Delta p/\mu\beta}{\dfrac{1}{2\bar{h}}\left[2 \log \dfrac{4h}{r_w} - \log \dfrac{\Gamma(0.875\bar{h})\,\Gamma(0.125\bar{h})}{\Gamma(1 - 0.875\bar{h})\,\Gamma(1 - 0.125\bar{h})}\right] - \log \dfrac{4h}{r_e}}, \tag{6}*$$

* J. Kozeny [*Wasserkraft und Wasserwirtschaft*, **28**, 101 (1933)] indicates that the resulting fluxes plotted in Fig. 5.11 can be represented by the still simpler formula

$$Q = \frac{2\pi k h \bar{h} \ \Delta p/\mu\beta}{\log r_e/r_w}\left(1 + 7\sqrt{\frac{r_w}{2h\bar{h}}} \cos \frac{\pi\bar{h}}{2}\right).$$

A direct test, however, shows that for sand thicknesses exceeding 75 ft this formula will give values for Q which are too low by 3 to 4 per cent.

where $\bar{h}$ is the well penetration expressed as a fraction of the stratum thickness h, k is its permeability, μ is the liquid viscosity, β its formation-volume factor, and Q is now expressed in surface measure.

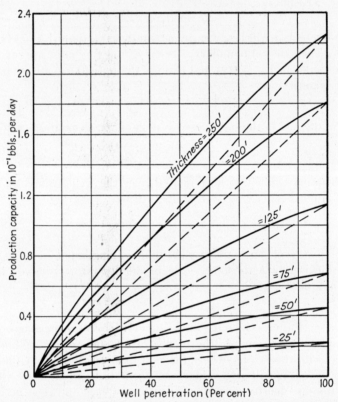

Fig. 5.11. The calculated steady-state homogeneous-fluid production capacities of partially penetrating wells as functions of the well penetration, for various thicknesses of the producing stratum. Straight dashed lines given the production capacities for strict radial flow into the exposed well section. $k \, \Delta p / \mu \beta$ is taken as unity, with k in millidarcys, Δp in psi, and μ in centipoises; well radius $= \frac{1}{4}$ ft; external-boundary radius $= 660$ ft.

In Fig. 5.11, Q, as calculated by the above methods, by using either Eq. (5) or Eq. (6), is plotted as a function of the percentage well penetration $(100\bar{h})$ for various sand thicknesses, r_e/r_w being taken as 2,640, and $k \, \Delta p / \mu \beta$ as unity[1] (the units of k, Δp, and μ being in millidarcys, psi, and centipoises, respectively).

The straight lines in Fig. 5.11 represent the production rates that would

---

[1] In a practical situation where k is given in millidarcys, μ in centipoises, and Δp in psi the values of barrels per day plotted in Fig. 5.11 and also in Figs. 5.12 to 5.14 should be multiplied by the numerical factor $k \, \Delta p / \mu \beta$.

be obtained if the flow into the partially penetrating wells were strictly radial. It is seen from these that as the penetration decreases the excess of the actual production capacity over that for strict radial flow continually increases until at penetrations of about 20 per cent the excess may even

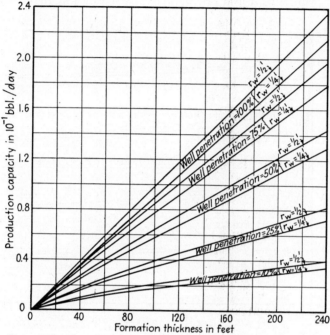

FIG. 5.12. The calculated steady-state homogeneous-fluid production capacities of partially penetrating wells, as functions of the formation thickness. r_w = well radius; external-boundary radius = 660 ft; remaining conditions and notation as in Fig. 5.11.

exceed 50 per cent of that for strict radial flow. The approximation of strict radial flow for low penetrations will therefore lead to large errors. Further, these results also show that one cannot consider the actual system as being equivalent to a simple superposition of a radial flow into the well proper and a semispherical flow into the well extremity, since the contribution of the latter is only of the order of 2 to 3 per cent of the radial flow.

In Fig. 5.12 the results are presented in a somewhat different manner. Here the production capacities are plotted against the thickness of the formation for various penetrations and two well radii. It is of interest to note that for formation thicknesses greater than about 60 ft the variation of the production capacity with thickness is almost exactly linear even for the partially penetrating wells.

The effect of adding formation thickness below the bottom of a well bore also can be readily derived by comparison of the data plotted in the several curves of Fig. 5.12. Thus it is found that beyond the first few feet of pay below the well bottom the additional layers of formation give successively decreasing contributions to the production capacity of the well. For example, for a 25-ft well penetration, increasing the pay thickness from 125 to 200 ft will increase the production capacity by less than 2 per cent,

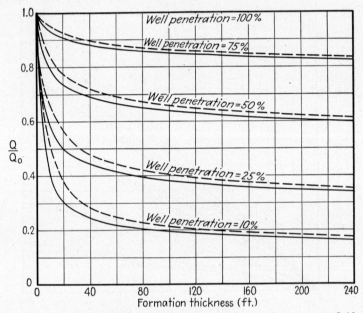

FIG. 5.13. The calculated variation of the relative steady-state homogeneous-fluid production capacities of partially penetrating wells with the formation thickness. Q/Q_o = (production capacity of partially penetrating well)/(production capacity of completely penetrating well). For solid curves, well radius = $\frac{1}{4}$ ft. For dashed curves, well radius = $\frac{1}{2}$ ft. External-boundary radius = 660 ft in all cases.

whereas the first 75 ft below the well bottom (from 25 to 100 ft) increases the production capacity by more than 50 per cent.

The relative production capacities, as compared with those for strict radial flow for the whole pay section, are plotted in Fig. 5.13 as functions of the formation thickness for two well radii.[1] It will be seen from these, as well as Fig. 5.12, that for the large penetrations the production capacities vary approximately logarithmically with the well radius, as they do exactly for the radial-flow case. They increase in their rate of variation, with decreasing penetration, until in the limit of the nonpenetrating well the production capacities vary directly as the well radius, as in the case of spherical flow [cf. Eq. 5.3(10)].

[1] The ratios Q/Q_o plotted in Fig. 5.13 give the values of the correction factor F referred to in Sec. 5.8.

5.6. Wells Producing through Casing Perforations.—An effect on the idealized well productivity, as predicted by the simple radial-flow formula [Eq. 5.1(10)], analogous to that caused by incomplete well penetration is that resulting from the use of perforated casing. The completion of wells by setting casing opposite the producing section and subsequently perforating the casing to permit fluid entry into the well bore is becoming standard practice in many oil-producing districts. While many practical arguments, such as the prevention of formation caving, the elimination of water production from stray water-bearing strata, and easier control of the sequence of depletion from multizoned formations, make such a completion procedure highly desirable, it is clear that its use will lead to greater resistance to fluid production than if the well be completed with an open hole. This increase may be computed as follows:[1]

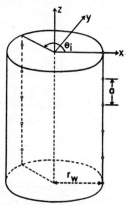

It will be assumed that the perforations are distributed along the casing in a spiral pattern formed of m vertical lines of perforations, of angular separation $2\pi/m$ and spacing a along the lines (cf. Fig. 5.14). Each perforation, of radius r_p, will be considered as a fluid sink, of strength q. Each vertical line of perforations will thus represent an infinite array of fluid sinks similar to those considered in Sec. 5.4 in the treatment of the partially penetrating well problem. Hence, by analogy with Eq. 5.4(9), the ith line will

FIG. 5.14. A diagrammatic representation of a spiral-perforation pattern.

contribute to the resultant pressure distribution in the system a term p_i, as

$$p_i = -\frac{4q}{a}\left[2\Sigma K_0(2n\pi\rho_i)\cos 2n\pi\left(w - \frac{i}{m}\right) + \log\frac{2}{\rho_i}\right], \tag{1}$$

where

$$\rho_i^2 = \frac{(x - x_i)^2 + (y - y_i)^2}{a^2}; \qquad w = \frac{z}{a};$$

$$r_w^2 = x_i^2 + y_i^2; \qquad \tan\theta_i = \frac{y_i}{x_i}; \qquad \rho_w = \frac{r_w}{a}, \left.\vphantom{\begin{array}{c}a\\b\end{array}}\right\} \tag{2}$$

and r_w is the casing or well radius.

The composite pressure distribution will be

$$p = \sum_0^{m-1} p_i. \tag{3}$$

At distant points, such as at the external-boundary radius r_e, the series term in Eq. (1) will become negligible. The value of p will then reduce to

$$p(r = r_e) = p_e = -\frac{4mq}{a}\log\frac{2}{\rho_e}; \qquad \rho_e = \frac{r_e}{a}. \tag{4}$$

[1] M. Muskat, *AIME Trans.*, **151**, 175 (1943).

The well pressure p_w, associated with the pressure distribution of Eqs. (1) and (3), is obtained by evaluating p at any one of the perforations, such as that along the x axis ($i = 0$). In doing so, it is to be noted that in p_0, ρ_0 will assume the value $\rho_p = r_p/a$. For the other p_i contributions the ρ_i's will be given by $2\rho_w \sin \theta_i/2$, where θ_i is the angle to the ith perforation line (cf. Fig. 5.14). It is so found that

$$p_w = -\frac{4q}{a}\left[2\Sigma K_0(2n\pi\rho_p) + 2\sum_i\sum_n K_0\left(4n\pi\rho_w \sin\frac{\theta_i}{2}\right)\cos\frac{2n\pi i}{m} + \log\frac{2}{\rho_p}\right.$$
$$\left. - (m-1)\log\rho_w - \sum_1^{m-1}\log\sin\frac{\theta_i}{2}\right]. \quad (5)$$

On subtracting Eq. (5) from Eq. (4) the over-all pressure differential Δp is readily obtained. In addition to the geometrical parameters m, a, ρ_e, ρ_p, and θ_i, it will involve the sink "strength" q. To eliminate the latter it is observed that the total flux through the system per unit thickness will be given by

$$Q = \frac{2\pi k}{\mu\beta}\left(r\frac{\partial p}{\partial r}\right)_{r_e} = \frac{8\pi mkq}{a\mu\beta}. \quad (6)$$

Upon replacing q by Q, according to this relation, in the equation for Δp, it is readily found that Q can be ultimately expressed as

$$Q = \frac{2\pi k\,\Delta p/\mu\beta}{C + \log r_e/r_w}, \quad (7)$$

where

$$mC = 2\Sigma K_0(2n\pi\rho_p) + 2\sum_i\sum_n K_0\left(4n\pi\rho_w \sin\frac{\theta_i}{2}\right)\cos\frac{2n\pi i}{m} + \log\frac{\rho_w}{m\rho_p}. \quad (8)$$

Thus it is seen that the effect of the perforations in increasing the flow resistance is expressed simply by the value of the constant C, as added to the term $\log r_e/r_w$. Or it may be considered as decreasing the effective well radius by the factor e^{-c}, that is, to the value $e^{-c}r_w$. Expressing the flux Q, given by Eq. (7), as a ratio to that for perfect radial flow Q_o, we have

$$\frac{Q}{Q_o} = \frac{\log r_e/r_w}{C + \log r_e/r_w}. \quad (9)$$

Before exhibiting the numerical implications of Eq. (9), it may be noted that formally the same equation applies if the perforations in the several lines lie in parallel planes normal to the axis of the well bore, rather than in a spiral pattern. For this case C is still given by Eq. (8), except that the factors $\cos 2n\pi i/m$ are all replaced by unity. However, this change will have but a negligible effect on the value of C, and hence on Q, unless the vertical perforation spacing a appreciably exceeds 12 in., for values of the other variables of practical interest. This, and direct computations of the ratio Q/Q_o for various numbers of perforation lines m and perforation spacings a, shows that the resultant values of the production capacity are essentially independent of the detailed pattern and distribution of the perforations but are largely determined by the over-all linear perforation density, *i.e.*, the total number of perforations per foot of casing. Moreover the effect of the perforation radius is such that to a close approximation the value of Q is a function only of the product of the perforation radius and density, *i.e.*, of $\rho_p{}^*$, so that 6 $\frac{1}{4}$-in.-radius perforations per foot will give the same production capacity as 12 $\frac{1}{8}$-in.-radius perforations per foot.

The values of Q/Q_o, calculated from Eqs. (8) and (9), are plotted in Fig. 5.15. For the reason just indicated the abscissa variable has been taken as the product of the perforation radius, in inches, and over-all density, in number of perforations per foot of casing.[1] On considering the former as fixed, the effect of the latter is given directly by the trend of the curves, of which the solid curve refers to a casing radius of 3 in. and the dashed curve to a casing radius of 6 in. It will be observed that whereas the relative production capacity increases linearly for low perforation densities, the rate of increase necessarily tapers off at higher densities. For example, for a perforation radius of $\frac{1}{4}$ in. and casing radius of 3 in., Q/Q_o will be 0.24 for a density of 1 perforation per foot, 0.41 for a density of 2, and only 0.59 for a density of 4 perforations per foot. Moreover, to achieve the same relative production capacities with $\frac{1}{8}$-in. perforations the corresponding densities will have to be doubled. It is thus seen that, while low perforation densities can materially restrict the flow capacity of a well, a point of diminishing returns will ultimately be reached as the density is continually increased. In particular the reduction in strength of the pipe as the density is increased may well counterbalance the small increases in production capacity thus gained. On the other hand, smaller radius perforations, with greater densities, may be advantageous in minimizing the loss in strength of the perforated pipe.

As indicated in Fig. 5.15, the effect of the casing size on the relative production capacity is minor. As is to be expected, the perforations represent a relatively greater part of the flow resistance of the composite sand

---

* These approximations break down, of course, in the limits of very small vertical perforation separations, where they simulate continuous slot systems (Sec. 5.7).

[1] The curves of Fig. 5.15 have been drawn through the average values of Q/Q_o for different combinations of m, a, and r_p, giving the same abscissa parameter. While these individual values fluctuate somewhat—to 2 per cent—with the exact combination of m, a, and r_p, the average curves, as drawn, should suffice for all practical purposes.

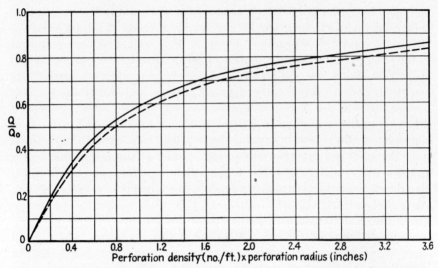

Fig. 5.15. The calculated effect of casing perforations on the steady-state homogeneous-fluid well productivity. Q/Q_o = (production capacity of cased and perforated well)/(production capacity of completely penetrating uncased well). For solid curve, casing radius = 3 in. For dashed curve, casing radius = 6 in.

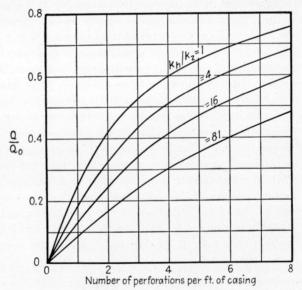

Fig. 5.16. The calculated effect of casing perforations on the steady-state homogeneous-fluid well productivity, as a function of the perforation density, in anisotropic formations. Q/Q_o = (production capacity of cased and perforated well)/(production capacity of completely penetrating uncased well). k_h/k_z = (horizontal permeability)/(vertical permeability); well radius = 3 in.; perforation radius = $\frac{1}{4}$ in. (From *AIME Trans.*, 1943.)

and casing system for wells with larger casing. The reduction in flow capacity due to the casing-perforation type of completion will therefore be somewhat greater in the larger well bores.

Finally it may be noted that Eqs. (7) and (8) formally remain valid if the producing section is anisotropic, provided that the dimensionless radii ρ_w, ρ_p are redefined as

$$\rho_w = \frac{r_w}{\alpha a}; \qquad \rho_p = \frac{r_p}{\alpha a}; \qquad \alpha = \sqrt{\frac{k_h}{k_z}}, \tag{10}$$

where k_h, k_z are the horizontal and vertical permeabilities. The corresponding constant-pressure perforation cavity is a spheroid of semiaxis r_p in the horizontal plane and r_p/α vertically. Upon taking the geometric mean of these as the radius of the equivalent circular perforation, the resistance constant C takes the form

$$mC = 2\Sigma K_0\left(\frac{2n\pi r_p}{a\sqrt{\alpha}}\right) + 2\Sigma\Sigma K_0\left(\frac{4n\pi r_w}{a\alpha}\sin\frac{\theta_i}{2}\right) + \log\frac{r_w}{mr_p}. \tag{11}$$

By using this equation for C and taking $m = 2$, Q/Q_o is affected by the permeability ratio as shown in Fig. 5.16. It will be seen that the effect of the anisotropy is not large unless the vertical permeability is but a small fraction of the horizontal permeability.

5.7. Wells Completed with Slotted Liners.—By considering the fluid sinks, representing the perforations discussed in the last section, to be distributed continuously along the surface of the casing the type of analysis outlined above will give the effect of completing a well with a slotted liner. To carry through this method one need only integrate Eq. 5.6(1), expressed in differential form, over the preassigned slot position and length to obtain the pressure-distribution contribution of the individual vertical lines of slots. Treating these pressure-distribution terms as before, one finally obtains a value for the production capacity identical with Eq. 5.6(7), except that in the equation for C [Eq. 5.6(8)] the Hankel functions are multiplied by the factors $\sin 2n\pi h/2n\pi h$, where h is half the slot length and the ρ_p's refer to their half width, divided by their vertical spacing.

A still simpler form for C that has been derived[1] for the slotted liner system is

$$C = \frac{2}{m}\log\frac{2}{\pi\Omega}, \tag{1}$$

where m is the number of slot columns and Ω is the open fraction of the pipe. This formula was developed on the assumption that the slots extend continuously along the whole casing ($a = 0$), so as to permit a two-dimensional analysis with the aid of conjugate-function transformations. How-

---

[1] C. R. Dodson and W. T. Cardwell, Jr., *AIME Trans.*, **160**, 56 (1945).

ever, it was found by both electrical-model experiments and a more rigorous analysis, similar to that mentioned above, that Eq. (1) gives a good approximation even when the individual slots are separated vertically by the casing wall, provided that $\Omega \leqslant 0.3$.*

The relative production capacities of wells completed with slotted liners, as given by Eq. 5.6(9), and with values of C computed by Eq. (1), are plotted in Fig. 5.17 for $r_e/r_w = 2{,}640$. As in the case of perforated pipe,

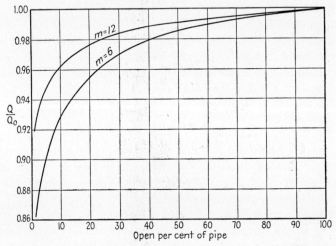

FIG. 5.17. The calculated effect of slotted liners on the steady-state homogeneous-fluid production capacity. $Q/Q_o =$ (production capacity of well with slotted liner)/(production capacity of completely penetrating open-hole well); $m =$ number of slot columns. (External-boundary radius)/(well radius) $= 2{,}640$.

for a given open area, the production capacity will be greater for that pattern having the larger *number* of openings. It will also be seen that the effects of the slotted liners are so small as to be quite negligible from a practical point of view, except when the open area is less than 5 per cent of the total liner surface. Hence, in choosing slot sizes and distributions for an actual installation, factors such as mechanical strength, sand retention, and clogging should be given primary consideration.

5.8. Flow Capacities of Wells Producing by Gravity; Free Surfaces.—If the productive capacity of a well should be so low that the pumping equipment can maintain the fluid level below the top of the producing section, the effect of gravity may become of importance even if the well completely penetrates the section. Such a situation may result entirely from the tightness of the pay and arise before the reservoir pressure has suffered

* Eq. (1) must obviously break down completely for $\Omega > 2/\pi$, as C must be positive. However, as the Q/Q_o for $\Omega = 0.3$ are already so close to 1, a smooth extrapolation for greater values (using a logarithmic scale) should be quite accurate.

appreciable depletion.   More commonly, however, it appears when the production has declined because of reservoir pressure depletion, and gravity is the remaining major source of driving energy for bringing the oil into the well bore.

For the latter case the fluid distribution in the system may be visualized by reference to Fig. 5.18.   Here h_w represents the fluid level in the well bore,

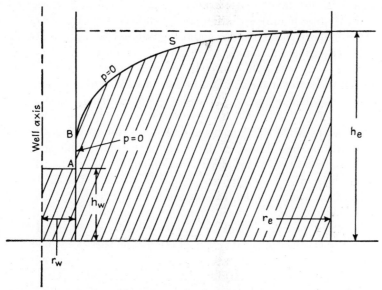

FIG. 5.18.   A diagrammatic representation of a well producing by gravity flow.

above the bottom of the pay, maintained by the pumping equipment.   On the assumption that the original reservoir pressure has been largely dissipated the fluid head at the external-boundary radius may be taken as equivalent to a value h_e, which does not exceed the pay thickness.[1]   The driving head acting on the system is thus $h_e - h_w$.

In proceeding further, however, the unique characteristic of all systems that are largely controlled by the action of gravity is immediately encountered.   This is the presence of a "free surface" connecting the inflow and outflow boundaries and providing the upper boundary of the flow system. This feature constitutes the basic difficulty in the analysis of such gravity-flow systems.   For in principle the exact shape of this boundary surface is initially unknown, and indeed it must be determined simultaneously with

---

[1] For purposes of clarity the capillary layer, which will always be spread over an unconfined free surface, has been omitted from Fig. 5.18.   Moreover it will generally be of minor significance in determining the production capacities of practical gravity-flow systems.

the potential distribution in the interior of the flow system and the rate of flow through it. In contrast to this situation it will be recalled that in the flow problems thus far treated the geometry of all the boundaries of the fluid-saturated porous medium were preassigned and fixed. The analysis was directed toward developing such potential distributions in the interior of the medium as would satisfy the preassigned conditions on the boundaries. And from these distributions the rate of fluid flow was readily obtained by suitable differentiation and integration. Here, however, only the bottom impermeable boundary and the internal- and external-cylindrical-boundary surfaces are subject to choice in advance. As the fluid level in the well bore is below the top of the original porous stratum, it is clear that not all the latter will carry and be saturated with the liquid. Rather, the liquid-saturated region will be delimited by a surface such as is indicated by S in Fig. 5.18, that is, the free surface. While this surface cannot be chosen arbitrarily, there is no a priori rule for locating its position independently of the solution of the whole of the flow problem.

From a mathematical point of view the difficulty created by the appearance of the free surface is apparent rather than real. The lack of initial definition of the shape of the free surface is compensated by imposing on it the twofold requirement that it must be both a streamline surface and one of constant pressure. The first is merely an expression of the fact that it represents the upper boundary of the mobile part of the flow system. The second takes note of the necessary existence of pressure equilibrium along the free-surface boundary and with the atmosphere above the fluid level in the well bore. The requirement that the upper boundary must simultaneously satisfy both conditions in effect serves to localize its position to a unique surface, namely, the actual free surface for the system in question.

Unfortunately, however, while the problem is really tractable in principle, no rigorous explicit solutions for three-dimensional systems have thus far been derived. A number of two-dimensional gravity-flow systems, with free surfaces, such as the seepage of water through dams of length great compared with their thickness, have been analyzed in detail.[1] But even these generally involve either specialized and ingenious analytical manipulations or rather laborious numerical computations. For three-dimensional problems, such as are represented by Fig. 5.18, one is forced to resort to approximation procedures.

Before considering the details of an approximation theory another significant feature of most gravity-flow systems is to be noted. This is the "surface of seepage" extending along the well bore from A to B in Fig. 5.18.

[1] Cf. M. Muskat, "The Flow of Homogeneous Fluids through Porous Media," Chap. VI, McGraw-Hill Book Company, Inc., 1937.

The surface of seepage is the part of the boundary of the porous medium where the liquid leaving the system enters a region free of both liquid and porous medium. As in the case of the free surfaces the pressure is constant along these surface elements. However, they do not represent streamline surfaces. They span and connect the top of the fluid level at the outflow boundary, the well bore, and the intersection with the latter of the free surface. We shall not enter here into the detailed reasons necessitating the existence of the surface of seepage. However, it may be observed that the free surface will obviously terminate above the top of the fluid level in the well bore for the special case where that fluid level is zero. From considerations of continuity it may be anticipated that such discontinuities and intervening surfaces of seepage will persist even when the fluid level in the well bore is made nonvanishing. On the other hand, it should be emphasized that the upper limits of the surfaces of seepage will not be known a priori, as they are the terminal intersections with the well bore of the free surfaces, which are themselves of unknown shape initially, as previously discussed.

To proceed now to an approximate treatment of the flow problem represented by Fig. 5.18, the analysis may be carried through as follows:[1] On the external and internal fixed boundaries and on the basal plane the boundary conditions are those which would be required in a rigorous treatment. These are (1) a constant potential over the external boundary $(r = r_e)$; (2) zero flow over the basal plane $(z = 0)$; and (3) a constant potential along the internal (well) boundary, at $r = r_w$, up to the fluid level in the well, and a constant pressure above that level. However, the presence of the free surface is ignored, and the upper boundary is taken as a plane parallel to the basal plane and at a height equal to the fluid level at the external boundary, assuming that the latter does not exceed the sand thickness. While this approximation system implies fluid flow above the free surface in the actual gravity-flow system, the magnitude of this additional contribution turns out to be negligible.

Expressed analytically, these boundary conditions are

$$
\left.
\begin{aligned}
r = r_e \quad &: \quad \Phi = \bar{k} h_e, \\[6pt]
z = 0, h_e \quad &: \quad \frac{\partial \Phi}{\partial z} = 0, \\[6pt]
r = r_w \quad &: \quad \bar{\Phi} = \bar{k} h_w; \quad 0 \leqslant z \leqslant h_w \\[4pt]
&\quad\;\; = \bar{k} z; \quad\;\; h_w \leqslant z \leqslant h_e,
\end{aligned}
\right\}
\tag{1}
$$

where $\bar{k} = k \gamma g / \mu$, k being the permeability of the stratum, γ the liquid density, μ its viscosity, g the acceleration of gravity, h_e the fluid head or

---

[1] Cf. M. Muskat, *AGU Trans.*, 1936, p. 391.

thickness of fluid-saturated section at the external boundary, and h_w the fluid head in the well bore. The potential function Φ is defined by

$$\Phi = \frac{k}{\mu}\,(p + \gamma g z), \tag{2}$$

and must satisfy Eq. 5.4(1). Its solution, satisfying the conditions of Eq. (1), is readily verified to be

$$\Phi = \bar{k}h_e + \frac{\bar{k}(h_e^2 - h_w^2)\log\dfrac{r}{r_w}}{2h_e\log\dfrac{r_e}{r_w}}$$
$$+ \frac{2\bar{k}h_e}{\pi^2}\sum_1^\infty \frac{\left[(-1)^n - \cos\dfrac{n\pi h_w}{h_e}\right]U(\alpha_n r)\cos\dfrac{n\pi z}{h_e}}{n^2 U(\alpha_n r_w)}, \tag{3}$$

where

$$U(\alpha_n r) = I_0\!\left(\frac{n\pi r}{h_e}\right)K_0\!\left(\frac{n\pi r_e}{h_e}\right) - I_0\!\left(\frac{n\pi r_e}{h_e}\right)K_0\!\left(\frac{n\pi r}{h_e}\right), \tag{4}$$

I_0, K_0 being zero-order Bessel functions of the third kind.

The rate of production of the well will be

$$Q = 2\pi r\int_0^{h_e}\frac{\partial\Phi}{\partial r}\,dz = \frac{\pi k\gamma g(h_e^2 - h_w^2)}{\mu\log r_e/r_w}. \tag{5}$$

While this simple formula has been derived without reference to the existence of the free surface, direct sand-model experiments[1] have shown it to predict the empirical results within experimental errors, after applying corrections for the effect of the additional flow in the capillary layer overlying the fluid-saturated zone. Moreover, when a similar approximation treatment is applied to two-dimensional gravity-flow systems, the corresponding flow-rate formula agrees almost exactly with that computed by a rigorous analysis.[2] On the other hand, it may be noted that Eq. (5) can also be derived by a generalization of the theorem, demonstrated in

[1] R. D. Wyckoff, H. G. Botset, and M. Muskat, *Physics*, **3**, 90 (1932). In a more recent study H. E. Babbitt and D. H. Caldwell (*Univ. Illinois Eng. Expt. Sta. Bull. Series* 374, Jan. 7, 1948) have again confirmed the validity of Eq. (5) by both sand and electrical-model experiments. They have also determined by both methods the shapes of the free surfaces and extents of the surfaces of seepage in the radial-flow systems.

[2] M. Muskat, *Physics,* **6**, 402 (1935). It should be noted that Eq. (5) was derived as long ago as 1863 by J. Dupuit by making the assumption that the radial-velocity component is proportional to the slope of the free surface. While Dupuit's work was subsequently extended by Forchheimer ("Hydraulik," 3d ed., B. G. Teubner, 1930) and has been rather widely used by hydrologists, the derivation of the production capacity [Eq. (5)] as given above is felt to be more satisfactory than that of Dupuit-Forchheimer (cf. M. Muskat, "The Flow of Homogeneous Fluids through Porous Media," Chap. VI, McGraw-Hill Book Company, Inc., 1937).

Sec. 5.2, giving the production capacity of a well in terms of the average pressures over the boundaries. The latter, expressed by Eq. 5.2(7), referred to strictly two-dimensional radial flow. Its analogue for the present type of three-dimensional-flow system may be written

$$Q = \frac{2\pi h(\overline{\Phi_e} - \overline{\Phi_w})}{\log r_e/r_w},$$ (6)

where h is the gross thickness of the fluid-saturated zone and $\overline{\Phi_e}$, $\overline{\Phi_w}$ are the averages of the potential over the external and internal cylindrical boundaries of radii r_e, r_w. For the present problem, h evidently equals the fluid head h_e at r_e. $\overline{\Phi_e}$ is $k\gamma g h_e/\mu$. And $\overline{\Phi_w}$ is readily found from Eq. (1) to be $k\gamma g(h_e^2 + h_w^2)/2\mu h_e$. On inserting these values into Eq. (6), Eq. (5) is immediately obtained. If k is expressed in millidarcys, γ in grams per cubic centimeter or as a specific gravity, μ in centipoises, and h_e, h_w in feet, Eq. (5) may be rewritten as

$$Q = \frac{6.667 \times 10^{-4} k\gamma(h_e^2 - h_w^2)}{\mu \log_{10} r_e/r_w} \text{ bbl/day.}$$ (7)

Equation (6) also permits a simple derivation of the appropriate formula for the production capacity when a pressure head is superposed on the gravity head. The fluid head at the external bounday, h_e, is then greater than the pay thickness h. $\overline{\Phi_e}$ is here again $k\gamma g h_e/\mu$. And $\overline{\Phi_w}$ will be $k\gamma g(h^2 + h_w^2)/2\mu h$, where h_w denotes here, too, the fluid level in the well bore and is less than h. On applying these expressions to Eq. (6), it is found that

$$Q = \frac{\pi k\gamma g(2hh_e - h^2 - h_w^2)}{\mu \log r_e/r_w}.$$ (8)

When Q, k, γ, μ, and h, h_e, h_w are expressed in the units used in Eq. (7), the numerical coefficient will be the same as in Eq. (7).

Another useful application of this method may be made to the gravity flow into partially penetrating wells. Here again, regardless of the details of the potential distribution over the well and external-boundary surface, the production capacity of a partially penetrating well in a sand of total thickness h may be expressed as

$$Q = \frac{2\pi h F \,\overline{\Delta\Phi}}{\log r_e/r_w},$$ (9)

where F is the geometrical factor taking into account the partial penetration of the well, as derived in Sec. 5.5 and expressed by the curves of Fig. 5.13, and $\overline{\Delta\Phi}$ is the average potential difference between the well and external-boundary surfaces. Now if h_w is the fluid head in the well and the zero of the potential is taken at the bottom of the well, the average potential over the well surface will be $\overline{\Phi_w} = k\gamma g(b^2 + h_w^2)/2b$, where b is the actual depth of penetration. The uniform and average potential over the external boundary is evidently $\overline{\Phi_e} = k\gamma gb/\mu$. Upon evaluating $\Delta\overline{\Phi}$ and inserting it into Eq. (9), Q becomes

$$Q = \frac{\pi k\gamma gF(b^2 - h_w^2)}{\mu \log r_e/r_w}. \tag{10}$$

Comparison of this expression with Eq. (5) shows that the effect of partial penetration in a well flowing by a gravity drive can be taken care of by the same correction factor (Fig. 5.13) as applies to normal pressure-drive wells. And in a similar manner it may be shown that Eq. (8), combined with the same correction factors F, will give the production capacities of partially penetrating wells producing under the composite action of gravity and pressure drives.

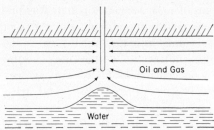

Oil and Gas

Water

Fig. 5.19. A diagrammatic representation (not to scale) of the elevated water table under a partially penetrating well producing oil and gas.

5.9. Water Coning.—"Water coning" is a term given to the mechanism underlying the entry of bottom waters into oil-producing wells. Such entry, when considered as true coning, is generally associated with high production rates. It is basically a rate-sensitive phenomenon and develops only after certain equilibrium conditions, to be discussed below, are unbalanced by increasing the pressure differential over the producing section beyond critical limits.

In its broadest sense the general water-coning phenomenon constitutes one of the most complex problems pertaining to oil production. To develop a quantitative treatment, however, it is necessary to limit the problem to the rather idealized situation where the bottom waters themselves do not directly participate in the production mechanism. The latter will be basically of the gas-drive type,[1] providing an essentially radial flow into the well except for the local distortions arising from the incomplete well penetration, as indicated in Fig. 5.19.

The water, being of greater density than the oil, will, under static conditions, remain at the bottom of the producing section.[2] Its rise, as shown in

---

[1] Cf. Chap 9 for a general discussion of oil-production mechanisms.

[2] The initial water-oil transition zone (cf. Sec. 7.9) as well as that rising with the water table will be neglected here, and the water-oil boundary will be considered as a sharp geometrical surface.

Fig. 5.19, therefore represents a dynamic effect in which the upward-directed pressure gradients associated with the oil flow are able to balance the hydrostatic head of the elevated water column. When and if equilibrium is established at any fixed production rate from the well, or producing bottom-hole pressure, with the water lying statically as the lower boundary of the oil zone, the pressures at the water-oil interface, $p(r,z)$, will satisfy the equation

$$p(r,z) + \gamma_w g(h - z) = p(r,z) + \gamma_w gy = P_b, \tag{1}$$

in the notation of Fig. 5.20. P_b is the formation or reservoir pressure as measured at the bottom of the oil zone at a point remote from the well. γ_w is the density of the water.

Equation (1) represents only a necessary condition for static equilibrium. To maintain *dynamic* equilibrium the water-oil interface must also coincide with a limiting streamline surface in the oil zone. Since the detailed structure of the pressure-distribution function $p(r,z)$ will depend on the shape and location of the water-oil interface, the requirement that the latter be a streamline surface, together with Eq. (1), should suffice in principle to determine the form of the interface.[1] From a practical point of view, however, this is not feasible. To the extent that the movement of oil through the oil zone is that of heterogeneous-fluid flow, the derivation of the pressure-distribution function $p(r,z)$ is as yet an unsolved problem for three-dimensional systems, even if the shape of the interface or bounding surface be considered as known.[2] And if the heterogeneous character of the oil and gas flow in the oil zone be ignored, the treatment of the three-dimensional potential-flow problem is still virtually impossible analytically,[3] from a practical standpoint, unless all the boundaries are both fixed and of reasonably simple geometry. Accordingly the dis-

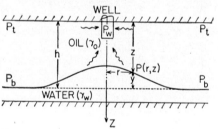

FIG. 5.20. The coordinate and pressure scale systems used in constructing the equilibrium equation for a water cone.

---

[1] This is analogous to the problem of determining the free surfaces in gravity-flow systems (cf. Sec. 5.8). In fact, the oil-water interface in a water-coning system will actually be a type of inverted free surface, although the flow conditions in the oil zone will not be strictly of the potential type.

[2] The solution of heterogeneous-fluid-flow problems will be considered in Chap. 8.

[3] It would, of course, be possible to develop trial-and-error solutions by such numerical procedures as the Southwell relaxation-oscillations method. However, in view of the basic approximation of potential flow, such extended numerical effort is hardly warranted.

cussion to be given here will be based[1] on two simplifying assumptions, namely, that the flow in the oil zone is governed by Laplace's equation as if it were a homogeneous-incompressible-fluid system and that the pressure function $p(r,z)$ in Eq. (1) is that representing the pressure distribution in the corresponding oil zone without any water cone. The latter assumption also implies that the water-cone surface, the oil-water interface, as determined by the application of Eq. (1), will not be an exact streamline surface. Because of these assumptions the quantitative results of the following treatment will have no absolute significance. The relative magnitudes, however, when the conditions are varied should be less seriously affected by these assumptions.

If the water cone is to lie statically below the oil zone, it is clear that at the top of the cone the vertical pressure gradient in the oil zone must balance the differential-gravity gradient in the water, $(\gamma_w - \gamma_o)g$, where γ_o is the density of the reservoir oil. As the flow of oil into the bottom of the well is highly convergent, the pressure gradient will increase rapidly on approaching the well bore (cf. Fig. 5.9). There will thus be an upper limit to the height of rise of the cone at which it can remain in static equilibrium. Above this height the upward pressure gradient in the oil zone will exceed the gravity gradient, and the water will be pulled into the well. Of course, if the total pressure drop across the oil zone is less than the total differential head of the column extending from the bottom of the well to the original oil-water contact the cone will not of itself rise into the unstable region. On the other hand, it may be shown[2] that if the pressure differential is continually increased by lowering the well pressure, the cone will rise to a critical height below the bottom of the well and then suddenly break into the well. The production rate corresponding to this critical pressure differential will thus be the maximum at which water-free oil production can be obtained.

To apply Eq. (1) to the determination of these critical cone heights, pressure differentials, and production rates, as well as the general shapes of the water-oil interfaces, it is convenient to introduce the following transformation: Denoting the potential function and total pressure drop over the system, as measured at the top of the oil zone, by

$$\left. \begin{aligned} \phi &= \frac{k}{\mu}\,(p - \gamma_o g z), \\ \Delta P &= P_t - p_w = P_b - \gamma_o g h - p_w, \end{aligned} \right\} \tag{2}$$

---

[1] This treatment follows that of M. Muskat and R. D. Wyckoff, *AIME Trans.*, **114**, 144 (1935).

[2] Cf. M. Muskat, "The Flow of Homogeneous Fluids through Porous Media," Sec. 8.10, McGraw-Hill Book Company, Inc., 1937.

the potential drop $\Delta\phi$ between any point (r,z) on the surface of the cone and the well will be

$$\Delta\phi = \phi(r,z) - \phi_w(z = 0) = \frac{k}{\mu}[\Delta P - g(h - z)\Delta\gamma], \tag{3}$$

where $\Delta\gamma = \gamma_w - \gamma_o$. On noting that, at large r, $\Delta\phi = (k/\mu)\Delta P \equiv (\Delta\phi)_e$, Eq. (4) can be rewritten as

$$\frac{\Delta\phi(r,z)}{(\Delta\phi)_e} = 1 - \frac{gh\,\Delta\gamma}{\Delta P}\left(1 - \frac{z}{h}\right). \tag{4}$$

Equation (4) is the basic equation determining the shape of the water-oil interface $z = z(r)$. As previously indicated, the potential function

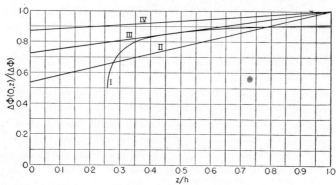

FIG. 5.21. An illustration of the graphical solution of Eq. 5.9(4) to obtain the equilibrium heights of a water cone. (*After Muskat and Wyckoff, AIME Trans., 1935.*)

$\phi(r,z)$, and hence $\Delta\phi(r,z)$, will be assumed to be that corresponding to the flow of an incompressible liquid into partially penetrating wells in the absence of the water cone. These functions are those developed and discussed in Sec. 5.4. If they are considered as known, Eq. (4) can readily be solved graphically by finding the intersections of the curves for the left-hand side, for fixed values of r, as functions of z or z/h, with the straight lines representing the right-hand side. The slopes of the latter, for fixed $gh\,\Delta\gamma$, will be inversely proportional to the pressure differential ΔP imposed on the particular system of interest, as shown in Fig. 5.21, where curve I gives the value of $\Delta\phi/(\Delta\phi)_e$ along the axis of a 25 per cent penetration well in a 125-ft formation, with an external radius of 500 ft.

It will be seen from Fig. 5.21 that the straight line, representing the right-hand side of Eq. (4), of highest slope (line II) gives two intersections with curve I. Since its slope exceeds that of curve I at the higher intersection z/h, the latter represents a stable cone height, along the well axis, for the corresponding value of ΔP. Conversely the intersection at the lower z/h (higher cone elevation) is unstable and will not occur in practice.[1] It will also be clear that the point of tangency, at $z/h = 0.48$, with line III gives the maximum and critical cone height. The slope of line III determines the maximum pressure differential permitting a stable cone, and the corresponding production rate will be the maximum for water-free production. For line IV, corresponding to still higher ΔP, there is no intersection with curve I and hence no stable cone height.

By using this type of graphical solution of Eq. (4) the cone height vs. ΔP variations so obtained are plotted in Figs. 5.22 and 5.23 for three well penetrations. The numerical values in these figures were derived on the assumptions that the formation-thickness = 125 ft, $\Delta\gamma = 0.3$ gm/cc, the well radius = $\frac{1}{4}$ ft, and the external-boundary radius = 500 ft. The calculated shapes of the cones for several pressure differentials below the critical for the 50 per cent well penetration are plotted in Fig. 5.24. While the effect of the cone on the potential distribution has been neglected in these computations, the general characteristics of Figs. 5.22 to 5.24 have been confirmed by an electrical-model study[2] in which the shape of the cone was determined simultaneously with its perturbation on the potential distribution.

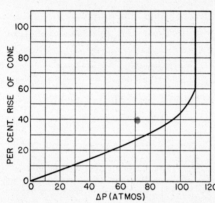

FIG. 5.22. The calculated variation of the water-cone height below a "nonpenetrating" oil well with the pressure drop ΔP across the producing formation. The vertical segment represents unstable cone heights. Assumed oil-zone thickness = 125 ft; well radius = $\frac{1}{4}$ ft; formation-boundary radius = 500 ft; water-oil density contrast = 0.3 gm/cc. (*After Muskat and Wyckoff, AIME Trans., 1935.*)

The critical, or maximal, pressure differentials without cone break-through, as calculated by the above-outlined procedure, are plotted in Fig. 5.25 as a function of the well penetration for various values of oil-zone thickness. The corresponding maximal production rates, as obtained by the application of Figs. 5.11 and 5.12 to Fig. 5.25, are plotted in Fig. 5.26.

It will be seen from Fig. 5.25 that the critical pressure differentials increase sharply as the well bottom is first removed from the original oil-water table. The rise becomes rapid again as the region of low well penetrations is approached, especially for the larger oil-zone thicknesses. For the latter

---

[1] On the assumption that the functions $\Delta\phi$ have significance, these solutions of Eq. (4) would represent positions of unstable equilibrium for cones dropping down from higher elevations after a reduction in the pressure differential.

[2] Cf. Muskat and Wyckoff, *loc. cit.*

conditions the absolute values of ΔP evidently are much larger than the
total differential fluid heads of the formation thickness.   This arises from
the fact that at low well penetrations a very large part of the total pressure

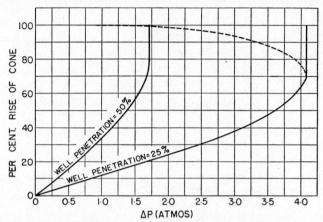

FIG. 5.23.   The calculated variation of water-cone elevations below oil wells of 25 and
50 per cent penetration with the pressure drop ΔP across the producing formation.   One
hundred per cent rise of cone represents the total distance between the bottom of the well
and oil zone.   The vertical segments represent unstable cone heights.   The dashed segment
corresponds to the lower intersections of the curves of Fig. 5.21.   Assumed data are the same
as in Fig. 5.22.   (*After Muskat and Wyckoff, AIME Trans., 1935.*)

drop is concentrated immediately below the bottom of the well and only
a small fraction remains for balancing the differential head of the water
cone.

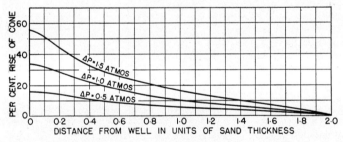

FIG. 5.24.   The calculated cross-sectional shapes of stable water cones below an oil well
of 50 per cent penetration for various pressure drops ΔP across the producing formation.
Assumed data are the same as in Fig. 5.22.   (*After Muskat and Wyckoff, AIME Trans., 1935.*)

Figure 5.26 shows that while the maximum water-free oil-production
rates increase continuously as the well penetration decreases, there is a
marked flattening of the curves at low penetrations.   This represents the
compensating effect of the increased flow resistance as the well penetration

is decreased against the increasing critical pressure differentials indicated
by Fig. 5.25.   The absolute values of the production rates shown in

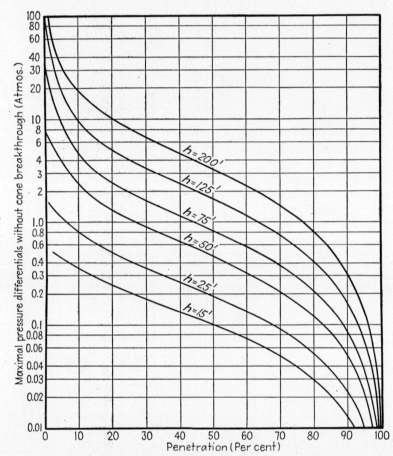

FIG. 5.25.   The calculated maximal producing pressure differentials under steady-state
homogeneous-fluid flow without water-cone break-through, as functions of the well penetration,
for various oil-zone thicknesses h.   Well radius = $\frac{1}{4}$ ft; external-boundary radius = 500 ft;
water-oil density contrast = 0.3 gm/cc.   (*After Muskat and Wyckoff, AIME Trans., 1935.*)

Fig. 5.26, however, should not be considered as quantitatively accurate.
Because of the higher concentrations of the pressure drop about the well
bore in heterogeneous-flow systems (cf. Sec. 8.2), the critical pressure differentials for actual gas-drive oil-producing wells should be higher than
those shown in Fig. 5.25.   And in estimating the corresponding maximal
production rates from Fig. 5.26 account should be taken of the reduced
values of $k/\mu\beta$, taken as 1 in Fig. 5.26, because of the effect of connate

water and free gas on k and of the gas evolution on μ.* Moreover the
inherent flow capacities of the well will be less than those indicated by

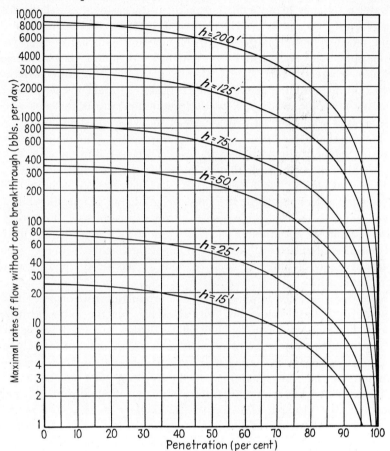

FIG. 5.26. The calculated maximal rates of steady-state homogeneous-fluid flow without
water-cone break-through, as functions of the well penetration, for various oil-zone thicknesses h. Well radius $= \frac{1}{4}$ ft; external-boundary radius $= 500$ ft; water-oil density contrast
$= 0.3$ gm/cc. $k/\mu\beta$ assumed $= 1$; $k =$ permeability; $\mu =$ viscosity; $\beta =$ formation-volume
factor. (*After Muskat and Wyckoff, AIME Trans., 1935.*)

Figs. 5.11 and 5.12 because of the multiphase character of the flow system
(cf. Sec. 8.5).

While the quantitative features of the theory of water coning presented
above are thus subject to much improvement, there is little reason to

* Aside from these effects on $k/\mu\beta$ the maximal production rates plotted in Fig. 5.26
should be multiplied by the actual value of $k/\mu\beta$ even if the flow were strictly that of a
homogeneous fluid. The basic criterion for cone break-through is represented by the
pressure differentials of Fig. 5.25, rather than by the production rates.

doubt the correctness of the basic physical aspects of the discussion. The desirability of minimizing the well penetration into an oil zone underlain by water is, of course, quite obvious on intuitive grounds as well as on the basis of Figs. 5.25 and 5.26. Hence if, at a certain well penetration, water coning has developed, plugging back the hole and reducing the well penetration should be an effective remedial measure unless the net oil-zone thickness is already very small. Reducing the pressure differential and production rates should, of course, be tried first. In fact the sensitivity of the entry of water into a well to the production rate is the most direct evidence that water coning is the cause of the water production.

The reduction of well penetration will be especially effective in eliminating water coning if the bottom of the well can be set above a shale lens or an equivalent body of rock that is much less permeable than the main producing formation. Such permeability barriers, even if of limited areal dimensions, will concentrate the pressure gradients about the well and above the barrier so that the pressures below will be raised. The pressure drop available for maintaining the cone against its gravity head will thus be reduced. Both electrical-model experiments and field experience confirm the effectiveness of plugging back to shale breaks in eliminating water-coning difficulties.

The above discussion implies that if the production rate or pressure differentials are reduced below their critical values after the water has broken through into the well, the water cone will settle down to a new stable level. The details of this process are too complex to describe quantitatively, and the time required for reestablishment of a stable cone is difficult even to estimate. On the other hand, none of the so-called "stable-cone heights" considered above will represent states of dynamical equilibrium persisting throughout the production history of the well or reservoir. Since it is tacitly assumed that the oil is being produced by gas drive, the reservoir pressure in the oil zone will undergo a continuous decline, unless gas is injected to replace the oil and gas withdrawals. A pressure differential will therefore develop between the water and oil zones, and the bottom water will gradually rise and encroach into the oil reservoir, simulating the bottom-water-drive reservoir mechanism.[1] The net residual-oil-zone thickness will thus continually decrease, and the conditions for preventing the cone break-through will become increasingly difficult to maintain at economical production rates. Ultimately it will become simply impossible to obtain sustained "clean" oil production[2] at rates above the economic limit

[1] For a discussion of bottom-water-drive systems cf. Sec. 11.13.

[2] The term "clean" oil refers here to oil entering the well bore without a free-water phase, although in practice the oil production is termed clean as long as extraneous components, including free water and sediments, do not exceed 2 per cent, when measured at the surface.

for profitable operation, even though the well be plugged back as the general water-oil contact rises.

Under the same basic approximations as those used above for the discussion of the coning of underlying bottom waters a treatment can be given of the problem of the downward coning of gas overlying an oil-saturated zone. If the well completely penetrates the oil zone but is open only over its lower part, the graphical and numerical procedure outlined for the water-coning problem can be applied directly with but two minor changes. The effective penetration is taken as the open fraction of the total oil-zone thickness; and the gravity difference $\Delta\gamma$ in Eq. (4) is interpreted as the difference between the oil density and that of the reservoir free-gas phase. The latter change will imply a proportionate increase in the maximal differential pressure and production rates for the gas cone to break into the well. Thus, for example, if the total oil-zone thickness is 75 ft and the casing is set 15 ft below the top of the gas-oil contact,[1] the effective well penetration would be 80 per cent. If the density difference between the oil and gas is 0.6, the critical pressure differential without gas coning will therefore be $2 \times 0.21 = 0.42$ atm (cf. Fig. 5.25). The corresponding maximal production rates, under the same assumptions and approximations that underlie Fig. 5.26, will then be 400 bbl/day.

If a well subject to gas coning does not completely penetrate the oil zone, a further approximation must be made to obtain the proper potential function $\phi(r,z)$ to be used in Eq. (4). This may be based on a suitable choice of the horizontal plane intercepting the well which divides the fluid having an upward component from that having a downward component.[2] The region above this plane may then be considered as isolated, as if the plane itself formed a lower impermeable boundary, and the gas-coning problem will thus be reduced to that discussed in the preceding paragraph. The choice of the dividing plane may be based on the requirement that the well surface potential, as calculated for example by the method of Sec. 5.4, be the same both above and below the plane. Once this plane is determined, the fractional penetration of the open part of the well bore in the total oil-zone thickness above the plane may be used in working out the limiting pressure differential without gas coning. The corresponding production rate into the exposed well bore both above and below the plane will then be the maximum possible without gas-cone break-through.

The above-outlined method of treating partially penetrating wells that

---

[1] Under modern completion practice the well would probably be cased through the whole producing section and the lower 60 ft perforated. Such perforations will not affect the basic equilibrium equation [Eq. (4)], though the effective radius of the well used to determine the potential function $\phi(r,z)$ will be reduced to $e^{-c} r_w$ [cf. Eq. 5.6(8)] and the corresponding production rates will be reduced by the ratios Q/Q_o of Fig. 5.15.

[2] Cf. M. G. Arthur, *AIME Trans.*, **155**, 184 (1944).

have open sections extending neither to the top nor to the bottom of the original oil-zone stratum may also be applied to the case where at the same time the oil-bearing formation is overlain by a free-gas zone and underlain by communicating bottom water. The limiting pressure drop for gas coning is then derived for the part of the oil pay above the dividing plane, and that for water coning for the section below the dividing plane.[1] These will in general be different, depending on the total oil-zone thickness, the total open interval, and the location of that interval. The smaller of the two will represent the maximal pressure differential without break-through of either the gas or the water zone.

Finally it should be noted that the elevation of the water-oil interface induced by upward pressure gradients in an overlying oil zone will also occur as a regional type of coning. Areal concentrations of fluid withdrawal from oil-producing formations in communication with mobile water reservoirs will induce a tendency for the water to rise toward the low-pressure regions, even though the water reservoir is not exerting directly a water-drive action on the field. The water-oil contact may thus develop regional contours suggestive of the conelike surface formed locally underneath individual wells. Of course, if there should be an appreciable amount of water encroachment into the oil reservoir, the effect of coning on the water-oil interface may well be masked by extraneous factors such as differential movements associated with permeability stratification in the producing formation.

5.10. Summary.—The production capacities (in surface measure) of individual wells producing homogeneous liquids under steady-state conditions are given by

$$Q = \frac{2\pi kh(p_e - p_w)}{\mu\beta \log r_e/r_w} = 0.003076 \times \frac{kh(p_e - p_w)}{\mu\beta \log_{10} r_e/r_w} \text{ bbl/day}, \qquad (1)$$

where k is the permeability of the pay, of thickness h; μ is the liquid viscosity; p_e, p_w are the pressures at the external boundary, of radius r_e, and well bore, of radius r_w, respectively; β is the formation-volume factor of the liquid; and the well completely penetrates the pay. In the first part of the equation, Q will be expressed in cubic centimeters per second, if k is in darcys, h in centimeters, μ in centipoises, and p_e, p_w in atmospheres. In the second expression, Q will have a value in barrels per day, provided that k is given in millidarcys, h in feet, p_e, p_w in psi, and μ in centipoises.

Equation (1) will be applicable, to a high degree of approximation, even

---

[1] This method of treating such problems and examples of its application are discussed in detail by Arthur, *loc. cit.*, who also treats a case of upstructure edgewater fingering into a well producing by simple radial flow and gives charts for taking into account different well and external-boundary radii.

if the well bore and external boundary do not form two exactly concentric cylindrical surfaces, provided that r_e is taken as any reasonable average distance of the well from the external boundary or is interpreted as an effective drainage radius.[1]   Likewise Eq. (1) may be used when the well and external-boundary pressures are not strictly uniform, provided that their average values are substituted for p_w, p_e.

If the well does not completely penetrate the producing section, the production capacity can be computed from the equation

$$Q = \frac{2\pi k h F (p_e - p_w)}{\mu\beta \log r_e/r_w}, \tag{2}$$

where F is a coefficient depending on the fractional well penetration and absolute pay thickness, plotted in Fig. 5.13.   Here, too, the numerical coefficient given in Eq. (1) will apply for corresponding units of k, h, p_e, p_w, and μ, with the resultant value of Q expressed in barrels per day.   Moreover, it will apply to systems where the well bore is not exactly concentric with the external boundary, and if the boundary pressures are not strictly constant, on making the same interpretations of r_e, and p_e, p_w as in the case of complete well penetration.

For the special case where the well just barely penetrates the pay, Q assumes the form (to a close approximation)

$$Q = \frac{2\pi k}{\mu\beta}(p_e - p_w)r_w. \tag{3}$$

It is thus seen to vary directly as the well radius and is essentially independent of the external-boundary radius, whereas in the case of appreciable or complete well penetration it varies logarithmically with the boundary radii.

If the well is completed with a perforated casing or slotted liner opposite the producing section, the production capacity may still be expressed by Eq. (1), provided that the well radius is considered as being reduced by a factor e^{-C}, where C is given by Eq. 5.6(8) or 5.7(1).   The resultant effect on Q is given directly by Figs. 5.15 and 5.17.   In the case of perforated-casing completions, only the perforation density, rather than pattern or distribution, and radius are of importance.   When a slotted liner is used, the determining factors are the number of slot columns and open fraction of the pipe.

When the driving force causing the flow into the well bore is gravity, rather than a pressure differential, the production capacity will still be

---

[1] This term is not to be interpreted as physically limiting the region of actual flow through the formation; it refers only to that part for which the analytical expressions for the pressure distributions are directly applicable.

given by a formula essentially equivalent to Eq. (1). Of course, such a situation will arise only if the bottom-hole pressure is so reduced by pumping or by virtue of the natural depletion of the reservoir pressure that the fluid level in the well lies below the top of the pay. From the bottom of the well to the fluid level the potential will be constant, and from there to the top of the pay the pressure will be constant. Within the formation the fluid-saturated region will be bounded by a free surface, which at the same time is a streamline and constant-pressure surface. It intersects the well above the fluid level, the intervening part of the well bore being a "surface of seepage." If the fluid level at the external boundary is h_e, which may be less than the pay thickness, and that in the well is h_w, the production capacity will be

$$Q = \frac{\pi k \gamma g (h_e^2 - h_w^2)}{\mu \log r_e/r_w} = \frac{6.667 \times 10^{-4} k \gamma (h_e^2 - h_w^2)}{\mu \log_{10} r_e/r_w} \text{ bbl/day}, \qquad (4)$$

where, in the right-hand side, k is expressed in millidarcys, h_e, h_w in feet, μ in centipoises, and the liquid density γ in grams per cubic centimeter or as the specific gravity. If the fluid head at r_e exceeds the pay thickness h,

$$Q = \frac{\pi k \gamma g (2hh_e - h^2 - h_w^2)}{\mu \log r_e/r_w}. \qquad (5)$$

Both Eqs. (4) and (5) can be derived by merely replacing p_e, p_w in Eq. (1) by the corresponding values of the average boundary potentials.[1]

It should be understood that the results of this chapter have been derived on the assumption that the fluid flow is homogeneous. In the case of practical oil production, they will therefore be valid only if the oil is produced above the saturation pressure of its dissolved gas. Moreover the treatment has been based on steady-state conditions of flow. While in a strict sense all oil-producing systems are inherently time varying, the transient[2] effects due to general changes in reservoir pressure will be such as to permit a representation in terms of a succession of steady states. Even so, however, the values of the permeability used in the above formulas must take into account the presence of the connate water in reducing the permeability as usually measured with air on dry cores. Of course, in the case of gravity flow, where the pressure has been depleted and gas evolution has taken place in the pay, the permeability to the oil, when used in Eq. (4), must be further corrected for the additional resistance caused by the free gas.

---

[1] The interchange of pressures and potentials in the various flow equations must, of course, be made with due account of the associated factors k/μ.

[2] These refer to well behavior within the oil zone. For the water reservoir it will in most cases be necessary to take into account directly the compressibility of the water and treat it as a compressible-flow system throughout.

Although the explicit flux formulas and graphs developed here have been referred to as production capacities, they will all—except only those pertaining to gravity-flow and water-coning systems—be equally valid as expressions for injection capacities. The flow resistances are determined only by the geometry of the bounding surfaces. As long as the same average pressure or potential difference is applied between the inflow and outflow boundaries, the rate of flow will be the same whether the "external" boundary be taken as the inflow or outflow surface. While an injection capacity for oil will seldom be of practical interest, it is useful to note that the steady-state capacity of wells to take water—which is of importance in water-injection and -flooding operations—will be given by the same formulas.

A problem that has been treated by a steady-state homogeneous-fluid approximation, simply because the proper heterogeneous-fluid analysis has not been developed as yet, is that of water coning. The phenomenon of water coning arises when a normal gas-drive producing system suffers from premature entry of bottom waters because of excessive producing rates or pressure differentials. If the latter be restricted below certain maximal or critical values, the bottom waters would lie statically below the bottom of the well with an elevated conelike interface with the oil zone without participating directly in the oil-production mechanism. The hydrostatic head of the local water elevations above the original water-oil contact plane is held in balance by the upward gradients in the oil zone driving the oil into the bottom and lower parts of the well bore. When these gradients are increased by reducing the well pressure, the water cone rises. Ultimately the tip of the cone will enter a region of such steep pressure gradients as to overcome the downward differential-gravity gradient, between the oil and water, and the cone will become unstable and break into the well bore. It is by virtue of this physical mechanism that the cone rise is sensitive to the production rate.

To describe accurately the development of water coning it would be necessary to establish first the proper pressure distribution in the oil zone associated with the flow of oil into the partially penetrating well. Moreover this distribution would have to be adjusted for the presence of the underlying water cone distorting the lower oil-zone boundary. The shape of the latter, in turn, will be affected by and determined simultaneously with the pressure distribution. This type of analysis, however, is as yet intractable, from a practical standpoint. To obtain at least an order-of-magnitude solution of the problem the pressure distribution has been approximated by that for a steady-state homogeneous-fluid drainage by a partially penetrating well, with no correction due to the presence of the water cone. The shape and height of rise of the water cone at any radial distance from the well were obtained by determining the heights above the main water-oil contact at which the differential head of the corresponding water column equaled the pressure drop at the top of the column below the external-boundary pressure [cf. Eq. 5.9(4)]. No attempt was made to apply the requirement that the water-oil interface must represent a bounding-stream-line surface for the oil zone. By a graphical solution (cf. Fig. 5.21) of the basic hydrostatic-equilibrium equation so obtained the critical cone heights (Figs. 5.22 and 5.23) and maximum pressure differentials (Fig. 5.25) were derived for water-free oil production for different well penetrations and oil-zone thicknesses. The corresponding production rates (Fig. 5.26) represent the maximal water-free values.

As would be expected from general physical considerations, the maximal pressure differentials and water-free production rates are found to increase with decreasing well penetration and increasing oil-zone thickness. At the lower penetrations the critical pressure differentials even exceed the total differential fluid heads corresponding to the total oil-zone thickness, and at penetrations of 5 to 10 per cent may be 2 to 5 times as great as the fluid head.

These considerations have the practical implication that a reduction in production rate or back plugging a well in which a water cone has broken through should be tried as remedial measures for reducing or eliminating the water production. Back plugging will be especially effective if the well can be bottomed above a shale break or other impermeable barrier. On the other hand, as the pressure in the oil zone suffers its normal decline, the general water level will automatically rise and make it increasingly difficult to prevent water coning at economically practicable production rates.

The same approximate type of treatment developed for the water-coning problem can be applied to the problem of coning and break-through of gas from a free-gas zone overlying an oil-producing formation. The effective well penetration is then expressed as the ratio of the open section of the well bore to the total oil-zone thickness, on the assumption that the well is actually open to the bottom of the pay. The differential gravity gradient tending to prevent the break-through of the gas cone is that due to the difference between the density of the reservoir oil and free-gas phase. With these changes the basic equilibrium equation governing the water-coning problem can be applied directly to determine the critical pressure differentials and production rates for gas coning. This treatment can be still further extended to the case where the oil-producing stratum is both overlain by a free-gas zone and underlain by mobile bottom waters.

CHAPTER 6

STRATA OF NONUNIFORM OR ANISOTROPIC PERMEABILITY WITH HOMOGENEOUS-FLUID FLOW

In view of the accumulated experience in the analyses of cores from actual underground strata, it appears extremely unlikely that such strata will be of strictly uniform permeability over distances or areas associated with oil-producing reservoirs. However, as such random lateral variations as undoubtedly occur are literally impossible of exact determination, they must be considered as averaged to give an equivalent uniform-permeability stratum. Moreover, even if the nature of these variations were known, the difficulties of exact analytical[1] treatment would still force the use of an averaging procedure and reduction to an equivalent uniform-permeability system.[2] On the other hand, systematic permeability variations that are either continuous or discontinuous do permit analysis in many cases. While the establishment of even systematic permeability variations will seldom be feasible, it is of value to study certain types of variations so as to aid in the interpretation of such anomalous well characteristics as may be caused by them. It is for this reason, rather than for direct application, that the following discussion of steady-state homogeneous-fluid flow into wells producing from strata of nonuniform permeability is presented.[3] And for the same reason several problems involving anisotropic media will be treated.

6.1. Continuous Permeability Variations.—If the permeability changes continuously throughout an isotropic producing stratum, one may still formally consider the Darcy equation

$$v = -\frac{k}{\mu} \nabla (p - \gamma gz), \tag{1}$$

[1] It is possible, however, to treat nonuniform-permeability systems by the electric analyzer (cf. Sec. 11.8) and potentiometric models (cf. Sec. 13.6).

[2] The effective hydrodynamic average permeability of a heterogeneous system may be shown to be less than the arithmetic mean and greater than the harmonic mean [cf. W. T. Cardwell, Jr., and R. L. Parsons, *AIME Trans.*, **160**, 34 (1945)]. These limits are essentially equal to the "parallel" and "series" resultant permeabilities indicated in the footnote on p. 242.

[3] Except for Sec. 6.6 the problems considered in this chapter will involve only areal variations in permeability in single producing zones. Parallel superposition treatments of vertical stratifications in permeability will be given in Chaps. 12 and 13.

if $+z$ is directed downward, as applicable, except that k is to be expressed explicitly as a function of the coordinates. Upon applying the equation of continuity, it follows that for steady-state flow

$$\nabla \cdot \bar{v} = \frac{\partial}{\partial x}\left(k\frac{\partial p}{\partial x}\right) + \frac{\partial}{\partial y}\left(k\frac{\partial p}{\partial y}\right) + \frac{\partial}{\partial z}\left(k\frac{\partial p}{\partial z}\right) - \gamma g \frac{\partial k}{\partial z} = 0. \tag{2}$$

This equation is, in principle, integrable provided that $k(x,y,z)$ is known. For practical purposes, however, solutions of Eq. (2) are difficult to construct except for systems of simple geometry. One such is that of radial flow where $k = k(r)$. The equivalent of Eq. (2) then becomes

$$\frac{\partial}{\partial r}\left[rk(r)\frac{\partial p}{\partial r}\right] = 0. \tag{3}$$

This is readily integrated to give

$$p(r) = \frac{Q\mu}{2\pi}\int_{r_w}^{r}\frac{dr}{rk(r)} + p_w, \tag{4}$$

where Q is the rate of flow in the reservoir per unit thickness and p_w is the pressure at $r = r_w$. Once the form of $k(r)$ is chosen, the pressure distribution can thus be determined by simple quadrature.

It is interesting to note from Eq. (4) that if k varies directly with the radius, that is, $k = k_0 r$, the pressure will vary inversely with the radius, as in the case of spherical flow (cf. Sec. 5.3). On the other hand, if k varies inversely as the radius, that is, $k = k_0/r$, the pressure will vary directly with the radius, the permeability increase at small radii compensating for the converging area for flow.

6.2. Discontinuous Radial Variations in the Permeability.[1]—While, as previously indicated, the extremely localized information provided by an analysis of cores will in itself give but meager indications of the lateral uniformity or variability of a producing formation, it is of value to see what the effect on the flow characteristics would be if the well bore happened to penetrate a region either more or less permeable than the main body of the pay. The very process of drilling[2] and well completion may lead to a localized plugging of the formation, and such plugging may also develop during the course of production. Moreover the process of "acid treatment" in limestone formations may lead to an increase in the perme-

[1] The case of linear systems composed of sections of different permeability can be simply treated by analogy with series or parallel combinations of electrical resistances. Thus a linear "series" combination of sections of sand of permeabilities k_i and thicknesses h_i will have a resultant permeability $\Sigma h_i/\Sigma(h_i/k_i)$. On the other hand, if the different sections are of the same areal geometry but are assembled in "parallel," the resultant permeability will be $\Sigma h_i k_i/\Sigma h_i$.

[2] While direct plugging under modern well-completion practices is probably rather uncommon, equivalent localized effects may result from drilling-water invasion in "dirty" sands (cf. Sec. 3.7), causing what are often termed "water blocks."

ability in the immediate vicinity of the well bore, giving rise to a discontinuous and approximately radial permeability distribution.

All these problems, when idealized, may be described analytically as follows: Find the pressure distributions in regions 1 and 2 of permeabilities k_1 and k_2 such that (cf. Fig. 6.1)

$$p = p_w \ : \ r = r_w; \qquad p = p_e \ : \ r = r_e. \tag{1}$$

The essence of this problem lies in the observation that separate pressure distributions must be constructed for the regions 1 and 2 and then "adjusted" at the boundary of contact between the two regions, the "surface of discontinuity," so that the pair of distributions when considered together correspond to the composite fluid system. This "adjustment" consists in imposing the conditions

$$\left. \begin{array}{c} p_1 = p_2, \\[2ex] k_1 \dfrac{\partial p_1}{\partial r} = k_2 \dfrac{\partial p_2}{\partial r} \end{array} \right\} \text{at } r = r_o, \tag{2}$$

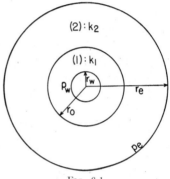

Fig. 6.1

where p_1, p_2 represent the individual pressure functions in the regions 1 and 2 joining at the radius r_o. The first of Eqs. (2) expresses the requirement of pressure continuity at the surface of discontinuity. The second ensures continuity of the normal fluid velocity across the interface.[1]

To carry through this procedure it is necessary only to choose the constants a and b in the generalized radial-flow pressure distributions [cf. Eq. 5.1(2)], namely,

$$\left. \begin{array}{c} p_1 = a_1 \log r + b_1 \\[1ex] p_2 = a_2 \log r + b_2, \end{array} \right\} \tag{3}$$

so as to satisfy Eqs. (1) and (2). It is so found that

$$\left. \begin{array}{ll} p_1 = p_w + \dfrac{p_e - p_w}{\log r_o/r_w + \alpha \log r_e/r_o} \log \dfrac{r}{r_w} & : \quad r_w \leqslant r \leqslant r_o, \\[3ex] p = p_e + \dfrac{\alpha(p_e - p_w)}{\log r_o/r_w + \alpha \log r_e/r_o} \log \dfrac{r}{r_e} & : \quad r_o \leqslant r \leqslant r_e, \end{array} \right\} \tag{4}$$

where $\alpha = k_1/k_2$. These represent the resultant pressure distribution in the system.

To illustrate the effect of radial permeability discontinuities on the pressure distribution, Eqs. (4) have been evaluated for two cases, namely, $\alpha = \frac{1}{5}$ and 5, with $r_o = 50$ ft, $r_w = \frac{1}{4}$ ft, $r_e = 660$ ft, and $p_w = 0$, p_e

---

[1] These general physical requirements must be applied at all surfaces of discontinuity in the medium in homogeneous-fluid systems.

= 100 psi. The results are plotted in Fig. 6.2. For comparison the distribution for a uniform permeability ($\alpha = 1$) has been plotted as the dashed curve. It will be seen that when the region about the well bore is relatively tight ($\alpha = \frac{1}{5}$), the pressure gradients near the well bore are accentuated, whereas those in the distant parts of the medium are reduced; and conversely for a local high-permeability zone about the well ($\alpha = 5$). In both cases the pressure is continuous at the boundary radius $r_o = 50$ ft. On the other hand, the slopes at r_o are discontinuous, as required by Eq. (2).

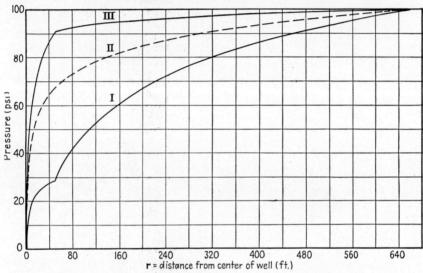

FIG. 6.2. The calculated pressure distributions in nonuniform steady-state homogeneous-fluid radial-flow systems. Curve I, $\alpha = 5$. Curve II, $\alpha = 1$ (uniform system). Curve III, $\alpha = \frac{1}{5}$. $\alpha =$ (permeability to radius of 50 ft)/(permeability between 50 and 660 ft); well radius $= \frac{1}{4}$ ft; total pressure drop $= 100$ psi.

The reservoir fluid-production capacity corresponding to the pressure distributions of Eqs. (4) is readily found to be

$$Q = \frac{2\pi k_1 h r}{\mu} \frac{\partial p_1}{\partial r} = \frac{2\pi k_2 h r}{\mu} \frac{\partial p_2}{\partial r} = \frac{2\pi k_1 h (p_e - p_w)/\mu}{\log r_o/r_w + \alpha \log r_e/r_o}. \tag{5}$$

In comparison with that of a similar system of uniform permeability k_2, Q bears a ratio [cf. Eq. 5.1(6)]:

$$\frac{Q}{Q_o} = \frac{\alpha \log r_e/r_w}{\log r_o/r_w + \alpha \log r_e/r_o}, \tag{6}$$

where Q_o is the capacity of the uniform system.

Equation (6) is plotted in Fig. 6.3 for several values of r_o and for $r_w = \frac{1}{4}$ ft, $r_e = 660$ ft. The curves of Fig. 6.3 show that even highly localized permeability anomalies will have considerable influence on the production capacity. For example, a zone about the well of only 5 ft radius—0.006 per cent of the total pay volume—will increase the production capacity by 43 per cent if its permeability is 5 times as great as the remainder of the formation or cut it to 41 per cent of normal if its permeability is low by a factor of 5. These effects may be readily understood on consideration of

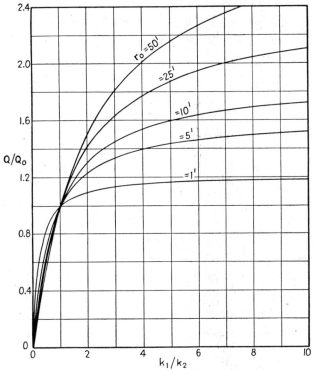

FIG. 6.3. The calculated variation of the production capacity of a well, producing a homogeneous fluid under steady-state conditions, as a function of k_1/k_2 = (permeability within the annulus of radius r_o)/(permeability from r_o to 660 ft). Q/Q_o = (production capacity of well in stratum where $k_1/k_2 \neq 1$)/(production capacity of well in stratum with uniform permeability k_2); well radius = $\frac{1}{4}$ ft.

the nature of the pressure distributions in such nonuniform-permeability systems as shown in Fig. 6.2.

For permeabilities about the well bore higher than the rest of the formation $(\alpha > 1)$ the maximum increase in production capacity which can result is evidently that corresponding simply to an enlargement of the well bore to a radius r_o. This is given by Eq. (6) on letting α approach infinity. Because of this finite limitation the increase in Q/Q_o rapidly tapers off with increasing α, as shown in Fig. 6.3. On the other hand, if the zone about the well bore is abnormally tight $(\alpha < 1)$, the physical limit is an absolute vanishing of the production capacity, as required by Eq. (6) $(\alpha = 0)$ and indicated in Fig. 6.3.

6.3. Adjacent Beds of Different Permeability; Fractured Limestones.—
A practical oil-production problem involving systems of nonuniform permeability is that of describing the flow conditions in fractured limestones. Such formations are characterized by distributions of fractures, or crevices,[1] throughout the body of the rock. The permeability of the limestone proper is often extremely low and would not alone provide for production rates of commercial importance. However, when such a rock is permeated by fractures, it can serve as a reservoir feeding oil into the fractures, which in turn carry the production, directly or through interconnection with other similar channels, to the wells. Because these fractures have very high effective permeabilities, even when of extremely small width, the composite system of limestone body and fractures may have a resultant flow capacity quite comparable with and often greater than the average sandstone.

The treatment of a general fractured limestone system containing a number of fractures distributed at random throughout the mass of limestone proper is not as yet practical mathematically. Nevertheless it is possible to get an idea as to the significance of the fractures as fluid carriers by considering the simplified case of a single fracture opening into the well bore and extending for some distance into the body of limestone. For this purpose it is convenient to regard the fracture as the equivalent of a porous medium of high permeability bounded on either side by the limestone proper. The effective permeability of the fracture will be taken as that corresponding to the classical hydrodynamics fluid-carrying capacity of narrow linear channels. The latter, per unit length, is given by[2]

$$Q = \frac{w^3 \, \Delta p}{12\mu},$$ (1)

where w is the width of the channel. The equivalent permeability of the channel is therefore

$$k = \frac{w^2}{12} = \frac{10^8 w^2}{12} \text{ darcys},$$ (2)

if w is expressed in centimeters.

For analytical purposes the problem may be defined by the quadrant shown in Fig. 6.4, with the indicated boundary conditions.[3] The condition

---

[1] It is assumed here that the fractures are essentially vertical. If the fractures were horizontal and had areal extensions comparable with that of the main pay, they would simply represent an extreme form of the common type of permeability stratification, which can be represented by a "parallel" system of producing strata (cf. also footnote on p. 242).

[2] Cf. H. Lamb, "Hydrodynamics," 6th ed., p. 582, Cambridge University Press, 1932.

[3] The equivalent problem, in which the limestone extends uniformly and indefinitely from the well center, has been treated by M. Muskat and R. D. Wyckoff, *Physics*, **7**, 106 (1936); (cf. also M. Muskat, "The Flow of Homogeneous Fluids through Porous Media," Sec. 7.4, McGraw-Hill Book Company, Inc., 1937).

of zero gradient along $y = 0$ and $x = 0$ arises from symmetry. The fracture width is taken as the unit of length, and its permeability as k_1, that of the limestone proper being denoted by k_2. The lateral extensions of the pay will be taken as x_o and b, parallel and perpendicular to the fracture, and the pressure at these boundaries will be assumed to have the uniform value p_e. Finally the flux into the well from the fracture will be assumed such as corresponds to a pressure gradient a.

To solve this type of problem we must find pressure-distribution functions p_1 and p_2, pertaining, respectively, to the fracture and limestone proper, each a solution of

$$\frac{\partial^2 p}{\partial x^2} + \frac{\partial^2 p}{\partial y^2} = 0, \qquad (3)$$

and such that they satisfy not only the boundary conditions indicated above

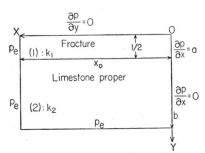

Fig. 6.4. A diagrammatic representation of a quadrant of a section of a limestone bisected by a fracture, showing the assumed boundary conditions.

and in Fig. 6.4 but in addition the continuity requirements at their common boundary, namely,

$$\left. \begin{aligned} p_1 &= p_2 \\ k_1 \frac{\partial p_1}{\partial y} &= k_2 \frac{\partial p_2}{\partial y} \end{aligned} \right\} \, y = \tfrac{1}{2}. \qquad (4)$$

This may be accomplished by choosing p_1 and p_2 in the forms

$$\left. \begin{aligned} p_1 &= p_e + a(x - x_o) + \sum_{\text{odd}} A_n \cos \frac{n\pi x}{2x_o} \cosh \frac{n\pi y}{2x_o}, \\ p_2 &= p_e + \sum_{\text{odd}} B_n \cos \frac{n\pi x}{2x_o} \sinh \frac{n\pi (b - y)}{2x_o}, \end{aligned} \right\} \qquad (5)$$

and adjusting the A_n's and B_n's so that Eqs. (4) are satisfied.[1] Upon so determining the A_n's and B_n's and evaluating the pressure drop across the system, it is found to be

$$p_e - p_w = \frac{8ax_o}{\pi^2} \sum_{\text{odd}} \frac{1}{n^2 \left[1 + \dfrac{k_2}{k_1} \coth \dfrac{n\pi}{4x_o} \coth \dfrac{n\pi (b - \frac{1}{2})}{2x_o} \right]}, \qquad (6)$$

where

$$p_w = p_1(0, \tfrac{1}{2}) = p_2(0, \tfrac{1}{2}), \qquad (7)$$

and represents the well pressure.

[1] It will be readily verified that Eqs. (5) directly satisfy all the other boundary conditions of the problem.

As x_o is measured in units of the fracture width, it will be a very large number. Moreover, by taking b to be the order of $2x_o$, Eq. (6) may be simplified to the form

$$p_e - p_w = \Delta p = \frac{8ax_o}{\pi^2} \sum \frac{1}{n[n + (4k_2x_o/\pi k_1)]}. \tag{8}$$

Now the flux in the system, from both sides of the y axis, will be

$$Q = \frac{2k_1}{\mu}\left(\frac{\partial p_1}{\partial x}\right)_{x=0} = \frac{2ak_1}{\mu} = \frac{\pi^2 k_1 \,\Delta p}{4\mu x_o \sum_{\text{odd}} n[n + (4k_2x_o/\pi k_1)]}. \tag{9}$$

Upon evaluating the series, Eq. (9) may be rewritten as

$$\frac{Q\mu}{k_2\,\Delta p} = \frac{2\pi}{0.5772 + 2\Psi(2s) - \Psi(s)}; \qquad s = \frac{2x_ok_2}{\pi k_1}, \tag{10}$$

Ψ being the logarithmic derivative of the Γ function [cf. Eq. 5.4(6)].

The values of Q as given by Eq. (10) are plotted in Fig. 6.5, as a function of the fracture width, as fractions of the simple radial-flow capacity of the limestone proper, Q_o, that is, as

$$\frac{Q}{Q_o} = \frac{\log r_e/r_w}{0.5772 + 2\Psi(2s) - \Psi(s)}, \tag{11}$$

where the equivalent radius r_e, as well as the absolute linear extension represented by x_o, is taken as 300 ft. It will be seen that for fracture widths exceeding 0.035 mm the production capacity of the fractured formation exceeds that of the simple radial-flow system of well radius r_w without the fracture. At fracture widths greater than 1 mm the production capacity increases as the cube of the width, indicating that the fracture is the controlling factor [cf. Eqs. (1) and (2)]. Moreover the apparent fall of Q below Q_o for widths less than 0.035 mm does not imply that such narrow fractures will in some manner reduce the production capacity in actual producing formations. For it should be noted that, whereas Q_o refers to a system with a circular fluid outlet of radius r_w ($\frac{1}{4}$ ft), in the idealized representation on which Eq. (10) is based the fluid outlet is essentially limited to a vertical slit of length w. Evidently such a system will inherently have a much higher flow resistance than one provided with an outlet simulating a well bore. The ratios Q/Q_o, as plotted in Fig. 6.5, thus represent measures of Q in terms of a simple arbitrary unit Q_o, rather than absolute comparisons between the fractured and radial-flow systems on a physically comparable basis.

In Fig. 6.5 are also plotted the fractions of the total capacity of the limestone-fracture system that would have been contributed by the frac-

ture itself if it alone were providing the flow. The latter was calculated
by [cf. Eq. (1)]:

$$Q_f = \frac{2 \times 10^8 w^3 \, \Delta p}{12 \mu r_e}.$$ (12)

It will be seen that even for fracture widths of 0.5 mm the fracture alone
could carry more than 90 per cent of the total capacity of the composite
limestone-fracture system. At widths of 1 mm or greater the limestone
proper, within the 300 ft from the well bore, will feed into the fracture but

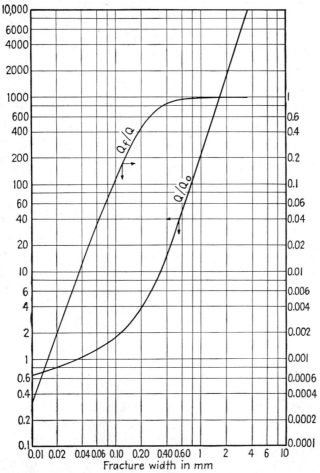

Fig. 6.5.   The calculated steady-state homogeneous-fluid production characteristics of
limestone-fracture systems, as functions of the fracture width. Q/Q_o = (production capacity
of limestone-fracture system)/(production capacity of corresponding radial-flow system);
Q_f/Q = (production capacity of a single fracture)/(production capacity of composite limestone-fracture system).   Permeability of limestone proper = 10 md; linear or radial extent
of system = 300 ft.

a negligible fraction of the total oil carried by the fracture. On the other hand, it is to be noted that the fractions Q_f/Q plotted in Fig. 6.5 will not give exactly the relative contributions to the total flow of the fracture and limestone body in the composite system, for the pressure distribution in the fracture will be changed from the strictly linear variation underlying Eq. (12).

6.4. The Theory of Acid Treatment of Wells Producing from Limestone or Dolomite Reservoirs.—In the completion of wells drilled into oil-bearing limestone or dolomite reservoirs it is common practice in many districts to "acidize" the wells before putting them on production. Often, preliminary to acidizing, the wells are "shot" with nitroglycerin. The well-established success of these practices may be explained on the basis of the developments of the last several sections.

Acid treatment is the process of injecting acid[1] into the oil-bearing rock so as to react with and dissolve the carbonates of the limestone or dolomite. Hydrochloric acid, in suitable concentration, generally 15 per cent by weight, is the acid used. Usually it has added to it inhibitors against reaction with the pipe in the hole, and often other agents to control or facilitate the chemical reaction[2] with the reservoir rock. The amount of acid used may vary from 500 to 10,000 gal, and frequently a well may be given several "treatments." The purpose is simply to increase the production capacity of the well. Often, when a well appears to be "dry," it will be acidized in an attempt to create channels into the well from the surrounding rock. Or if a producing well seems to have lost its production capacity prematurely, acid treatment may be used to "rejuvenate" the well. On the other hand, in some oil-producing districts such treatments are given the wells as a matter of routine, even without prior testing as to the need for it in each individual case, after their value has been established for the first few wells in the field. As a whole there is no doubt that when the oil-bearing rock contains substantial amounts of acid-soluble constituents, an acid treatment will increase its production capacity.

The "shooting" of wells is intended to accomplish the fracturing of the rock or a sloughing off of the surface layers of the rock exposed by the well bore by creating mechanical stress failure. From the nature of their structure, limestones and dolomites appear to be inherently more suscepti-

---

[1] In actual practice the acid is not left permanently in the formation. In the earlier development of such treatments the acid was allowed to remain 48 to 60 hr in the pay. Now, however, it is generally removed after a few hours contact with the rock, and often reverse circulation and removal are begun as soon as the injection is completed.

[2] The material balances of these reactions may be computed from the controlling equations, namely, $2HCl + CaCO_3 = CaCl_2 + CO_2 + H_2O$ for limestone and $4HCl + CaMg(CO_3)_2 = CaCl_2 + MgCl_2 + 2H_2O + 2CO_2$ for dolomite.

ble to such mechanical breakdown by shooting than consolidated sands.[1] Moreover, it provides a natural preliminary preparation for acidization, in exposing or creating fresh rock surfaces or fractures to be attacked by the acid.

While it is quite possible that both shooting and acidizing may open up to the well sources of oil otherwise sealed off and hence not available for production, the primary function and accomplishment of these practices is generally an increase in the production *capacity*, rather than in the physical ultimate recovery. On the other hand, from an economic point of view, increased production rates will often be equivalent to increased *economic* ultimate recoveries of considerable importance.[2] However, for the purposes of the present considerations it will suffice to examine only the matter of production capacity.

To increase the production capacity of a well it is evidently necessary to increase the permeability of the formation either locally or throughout. The latter can be eliminated here because of the limited volumes of acid used in acidizing. For example, even 10,000 gal of acid will penetrate a 25-ft 15 per cent porosity pay only 10.6 ft. As to the former, however, the permeability increase may occur in any one or combination of several forms, namely: (1) raising the permeability in a small zone about the well from its normal value to a much greater permeability; (2) eliminating the plugging from a zone immediately about the well bore and raising its permeability to that of the main body of pay; (3) the creation or widening of extended fractures. The quantitative features of these possibilities may be evaluated from the results of the last two sections.

Case 1 above will not lead to large increases in production capacity. If the acid-penetrated zone is of radius r_o, the maximum increase possible would evidently be that corresponding to removing completely the pay to radius r_o, *i.e.*, to increasing the well radius to r_o. This, as was seen in Secs. 5.1 and 6.2, is not very effective. The increases to be expected if the acid-affected zone of radius r_o is not completely removed but has its permeability raised to k_2—that of the main body of pay—from an

---

[1] It is to be noted, however, that shooting is also a common practice in well preparation or completion for water-flooding operations in the sandstone reservoirs of Pennsylvania and the Mid-Continent district.

[2] Specific illustrations of the production histories of individual wells and fields following acid treatment may be found in R. F. Heithecker, *U.S. Bur. Mines Rept. Inv.* 3445 (April, 1939), dealing with the limestone fields of Kansas. And at Noodle Creek, Tex., an acidization program instituted after the field had been producing for 7 years led to an increase in ultimate recovery, above the previous decline curve, of 12 per cent ([N. W. Imholz, "Stratigraphic Type Oil Fields," *AAPG Bull.*, p. 698 (1941)]; cf. also A. S. Bunte, *AAPG Bull.*, **23**, 643 (1939), for similar examples from the Greenwich pool, Kansas.

initial value of k_1, may be read off the curves of Fig. 6.3, to the right of the point of convergence. These, of course, are still smaller than if the well radius were increased to r_o.

For case 2 the increases that may be expected are plotted in Fig. 6.6.[1] Here it will be seen that virtually any degree of increase may be achieved if the zone immediately about the well is sufficiently tight initially. Of course, in the limit where the well is originally completely plugged with an acid-soluble material and time is given for the acid to dissolve and remove

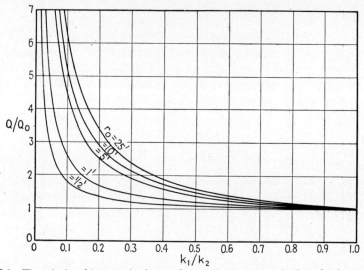

FIG. 6.6. The calculated increase in the steady-state homogeneous-fluid production capacity of a radial-flow system due to acid treatment, if the acid-affected zone, of radius r_o, is initially of lower permeability, k_1, than the rest of the limestone, k_2, and is raised to k_2 by the acid. Q/Q_o = (production capacity after treatment)/(production capacity before treatment); well radius = $\frac{1}{4}$ ft; external-boundary radius = 660 ft.

the plugging material, the apparent increase in production capacity will be infinitely large.

If the increase in production capacity is due to the creation or widening of extended fractures, the magnitude may be estimated from Fig. 6.5. The latter gives directly, as discussed in Sec. 6.3, a measure of the effect of creating a fracture and thus forming a composite limestone- (or dolomite-) fracture system, as compared with an original radial-flow system.[2] If the immediate result of the acid treatment is a widening of previously

[1] The ordinates in Fig. 6.6 are the reciprocals of those in Fig. 6.3 for $k_1/k_2 < 1$. They may be calculated directly by taking for Q/Q_o the reciprocals of the right-hand side of Eq. 6.2(6).

[2] Note should be taken, however, of the difference in fluid outlets involved in the formulas for Q and Q_o, whose ratio is plotted in Fig. 6.5 (cf. p. 248).

existing fractures, the effect on the production capacity may be obtained by comparing the value of Q/Q_o in Fig. 6.5 for the new fracture width with that for the original width. The ratios of these ordinates are plotted in Fig. 6.7 as a function of the increase in fracture width.

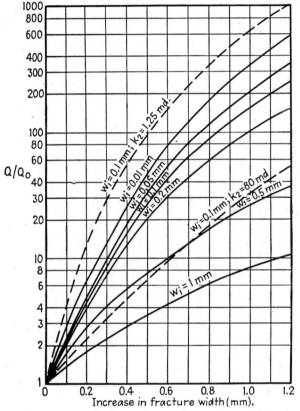

FIG. 6.7. The calculated increase in the steady-state homogeneous-fluid production capacity of a fractured limestone due to acid treatment. Q/Q_o = (production capacity before treatment)/(production capacity after treatment). w_i = initial fracture width. For solid curves, limestone permeability (k_2) = 10 md.

It will be evident from Fig. 6.7 that the theory of fracture widening has sufficient scope to explain virtually any increase of production capacity following acid treatment. The quantitative values of Fig. 6.7 cannot be safely used in actual field applications because of the assumptions that the producing formation contains but a single fracture and that this fracture will be uniformly widened over its whole assumed length of 300 ft. While even a moderate treatment of 1,000 gal of 15 per cent HCl could dissolve a layer 0.4 mm thick for a distance of 300 ft along each side of a fracture in a 25-ft limestone pay, that this will actually occur in the normal 1,000-gal treatment is, of course, quite unlikely. Nevertheless, there is no reason to doubt that Fig. 6.7 indicates the correct order of magnitude of the results of fracture widening, as well as the relative effects for different initial fracture widths.

As would be expected, a given increase in fracture width will lead to greater relative increases in production capacity for the smaller initial fracture widths and production capacities. Likewise, as indicated by the dashed curves in Fig. 6.7, if the pay proper is initially tight the effect of the fracture widening will be more pronounced than for higher permeability limestones or dolomites. Moreover Fig. 6.6 implies that greater improvement may be achieved by acidizing in wells of low initial production capacity also in situations where the acid serves mainly to eliminate a localized plugging about the well bore. On the other hand, for a fixed permeability of the zone affected by plugging, the higher the normal permeability of the limestone (or dolomite) proper and the initial capacity, the greater the improvement on removing the plugging.

In the light of these results, observations on acid treatment may be interpreted as follows:

1. Small increases in production capacity (less than 100 per cent) due to acid treatment *may* be explained on the assumption that the permeability of a small radial zone about the well bore has been increased from normal to higher values, as well as by removal of radial plugging or widening of extended fractures fed laterally by the main body of the pay. Unless the formation does have extended fractures or is appreciably plugged near the well bore, acid treatment should be relatively ineffective in stimulating the production.

2. Moderate increases in production capacity, of the order of one- to tenfold, can be explained on a radial-flow basis only on the assumption that the wells were initially plugged, the extent of the plugging being the principal factor in determining the initial production capacity, so that small producers will show larger responses. They can also be explained equally well by the assumption of extended-fracture flow.

3. Increases in production capacity appreciably larger than 1,000 per cent,[1] for wells of initially moderate production rates, can be explained only on the assumption that there are extended fractures in the limestone or dolomite which are penetrated and widened by the acid. Here the smaller producers should show the greater responses, whether their initially small production rates are due to low permeabilities of the main body of pay or

---

[1] Examples of increases in production capacity of this order of magnitude by acid treatment are given by Heithecker (*op. cit.*).

small widths of the fractures. Similar increases resulting from the radial-flow mechanism will occur only if there were initially a condition of *almost complete plugging* near the well bore and hence a very low initial well capacity.

In order to be able to anticipate the effect of the acid treatment it is necessary to make a detailed geological inspection of cores of the formation. Fracturing or jointing of limestones or dolomites can generally be detected by complete coring or sampling of the producing rock. Moreover, special core tests on the solubility of the rock in acid and the effect of acid in increasing the permeability of cores will also serve to indicate the type of reaction to be expected from acid treatment.[1] On the other hand, simple flow tests on the well will in themselves merely give the resultant flow resistance of the formation. By suitable adjustment of the many physical and geometrical constants involved in defining the details of the producing system it will usually be possible to construct either a radial-flow or a radial-fractured type of producing system such as will have any preassigned resultant flow resistance.

Finally it should be observed that the quantitative conclusions developed above will apply only if the flow conditions simulate those of a steady-state homogeneous-fluid system. In actual oil-producing reservoirs it will be necessary to correct the permeabilities for the effects of connate water even if the oil is undersaturated. And if free gas is flowing in the pay, the details of the pressure distribution will also be changed (cf. Sec. 8.2). If the permeability-saturation relationship (cf. Sec. 7.1) is substantially uniform throughout the pay, the multiphase character of the flow should make the production capacity of the well even more sensitive to the absolute permeability of the region immediately surrounding the well bore. Accordingly the improvements in production capacity due to localized increases in permeability should be even greater than those derived on the basis of the homogeneous-fluid assumption.

6.5. The Effect of a Sanded Liner on the Production Capacity of a Well.—Another problem involving porous media of different permeabilities within the same flow system concerns the effect of a sanded liner on the production capacity of a well. At first glance it might appear that the presence of a column of sand at the bottom of a well bore could have but little effect upon the production capacity of the well, since the permeability of such a column of loose sand should be much higher than that of the original pay. Nevertheless, it is frequently observed that a flowing or pumping well will suffer a considerable decrease in production rate during the accumulation of sand in the well bore. That the loss in production rate

[1] Cf. L. C. Chamberlain, Jr., *Oil Weekly*, **88**, 20 (Feb. 28, 1938).

is in many cases due to the presence of the sand within the liner, and not to the "mudding off," or clogging, of the screen itself, is proved by the fact that complete recovery often results merely from the removal of the accumulated debris. It is therefore of practical interest to analyze this phenomenon from the point of view of fluid flow.

The physical system may be represented diagrammatically as in Fig. 6.8. It will be assumed that the sand fills the well bore to the top of the pay

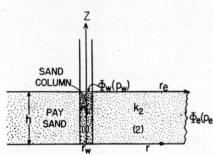

and is of uniform permeability k_1, while the permeability of the pay itself is k_2. Although the sand in the well bore is maintained there by gravity alone, so that near the top of the column the upward-flowing liquid will tend to loosen the packing, it will suffice for the present purpose to assume that the sand column has a uniform permeability from top to bottom. Moreover the presence of the liner as a

FIG. 6.8. A diagrammatic representation of a well with a sanded liner.

flow barrier between the well bore and pay will be ignored. The problem may then be analytically formulated as follows: Find potential functions Φ_1 and Φ_2,* pertaining to the sand column within the well bore and the surrounding formation, respectively, *i.e.*, solutions of Eq. 4.5(4), such that

$$\left.\begin{array}{ll} \dfrac{\partial \Phi_1}{\partial z} = \begin{cases} 2c & : \ z = 1 \\ 0 & : \ z = 0 \end{cases}; & \dfrac{\partial \Phi_2}{\partial z} = 0 \ : \ z = 0, 1; \\[4mm] \dfrac{\Phi_1}{k_1} = \dfrac{\Phi_2}{k_2}; \quad \dfrac{\partial \Phi_1}{\partial r} = \dfrac{\partial \Phi_2}{\partial r} \ : \ r = r_w; & \Phi_2 = \Phi_e, \ \text{for } r = r_e \gg r_w, \end{array}\right\} \quad (1)$$

where the unit of length has been taken as the formation thickness h. The uniform flux condition at the top of the sand in the well bore, *i.e.*, the first of Eqs. (1), has been chosen instead of the requirement that the well potential be uniform, in order to simplify the analysis. However, in view of the small value of the well radius r_w, as compared with the other dimensions of the systems, the two conditions are essentially equivalent.

The vanishing of the gradients of Φ_1 and Φ_2 with respect to z expresses the assumption that the producing stratum is isolated by impermeable boundaries at top and bottom, $z = 1, 0$, respectively. The next boundary conditions impose the requirements of pressure and velocity continuity at the surface of discontinuity, the original well bore (cf. Sec. 6.2). The final condition in Eqs. (1) simply specifies the external-boundary potential or pressure.

* The potential functions used here are defined as $k(p + \gamma g z)/\mu$.

It may be readily verified that potential functions satisfying the conditions of Eqs. (1) are

$$
\left.
\begin{aligned}
\Phi_1 &= \Phi_w + c\left(z^2 - \frac{r^2}{2} - 1\right) + \sum_1^\infty B_n[I_0(n\pi r)\cos n\pi z - (-1)^n] \\[2mm]
\Phi_2 &= \Phi_e + cr_w^2 \log \frac{r_e}{r} + \sum_1^\infty A_n K_0(n\pi r)\cos n\pi z,
\end{aligned}
\right\}
\tag{2}
$$

where I_0 and K_0 are zero-order Bessel functions of the third kind, provided

$$
A_n = -\frac{I_1(n\pi r_w)B_n}{K_1(n\pi r_w)}; \qquad B_n = \frac{-4c(-1)^n/n^2\pi^2}{I_0(n\pi r_w) + \dfrac{\alpha I_1(n\pi r_w)K_0(n\pi r_w)}{K_1(n\pi r_w)}}; \qquad \alpha = \frac{k_1}{k_2}, \tag{3}
$$

and the well potential Φ_w is taken as that at the center of the sand column and is given by:

$$
\Phi_w = \alpha\Phi_e + c\left[\frac{2}{3} + \frac{r_w^2}{2} + \alpha r_w \log\frac{r_e}{r_w}\right] + \sum_1^\infty (-1)^n B_n. \tag{4}
$$

Introducing the notation

$$
\overline{B}_n = -\frac{(-1)^n \pi^2 B_n}{4c}, \tag{5}
$$

and observing that the flux rate from the well is related to the constant c by the relation

$$
Q = -\pi r_w^2 h \frac{\partial\Phi_1}{\partial z}\bigg|_{z=1} = -2\pi chr_w^2, \tag{6}
$$

it is found that the fluid conductivity of the system may be written in the form

$$
\frac{Q}{\Delta p} = \frac{2\pi k_1 h r_w^2/\mu}{\dfrac{2}{3} + \dfrac{r_w}{2} + \alpha r_w^2 \log\dfrac{r_e}{r_w} - \dfrac{4}{\pi^2}\sum_1^\infty \overline{B}_n}, \tag{7}
$$

where $\Delta p = p_e - p_w$, and p_e, p_w are the pressures at the top of the pay corresponding to Φ_w and Φ_e. Upon denoting by Q_o the flux rate for the sand-free well bore the effect of the sand in the well bore may be expressed by the ratio

$$
\frac{Q}{Q_o} = \frac{\alpha r_w^2 \log r_e/r_w}{\dfrac{2}{3} + \dfrac{r_w^2}{2} + \alpha r_w^2 \log\dfrac{r_e}{r_w} - \dfrac{4}{\pi^2}\sum_1^\infty \overline{B}_n}. \tag{8}
$$

The values of Q/Q_o for general values of k_2/k_1 with $r_w/h = 0.01$* and $r_e/r_w = 2{,}000$ are plotted in Fig. 6.9. The tremendous effect of even very

---

* The values of r_w, r_e in Eqs. (1) to (8) are expressed as fractions of the sand thickness h.

permeable sand columns is shown in this figure by the almost vertical descent of the curve for small values of k_2/k_1. Thus a sand column of permeability as high as 200 times that of the sand pay will reduce the production capacity of the well to only 34 per cent of its original value. Furthermore, if the permeability of the sand column is of the same order as that of the pay, the production capacity is effectively that of a non-penetrating well regardless of the exact value of the ratio of the permeabilities.

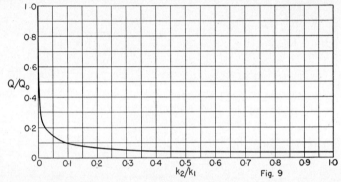

FIG. 6.9. The calculated effect of a sanded liner on the steady-state homogeneous-fluid production capacity of a well. Q/Q_o = (production capacity of well with sanded liner)/(production capacity of sand-free well); k_2/k_1 = (permeability of pay sand)/(permeability of sand column in liner); sand thickness = 25 ft; well radius = ¼ ft; external-boundary radius = 500 ft.

The physical reason for the marked effects of sand columns even of high permeability becomes clear when one analyzes the potential distribution in the well bores when filled with sand. It is then found that the potential rises very rapidly in passing down the sand column. For example, if the permeability of the sand column is 100 times that of the main pay, 88 per cent of the total potential drop will be dissipated within the sand column. And if the permeability ratio is 10, more than 94 per cent of the total potential drop occurs between the top and bottom of the sand in the well bore. Thus the lower parts of the sand are effectively producing against back pressures that are very high as compared with the well pressure at the top of the sand column and hence cannot contribute appreciably to the production from the well. The penetration of the well is therefore reduced from physical completeness to an effective value in homogeneous formations of less than 25 per cent of the thickness of the pay, if the sand-column permeability is less than 200 times that of the main body of producing rock.

All these effects will evidently be greatly aggravated if the sand column extends appreciably above the top of the pay.[1] It is therefore of significant

[1] For an approximate treatment of such cases, cf. Muskat, op. cit., Sec. 7.10.

practical importance to keep the well bore free of sand.   During the flush production stage the velocities of flow will probably be sufficient to flush out any sand entering the well bore.   Sand accumulation may develop, however, in the later stages of production and will rapidly tend toward an aggravated case of "sanding," unless remedial measures are applied promptly.   Of course, any plugging of the screen will act as an added choke beyond that caused by the sand within the liner.

6.6. Partially Penetrating Wells in Stratified Horizons.—Another three-dimensional problem of practical interest, involving regions of different permeability, is that in which a partially penetrating well is drilled into a producing section composed of several strata of different permeability.   Although in practice, where this problem arises,[1] the well penetration will be of nonvanishing magnitude and the individual strata will be of finite thickness and may be of variable permeability, it will be assumed here for simplicity that the well is nonpenetrating and that the producing section is composed simply of a layer (1) of permeability k_1 overlying an infinite homogeneous stratum (2) of permeability k_2, as indicated in Fig. 6.10.

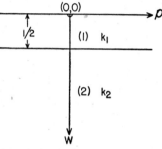

FIG. 6.10.   A diagrammatic representation of a "nonpenetrating" well tapping a stratified producing section.

The physical boundary conditions defining this particular flow system are that (1) the plane $w = 0$ be impermeable to flow, (2) the pressures be continuous as one passes across the interface between the two strata, and (3) the normal velocity be continuous at the interface.   Upon assigning potential functions Φ_1 and Φ_2 to the two strata, and expressing the vertical coordinate w (and also the radial coordinate ρ) in units of twice the upper zone thickness (cf. Fig. 6.10), these conditions can be stated analytically as

$$\left.\begin{array}{ll} \dfrac{\partial \Phi_1}{\partial w} = 0 & : \quad w = 0; \\[2mm] \dfrac{\Phi_1}{k_1} = \dfrac{\Phi_2}{k_2} & : \quad w = \dfrac{1}{2}; \\[2mm] \dfrac{\partial \Phi_1}{\partial w} = \dfrac{\partial \Phi_2}{\partial w} & : \quad w = \dfrac{1}{2}. \end{array}\right\} \qquad (1)$$

[1] Since, as seen in Chap. 3, producing strata are virtually always comprised of a multitude of layers of different permeability, the two-layer system considered here actually represents an extreme idealization of practical systems.   The purpose of this discussion, however, is only to demonstrate the rather small effect of variations in the exact value of the permeability of the strata lying below the well, and it is not intended as a treatment of the complex multilayer producing zones occurring in practice.

The development of solutions satisfying these conditions is quite similar to that involved in the problem of the preceding section, except that, because of the infinite extent of the region of interest in the present case, infinite integral syntheses of elementary solutions will be used here, in contrast to the infinite series of Sec. 6.5. Thus it is found that on introducing the notation

$$\delta = \coth \epsilon \quad : \quad \frac{k_2}{k_1} = \delta > 1,$$

$$= \tanh \epsilon \quad : \quad \frac{k_2}{k_1} = \delta < 1, \tag{2}$$

Φ_1 and Φ_2 may be expressed as follows:
For $\delta > 1$,

$$\left. \begin{aligned} \Phi_1 &= -\int_0^\infty \frac{J_0(\rho\alpha)\,\sinh\,[\epsilon + (\alpha/2) - w\alpha]}{\cosh\,[\epsilon + (\alpha/2)]}\,d\alpha, \\ \Phi_2 &= -\cosh\epsilon \int_0^\infty \frac{J_0(\rho\alpha)e^{-(w-\frac{1}{2})\alpha}}{\cosh\,[\epsilon + (\alpha/2)]}\,d\alpha. \end{aligned} \right\} \tag{3}$$

For $\delta < 1$,

$$\left. \begin{aligned} \Phi_1 &= -\int_0^\infty \frac{J_0(\rho\alpha)\,\cosh\,[\epsilon + (\alpha/2) - w\alpha]}{\sinh\,[\epsilon + (\alpha/2)]}\,d\alpha, \\ \Phi_2 &= -\sinh\epsilon \int_0^\infty \frac{J_0(\rho\alpha)e^{-(w-\frac{1}{2})\alpha}}{\sinh\,[\epsilon + (\alpha/2)]}\,d\alpha. \end{aligned} \right\} \tag{4}$$

These equations contain the whole description of the potential distribution in the system. However, we shall not enter into the details of the evaluation of the integrals[1] but shall limit the treatment to obtaining the relation between the effective resistance of the system and the permeability ratio k_2/k_1. This may be expressed in a practical form by computing the production capacity of a well for a unit pressure or potential difference over the system, as a function of k_2/k_1. In making the computations it is noted first that the potential expressions of Eqs. (3) and (4) correspond to a well production rate of $4\pi h$, where h is the upper zone thickness, and well potentials given by

$$\Phi_w = -\frac{1}{\rho_w} + 2\log\frac{2k_2}{k_1 + k_2}, \tag{5}*$$

where ρ_w is the well radius, in units of twice the upper zone thickness For the potentials at distant points, and at the top of the upper zone, an

---

[1] The complete derivation of Eqs. (3) and (4) and evaluation of the integrals are given in Muskat, *op. cit.*, Sec. 7.9.

* Terms of the order of ρ_w^2 or smaller are here neglected.

evaluation of the integrals of Eqs. (3) and (4) leads to the asymptotic expansion for Φ_1,

$$\Phi_1(w = 0) = \frac{-1}{\delta\rho}\left[1 + \frac{1}{4\rho^2\delta^2}(\delta^2 - 1) - \frac{3}{16\rho^4\delta^4}(\delta^2 - 1)(3 - 2\delta^2) + 0\left(\frac{1}{\rho^6}\right)\right]. \quad (6)$$

In evaluating these equations numerically it has been assumed that $\rho_w = 0.005$ and $\rho_e = 10$, where ρ_e is the external radius, also divided by twice the upper zone thickness, at which the potential is Φ_e. Thus, for

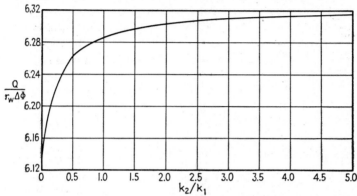

FIG. 6.11. The calculated steady-state homogeneous-fluid production capacity Q of a well just tapping a 25-ft stratum of permeability k_1, overlying an infinitely thick formation of permeability k_2. $\Delta\Phi$ = potential drop; r_w = well radius = $\frac{1}{4}$ ft; external-boundary radius = 660 ft.

example, for a well radius of $\frac{1}{4}$ ft, $\rho_w = 0.005$ corresponds to an upper stratum 25 ft thick, and $\rho_e = 10$ corresponds to an external radius of 500 ft.

The quantity $Q/r_w \Delta\Phi$,* or the flux into the well per unit well radius per unit potential drop across the pay (or unit pressure drop for a value of $k/\mu = 1$ in the upper zone), as calculated from the above expressions, is plotted against k_2/k_1 in Fig. 6.11. It will be seen from this curve that the effect of the lower zone may be considered merely as a small correction to the main flow in the upper pay. Thus, over the infinite[1] range of permeability, k_2, for the lower zone the total change in Q is only 3 per cent.

Although this result was derived on the explicit assumption that the well just taps the upper stratum, it is not difficult to see in a qualitative way that the increase in the effect of the lower zone when the well penetration is nonvanishing also will not be large. Assuming for the moment that the lower pay is of the same permeability as the upper one, we may apply the results of Secs. 5.4 and 5.5 and Fig. 5.12. These show that for a

---

* $\Delta\Phi$ is simply the difference between $\Phi_1(10)$, as given by Eq. (6), and $\Phi_w(0.005)$, as given by Eq. (5).

[1] In the limit of $k_2 \to \infty$ the ordinate of Fig. 6.11 approaches 6.32, and for $k_2/k_1 = 1$ it becomes exactly 2π, as it should for strict spherical flow (cf. Sec. 5.3).

uniform stratum with a partially penetrating well the lower layers of pay below the bottom of the well give rapidly decreasing contributions to the total flow as these layers lie at increasing depths in the sand. In fact, unless the penetration of the well in the upper zone exceeds 50 per cent and is in numerical value less than 25 ft, the addition below it of a very thick stratum of the same permeability will probably increase the total production capacity of the well by not more than 15 per cent. Returning to Fig. 6.11, it is seen that, at least for a nonpenetrating well, more than three-fourths the increase in production capacity which would result from a lower zone of infinite conductivity is already attained if the lower pay has a permeability no greater than that of the upper layer. Hence it may be concluded that the production capacity of a well partially penetrating a formation overlying and contiguous with a deeper layer will probably not be increased by the presence of the lower stratum by more than 20 per cent for penetrations up to 50 per cent, whereas in cases of small well penetrations and lower permeability deeper layers the increase will probably not exceed 10 per cent.[1]

6.7. Anisotropic Media.—As mentioned in Chap. 3, measurements on cores have shown that not infrequently the permeability normal to the bedding plane of a sand is appreciably different from that parallel to the bedding plane. In such cases the formation may be considered as an anisotropic medium with the permeability depending on the direction of flow. While this anisotropy is of no great significance in many practical problems, especially those in which the velocities are essentially confined to planes parallel to the bedding planes, it is of interest to see how the anisotropy may be taken into account when it does have to be considered.

Since the present discussion will be concerned only with the anisotropic character of the medium, it will suffice to assume here that the permeability, though different for directions of flow along the different coordinate axes, is otherwise uniform and independent of the coordinates. Upon recalling Eqs. 3.4(2), Darcy's law for a homogeneous but anisotropic medium can be written as

$$v_x = -\frac{k_x}{\mu}\frac{\partial p}{\partial x}; \qquad v_y = -\frac{k_y}{\mu}\frac{\partial p}{\partial y}; \qquad v_z = -\frac{k_z}{\mu}\frac{\partial p}{\partial z}+\frac{k_z}{\mu}\gamma g, \qquad (1)$$

where $+z$ is directed downward.

Upon assuming that the individual permeabilities k_x, k_y, k_z are uniform and applying the equation of continuity, it is found that the pressure dis-

---

[1] A close approximation to a quantitative solution when the well penetration is nonvanishing should be given, as in Sec. 5.5 for the case of a uniform stratum, by assuming a uniform flux along the well surface and then taking the well potential as that at three-fourths the depth of penetration from the top of the pay. However, for most practical purposes the qualitative results given above should suffice.

tribution $p(x,y,z)$ for steady-state homogeneous-liquid flow is no longer given by Laplace's equation but rather by

$$k_x \frac{\partial^2 p}{\partial x^2} + k_y \frac{\partial^2 p}{\partial y^2} + k_z \frac{\partial^2 p}{\partial z^2} = 0. \tag{2}$$

However, a slight change will reduce this to Laplace's equation. For by transforming the coordinate system to that of $(\bar{x}, \bar{y}, \bar{z})$, defined by

$$\bar{x} = \frac{x}{\sqrt{k_x}}; \qquad \bar{y} = \frac{y}{\sqrt{k_y}}; \qquad \bar{z} = \frac{z}{\sqrt{k_z}}, \tag{3}$$

it follows at once that

$$\frac{\partial^2 p}{\partial \bar{x}^2} + \frac{\partial^2 p}{\partial \bar{y}^2} + \frac{\partial^2 p}{\partial \bar{z}^2} = 0. \tag{4}$$

It thus appears that the effect of anisotropy in the permeability can be replaced by an equivalent shrinking or expansion of the coordinates. Hence, to find the pressure at (x,y,z) according to Eq. (2), one need only transform the boundaries by the transformation of Eq. (3), solve Laplace's equation (4) for these new boundaries, and then compute the pressure at $(x/\sqrt{k_x}, \, y/\sqrt{k_y}, \, z/\sqrt{k_z})$. Applications of this method of analysis will be made in Secs. 6.8 and 11.13.

With regard to the fluid motion in the original physical system it may be noted that in general the streamlines will not be normal to the equipressure curves. In fact the angle θ between these curves is readily seen to be given by the equation

$$\cos \theta = \frac{\bar{v} \cdot \nabla p}{|\bar{v}| \cdot |\nabla p|} = \frac{k_x(\partial p/\partial x)^2 + k_y(\partial p/\partial y)^2 + k_z(\partial p/\partial z)^2}{\mu \, |\bar{v}| |\nabla p|}. \tag{5}$$

The effective permeability along the streamline will therefore be

$$k_e = \frac{\mu \, |\bar{v}|}{|\nabla p| \cos \theta} = \frac{\bar{v}^2 \mu}{\bar{v} \cdot \nabla p} = \frac{1}{\dfrac{\cos^2 \theta_x}{k_x} + \dfrac{\cos^2 \theta_y}{k_y} + \dfrac{\cos^2 \theta_z}{k_z}}, \tag{6}$$

where θ_x, θ_y, θ_z are the angles between the vector $\bar{v}$ and the coordinate axes (x,y,z).

In solving specific problems of flow in anisotropic media, account must be taken not only of the modified form of the basic potential functions but also of the flux coefficients associated with them. Thus, whereas for two-dimensional radial isotropic flow, the fundamental pressure function corresponding to a flux Q is

$$p = \frac{\mu Q}{2\pi k} \log r, \tag{7}$$

that for a two-dimensional flow with permeabilities k_x, k_y is

$$p = \frac{\mu Q}{4\pi \sqrt{k_x k_y}} \log \left(\frac{x^2}{k_x} + \frac{y^2}{k_y} \right). \tag{8}$$

For three-dimensional anistropic flow with axial symmetry the pressure function for a point sink at $r = z = 0$, and a flux Q (positive) from the lower half space is

$$p = \frac{-\mu Q/2\pi}{k_r \sqrt{k_z[(r^2/k_r) + (z^2/k_z)]}} + \gamma gz, \tag{9}*$$

where γ is the fluid density and g the acceleration of gravity.

Finally for a general three-dimensional anisotropic flow system,

$$p = \frac{-\mu Q/2\pi}{\sqrt{k_x k_y k_z[(x^2/k_x) + (y^2/k_y) + (z^2/k_z)]}} + \gamma gz. \tag{10}$$

6.8. Partially Penetrating Wells in Anisotropic Strata.—An example of a problem which can be treated by the method of the preceding section is that of wells partially penetrating an anisotropic oil-bearing formation, producing under steady-state homogeneous-fluid-flow conditions. For this purpose it is convenient to use a potential function defined by

$$\Phi = \frac{1}{\mu}(p - \gamma gz). \tag{1}$$

The potential equation that Φ must satisfy is therefore

$$\frac{k_h}{r} \frac{\partial}{\partial r}\left(r \frac{\partial \Phi}{\partial r}\right) + k_z \frac{\partial^2 \Phi}{\partial z^2} = 0, \tag{2}$$

where k_h, k_z are the horizontal and vertical permeabilities, respectively. This can be reduced to the standard Laplace form by the transformations

$$r = r'\sqrt{\frac{k_h}{k_z}}; \quad z = z'; \quad \text{or} \quad : \quad r = r'; \quad z = z'\sqrt{\frac{k_z}{k_h}}. \tag{3}$$

By analogy with the partially penetrating well problem for isotropic media (cf. Sec. 5.4) the dimensionless variables

$$\rho' = \frac{r'}{2h'}; \quad w' = \frac{z'}{2h'}; \quad x' = \frac{b'}{2h'} = \frac{b}{2h} = x, \tag{4}$$

are introduced. In these h' and b' are the thickness of the stratum, h, and well penetration, b, as transformed by Eq. (3). In terms of these variables the boundary conditions become (cf. Fig. 6.12)

$$\frac{\partial \Phi}{\partial w'} = 0; \quad w' = 0, \frac{1}{2},$$

$$\Phi = \text{const} \quad : \quad \rho_w' = \sqrt{\frac{k_z}{k_h}} \rho_w; \quad w' \leqslant x' = x$$

$$\Phi = \text{const} \quad : \quad \rho_e' = \sqrt{\frac{k_z}{k_h}} \rho_e; \tag{5}$$

---

* It may also be inferred from Eq. (9) that such systems will be equivalent to isotropic formations with an effective permeability equal to $\sqrt{k_r k_z}$ and vertical coordinates $z' = z\sqrt{k_r/k_z}$.

where ρ_w, ρ_e are the dimensionless well and external radii in the original untransformed system. The potential problem is thus identical with that for a well with the same fractional well penetration in an isotropic formation but having well and external radii equal to $\sqrt{k_z/k_h}$ times those in the actual physical system. Accordingly the production capacities in the latter can be computed by the method of Sec. 5.5 for the isotropic system. The values so found for a stratum thickness of 125 ft are plotted in Fig. 6.13 as a function of k_z/k_h, and for $k_h \, \Delta p/\mu\beta = 1$, with units of millidarcys for k_h, psi for Δp, and centipoises for μ.*

The extreme values of $k_z/k_h = 1$ and $k_z = 0$ correspond, respectively, to the case of an isotropic sand, as treated in Secs. 5.4 and 5.5, and to the case of strict radial flow confined to the part of the sand actually penetrated by the well.[1] Although the effect of the variation in the vertical permeability may not appear large from Fig. 6.13, a comparison of the ratios of the extreme values of the production capacities, for the isotropic system and that for zero vertical permeability, as may be read from Fig. 6.13 or 5.11, shows that the addition to the flux due to the vertical

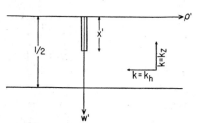

FIG. 6.12. A diagrammatic representation, in the transformed dimensionless coordinate system, of a partially penetrating well in an anisotropic sand.

flow is after all quite considerable, especially for the smaller well penetrations. Thus, for a well penetration of 20 per cent the addition to the strictly radial flow is 55 per cent on introducing an average—as a "series" resultant—vertical permeability equal to the horizontal permeability; and for a 10 per cent well penetration the addition is approximately 80 per cent.

It is to be noted, however, that the increase in production capacity due to the vertical permeability is not simply proportional to k_z/k_h. If that were so the curves of Fig. 6.13 would be straight lines connecting the terminal points. Rather the increase in capacity with k_z/k_h sets in rapidly even for small k_z/k_h and changes but slowly for $k_z/k_h > 0.1$. This result is of practical significance in that it shows that, as long as k_z/k_h is 0.1 or greater, it need not be determined with great precision, while still permitting a reasonably accurate estimate of the value of the flux into the well. On the other hand, unless k_z is an appreciable fraction of the horizontal permeability, the anisotropy[2] of the sand may cause a very material

---

* Further details of the analysis are given in Muskat, *op. cit.*, Sec. 5.5.

[1] When $k_z = 0$, the boundary conditions of Eq. (5) become meaningless owing to the breakdown in the transformations of Eq. (3). This case, however, can be treated directly as a simple radial-flow system.

[2] If there should be a continuous zone, as a shale break, of effectively zero permeability lying below the bottom of the well, the pay below it will be completely isolated without drainage, no matter how thin the barrier may be.

diminution in the rate at which oil enters a well that only partly penetrates the anisotropic sand, and hence it should be taken into account whenever possible.

6.9. Summary.—The virtually infinitesimal sampling of underground strata provided by cores several inches in diameter taken at lateral separations of several hundred feet can hardly suffice to establish the details of permeability variations, if any, between producing wells. On the other

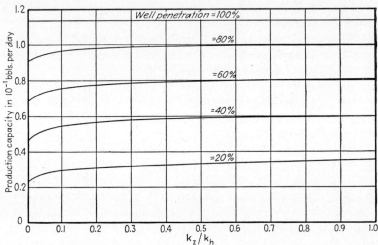

Fig. 6.13. The calculated steady-state homogeneous-fluid production capacities of partially penetrating wells as functions of the permeability ratio k_z/k_h = (vertical permeability)/(horizontal permeability). Pay thickness = 125 ft; well radius = $\frac{1}{4}$ ft; external-boundary radius = 660 ft; $k_h \Delta p/\mu\beta$ is taken as 1, with k in millidarcys, Δp in psi, and μ in centipoises.

hand the very fact that the permeabilities of cores at equivalent stratigraphic depths generally do vary from well to well and sometimes can be correlated in contour form shows that an assumption of strict uniformity may have as little basis as any assumed type of variation. Although there is no means now available for a satisfactory resolution of this dilemma, it is possible to estimate the range of effects of permeability variations by a study of specific types of such variations.

While it is possible to solve certain problems with continuous permeability distributions, it is generally simpler to visualize the flow system as being comprised of adjacent parts of uniform but different permeabilities. A particular case of this type is that in which the liquid-bearing rock may be considered as formed by two adjacent concentric annular regions of different permeabilities. Such a system may represent a well which has been drilled into a region having a permeability higher or lower than that of the producing formation as a whole. It will also correspond to a well that was initially drilled into a homogeneous stratum, the inhomogeneity having been caused by a partial plugging or mudding off of the region immediately adjacent to the well bore during the course of production or drilling, or conversely by an increase in the permeability about the well bore as the result of acid treatment. The analysis for this problem shows that, as is to be expected in view of the highly localized character of the pressure drop in a radial-flow system about the well center (cf. Fig. 6.2), the production capacity of a well is quite sensitive to the value of the permeability of the zone immediately surrounding the well bore. Thus if the annular zone adjacent to the well bore has a permeability 2.5 times that of the remainder of the rock and is only 5 ft in radius, occupying less than 0.01 per cent of the total volume of the system, the production rates for given total pressure differentials will be 29 per cent higher than if the permeability were uniform. Likewise, if this zone has a permeability one-fourth that of the main body of rock, the production capacity of the well will be reduced to only 48 per cent of its normal value. On the other hand, these effects do not increase in proportion to the radius of the zone of abnormal permeability; rather, the major effect is caused by the first few feet about the well bore, and additions to the zone give successively smaller changes in the production capacity. Similarly for zones with relatively high permeabilities, the increase in production capacity (decrease in total resistance of the system) as the permeability of the inner zone is increased rapidly approaches the limiting value corresponding to that of a well in which the inner zone has been removed entirely, and hence which has a radius equal to that of the inner zone (cf. Fig. 6.3).

These results provide a possible explanation for the large variations in the production capacities that are frequently observed for neighboring and apparently identical wells. For if the formation is not strictly homogeneous in detail, it is not unlikely that neighboring wells may penetrate zones of appreciably different local permeabilities and hence show markedly different production capacities, even though the producing stratum as a whole may be considered as homogeneous.

Another problem involving systems composed of regions of different permeability arises in the study of the flow of fluids in limestone or dolomite reservoirs. Except when oölitic or cavernous, the limestone or dolomite rock itself is usually of very low permeability, and high production capacities of wells penetrating such tight reservoir formations must be attributed to the fractures and crevices that permeate and are disseminated through the limestone or dolomite proper. When such fractures are of limited extent and uniformly distributed through the pay, they will give a resultant effect equivalent to that of a homogeneous porous medium. However, when they are of extended length and limited in number, they may be considered separately as linear channels that are fed laterally with the fluid issuing from the main body of rock. The fractures themselves may then be represented as distinct zones of the porous medium with permeabilities equal to the effective permeability of a linear free channel carrying a liquid under viscous-flow conditions. From classical hydrodynamics it follows that a linear free channel of width w (in centimeters) has an effective permeability of $10^8 w^2/12$ darcys.

With this representation for a limestone-fracture system it is possible to calculate the inherent production capacities of wells penetrating such formations. It is thus found that these production capacities are very sensitive to the fracture width. For example, an increase of the fracture width from 0.1 to 0.3 mm will quadruple the capacity. Moreover for fracture widths greater than 0.22 mm a single fracture, assuming that it extends for at least 300 ft, will alone have half as much flow capacity as the composite limestone-fracture system (cf. Fig. 6.5).

These analyses of formations with radial permeability variations and the limestone-fracture systems may be applied in interpreting the effect of acid treatment of oil wells producing from limestone or dolomite reservoirs. Thus, if the flow into the well is essentially radial, owing to an approximately uniform distribution of fractures of limited extent or to an absence of fractures, and the pay immediately surrounding the well bore is of normal permeability as compared with the main body of rock, an increase of the permeability in the zone adjacent to the well bore due to the introduction of acid will give a relatively small increase in the production capacity of the well. If, however, the well is plugged or if the rock immediately surrounding the well bore is of an abnormally low permeability, the introduction of acid should be considerably more effective in increasing the production capacity of the well. If, for example, the zone about the well bore has a permeability one-tenth that of the main body of rock and is raised to that of the latter by the acid, the production capacity of the well will be increased by 70 per cent even if the acid-affected zone is only 3 in. thick, and by as much as 345 per cent, if the effect of the acid penetrates to a radius of 5 ft (cf. Fig. 6.6).

Still larger effects of acid treatment will result if the limestone or dolomite is permeated by extended fractures that are penetrated and widened by the acid. Thus if a limestone of permeability 10 md is cut by only a single extended fracture 0.1 mm in width, a 0.5-mm addition in width will result in a 2,600 per cent increase in production capacity (cf. Fig. 6.7). Furthermore, smaller wells will in general show larger responses than good producers, a result often verified by field experience. Thus the theoretical analysis of the flow in porous media of nonuniform permeability is fully capable of explaining the wide range of field observations as to the effects of acid treatment in increasing the production capacities of wells producing from limestone or dolomite reservoirs, simply on the basis of the differences in detail of the mechanism of production among individual wells. In fact the theory indicates that, since fractured limestones or dolomites will show greater responses to acid treatment than such media as will give an essentially radial flow, "shooting" wells penetrating reservoirs of the latter type should be an effective preliminary to acid treatment.

Another practical problem involving a system composed of parts of different permeability is that of the effect of a sanded liner on the production capacity of a well. For although one would hardly expect that a column of clean sand at the bottom of a well bore would appreciably reduce the production capacity of the well, it is a common field observation that both flowing and pumping wells suffer very considerable decreases in production rate upon the entry of sand into the well bore. A closer analysis of this question, however, fully explains the observed effects. For as all the production from the well must pass through the narrow well bore, the pressure gradients in the latter, considered as a porous medium, will be very high and will be very sensitive to its effective permeability. In fact, even if the sand-column permeability is 100 times as great as that of the main sand body, 87 per cent of the total pressure drop in the system will take place in simply passing along the well bore from the top to the bottom of the pay. This situation is evidently equivalent to one in which a high back pressure is imposed on the lower parts of the sand, thus cutting down the flow coming from these parts. The effective well penetration is thus reduced, and with it the associated production capacity of the system. In particular the analysis shows that if the sand column extends to the top of a pay of 25 ft thickness and has a permeability 100 times as great as that of the pay, the production capacity will be cut to 26 per cent of that of the same system with a sand-free well bore (cf. Fig. 6.9). Furthermore these effects are greatly accentuated if the sand column extends beyond the top of the pay. An additional sand-column height of only 5 ft above the top of the pay will cut the production capacity of the above system to only 10.9 per cent of that with a sand-free well bore.

Analyses can also be made of systems in which the nonuniformity is due to a vertical stratification in the producing section. If the wells completely penetrate the full section, the group of different strata, under steady-state conditions, may be simply represented as a "parallel" superposition of the individual uniform strata. For wells that penetrate only the uppermost layers of the composite section the three-dimensional character of the flow must be taken into account. However, when this is done, it is found that the contribution of the nonpenetrated strata to the production capacity will be quite small unless the well has a high penetration in the upper zones.

The assumption of complete isotropy of oil-producing strata is as much an idealization as that of strict permeability uniformity. While there is only meager evidence that permeabilities vary in different directions in the planes of bedding, core analyses show that exact equality between the permeability in the bedding plane and normal to it is the exception rather than the rule. Fortunately, however, when the flow is two-dimensional in the bedding plane, as in the case of completely penetrating wells, the permeability normal to that plane does not enter the problem, whether or not the medium is isotropic. On the other hand this permeability must be taken into account when there is a flow component normal to the bedding planes. In such circumstances the different terms in the differential equation for the pressure distribution must be given coefficients equal to the respective permeabilities, rather than the value unity as in the case of isotropic systems. This complication can be removed by simple transformations of the variables. While the flux coefficients in the pressure-distribution functions will be modified, the formal analyses in the transformed system can be carried through as if it were isotropic.

An illustration of this type of treatment is provided by the problem of determining the production capacity of partially penetrating wells in anisotropic strata. By following the transformation procedure indicated above, it is found that the potential distribution in the anisotropic system will be the same as for an isotropic-medium problem with the same fractional well penetration, but with the dimensionless radial coordinates, including the well and external radii, equal to those in the original system multiplied by the square root of the ratio of vertical to horizontal permeability. The result of the analysis is, as would be expected, that the effect of the vertical permeability depends on the extent to which the vertical-flow component normally contributes to the production capacity. Thus in the case of a well penetration of 80 per cent in a 125-ft formation, the isotropic-stratum production capacity will be reduced by 9.3 per cent if the vertical permeability is made vanishing, whereas the reduction for a 20 per cent penetration well will be 35 per cent (cf. Fig. 6.13). Another important problem involving anisotropic media will be considered in Sec. 11.13.

CHAPTER 7

DYNAMICAL FOUNDATIONS FOR HETEROGENEOUS-FLUID-FLOW PHENOMENA

7.1. The Generalized Permeability Concept.—The treatments of specific problems in the previous chapters have been based on the assumption of homogeneous-fluid flow. This assumption will be generally satisfied when the oil is undersaturated or is produced without gas evolution within the pay. Such producing conditions do occur, as, for example, in the East Texas field, some of the reservoirs in the Thompson field, Texas, and many of the Kansas fields producing from the Arbuckle Limestone. Accordingly their quantitative description[1] can be made in terms of the homogeneous-fluid theory. Similarly the production characteristics of gas and condensate[2] fields can be quantitatively interpreted in terms of the previously developed homogeneous-fluid concepts and analytical methods.

In the majority of oil-producing fields thus far discovered, however, both gas and oil flow through the porous formations at some period, at least, during the course of production. In these reservoirs the flow of the oil and its expulsion from the oil pay are intimately associated with the presence and flow of free gas in the pores of the rock. It is true that on lowering the pressure at the exposed face of the rock, by opening the well to production, pressure gradients are created in the formation which will directly induce flow of oil toward the well bore. If, however, there were no gas evolution in the pay or a bounding mobile water body free to replace the oil leaving the formation, both the reservoir pressure and production rate would drop with extreme rapidity. For the total production would be merely that due to expansion of the oil and connate water as a result of the pressure decline. As the compressibility of crude oils is of the order of 10^{-4} per atmosphere, a reservoir at an initial pressure of 200 atm ($\sim 3,000$ psi) would yield only about 2 per cent of the volume of the oil[3] initially in place by the time the pressure is depleted, assuming that the

---

[1] An example of such an analysis, as applied to the East Texas field, will be given in Sec. 11.9.

[2] As will be seen in Chap. 13, the theory of cycling patterns in condensate reservoirs may be based on the homogeneous-fluid theory.

[3] The connate-water expansion will generally provide an expulsion of the order of 10 per cent that due to the oil.

oil contains no dissolved gas whatever. This would imply a recovery of only 31.0 bbl/acre-ft of a 25 per cent porosity pay having a connate-water saturation of 20 per cent.

It is clear, therefore, that recoveries of 200 to 600 bbl/acre ft, such as are obtained in practice, cannot be explained simply as the result only of the direct compressibility expansion of the reservoir oil. The production of large volumes of gas with the oil, generally in excess of that currently dissolved in the oil, except in complete water-drive reservoirs (cf. Chap. 9), also makes it obvious that the natural gas will often play an important role in the production. Moreover, when the oil is initially saturated with gas at the reservoir pressure, the gas must of necessity come out of solution within the formation as the pressure falls, if equilibrium behavior is to prevail.[1]

The coexistence of two phases in a porous medium, such as gas and oil or oil and water, does not of itself invalidate the concept of homogeneous-fluid flow. As previously indicated, formations producing at pressures above the saturation pressure of the oil may generally be treated as homogeneous-fluid systems even though they may contain 10 to 30 per cent of connate water. The same is true of free-gas-producing formations. The reason is that in these cases the connate water is *immobile*, within the oil-productive area, so as to leave only one mobile phase in the system.

While the formal analytical structure describing the motion of the single mobile phase remains the same as if the immobile phase were not present at all, there is one basic numerical factor that enters the problem because of the presence of the immobile phase. This is a change in the permeability of the porous medium to the mobile phase. It will be recalled that the permeability, as considered thus far, has referred to measurements or conditions of flow in which the flowing phase occupies the whole of the void space represented by the effective porosity of the medium. If part of this space were to be occupied by another phase, it is clear that the resistance to flow of the mobile phase would be increased. That is, the permeability to this fluid would be less than if it alone filled the whole of the void space of the rock. The magnitude of the reduction in the homogeneous-fluid permeability will evidently depend on the amount of the immobile phase present. Moreover, it will be different if the immobile phase wets the internal solid boundaries of the rock and hence tends to concentrate in the smaller pores and sharp-angled recesses of the rock than if it is non-

---

[1] While there occasionally have been reports of evidence indicating the occurrence of supersaturation within oil-bearing reservoirs, such evidence presumably refers only to a lag in equalization between the solution and free-gas-phase pressures, rather than a permanent and complete absence of gas evolution. On the other hand, there are conditions under which supersaturation will occur in laboratory flow experimentation.

wetting and is distributed in separated elements occupying the central regions of the individual pores.

In either case, when the producing rock contains more than a single fluid phase, the permeability concept, as heretofore developed, must be generalized. It can no longer be considered as an invariable quantity, uniquely and completely fixed by the nature and structure of the rock. Cognizance must be taken of the fact that its magnitude, when referring to the mobile phase, will be affected by the presence of other fluids in the pores of the rock, even though they may remain immobile. And it is to be anticipated that the effect of the immobile phase will vary with its nature, distribution, and the amount present.

These same considerations suggest the further generalization that when more than one fluid phase is present in a porous medium, the term "permeability" must be specifically associated with the individual phases involved. The rock itself will still have a permeability referring to its flow capacity for a single-phase, or homogeneous, fluid. And, as before, this will be independent of the nature of the fluid, as long as the latter does not interact with the porous medium. However, in referring to the composite system of the porous medium and its fluids the flow capacity must be expressed in terms of permeabilities to the separate fluid phases present. Their absolute magnitudes may be termed "effective" permeabilities. Or, more conveniently, they may be expressed as fractions of the homogeneous-fluid permeability, *i.e.*, as "relative" permeabilities. For example, if a rock of 500-md homogeneous-fluid permeability containing 20 per cent connate water produces "clean" (water-free) oil at a rate corresponding to a permeability of 400 md, the situation may be described by saying that the *relative* permeability of the water is zero and that of the oil is 80 per cent or that the effective permeability to the water is zero and that to the oil 400 md. Likewise, if the 500-md pay is producing both free gas and oil and if the permeabilities as calculated for each phase separately, as if each alone were flowing through the rock, are 200 md for the gas and 50 md for the oil, the results would be expressed as 40 per cent relative permeability for the gas and 10 per cent relative permeability for the oil.

An alternative but equivalent physical representation of these basic phenomena may be constructed by considering the porous medium itself to have a *local* structure determined by the fluid-saturation distribution, which is superposed on the grain structure. While the latter is represented by the homogeneous-fluid permeability, the former determines the permeabilities to the individual phases of a multiphase fluid. That is, the porous medium may be assigned a set of local permeabilities varying from point to point, and with time, in accordance with the variations in the local volumetric fluid distribution. If the latter be constant throughout the medium, the flow system may be treated simply as a superposition "in parallel" of individual homogeneous-flow systems, with permeabilities for the separate phases reduced from the homogeneous-fluid value by constant factors. From this point of view it appears that the unique feature requiring a generalized treatment of multiphase-flow systems is the *variability* in the phase distributions as the fluid progresses along its macroscopic streamlines. The local permeabilities for the individual phases then become variable, even though the homogeneous-fluid permeability is strictly uniform, and their variation, together with that in the phase saturations, must be determined simultaneously with the pressure and velocity distributions in the system.

From either point of view, it follows that as the carrier of a heterogeneous fluid a porous medium is characterized by a set of relations—curves or equations—between the local permeabilities for the individual fluid phases and the local saturation distribution. These have been termed the "permeability-saturation relationship." Their specific analytical significance is that they provide the coefficients, now variable, in the *set* of Darcy equations[1]

$$\left.\begin{aligned}
\bar{v}_o &= -\frac{k_o}{\mu_o}\nabla(p - \gamma_o gz), \\
\bar{v}_g &= -\frac{k_g}{\mu_g}\nabla(p - \gamma_o gz), \\
\bar{v}_w &= -\frac{k_w}{\mu_w}\nabla(p - \gamma_w gz),
\end{aligned}\right\} \tag{1}$$

where the subscripts o, g, w refer to oil, gas, and water, $\bar{v}$ is the vector velocity (volume flux per unit gross area), k the permeability, μ the viscosity, p the pressure, γ the fluid density, g the acceleration of gravity, and z the vertical coordinate (directed downward). It is the variation of k_o, k_g, k_w with the fluid distribution that is the key to the behavior of heterogeneous-fluid systems. It will be instructive to consider further these relationships before proceeding with the formulation of the final hydrodynamic equations.

7.2. The Permeability-Saturation Relationship for Two-phase Systems; Gas-Liquid Mixtures.—The first experimental study of two-phase systems from the point of view discussed in the preceding section, so as to obtain explicitly permeability vs. saturation curves for both phases, was that of

[1] For simplicity the capillary pressures associated with the curvatures of the fluid interfaces, which in turn may be related to the fluid saturations, are neglected here. They will be discussed in Secs. 7.8 to 7.10.

Wyckoff and Botset.[1] These experiments were made with a 10-ft-long tube packed with unconsolidated sand. Four sands were studied, varying in permeability from 17.8 to 262 darcys. The fluids used were water and carbon dioxide. Sodium chloride, NaCl, was added to the water to make it slightly conducting. Cylindrical electrodes spaced along the bakelite flow tube permitted measurement of the electrical conductivity along the tube. On the basis of previous calibration experiments these conductivities were translated into equivalent liquid saturations. For the study of the region of high liquid saturation and low gas-liquid ratios the inflowing water was saturated with carbon dioxide, CO_2, and the latter was allowed to evolve within the sand by virtue of the pressure drop. For higher gas-liquid ratios and lower liquid saturations, additional free gas, CO_2, was injected at the inflow end of the tube. The pressure distribution along the tube was determined by means of piezometer rings, located at and combined with the electrodes, connected to manometers. From simultaneous measurements of the conductivity and pressure drop across several (eight) interior sections of the tube, together with observations on the flow rates, the permeabilities were calculated and plotted against the corresponding liquid saturations.

The data obtained on the four sands are plotted in Fig. 7.1. The multiplicity and distribution of the points show at the same time the natural dispersion of the data and the basic equivalence of the results for the different sands. While it will be presently seen from additional data that the average curves of Fig. 7.1 are not at all universally applicable to other porous media and fluids from a quantitative point of view, their significant qualitative features are common to all data of this type thus far obtained in laboratory[2] studies. These are (1) a rapid fall in the permeability to the liquid as the liquid saturation first decreases from 100 per cent, (2) its approach to zero at saturations appreciably greater than zero, (3) the rapid rise of the permeability to the gas as the liquid saturation decreases, and (4) the virtual attainment of 100 per cent permeability to gas before all the liquid is completely removed. Still another very important feature of the data shown in Fig. 7.1 is that the permeability to the gas apparently does not rise appreciably above zero until the liquid saturation has fallen to approximately 90 per cent. This, however, will be discussed in detail

---

[1] R. D. Wyckoff and H. G. Botset, *Physics*, **7**, 325 (1936). While not emphasizing the multiphase features of simultaneous wetting- and nonwetting-phase flow, the basic variability of the permeability of the wetting phase with its saturation in a porous medium was recognized and experimentally determined for two soils and a ceramic clay 5 years earlier by L. A. Richards [*Physics*, **1**, 318 (1931)].

[2] Related data derived from field observations and referring, in particular, to the ratio of the gas to oil permeabilities will be discussed in Sec. 10.12.

separately later, as will the physical interpretation and significance of the other four characteristics of Fig. 7.1.

Figure 7.1, as well as the similar curves to follow, show the value and significance of expressing the permeabilities of the individual phases in terms of *relative* permeabilities, *i.e.*, as ratios to the homogeneous-fluid value. In this way a spread in absolute permeabilities of a factor of 15 is

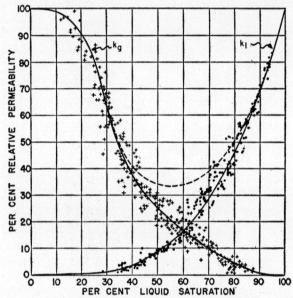

FIG. 7.1. Permeability-saturation data for four sets of unconsolidated sands. k_l, k_g refer to liquid and gas phases. The dashed curve represents the sum of the gas and liquid permeabilities. (*After Wyckoff and Botset, Physics, 1936.*)

reduced to a dispersion that is hardly greater than the inherent experimental errors of the measurements. For this set of unconsolidated sands the relative or percentage effect of the phase distribution is thus seen to be independent of the absolute permeability.[1] That is, for example, a gas saturation of 18 per cent will cut the liquid permeability in approximately half whether the homogeneous-fluid permeability is 17.8 or 262 darcys. Likewise, a 35 per cent liquid saturation will reduce the gas permeability to approximately half its homogeneous-fluid value even though the latter varies from 17.8 to 262 darcys. Hence the variations in the relative-permeability curves that are established as real must reflect such differences

[1] While the expression of the effective permeabilities to the individual fluid phases relative to the homogeneous-fluid (air) permeability will be used hereafter in this work, as is now common practice in the oil industry, the proper base to be used in dirty sands or other rocks that are affected by the fluids carried is still uncertain. Fortunately, however, it always suffices for actual applications to use the effective permeabilities.

in the structure of the porous medium as grain-size distribution, grain-shape factors, and the character and degree of cementation, which may have a greater effect on the multiphase permeabilities than on the gross resultant homogeneous-fluid permeability.

By using techniques very similar to that described above for unconsolidated sands the results shown in Fig. 7.2 have been obtained[1] for a Nichols

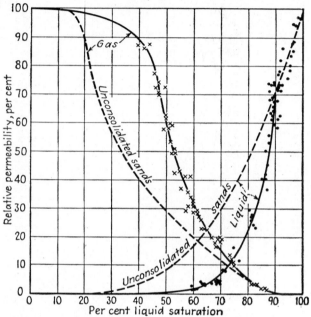

Fig. 7.2. Permeability-saturation data for a consolidated sand and the average curves for a 17.8-darcy unconsolidated sand. (*After Botset, AIME Trans., 1940.*)

Buff Sandstone column, 4.5 ft long and 4 in. in diameter. Its permeability —quite uniform along its length—averaged 495 md, and its porosity was 21.8 per cent. Again, the gas used was CO_2, and the liquid was water, made conducting by the addition of K_2SO_4. For ready comparison with the corresponding data for unconsolidated sands the average curves for the unconsolidated sand of 17.8 darcys are also plotted in this figure. It will be observed that while there are qualitative similarities between the curves and complete overlapping for liquid saturations above 90 per cent, the quantitative differences are readily apparent. Thus the liquid-permeability curve drops more sharply, as the liquid saturation decreases, for the consolidated sand. At the same time the gas permeability rises more steeply and approaches the homogeneous-fluid value more rapidly as the gas saturation increases.

[1] H. G. Botset, *AIME Trans.*, **136**, 91 (1940).

In these experiments both the liquid and gas phases were continually flowed through the porous medium, so as to achieve steady-state conditions. In other investigations,[1] however, air was flowed through Bradford Sand cores containing "dead" (gas-free) oil, and the permeability to the air was

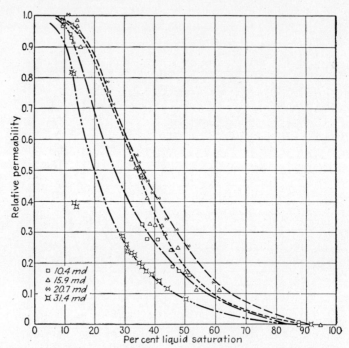

Fig. 7.3. Data on the relative permeability to air in four cores of Bradford Sandstone. (*After Hassler, Rice, and Leeman, AIME Trans., 1936.*)

observed as the oil was swept out of the cores. The oil saturation was measured by removing the core from its holder and weighing, at various stages of the experiment. The behavior of the liquid phase was not determined directly as a permeability but expressed in terms of oil-air ratios obtained from the rate of oil depletion in the cores. Measurements were made with six cores varying in permeability from 8.3 to 31.4 md and five oils ranging in viscosity from 4.4 to 66 cp. For the individual oils there was no systematic variation of the curves with the permeability of the cores. As a whole the relative permeability to the gas was found to be lower with the lower viscosity oil, though this trend could not be expressed by a simple factor varying only with the viscosity. In all cases the curves lay below and to the left of the consolidated sand curve for the gas phase shown in Fig. 7.2. The data on four cores for the oil of 8.66 cp viscosity

[1] G. L. Hassler, R. R. Rice, and E. H. Leeman, *AIME Trans.*, **118**, 116 (1936); cf. also G. L. Hassler, *Pennsylvania State Coll. Min. Ind. Bull.* 20, p. 19 (1936).

are plotted in Fig. 7.3. It will be noted that the curves of Fig. 7.3 more nearly approximate those obtained on long columns of unconsolidated sands (Fig. 7.1) than the long consolidated core (Fig. 7.2).[1]

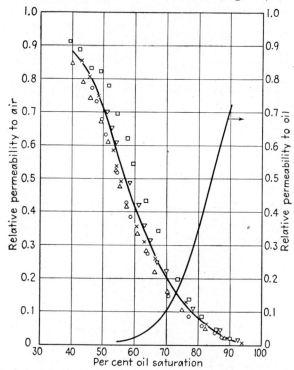

FIG. 7.4. Data on the relative permeability to air measured in a long 522-md sandstone core at different pressure gradients. Input pressures in centimeters mercury were ∇, 10; o, 20; x and Δ, 30; □, 40. (*After Krutter and Day, Petroleum Technology, 1943.*)

An explicit verification of Darcy's law, at least for the gas phase, was also reported in this study, by showing that the variation of the rate of flow with the pressure gradient was linear for a fixed oil saturation. The latter, 23.4 per cent, was so low that the oil permeability was probably zero, so that the gas flowed essentially as a homogeneous fluid. However, additional data showed the complete gas permeability vs. saturation curve, as determined with a pressure differential of 6.7 psi, to be identical, within experimental errors, with that obtained using a 42-psi differential.[2]

---

[1] The early relative-permeability data on short cores, as shown in Figs. 7.3 and 7.5, which were obtained by continuous liquid-phase desaturation, are of doubtful quantitative significance because no attempts were made to eliminate "end effects." These may be especially serious at high wetting-phase saturations and high pressure gradients.

[2] On the other hand recent experiments by J. H. Henderson and S. T. Yuster [*Producers Monthly*, **12**, 17 (January, 1948)] indicated the permeability to oil in Bradford cores flowing both brine and oil to be appreciably higher at 42 psi differentials than at 21 psi.

Similar experimental data have been reported[1] on long consolidated sandstone cores, in which the liquid phase, oil, was not fed into the flow system together with the gas (air) but rather was continuously displaced from the core by the air drive. Thus only the gas permeability was measured directly. That for the oil was computed from the observed air-oil ratio in the outflow stream. The results obtained for a core of 522 md average permeability are reproduced in Fig. 7.4. The various types of points refer to experiments with different input air pressures. While there appears to be some separation between the several sets of data, the correlation with the input pressure is not monotonic and perhaps of no real significance. In any case a parallel variation was observed in the air-oil ratios, so that the computed oil permeability was found to be essentially independent of the input pressure. It is therefore indicated as a single curve in Fig. 7.4.

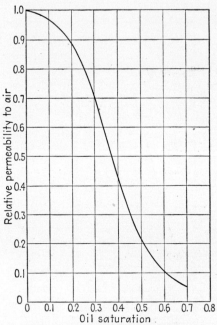

FIG. 7.5. The relative permeability to air, averaged for 36 core samples, as a function of the fractional oil saturation. (*After Krutter, Producers Monthly, 1941.*)

In further experiments on small cores,[2] in which air was injected into oil-saturated cores and the permeability to the air was measured as a function of the residual-oil saturation, no systematic or significant variations in the permeabilities were observed in changing the pressure drop over the core over the range of 5 to 30 cm Hg. Oils ranging in viscosity from 2 to 100 cp at 25°C gave substantially the same permeability data. The permeability of the cores studied ranged from 46 to 1,180 md. An average curve obtained from data in 36 experiments is shown in Fig. 7.5.

The effect of the liquid viscosity was also studied in the first reported permeability-saturation investigation[3] by adding sugar solution to the water to raise its viscosity to 3.4 cp. Within experimental errors the permeability curve for the liquid was identical with that for the original water of 0.9 cp viscosity. Still other experiments, using CO_2 and alcohol, with a

[1] H. Krutter and R. J. Day, *Petroleum Technology* **6**, 1 (November, 1943).

[2] H. Krutter, *Producers Monthly*, **6**, 25 (June, 1941).

[3] Wyckoff and Botset, *loc. cit.*

surface tension of 27 dynes/cm, gave results agreeing, within the experimental errors, with those for CO_2 and water (the surface tension is equal to 72 dynes/cm).

Of interest with respect to the behavior of limestones as carriers of heterogeneous fluids are the curves shown in Fig. 7.6. These represent the average relative permeabilities for 26 cores from three Permian Dolomites in West Texas.[1] While core measurements on small samples from fissured

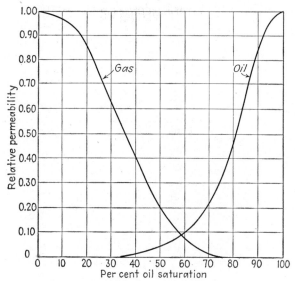

Fig. 7.6. Average relative permeabilities to gas and oil for 26 cores from three Permian Dolomites in West Texas. (*After Bulnes and Fitting, AIME Trans., 1945.*)

or cavernous limestones give highly erratic results and are virtually meaningless, the data plotted in Fig. 7.6 show that intergranular limestones of uniform texture possess permeability-saturation relationships quite similar to those of sands and sandstones.

These results comprise most of the published laboratory-determined information on gas-liquid-mixture flow that includes data on the permeabilities of the individual phases.[2] A number of experimental studies have been

---

[1] A. C. Bulnes and R. U. Fitting, Jr., *AIME Trans.*, **160**, 179 (1945).

[2] Several of the papers cited in this and the following two sections include still other data in numerical or graphical form. However, the figures reproduced here give the typical as well as the most important empirical results. A number of additional results have been recently reported by R. A. Morse, P. L. Terwilliger, and S. T. Yuster, *Producers Monthly*, **11**, 19 (August, 1947), and Henderson and Yuster, *op. cit.*, p. 130. An example of these (Fig. 7.12) will be discussed in Sec. 7.5.

reported on the general behavior of two-phase mixtures,[1] and these could be interpreted in terms of the permeability-saturation characteristics of the systems investigated. However, as the latter were not determined directly, they will not be discussed further here.

7.3. Permeability-Saturation Data for Two-phase Systems; Immiscible Liquids.—Several studies have been reported on the flow of two immiscible-liquid phases, oils and water, through unconsolidated sands.[2] These include a systematic investigation[3] of the effects of the liquid viscosities, pressure gradients, and interfacial tension. The viscosity of the oils used varied from 0.31 to 76.5 cp. In one set of experiments, by the addition of glycerol, the viscosity of the water was increased to 32.2 cp. By suitable combinations of these liquids the ratio of oil to water viscosity ranged from 90.0 to 0.057. The unconsolidated sands were of 40 to 42 per cent porosity and of permeability 3.2 to 6.8 darcys. They were initially saturated with water and subsequently subjected to streams of oil-water mixtures of constant composition.

The results are replotted in Fig. 7.7 for the four oil-water mixtures of different viscosity ratios. It will be seen that in comparison with the natural spread of the data there is no significant variation of the permeabilities with the relative viscosities of the two liquids. The striking feature of these curves is their marked similarity to those obtained for gas-liquid mixtures, illustrated in the last section. The significance and physical interpretation of this basic fact will be discussed in Sec. 7.5.

The variation of the permeability measurements with the pressure gradient was negligible at the higher pressure gradients (of the order of 1 cm Hg per centimeter), but at gradients of the order of 0.1 cm Hg per centimeter the permeabilities were somewhat lower. The interfacial tension was reduced from the values of 24 to 34 dynes/cm for the previously used oil-water systems to 5 dynes/cm by replacing the oil with amyl alcohol. The relative permeabilities for the latter were higher than those shown in

---

[1] Among these are W. F. Cloud, *AIME Trans.*, **86**, 337 (1930); L. C. Uren and E. J. Bradshaw, *AIME Trans.*, **98**, 438 (1932); L. C. Uren and M. Domerecq, *AIME Trans.*, **114**, 25 (1937); I. I. Gardescu, *Petroleum Engr.*, **4** (November, December, 1932, January, February, 1933); F. B. Plummer, J. C. Hunter, Jr., and E. H. Timmerman, *API Drilling and Production Practice*, 1937, p. 417; L. S. Reid, and R. L. Huntington, *AIME Trans.*, **127**, 226 (1938).

[2] In reporting on multiphase steady-state flow of relative-permeability measurements, illustrative examples have recently been given of the permeability-saturation curves for the flow of oil-brine mixtures in consolidated sands and synthetic cores [cf. Morse, Terwilliger, and Yuster, *loc. cit.*; Henderson and Yuster, *op. cit.*, p. 13; and J. H. Henderson and A. H. Meldrum, *Producers Monthly*, **13**, 12 (March, 1949)]. Examples from the first paper will be discussed in Sec. 7.5.

[3] M. C. Leverett, *AIME Trans.*, **132**, 149 (1939).

Fig. 7.7. The difference increased with the relative permeability for either phase, being approximately 0.1 at relative permeabilities of 0.6 for the normal oil-water mixtures. On varying the grain-size distribution of the sand a small but rather systematic change was observed in the position of the permeability curves. The effects of these several factors were correlated[1] with the dimensionless quantity $(P_d/D)/(dp/dx)$, where dp/dx is the

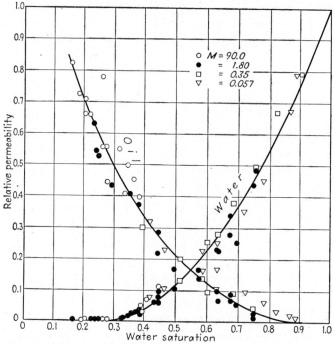

FIG. 7.7. The effect of the viscosity ratio M (oil to water) on the relative permeabilities in a 100- to 200-mesh sand. (*After Leverett, AIME Trans., 1939.*)

pressure gradient, P_d the "displacement pressure" (cf. Sec. 7.8), and D the average pore diameter. The quantity $(P_d/D)/(dp/dx)$ may be shown to be a measure of the ratio of the capillary interfacial forces opposing the fluid motion to the fluid pressure gradient driving the mixture through the sand.

A few experiments on oil-water mixtures, similar to those discussed above, were reported with the first work done on gas-liquid-mixture flow.[2] When the unconsolidated sand columns were originally wet and saturated with water, the results were similar to those indicated in Fig. 7.7, although

---

[1] While doubt has been cast on the reality of these effects and correlations, since "end effects" were not completely eliminated in the experiments [cf. M. C. Leverett, *AIME Trans.*, **142**, 152 (1941)], they are nevertheless of interest in illustrating the physical factors that may affect the permeability-saturation relationship.

[2] Wyckoff and Botset, *loc. cit.*

only the permeability to the water was plotted. However, when flow of a water-oil mixture through an oil-wet and oil-saturated sand was attempted, the water quickly channeled through even at very low water saturations and consistent data could not be obtained.

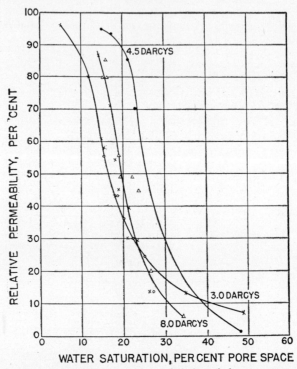

Fig. 7.8. The relative permeability to oil, as a function of the water saturation, in three unconsolidated sands. (*After Dunlap, AIME Trans., 1938.*)

Nonsteady-state experiments with oil and water, on several unconsolidated sand columns ranging in permeability from 2.9 to 13 darcys, have been made[1] similar to those using gas-liquid mixtures. In these studies oil (kerosene) was flowed into the water-saturated cores, and the permeability to the oil was measured as a function of the residual-water saturation. The results for three of the sands are plotted in Fig. 7.8. No definite correlation with permeability was observed. Qualitatively the curves of Fig. 7.8 are evidently similar to those of Fig. 7.7 obtained by flowing mixtures of oil and water through the sand and measuring the permeabilities under steady-state conditions.

7.4. Permeability-Saturation Data for Three-phase Systems.—The results of only one study of permeability-saturation characteristics of three-

[1] E. W. Dunlap, *AIME Trans.*, **127**, 215 (1938).

phase systems have thus far been published.[1] This was made on a series of five unconsolidated sands, varying in permeability from 5.4 to 16.2 darcys. The fluids comprising the three phases were nitrogen, water, and two kinds of oil. The latter were either kerosene, of viscosity 1.67 cp

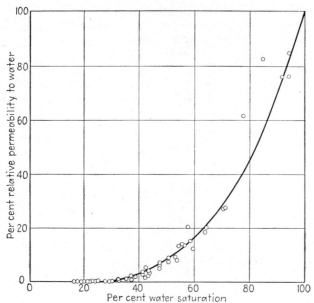

FIG. 7.9. The relative permeability to water in three-phase systems as a function of the water saturation. The solid curve is that determined for CO_2-water mixtures. (*After Leverett and Lewis, AIME Trans., 1941.*)

at 77°F, or a kerosene motor oil, of viscosity 18.2 cp at 77°F. To measure the water saturation an electrical-resistivity determination was made, as in the case of earlier studies of two-phase systems. The free-gas saturation was determined by expanding the gas in the sand to atmospheric and computing the initial free-gas volume from the volume at atmospheric pressure. The oil saturation was obtained by difference from unity of the sum of the water and free-gas saturations.

The relative permeability to the water was found to be independent of the distribution of the oil and gas and determined only by the water saturation, as shown in Fig. 7.9. In fact, as indicated, its variation with the latter was identical, within experimental errors, with that observed for CO_2-water mixtures. However, the relative permeabilities to the gas and oil were found to vary with the distribution of the other two phases and hence cannot be represented simply by single curves, as that for the water permeability. A convenient representation is provided by the triangular dia-

[1] M. C. Leverett and W. B. Lewis, *AIME Trans.*, **142**, 107 (1941).

grams (of widespread use for alloy phase diagrams) on which are plotted "isoperms," or curves of constant relative permeability. Such isoperms are shown in Figs. 7.10 and 7.11. The deviation[1] of these isoperms from straight lines parallel to the sides of the triangles is a measure of the variation of the permeability with the distribution of the other phases. Accordingly in this type of plotting the isoperms for water would be straight segments parallel to the zero per cent water base line.

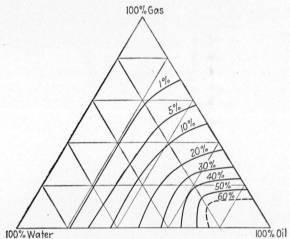

Fig. 7.10. Isoperms (curves of constant relative permeability, in per cent) for oil, as functions of the fluid saturations, in oil, gas, and water multiphase flow through unconsolidated sands. (*After Leverett and Lewis, AIME Trans., 1941.*)

These data are all that have been reported for three-phase systems. The experiments with the kerosene and kerosene motor oil gave the same results, thus indicating that the relative permeabilities are essentially independent of the viscosity of the oil phase. However, as the various sands used in the study were all of approximately the same grain size, the experiments do not show the extent to which the relative permeabilities may vary with the character of the sand. Moreover, no attempt was made to check the constancy of the permeabilities with variations in the pressure gradient.

7.5. The Physical Interpretation of the Permeability-Saturation Curves.—The admittedly meager quantitative data reviewed in the previous sections evidently do not suffice to provide means for predicting the permeability-saturation characteristics for all types of porous media and

[1] In contrast to the symmetry of the isoperms for the gas (cf. Fig. 7.11), which remains a nonwetting phase regardless of the relative saturations of the oil and water, the asymmetry of the isoperms for oil (cf. Fig. 7.10) reflects the change in the role played by the oil from a wetting phase at zero water saturation to that of a nonwetting phase as it is displaced from the sand surface by the increasing saturations of water.

heterogeneous fluids.   Nevertheless they do serve definitely to establish the basic physical features characterizing the flow of multiphase fluids through porous media.   In the discussion of these a fundamental distinction must be made between the fluid phase that wets the porous medium and the nonwetting phase or phases.

In gas-liquid-mixture flow, discussed in Sec. 7.2, the liquid is evidently the wetting phase.   The liquids used in these experiments included both

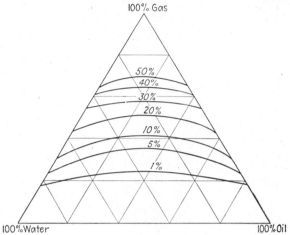

Fig. 7.11.   Isoperms (in per cent) for gas, as functions of the fluid saturations, in oil, gas, and water multiphase flow through unconsolidated sands.   (*After Leverett and Lewis, AIME Trans., 1941.*)

water and oil.   For the immiscible-liquid experiments, reviewed in Sec. 7.3, the wetting phase was water.   And in the three-phase systems, referred to in Sec. 7.4, water again was the wetting phase.   Referring now to Figs. 7.1, 7.2, 7.4, 7.6, 7.7, 7.9, and 7.10, it will be clear that there is a basic similarity among all the permeability curves for the wetting phases in the various multiphase fluids and for the different porous media used in these experiments.   The significant points of similarity are (1) the rapid fall in permeability as the saturation of the wetting phase first falls from unity, and (2) the almost complete vanishing of the permeability to the wetting phase by the time its saturation has fallen to 15 to 35 per cent.   As indicated, these features obtain whether the wetting phase is water and the nonwetting phase is gas, oil, or a combination of the two or whether the wetting phase is oil and the nonwetting phase is gas.

These observations can readily be given a physical interpretation.   From the nature of the surface wetting phenomenon it is clear that the most easily displaced part of the fluid content of a porous medium is comprised of the open and central regions of the pores between the grains.   It is therefore these parts of the pore space that are first occupied by a non-wetting phase. Moreover these same regions of the pore space provide the channels of least resistance for fluid flow, although they may constitute only a rather small fraction of the whole pore volume. Hence the blocking off of this portion of the pore space by a nonwetting phase will lead to an increase in flow resistance or fall in permeability to the wetting phase considerably greater, in proportion, than the direct volumetric displacement of the wetting phase.[1] From the curves cited above it appears that in sands the 10 per cent of the pore space first displaced by the nonwetting phase will result in a reduction of the permeability of the order of 15 to 30 per cent.

Further increments in the nonwetting-phase saturation must of necessity displace the wetting phase from parts of the pore space of continually decreasing effectiveness with respect to flow. The rate of fall in permeability of the wetting phase will therefore taper off with decreasing saturation. Ultimately, however, the state will be reached in which the wetting-phase saturation does not suffice to provide continuity throughout the porous medium except for a very thin layer adsorbed on the individual grains. Under such conditions the permeability to the wetting phase will be zero, except for the flow that may be carried by the film adsorbed on the sand grains. The residual wetting phase will be distributed in toroidlike rings about the contact points between the grains,[2] but the individual rings will no longer be connected. On the other hand, such a distribution may still occupy an appreciable fraction of the pore space, the exact amount depending upon the nature and shapes of the individual grains, their size distribution, and the kind and degree of cementation. The important point is that the break in continuity of the wetting phase[3] and vanishing

[1] While the terminology of this discussion is suggestive of conditions of simultaneous flow of both wetting and nonwetting phases within the same pores, this does not imply that such a simple picture can fully describe the extremely complex microscopic behavior of multiphase-flow systems. It is possible that some degree of local phase segregation among pores of various sizes may also play a role, although complete equivalence to extended bundles of capillaries is most unlikely.

[2] This has been termed a "pendular" state of saturation by J. Versluys [*AAPG Bull.*, **15**, 189 (1931)] and W. O. Smith [*Physics*, **4**, 425 (1933)] (cf. Fig. 7.17).

[3] While the above terminology and physical pictures for convenience follow those commonly accepted in visualizing multiphase fluid distributions in static systems, the quantitative equivalence of the latter to those in dynamic systems is not to be inferred. Thus the break in continuity and vanishing of the wetting-phase permeability under dynamical-flow conditions generally appear to develop at higher saturations than when the porous material is desaturated under continuous and static capillary-pressure application. Although such differences may reflect in part difficulties of permeability experimentation in the range of low-wetting-phase saturation, they may also be due in part to fundamental hysteretic effects in the dynamics of multiphase flow similar to those known to occur under static capillary-pressure interactions (cf. footnote 2, p. 320.

of its permeability will develop before it has been completely stripped from the sand. Rather, it will occur while the liquid saturation still occupies a substantial part of the total pore space, as expressed by the conventionally measured porosity.

In the case of the nonwetting phase or phases there are likewise several significant characteristics of the permeability-saturation curves that are common to all thus far obtained in reported laboratory experiments. These are (1) the development of a measurable permeability only after the saturation from the nonwetting phase has reached a definite or critical value, ranging from 5 to 20 per cent, (2) a rapid rise in permeability as the nonwetting-phase saturation increases beyond the critical value, and (3), the attainment of almost complete homogeneous-fluid permeability before the wetting-phase saturation falls completely to zero. The generality of these characteristics may be verified by references to Figs. 7.1 to 7.8 and 7.10 and 7.11, although the prominence with which they appear differs among the various systems studied. Here again these empirically observed features can be given a simple physical interpretation, although the quantitative aspects of the behavior cannot be readily predicted in advance.

From a purely geometrical point of view, it is clear that the nonwetting phase in a porous medium must of necessity be distributed throughout the medium in a discontinuous manner unless its saturation exceeds a definite minimal value. That is, the nonwetting phase will be broken up, as it were, in separate bubbles or globules confined to individual pores or small groups of adjacent pores. Such a distribution will automatically develop if the nonwetting phase represents a gas coming out of solution from a liquid within the porous medium and while the total free gas so evolved still constitutes a small fraction of the pore space. Such was the situation obtaining in the experiments giving the data of Figs. 7.1, 7.2, 7.10, and 7.11 on gas-liquid mixtures, reviewed in the previous sections. In the case of the experiments represented by Figs. 7.3 to 7.5, where gas was injected into a fully saturated medium containing a wetting phase, or in the similar studies of two immiscible-liquid phases (Fig. 7.8), there was no automatic development of a discontinuous nonwetting-phase distribution. However, here, too, measurable nonwetting-phase permeabilities could not be established until limiting saturations were built up to provide continuous-flow paths. While the latter saturations would be expected to be lower than those limiting the mobility of a gas phase developed by internal gas evolution, comparative measurements have not yet been reported, and the data on the different porous media and flow systems suggest that they are at least comparable in order of magnitude. In any case, it appears that some minimal saturation will be required for nonwetting-phase mobility regardless of the manner of experimentation, although the hysteretic effects giving rise to the differences obtained by different types of experiment may be of importance with respect to oil-reservoir performance.

Considering further the region of nonwetting-phase saturation below the limiting minimal value, where the distribution is discontinuous, the vanishing of the permeability can be predicted from physical considerations. For if the nonwetting phase is distributed as individual bubbles or globules, of diameter exceeding that of the pore constrictions[1] in the direction of flow, their passage through these constrictions will be resisted by interfacial forces, as well as the normal viscous shearing resistances acting on the continuous wetting phase. These interfacial forces are those involved in distorting the bubbles or globules so that they can pass through the narrow passageways between the grains of the porous medium. They arise because of the increased surface area and the energy of the distorted fluid particles, as compared with their essentially spherical shape when at rest in the central parts of the pores. Their magnitude is determined by the effective radius of the constriction, r_1, that at the rear surface of the fluid particle, r_2, and the interfacial tension σ between the wetting and nonwetting phases, according to the equation

$$\Delta p = 2\sigma \cos \theta \left(\frac{1}{r_1} - \frac{1}{r_2} \right), \qquad (1)*$$

where Δp is the pressure difference between the forward and rear surfaces of the fluid particle required to force it through the constriction and θ is the contact angle with the walls. Δp is given in dynes per square centimeter if r_1, r_2 are expressed in centimeters and σ in dynes per centimeter. Thus if $r_1 = 0.002$ cm, $r_2 = 0.005$ cm, $\sigma = 20$ dynes/cm, and assuming a 0 contact angle, $\Delta p = 12,000$ dynes/cm^2.

Such an application of Eq. (1), with r_2 taken as considerably larger than r_1, refers strictly only to a single bubble trapped against a constriction leading into a pore which is substantially filled with liquid, as occurs in the process of overcoming the displacement pressure (cf. Sec. 7.8) during the initial advance of a nonwetting phase into a saturated rock. It therefore gives the maximum pressure drop which can be sustained without break-through by isolated bubbles or globules in individual pores. When the latter grow, as the average nonwetting-phase saturation in-

---

[1] If the nonwetting-phase particles are of colloidal dimensions and much smaller than the pore radii, they will be carried along by the continuous phase as if they were an integral part of the latter.

* Strictly speaking the contact angle associated with $1/r_2$ should include the inclination to the axis of the constriction of the surface at the ring of contact. However, Eq. (1) should suffice for order-of-magnitude estimates.

creases, so as to be in contact with both the forward and rear constrictions of the pores, Eq. (1) will still apply. But r_2 will then become comparable to r_1, and Δp will be correspondingly reduced. It is such lower values of Δp—the static analogues of which are the "threshold" pressures (cf. Sec. 7.8)—which would have to be overcome to cause motion of a non-wetting phase when its saturation is below the minimum for mobility. However, even when r_2 is only 10 per cent greater than r_1, Δp will still be 1,800 dynes/cm² for $r_1 = 0.002$ cm. If this pressure drop is to be provided by the parallel flow of the surrounding liquid phase, and the pore length is taken as 0.01 cm, the equivalent local pressure gradient would be about 5.4 atm/ft. While such pressure drops over the individual pores would not be simply cumulative and additive if flow of the non-wetting phase were actually established, it is clear that the required pressure gradients nevertheless will still be so high that substantial mobility is not to be expected under laboratory or reservoir conditions, except possibly in the immediate vicinities of well bores or when coalescence of the individual bubbles and establishment of continuity is already imminent. Although the lack of mobility of a discontinuous nonwetting phase in a strict sense would thus appear to be related both to the local pressure gradients in the associated flowing phase and the absolute value of its mean saturation, it still represents a characteristic of multiphase-fluid phenomena of physical significance in actual oil-producing processes.

As the gas saturation builds up in the range of the minimum value for mobility, the separated bubbles in the individual pores will overflow, as it were, and coalesce with those in adjoining pores. The pressure drop, given by Eq. (1), required to force these large masses through the constrictions will be extended over the several pores spanned by the fluid particle and will be equivalent to a lower pressure gradient. Moreover, because of the natural distribution of pore sizes and dimensions of the pore constrictions, there will develop a gradual diffusion through the medium of local regions where the nonwetting phase has become mobile. As its total saturation increases, these regions will spread and encompass an increasingly larger number of pores, until the medium as a whole is permeated by continuous-flow channels for the nonwetting phase. It is during this development that the permeability of the nonwetting phase rises from zero and rapidly increases and exceeds that for the wetting phase. Ultimately, when the wetting-phase saturation falls into the "pendular" state, so that it loses its own mobility, its interference with the flow of the nonwetting phase likewise becomes small. The remnants of the wetting phase that are confined to the smallest pores or the almost inaccessible recesses of some of the larger pores would not in any case contribute greatly to normal fluid flow, and the nonwetting phase develops a permeability virtually equal to that for a homogeneous fluid.[1]

These same general considerations will apply in the interpretation of the permeability characteristics of systems containing two nonwetting phases. As Fig. 7.9 shows, the wetting phase of such systems has a permeability curve independent of the saturation distribution among the other two and the same as if there were only one nonwetting phase. In the case of the individual nonwetting phases the saturation distribution of the remaining two does affect the permeability values, as indicated by the nonlinearity[2] of the isoperms of Figs. 7.10 and 7.11. Nevertheless they still show regions of very low permeability till some 20 per cent of the saturation of the phase is reached, which is suggestive of the "equilibrium" saturation (cf. Sec. 7.6), and the rapid rise to rather high permeabilities while the medium still contains appreciable amounts of the other phases. The interaction between the nonwetting phases is, of course, a new factor in the behavior of the three-phase systems. However, interactions between the wetting phase and the individual nonwetting phases are essentially similar to those involved in the simpler two-phase mixtures.

While there can be little doubt that the physical mechanism underlying the permeability-saturation curves described here is substantially correct, it must be emphasized that it is essentially of a qualitative nature. Porous media are statistical ensembles of elements differing in detailed geometry over considerable ranges. Hence the development of flow of the nonwetting phase will disseminate gradually from the larger pores with larger interconnecting channels to the smaller pores with narrow constrictions, rather than as a sharp break from the state of no flow. Likewise the immobilization of the wetting phase will diffuse from the smaller and locally tighter groups of pores to the larger and less cemented grain clusters as its saturation is decreased. It will not suddenly crystallize throughout the medium just when a critical-saturation limit is reached. This same observation makes it clear that while similar groups of porous media, such as unconsolidated sands of limited grain sizes, may give nearly the same permeability-saturation curves, as in Fig. 7.1, variations in grain sizes, shapes, and cementation, as well as the detailed surface and interfacial characteristics, may well lead to substantial modifications in the curves, from the quantitative point of view.

It should be noted that, while the distinction between the gross wettabil-

---

[1] The details of the approach to 100 per cent relative permeability as the wetting-phase saturation is reduced will evidently depend on the microscopic pore-space geometry, amount and character of the cementing material, etc. Intuitively, however, it appears plausible to assume that the 100 per cent relative-permeability limit at zero wetting-phase saturation will be approached with a zero slope in actual consolidated sands.

[2] Because of the small magnitude of the curvatures of the gas isoperms of Fig. 7.11, further evidence is needed to establish their reality in general porous media.

ity of the phases suffices to determine the broad features of their permeability-saturation characteristics, the quantitative aspects of the latter depend
both on the microscopic geometry and surface properties of the rock.   And
even for the same porous medium the permeability-saturation curves are
not necessarily independent of the nature of the wetting or nonwetting
fluids, although in the case of unconsolidated sands or synthetic sand

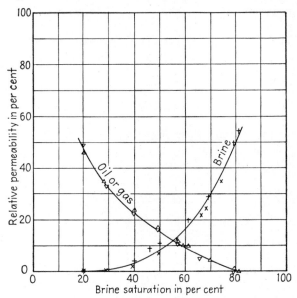

Fig. 7.12.   Relative permeabilities in a synthetic sandstone for two-phase systems comprised
of brine and air, and brine and oil. $+$, $\times$, permeabilities to brine when the nonwetting phases
are oil and air. $\triangle$, permeability to air. $\triangledown$, permeability to oil.

samples such dependence appears to be of minor importance.   Thus, as
previously noted, Fig. 7.9 shows that the permeability to water, as a
wetting phase, is independent of the distribution of the nonwetting-phase
saturation between oil and gas.   Moreover it may be verified by reference
to Fig. 7.10 that the permeability curve to oil as a wetting phase in the
unconsolidated sands, with gas as the nonwetting phase, is virtually the
same as the curve for water of Fig. 7.9.   The symmetry of the gas-permeability curves of Fig. 7.11 shows the gas permeability to be the same for
oil as the complementary wetting phase as when water is the wetting phase.
Substantially equivalent results have been reported for two-phase flow in
a synthetic sandstone, as shown in Fig. 7.12.[1]   On the other hand, for a

---

[1] Figs. 7.12 and 7.13 are replotted from separate graphs given by Morse, Terwilliger,
and Yuster, *loc. cit.*   While there is some question whether or not the techniques used
in obtaining Figs. 7.12 and 7.13 ensure quantitative accuracy, they nevertheless serve
to illustrate the possible dependence of the relative permeabilities on the exact nature
of the fluid components (cf. also Henderson and Meldrum, *loc. cit.*).

sample from an actual producing sand, the Second Venango, of Pennsylvania, the permeability curves for oil and gas as nonwetting phases are quite different, and even that for the same wetting phase, brine, shows some variation with the nature of the nonwetting phase (cf. Fig. 7.13). These differences reflect the microscopic nonuniformity in wettability and interfacial forces of the internal solid surfaces, as would be expected in

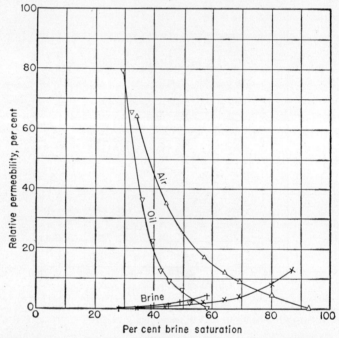

FIG. 7.13.   Relative permeabilities in a sample of Second Venango Sandstone for two-phase systems comprised of brine and air and brine and oil. $+$, $\times$, permeabilities to brine when the nonwetting phases are oil and air. $\triangle$, permeability to air. ∇, permeability to oil.

cemented sandstones, and especially in argillaceous sands, and for which there is also independent evidence.   For quantitative applications of permeability-saturation data the latter should therefore be determined by using the actual rock and fluids of interest.

Since the experimental techniques for determining the relative permeability-saturation relationships are not yet established as routine techniques comparable with those for measuring the homogeneous-fluid permeability, it is of interest to note that considerable progress has recently been made[1] in deriving analytical expressions describing the permeability-saturation curves.   This development is based on an analysis of the physical interrelationship between the fluid-flow phenomena in porous media and static

[1] W. D. Rose, *Jour. Petroleum Technology*, **1**, 111 (1949).

interfacial characteristics described by capillary-pressure properties. The latter, to be discussed in detail in Sec. 7.8 and also referred to in Sec. 3.11, relate the pressure discontinuities across the fluid interfaces in a porous medium, termed the "capillary pressures," and the magnitude of the fluid saturations. This type of analysis leads first to a relation between the relative permeability to the wetting phase k_{rw}, the wetting-phase saturation ρ_w, and the capillary pressure p_c between the wetting phase and its bounding nonwetting phase, given by

$$k_{rw} = \rho_w \left(\frac{p_d}{p_c}\right)^2, \tag{2}$$

where p_d, the displacement pressure, is the limiting value of p_c at 100 per cent wetting-phase saturation, i.e., at $\rho_w = 1$. A study of the physical factors controlling the capillary-pressure function p_c and empirical data on its variation with ρ_w provided means for expressing k_{rw} explicitly as a function of ρ_w, the final result being

$$k_{rw} = \frac{16\rho_w^2(\rho_w - \rho_{wm})^3(1 - \rho_{wm})}{[2\rho_w^2(2 - 3\rho_{wm}) + 3\rho_w\rho_{wm}(3\rho_{wm} - 2) + \rho_{wm}(4 - 5\rho_{wm})]^2}, \tag{3}$$

in which ρ_{wm} is the limiting value of ρ_w for wetting-phase mobility. This is the only empirical parameter determining the complete family of wetting-phase relative-permeability curves. Although the derivation of Eq. (3) implied that ρ_{wm} should be the "irreducible" minimum saturation measured in capillary-pressure experiments, such as are made for connate-water saturation determinations (cf. Sec. 3.11), the latter are generally lower than the limiting values found in actual relative-permeability measurements. The complete explanation for this difference is not yet clear. It may reflect basic hysteretic effects in interfacial forces associated with the detailed dynamics of multiphase-fluid motion and the dependence of the microscopic structure of the fluid-phase distribution on the manner in which the relative saturations are established.[1] Thus ρ_{wm} may be considered as a dynamical equivalent of the "irreducible" static wetting-phase saturation, and possibly depending on the detailed flow conditions. In any case, by suitable choice of ρ_{wm} in Eq. (3) it is possible to reproduce virtually all accurately measured wetting-phase relative-permeability curves over most of the saturation range with an accuracy comparable with that of the experiments.

The physical theory leading to Eqs. (2) and (3) does not permit direct generalization or application to the nonwetting phase of a multiphase system. However, it is possible to construct an analogue of Eq. (3), which appears to approximate measured relative-permeability curves for the nonwetting phase in two-phase systems, by simply introducing the concept

---

[1] Cf. footnote 2, p. 320.

of an effective freezing of that part of the wetting-phase saturation, Ψ_w, which presumably does not interfere with the nonwetting-phase mobility. That is, the effective porosity for the nonwetting phase is considered to be $f(1 - \Psi_w)$, where f is the true porosity, and the actual nonwetting-phase saturations ρ_n are introduced in Eq. (3) as $\rho_n/(1 - \Psi_w)$. The resulting equation for the nonwetting-phase relative permeability k_{rn} is then

$$k_{rn} = \frac{16\rho_n^2(\rho_n-\rho_{nm})^3(1-\Psi_w-\rho_{nm})}{[2\rho_n^2(2-2\Psi_w-3\rho_{nm})+3\rho_n\rho_{nm}(3\rho_{nm}-2+2\Psi_w)+\rho_{nm}(1-\Psi_w)(4-4\Psi_w-5\rho_{nm})]^2}, \quad (4)$$

where ρ_{nm} is the equilibrium nonwetting-phase saturation. While this relation is of questionable significance from a fundamental physical standpoint, it may serve as a semiquantitative guide in estimating or extrapolating nonwetting-phase relative-permeability data in two-phase systems.

Further generalizations have been suggested for computing the permeability to oil in oil, gas, and water systems.[1] However, these will not be reproduced here, since the available experimental data are too meager for evaluating their accuracy and significance. As to the relative permeability to gas in three-phase systems, the reported empirical data are again limited to the single study represented by Fig. 7.11. As an approximation until further information is developed it may suffice to ignore the curvature in the isoperms of Fig. 7.11 and assume the relative permeability to the gas to depend only on its own saturation and to be the same function of the gas saturation in three-phase as in two-phase systems.

7.6. The Direct Implications of the Permeability-Saturation Curves; The Equilibrium Saturation.—The fundamental data on the permeability-saturation relationship, presented in Secs. 7.2 to 7.5, and their physical interpretation have a number of significant implications. Indeed, in a broad sense they provide the basic physical mechanism underlying the whole complex of processes involved in the expulsion of oil and gas from oil-bearing rocks. It will be one of the purposes of the following chapters to develop such applications. Here, however, some of the simpler and qualitative implications will be considered.

The first of these follows from the observation that the nonwetting-phase permeability remains substantially zero until its saturation attains a definite nonvanishing value. If it is less than this value, it remains effectively locked in the pores. On the other hand, as this saturation is increased from zero the permeability to the wetting phase continually decreases. Evidently, then, if one attempts to flow a gas-saturated liquid through a porous medium so that the pressure in the medium falls below the saturation pressure, the gas first evolved in the pores will be trapped and accumulate as

[1] Cf. *ibid.*

flow continues until its saturation reaches the value at which the permeability becomes nonvanishing. During this process of gas accumulation the permeability to the liquid will continue to fall along a curve such as those shown in Figs. 7.1, 7.2., 7.6, and 7.9. The significant point about this situation is that the process described is inherently transient. Steady-state flow will not be achieved, nor can it be maintained, as long as the free-gas saturation is such that the gas permeability is zero, provided, of course, that there is available a source for additional free gas. In other words, the system will not reach equilibrium until the liquid saturation has fallen and the free-gas saturation has increased to the state where the gas-phase permeability becomes nonvanishing. Accordingly these complementary saturations of the wetting and nonwetting phases will be termed "equilibrium saturations." And the permeability of the wetting phase at these saturations is termed the "equilibrium permeability." They represent the limiting states of saturation and permeability at which equilibrium, or steady-state, conditions can be established and maintained.

At saturations of the wetting phase below the equilibrium value steady-state conditions can be maintained provided that there is a steady source of a constant-composition mixture. The fluid saturations for each such steady-state condition of flow will be determined by the composition of the flow stream. That is, the saturations will adjust themselves so that the corresponding permeabilities will just suffice to carry the respective components in the flow stream in the same proportions as they are being supplied. This adjustment will involve a transient period if the initial-fluid saturations do not correspond to those of the steady-state condition or if the composition of the flow stream should be changed after an initial steady state has been established. Moreover, if along the flow path the volumetric composition of the composite flow stream should change, the saturation distribution in the rock will likewise adjust itself to the changed permeability requirements for transmitting the modified flow-stream composition. In particular, if owing to entry into regions of lower pressure the free-gas component of a gas-liquid mixture should be augmented by gas evolution and expansion, the free-gas saturation in these regions will increase so as to accommodate a greater flow rate and permeability for the free-gas phase as compared with the liquid phase. As such changes will be distributed continuously throughout the porous medium, even though the system as a whole will be in a steady-state condition, the medium will display a corresponding continuous distribution of fluid saturations (cf. Secs. 8.2 and 8.4). It is in this sense that the rock may be described as possessing a continuously variable but localized structure, with which are associated continuously varying local fluid saturations and permeabilities. And if the system as a whole is undergoing a transient history, as that of an actual oil-producing reservoir, these spatial distributions of saturation and permeability themselves experience continual changes to correspond to the variations in total fluid content and pressures of the composite and integrated fluid reservoir. Indeed, it is the determination of these space and time variations in the fluid saturations, permeabilities, and pressures that constitutes the basic problem of describing and predicting the behavior of heterogeneous-flow systems, such as oil-producing reservoirs.

As will be considered more fully in later chapters, the changes in fluid saturations required for adjustment to variations in the volumetric composition of the flow stream may be readily inferred from the basic permeability-saturation curves. For the ratios of the relative permeabilities for the individual phases, at a fixed saturation distribution, when divided by the ratios of the viscosities of the same phases, will evidently give the relative local rates of flow for the corresponding phases. In particular, for a gas-liquid mixture, such composite ratios derived from the gas- and liquid-permeability curves directly represent the local free-gas–oil ratio in the flow stream, as a function of the local oil or gas saturation. When to these are added the dissolved gas carried along in the oil phase, the total values of the local gas-oil ratio are readily obtained. An inspection of such data as are plotted in Figs. 7.1 and 7.2 shows that the free-gas–liquid ratios will be zero until the equilibrium liquid saturation is reached and will then rise rapidly as the liquid saturation is further reduced. It is in this observation alone that some of the most important performance characteristics of oil fields producing by the gas-drive mechanism find their basic explanation, as will be seen in Chap. 10.[1]

It may be noted, from a historical point of view, that the fundamental representations and concepts thus far developed in this chapter used to be described as manifestations of a "Jamin effect." The latter referred to the resistance offered by individual gas bubbles to attempts to force them through the pore constrictions. As explained in the last section, such resistances are quite real and are associated with the capillary forces at the interfaces of the immiscible phases. Because of this connection with the capillary phenomena, flow systems, such as oil reservoirs, in which they occurred were often described as being under "capillary control."

While such terminology may not be inappropriate per se, it apparently

---

[1] It should be pointed out, however, that, while the vanishing of the gas permeability until the equilibrium saturation is reached does play an important role in determining the early history of solution-gas-drive reservoirs from a theoretical standpoint, the existence of an appreciable free-gas saturation without mobility is often masked in reported field observations by other factors associated with actual reservoir and producing conditions (cf. Sec. 10.12).

diverted an attack on the basic problem of multiphase-fluid flow from the point of view presented here into one or the other of two rather fruitless channels: (1) a completely generalized and qualitative proposition that fluid mixtures systems were controlled by the Jamin effect, with the virtual implication that by so naming the phenomenon it was fully described and quantitatively specified; (2) a disproportionate emphasis on a microscopic analysis of gas-bubble behavior under such extremely simplified conditions as substantially to vitiate its pertinence to the problem of flow in actual porous media. This second point of view led to the discussion of chains of bubbles in single capillary tubes, individually trapped between well-defined constrictions, as analogues of the trapping of gas bubbles in porous media. The demonstration that such a bubble chain, when aligned in phase against the constrictions, can withstand very appreciable pressures before breaking through was interpreted to imply that actual oil-bearing strata under such "capillary control" will yield oil and gas only from a region about the well extending to a radius at which the pressure exceeds the equivalent of the hypothetical individual bubble resistances acting in series between the well and that radius. Beyond this radius the fluids were considered to be immobile. As a corollary it was deduced that this limiting radius, or "radius of drainage," would shrink as production proceeds and the pressure declines. And as an underlying hypothesis it was supposed that gas bubbles are uniformly distributed throughout the reservoir rocks initially.

These unwarranted extrapolations of the validity of the primary Jamin effect experiments do not, of course, invalidate the basic significant features of the bubble-resistance phenomena. For it is true that separated individual bubbles or globules will require finite pressure differentials to force them through the pore constrictions. Indeed, as previously discussed, this observation is fundamental to an understanding of the existence of the equilibrium-saturation limit required for free-gas flow. But, contrary to the behavior of the bubble chains in closed capillaries, the multiplicity of lateral pore interconnections in actual rocks will permit continued flow of the *liquid* phase even though the free-gas phase may be locked in place. On the other hand, even when the nonwetting-phase permeability is nonvanishing, the resultant flow stream experiences greater flow resistance than the homogeneous fluids. This is evident on noting from the various sets of permeability-saturation curves of Secs. 7.2 to 7.4 that the sum of the individual relative permeabilities is very materially less than 100 per cent. As indicated by the curves of Fig. 7.2, the sum of the gas and liquid permeabilities for the consolidated sand falls to only about a fifth of that for homogeneous fluids. And in the case of three immiscible phases, as gas, oil, and water, the sum of the individual permeabilities in unconsolidated sands may fall to only 10 per cent of that for a single homogeneous phase. The reason undoubtedly lies in the interfacial capillary forces between the phases and is basically of the same nature as those involved in the idealized Jamin effect experiments. However, the latter do not provide a basis for a quantitative evaluation of the phenomena.

It should be observed that, while the Darcy law representation of Eqs. 7.1(1) is a natural generalization to multiphase-flow systems of the original homogeneous-fluid "law of force" [cf. Eq. 4.3(3)], its quantitative validity is not so well established as in the latter case. Such validity implies that the permeability functions k_o, k_g, and k_w are determined only by the saturation distribution[1] but are independent of the fluid viscosity and pressure gradient. With respect to the former, it will be recalled from the discussion of the basic experimental data on the permeability-saturation relationship that in virtually all multiphase-fluid combinations thus far studied no significant effect of the fluid viscosity was found. Some variations were observed in several of the experiments, but these either were within the range of the experimental errors or were not subject to unique correlation with the viscosity. Until such correlation, if any, be established, the separation of the viscosity from the permeability in the generalized Darcy equations may be considered as reasonably justified.

As to the dependence of the permeabilities on the pressure gradient, the situation is rather less satisfactory. The basic reason lies in the paucity of available data, because of the complexity of the techniques required for precise experimentation. For gas-liquid mixtures the reported[2] measurements with different pressure gradients showed no definite variation of the permeability that could be correlated with the pressure gradient. In the case of oil-water mixtures an indication of a definite variation of the relative permeabilities with the pressure gradient has been observed in one study.[3] While there is some doubt as to the reality of the effect, the magnitude of the variations was small. In any case, from a practical standpoint, it appears reasonable to accept Darcy's law for the present,

---

[1] While there is little doubt that the saturation *distribution* within, as well as between, the various phases is the primary variable controlling the permeability functions, there is little evidence that the *magnitudes* of the saturations uniquely fix the distributions. In fact there is reason to believe that further study of the microscopic aspects of multiphase-flow phenomena will reveal dynamical hysteresis effects and some degree of variation of the phase distributions with the history of the flow system. Ultimately, therefore, it may turn out that the multiphase permeabilities will have to be defined not only as functions of the saturation magnitudes but also as functions of the past history of the system (cf. Henderson and Meldrum, *loc. cit.*).

[2] Hassler, Rice, and Leeman, *loc. cit.;* and Krutter, *loc cit.*

[3] M. C. Leverett, *AIME Trans.*, **132**, 149 (1939). Cf. also Henderson and Yuster, *loc. cit.*

at least as a first approximation and until conclusive information to the contrary is developed.

There are no published data on the possible dependence of the equilibrium saturation and permeability on the pressure gradient. From physical considerations, however, it is to be anticipated that such a dependence should appear at least at very high gradients. This follows from the physical interpretation of the equilibrium-saturation phenomena, as discussed in Sec. 7.5. Evidently, if the pressure gradient be made sufficiently high, separated bubbles or globules of the nonwetting phase will be forced out of pores in which they might remain trapped at lower gradients. The equilibrium free-gas saturation should therefore decrease as the pressure gradient is increased. On the other hand, from the order of magnitude of the capillary forces involved, as indicated in Sec. 7.5, it would appear that in actual oil-bearing rocks the equilibrium values should correspond to those determined in the laboratory at moderate gradients, except possibly in the immediate vicinity of well bores. Until definite evidence to the contrary appears, both the equilibrium saturations and general permeability values will be considered to be independent of the pressure gradient.[1]

A final important inference that may be drawn directly from the permeability-saturation curves, especially those for the three fluid phases discussed in Sec. 7.4, is that there is only a small range of fluid saturations in which all three phases will simultaneously have appreciable permeability. Hence it is only in this limited range that the three phases can be produced simultaneously at comparable rates, per unit viscosity. Conversely, small variations in the saturation distributions in this region would lead to relatively large changes in the composition of the fluid stream being produced. Thus, for example, at a saturation distribution of 40 per cent water, 30 per cent oil, and 30 per cent gas, the relative permeabilities are approximately 3 per cent, 5 per cent, and 3 per cent, respectively, on the basis of Figs. 7.9 to 7.11. Increasing the water saturation to 45 per cent and decreasing the gas saturation to 25 per cent increases the water permeability to 6 per cent and reduces the gas permeability to about 1.5 per cent, so that these two now have a ratio of 4 as compared with 1 previously. On the other hand, if the water saturation is decreased to 35 per cent and the gas saturation is raised to 35 per cent, the new relative permeabilities will be about 1.5 per cent and 5 per cent, respectively, or a ratio of 0.3 as compared with 1. Similarly, if the oil saturation is increased to 35 per cent and the gas

----

[1] With respect to the effect of slippage phenomena on the relative permeability to gas, recent studies (Rose, API meetings, Chicago, Ill., November, 1948) indicate that if the same mean pressure is used for the relative as for the homogeneous-fluid gas permeability, no further correction need be made for the mean flow pressure to obtain values sufficiently accurate for practical purposes.

saturation reduced to 25 per cent, their respective relative permeabilities will change to about 8 per cent and 1 per cent, with a ratio of 8 as compared with 1.7 previously. These observations have a direct bearing on the performance of actual oil-producing reservoirs, as will appear from the discussion in later chapters.

7.7. The Equations of Motion.—Returning now to the problem of formulating the final equations of motion for the flow of heterogeneous fluids through porous media, one may follow the procedure applied in Chap. 4 for homogeneous-fluid systems. That is, one merely combines with the "law of force" [Eqs. 7.1(1)] the equations of state for the fluid phases and their respective equations of continuity. The former are treated implicitly, in the sense that the relationships between the specific volumes of the phases and the pressure (and temperature if considered as variable) are assumed to be independently preassigned and available for insertion in the equations when they are being solved. The latter are applied directly for the individual phases to the corresponding Darcy equations of Eqs. 7.1(1). One thus readily obtains the three equations[1]

$$
\left.\begin{aligned}
\nabla \cdot \left[\frac{S_o k_o}{\beta_o \mu_o} \nabla(p - \gamma_o gz) + \frac{S_w k_w}{\beta_w \mu_w} \nabla(p - \gamma_w gz) + \frac{\gamma_g k_g}{\mu_g} \nabla(p - \gamma_g gz) \right] \\
= f \frac{\partial}{\partial t} \left(\frac{S_o \rho_o}{\beta_o} + \frac{S_w \rho_w}{\beta_w} + \gamma_g \rho_g \right), \\
\nabla \cdot \left[\frac{k_o}{\beta_o \mu_o} \nabla(p - \gamma_o gz) \right] = f \frac{\partial}{\partial t} \left(\frac{\rho_o}{\beta_o} \right), \\
\nabla \cdot \left[\frac{k_w}{\beta_w \mu_w} \nabla(p - \gamma_w gz) \right] = f \frac{\partial}{\partial t} \left(\frac{\rho_w}{\beta_w} \right), \\
\rho_o + \rho_w + \rho_g = 1,
\end{aligned} \right\} \quad (1)
$$

where again the subscripts o, w, g refer to the oil, water, and gas phases, S is the volume (at standard conditions) of the gas in solution per unit volume of the liquid phase at standard conditions, k is the permeability, β the formation volume of the liquid phase, μ the viscosity, γ the phase density, p the pressure, f the porosity, t the time, and ρ the phase saturation expressed as a fraction of the pore space.

In these equations the functions S, β, μ, and γ are to be considered as known functions of the pressure. The k's are to be expressed analytically or numerically as functions of the saturations ρ, according to the permeability-saturation relationship of the porous medium of interest. Thus the basic dependent variables of Eqs. (1) reduce to the pressure p and the

[1] M. Muskat, *Proc. Sec. Hydraulics Conference, Univ. Iowa*, 1943, p. 130. The basic equivalent of these, but written out explicitly under several simplifying assumptions, were given by M. Muskat and M. W. Meres, *Physics*, **7**, 346 (1936).

three phase saturations ρ. In principle the four equations should suffice for a complete determination of p and the ρ's, as functions of time and space, after the assignment of the physical-boundary and initial conditions defining the particular flow system under consideration.

Equations (1) do not include the effect of the variation in fluid saturations in giving rise to capillary-pressure gradients superposed on the gradients in p. As virtually no quantitative evaluations of this effect in general flow systems have thus far been made, their discussion will be deferred to the next several sections, where the general subject of capillarity will be given separate consideration. As will be seen in Sec. 7.10, however, the inclusion of the capillary-pressure terms in Eqs. (1) is actually not justified for the great majority of dynamical systems of practical interest.

From a practical point of view Eqs. (1) have a far greater generality than need be considered in most applications. For example, in the great majority of oil-field reservoirs the solubility of the gas in the water phase, S_w, can be neglected as compared with that in the oil, S_o, except as one may be interested directly in the exact behavior of a contiguous and mobile water reservoir. Moreover, within the oil-producing section itself, and above water-oil transition zones, ρ_w will generally represent the connate-water saturation, which will usually remain immobile[1] ($k_w = 0$) until the pay has been invaded by external waters. In such cases one may therefore drop out the second term in each side of the first of Eqs. (1), and the whole of the third equation. Finally, while the effect of gravity undoubtedly plays an important role in a number of phases of oil production, it is extremely difficult to treat quantitatively. Such gravity-controlled systems as will be considered in this work will therefore be based on independently formulated representations, rather than as special applications of the generalized equations (1).

Even with these simplifications Eqs. (1) are unfortunately of such a high degree of complexity that the derivation of general solutions is literally impossible. They are basically nonlinear in both the pressure and saturations. In addition, the various coefficients, as S, β, μ, γ_g, and k, will in general be empirical functions of the pressure or saturation rather than simple analytical expressions. At the present time a direct analytical treatment of nonsteady states on the basis of Eqs. (1) is simply not practical. In fact, only one attempt has thus far been reported for the solution of a transient system governed by Eqs. (1), and that[2] was carried out by a numerical

---

[1] The frequently observed small percentages of water production immediately on opening a well to production, even in virgin fields, may be due to the presence of exposed thin streaks of "wet" strata, rather than a flow of the connate water in the main oil-producing section.

[2] Muskat and Meres, *loc. cit.*

integration of a simple linear system and applied only to a highly limited special case. Hence, instead of attempting to apply the basic physical concepts underlying Eqs. (1) to the prediction of the performance histories of oil reservoirs through the medium of exact solutions of these equations, it will be necessary to devise approximation procedures.[1] On the other hand, for steady-state conditions, it will be possible formally to integrate Eq. (1) for several special cases. While these will be of limited scope, they will serve to show some of the significant differences between the characteristics of multiphase- and homogeneous-fluid systems. Such steady-state solutions and approximation procedures based on Eqs. (1) will be considered in Chaps. 8 and 10.

7.8. Capillary Phenomena; Capillary, Displacement, and Threshold Pressures.—Because oil-bearing reservoirs universally contain more than one fluid phase, interfacial forces and pressures are ever present in influencing both static and dynamical states of equilibrium. These arise from and are associated with the curvatures of the interfacial surfaces. The reason they are of a magnitude of practical interest is the inherently small pore sizes in oil-producing porous media. These, in turn, necessitate that when the interfacial surfaces are not strictly plane by mere accident, they will have small radii of curvature and high "curvatures."

It may be shown by application of the principles of mechanics or thermodynamics that across an interface between two fluid phases there will exist a pressure difference given by

$$\Delta p (\text{dynes/cm}^2) = \sigma \left(\frac{1}{R_1} + \frac{1}{R_2} \right),$$ (1)

where σ is the interfacial tension, in dynes per centimeter, and R_1 and R_2 are the two principal radii of curvature of the surface, in centimeters. The sign of Δp is determined by the algebraic signs of R_1 and R_2 and is such that the pressure is greater on the concave side of the interface. In its broadest sense Δp, as defined by Eq. (1), is termed the "capillary pressure."

The physical significance of Eq. (1) may be conveniently illustrated by deriving the height of rise of liquid in a capillary tube above the free-liquid surface in which the capillary is partly submerged. Assuming the liquid to wet the internal wall of the capillary tube and the meniscus to have a semispherical shape, so that $R_1 = R_2 = r$, the capillary radius, it follows from Eq. (1) that the pressure on the underside of the meniscus is lower

---

[1] While there is little hope for ever deriving general analytic functional solutions to Eqs. (1), the development of large-scale digital computing machines during the last war offers promising possibilities of treating numerically specific problems. In fact, the transient histories of a number of solution-gas-drive systems, as governed by Eqs. (1), are currently being derived by the author, in collaboration with the International Business Machines Corp., using the IBM Selective Sequence Electronic Calculator.

than atmospheric by $2\sigma/r$, where σ is now the surface tension of the liquid. This pressure deficiency must be in equilibrium with the hydrostatic pressure at the level of the meniscus. Denoting the height of rise by h, the liquid density by γ, and atmospheric pressure by p_a the pressure equilibrium at the underside of the meniscus requires that

$$p_a - \frac{2\sigma}{r} = p_a - \gamma gh, \tag{2}$$

where g is the acceleration of gravity and the change in atmospheric pressure across the height h is neglected. It follows at once that

$$h = \frac{2\sigma}{\gamma gr}, \tag{3}$$

which is the well-known expression for the height of rise of a wetting liquid in a capillary tube.[1]

As Eq. (3) shows explicitly and Eq. (1) implies generally, the interfacial pressure discontinuities will vary inversely as the linear dimensions of the vessel confining the fluid system. It is only when the latter are of capillary dimensions that the order of magnitude of the pressure differentials becomes large enough to be of significance in many practical problems. It is for this reason that the various implications of Eq. (1) are usually termed "capillary" phenomena.

In porous media the radii of curvatures of the fluid interfaces will evidently be of an order of magnitude comparable with the grain or pore radii of the medium. For similar grain-shape- and grain-size-distribution characteristics the interfacial pressure differentials in porous media should therefore vary inversely as the square root of the permeability. Hence capillary phenomena generally will be associated with greater magnitudes and may be expected to play a more significant role in low permeability and tight rocks than in systems of high permeability.

The origin of the "displacement" pressure can be readily explained on the basis of Eq. (1). This pressure may be defined as the minimum required to force the entry of a nonwetting fluid into a porous medium saturated with a wetting liquid. Since the entry of the foreign fluid will evidently be associated with a multiply connected nonplanar interfacial surface at the contact with the native wetting fluid, there will exist a pressure drop across each point of the interface given by Eq. (1). In fact, under static equilibrium, before actual break-through at any surface pore has developed, Eq. (1) defines the exact nature of the interface, as a statement of the requirement that the sum of the principal curvatures is every-

---

[1] It is clear that Eq. (3) will also give the depth of meniscus depression when the liquid is completely nonwetting with respect to the capillary-wall surface.

where constant and equal to $\Delta p/\sigma$. As the external pressure and Δp are increased, the curvatures of the interface will increase, until at some pore the protruding forefront of the displacing fluid will just be able to pass through the constriction leading to the adjoining pore. At this pressure there will be at least a local forward advance of the nonwetting fluid, and such advance will continue into the porous medium if the latter is substantially uniform in microscopic structure. It is this initial driving pressure difference, or the maximum that can be maintained at the saturated face of the medium without mass displacement of the saturating liquid, that is the displacement pressure.

As the applied pressure is increased beyond the displacement value, the interface advance will be accelerated and break-through will develop at additional surface pores of smaller effective radius than those first penetrated. In fact by reversing the direction of flow and observing the sequence of break-through of a nonwetting fluid from the porous medium into a surrounding wetting fluid, the distribution of effective pore sizes at the surface of the porous material can be determined.

In a partially saturated porous medium the interfacial curvatures that, by Eq. (1), determine the capillary pressures will evidently depend on the fluid saturation. Once the displacement pressure has been exceeded and the nonwetting phase has established intercommunicating channels throughout the porous medium, the average interfacial curvature will increase as the wetting-phase saturation decreases. It is this relationship, between the interfacial curvatures and the fluid saturations, that constitutes the empirical description of the capillary properties of a porous medium.

An understanding of the physical significance of capillary phenomena in porous media and means for their empirical study were first developed by those interested in the soil sciences.[1] And a number of investigations[2] to calculate theoretically the relation between capillary pressures and fluid saturations were made long before its applicability to oil reservoirs was widely appreciated. It is only recently, however, that serious attempts have been made to apply and extend this work from the point of view of its bearing on oil reservoirs and oil production.[3] By methods similar or equivalent to that outlined in Sec. 3.11 for connate-water determinations,

---

[1] Cf. F. H. King, *USGS 19th Annual Report*, pt. II, p. 59 (1897–1898); J. Versluys, *Inst. Bodenkunde*, **7**, 117 (1917); B. A. Keen, "Physical Properties of the Soil," Longmans, Green & Co., Inc., 1931.

[2] Cf. W. O. Smith, P. D. Foote, and P. F. Busang, *Physics*, **1**, 18, (1931); W. O. Smith, *Physics*, **3**, 139 (1932), **4**, 184, 425 (1933).

[3] M. C. Leverett, *AIME Trans.*, **142**, 152 (1941); G. L. Hassler, E. Brunner, and T. J. Deahl, *AIME Trans.*, **155**, 155 (1944); W. D. Rose and W. A. Bruce, *Jour. Petroleum Technology*, **1**, 127 (1949).

numerous measurements have been made and reported[1] on the variation of the capillary pressure in actual oil-productive sandstones and limestones with the saturation of the wetting phase.   Figures 7.14 to 7.16 show groups

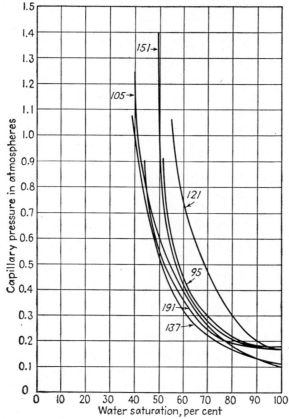

FIG. 7.14.   Capillary-pressure curves, obtained with water, on benzene-extracted sandstone samples from a California field.   (*After Hassler, Brunner, and Deahl, AIME Trans., 1944.*)

of curves obtained for different samples from a sandstone reservoir in California, a cavernous dolomitic limestone in a Texas field, and various Mid-Continent oil fields, respectively.[2]   The numbers associated with the different curves give the permeabilities of the samples in millidarcys.

Figures 7.14 to 7.16 have several broad features in common.   They show nonvanishing values of capillary pressure at 100 per cent water saturation.

[1] J. J. McCullough, F. W. Albaugh, and P. H. Jones, *API Drilling and Production Practice*, 1944, p. 180; O. F. Thornton and D. L. Marshall, *AIME Trans.*, **170**, 69 (1947); W. A. Bruce and H. J. Welge, *Oil and Gas Jour.*, **46**, 223 (July 26, 1947).

[2] Hassler, Brunner, and Deahl, *loc. cit.*

These evidently represent the displacement pressures previously discussed. The initial rise in capillary pressure as the water saturation is decreased is generally quite slow. Beyond the region of rapid fall in saturation as the capillary pressure is increased the latter rises sharply in a manner indicating

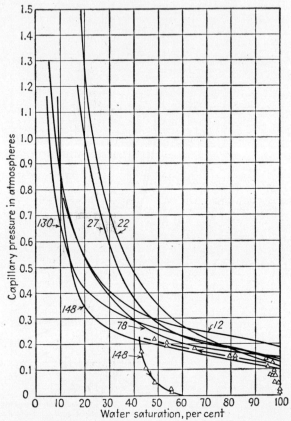

FIG. 7.15. Capillary-pressure curves, obtained with water, on benzene-extracted dolomite samples from a West Texas field. (*After Hassler, Brunner, and Deahl, AIME Trans., 1944.*)

an approach to a vertical asymptote. An order-of-magnitude interpretation of the ordinate scale used in Figs. 7.14 to 7.16 is obtained by noting that, if the water-air interfaces are assumed spherical [$R_1 = R_2$ in Eq. (1)], a capillary pressure of 0.1 atm implies a radius of curvature of 1.4×10^{-3} cm and 1 atm corresponds to 1.4×10^{-4} cm for the radius of curvature.

The rates of variation or slopes of the curves of Figs. 7.14 to 7.16 in the region of high liquid saturation largely reflect the pore-size distribution within the rock. A slow rise in the curve, as observed with the 90- and 218-md specimens in Fig. 7.16, indicates that a large fraction of the pores have substantially the same effective radius and geometry, such as would be found in clean sands with well-sorted and rounded grains.   A rather rapid initial rise[1] in the capillary-pressure curve, as found for most of the

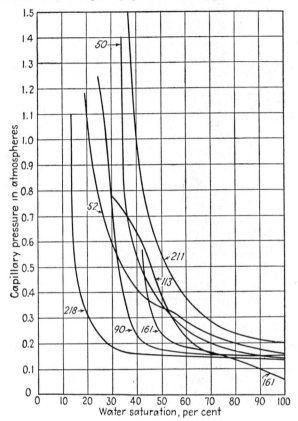

FIG. 7.16.   Capillary-pressure curves, obtained with water, on benzene-extracted sandstone samples from various Mid-Continent oil fields.   (*After Hassler, Brunner, and Deahl, AIME Trans., 1944.*)

samples of Fig. 7.14, suggests a continuous gradation over an extended range in pore sizes.   This may well be expected in the case of the argillaceous sands in most California fields.

[1] As noted in Sec. 7.9, theoretical considerations suggest that, except when the rock specimen has such continuous channels as would lead to a substantially zero-equilibrium gas saturation, the initial desaturation segment should always be strictly horizontal. If the nonwetting phase be considered as entering and completely occupying pores of continually decreasing size as the capillary pressure is increased, the corresponding pore-size distribution function will be proportional to the reciprocals of the slopes of the capillary-pressure curves [cf. H. L. Ritter and L. C. Drake, *Ind. and Eng. Chemistry, Anal. Ed.*, **17**, 782 (1945)].

The apparently asymptotic limits in water saturation approached by the curves as the capillary pressure increases represents the irreducible water saturations to be discussed in the next section. It is these limiting values that are considered as equivalent to the connate-water saturations that would be found in the rock samples if they were located at an appreciable elevation above a water-saturated zone.

While the general subject of capillary phenomena in porous media has in no sense been thoroughly explored, studies thus far made already suffice to disclose a number of complications beclouding the rather simple outline presented above. First of these is that of hysteresis. The very shapes and continuity of the curves of Figs. 7.14 to 7.16 resulted from the fact that the rock samples were investigated by exposing them to a continuous liquid-desaturation process. Such a process would occur by gravity drainage if a long liquid-saturated column of rock were mounted vertically with its bottom end in contact with the free-liquid surface. To obtain an equivalent result with short rock specimens, suction may be applied at the bottom of the sample in contact with a liquid-saturated capillary diaphragm. This type of desaturation process may be accelerated by subjecting the specimen to a centrifugal-force field. Or, as is the more common practice in making connate-water determinations, pressure may be applied to the exposed surface of the sample resting on a liquid-saturated capillary diaphragm (cf. Fig. 3.11). Such procedures give "drainage" capillary-pressure curves, which are essentially reproducible and represent well-defined relationships between the interfacial curvatures and saturations.

If, however, initially, or at any state of partial saturation, the liquid content is allowed to increase by imbibition of liquid, the capillary-pressure curve will generally not reproduce that obtained under desaturation processes. A typical curve segment so obtained is indicated by the points plotted in Fig. 7.15 for the 148-md sample, beginning at a saturation of 42 per cent and terminating at 60 per cent. These points, together with the upper drainage curve, form a hysteresis loop giving an apparent ambiguity in the capillary-pressure–saturation relationship. The exact nature of the hysteretic effect will depend on the rock sample and the starting point or past saturation history of the specimen.[1] Generally the saturation developed by imbibition at zero capillary pressure will not be complete, so that a displacement pressure will not be necessary to initiate desaturation processes. This feature appears to be of special importance in the development of interfluid transition zones in actual reservoirs, as will be discussed in the next section.

---

[1] While desaturation capillary-pressure curves are also inherently subject to hysteretic effects, depending on the past saturation history and starting point, such curves are generally determined with cores initially fully saturated with the wetting phase and hence are unique and reproducible.

If both the nonwetting and wetting phases are continuous, the multiple hysteresis states of saturation for fixed capillary pressures are individually in stable equilibrium. In addition to these, however, there are states of metastable equilibrium in which the associated capillary pressure and fluid saturation have no systematic relationship and represent little more than accidental points. These correspond to separated or discontinuous concentrations of the nonwetting phase in the form of bubbles or globules, which are trapped and immobile. The curvatures and total fluid saturations are determined by the size and geometry of the individual bubbles or globules and are not changed by variations in the interfacial pressures except through the processes of solution and diffusion.

As previously indicated, a fully saturated porous medium will not permit the entry of a nonwetting phase until the displacement pressure has been exceeded. On the other hand, as discussed in Sec. 7.6, when the liquid saturation is reduced below its "equilibrium saturation," the nonwetting or gas phase will be continuous and it will move even under an infinitesimal pressure gradient. At intermediate wetting-phase saturations (between the equilibrium value and 100 per cent) the gas phase will be discontinuous and will tend to remain locked in the pores of the medium. However, it is still subject to displacement if sufficient pressures are imposed on the individual bubbles to force them through the pore constrictions. Such pressures are "threshold pressures." Their maximum is the displacement pressure in the limit of 100 per cent wetting-phase saturation. They will decrease as the gas saturation increases. And they will vanish at the limit of the equilibrium gas or liquid saturations. In fact the latter may be considered as that saturation at which the threshold pressure becomes zero. If the gas-phase distribution is of the type developed by the continual pressure decline associated with the solution-gas-drive production mechanism (cf. Sec. 9.4), the zero threshold-pressure saturation will give the maximum possible "equilibrium" wetting-phase saturation. It is this value which is of importance in solution-gas-drive reservoir performance. However, it is possible to have still lower metastable wetting-phase saturations in which the nonwetting phase is in the hysteresis range and remains immobile owing to its accidental discontinuous distribution. Such states of fluid distribution may occur during imbibition processes, as the migration of oil into a dry-gas zone or partially depleted oil zone.

7.9. The Fluid Distribution in Virgin Reservoirs.[1]—An important application of the concept of capillary pressures pertains to the fluid distribution in an oil- or gas-bearing formation prior to its exploitation. It is a generally accepted theory that virtually all oil- or gas-bearing reservoirs were saturated with water prior to the entry of the petroleum fluids. Both

---

[1] The discussion of this section follows that of M. Muskat, *Petroleum Technology*, **11**, 1 (July, 1948); cf. also Leverett, *loc. cit.*

geological evidence and the universally found occurrence of connate water in oil- and gas-productive formations strongly support this theory. The details of the dynamical processes of displacement of the water from the reservoir during its invasion by the migrating oil are still subject to much speculation and conjecture. It is clear, however, that the ultimate equilibrium fluid distribution can be developed only in relation to the gross thickness of the oil-bearing section and the final position of the water-oil contact, which depend on the total volume of oil available from the source bed. The oil and water saturations resulting from the initial entry of oil into the trap will therefore require some readjustment during the later stages of the oil accumulation or after the latter is completed. While the fluid motion will be subject to gravity forces throughout the accumulation process, it is in the establishment of the final equilibrium distribution that gravity will play its most important role. Now if the effect of gravity forces were entirely unrestricted, all the water lying above the undisturbed water-saturated zone would have drained down to the bottom of the reservoir, forming a sharp horizontal plane of demarcation with the overlying oil mass. And the latter, in turn, would have been overlain, above a sharply defined horizontal plane, by whatever free-gas phase may have been associated with the invading oil mass. Such would have been the fluid distribution found in oil reservoirs at the time of discovery.

It is because of capillary forces that actual oil reservoirs would not be expected and are not found to show the above-described fluid distribution. The actual fluid segregations as observed are neither complete nor sharp. The failure to develop complete fluid segregation may be considered as simply due to a vanishing of the permeability to the water phase long before its saturation falls to zero, as found experimentally and discussed in Sec. 7.5. Accordingly the mass downward drainage of water overlying the saturated water zone would of necessity be terminated before complete segregation could be achieved. But this vanishing of the permeability is in itself the result of capillary forces. For at saturations of vanishing permeability the water may be visualized as confined largely to separated rings enveloping the points of areas of contact between the grains, their interconnection, if any, being limited to films over the intervening grain surfaces of a small number of molecular layers in thickness. Such a distribution has been termed "pendular," in contrast to the "funicular" distribution, where there is a continuous mass of wetting phase covering the solid surfaces (cf. Fig. 7.17). The localization of the liquid in these regions is a consequence of the thermodynamic equilibrium requirement that the total surface energy of the free and interfacial surfaces be a minimum for the given volume of liquid associated with each grain cluster. Because of the increasing surface energy resulting from a distortion and displacement of these pendular rings, they will resist mass movement even aside from the viscous-friction forces. As a result there will be an "irreducible water saturation" below which the natural force of gravity will be unable to drain the porous medium at any elevation[1] above the water-saturated zone.

This irreducible water saturation represents only a limiting minimum for the residual-water content. For, as in the case of the capillary-tube system discussed in Sec. 7.8, the capillary pressures can sustain a height of

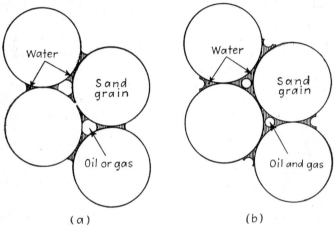

(a)   (b)

FIG. 7.17.   A diagrammatic and idealized representation of wetting- and nonwetting-phase distributions about the contacts between spherical sand grains. (a) Pendular-ring distribution. (b) Funicular distribution.

liquid against the force of gravity even when the liquid is mobile and could drain downward if free of upward retentive forces. Hence the condition of equilibrium approached by the drainage of the water toward the free-water zone will be established by the balance between the average upward component of the capillary-pressure forces and the downward force of gravity.

While there can be little doubt that, in principle, the equilibrium fluid distribution in virgin reservoirs must be determined by the nature of the capillary-pressure vs. saturation curve, the quantitative calculation of this distribution is still subject to a number of uncertainties. Furthermore, several assumptions must be introduced in constructing the equations governing the calculations. For convenience it will be assumed[2] that the interfacial-curvature vs. saturation relationship in a particular porous medium may be represented by a universal function or curve, independently

---

[1] Interstitial water has been found in oil reservoirs as much as 2,000 ft above the water-oil contact (cf. Bruce and Welge, *loc. cit.*).

[2] While the validity of this assumption has been confirmed for some rock and fluid systems, in others it definitely is not valid (cf. J. C. Calhoun, Jr., M. Lewis, Jr., and R. C. Newman, AIME meetings, San Francisco, Calif., February, 1949).

of the nature of the fluids.   Denoting the total liquid saturation bounded by the interface by ρ and the curvature by $C(\rho)$ the latter will then be independent of the manner in which ρ is composed.   When the interfaces are formed by a desaturation process, it will be tentatively assumed to have a form such as is indicated by the solid curve in Fig. 7.18.[1]   This curve is to be considered as representing the directly measured capillary-pressure curve, as plotted in Figs. 7.14 to 7.16, with the capillary-pressure ordinates divided by the surface tension of the liquid used in the experiment.

Representing diagrammatically the water-oil and gas distribution in a vertical section of a formation as shown in Fig. 7.19 the capillary equi-

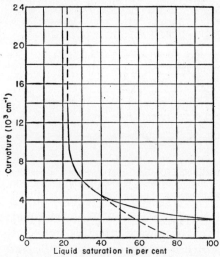

FIG. 7.18.   A hypothetical curvature-saturation relation in a porous medium.   Solid curve, drainage.   Dashed lower segment, imbibition.

librium in the two transition zones, above the water- or oil-saturated sections, will be given by

$$p_b - \gamma_1 gh = p_b - \gamma_2 gh - \sigma_{12} C(\rho),\tag{1}$$

where p_b denotes the base pressure at the beginning of the transition zone, γ_1, γ_2 the densities of the two phases under consideration, σ_{12} their inter-

[1] The solid curve of Fig. 7.18 has been drawn without a zero slope at 100 per cent saturation to simulate the characteristics of those most commonly reported from laboratory measurements.   However, a horizontal section extending to the equilibrium wetting-phase saturation would be expected theoretically, unless the sample has localized but continuous channels of abnormally large pore radii, which comprise only a small fraction of the total pore volume of the specimen.   A nonvanishing initial slope would imply both mobility and pressure equilibrium under conditions where the nonwetting phase is dispersed and discontinuous.

facial tension, ρ the liquid saturation enclosed within the gross interface, and g the acceleration of gravity. It follows at once from Eq. (1) that

$$C(\rho) = \frac{g(\gamma_1 - \gamma_2)h}{\sigma_{12}}. \tag{2}$$

Before attempting to apply Eq. (2) it is well to make a closer examination of the fundamental mechanism that was probably involved in the process

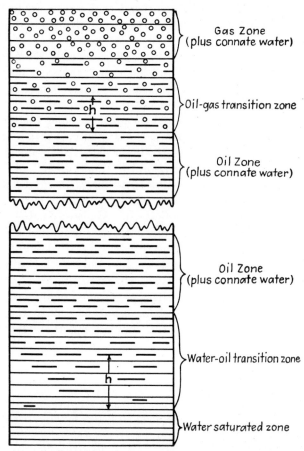

FIG. 7.19. An idealized representation of the transition zones in a vertical section of a reservoir with a free-gas zone and in contact with an intercommunicating water table. $h = $ height above the bottom of the transition zone.

of oil accumulation and subsequent fluid redistribution. As noted previously, there is virtually universal agreement that immediately prior to the appearance of the oil in its trap the latter was filled with water. The accumulation of the oil within the reservoir trap therefore involved partial displacement of the water. It appears probable[1] that the residual water left by this original displacement process of "oil flooding" represented a saturation appreciably higher than that which may be considered as its normal connate-water content. The upper part of the oil zone was thus left with an excess of water. During the subsequent[2] geologic time before discovery it was therefore draining downward so as to establish a capillary-pressure–gravity-head equilibrium by a wetting-phase drainage or desaturation process. This equilibrium, expressed by Eq. (2), would then be determined by a curvature-saturation relation as represented by the solid curve of Fig. 7.18.

Associated with the downward drainage of water in the upper parts of the pay there must evidently have been a countercurrent upward flow of oil to replace the removed water. This, in turn, required an increase in water saturation at the lower levels suffering oil depletion. Thus, whereas the main oil-producing pay itself and the upper layers of the zone of transition to the water-saturated section were being subjected to a drainage and desaturation of the wetting phase, the lower part of the water-oil transition zone was experiencing a resaturation, or "imbibition," of the wetting phase. Accordingly in applying Eq. (2) to the region immediately above the free-water zone an imbibition curvature-saturation curve, such as the dashed curve in Fig. 7.18, should be used. As also indicated by the lower segment of the 148-md curve in Fig. 7.15, the imbibition curves do not, in general, have a displacement-pressure type of discontinuity at 100 per cent saturation. In fact they usually show only partial saturation even at vanishing capillary pressure. Hence the oil saturation at the beginning of the transition zone will have the nonvanishing value characteristic of the capillary-imbibition process. The uncertainty as to the exact point of change from the imbibition to the drainage curves is not serious, since in the region of moderate and low wetting-phase saturations the two curves generally tend to merge. In fact it will suffice for illustrative and most practical purposes to use the imbibition curve for the whole of the transi-

---

[1] In fact, if the invading driving pressure on the oil is just equal to the displacement pressure, it would be expected that the water saturation would be reduced only to its equilibrium value. If this were the case, there would be no imbibition phase, as discussed below. However, since the buoyant forces will continually increase as the oil migrates upstructure toward its trap, the invasion pressure differential will become increasingly greater than the displacement pressure, except for the friction head, and the residual-water saturation immediately following the initial oil entry should be less than the equilibrium value, although higher than the "irreducible" saturation.

[2] For purposes of simplicity it is tacitly assumed in this discussion that the processes of oil accumulation in the reservoir are rapid compared with those of establishing capillary and gravity equilibrium. This, however, is not essential to the argument that the imbibition type of curve controls the lower part of the water-oil transition zone.

tion-zone calculation. On the other hand, since the imbibition curve is quite sensitive to hysteretic effects, it will not be feasible to establish the quantitative character of the curve with any degree of certainty.

On applying the imbibition curve to Eq. (2) one obtains the water-oil saturation-phase boundary curve of Fig. 7.20, assuming $\gamma_w - \gamma_o = 0.3$ gm/cc and $\sigma_{wo} = 30$ dynes/cm. In contrast to previously published transition-

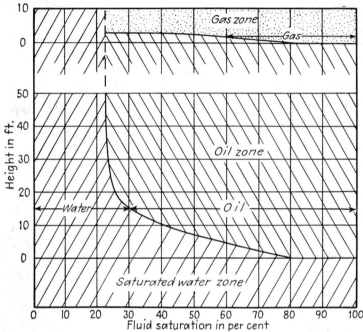

Fɪɢ. 7.20. The calculated saturation distributions in water-oil and oil-gas transition zones using the curvature-saturation functions of Fig. 7.18. Water-oil and oil-gas density differences assumed = 0.3 and 0.5 gm/cc; water-oil and oil-gas interfacial tensions assumed = 30 and 20 dynes/cm.

zone diagrams, in which the oil saturation has been assumed or calculated to build up continuously from zero, it will be seen that in Fig. 7.20 it starts abruptly at a saturation of 20 per cent, the assumed zero-curvature intercept of the imbibition curve of Fig. 7.18. Unfortunately there are no satisfactory published quantitative field data on the water-oil transition zone. The reality of the curves indicated in Fig. 7.20 must for the present therefore be judged on the basis of physical considerations.

The situation at a gas-oil contact, if such occurs, should be similar to that obtaining at the water-oil contact. Although the oil is a nonwetting phase with respect to the rock, it may be considered as a wetting phase in relation to the gas at the high oil saturations at the gas-oil contact. Accordingly it may be expected that adjustment to equilibrium will also involve here a countercurrent flow, of oil and gas, resulting in an imbibition process with respect to the oil phase in the lower part of the transition layer, and drainage or desaturation of the oil in the part adjoining the free-gas zone. While the details of the imbibition curve will undoubtedly be different for the three-phase system of oil absorption in a water-wet rock containing a dispersed gas phase from those for water absorption in the same medium when partly saturated with oil, they will be taken here as the same for illustrative purposes. Upon choosing the oil-gas density difference as 0.50 and the surface tension of the oil in equilibrium with the gas as 20 dynes/cm, an application of Eq. (2), with the above changes in constants, gives a fluid distribution in the lower part of the oil-gas transition zone as plotted in Fig. 7.20.

The determination of the fluid distribution in the upper part of the gas-oil transition zone requires further consideration. If the basic curvature-saturation relations of Fig. 7.18 be assumed to apply to the total liquid saturation, the application of Eq. (2) would formally describe the reduction in total liquid content with increasing height until the irreducible saturation corresponding to the connate water is reached. This would imply that the oil saturation at the same time would continuously decline to zero, in spite of the loss in mobility of the necessarily dispersed oil phase and the associated breakdown of the applicability of the hydrostatic-equilibrium requirement, implied by Eq. (2).

This phase of the problem can be treated by introducing the assumption that in the range of low total liquid saturations the oil and gas invert their roles. The gas becomes the continuous apparent wetting phase, and, relative to it, the oil will behave as the single nonwetting phase. The oil may therefore retain any distribution associated with an imbibition development of the gas saturation. The initial growth of the continuous oil phase below the upper limit of the transition zone is then to be calculated by Eq. (2), with h representing the depth below this upper limit and with the argument of the curvature function referring to the sum of the gas-phase and water saturations. The density difference will again refer to oil and gas, and $\sigma_{1,2}$ to the surface tension of the oil, in equilibrium with the gas. Assuming the same imbibition curve the gas-oil distribution in the upper part of the transition zone will thus simply be a symmetrical inversion of that near the oil-saturated zone. The region between will depend on the details of the change-over between the oil and gas phases as the apparent wetting phase and on the interfacial geometry in this intermediate saturation range. As

[1] The difficulty could also be resolved by considering the connate water as "frozen" and the oil as a similar wetting phase with its own curvature-saturation relation and "irreducible saturation." There is no evidence, however, to support such hypotheses.

very little is experimentally established regarding this region[1] in three-phase systems, the illustrative curve in Fig. 7.20 was computed by assuming complete symmetry about the 61 per cent saturation.

The magnitude of the initial nonwetting-phase saturations has been deliberately made high in Figs. 7.18 and 7.20 to emphasize the possibility of occurrence of such distributions. It is conceivable that the imbibition process may displace all the nonwetting phase, and the transition zones will correspondingly show a build-up of this phase from a strictly zero saturation. This, however, is unlikely. In any case, since the first layers of the transition zones are apparently the result of wetting-phase absorption processes, the initial nonwetting-phase saturations need not necessarily build up from zero but may have any starting saturation,[2] determined by the past history and the details of the capillary-pressure imbibition curve. On the other hand, it must be recognized that any discontinuous nonwetting phases that might be left in the transition zones by the imbibition processes would be unstable thermodynamically. Solution and diffusion processes will tend to remove such dispersed-phase inclusions. In the ultimate equilibrium state only continuous nonwetting phases will occur.

The break drawn in Fig. 7.20 indicates that the absolute thickness of the oil zone is not directly dependent on the capillary phenomena. It is determined simply by the total oil content of the reservoir and the gross geometry and porosity of the latter. The transition-zone distributions are to be considered as superposed and joined to the main body of the oil-saturated section, the capillary effects in the latter being mainly restricted to the fixing of the connate-water content. Of course, if the total oil content is very limited, the transition zones may comprise an appreciable part of the total oil-productive section and, in extreme cases, may even span the whole of the so-called "oil zone."

The water-saturation area of Fig. 7.20 has been abruptly limited by a reduction to an asymptotic limit within the oil zone. This is in accord with the manner in which the curvature-saturation curves were drawn in Fig. 7.18. Because of experimental uncertainties, measured capillary-

---

[1] A change in the role played by the gas and oil, as their mutual saturations are varied in a water-containing sand, is also suggested by their relative permeabilities, as plotted in Figs. 7.10 and 7.11. It is to be emphasized, however, that the suggested inversion of the gas and oil as apparent wetting phases is a purely hypothetical artifice for avoiding the continuous decrease of the oil saturation at the upper part of the gas-oil transition zone.

[2] It should be noted that the equilibrium nonwetting-phase saturations at the beginnings of the transition zones may extend downward into the oil- or water-saturated zones below, if the readjustment of the initial fluid distributions and upward flow of the oil or gas may have led to a shrinkage in the gross thicknesses of the oil or gas zones.

pressure curves, such as those reproduced in Figs. 7.14 to 7.16, do not answer the question whether the wetting-phase saturation can be continuously reduced to an asymptotic limit, or even zero, as the capillary pressure is increased indefinitely, or whether an "irreducible" water saturation develops at a finite and possibly sharp limiting value of the capillary pressure. The extremely complex geometry of the wetting-phase surfaces, as the pendular region is approached, makes it difficult to predict the detailed nature of the development of the pendular wetting-phase saturation. It seems likely, however, that, at least locally, the latter state results from sharp breaks in the continuous funicular wetting phase as a critical capillary pressure is exceeded.[1] This has been assumed in drawing the curves of Fig. 7.18. At heights corresponding to still greater capillary pressures, hydrostatic equilibrium in the wetting phase will no longer obtain. It is for this reason that no provision was made in the above discussion for pressure equilibrium between the gas phase in the oil-gas transition zone or in the overlying free-gas section and its water phase.

Much of the above discussion has been, of necessity, hypothetical, and even conjectural. However, there is little reason to doubt that the order of magnitude of the thickness of the transition zones, as determined by the above-outlined procedure, is substantially correct. Moreover, it is clear from Eq. (2) that, since the density difference will generally be greater between oil and gas than water and oil and the surface tension for the former will be lower than the water-oil interfacial tensions (cf. Sec. 2.12), the height of the oil-gas transition zone will be smaller than for the oil-water zone. On the other hand, the numerical values indicated in Fig. 7.20 should not be considered as applying to actual reservoirs. Aside from the different magnitudes of the coefficients in Eq. (2) occurring for different reservoir-fluid combinations from those assumed here, the basic curvature-saturation function will vary with the detailed nature of the rock, as is evident from Figs. 7.14 to 7.16.

That the lower limiting value for the wetting-phase saturation, indicated by the capillary-pressure curves, should represent the actual connate-water saturation at appreciable distances above the water-saturated zone is also a definite[2] implication of the above considerations, regardless of the

[1] A number of published capillary-pressure curves show no change whatever in the residual-water saturations for the last several measurements (cf. Thornton and Marshall, *loc. cit.*).

[2] Note should be taken, however, of the still unexplained differences which have been sometimes observed between the "irreducible" or connate-water saturations as obtained when air is used as the displacing medium as compared with those obtained when oil is used (cf. Sec. 3.11). On the other hand, it should be observed that while the concept of "irreducibility" of the connate-water saturations above transition zones in actual reservoirs is supported by the apparent immobility of the water during actual oil-

detailed fluid distributions within the transition zones. And from a practical point of view it is immaterial whether this saturation be considered as an asymptotic limit or an irreducible saturation developed by the sudden break-over from the funicular to the pendular distribution. Moreover, owing to variations in the permeability of virtually all producing formations, no constant or monotonically changing connate-water saturation is to be expected in actual reservoir rocks. Except immediately above the water-saturated zone the values will generally reflect the local capillary structure of the rock more than its location above the water-oil contact.

7.10. Dynamical Effects Associated with Capillary Phenomena.—The two preceding sections have dealt with capillary effects that are basically of an equilibrium and static nature. There are also various dynamical aspects to capillary phenomena, although they have been crystallized even less completely into quantitative implications. These will briefly be discussed here.

One of the basic dynamical implications of the capillary-pressure concept is that the actual pressures in the various mobile phases at equivalent positions along their streamlines will not be identical. That is, the pressures associated with the individual phases are, in principle, distinct rather than identical variables. Thus in considering the statement of Darcy's law for the gas and oil phases of a multiphase system the form given in Eq. 7.1(1) would have to be generalized to

$$\left.\begin{aligned} \bar{v}_o &= -\frac{k_o}{\mu_o}\nabla(p_o - \gamma_o gz), \\ \bar{v}_g &= -\frac{k_g}{\mu_g}\nabla(p_g - \gamma_g gz), \end{aligned}\right\} \tag{1}$$

where p_o, p_g refer to the oil and gas phases, and are not assumed equal[1] a priori. To proceed further it would be assumed[2] that the static interfacial-pressure discontinuities, as a function of the interfacial curvature, will also obtain under flow conditions, so that

$$p_g - p_o = p_c, \tag{2}$$

---

producing operations, the magnitude of the irreducible water saturation may be affected by hysteresis phenomena and the processes leading to its development. Relative-permeability experimentation indicates that under dynamical conditions significant wetting-phase mobility usually ceases at saturations higher than the limit obtainable by capillary-pressure desaturation (cf. Sec. 7.5), and oil-base cores which presumably are taken above transition zones often also show higher water contents than indicated by capillary-pressure experiments. In the above discussion the static capillary-pressure type of data have been used only in lieu of unavailable data related to the actual reservoir processes.

[1] A similar equation will apply to the water phase, if mobile.

[2] This is an unverified assumption and could conceivably be greatly in error.

or, on referring to a mean pressure $\bar{p}$,

$$p_g = \bar{p} + \frac{p_c}{2}; \qquad p_o = \bar{p} - \frac{p_c}{2}, \tag{3}$$

where p_c is the capillary pressure. Equation (1) will then take the more symmetrical form

$$\left. \begin{aligned} \bar{v}_o &= -\frac{k_o}{\mu_o}\left(\nabla\bar{p} - \frac{1}{2}\nabla p_c - g\nabla\gamma_o z\right) \\ \bar{v}_g &= -\frac{k_g}{\mu_g}\left(\nabla\bar{p} + \frac{1}{2}\nabla p_c - g\nabla\gamma_g z\right). \end{aligned} \right\} \tag{4}$$

The conventional form of Darcy's law will thus be modified if the capillary pressure varies along the streamlines. Such variations, however, in a uniform medium, will result only from corresponding changes in the saturation distribution. It is therefore the magnitude of the latter that will determine the dynamical importance of the capillary-pressure effects. An order-of-magnitude estimate of the capillary-pressure gradients, in comparison with the mean pressure gradients, may be derived from the relation

$$\frac{\nabla p_c}{\nabla \bar{p}} \sim \frac{dp_c}{d\rho}\frac{d\rho}{d\bar{p}}, \tag{5}$$

where ρ denotes the total liquid saturation. The term $d\rho/d\bar{p}$ may be approximated by that characterizing the normal gas-drive depletion process, as discussed in Chap. 10 (cf. Fig. 10.3). An upper limit to the value so obtained is of the order of 0.02 per cent/psi. Values of the same order of magnitude are obtained from actual saturation-distribution curves in steady-state gas-liquid-mixture flow, except in the immediate vicinity of the well bore (cf. Fig. 8.5). As is clear from Figs. 7.14 to 7.16, $dp_c/d\rho$ has no fixed value. However, in the range of liquid-phase saturations obtaining in actual producing gas-drive reservoirs a value of 10^{-2} atm/per cent will usually represent an upper limit for water-gas interfaces and would be even smaller by a factor of 2 to 3 for the oil-gas interfaces. In any case an upper limit to the above ratio would appear to be a magnitude of the order of 10^{-3}. There appears to be little reason, therefore, from the standpoint of practical significance, to introduce such capillary-pressure effects into Darcy's law as indicated in Eq. (4), as a correction to the mean-pressure-gradient term.

The relative magnitude of the capillary-pressure and gravity terms can be similarly estimated from the relation

$$\frac{\nabla p_c}{g\,\nabla\gamma z} \sim \frac{(dp_c/d\rho)(d\rho/dz)}{g\,\Delta\gamma}, \tag{6}^*$$

---

* The right-hand side of Eq. (6) expresses the fact that the actual gravity force of significance is $g\,\Delta\gamma$, neglecting the variations of the individual fluid densities, rather than a formal equality between $\nabla\gamma z$ and $\Delta\gamma$.

where z denotes the vertical coordinate and $\Delta\gamma$ the fluid-density difference. The right-hand side of Eq. (6) will be of the order of 20 $d\rho/dz$ cm/per cent. It is doubtful that outside of transition zones between different-density fluid layers $d\rho/dz$ would be greater than 0.01 per cent/cm in a uniform medium, so that the capillary pressures will again be small as compared with the gravity gradients. Within gas-oil transition zones, however, $d\rho/dz$ may become of the order of 1, and the capillary pressure will exceed considerably the gravity term. Hence, if one were to study the dynamical histories of transition zones, where the pressure gradients are inherently small, it would be necessary to consider both the capillary-pressure and gravity effects. On the other hand, because of the tremendous analytical complexity even of problems of multiphase flow in which both these effects are neglected, no attempt will be made in this work formally to take these terms into account.

In nonuniform media or at the contact planes between layers or regions of different permeability the fluid saturations may change rapidly or even abruptly. However, because of the continuity of pressure across rock interfaces the capillary pressures will also be continuous, with respect to phases that are themselves continuous.

In the extreme case where the porous medium terminates and is bounded by a free volume, as at the exposed face of the rock along a well bore, the effect of the capillary forces on the fluid distribution will be a maximum. Such exaggerated modifications of the fluid saturations are known as "end effects." They are of especial importance in multiphase-flow experimentation with short rock specimens. They tend to concentrate excessive wetting-phase saturations near the outflow faces when the major flow component is the nonwetting phase,[1] and reduce the apparent nonwetting-phase permeability. In producing oil reservoirs, however, the end effects appear to be of minor importance,[2] except when the reservoir pressure is substantially depleted, and hence will not be considered further here.

A different type of effect caused by capillary phenomena in nonuniform reservoirs is related to the differential fluid movement in contiguous strata of different permeability. Aside from capillary effects, it would, of course, be expected that the rates of invasion of an extraneous fluid, as water, into the individual members of a stratified formation would be essentially proportional to their effective permeabilities. If there is a great range of variation in the permeabilities, there will be a correspondingly wide range

---

[1] It is quite possible that the rather low gas permeabilities measured on short consolidated sandstone cores, such as are shown in Figs. 7.3 and 7.5, may have been appreciably reduced by such end effects.

[2] Qualitatively the end effects will probably tend to reduce the gas-oil ratios in gas-drive reservoirs, although a quantitative evaluation of this reduction is impossible at present without introducing many simplifying assumptions.

of water intrusion, such as is usually described as a "fingering" or "irregular" water advance. It has often been suggested that if the reservoir withdrawal rates be restricted so as to retard the total rate of water influx, the water advance will be made "regular" and the fingering will be substantially reduced. These assumed effects, although never explained in detail, have been attributed mainly[1] to capillarity, and their practical importance has been considered as intuitively obvious and axiomatic.

In principle, capillary-pressure forces will have a resultant tendency to equalize the rate of oil displacement between tight and permeable parts of a pay. If continuity and equality be assumed for the pressure in the water phase[2] in the flooded part of a highly permeable stratum and that in an adjacent and intercommunicating tighter stratum, which has not yet been invaded by the encroaching edgewater, the oil-phase pressure in the latter should exceed that in the former. This pressure differential would tend to move the oil from the tighter into the more permeable stratum and thus accelerate the water entry into the former. It is this type of mechanism that can cause an interchange of oil and water in adjacent and communicating bodies of porous media of different permeability that initially have the same fluid distribution. If suitably disposed geometrically the capillary pressures will give rise to a flow of water from the permeable to the tight rock, and a countercurrent flow of oil.

The practical importance, however, of effects of this type is subject to serious question. Whatever interstratum flow would be induced by them would depend on the cross-bedding permeability, which often is very small as compared with that parallel to the bedding planes. And if any oil should migrate into the water-invaded higher permeability strata, it would be carried along in the latter at high water-oil ratios. Moreover if the permeability-to-viscosity ratio in the water-invaded rock should be greater than in the uninvaded region, the pressure distributions will be so modified during the water intrusion as to oppose cross flow into the lower permeability strata.

If the withdrawal rates from a nonhomogeneous water-drive reservoir be restricted, the time for the capillary cross flow between the tight and permeable strata would be lengthened and its cumulative magnitude would

---

[1] The other factor commonly assumed to be effective in reducing "fingering" at slow withdrawal rates is the gravity head between the water and oil (or oil and gas in the case of gas-cap expansion). Here, too, this effect is in the "right direction," but its magnitude will usually be small under conditions where the average water-intrusion rate is sufficiently high to control the reservoir performance.

[2] If pressure equality be assumed for the oil phase, an opposite differential pressure would arise in the water. The resultant cross-flow effects and tendency for equalization of oil displacement in the tight and permeable strata would be substantially the same as in the reverse case.

be accentuated. This would presumably be equivalent to a more "regular" advance of the encroaching waters. While the large contact areas along bedding planes would tend to accelerate the cross flow, such areas will be proportional to the degree to which the water invasion has already suffered "fingering." Furthermore the total equivalent cross-bedding pressure differential will be limited to the normal capillary-pressure discontinuity in the tighter strata, whereas the driving pressure differential causing the general water intrusion will usually be of the order of several hundred psi in reservoirs where the water drive is the controlling producing mechanism. While under extreme conditions of severe restrictions in the production rates the capillary effects may exert some influence on the differential rates of water intrusion, it is very doubtful that it will generally be economically feasible to operate at such low withdrawal rates as would be required to induce significant recovery contributions from capilary cross flow, if this were to be the only advantage to be gained from such operating control.

As a whole, capillary forces will represent a minor factor in the dynamics of oil recovery, beyond their direct effect in determining the basic characteristics of multiphase-fluid flow, as expressed by the permeability-saturation relationship. They may influence appreciably the role played by gravity forces, especially in transition zones. They will also be of importance locally wherever the pressure gradients are low or the saturation gradients are high. However, there is little evidence that such control of the capillary phenomena as may be practical under actual operating conditions will often be of great significance with respect to gross reservoir performance and recovery.

7.11. Summary.—As actual oil-producing rocks universally contain both water and gas, as well as oil, the homogeneous-fluid permeability, as measured with a single-phase fluid, requires modification or correction if it is to be applied to practical systems. If the gas associated with the oil is in solution and remains so throughout the course of travel of the oil to the producing wells, its only effect will be to reduce the viscosity of the oil, as discussed in Sec. 2.11. However, if the pressure within the rock falls below the saturation pressure of the gas and free gas is liberated in the porous medium, the resistance to flow of the oil will increase and the effective permeability to the oil will decrease. Likewise, the connate water in oil-bearing rocks will also reduce the permeability to the oil and gas phases.

As long as other phases are immobile, the flow of the oil phase may be considered as equivalent to that of a homogeneous fluid, except that the numerical value of the permeability coefficient must be reduced to take account of the effect of the other phases. However, when two or more of the phases are mobile, the basic concept as well as the numerical value of the permeability must be generalized. In particular a permeability must be associated with each fluid phase as if the individual phases were flowing separately in parallel channels. And their interaction is expressed by the fact that the numerical values of the permeability for the separate phases are determined by the volumetric distribution of the fluid saturation of the rock among all the phases. The permeability is no longer a constant but is a separate function for each phase of the local phase distribution within the porous medium. Thus, superposed on its granular structure, which is dynamically characterized by its homogeneous-fluid permeability, a porous medium carrying a heterogeneous fluid may be considered as possessing a local structure defined by the saturation distribution of the several fluid phases, which in turn determine the local permeabilities for the individual phases.

The empirical basis for this generalized concept lies in actual measurements of the permeabilities, computed by individual Darcy equations [cf. Eqs. 7.1(1)], for the individual phases as functions of the fluid saturations. The latter are conveniently expressed as fractions or percentages of the pore space occupied by the individual phases. And the permeabilities are similarly expressed as fractions or percentages of the homogeneous-fluid permeability and termed "relative permeabilities."

Qualitatively the experimental data on the variation of the permeabilities with the fluid saturations (the "permeability-saturation relationship") show several basic characteristics regardless of the detailed nature of the multiphase system or rock (cf. Figs. 7.1 to 7.13). The relative permeability for the wetting phase (that preferentially wetting the solid rock structure) rapidly falls from 100 per cent as the wetting-phase saturation is reduced below 100 per cent. It is of the order of 50 per cent in clean sands at wetting-phase saturations of 75 to 85 per cent and becomes negligibly small at saturations of 25 to 35 per cent. The nonwetting phase or phases generally show a zero or negligible permeability until its saturation has been built up to 5 to 15 per cent, then rises rapidly on further increase in saturation, and often approaches 100 per cent at saturations of only 80 to 90 per cent. The numerical sum of the individual permeabilities is less than 100 per cent, except at the limits of 100 per cent saturation for the separate phases. For two-phase systems the sum may fall to 33 per cent (cf. Fig. 7.1) and for three phases to as low as 10 per cent of the homogeneous-fluid value for unconsolidated sands (cf. Figs. 7.9 to 7.11).

These broad features obtain whether the wetting phase is oil and the nonwetting phase is gas or whether the wetting phase is water and the nonwetting phase is oil. They may be interpreted in terms of the natural distribution of immiscible phases in the pores of a porous medium. Thus, as the nonwetting phase will tend to occupy the larger pores and central parts of the pore space as a whole, a small amount of this phase will suffice to block off the most conductive part of the medium and lead to a large reduction in permeability to the wetting phase. And when the nonwetting phase occupies 65 to 75 per cent of the pore space, that left for the wetting phase is largely distributed among the finer pores and narrow recesses between the grains, which offer high flow resistance and provide but a negligible permeability. At the same time such a distribution of the wetting phase will not interfere seriously with the flow of the nonwetting phase, which therefore retains a high permeability in spite of the presence of appreciable amounts of wetting phase in the medium. On the other hand, when the saturation of the nonwetting phase is so reduced that it becomes distributed as separated bubbles or globules confined within individual pores or small groups of adjoining pores, it may lose its mobility completely, unless the driving pressure gradients exceed the equivalent of the interfacial capillary forces required to force the disconnected bubbles or globules through the pore constrictions. Under such conditions the nonwetting phase may be expected to show a zero or negligible permeability.

These general characteristics of the permeability-saturation curves are exhibited by all those thus far determined and can be interpreted in terms of the gross wetting properties of the phase in question. Quantitatively, however, they vary widely with the internal microscopic geometry of the porous medium. Moreover, they are generally also dependent on the nature of the fluids (cf. Fig. 7.13), although for clean unconsolidated sands and synthetic porous media even the quantitative aspects usually appear to be fixed regardless of the type of fluids used, provided that the broad distinction is made simply on the basis of gross wettability (cf. Figs. 7.9 and 7.12). In practical applications it will be necessary to use the curves as determined on the actual rock and fluid samples of interest, if the results are to have quantitative significance.

The tendency for the nonwetting phase to be immobile when it is distributed in individual bubbles or globules of dimensions comparable with those of the pores means that in a flowing system in which this phase is continually created (as gas coming out of solution) its saturation will continue to build up until it does develop mobility. As a result, as long as the nonwetting phase is continuously supplied, equilibrium conditions cannot be maintained until its saturation has built up to the minimum value for mobility. The limiting conditions for which equilibrium conditions can first be maintained are termed the equilibrium saturation and permeability (for the wetting phase).

The ratios of the permeabilities for the several phases at fixed saturations, when corrected for their viscosities, give the flowing ratios of these phases. If, in the case of gas-oil systems, one adds to the latter the gas in solution, one obtains the value of the local gas-oil ratio as a function of the oil or gas saturation. From the nature of the individual permeability-saturation curves, it can thus readily be seen that, while the gas-oil ratio will be only the solution ratio for oil saturations above the equilibrium value, it will rise very rapidly as the oil saturation is decreased below the limiting equilibrium value. The quantitative application of this simple but basic observation directly provides predictions of the ultimate recoveries of oil reservoirs produced by virtue of the gas dissolved in the oil, as will be seen in Chap. 10.

Prior to the development of the generalized permeability concept and the permeability-saturation relationship the phenomena of gas-liquid-mixture flow through porous media were generally described as manifestations of the "Jamin effect." The latter denoted the general phenomenon of the resistance of individual gas bubbles to motion through narrow constrictions, as the pore passageways between the sand grains. While such resistance effects play a fundamental role in determining the permeability-saturation characteristics of porous media, they were not previously expressed in fundamental quantitative terms. On the other hand, they were erroneously interpreted as implying that in gas-drive reservoirs—under "capillary control"—there would be no flow of either liquid or gas beyond a distance from a well at which the total pressure increment equals the linearly superposed equivalent static resistances of the bubbles in the individual pores extending radially from the well to that distance. This limiting region of flow to a well was thus bounded by a "radius of drainage." As the generalized permeability concept and its expression by the permeability-saturation relationship both retain the valid substance of the Jamin effect principle and avoid the pitfalls of the extremely idealized extrapolations leading to the theory of the radius of drainage, no further consideration need be given to this earlier approach to the problem of heterogeneous-fluid flow.

The computation of a permeability value simply by inverting a Darcy equation does not of itself prove the validity of the latter. Such proof requires evidence that the calculated permeabilities are independent of both the fluid viscosity and the pressure gradient. Experimental data covering wide ranges of viscosity of the liquid phase show that within the experimental errors there is no systematic variation of the permeability with the viscosity. With regard to the effect of the pressure gradient the available data need to be considerably augmented before definite conclusions can be reached. Studies on oil-water mixtures have indicated that the permeabilities, at least to the nonwetting phase, are higher at higher pressure gradients. On the other hand, none of the investigations of this question when using gas-liquid mixtures gave variations, within the ranges of pressure gradients studied, that were definitely larger than the experimental errors or that could be correlated with the pressure gradient. While physical considerations suggest that the equilibrium-permeability and saturation values should not be entirely independent of the pressure gradient, it appears that at present the neglect of such effects is warranted, at least as a first approximation.

Once the validity of the generalized Darcy "law of force" is accepted, the equations of motion are readily obtained by the application of the equations of continuity for each phase. One thus obtains a set of three basic differential equations in the pressure and phase saturations with coefficients that are functions of these [cf. Eqs. 7.7(1)]. The latter include the gas solubility in the oil and water, the formation-volume factors of the oil and water, the gas density, and the viscosities of each phase, all depending on the pressure, and the permeabilities for the three phases, which are to be considered as known functions of the phase saturations. In addition the porosity of the medium enters explicitly in the equations, and the homogeneous-fluid permeability is involved implicitly to the extent that the permeability-saturation relationships are expressed in terms of the *relative* permeabilities. In principle these equations, if solved with the proper boundary and initial conditions, will describe the dynamical behavior of all types of heterogeneous-fluid systems.

From a practical point of view a number of simplifications of the most general form of the equations can be made in considering specific situations. Under most circumstances the gas solubility in the water phase, within the oil reservoir, can be neglected as compared with that in the oil. And until there has been encroachment of external water into the oil section, it will generally suffice to consider the connate water as immobile within the individual oil-productive strata, thus eliminating the differential equation for the water saturation. Moreover the effect of gravity will usually be of minor significance unless the pressure gradients are comparable with the differential gravity gradients.

Unfortunately, however, even if all such simplifications are introduced, the fundamental equations are still of such a high order of complexity that virtually no direct application of them has yet been made except in the treatment of a few special steady-state systems. The problem is basically mathematical, and serious study of it has been hardly even begun. On the other hand, some progress has been made in developing approximate treatments of certain classes of practical systems, in which the basic concepts of multiphase-fluid flow are preserved in formulations that emphasize the gross behavior of the oil-producing reservoirs rather than the detailed local variations about well bores or external boundaries.

Although the whole microscopic complex of physical interactions underlying multiphase-fluid dynamics and permeability-saturation phenomena reflect the dynamical balance between viscous and interfacial forces, it is not yet practicable to synthesize the microscopically observed relationships from a first-principle analysis of the capillary-force distribution. Rather it is necessary to discuss capillary phenomena as supplementary considerations and as an aid in the general understanding of fluid-phase interactions.

The simplest measure of capillary forces is the "capillary pressure," which is the pressure difference sustained by an interface separating two fluid phases. It is given by the product of the interfacial tension and the sum of the reciprocals of the principal radii of curvature of the interface [cf. Eq. 7.8(1)]. The latter sum, which is termed the "curvature" of the surface, is a basic geometrical property of the interface and is related to the total liquid saturation between the interface and rock walls. It is because the radii of curvature of the interfaces in a porous medium are necessarily of the order of magnitude of the pore dimensions that the curvatures and associated capillary phenomena are relatively much larger in porous materials than in free vessels.

The variation of the capillary pressure, or curvature, with the liquid saturation represents the composite empirical description of the microscopic structure and capillary forces that can be developed in a porous medium. When the sample saturation is continually reduced from 100 per cent by the application of suction or pressure, it is found that the desaturation does not begin until a definite pressure is exceeded, of the order of 0.1 atm. The desaturation first proceeds rapidly as the capillary pressure is increased. The rate of desaturation, however, continually declines, and ultimately a saturation limit is approached that cannot be lowered further even if the capillary pressure is raised to several atmospheres (cf. Figs. 7.14 to 7.16).

The pressure required to begin the desaturation process and permit the first entry of the nonwetting displacing phase (oil or gas) into the saturated specimen is termed the "displacement pressure." The irreducible lower limit of desaturation is considered as the equivalent of the connate-water saturation for a sample lying in an oil or gas zone above the transition zone adjoining the water-saturated section, when applying the capillary-pressure method for connate-water determination (cf. Sec. 3.11).

While the capillary-pressure curve obtained by continuous drainage or desaturation of a saturated rock sample is a reproducible property of the rock, a different curve is obtained if the dry or partially saturated specimen is allowed to absorb the wetting phase. A cyclical process of desaturation and imbibition will thus lead to a hysteresis loop (cf. Fig. 7.15). Moreover the imbibition curve is not inherently unique but depends on the starting point of the imbibition phase and the previous saturation history. An important feature of these curves is that complete saturation is usually not obtained even at vanishing capillary pressure. These curves thus do not show displacement pressures. On the other hand the nonwetting-phase saturation at intermediate values between 100 per cent and that left after completion of the imbibition process, which is generally in a dispersed and discontinuous distribution, will develop mobility if a finite pressure is applied. These pressures are termed "threshold pressures" and increase from zero at the maximum imbibition saturation to the displacement pressure at 100 per cent saturation. The maximum imbibition saturation at which the threshold pressure vanishes corresponds to the "equilibrium saturation," determined by the permeability-saturation curve as that at which the nonwetting phase first develops mobility.

The transition zones between the water-saturated and oil-productive sections of a formation, as well as those between the oil pay and overlying free-gas caps (cf. Fig. 7.19), are determined by a hydrostatic balance between the differential gravity heads and the capillary pressures [cf. Eq. 7.9(1)]. If the curvature-saturation relationship is known for the rock, the saturation distribution can be formally calculated as a function of the height within the transition zone [cf. Eq. 7.9(2)]. Although the ultimate equilibrium fluid saturation in the oil-productive section is undoubtedly developed by a downward drainage of the excess water left by the oil-accumulation process, that in the lower part of the transition zone probably results from a wetting-phase imbibition mechanism. The use of the imbibition curve in calculating the fluid-saturation distribution automatically avoids a number of difficulties that would arise if the drainage curve were assumed to apply. Hence the oil saturation should start, not from zero, but rather with a value corresponding to the zero-capillary-pressure intercept of the imbibition curve (cf. Fig. 7.20). A similar situation obtains in the case of the oil-gas transition zone, and the initial-gas saturation likewise is the nonvanishing intercept given by the imbibition curve. On the other hand, whereas the water-oil transition zone merges into the typical oil-pay fluid distribution, comprised only of connate water and oil, the uppermost levels of the oil-gas transition zone may retain a residual-oil saturation such as is left by the imbibition process, as well as its complement of connate water.

The over-all thicknesses of the transition zones will be proportional to the interfacial tension of the phases involved and inversely proportional to their density difference. The gas-oil transition zone will therefore be usually thinner than the oil-water transition zone by several fold. If the total oil content of the productive formation is quite limited and if the capillary-pressure curve rises gradually to an asymptotic connate-water saturation, as is common in clay-containing sands (cf. Fig. 7.14), the oil-water transition zone may span virtually the whole of the oil-bearing section. Such conditions apparently obtain in some of the reservoirs in the Rocky Mountain states.

The role played by capillary phenomena in the dynamics of oil production, beyond their direct effect in determining the nature of the permeability-saturation relationship, has not been definitely established. If the capillary-pressure interfacial discontinuities are assumed to apply to the continually changing and moving interfaces under flow conditions, the pressures in the separate phases will be different. Darcy's equations will then have to be generalized so as to distinguish among these pressures [cf. Eq. 7.10(1)]. The change so introduced can be expressed as a term proportional to the gradient in the capillary pressure added to or subtracted from that in the mean-phase pressure [cf. Eq. 7.10(4)]. However, as an order-of-magnitude evaluation of this correction term indicates that in uniform media it will usually be less than 1 per cent of the primary pressure-gradient term, it cannot be of great importance in most production problems. On the other hand, in fluid-transition zones it may be of the same order of magnitude as or may even exceed the gravity term and may therefore appreciably affect the fluid dynamics within the transition zones.

In nonuniform media or at contacts between porous materials of different microscopic structure the variations in fluid saturations associated with the different capillary-pressure characteristics will be accentuated. They will be especially exaggerated at free boundaries, such as the exposed rock face in a well bore. These lead to "end effects," which may be quite troublesome in multiphase-flow experimentation with cores of limited length. It is doubtful, however, that these localized distortions of the fluid saturations will seriously affect the gross performance of oil-producing fields, at least until they have become substantially depleted in reservoir pressure.

A final effect of capillary-pressure phenomena that could possibly influence reservoir performance and recovery arises from the encroachment of edgewater in water-drive fields. By the indirect process of encouraging cross flow of oil from strata of low permeability into adjacent strata of higher permeability, and a countercurrent flow of water, the capillary forces tend to equalize the resultant water invasion in the zones of different permeability. It does not seem, however, that under practical operating conditions processes of this type will play an important role in oil recovery.

CHAPTER 8

STEADY-STATE HETEROGENEOUS-FLOW SYSTEMS;
THE PRODUCTIVITY INDEX

From a physical point of view strictly steady-state conditions of hetero-geneous-fluid flow in oil-producing systems are virtually never encountered. The mechanism of oil production is inherently one of a continual change in the volumetric contents of the oil-producing section. Of necessity, the oil expelled must be displaced either by gas or by water.[1] Accordingly, as production proceeds, the average oil saturation in the original reservoir monotonically decreases—except when the production is the result of reservoir liquid expansion—while the resultant saturation of the displacing phase of gas or water, or both, simultaneously increases. Nevertheless there is value in a consideration of the theoretical steady-state behavior of multiphase-fluid systems, for two reasons.

The first is that, as pointed out in the preceding chapter, a rigorous analysis of time-varying systems by means of Eqs. 7.7(1) is virtually im-possible because of the complexity of the nonlinear equations. This un-fortunate circumstance in itself does not, of course, make the steady-state analogue of a particular transient system a practical equivalent of the latter. On the other hand, such steady-state analogues will provide a guide to the qualitative interpretation and understanding of the behavior of the corresponding time-varying systems in lieu of quantitative treatments of the transient problem. They will serve to give a physical picture of the phenomena associated with heterogeneous-fluid flow by which their basic characteristics may be visualized even when the conditions change with time.

Second, in many cases the steady-state prototypes will actually repre-sent physically reasonable approximations to the corresponding practical nonsteady-state systems. That is, the rates of change in oil-producing reservoirs, when considered as a whole, will often be so slow that one may approximate the changing conditions by a continuous succession of steady states. Of course, rapid transients about well bores following artificial changes in their production rates cannot be so approximated. However,

[1] In highly undersaturated reservoirs much of the initial oil production may be replaced by expansion of the residual-liquid content. This too, however, involves a transient decline in pressure and changes in the mass of the reservoir oil (cf. Sec. 11.2).

the flow conditions about a well whose production rate or flowing pressure may be changing only as a consequence of the changes in the reservoir as a whole should be subject to satisfactory representation by sequences of steady states for certain practical applications.

8.1. Linear Systems.—While linear systems in themselves have no immediate practical analogue in oil-producing reservoirs, they serve to illustrate the nature of the heterogeneous-flow characteristics in their most simple form. For this case Eqs. 7.7(1) (under steady-state conditions) reduce to

$$\left. \begin{array}{c} \dfrac{\partial}{\partial x}\left(\dfrac{S_o k_o}{\mu_o \beta_o}\dfrac{\partial p}{\partial x}\right) + \dfrac{\partial}{\partial x}\left(\dfrac{S_w k_w}{\mu_w \beta_w}\dfrac{\partial p}{\partial x}\right) + \dfrac{\partial}{\partial x}\left(\dfrac{\gamma_g k_g}{\mu_g}\dfrac{\partial p}{dx}\right) = 0; \\[4mm] \dfrac{\partial}{\partial x}\left(\dfrac{k_o}{\mu_o \beta_o}\dfrac{\partial p}{\partial x}\right) = 0; \qquad \dfrac{\partial}{\partial x}\left(\dfrac{k_w}{\mu_w \beta_w}\dfrac{\partial p}{\partial x}\right) = 0, \end{array} \right\} \tag{1}$$

where S is the gas solubility, k the phase permeability, μ the viscosity, β the formation-volume factor, γ_g the gas density, p the pressure, and o, w, g indicate the oil, water, and gas phases.

The first integrals of these equations are evidently

$$\left. \begin{array}{c} \left(\dfrac{S_o k_o}{\mu_o \beta_o} + \dfrac{S_w k_w}{\mu_w \beta_w} + \dfrac{\gamma_g k_g}{\mu_g}\right)\dfrac{\partial p}{\partial x} = \text{const} = Q_g; \\[4mm] \dfrac{k_o}{\mu_o \beta_o}\dfrac{\partial p}{\partial x} = \text{const} = Q_o; \qquad \dfrac{k_w}{\mu_w \beta_w}\dfrac{\partial p}{\partial x} = \text{const} = Q_w, \end{array} \right\} \tag{2}$$

where Q_g, Q_o, Q_w are the rates of flow of the gas, oil, and water, respectively, as measured under surface or standard conditions. It follows readily from these that

$$R = \frac{Q_g}{Q_o} = S_o + \frac{S_w k_w \mu_o \beta_o}{k_o \mu_w \beta_w} + \frac{\gamma_g k_g \mu_o \beta_o}{k_o \mu_g}; \qquad R_w = \frac{Q_w}{Q_o} = \frac{k_w \mu_o \beta_o}{k_o \mu_w \beta_w}, \tag{3}$$

where R is the gas-oil ratio and R_w the water-oil ratio. R can also be expressed as

$$\left. \begin{array}{c} R = S_o + S_w R_w + \alpha(p)\Psi(\rho), \\[4mm] \text{where} \\[4mm] \alpha(p) = \dfrac{\gamma_g \mu_o \beta_o}{\mu_g}; \qquad \Psi(\rho) = \dfrac{k_g}{k_o}, \end{array} \right\} \tag{4*}$$

and the argument ρ indicates the fluid-saturation distribution.

---

* The strict validity of Eqs. (3) and (4) and the conclusions drawn from them depend on the tacit assumption that capillary-pressure phenomena may be neglected, which has been made in writing Eqs. (1) and (2) without distinguishing between the pressure gradients in the individual phases.

To obtain the pressure distribution one may integrate any of Eqs. (2). In particular the pressure distribution may thus be expressed formally as

$$\frac{x}{L} = \frac{\int_{p_w}^{p} (k_o/\mu_o\beta_o)dp}{\int_{p_w}^{p_e} (k_o/\mu_o\beta_o)dp},$$
(5)

where the length of the system is L and p_w, p_e are the terminal pressures. The relation between the rate of oil flow, L, and p_e, p_w is

$$Q_o = \frac{1}{L} \int_{p_w}^{p_e} \frac{k_o}{\mu_o\beta_o} dp.$$
(6)

Similar expressions are readily constructed for the gas and water phases, from their corresponding equations in Eqs. (2).

Equations (5) and (6) evidently cannot be integrated directly, as the integrands involve functions of both the pressure and fluid saturations. To carry out the integration one must therefore add a relationship between these variables. Such is provided by the fluid-ratio equations, Eqs. (3) or (4). For example, as in steady states the ratios R and R_w will be constants, Eq. (4) permits the computation of $\Psi(\rho)$, and hence k_g/k_o, as a function of the pressure. Likewise the second of Eqs. (3) gives k_w/k_o as a function of p. From both of these and the permeability-saturation relationship for the porous medium the phase distribution may be determined, and hence ultimately the value of k_o to be used in Eqs. (5) and (6).

As this procedure requires considerable numerical manipulation, its numerical illustration will be reserved for the case of radial flow, which is more practical but for which the formal analysis is quite similar. Some general observations, however, can be deduced directly from the structure of the above equations. For example, upon rewriting the second of Eqs. (2) as

$$\frac{\partial p}{\partial x} = \frac{\mu_o\beta_o Q_o}{k_o},$$

it becomes evident that on proceeding to the outflow terminal the pressure gradient will increase owing to the viscosity increase with decreasing pressure, as generally observed for gas-saturated oils (cf. Sec. 2.11). In addition the permeability k_o may be expected to fall on approaching low-pressure regions, so as to lead to a still more rapid rise in the pressure gradient. These effects will usually more than counterbalance the decrease of β_o with decreasing pressure. It will be recalled that in linear homogeneous-fluid systems the pressure gradient is uniform along the direction of flow.

If the water phase is mobile, similar considerations as for the oil flow show that the permeability to the water must drop along the direction of flow even more rapidly than that to the oil. On the other hand an inspection of Eq. (4) shows that $\Psi(\rho)$ will be greater in the regions of lower pressure. As is to be expected, this implies increasing values of the gas-phase saturation on approaching the outflow terminal.

While the rates of flow of the various phases are inversely proportional to the length of the flow channel [cf. Eq. (6)], they are not directly proportional to the total pressure differential. Moreover their absolute values for given pressure drops will depend on the mutual ratios R and R_w. In fact, even for such simple systems as those of linear steady-state flow, it is necessary to carry out detailed numerical procedures to evaluate quantitatively all features of the flow behavior.

8.2. Radial Flow; Immobile Water Phase.[1]—The analytical solution of Eqs. 7.7(1) for the steady-state radial flow of multiphase fluids is virtually identical with that developed in the last section for linear flow except for simple changes in the geometrical factors. Equations 8.1(3) and (4) for the flow ratios will also apply here. And for the pressure distribution one need only replace x/L in Eq. 8.1(5) by $(\log r/r_w)/(\log r_e/r_w)$, so as to obtain

$$\frac{\log r/r_w}{\log r_e/r_w} = \frac{\displaystyle\int_{p_w}^{p} (k_o/\mu_o\beta_o)dp}{\displaystyle\int_{p_w}^{p_e} (k_o/\mu_o\beta_o)dp}, \tag{1}$$

where r_e, r_w are the bounding radii at which the pressures are p_e, p_w. Similarly the rate of oil flow, by analogy with Eq. 8.1(6), is readily found to be

$$Q_o = \frac{2\pi h k}{\log r_e/r_w} \int_{p_w}^{p_e} \frac{k_o/k}{\mu_o\beta_o} \, dp, \tag{2}$$

where h is the thickness of the producing stratum and k is its homogeneous-fluid permeability.

The same general observations noted in Sec. 8.1 regarding the pressure gradients and phase-saturation variations will also apply to radial systems. And as in the case of Eq. 8.1(6), Eq. (2) shows that here, too, the rate of

---

[1] The dicussions of this and the following two sections follow the treatment of H. H. Evinger and M. Muskat [*AIME Trans.*, **146**, 126, 194 (1942)]; cf. also M. Muskat and M. W. Meres, *Physics*, **7**, 346 (1936). The assumed immobility of the water phase implies either that the water saturation is so low as to have strictly zero permeability or that its saturation builds up on approaching the outflow surface to maintain hydrostatic pressures in the water phase in equilibrium with the declining pressures in the oil and gas by virtue of the capillary-pressure interfacial discontinuities. To emphasize the multiphase character of the gas and oil flow the first condition is tacitly assumed to obtain in the system under consideration here.

flow does not vary linearly with the total pressure drop, $p_e - p_w$, although formally Eq. (2) can be expressed in the conventional radial-flow form

$$Q = \frac{2\pi h k (p_e - p_w)}{\log r_e/r_w} \left(\frac{k_o/k}{\mu_o \beta_o}\right)_{\mathrm{av}} \tag{3}$$

As a first illustration of these considerations it will be assumed that the connate-water saturation is uniformly 20 per cent and has a zero permea-

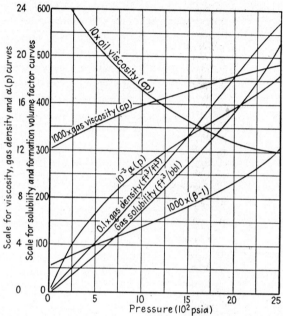

FIG. 8.1.   The physical characteristics of oil and gas assumed in the calculation of steady-state radial heterogeneous-flow characteristics. β = formation-volume factor of oil. $\alpha(p)$ is function defined by Eq. 8.1(4).

bility.   As the water-oil ratio R_w is therefore 0, the gas-oil ratio R will be given by [cf. Eq. 8.1(4)]

$$R = S_o + \alpha(p)\Psi(\rho). \tag{4}$$

For the physical properties of the reservoir fluids, the data shown graphically in Fig. 8.1 will be assumed.[1]   It will be observed that the solubility at the maximum pressure of 2,500 psia is 534 ft³/bbl.   To facilitate the application of Eq. (4) the composite function $\alpha(p)$ [cf. Eq. 8.1(4)] is also plotted in Fig. 8.1.

[1] These are substantially the same as determined for 190°F by B. H. Sage and W. N. Lacey (*API Drilling and Production Practice*, 1935, p. 141) on the Dominguez field oil (33.9° API) and gas, except that the viscosities have been taken as approximately twice those given there.   Here, as in subsequent discussions of heterogeneous-flow systems,

For the permeability-saturation relationship the curves of Fig. 8.2 will be used. As the flow conditions are here considered to be steady, an equilibrium free-gas saturation of 10 per cent has been assumed. Accordingly the gas permeability remains zero until the oil saturation falls below 70 per cent.

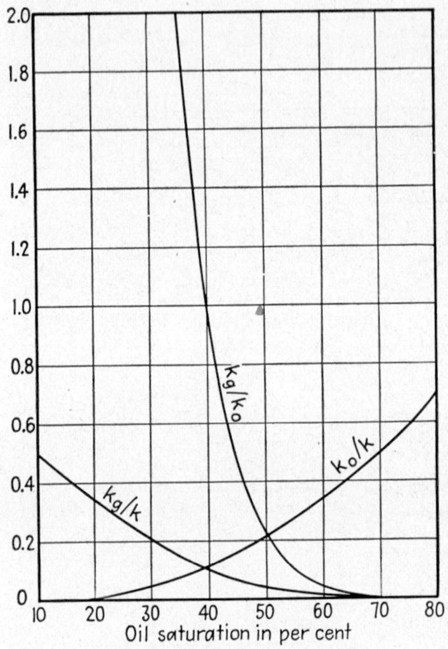

FIG. 8.2. The permeability-saturation relationship assumed in steady-state heterogeneous-flow calculations. k_g, k_o, k are gas, oil, and homogeneous-fluid permeabilities.

To proceed with the determination of the flow conditions in a radial system (into a well bore), the terminal pressures p_w, p_e must be specified. These will be taken as 100 and 2,500 psi, respectively. Next the assumed gas-oil ratio is fixed. This may be chosen as any value equal to or exceeding the solubility at 2,500 psi, that is, 534 ft^3/bbl. The $\Psi(\rho)$ are then calculated as a function of pressure by inverting Eq. (4) as

$$\Psi(\rho) = \frac{R - S_o(p)}{\alpha(p)} = \frac{k_g}{k_o}. \qquad (5)$$

Such curves for values of $R = 534$, 1,500 and 5,000 ft^3/bbl are shown in Fig. 8.3. By reference to Fig. 8.2 the corresponding values of ρ_o are obtained, and then the values of k_o/k. Having thus determined k_o/k as a function of the pressure, the integrals of Eqs. (1) and (2) can be evaluated numerically or graphically. While Eq. (1) gives the pressure distribution, that is, p as a function of r, the previously determined relationship between k_o/k and ρ_o and the pressure may be readily replotted to give k_o/k and ρ_o also as functions of r.

In Fig. 8.4 are plotted the results so obtained for the relative production capacities: $\overline{Q} = Q_o (\log r_e/r_w)/2\pi hk$, that is, the integrals of Eq. (2),[1] vs.

isothermal-flow conditions will be assumed, and the physical properties of the fluids will be considered as referring to a fixed "reservoir temperature." While the validity of this assumption from a practical standpoint may be anticipated intuitively, calculations similar to those given here and for systems of the type which might occur in field operations have shown that even if the flow be completely adiabatic, the temperature drop will be only several degrees Fahrenheit [cf. M. P. O'Brien and J. A. Putnam, *Petroleum Technology*, **4**, 1 (July, 1941)].

[1] These relative production capacities will also apply to linear steady-state systems [cf. Eq. 8.1(6)] with the same fluid and rock properties.

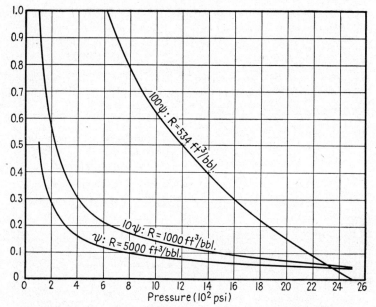

FIG. 8.3. The calculated variation with pressure of the gas-oil permeability ratio in hypothetical steady-state heterogeneous-fluid systems, for fixed gas-oil ratios R. $\Psi = k_g/k_o =$ (permeability to gas)/(permeability to oil).

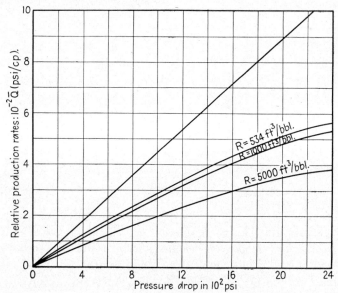

FIG. 8.4. The calculated relative steady-state heterogeneous-fluid-production rates vs. the pressure drop, for fixed gas-oil ratios R. For the straight line the permeability and viscosity were assumed constant. $\overline{Q} = (Q_o \log r_e/r_w)/2\pi hk$, where $Q_o =$ actual production rate; h, k = thickness and permeability of formation; p_e, p_w = pressures at r_e, r_w; p_e is assumed fixed at 2,500 psi.

$p_e - p_w$ for the reservoir pressure of 2,500 psi. The corresponding pressure, permeability, and oil-saturation distributions for the total pressure differential of 2,400 psi, as computed by Eq. (1), are plotted in Fig. 8.5, taking $r_e = 660$ ft and $r_w = \frac{1}{4}$ ft.

It will be seen from Fig. 8.4 that the relative production capacities do not increase linearly with the pressure differential. This is, of course,

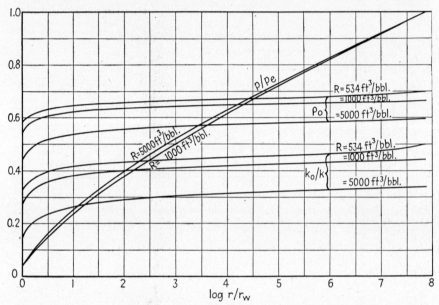

Fig. 8.5. The calculated pressure, permeability, and oil saturation distributions in steady-state radial heterogeneous-fluid systems with fixed gas-oil ratios R. $p/p_e =$ (pressure at radius r)/(pressure at r_e). $p_e = 2,500$ psi; $r_e = 660$ ft; $\rho_o =$ oil saturation; $k_o/k =$ relative permeability to oil; $r_w =$ well radius $= \frac{1}{4}$ ft. Connate-water saturation assumed $= 20$ per cent.

due to the decreased average oil saturation and oil permeabilities as the well pressure is lowered. Moreover the capacities become markedly smaller for the higher gas-oil ratios. The reason evidently lies again in the lower oil saturations and higher free-gas saturations required for carrying the higher gas-oil-ratio fluid streams. This will be verified by reference to Fig. 8.5. The straight line in Fig. 8.4 gives the homogeneous-fluid relationship between the relative production capacity and pressure differential, assuming that throughout the radial system β and μ maintained their values as at r_e, that is, 1.31 and 1.2 cp, and that k_o/k was reduced only by the connate water, that is, $k_o/k = 0.7$. The deviations of the curves from this line show directly the effect of changes in fluid properties and oil saturation due to the heterogeneous character of the actual system.

The curves of Fig. 8.5 show in detail the nature of the radial variations of oil saturation, oil permeability, and pressure. Because of the assumed steady-state character of the flow and the 10 per cent equilibrium free-gas saturation for the rock, the oil saturations never exceed 70 per cent even for the case when there is no free gas entering at the external boundary. On the other hand, neither the oil saturation nor oil permeability falls rapidly from its entrance value until the immediate vicinity of the well bore is reached. Likewise the pressure distribution remains substantially linear, on the logarithmic scale, until the region near the well is reached, where the oil saturation and permeability drop rapidly.

If the reservoir pressure for the same fluid and rock system is less than 2,500 psi, the relative production capacities can also be readily obtained from Fig. 8.4. If these be denoted by $\overline{Q}$, it readily follows from Eq. (2) that

$$\overline{Q}(p_e - p_w) = \overline{Q}(2,500 - p_w) - \overline{Q}(2,500 - p_e), \tag{6}$$

where p_e, p_w are the reservoir and well pressures of the new system and the differences in parentheses represent the actual pressure differentials to be read as abscissas in Fig. 8.4. On computing relative production capacities in this manner for lower reservoir pressures it will be found that for the same gas-oil ratios the production capacity per unit pressure drop decreases with decreasing reservoir pressure. While such comparisons refer to systems producing with the same gas-oil ratio, it will be seen in Chap. 10 that the productive capacities of wells in actual gas-drive oil-producing reservoirs will also generally fall as the reservoir pressure declines.

In oil-field practice the production capacity of producing wells is generally measured as the rate of production per unit pressure drop. When expressed as barrels per day per psi drop, it is termed the "productivity index" and usually denoted by the symbol PI. In this definition no cognizance is given to the possible variation of the productivity index with the pressure differential. However, as is implied by the above discussion and expressed in Fig. 8.4, the productivity index should decrease with increasing pressure differential. Hence, to give this term meaning from a theoretical standpoint, it is necessary to specify the conditions of measurement more precisely. While no such specification has been adopted in the oil industry, a convenient and simple limitation to fix a unique value of the productivity index is the requirement that it refer to the limiting condition of zero pressure differential[1], and it will be so used hereafter with reference to theoretical predictions. This productivity index will evidently be proportional to the slopes at the origin of such curves as those of Fig. 8.4. Analytically they will be defined by

$$PI = \frac{0.007082 k_e h}{\mu_e \beta_e \log r_e/r_w} \text{ bbl/day/psi}, \tag{7}$$

[1] Cf. Evinger and Muskat, *loc. cit.*

where k_e, μ_e, β_e refer to the oil permeability (in millidarcys), viscosity, and oil formation-volume factor at the reservoir pressure, h is the pay thickness (in feet), and r_e, r_w are the effective external boundary and well radii. To use Eq. (7), one does not need to evaluate the integral of Eq. (2). On the other hand, just because it refers to the limiting condition of zero pressure differential, it should not be used to predict production rates at high differentials by simply multiplying the PI by the pressure drop of interest.

8.3. Radial Two-phase-liquid Flow—No Free-gas Flow.—The absence of free-gas flow in heterogeneous-fluid systems would in itself imply either that there simply is no free-gas-phase saturation or that nowhere in the system does it exceed the equilibrium gas saturation. The latter condition, however, cannot be maintained in steady-state flow. For if $\Psi(\rho_o) = 0$, Eq. 8.1(4) requires that the local gas-oil ratio R will decrease in the direction of decreasing pressure. Of course, under transient conditions, during the build-up of the equilibrium free-gas saturation, there will be no free-gas flow until somewhere in the system the equilibrium gas saturation has been reached. However, if the flow is to be strictly of the steady-state type, free-gas flow cannot be avoided unless there is no free-gas phase whatever. In such systems there will be simultaneous flow[1] of oil and water under pressures that are everywhere above the bubble point, so that S_o and S_w are constant over the pressure differential inducing the flow.

The characteristics of steady-state oil-water-flow systems are determined by the water-oil ratio R_w, related to the permeability ratios by

$$\frac{k_w}{k_o} = \frac{\mu_w \beta_w}{\mu_o \beta_o} R_w. \tag{1}$$

In principle Eq. (1) will give k_w/k_o as a function of the pressure, so that from the relation between k_w/k_o and the liquid saturations the latter may be computed as a function of pressure. Actually, however, since, as discussed above, the pressures must everywhere be above the bubble point, the coefficient $\mu_w\beta_w/\mu_o\beta_o$ may be considered as independent of the pressure. Hence k_w/k_o and the fluid distribution will be uniform throughout the flow system.[2] From Eq. 8.2(1), which also applies to the present case, it

[1] Although a steady-state simultaneous flow of oil and water in both two-phase and three-phase systems is inherently possible in limited ranges of fluid-saturation distribution (cf. Sec. 7.6), it is doubtful that flows of this type are of importance in actual oil-reservoir performance (cf. Sec. 14.13). The discussions of this and the next sections are presented mainly to illustrate the relationships between fluid-saturation distributions and the volumetric composition of the flow stream, as implied by the permeability-saturation characteristics of porous media.

[2] Here, too, this constancy is a requirement imposed by the neglect of capillary pressures. If the latter be taken into account, both constant as well as variable saturation distributions will be found permissible.

therefore follows that the pressure will have a logarithmic distribution just as in a homogeneous-fluid system, *i.e.*,

$$p = p_w + \frac{(p_e - p_w) \log r/r_w}{\log r_e/r_w}, \tag{2}$$

and the rates of production for oil and water, Q_o and Q_w, will be given by

$$Q_o = \frac{2\pi h k_o (p_e - p_w)}{\mu_o \beta_o \log r_e/r_w}; \qquad Q_w = \frac{2\pi h k_w (p_e - p_w)}{\mu_w \beta_w \log r_e/r_w}, \tag{3}$$

which are also formally identical with that for homogeneous-fluid flow.

To illustrate the possible variation of the water-oil ratio with the fluid distribution, permeability-ratio curves will be taken as those plotted in Fig. 8.6[1] for $\rho_g = 0$. These correspond approximately to the individual permeability curves of Fig. 7.7. Assuming also that $\mu_w = 1$ cp, $\beta_w = 1$, $\mu_o = 1.2$ cp, $\beta_o = 1.2$, and applying the curves of Fig. 8.6 to Eq. (1), one obtains the results plotted in Fig. 8.7. It will be seen that the water-oil ratio falls rapidly as the oil saturation increases, as would be expected simply from the nature of the curves of Fig. 8.6. This sensitivity of the water-oil ratio to the oil saturation is quite analogous to that of the gas-liquid ratio for gas-liquid-flow systems. In both cases this phenomenon results from the rapid rise of the

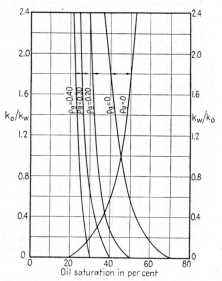

FIG. 8.6. Illustrative permeability-ratio curves for oil and water, as functions of the oil saturation, for fixed values of the free-gas saturation ρ_g. $k_w/k_o = $ (permeability to water)/(permeability to gas).

permeability to the nonwetting phase,[2] as its saturation increases beyond its equilibrium value, and associated decline in permeability to the wetting phase.

[1] It may be noted that the abscissa intercepts in Fig. 8.6 may be interpreted as implying that the porous medium would have an "equilibrium" oil saturation, or a residual water-flooding oil saturation of 20 per cent, and a connate-water saturation of 30 per cent, to the extent that the permeability-saturation phenomena are completely independent of pressure gradients.

[2] In the systems discussed in the preceding section, where the immobile water phase is the true wetting phase, the oil may be considered for some purposes as a wetting phase at the higher oil saturations, even though it, too, exhibits an "equilibrium" saturation characteristic of nonwetting phases.

As implied by Eq. (3), the productivity index for the oil in steady-state water-oil-flow systems will be given by

$$\text{PI} = \frac{0.007082k_o h}{\mu_o \beta_o \log r_e/r_w} \text{ bbl/day/psi,} \qquad (4)$$

where k_o is in millidarcys, h in feet, and μ_o in centipoises. It will be a true index, independent of the pressure differential, though it will vary with the water-oil ratio. The latter va-

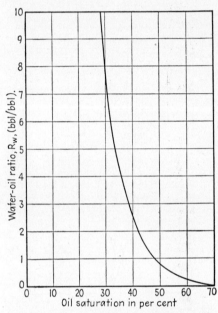

riation may be determined by combining curves such as that in Fig. 8.7 with the oil-permeability curve, such as that shown in Fig. 7.7.

8.4. Radial Three-phase Flow. —While in principle the treatment of steady-state three-phase flow is quite similar to that presented for two-phase systems, the details of the calculations are much more involved. Moreover they require a knowledge of the complete system of permeability-saturation relations, as discussed in Sec. 7.4.

The limiting productivity index and interrelationships between the gas-oil ratio, water-oil ratio, and fluid saturations at the external inflow boundary may be deter-

FIG. 8.7. The calculated variation of the water-oil ratio with the oil saturation in steady-state water-oil-flow systems, using the permeability-ratio curves of Fig. 8.6 and assuming water and oil viscosities of 1.0, 1.2 cp, and water and oil formation-volume factors of 1.0 and 1.2.

mined as follows: fixing the inflow pressure and choosing R and R_w, k_g/k_o can be calculated from the first of Eqs. 8.1(4). Similarly k_w/k_o at the inflow boundary is computed from the second of Eqs. 8.1(3). By reference to the permeability-saturation curves the fluid distribution is found that will give the computed values of both k_g/k_o and k_w/k_o.

By assuming for simplicity that the gas solubility in the water, S_w, may be neglected and taking for the physical properties of the oil and gas those shown in Fig. 8.1, the value of k_g/k_o for an inflow pressure of 2,500 psi is found [by application of Eq. 8.2(5)] to vary with the gas-oil ratio as shown in Fig. 8.8. The corresponding curve for k_w/k_o as a function of the water-oil ratio R_w is also shown in Fig. 8.8, assuming $\mu_w = 1$ and $\beta_w = 1.05$.

The fluid saturations corresponding to the k_w/k_o's and k_g/k_o's implied by Fig. 8.8 may be determined by reference to curves such as are shown in Figs. 8.6 and 8.9, in which the free-gas saturation is used as the parameter fixing the individual curves.[1]   For example, if the water cut is 13 per cent ($R_w = 0.15$) and the gas-oil ratio is 5,000 ft³/bbl, Fig. 8.8 shows $k_g/k_o = 0.043$, $k_w/k_o = 0.10$.   In Figs. 8.9 and 8.6 it is seen that both these values fall on the curves for $\rho_g = 0.20$ at $\rho_o = 0.45$.   Accordingly these, together with $\rho_w = 0.35$, represent the fluid distribution that will give a flow stream with a water cut of 13 per cent and gas-oil ratio of

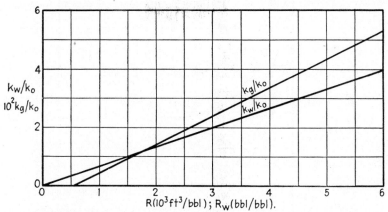

FIG. 8.8.   The calculated values of the gas-oil and water-oil permeability ratios, k_g/k_o and k_w/k_o, required to give the abscissa values of gas-oil and water-oil ratios, using the gas and oil properties plotted in Fig. 8.1.   Assumed viscosity and formation-volume factor of water = 1 cp and 1.05.

5,000 ft³/bbl.   From Fig. 7.10 it is seen that for the above fluid saturations the relative permeability to oil is 0.19.   This means that the limiting productivity index will be 19 per cent as great as if the rock were completely saturated with oil and it were flowing as a homogeneous fluid.   The absolute value of the rate of oil flow or the productivity index near the limit of vanishing pressure differential can be computed by Eq. 8.3(3), and the rate of water production will be 15 per cent that of the oil.

The same procedure can be followed to determine the fluid-saturation distribution at the producing well.   Thus, if the flowing pressure at the well bore is 250 psi, the value of k_g/k_o at the well bore will be 0.235, as may be read from Fig. 8.3.   Noting from Fig. 8.1 that, at 250 psi, $\mu_o = 2.37$ cp

---

[1] It is to be understood that the curves of Figs. 8.6 and 8.9, which were obtained by smoothing the data of Leverett and Lewis [*AIME Trans.*, **142**, 107 (1941)] on unconsolidated sands, are of illustrative significance only.   In practical applications, data should be used that refer to the particular producing stratum of interest.   Moreover they may be plotted with any of the saturations as the parameter or preferably as contours in a triangular diagram of constant values of k_g/k_o and k_w/k_o.

and $\beta_o = 1.077$, Eq. 8.3(1) gives for k_w/k_o the value 0.062. Reference to Figs. 8.6 and 8.9 shows that $k_g/k_o = 0.235$ and $k_w/k_o = 0.062$ both lie on the corresponding curves for $\rho_g = 0.30$ at $\rho_o = 0.385$. Accordingly these values, together with $\rho_w = 0.315$, represent the fluid saturations at the

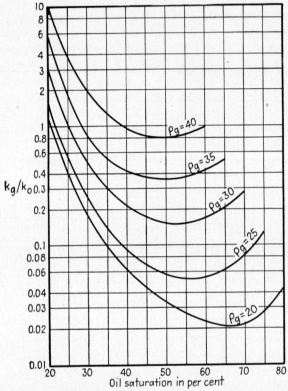

FIG. 8.9. Illustrative curves showing the variation of the gas-oil permeability ratio k_g/k_o in three-phase systems as a function of the oil saturation, for fixed free-gas saturations ρ_g (in per cent).

well bore when producing with a water cut of 13 per cent and gas-oil ratio of 5,000 ft³/bbl at a flowing pressure of 250 psi. Thus between the external boundary (at 2,500 psi) and the well the gas saturation will have increased by 10 per cent, whereas the oil and water saturations will have dropped 6.5 and 3.5 per cent, respectively.

In principle this procedure can be applied at intermediate pressures and ultimately used to determine the complete pressure and fluid-saturation distribution, by means of Eq. 8.2(1). Unfortunately, however, the above examples show that even for unconsolidated sands the necessary permeability-saturation data (Figs. 8.6 and 8.9) are hardly well defined enough to determine such distributions with any precision. And for consolidated rocks there are available no published data whatever describing the full ranges of mobility of three-phase systems. This situation should serve to emphasize the urgent need for the accumulation of such data. On the other hand the considerations of this and previous sections of this chapter show that the physical theory is available for applying the permeability-saturation curves to steady-state heterogeneous-flow systems. But until the specific data pertinent to particular rocks of practical interest are obtained, their implications, as developed here, will be at best only of semiquantitative significance.[1]

If the fluid saturations and associated pressures are known, the gas-oil ratio and water-oil ratios can be determined by simply inverting the above-outlined procedure or by direct reference to the permeability-saturation curves. And the variation of the composition of the flow stream with the fluid saturation can be readily inferred from curves such as plotted in Figs. 8.6 and 8.9. Thus the curves of Fig. 8.6 show the extremely rapid rise in k_w/k_o, and hence water-oil ratio, as the oil saturation decreases, for fixed free-gas saturation. And even if the oil saturation is kept fixed, k_w/k_o rapidly increases as the free-gas saturation decreases and the water saturation increases.

The relative positions of the different curves in Fig. 8.9 show the direct effect of the free-gas saturation on the gas-oil ratio, which is largely determined by k_g/k_o. The rapid rise to the left of the plot is due mainly to the decline in oil permeability as the oil saturation decreases. The initial decline in the k_g/k_o curves, as the oil saturation decreases, is due to the curvature of the gas-permeability curves of Fig. 7.11, which imply a fall in gas permeability, even at constant gas saturations, as water begins to displace the oil.

8.5. The Productivity Index—Theoretical Considerations.—From a practical point of view the productivity index referred to in the previous sections is the most direct measure of the actual productive capacity of an oil-bearing rock. Theoretically, however, it is a quantity dependent on so many factors that a quantitative interpretation of specific numerical values in terms of known physical parameters is often virtually impossible. From its definition

$$\text{Productivity index} = \text{PI} = \frac{\text{production rate (bbl/day)}}{\text{pressure drop (psi)}}, \qquad (1)*$$

---

[1] The neglect of capillary phenomena is another approximation of the treatment of this section. However, attempts to take these into account also will not be quantitatively significant unless capillary-pressure data for actual three-phase systems are used.

* For some purposes it is more convenient to use the "specific" productivity index (SPI), which is simply the PI per unit pay thickness.

it follows that for homogeneous-fluid systems it should have the value

$$PI = \frac{0.003076kh}{\mu\beta \log_{10} r_e/r_w} \text{ (bbl/day)/(psi)}. \tag{2}$$

where k is the homogeneous-fluid permeability in millidarcys, h is the effective pay thickness in feet, β is the formation-volume factor of the liquid phase, and μ is the viscosity in centipoises.

To apply this formula, even accepting its basic limitations, requires a knowledge of k, h, μ, β, and r_e. The value of k must evidently represent an average over individual sample measurements, which often vary by factors of 10 to 100 in single geologically identifiable producing strata. μ and β refer to the oil viscosity and formation-volume factor at reservoir temperature and pressure. These can be determined by appropriate laboratory measurement. r_e, it will be recalled, represents that radius from the axis of the well bore at which the pressure is known to be p_e, which is the base for computing the pressure drop inducing the flow. While the latter is generally taken as the "reservoir pressure," the corresponding value of r_e is not readily determined and is often chosen as half the distance to the nearest neighboring well that is also producing. This arbitrariness is fortunately mitigated by the fact that r_e enters Eq. (2) logarithmically, so that the calculated value of the PI will be rather insensitive to the absolute value chosen for r_e.

To proceed further it must be recognized that there are no strictly homogeneous-fluid systems among actual oil-producing reservoirs. On the other hand, as was seen in Sec. 8.3, when there is no free gas in the rock, as will be the case when the pressures are above the bubble point of the oil, the equations governing the steady-state oil flow will be formally identical with those for a homogeneous fluid, except that the permeability to the oil must be corrected for the water saturation, whether or not the water phase is mobile.[1] Under such conditions Eq. (2) should still be applicable, with k representing the permeability to the oil phase.

Unfortunately the problem is not thus completely solved. For even when flowing above the bubble point, the oil will possess a measurable compressibility, of the order of 15×10^{-6} per psi. Accordingly the assumption of steady-state flow will be only approximately true at the best. It is a well-established observation that when a well is first brought in or is opened after an extended shut-in period, the initial production rates are generally much higher than the subsequently established "settled" rate. This transient behavior is observed both with saturated and undersaturated

---

[1] The flow of condensate-containing single-phase fluids will likewise be subject to an essentially homogeneous-fluid description, if the pressures are maintained above the dew point.

oils and arises from the compressibility of the oil and the free gas, if the latter is present. Until these transients pass and the effect of opening the well or changing the well pressure has been transmitted to the distant parts of the flow system, so that at least the region near the well assumes a substantially steady-state pressure distribution, the productivity index as computed by Eq. (1) will not approximate a constant of physical significance.

When the pressures are below the original bubble point and the fluid stream is inherently heterogeneous owing to gas evolution, the productivity index computed by Eq. (1) becomes even less well defined. The transients, previously mentioned, will be of longer duration, and the steady-state approximation underlying Eq. (2) may have only asymptotic validity. Moreover, as seen in Sec. 8.2, even under steady-state conditions, the value of the PI calculated by Eq. (1) will theoretically depend upon the absolute magnitude of the pressure differential.[1] In particular it should decrease as the pressure differential increases. Completely to specify a computed or measured PI it is therefore necessary to indicate the value of the well pressure or pressure differential to which the PI refers. As an alternative the PI can be redefined as the limiting value of Eq. (1) as the pressure differential is made to vanish, as suggested in Sec. 8.2. Analytically the PI so defined will be given by

$$\text{PI} = \frac{0.003076h}{\log_{10} r_e/r_w} \left(\frac{k}{\mu\beta}\right)_e (\text{bbl/day})/(\text{psi}), \tag{3}$$

where the subscript e denotes that the values of k, μ, β should each refer to those at r_e, or at the reservoir pressure. Hence, as seen in the previous sections, the PI will depend on the composition of the total fluid production as well as on the basic characteristics of the porous medium. And generally, it will be considerably smaller than would be implied by Eq. (2) for a homogeneous-fluid system, because of the reduction in k, as the following illustrations will show:

In Fig. 8.5 it will be seen that, if the water phase is immobile and its saturation is 20 per cent, the relative permeabilities to oil, k_o/k, at the external boundary of the system treated in Sec. 8.2 will be 0.50, 0.44, and 0.34 for gas-oil ratios of 534, 1,000, and 5,000 ft³/bbl. These will also be the ratios of the corresponding PI's implied by Eq. (3) to those predicted by Eq. (2). On the other hand, if only oil and water were being produced as free phases and at equal rates, Fig. 8.7 shows that the oil saturation would be 0.48. Figure 7.7 then implies that the relative permeability to the oil would be 0.20, and the PI would be but 20 per cent of the homo-

---

[1] This dependence is caused by changes in viscosity and formation-volume factor, due to variations in bottom-hole flowing pressure, as well as those in the permeability to oil, although the latter effect will often predominate under multiphase-flow conditions.

geneous-fluid value.   Finally it will be recalled from the preceding section
that, if the gas-oil ratio in a hypothetical steady-state system is 5,000 ft³/bbl
and the water cut is 13 per cent, the saturation distribution at the inflow
boundary (2,500 psi) would be 0.20, 0.45, and 0.35 for the gas, oil, and
water, respectively.   The relative oil permeability for this distribution as
read from Fig. 7.10 is 0.19, so that the PI would be 19 per cent of the
homogeneous-fluid value.

These numerical examples are based, of course, on the particular set of
fluid properties and permeability-saturation characteristics used for illustration in the discussions of the previous sections.   Moreover, they refer
to a reservoir pressure of 2,500 psi.   Thus they have no absolute significance.
However, they should serve to show the order of magnitude of the effect
of the multiphase character of the flow stream on the PI.

To obtain the PI corresponding to Eq. (3) from field data one need
only determine the slope at the origin of a plot of the production rate
vs. the pressure differential.   As previously pointed out, such data should
be taken only after the production rates and bottom-hole flowing pressures
have become stabilized, so as to give at least some approximation to
localized steady-state flow conditions.   The time required for stabilization
will evidently depend on the permeability, oil viscosity, fluid composition,
and magnitude of change from the previous state of flow.   On the basis of
general field experience it appears that conditions will seldom stabilize in
less than 1 hr, that usually 4 to 24 hr is required, and that, in very tight
pays even several days may not suffice.

8.6. Field Measurements of Productivity Indexes.[1]—From the point of
view of actual field experience in the determination of productivity indexes
the situation is perhaps no better crystallized than the theoretical status
of the problem.   Many examples may be cited of all manner of variations
with the production rate.   Such data are given in Table 1 for two wells in
the Section # 28 field, Louisiana, and one well in the Delhi field, Mississippi.

Thus, whereas the data for the first case would show an essentially linear
relation between production rate and pressure differential, a plot of those
for the B well would be a curve, concave upward, whereas the curve for the
third example would be convex upward.[2]   While in individual instances the

---

[1] For discussion of the practical aspects of the field determination of productivity
indexes see W. S. Walls (*API Drilling and Production Practice*, 1938, p. 146).

[2] Similar irregularities have been reported by M. L. Haider [*API Drilling and Production Practice*, 1936, p. 181, and *AIME Trans.*, **123**, 112 (1937)].   On the other hand,
under the more idealized conditions (undersaturated oil and high permeabilities) of
the East Texas field, C. E. Reistle and E. P. Hayes [*U.S. Bur. Mines Rept. Inv.* 3211
(May, 1933), especially Fig. 2] found the production rate to vary linearly with the pressure differential within the accuracy of the experiments.   The same applies to many of
the fields in Kansas, which are highly undersaturated and have but negligible gas
contents.

apparently anomalous behavior can often be rationalized in the light of
related observations regarding gas-oil-ratio variations, duration of test,
etc., in general it is difficult to evaluate the PI data except from an empirical and comparative point of view.   In fact it is not uncommon practice
to compute productivity indexes from individual pressure-differential and
production-rate measurements, as was done in Table 1, and simply to
average the values computed for two or three sets of such measurements
without regard to the implications of their variations.   Moreover such
values have been used, by linear extrapolation, to obtain the "open-flow
potentials" or production rates to be expected at zero bottom-hole flowing

TABLE 1

Well	Pressure, psi	Prod. rate, bbl/day	GOR	PI
A—Section # 28 field.........	4,510	0	—	—
	4,175	154	886	0.46
	3,805	260	962	0.49
	2,975	325	921	0.46
B—Section # 28 field.........	5,172	0	—	—
	5,096	211	1,012	2.76
	5,066	341	930	3.21
	5,046	536	891	4.25
C—Delhi field..............	1,524	0	—	—
	1,520	65	—	16.25
	1,504	244	—	12.20
	1,468	530	—	9.46

pressure.   Until the whole problem of productivity-index measurement is
established on a more satisfactory physical basis, these empirical procedures
offer the simplest means for at least the comparative evaluation of individual wells and the formations immediately surrounding them with respect to oil-production capacity.   However, because of the importance of
such data from both the economic and physical standpoints, the resolution
of the many complicating factors of this problem may well be considered
as an important task for future reservoir-engineering research.[1]

With regard to the numerical magnitudes of the productivity index as
obtained by field measurements, correlation between theory and practice
is also unsatisfactory in some respects.   From Eq. 8.5(3) it follows that,
in order of magnitude, the productivity index should be $0.001kh$, that is,
0.001 times the production capacity of the formation expressed as milli-

---

[1] Rather impressive consistency in correlation has been reported by H. C. Miller,
E. S. Burnett, and R. V. Higgins [*AIME Trans.*, **123**, 97 (1937)] when redefining the
PI as the rate of total *mass* flow of fluid (gas and oil) per unit pressure drop.   Many
of the vagaries of the conventional PI data appear to be eliminated on including the
gas in the flow rate.   While also essentially empirical, this method would be of value if
it were shown to have a wide range of applicability.

darcy-feet. Except for the extremes of very tight and thin or loose and
thick pays the numerical value of the index should therefore lie in the range
of 0.1 to 50. While the examples cited in Table 1 thus appear to be
reasonable, detailed comparisons between the field observations and the
predictions from core-analysis data show cases both of reasonably close
agreement and of ten- to a hundredfold discrepancies. Thus in Fig. 8.10

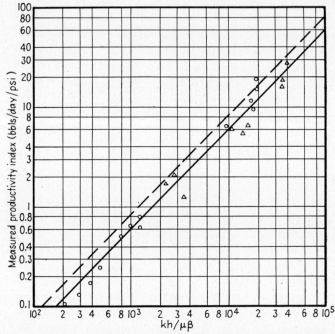

FIG. 8.10. The variation of productivity indexes observed in the Mid-Continent area with
the reservoir parameters. k = air permeability in millidarcys; h = exposed formation
thickness in feet; μ = reservoir oil viscosity in centipoises; β = formation-volume factor.
———, drawn according to Eq. 8.6(1). ———, relation predicted by steady-state radial-flow
formula with $r_e/r_w = 4,000$. o, all data determined experimentally. Δ, μ and/or k estimated.
(*After Lewis, Horner, and Stekoll, Petroleum Technology, 1942.*)

are shown data[1] on the PI, obtained in the Mid-Continent district, plotted
vs. the factor $kh/\mu\beta$. The wells, both pumping and flowing, for which
the data are shown had a range in kh from 400 to 150,000 millidarcy-feet,
reservoir oil viscosities from 0.5 to 3.4 cp, and values of β from 1.02 to 1.48.
For all but one of the wells with a PI less than 1.0, a correction was made
for an average estimated free-gas saturation of 11.5 per cent by using
for k 51 per cent of the measured permeability. For the other wells the
pressures during the tests were above or near the initial bubble point, so
that the effect of the free-gas phase was neglected.

[1] J. A. Lewis, W. L. Horner, and M. Stekoll, *Petroleum Technology*, **5,** 1 (March, 1942).

The solid line drawn through the data of Fig. 8.10 is defined by the equation

$$\text{PI} = 6.0 \times 10^{-4}\frac{kh}{\mu\beta}.\tag{1}$$

The dashed line is that implied by Eq. 8.5(3) with $r_e/r_w = 4,000$. It will be seen that within the limits of inherent errors in the data these observed productivity indexes differ from those implied by the core-analysis permeabilities and fluid constants [Eq. 8.5(3)] by an essentially constant factor of about 1.4. Since no correction was made for the effect of the connate water on k, the difference might well be accounted for by this factor alone.[1] In fact, in the light of the discussion of the last section, the data of Fig. 8.10 would appear to represent entirely satisfactory agreement with the simplified theoretical expectations.

Quite different are the findings derived from a study[2] of 141 wells in California, located in 16 fields. Among these wells the sand thickness ranged from 13 to 663 ft, the permeability from 10 to 8,500 millidarcys, the oil gravity from 13.5 to 44.0°API, the reservoir oil viscosity from 0.096 to 1,040 cp, the formation-volume factor from 1.03 to 1.77, the reservoir pressure from 76 to 4,850 psi, and the gas-oil ratio from 14 to 1,390 ft^3/bbl. The resultant data are plotted in Fig. 8.11. The ordinates are the specific productivity indexes multiplied by $\mu\beta$ $(\log_{10} r_e/r_w)/3.076$, for $r_e/r_w = 2,000$, and by a factor, $1 - C$, to correct for the water cut, C being the fraction of total fluid that is water. From Eq. 8.5(3) it will be seen that, except for the factor $1 - C$, the ordinates should theoretically equal $10^{-3}k$ in millidarcys. The straight line in Fig. 8.11 represents this idealized theoretical prediction, reduced by the factor 1.073.

While the mean deviation of the individual data, plotted in Fig. 8.11, from the straight line is represented by a factor of 31, there are 14 instances where the straight-line prediction is more than 64 times that observed. Obviously such large discrepancies cannot be explained merely by the heterogeneous-fluid character of the flow system, as discussed in Secs. 8.4 and 8.5, or to the increase in flow resistance when perforated-casing rather than open-hole well completions are used (cf. Sec. 5.6). In special circumstances these factors could result in over-all reductions in the PI by a

---

[1] Discrepancies of two- to fivefold, apparently accounted for by known deviations from the idealized steady-state homogeneous-fluid formula, have also been observed in the Gulf Coast (cf. T. V. Moore, *API Drilling and Production Practice*, 1941, p. 197). And even better agreement is reported in the study of data from a deep field in the San Joaquin Valley, California, in which the oil was undersaturated even at the flowing bottom-hole pressures [cf. N. Van Wingen, *AIME Trans.*, **146**, 63 (1942)]. Part of the discrepancy with respect to Eq. 8.5(3) implied by Fig. 8.10 may also be due to the assumption of the fixed value of 4000, for r_e/r_w.

[2] N. Johnston and J. E. Sherborne, *API Drilling and Production Practice*, 1943, p. 66.

factor of 20. But it is hardly reasonable that such cases should constitute the general rule. And if mudding off the sand face and general poor completion practice[1] were the major cause, a much more erratic distribu-

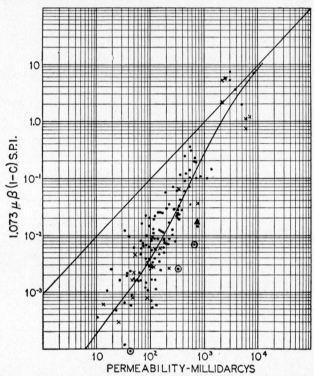

FIG. 8.11. The variation of specific productivity indexes, SPI, observed in California fields with the permeability. μ, β are the viscosity and formation-volume factors of the reservoir oil. C = water cut, as a fraction of the gross fluid. Straight line is the theoretical variation for homogeneous-fluid flow except for the factor $1.073\ (1 - C)$. (*After Johnston and Sherborne, API Drilling and Production Practice, 1943.*)

tion of the data would be expected. In spite of the scattering of the data plotted in Fig. 8.11 they show a definite trend, as represented by the curve.

While no quantitative interpretation has been made of these data, a factor that can and very likely does explain the very low values is the inherent effect of the connate water on the permeability of many of the producing formations of California. As discussed in Sec. 3.7, many of these formations have homogeneous-fluid permeabilities to water that are much lower than for air. The difference is due to the reaction with the

<hr>

[1] A detailed discussion of effects of completion practices on the productivity index is given by W. J. Travers, Jr., *Petroleum Technology*, **5**, (March, 1942); cf. also F. R. Wade, *API Drilling and Production Practice*, 1947, p. 186.

water of the clay and intergranular content of the rock. In some cases the permeability even to salt water has been found to be less than 1 per cent that to air. Under such conditions the use of the air permeability provides only a fictitious basis for comparison with observed PI data (cf. Table 3, Chap. 3). While only a few measurements have been reported on effective oil permeabilities of "dirty" sands either before or after exposure to fresh waters, special tests[1] on the Stevens Sand in California indicate that fresh-water-containing cores have effective oil permeabilities of the order of one-tenth the air permeabilities. It is therefore reasonable to suppose that if the permeability to oil in the presence of connate or fresh water, as measured for the individual cores or wells, were used, or if the permeability to the formation water were taken as the basic homogeneous-fluid permeability, a large part, if not all, of the apparent discrepancies with the measured PI's would be removed, provided that such other factors as the free-gas saturation and the effect of perforated-casing completions were taken into account.

The increasing divergence between the idealized predictions and observed productivity indexes as the permeability decreases also does not have a well-established explanation. It may be rationalized, however, by the supposition that the relative content of clay or argillaceous material is greater in the tighter than in the looser sands.

While the California data (Fig. 8.11) do show a rather definite and useful correlation between the observed PI and the air permeability, it should be noted that it is only empirical. The previously discussed observations from other producing districts show that correlations of this type cannot be safely carried over from one region to another, where the nature of the producing formations may be greatly different. Of course, if and when it is found that the California data will follow the simple theoretical relationship on using the proper fluid permeability, a common basis for predicting the productivity index will be a definite possibility.

A final point pertaining to productivity indexes concerns their variability with time. Of course, from the simple homogeneous-fluid point of view, the PI should be a fixed property of the well. And this may be expected in actual oil-producing systems in which the oil saturations and viscosity are maintained constant. It may occur in complete-water-drive systems in which the pressure is maintained above the bubble point or in reservoirs which may have initially produced by gas drive but have become subsequently stabilized by natural or artificial pressure and saturation-maintenance agencies. However, when a significant role in the production mechanism is played by a gas drive, the PI will naturally decline as the oil

---

[1] K. T. Miller, F. Morgan, and M. Muskat, *Producers Monthly*, **11**, 31 (November, 1946).

saturation decreases and the viscosity rises because of liberation of solution gas. As will be seen in Chap. 10, where the gross production performance of gas-drive reservoirs is treated, the PI may be expected to fall by factors of the order of 10 during the depletion history. In fact these slow declines in productivity index will simply reflect the rate and degree of depletion of the part of the reservoir drained by the particular well.

Aside from these slow variations, rapid transients are generally observed whenever any changes in well conditions—pressure or flow rate—are made. If productivity indexes are calculated from the pressure and production data obtained during such unstabilized periods, they will also show a time-varying behavior. The character and duration of these transients will depend on the past history or state of the system just prior to the change, the nature of the change in production rate or bottom-hole pressure, the compressibility of the fluid stream within the rock, and the permeability of the formation. In general their quantitative interpretation will be extremely difficult,[1] and for practical application the measurements should be made only after conditions have stabilized and at least some semblance to steady-state behavior has developed.

There is one type of transient behavior, however, that can, under ideal conditions, be analyzed to give the equivalent of a steady-state productivity index. This is that of the pressure build-up in a well after shutting in.[2] If during the build-up period the fluid entry in the well bore be assumed to be of the steady-state homogeneous-fluid character,[3] it can be expressed as

$$Q = c(p_e - p); \qquad c = \frac{2\pi kh}{\mu\beta \log r_e/r_w}, \tag{2}$$

where Q is the rate of flow, p_e the reservoir pressure, and p the instantaneous

---

[1] It has been reported [cf. C. V. Millikan and H. F. Beardmore, *AIME Trans.*, **160**, 248 (1945)] that many features of the gross reservoir behavior can be predicted from a study of these transients. Because no satisfactory explanation for these conclusions has been developed on the basis of the physical principles underlying reservoir performance, a detailed discussion of these developments will be deferred until they have been independently confirmed, in spite of their tremendous practical importance if they were to be established.

[2] M. Muskat, *AIME Trans.*, **123**, 44 (1937).

[3] If the flow is homogeneous, the transient history of the pressure build-up or drawdown on making definite changes in the well-production rate can be computed by taking into account the compressibility of the reservoir fluid [cf. T. V. Moore, R. J. Schilthuis, and W. Hurst, *Oil and Gas Jour.*, **32**, 58 (May 25, 1933); also M. Muskat, "The Flow of Homogeneous Fluids through Porous Media," Chap IX, McGraw-Hill Book Company, Inc., 1937]. However, as both the field measurements for determining the complete transient histories and subsequent handling of the data are considerably more involved than in using the steady-state approximation, tests of this type have been made in only a few instances.

bottom-hole pressure. Moreover, if on entry into the well bore, of area A, the fluid has a uniform density γ_o, Q may also be expressed as

$$Q = A \frac{\partial h}{\partial t} = \frac{A}{\gamma_o g} \frac{\partial p}{\partial t}, \tag{3}$$

where h is the instantaneous fluid height in the well bore[1] and g is the acceleration of gravity. By combining Eqs. (2) and (3) it follows that

$$\left. \begin{aligned} p &= p_i + (p_e - p_i)(1 - e^{-\gamma_o g c t / A}) \\ h &= h_i + (h_e - h_i)(1 - e^{-\gamma_o g c t / A}) \end{aligned} \right\} \tag{4}$$

where p_i, h_i are the initial bottom-hole pressure and fluid head and p_e, h_e represent the equilibrium shut-in values.

On rewriting Eqs. (4) as

$$\log \frac{p_e - p_i}{p_e - p} = \log \frac{h_e - h_i}{h_e - h} = \frac{\gamma_o g c t}{A}, \tag{5}$$

it is seen that a semilogarithmic plot of pressure or fluid rise vs. time should be linear. If such a linear plot is obtained, the slope will be $\gamma_o g c / A$, from which c can be computed. From the form of Eq. (2) it is seen that c is simply the theoretical productivity index. Because of the severe limitations to the validity of Eqs. (2) and (3), but little use has been made of pressure-build-up–time curves for determining productivity indexes. In several instances, however, where conditions were favorable,[2] their use has been found to give results comparable with those measured directly from rate-of-flow vs. pressure-differential tests or as predicted from core-analysis data.

8.7. The Application of Productivity-index Measurements.—In spite of the basic difficulties still outstanding regarding the quantitative interpretation of individual PI determinations, they are of considerable value from a practical point of view. If the measurements on a group of wells in a single field are made under similar conditions, their relative values should be of significance in reflecting the comparative thickness and permeability

---

[1] For practical purposes, Eq. (3) and the use of h as a measure of the bottom-hole pressure mean that the well is pumping.

[2] Van Wingen, *loc. cit.*, and C. C. Rodd [*AIME Trans.*, **151**, 48 (1943)]. Use has also been made (E. Kemler and G. A. Poole, *API Drilling and Production Practice*, 1936, p. 140) of Eq. (5) to determine the ultimate equilibrium build-up pressure p_e, in flowing wells, although such applications would appear to have only empirical significance, in view of the questionable validity of the underlying Eqs. (2) and (3) for the transients in flowing wells with aerated fluid columns in the flow string. While it is quite feasible to develop a generalized approximation theory for the pressure build-up transients in flowing wells with both free gas and oil flowing in the well bore, it would be of doubtful practical applicability, since its quantitative features would depend on the details of the fluid distribution in the flow string.

of the sections of the reservoir drained by the individual wells. Reasonable correlations have been reported[1] between the initial productivity index and the cumulative production, over a period of $3\frac{1}{2}$ years, for wells in the same field.

An application of practical importance in certain producing districts is the computation of the "open-flow potential" as the product of the PI and the reservoir pressure. These potentials have been used by state regulatory bodies[2] in formulas for the allocation of allowable production rates of individual wells in a field. The calculation of such potentials made it unnecessary to perform the actual open-flow tests,[3] which would be undesirable or impractical for a number of reasons.[4] While it is doubtful that these calculated potentials would agree closely with the results of actual tests, were they to be made, their relative magnitudes may well reflect the comparative production capacities with fair approximation.

The PI provides a means for evaluating well treatment or repair work. A comparison of the PI before and after treatment should give a better measure of the effect of the work than the absolute production rates at a fixed choke size or other arbitrarily defined specification. For example, before PI measurements became common practice, the effect of acid treatment in pumping wells was often underestimated, because the production rate after acidization was limited by the pump capacity, whereas the PI would have indicated a much greater production capacity.

To the extent that the PI is a physically significant measure of the productive capacity of a sand in terms of the factors entering the theoretical formulas of Sec. 8.5, it should be far more representative of these factors than the direct permeability measurements on core samples. Except for the previously discussed complications relating to heterogeneous fluid and nonsteady-state effects, the PI is an integrated resultant over a large body of the sand, whereas the cores constitute hardly more than an infinitesimal sample of the rock drained by a well. On the other hand the PI involves an averaging vertically as well as areally and hence can give no information regarding vertical variations in the character of the producing section within the zone tested.

[1] Travers (*op. cit.*). Essentially equivalent approximate and statistical correlations have been found in older fields between the cumulative recovery and the initial open-flow production rates.

[2] Such applications, now practiced mainly in Kansas, used to be quite common, although the allocation "formulas" were generally of arbitrary structure.

[3] The open-flow tests may in themselves also be of questionable physical significance, as the maximum obtainable production rates may be limited by the nature of the flow string or pumping equipment rather than by the production capacity of the formation.

[4] The use of flow-rate vs. pressure-differential plots for determining potentials in proration formulas was first proposed by T. V. Moore [*API Proc.*, **11**, No. IV, 27 (1930)].

As previously mentioned, the long-term variations in PI will reflect gross changes in the state and character of the reservoir fluids. Provided that extraneous effects of well plugging or drowning out of pays by water incursion have not developed, it should be possible to correlate the decline of PI in gas-drive fields with those in the reservoir pressure (increase of oil viscosity) and oil saturation (decrease in oil permeability). Although no quantitative relationship of this type has been established on the basis of field data,[1] qualitative correlations between reservoir depletion and declining PI have often been observed.[2]

When combined with gas-oil-ratio observations, the PI may be of value in interpreting abnormalities of well behavior in gas-drive reservoirs. As seen in the previous sections, both the gas-oil ratio and the PI depend mainly on the oil saturation taking into account, of course, the reservoir pressure, oil viscosity, gas solubility, and water saturation. Thus, by Eq. 8.2(5) an effective gas-oil permeability ratio can be computed from the gas-oil-ratio measurement (cf. Sec. 10.12). This will refer directly to the conditions at the well bore. But if the flow rate and pressure differential are low, the calculated k_g/k_o may be considered as an approximation to the average for the rock being drained by the well. The same will apply to the effective oil permeability k_o/k, computed by Eq. 8.2(7) or 8.5(3) from the PI measurement. Now both k_o/k and k_g/k_o imply definite values of the oil saturation, if the water saturation is known. If these oil saturations are entirely inconsistent, the action of extraneous factors on either the PI or the gas-oil ratio may be inferred. If the oil saturation implied by the gas-oil ratio is materially higher than that deduced from the PI, the possibility of a local plugging near the well bore should be investigated. Conversely, if the gas-oil ratio indicates an appreciably lower oil saturation than the PI, the cause may lie in the entry of free gas from a gas cap or a gas sand that is sealed off imperfectly.

Because of previously discussed uncertainties regarding the significance of the absolute values of the factors involved, this type of analysis of well

---

[1] As will be seen in Chap. 10, it is possible to predict such relations theoretically.

[2] Examples of such decline are given by R. V. Higgins, *U.S. Bur. Mines Rept. Inv.* 3657 (September, 1942). And at Midway, Ark., it has been observed (cf. W. L. Horner, *API Drilling and Production Practice*, 1945, p. 27) that in the region maintained (by water injection) above the bubble point the PI has remained substantially constant while at the same time the wells in the area where the pressures have fallen below the bubble point have suffered declines in PI by factors of the order of 5 to 10, although some of the latter reduction may have been due to the development of water production. Rather striking correlations of long-term declines in the productivity index with the pressure, by factors exceeding 10, have also been reported for the Kettleman Hills field, Kern and Fresno Counties, Calif., by E. W. McAllister, *AIME Trans.*, **142**, 39 (1941).

data will seldom be feasible, except by using ratios of gas-oil ratio or PI data taken at different times or by comparing changes in gas-oil ratio with those in the PI. For example, a stable gas-oil ratio with a declining PI will be indicative of plugging at the well bore. But if the gas-oil ratio has increased markedly without a corresponding large decline in PI, the entry of extraneous free gas may be inferred. The latter inference may also be drawn if the gas-oil ratio changes appreciably with varying rates of production from the well while the PI remains substantially constant.[1]

Similar considerations can be applied to the interpretation of the source of water production. A rapid growth in the water production should bring with it a decline in the PI if the water is entering the well through typical strata within the oil pay. If, however, the PI is maintained while the water production rises, it may be inferred that the water is coming in from entirely separate strata or has broken through sections of the oil pay which had not been contributing significantly to the oil production.

It should be understood that all such interpretations of individual well behavior should be considered as guides for more detailed investigation, rather than as proofs in themselves of particular mechanisms of well performance. Without complete data on the permeability-saturation relationships for the specific producing strata in question and a more satisfactory theoretical basis for evaluating PI determinations in multiphase systems it will not be possible to make satisfactory quantitative deductions from either individual PI or gas-oil-ratio measurements. This limitation should not, however, vitiate the practical significance of the comparative, qualitative, and even semiquantitative applications discussed above.[2]

8.8. Summary.—Because of the extreme complexity of the general hydrodynamic equations for heterogeneous-fluid flow no satisfactory solutions for time-varying systems have been developed thus far. To get some idea regarding the quantitative implications of these equations one is therefore forced to resort to the steady-state approximation. Although these will not correctly describe the behavior of rapidly varying systems and local well transients, they may serve as a basis for approximating the flow conditions when the time variations are slow, such as those associated with changes in pressure and fluid content in a reservoir as a whole, due to the normal producing and depletion processes.

While for systems of simple geometry the steady-state equations can be

---

[1] Applications of this type of PI data have been reported and discussed by Higgins (*op. cit.*).

[2] Especially interesting conclusions with regard to the effect of drilling time in the producing zone, the type of drilling mud, and type of perforations have been derived from a statistical study of PI data for wells in California fields (cf. Wade, *loc. cit.*).

integrated formally, their detailed implications can be derived only by extended numerical or graphical procedures. As is to be expected, these involve a knowledge of the specific data on the thermodynamic properties of the fluids and of the permeability-saturation relationship of the particular rock formation of interest. The former are generally readily available from the analysis of bottom-hole or surface recombination samples of the oil and gas. With respect to the latter, however, information is very meager, and data for three-phase systems have been reported only for a series of unconsolidated sands (cf. Sec. 7.4). Accordingly the numerical analyses of the steady-state systems given in this chapter must be considered as having only illustrative significance. On the other hand the qualitative features of the results should also be applicable to consolidated producing strata, in view of the general similarity in the permeability-saturation curves for two-phase flow for all types of porous media thus far studied.

A basic result of the integration of the steady-state equations is the expression of the fluid ratios—gas-oil and water-oil—in terms of functions of the pressure and permeability ratios of the phases in question [Eqs. 8.1(3) and 8.1(4)]. While these fluid ratios can be used directly as constant parameters of strictly steady-state systems, they should also apply *locally* under transient conditions, where the several phases are flowing simultaneously. They are directly proportional to the corresponding permeability ratios, which, in turn, are determined by the fluid saturations.

Without carrying through the numerical integrations the formal integrals of the equations of motion for steady-state flow [Eqs. 8.1(6) and 8.2(2)] show that the production rates will not be strictly proportional to the pressure differential but will decrease with increasing pressure differential. Likewise the pressure distribution in linear systems will not vary linearly with distance along the flow channel [Eq. 8.1(5)]. And in radial systems the pressure will not vary in a strictly linear manner with the logarithm of the radial distance [Eq. 8.2(1)].

The steady-state production rates in heterogeneous-fluid systems are, in general, lower than for homogeneous-fluid flow in the same porous medium. They decrease with increasing gas-oil ratio (Fig. 8.4) or water-oil ratio. Moreover for fixed fluid ratios and pressure differential the production rate decreases with decreasing inflow or reservoir pressure.

The oil saturation in systems producing gas and oil, but with an immobile water phase, continually decreases on approaching the outflow surface (Fig. 8.5). However, most of the decline in saturation is concentrated in the immediate vicinity of the outflow surface. The same applies to the permeability to the oil phase. On the other hand, if there is no free gas in the system, the steady-state flow of both the oil and the water will be of a homogeneous-fluid character, except that the production rates will be determined by the effective permeabilities for the respective phases [Eq. 8.3(3)]. The fluid distribution will be uniform and determined by the water-oil ratio.

If all three phases should be flowing simultaneously, the determination of the detailed behavior is considerably more complicated, although the method of calculation is basically the same as for gas-liquid-mixture flow. Both the gas-oil and the water-oil ratios will increase rapidly with decreasing oil saturations if the formation has a water content high enough to give it mobility. On approaching a well the free-gas saturation will in general increase. The increased gas saturation will be provided mainly by a drop in the oil saturation but also in part by a decline in the water saturation.

In the practical evaluation of the producing characteristics of individual oil wells the production rate per unit pressure drop (the productivity index PI) is generally used. This can be defined theoretically in terms of the fluid properties (the oil viscosity and formation-volume factor), the well and external radii, the pay thickness, and the permeability to the oil [Eq. 8.5(2)]. This, however, applies directly only to steady-state flow, to which the flow in actual wells can at best be but an approximation. Moreover, because of the heterogeneous character of the flow, the permeability to the oil phase will be sensitive to the gas-oil ratio and will also be affected by the differential as well as the absolute pressures. In addition the viscosity and formation-volume factors will give some variation with the pressure. Theoretically, therefore, the productivity index cannot be expected to be an absolute constant of the producing system, and its numerical value should depend on the conditions of measurement. One way of making its definition more unique is to express the PI as the limiting value of the production rate per unit pressure drop as the latter is made to vanish [Eq. 8.5(3)]. On the basis of such a definition its value can be computed for various conditions of flow, if the fluid and rock properties are known. Illustrative calculations indicate that owing to the multiphase character of the flow it may be reduced from the equivalent homogeneous-fluid value by factors of the order of 5.

While PI determinations are common field practice, actual field measurements are often difficult of detailed interpretation. Under idealized conditions of undersaturated oil flow and substantially constant gas-oil ratios the PI has been found to be essentially independent of the flow rate or pressure differential. Quite often, however, it shows either an increasing or a decreasing trend with increasing production rate (cf. Table 1). Moreover its absolute magnitudes, in some producing districts and under favorable circumstances, have agreed within factors of 2 to 5 with those anticipated from the permeability of the producing formation as determined from core analysis. Most data from California fields, however, indicate values of the PI much smaller than predicted from the air permeabilities of the producing sections (cf. Fig. 8.11). The discrepancies (on the average by a factor of 31) are far too large to be accounted for by the heterogeneous character of the flow. The reason appears to lie in a great reduction in both the homogeneous and the effective fluid permeability caused by the reaction between the connate water and the intergranular argillaceous material common in California producing pays, although it has not been confirmed by a quantitative analysis that the discrepancies are completely accounted for by these effects.

Even though the quantitative significance of the absolute magnitude of individual PI values may be in doubt, many applications of practical importance can be made of such data. Their relative values for wells in the same field should give a good measure of the comparative permeabilities and pay thicknesses of the areas drained by the wells. On multiplying them by the reservoir pressures they give calculated "open-flow potentials," which have been used in the allocation of allowable production rates. By comparing the PI before and after well repair or acid treatment the effectiveness of such operations can be better evaluated. Their decline during the course of production will reflect the general state of reservoir depletion. Such decline should parallel the growth in gas-oil or water-oil ratios. If it appears to be more rapid than would be anticipated from the latter, a plugging at the well bore would be indicated. On the other hand, if the rise in the fluid ratios does not bring with it a corresponding fall in the PI, an investigation into the possible extraneous entry of gas or water would be warranted. Of course, under all circumstances the PI should be measured only after the well has stabilized itself for each production rate and some approximation to a steady state has been established, even if no quantitative interpretation is to be made of the numerical value obtained.

CHAPTER 9

GENERAL RESERVOIR MECHANICS

9.1. The Classification of Reservoir Energy and Producing Mechanisms.—The general performance of oil-producing reservoirs is largely determined by the nature of the energy available for moving the oil to the well bore, and the manner in which it is actually used during production. These controlling factors are, of course, in turn determined by a multitude of other variables, such as the structural conditions defining the reservoir,[1] the nature of the oil, the gas in solution in the oil, the flow capacity of the rock, the mobility of contiguous water reservoirs, if any are present, and the rate of oil, gas, and water withdrawals. In practice, conditions will not often be such that an oil reservoir can be described throughout its producing history by any single sharply defined type of production mechanism. The definition of such mechanisms nevertheless will serve to classify the predominating factors that may influence, individually or in combination, the observed reservoir behavior.

The major types of energy available for oil production are (1) energy of compression of the oil and water within the producing section of the reservoir rock; (2) the gravitational energy of the oil in the upper parts of the formation, as compared with that at greater depths; (3) the energy of compression and solution of the gas dissolved in the oil (and also water) within the producing stratum or in free-gas zones overlying the oil-saturated section; and (4) the energy of compression of the waters in reservoirs that are contiguous to and in intercommunication with the oil-bearing rock.

As releases are provided for these forms of energy, by the drilling and operation of wells, the energy is expended by the action of forces or pressures exerted in the direction of lower energy levels or pressures. These forces serve to overcome the flow resistance of the rock to the fluids passing to the producing wells. The work done by these forces accounts for the loss in energy within the reservoir between the initial and final states (at the well bore) of the fluids involved in the production processes. Thus the energy of compression of the oil and water within the oil reservoir manifests itself in an expansion of these fluids, a dissipation of the pressure holding them in a compressed state, and the flow of the volume of expansion into the wells or low-pressure outlets inducing the expansion. The gravita-

[1] Cf. Sec. 1.5.

364

tional energy is made available through the action of the body force of gravity on the various fluid phases, in proportion to their density, tending to move them to the lower levels of the pay and thence into the wells. The differential magnitude of the gravity forces on the gas and liquid phases also leads to a relative residual upward force or buoyancy on the gas phase and a tendency for a segregation between the two phases.

The energy of the gas dissolved in the oil is made available as the gas is liberated from solution and expands in place or while proceeding to the low-pressure regions surrounding the producing wells. By virtue of the volume expansion of the gas phase this process leads directly to an equivalent volume expulsion of oil, and in the course of its flow through the rock it will also be accompanied by a flow of oil to the wells. Finally the energy of compression of contiguous water reservoirs becomes available for moving oil to the well in a manner quite similar to the energy of compression of the oil itself. The expansion volume of the fluids—the water itself or the gas evolved from the water—in the water reservoir will in effect overflow into the oil section and hence displace a corresponding volume of oil in the latter.

In addition to these major forms of energy that may control the reservoir performance, two others should be noted for the sake of completeness. The first is the differential energy of the internal surfaces of the porous rock for the different fluid phases, discussed in Secs. 7.8 to 7.10. Under favorable conditions these may lead to flow and changes in fluid distributions between various regions of a rock even though none of the other types of energy is active. For example, if a tight section of rock with a high oil saturation is in contact with a region of coarser structure but higher water saturation, there will generally be a tendency for some of the water to flow into the tighter rock independently of the action of gravity or pressure effects, assuming that the rock is preferentially wet by water. While the gravity and pressure forces predominate in most practical producing operations, under special conditions, such as extended shutdown periods, as well as during the establishment of the initial virgin-fluid distribution in the reservoir prior to exploitation, capillary energies and forces may well be of significance (cf. Sec. 7.9).

The final source of energy that, in principle, may play a role in oil production lies in the compression of the rock itself. On release of pressure the direct change in volume of the pore space or such changes as may be caused by a redistribution of the granular structure associated with subsidence or settling of the overburden will be superposed on the effect of the other energy transformations. However, such effects will be given no further separate consideration here, as there is no evidence of their importance in the great majority of actual oil-producing reservoirs. Moreover, as will be seen in Sec. 11.3, in the case of complete-water-drive reservoirs the effects of the rock compression, if any, can be formally combined with those due to the compressibility of the water.

Among the four primary energy sources listed above, the first is definitely of lesser significance than the others. The compressibility of the oil itself simply is not great enough to account for a major part of the total oil production derived from fields warranting commercial exploitation. Thus the compressibility of reservoir crudes will generally be of the order of magnitude of 10^{-5} per psi. Hence even if undersaturated by 1,000 psi, the oil will expand by only 1 per cent before reaching the bubble point. The compressibility of water (of the order of 3.10^{-6} per psi) will lead to an even smaller percentage expansion of the connate water in the oil pay, if it is also undersaturated. While the oil and water expansion may provide the major source of fluid withdrawals during the early history of complete water-drive fields (cf. Sec. 11.1), closed reservoirs would evidently not be of commercial value if the oil production were caused only by the simple expansion of the liquid-phase contents of the reservoirs.

If, however, the reservoir oil originally is undersaturated but still contains enough gas for significant gas-drive oil expulsion when the pressure is sufficiently reduced, the field will undergo an initial direct liquid-phase expansion history, which should be taken into account in the consideration of its over-all performance. This period of production will be characterized by a rapid pressure decline but may provide a substantial part of the fluid-withdrawal replacement during this period even if the reservoir is bounded by a mobile water body. On the other hand, in comparison with the role played by the energy of the gas dissolved in most[1] reservoir oils or the compression energy in adjoining water reservoirs, that due to direct liquid-phase expansion within the oil-saturated rock is of minor significance from a recovery standpoint.

The force of gravity is ever present in fluid-bearing underground rocks. And likewise the tendency for gravity segregation between the gas and liquid phases, and even between the oil and water phases, is always present wherever these phases occur and have different densities. These manifest themselves most prominently in the development of "gas caps," or zones of relatively high gas saturation, at the structural crests of the oil-saturated section, and in the long-continued oil flow into the producing wells after virtually all the dissolved gas pressure has been dissipated. The former

[1] In exceptional cases, however, such as the D-7 zone of the Ventura Ave. field, California, where, associated with an abnormally high reservoir pressure—the latter is 5,000 psi greater than the bubble point—the reservoir liquid expansion may lead to a recovery 40 per cent as great as is to be expected from the solution-gas drive [cf. E. V. Watts, *AIME Trans.*, **174**, 191 (1948)].

is a common occurrence in gas-drive reservoirs of appreciable structural relief, when production is not too rapid. However, the force of gravity will not represent a major factor in the direct mechanism of oil expulsion unless the pressure differentials over the pay are comparable with the equivalent head of an oil column of height equal to the oil section. Thus, from the point of view of maintaining production rates, gravity drainage will be of significance mainly under conditions where such energy as is related to fluid pressures is largely depleted and the producing formation is of sufficient thickness and permeability to give production rates justifying continued pumping operations. On the other hand the action of gravity, when given full play, in inducing gas segregation and gas-cap expansion with the associated downdip oil drainage may be of very great value in leading to much higher oil recoveries than can be obtained by the simple dissolved-gas-drive mechanism (cf. Secs. 10.16, 10.17, and 14.14).

The reservoir oils in all oil fields thus far discovered contain dissolved gas.[1] Many reservoirs contain more gas than can be held in solution in the oil even at the initial pressure before exploitation. The excess is then found overlying the oil-saturated section in a gas cap or free-gas zone. In the case of perhaps most oil-bearing reservoirs the oil is simply saturated with gas with no appreciable excess in the form of a gas cap. And, finally, there are many that are undersaturated in varying degrees. In some cases the bubble-point pressure may be as low as 100 psi, even though the initial reservoir pressure may exceed 1,000 psi. Under such conditions the energy of the gas will not become available until the reservoir pressure falls to the bubble point.

The amount of energy available in an element of free-gas phase associated with an oil reservoir is proportional to its volume, at standard conditions, and to the logarithm of the pressure. However, even when the oil is undersaturated by several hundred psi, there is still generally enough energy available for an appreciable degree of oil expulsion.[2] In fact, as will be seen in Sec. 10.4, the increased oil shrinkage associated with increasing amounts of gas in solution may lead to even lesser quantities of stock-tank-oil expulsion than if the oil contained relatively smaller volumes of dissolved gas (cf. Sec. 10.4).

As implied by the above discussion, a "gas-drive" reservoir is one in which the major source of energy being used to induce oil flow toward the

---

[1] While this is literally correct, the gas content of many Kansas fields is so low that the reservoir oil is justifiably considered as "dead" even initially.

[2] To provide for the creation of a nonvanishing equilibrium gas saturation and the corresponding oil expulsion, solubilities as low as 1 ft^3/bbl will suffice if the gas is liberated at substantially atmospheric pressure. From a practical point of view, however, the oil so expelled would not be obtained at commercially profitable rates.

producing wells is that associated with the gas dissolved in the oil or in a free-gas zone which may overlie the oil-saturated section. If there is no gas cap initially, the producing mechanism is termed a "solution gas drive." If the reservoir contains a gas cap of appreciable magnitude and the production is so controlled as to permit an expansion of the gas cap without a direct bleeding of the gas-cap gas, the mechanism is often referred to as a "gas-cap-expansion drive." To lead to significantly different reservoir performance and oil recovery, however, gravity drainage ahead of the expanding gas cap must contribute to the oil movement, as previously indicated. While in such cases the energy for the downstructure oil drainage may be largely derived from gravity forces, the gas-cap-expansion reservoirs will be treated here under the general classification of gas-drive systems, for reasons to be discussed in Sec. 9.2.

There are commercially productive oil fields that are effectively sealed, throughout their producing life, from contact or interaction with water-bearing strata. However, the majority thus far discovered are bounded by and in fluid intercommunication with water-bearing reservoirs. The existence of such water reservoirs is usually established by the "dry" holes—generally water productive only—delineating the oil-productive area. If the oil-bearing stratum has a steep dip, the contact plane with the water body will be of limited areal extent and provide an "edgewater" boundary, and corresponding "edgewater drive," if its entry into the oil reservoir is the major oil-replacement agent. For gently sloping formations, as in the East Texas field (cf. Fig. 1.12b), the water-oil plane of contact may underlie an appreciable part of the oil pay, and a proportionate number of the producing wells will be subject to a "bottom-water drive," if the water is mobile and is permitted to invade the oil reservoir at a sufficient rate to replace the fluid withdrawals. In either case the water reservoir contains energy of compression that will be released on lowering of the pressure in the oil-bearing formation by the withdrawal of fluid from the latter through the producing wells. Because of the lower compressibility of water its fractional expansion in volume on pressure release will be lower than for the oil. However, when a mobile contiguous water reservoir is present at all, its area will often be very much greater than that of the oil reservoir, so that in spite of its lower compressibility the total expansion volumes may exceed the whole of the original reservoir oil volume. Thus, whereas the great majority of known oil fields have areas less than 10 sq miles,* water reservoirs extending over 1,000 sq miles are not at all uncommon. Moreover, while there is little specific evidence on the sub-

---

* In fact a statistical analysis shows that in this country it is only in Pennsylvania and in the Texas Panhandle that a majority of the fields cover surface areas exceeding 1,000 acres.

ject, it may be anticipated that at least in some water reservoirs the decline in pressure may be followed by gas evolution, similar to that in the oil zone, and may thus result in an effective compressibility even larger than that of the bubble-point oil.

In addition to the expansion in volume of the fluid content of a water reservoir due to pressure release, an additional source of water that may be available for ultimate entry into the adjacent oil-bearing reservoir is provided by the drainage of surface waters into exposed outcrops of the formation. In general, however, the contribution, if any, due to surface waters is very small as compared with the expansion volume of the original water content and may be neglected, except when otherwise it is known to be of significance.

9.2. General Performance Characteristics of Oil-producing Reservoirs.—
The "performance" of an oil-producing reservoir is the composite history of the various physical parameters describing its current and past "behavior." The basic variable defining the state of the reservoir is the time since production was begun or the value of the cumulative production. The latter is generally of more fundamental significance in the case of gas-drive fields, though the time scale often provides a more convenient basis for analysis. For water-drive fields the time variable enters explicitly in the description of both short-period transients and the gross transient history of the reservoir performance. For some purposes, however, the use of the cumulative recovery as denoting the state of reservoir depletion even of water-drive fields is quite satisfactory and appropriate.

The reservoir characteristics whose variation with the cumulative production or time constitute the record of its "performance" are the pressure, gas-oil ratio of the production, water production, motion of the water-oil contact, and the development or expansion of gas caps. The long-term variation of the production rate or production capacity of the reservoir is also a significant component of the over-all history, although in recent years in most states the production rates have been generally maintained at constant values for extended time periods or deliberately varied by proration regulations.

In addition to the pressures, gas-oil ratios, and water-production rates of the reservoir as a whole the distribution of the individual well data over the areal extent of the reservoir will throw much light on the mechanism controlling the production. These are most conveniently represented in contour form drawn on a map of the field, at various time or cumulative production intervals during the reservoir history.

From a physical point of view it is axiomatic that the average reservoir pressure must decline from its initial value as oil or gas is produced. The oil and gas withdrawn must be replaced volumetrically. This can take place by (1) an expansion of the residual oil or water within the producing section, (2) creation of a free-gas phase, (3) expansion of an existing free-gas phase, or (4) the intrusion of water into the oil-producing area.[1] Any of these processes or several in combination may provide for the replacement of the volume of oil and gas removed. But all of them require a lowering of the pressure from its initial value.

The significant features of the decline in reservoir pressure are its magnitude, its reaction to production rate, and the character of its variation with the cumulative production. In fact the pressure reaction of a reservoir to continued production or changes in withdrawal rate generally provides the most direct indication regarding the producing mechanism.

It is to be emphasized that the type of producing mechanism, or "drive," is not necessarily an inherent property of a reservoir. If the latter is completely sealed off by faulting or permeability pinch-out from communication with water reservoirs, its natural production history will, of course, have to be of the solution-gas- or gas-cap-drive type. Likewise, if the producing formation is a fissured or cavernous limestone in contact with a mobile water reservoir, the natural performance characteristics may be anticipated to be those of the ideal complete-water-drive field. On the other hand, most sand reservoirs as well as noncavernous limestones or dolomites are in communication with contiguous water reservoirs that have limited capacities for water supply into the oil pay. The degree to which this water entry will completely replace the oil and gas withdrawal and retard the reservoir pressure decline will depend on the magnitude of the oil- and gas-withdrawal rates. It is the latter that will determine whether the reservoir will actually produce by a gas-drive or water-drive mechanism. Moreover, as these rates may be varied, so will the production mechanism, after appropriate lags. Accordingly the same reservoir may produce by either gas or water drive or by a combination of both as "partial" water drives, at different periods in its history depending on the manner in which the field is produced.

As in the case of the structural features of oil-reservoir traps, the classification of their producing mechanisms is inherently arbitrary. The limiting forms of strict solution-gas-drive and complete-water-drive mechanisms are so different in their characteristics that their differentiation has been generally accepted. But the intermediate and more common "partial" water drives and gas-cap-expansion reservoirs are subject to several types of classification. In fact their definitions are hardly even well established.

It will be noted from the discussion in this and the following sections that the emphasis has been placed on the "performance" characteristics

---

[1] Reservoir compaction is also a possibility. But this, too, requires the development of a pressure decline.

of the reservoirs during their producing lives. Their distinctions have been based on differences in *producing* mechanisms. An alternative basis would have been the *recovery* mechanisms. The latter would still require a differentiation between the strict solution-gas and water drives. However, since the partial water drive usually also involves ultimately a flushing of the oil by the invading waters and often leads to ultimate recoveries comparable with those associated with complete water drives, it would be considered as a special case of the latter. Moreover, as the gas-cap-expansion drive, when fully effective, derives its effectiveness from the gravity-drainage mechanism of oil depletion, it would accordingly be treated as a third fundamental type of producing system, from a recovery standpoint.

It is admittedly largely for reasons of personal preference that in this work, and in the following chapters, the mechanism determining the immediate and current reservoir performance rather than the ultimate recovery has been considered as the primary criterion of reservoir-behavior classification. While the broad goal of oil production is the achievement of maximum oil recovery at minimum cost, the ultimate recovery represents mainly the integrated resultant of the whole producing life history. It cannot be fixed in advance independently of the producing performance. And its absolute magnitude for each mechanism can cover a range overlapping that of the other recovery mechanisms. The production, pressure, and gas-oil-ratio data gathered during the producing life of a field reflect directly the current and local oil-displacement processes within the productive area, rather than the more remote agents, as invading waters or gravity drainage, which may ultimately determine the absolute recoveries.

It is in the light of these considerations that partial-water-drive systems will be treated here as generalized gas-drive reservoirs (cf. Sec. 10.18). The dynamics of water influx will indeed be determined by the characteristics of the water-supply reservoir in exactly the same manner as in complete-water-drive fields. However, since by its very nature part of the fluid-withdrawal replacement will be provided by gas evolution, the immediate oil-expulsion mechanism in the productive area will be governed by solution-gas-drive processes. The general performance characteristics will therefore be controlled by the gas-drive mechanism. Of course, they will be modified by the water intrusion. And new features will be introduced, such as the shrinkage of the productive area, resaturation of the producing formation, and producing-rate sensitivity. Nevertheless the quantitative interpretation of the gross reservoir performance can be expressed most naturally by considering the actual fluid-withdrawal area as producing primarily by the gas-drive mechanism.

The situation is somewhat similar with respect to gas-cap-expansion reservoirs. The process of gravity drainage, under ideal conditions, can lead to oil recoveries much higher than those commonly observed in solution-gas-drive reservoirs. In the extreme case where, by pressure maintenance, gas evolution in the oil zone is prevented and the oil withdrawals are replaced directly and completely by downstructure oil drainage, the gas-cap-expansion reservoir would indeed constitute a fundamentally distinct type of producing system, though it would be similar in many respects to the complete-water-drive mechanism. In the great majority of actual reservoirs, however, such conditions do not obtain. The downdip oil drainage provides only a supplement to the withdrawal replacement by gas evolution. The reservoir pressures decline, and the productive area below the gas cap exhibits the basic solution-gas-drive characteristics. Here, too, the gravity drainage and associated gas-cap expansion may modify the strict solution-gas-drive depletion performance quite appreciably. It is the study of these modifications that is the reason for giving special consideration to this type of reservoir. Yet the solution-gas-drive mechanism seems to afford the basic physical framework for the interpretation and prediction of the gross reservoir behavior during the producing life. In the detailed discussions to be given in later chapters of the *performance* histories of oil reservoirs the gas-cap-expansion reservoirs will therefore also be treated as a special type or generalization of the fundamental gas-drive system (cf. Sec. 10.15). On the other hand, in the consideration of recovery factors and ultimate recoveries, recognition will be given to the unique possibilities of the gravity-drainage mechanism, by discussing the gas-cap-expansion reservoirs separately (cf. Sec. 14.14).

9.3. Water Drives.—While there is no established definition of the term "complete water drive," it will be used here to denote the production mechanism in which the rate of water intrusion into the oil pay substantially equals the volumetric net rate of oil and gas withdrawal. This definition does not necessarily imply that, if and when volumetric equality between the oil- and gas-withdrawal and water-intrusion rates is established, no further decline in reservoir pressure will take place. On the contrary the pressure may continue to drop throughout the production history even though the rate of water entry is at all times substantially equal to the volumetric fluid withdrawals from the reservoir. For example, the pressure in the East Texas field in Gregg, Smith, Rusk, and Cherokee Counties, Tex., had declined by 600 psi after an oil production of 2.3 billion barrels, although 98 per cent of that production was replaced by water intrusion (cf. Sec. 11.9). The reason is that to maintain an influx rate, due to water expansion in the water reservoir, equal to the rate of net fluid withdrawal from the field the pressure at the field boundary has had to decline. While with such decline the fluids within the oil field also expand and provide for withdrawal replacement, its contribution during the over-all[1] producing life compared with that of direct water entry will generally be small, and the system may still be described as operating under a *complete* water drive.

Thus with respect to the pressure history of a reservoir the water-drive mechanism usually leads to a slow pressure decline with increasing cumu-

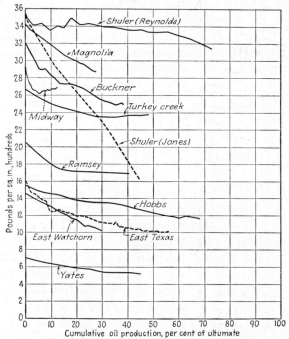

FIG. 9.1. The pressure vs. cumulative-production curves of various reservoirs that, except for the Schuler Jones Sand, have been subject to substantial or complete-water drives. (*After Elliott, AIME Trans., 1946.*)

lative recovery, after an initial rapid decline required to establish the pressure gradients that induce the water entry. A stabilization of the pressure as production continues at a fixed rate is definite evidence of a complete water drive. Likewise, if the average pressure increases on decreasing the production rate or on shutting in the field, an appreciable rate of water entry may be inferred. In all cases, when a water drive is a major component of the production mechanism, the reservoir pressure will be sensitive to the production rate. On the other hand, as the production rate is continually increased, the water-drive mechanism will ultimately lose control and the reservoir pressure decline will become subject to the gas-drive mechanism. Typical pressure-decline curves for several reser-

[1] Initially, however, the reservoir liquid expansion in undersaturated reservoirs will usually be the primary withdrawal-replacement medium (cf. Secs. 11.2 and 11.17).

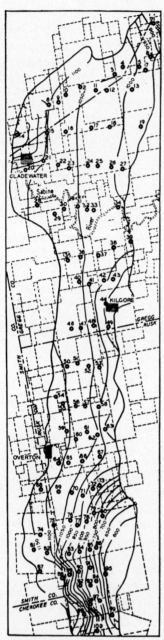

FIG. 9.2. The reservoir pressure contours in the East Texas field on July 1, 1945. (*After Meyer, Oil and Gas Jour., 1945.*)

voirs that were subject to substantial- or complete-water-drive action are shown in Fig. 9.1.[1]

Because under effective water-drive conditions the pressure declines but slowly, if at all, after the initial transients, and the growth of the free-gas phase is retarded or arrested, the gas-oil ratio will remain substantially constant as production continues. When the oil is undersaturated, as is usually the case in complete-water-drive systems, the gas-oil ratio will of necessity be constant until the bottom-hole flowing pressures fall to the bubble point. In any case, as long as the water drive remains effective, the gas-oil ratio will vary rapidly neither with the production rate nor with cumulative production, except as individual wells may be located near a gas-oil contact and gas coning is induced by excessive production rates.

The early appearance of water in oil-producing wells is not required as proof of the water-drive mechanism, unless the wells are known to be very near the original water-oil contact. However, if the water is actually entering at rates comparable with the rates of fluid withdrawal, it is to be expected that evidence of its intrusion will appear at the wells nearest the oil-water contact after a time required for the flushing of the intervening oil pay. On the other hand an early development of water cuts does not necessarily imply that a water drive is controlling the production of the reservoir as a

[1] These are taken from a paper of G. R. Elliott, *AIME Trans.*, **165**, 201 (1946). The decline curve for the Jones Sand of the Schuler field is included in Fig. 9.1 to show the contrast with a typical gas-drive pressure history, although there is still some question regarding the exact magnitude of the water intrusion into the Jones Sand reservoir (cf. Secs. 9.6 and 9.7.)

whole. Aside from the possibility that the water production may be the result of high local withdrawal rates and coning of bottom water, under a basically gas-drive mechanism (cf. Sec. 5.9) the water invasion may be restricted to a high-permeability zone representing a small part of the whole producing section.

In the case of the more common edgewater types of water drive the pressure distribution within the reservoir will reflect the existence and location of the external source of energy. That is, the pressures will be highest near the edgewater boundary and taper off in the regions most remote from the oil-water contact. This may lead to a continual decrease of pressure from one side of the field to the other, as in the case of the East Texas field, if the oil is confined by a stratigraphic trap or by faults and the water intrusion is limited to only one part of the reservoir boundary (cf. Fig. 9.2). Or if the structure is anticlinal and the oil is everywhere bounded by a water-oil contact, the pressure contours, under uniform areal withdrawal, will roughly parallel the reservoir boundary and the lower pressures will be found in the central part of the field, as in the case of the

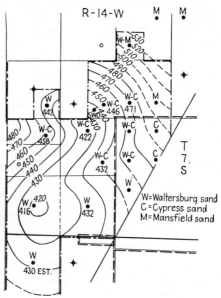

Fig. 9.3.   The reservoir pressure contours on July 1, 1943, in the Waltersburg Sand reservoir of the Mt. Vernon pool. (*After Barnes, Oil and Gas Jour., 1945.*)

Waltersburg Sand reservoir of the Mt. Vernon pool, Jefferson County, Ill. (cf. Fig. 9.3).[1] On the other hand, if the field as a whole is underlain by water and is producing entirely by a bottom-water drive, the pressures should tend to be substantially uniform, except as nonuniformities in character of the pay or local withdrawals may induce corresponding pressure variations.

Corollary to the slow rate of reservoir pressure decline, during most of the producing history, common to complete-water-drive reservoirs, is the maintenance of production capacity and rates from the individual wells.

---

[1] Figures 9.2 and 9.3 are to be considered as illustrations under almost ideal conditions. In many fields extraneous complicating factors virtually mask the simple pressure-contour patterns. Figure 9.2 is taken from L. J. Meyer, *Oil and Gas Jour.*, **44**, 67 (Dec. 1, 1945), and Fig. 9.3 from K. B. Barnes, *Oil and Gas Jour.*, **43**, 42 (Feb. 3, 1945).

Of course, envelopment of wells by the advancing or rising water front will result in severe reductions in oil-production capacity and often in rapid termination of the flowing life. But while the production remains "clean" (water-free), the producing characteristics of the wells will change but slowly.

9.4. Gas Drives.—As already noted in Sec. 9.1, "gas-drive fields" are those in which the major source of oil-expulsion energy is derived from a gas phase[1] and the oil produced is replaced by gas. If the oil-saturated formation is initially overlain by and in contact with a free-gas zone that expands as oil is withdrawn, the production mechanism is termed a "gas-cap drive." But if no such gas cap is present and the free-gas phase is that developed from pressure decline and evolution of dissolved gas and remains within the oil zone, the mechanism is termed a "dissolved-gas," "solution-gas," or "internal-gas" drive. The latter may change at least partly into the former if gas segregation and a gas cap develop during the course of the production. In either case, however, it is presumed that edgewaters, if present, do not enter the producing formation to an extent sufficient to provide an appreciable replacement of the gas and oil withdrawals.

The reservoir pressures in gas-drive fields will of necessity continually decrease as the fluids are withdrawn. For it is only by virtue of such pressure decline that the free gas already present can expand and additional solution gas be evolved to replace the volume occupied by the oil and gas produced. Moreover the magnitude of the pressure decline in a given reservoir, free of water intrusion, will be a function only of the total oil and gas withdrawal. The rate of oil production will not affect the pressure decline except indirectly, as it may influence the gas-oil ratio or degree of gas segregation and gas-cap formation. Even if the field be shut in, there will be no rise in average reservoir pressure, although pressure equalization within the field may appear to indicate a rise.

The pressure distribution within the field will reflect largely the local cumulative withdrawals, as compared with the local oil content of the formation. Pressures near a boundary may be low if at such boundary the pay actually pinches out. Or it may remain high if untapped sections of the reservoir extend beyond the region developed at the time. On the other hand, if the field has a gas cap, the part of the reservoir underlying the gas cap will tend to have uniform pressures due to the relatively easy fluid communication in the gas phase. The pressure contours shown in Fig. 9.4,[2] for a field in West Texas, illustrate many of the features of typical

---

[1] While this definition would encompass the broad class of condensate-producing fields, the latter are controlled by a basically different physical mechanism and will be treated separately (cf. Chap. 13).

[2] This figure was obtained through the courtesy of G. H. Fisher and C. N. Simpson.

gas-drive reservoirs with gas caps. In the southern part of the field, where there had been but little bleeding of the gas cap, the pressures were quite uniform. In the north, however, where the pay is tighter and the gas-oil

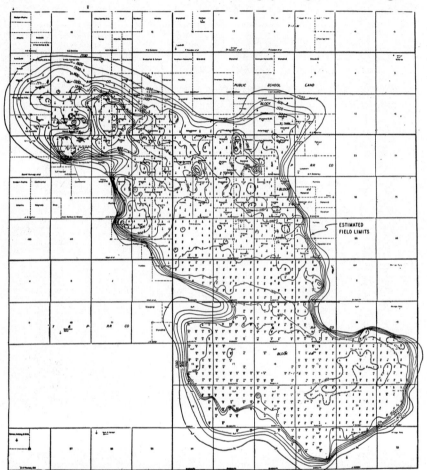

Fig. 9.4. The reservoir pressure contours in a West Texas field producing by gas drive.

ratios had been higher—in part owing to production of gas-cap gas—the average pressures were considerably lower and the contours mainly reflected variations in local withdrawal histories and pay characteristics.

Theory (cf. Sec. 10.3) indicates that for gas-drive fields the initial gas-oil-ratio history will be especially sensitive to the details of the permeability-saturation characteristics of the rock at high liquid saturations. If there is no equilibrium free-gas saturation, the gas-oil ratio will begin to rise from its initial solution value as soon as the field is put on production.

However, if there is a nonvanishing equilibrium free-gas saturation, the gas-oil ratio will first fall from the solution value. This decline will continue until the equilibrium gas saturation is developed, after which a rapid rise takes place. This rise continues for a period at an accelerated rate regardless of the initial behavior but ultimately reaches a maximum, after which it declines until the state of ultimate depletion.

In the great majority of cases in practice the gas-oil ratio has been found to rise from the very beginning of production,[1] without the initial decline, although the rate of rise varies from field to field. And where complete records have been kept throughout the whole production history, the theoretically predicted ultimate decline has also been observed. However, in recent years gas repressuring or pressure maintenance by gas injection has been applied to many gas-drive fields, thus modifying the normal depletion history and often leading to continued growth in gas-oil ratio until the time of abandonment of the gas injection. The presence of initial or developed gas caps also frequently results in an accentuated rise in the gas-oil ratio, as the upstructure wells become enveloped by the expanding gas cap. In fact, under modern production practices, controls over the operations or deliberate measures to improve the recovery efficiency are applied in virtually all fields at some period during the producing life. It is seldom now that a field is allowed to go through its whole history under a "natural" or "normal" depletion process.

It should also be noted that strict solution-gas-drive fields are perhaps the exception rather than the rule. Under conditions of controlled and limited production rates, such as generally obtain in this country, there usually develops some degree of water intrusion during the producing life, with the initial phases of gas drive changing into a partial-water-drive behavior, unless the reservoir is a strictly sealed and closed system.

9.5. Material-balance Principles—The Basic Equation.—Regardless of the nature of the production mechanism, all reservoirs must evidently obey the law of conservation of matter with respect to the gross fluid contents of the reservoir as a whole. The application of such conservation principles has generally been termed the "material-balance method" of reservoir analysis.[2] While the construction of the basic "material-balance equation" is itself a simple matter, the definition of the terms involved

---

[1] It should be noted, however, that gas-oil-ratio data are often subject to considerable uncertainty, especially during the early histories of fields before systematic gas-oil-ratio surveys are instituted.

[2] This method was first applied to oil-producing (gas-drive) reservoirs by S. P. Coleman, H. D. Wilde, Jr., and T. V. Moore [*AIME Trans.*, **86**, 174 (1930)]. Extensions and specific applications have since been reported by D. L. Katz, *AIME Trans.*, **118**,

requires careful consideration, especially with regard to the nature of the gas-liberation process occurring during the course of production. Unfortunately, as will be seen in the following sections, there are such serious limitations to the general applicability of the material-balance equation in the analysis of reservoir performance that a discussion of refinements and details of gas-liberation processes may appear wholly unwarranted. While this point of view may be justified from a strictly practical standpoint, it is still desirable to analyze thoroughly the physical principles involved, even though approximations must be made in applying the method to actual field problems.

As noted in Sec. 2.8, gas evolution from complex hydrocarbon systems may be of either the "flash" or "the differential" type. The former refers to a process in which the liberated gas is all maintained in contact with the liquid phase, so that the composition of the system as a whole remains constant as the pressure is reduced. In the latter the gas phase is continually removed as it is created by the declining pressure. While in the laboratory the gas liberation can be so controlled as to follow either process, the mechanism of oil production involves a complex composite of both types. The oil recovered in the stock tanks is the result of no single well-defined or fixed sequence of events. It is generally agreed that flash liberation prevails during the pressure decline up the flow string. The oil first produced will therefore undergo a complete flash-liberation process if it is flowed directly into the stock tanks at atmospheric pressure. But if it passes through one or more separators before proceeding to the stock tank, the flash liberation up the flow string will be followed by an approximate differential process before it becomes "stock-tank" oil. Moreover, an added similar complication will develop as soon as an appreciable pressure decline develops in the reservoir. For the reservoir oil and gas entering the flow string will not be the same at the bottom-hole pressure as that developed by flash liberation from the original pressure, except when the reservoir rock has an equilibrium gas saturation. If the equilibrium gas saturation is nonvanishing, then, during its build-up, the reservoir oil will undergo an approximate—though not exact—flash liberation. And the oil entering the flow string will simply be the liquid phase of the resulting gas-liquid mixture. But if there is no equilibrium gas saturation, the

---

18 (1936); R. J. Schilthuis, *AIME Trans.*, **118,** 33 (1936); R. L. Huntington, *Oil Weekly*, **83,** 25 (Oct. 12, 1936); B. H. Sage and W. N. Lacey, *API Drilling and Production Practice*, 1937, p. 81; W. Hurst, *AIME Trans.*, **151,** 57 (1943); R. E. Old, Jr., *AIME Trans.*, **151,** 86 (1943); C. B. Carpenter, H. J. Schroeder, and A. B. Cook, *U.S. Bur. Mines Rept. Inv.* 3720, part II (September, 1943); and R. W. Woods and M. Muskat, *AIME Trans.*, **160,** 124 (1945).

reservoir oil will follow approximately—though not exactly—a differential liberation process. In that case, or after an equilibrium gas saturation has been reached, the flash liberation in the flow string will be preceded by an approximate differential liberation during the passage of the oil to the well bore. Since the details of this preliminary gas-separation process are thus in themselves quite complicated and uncertain, it seems necessary, from a practical point of view, to accept the assumption generally made that the oil remaining in the reservoir at any pressure has undergone a simple differential liberation at the reservoir temperature. On the basis of this assumption the conventional procedure of bottom-hole-sample differential-liberation analysis should satisfactorily describe the behavior of the residual oil in the reservoir.

Aside from the inherent complexities associated with these variations in the different phases of gas liberation is the fundamental physical implication that the so-called "stock-tank oil" and "natural gas" are themselves no longer well-defined terms. For they are not fixed hydrocarbon systems. Strictly speaking the composition of the stock-tank oil and natural gas will vary during the life of the reservoir, especially with respect to their natural gasoline contents. This is to be expected, in view of the variation in gas-liberation processes experienced by successive increments of oil production if, and as, the reservoir pressure declines. If the well effluents were defined in terms of hydrocarbon composition or plant products, as is becoming common practice in the evaluation of condensate-producing reservoirs,[1] these difficulties would be obviated. But as long as stock-tank oil is merely defined in the conventional manner as the liquid product gathered in the stock tanks, and the natural gas as the sum of the separator gas-phase and stock-tank vapors, their inherently variable composition must be recognized.[2] And superposed on these essentially uncontrollable factors are those arising from deliberate changes in separator pressures and the natural fluctuations in separator and stock-tank temperatures.

As indicated before, the gas liberation within the reservoir itself may be reasonably approximated as one of differential liberation at reservoir temperature. It is this type of process that is also approximated by the conventional differential analysis of bottom-hole or surface recombined samples. If the finite pressure-decline increments used in the analysis are

---

[1] Cf. Sec. 13.3.

[2] These composition changes will generally be of minor importance in dealing with low-gravity (API) crudes. However, as the gravity is increased, the situation will merge into that pertaining to condensate-fluid systems, where the terms stock-tank oil and gas must be defined on a composition basis for quantitative reservoir and economic analyses.

not too great, it leads to a well-defined[1] terminal state of "residual" oil and cumulative development of the total gas liberation. This residual oil is not "stock-tank" oil. However, it provides a convenient base for the measurement of oil volumes in the construction of the material-balance equation.

While the nature (deviation factor) and total volumes of gas liberated in laboratory fluid-sample analyses can also be determined, it is not feasible to differentiate between the gas initially in the reservoir, that remaining in the reservoir after gas evolution, and the gas produced, unless their identifications are made on the basis of composition analyses, which will vary throughout the producing life. The initial free-gas phase could be accounted for by simply identifying it with the gas first evolved in the differential-liberation test. But the gas actually produced will not be equivalent, in a one-to-one ratio, to that obtained by the complete differential-liberation process. Hence, this phase of the problem will be treated by making the conventional assumption that "gas is gas," although this must be recognized as a basic weakness.

To proceed now to the construction of the material-balance equation, the following notation will be used:

$\overline{L}, L$ = initial reservoir content of residual oil and stock-tank oil, respectively, the former referring to the kind of oil left by a differential liberation of the gas in a reservoir oil sample, at reservoir temperature, from its initial pressure, and the latter to that remaining after a flash liberation of an equivalent sample to stock-tank conditions.

V_{gi} = initial reservoir volume of free-gas phase.

$\overline{S}, S$ = gas solubility by differential liberation per barrel of residual oil and per barrel of stock-tank oil, respectively.

γ = gas content at atmospheric conditions per unit volume of free-gas phase in the reservoir.

γ_j = value of γ for reservoir space occupied by injected gas.

G = surface (standard) volume of injected gas; Q_g = surface volume of produced gas.

$\overline{\beta}, \beta$ = formation volume of reservoir oil when pressure is declining by differential liberation in units of residual oil and stock-tank oil, respectively; β_f = formation volume of reservoir oil when flashed to stock-tank conditions.

W = reservoir volume of net water intrusion.

Q = stock-tank oil production; subscript i refers to initial values.

---

[1] In actual practice, further question may arise regarding the effect of the time during which the sample is held at reservoir temperature, after pressure reduction to atmospheric, before cooling to ambient stock-tank or standard temperature.

The composite statement of conservation of reservoir gas,[1] oil content.[2] and reservoir volume may be written as

$$\overline{S}_i\overline{L} + \gamma_i V_{gi} = \overline{S}\left[\overline{L} - Q\left(\frac{\beta_f}{\overline{\beta}}\right)_{\text{av}}\right] + \gamma\left\{V_{gi} + \overline{\beta}_i\overline{L} - \overline{\beta}\left[\overline{L} - Q\left(\frac{\beta_f}{\overline{\beta}}\right)_{\text{av}}\right]\right.$$
$$\left. - \frac{G}{\gamma_i} - W\right\} + Q_g . \quad (1)$$

In this equation the left-hand side represents the initial total gas content, no distinction being made between the nature of the gas to be liberated from solution $(\overline{S}_i)$ and that initially in the free-gas phase (γ_i). The first term on the right accounts for the gas dissolved in the unrecovered oil, and the second term on the right the gas in the free-gas phase.

Since the role played by the injected gas, if it remains trapped in the reservoir, is essentially one of volumetric displacement, it has been given a distinct density factor γ_j, which can be readily determined from its composition and deviation factor.

The term $(\beta_f/\overline{\beta})_{\text{av}}$ denotes an average of $\beta_f/\overline{\beta}$ according to the definition

$$Q\left(\frac{\beta_f}{\overline{\beta}}\right)_{\text{av}} = \int\frac{\beta_f}{\overline{\beta}}\,dQ. \quad (2)$$

It provides the means for converting the produced stock-tank oil to equivalent residual oil and could be calculated from the differential- and flash-liberation formation-volume-factor[3] data. It is assumed, of course, that the flashing simulates the actual composite process of the field operations if separators are used.

---

[1] As implied by the previous discussion, the change in character and effective solubility of the produced gas during the producing life, as the relative contributions of the differential- and flash-liberation processes vary, vitiates a strict applicability of conservation requirements with respect to the gas "phase," such as Eq. (1). The latter is in itself therefore basically an approximation, unless Q_g is adjusted by an averaging factor similar to that applied to the oil production Q.

[2] The oil content refers only to the residual oil from the initial-liquid phase. No account is taken here of the possible presence of condensable hydrocarbon components in the gas phase. Nor is a distinction made between the oil trapped in regions invaded by water and that in the unflooded part of the formation. Correction factors can be derived for the effect of segregation of the "occluded" oil and gas in the water-invaded part of the reservoir, as well as for the gas solubility in the water phase of the reservoir. However, these would represent refinements that hardly appear to be warranted in the light of the other approximations involved in the derivation and application of the material-balance equation.

[3] It should be noted that, while the differential-liberation data refer to the reservoir temperature, the flash-liberation tests for use in Eq. (2) would be preceded by a differential liberation at reservoir temperature to the pressure of interest, and these would be followed by flashing to atmospheric stock-tank conditions.

Equation (1) readily gives

$$\bar{L} = \frac{Q\left(\dfrac{\beta_f}{\bar{\beta}}\right)_{av} (\gamma\bar{\beta} - \bar{S}) + Q_g - \dfrac{\gamma}{\gamma_i} G - (\gamma_i - \gamma)V_{gi} - \gamma W}{\bar{S}_i - \bar{S} - \gamma(\bar{\beta}_i - \bar{\beta})}. \tag{3}$$

This is the general material-balance equation, giving the value of initial *residual* oil. All the *p-v-T* data refer to differential liberation except for the factors γ_i and $(\beta_f/\bar{\beta})_{av}$. To convert to the equivalent stock-tank oil L, of the type produced initially, it is noted that

$$\bar{L}\bar{\beta}_i = L\beta_{fi}. \tag{4}$$

Equation (3) can then be rewritten as

$$L = \frac{\dfrac{\bar{\beta}_i}{\beta_{fi}}\left(\dfrac{\beta_f}{\bar{\beta}}\right)_{av} Q(\gamma\beta - S) + Q_g - \dfrac{\gamma}{\gamma_i} G - (\gamma_i - \gamma)V_{gi} - \gamma W}{S_i - S - \gamma(\beta_i - \beta)}, \tag{5}$$

where S, β, S_i, β_i refer to unit volumes of *stock-tank* oil as produced initially, though the basic data still refer to the differential-liberation process.

If now the assumption or approximation is made that the coefficient of $Q(\gamma\beta - S)$ is unity and one introduces the cumulative gas-oil ratio R as measured conventionally, one obtains finally

$$L = \frac{Q(R - S + \gamma\beta) - \dfrac{\gamma}{\gamma_i} G - (\gamma_i - \gamma)V_{gi} - \gamma W}{S_i - S - \gamma(\beta_i - \beta)}. \tag{6}$$

There is no reason to expect the assumption that the coefficient of Q in Eq. (5) is unity to be strictly satisfied by actual natural-gas and crude-oil systems. However, in view of the other basic approximations involved in the development of Eq. (5) and the general parallelism of experimentally determined $\bar{\beta}$ and β_f functions, the simplified form of Eq. (6) should suffice for practical purposes. Moreover, if the injected gas should undergo mixing with the gas liberated from solution or the initial free-gas phase, the distinction between γ_i and γ will no longer be feasible. In that case the injected gas can be subtracted from that produced, and Eq. (6) re-written as

$$L = \frac{Q(\bar{R} - S + \gamma\beta) - (\gamma_i - \gamma)V_{gi} - \gamma W}{S_i - S - \gamma(\beta_i - \beta)}, \tag{7}$$

where $\bar{R}$ represents the *net* cumulative gas-oil ratio.

Equations (6) and (7) represent the basic "material-balance equations" applicable to fields producing crude oil. They involve several distinct sets of quantities. The first is comprised of the terms γ, β, and S, which are subject to direct determination by laboratory tests on samples of the reservoir fluids. γ may also be obtained by calculation, by using the equation

$$\gamma(p) = \frac{pT_a}{Zp_aT_r},\tag{8}$$

where p is the reservoir pressure at any time, T_r the reservoir temperature, T_a, p_a the ambient or standard surface temperature and pressure, and Z the deviation factor. The latter may be computed from the gas composition or gravity, if known, together with correlation charts (cf. Sec. 2.7).

The cumulative oil production Q, net cumulative gas-oil ratio $\overline{R}$, and the reservoir pressures associated with these are the basic operating data constituting the record of the production performance. In addition the total water production, which enters implicitly in the term W since the latter represents the *net* water intrusion into the reservoir, is to be considered as known at the same time intervals for which Q, $\overline{R}$ and the reservoir pressures are recorded.

In so far as Eq. (7) itself is concerned, the initial stock-tank oil in place, L, and volume of the original gas cap, if any, V_{gi}, are unknown reservoir constants. And W will be a progressively increasing but unknown function of time,[1] unless there is no natural water intrusion whatever in the reservoir. On the other hand, if in the latter case water is deliberately injected within the original confines of the reservoir, the amount of this injection will also be represented by the term W, which then becomes a known function.

Before considering the actual use and application of Eq. (7) it should be noted that in addition to the approximations and assumptions already discussed, it is fundamentally based on a representation of the reservoir as a continuous and uniform hydrocarbon-containing tank, throughout which the gas and oil are in continuous equilibrium.[2] The reservoir pressures involved in Eq. (7) refer to the strictly uniform pressures in such a tank. It is tacitly assumed that the fluid withdrawals, represented by Q and $\overline{R}$, are uniformly distributed throughout the reservoir. The water entry W, as well as the gas-injection term included in $\overline{R}$, also are to be

---

[1] An equivalent form of Eq. (7) in which the variable unknown function is the current reservoir free-gas content $\overline{V_g}$ in surface measure is

$$\overline{V_g} = \gamma_i V_{gi} + L[S_i - S + \gamma(\beta_{fi} - \beta_i)] - Q(\overline{R} - S),$$

which reduces in the case of no net water intrusion to $\overline{V_g} = \gamma[V_{gi} + L\beta_{fi} - \beta(L - Q)]$.

[2] The local thermodynamic equilibrium between the gas and liquid phases is a basic assumption underlying all quantitative treatments of reservoir phenomena. On the other hand in the use of the material-balance equation the additional assumption is made of pressure equilibrium and uniformity (or state of depletion) between various strata of different permeability that may be simultaneously drained by the producing well system.

visualized as uniformly affecting the whole of the reservoir. No account is taken of the use of wells as foci for fluid withdrawal or injection and of the natural pressure gradients associated with the flow of the fluid toward the wells. Nor is cognizance given to regional pressure variations over the reservoir, due to differences in local degrees of depletion or the action of nonuniformly distributed external sources of energy as edge-waters. With such and all pressure variations will be associated, of course, a nonuniformity in the values of the fluid parameters S, β, and γ.

The obvious implication of these considerations is that, in the application of Eq. (7) to actual producing reservoirs, *average* pressures must be used. And, for simplicity, equivalent averages of S, β, and γ are taken as the values corresponding to the average pressures. The latter, in turn, are not simply the arithmetical averages over the individual well measurements. Though it is not always feasible to carry through the procedure, they should be integrated averages weighted locally by the net reservoir void space[1] available for recoverable petroleum fluids. And, of course, they should refer to a common datum plane and the common date associated with the values of the cumulative oil and gas productions. While the use of such pressures will tend to bridge the gap between the hypothetical system on which Eq. (7) is based and the actual oil-producing reservoir, it should be recognized that in practice complete equivalence will virtually never be achieved.

9.6. The Application of the Material-balance Equation—No Water Influx.—In the early developments and applications of the material-balance method for interpreting reservoir performance, the main emphasis was on its use for the determination of the basic reservoir unknowns L, V_{gi}, and W. Upon neglecting for the moment the latter, and assuming that it is definitely known that there is no water intrusion, the procedure is essentially as follows: With the values of the pressure functions S, β, and γ known, the observed pressure and production data are introduced into the reduced form of Eq. 9.5(7), namely,

$$L = \frac{Q(\overline{R} - S + \gamma\beta) - V_{gi}(\gamma_i - \gamma)}{S_i - S + \gamma(\beta - \beta_i)} \qquad (1)^*$$

and the corresponding values of L are calculated for various fixed and assumed values of V_{gi}. Since both L and V_{gi} are inherently reservoir

---

[1] The further refinement in which variations in the local fluid saturations are taken into account can hardly be considered within the realm of practicability.

* The first term in the numerator of Eq. (1) divided by γ has sometimes been referred to as the "space voidage" created by the total oil and gas production, though it refers to the volume that would be occupied by the cumulative withdrawal if it all were subjected to the reservoir conditions at the current state of the continuous withdrawal process.

constants, that value of V_{gi} that leads to reasonably constant values of L, and the average of the latter, are presumably the correct magnitudes of these terms. If the size of the gas cap (V_{gi}) was established independently, or if it is known that there was no original gas cap ($V_{gi} = 0$), the average of the L's computed directly by Eq. (1) should give a reliable evaluation of the original oil in place, unless the individual values show such large fluctuations as to indicate serious inaccuracies in the basic

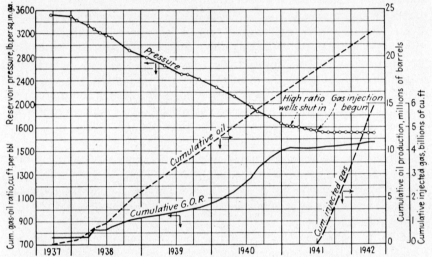

FIG. 9.5. The early pressure and production history of the Jones Sand reservoir of the Schuler field. (*After Old, AIME Trans., 1943.*)

data or in the assumptions regarding V_{gi} and W. Or, as is sometimes the case, if the ratio of the gas cap to initial reservoir oil volume is known, as expressed by the ratio m,

$$m = \frac{V_{gi}}{\beta_{fi}L},$$ (2)

L can be computed from the equation

$$L = \frac{Q(\overline{R} - S + \gamma\beta)}{S_i - S + \gamma(\beta - \beta_i) + m\beta_{fi}(\gamma_i - \gamma)}.$$ (3)

If the fluctuations are small, the average value of the calculated L's can again be accepted as trustworthy.

If, however, no prior information about the gas cap is available, the above-outlined procedure for calculating the L's for fixed values of V_{gi} will be subject to considerable uncertainty. For it will be found, in general, that approximate constancy in the L's will be obtained for a considerable range in the assumed values of V_{gi}, with corresponding appreciable differences in the average values of L. An example of this situation is provided by the data pertaining to the Jones Sand of the Schuler pool, Arkansas.[1] The production and pressure histories to April, 1942, are shown graphically in Fig. 9.5.[2] From these and the reservoir fluid data, represented by γ, S, and β, the corresponding values of L, denoted by L_j, are first calculated by Eq. (1), upon taking $V_{gi} = 0$. The results, together with the tabulated production data, are listed in the sixth column of Table 1. As indicated, the average value $\overline{L}_o = 116.05 \times 10^6$ bbl. The root-mean-square deviation of the individual values is 2.70×10^6 bbl, or 2.33 per cent of the average $\overline{L}_o$.

In view of the various assumptions and approximations underlying Eq. (1), as discussed previously, a rms mean deviation of only 2.33 per cent among the 30 individual values would certainly seem to justify faith in the average of 116.05×10^6 bbl and the associated specific assumption that $V_{gi} = 0$. Yet if V_{gi} be taken as 10^7 bbl, the constancy in the corresponding L_j's, as indicated in the seventh column in Table 1, is only slightly inferior—a rms deviation of 2.92×10^6 bbl, or 3.18 per cent of the average: 91.76×10^6 bbl. Indeed these latter deviations also appear to be no greater than might have been expected. And to these slight differences in constancy of the L_j's corresponds a difference of 24.3×10^6 bbl in the average calculated oil in place.

These considerations may be expressed analytically[3] on introducing the notation

$$L_{oj} = L_j(V_{gi} = 0) = \frac{Q(\overline{R} - S + \gamma\beta)}{S_i - S + \gamma(\beta - \beta_i)}; \qquad \lambda_j = \frac{\gamma_i - \gamma}{S_i - S + \gamma(\beta - \beta_i)}, \qquad (4)$$

so that Eq. (1) can be rewritten as

$$L_j = L_{oj} - \lambda_j V_{gi}. \qquad (5)$$

Upon denoting average values by bars[4] it follows that

$$\overline{L} = \overline{L}_o - \overline{\lambda} V_{gi}, \qquad (6)$$

[1] Although there still appears to be some doubt regarding the importance of water-drive action in this reservoir [cf. H. H. Kaveler, *AIME Trans.*, **155**, 58 (1944) and Sec. 10.13], this question will not affect the present considerations with respect to the role played by the gas-cap size and it will therefore be assumed for the present purposes that there simply is no water intrusion in the reservoir.

[2] Fig. 9.5, as well as the other basic data for Schuler, were taken from the work of Old, *loc. cit.* Since the purpose of the discussion given here of the Schuler Jones reservoir is illustrative only, no attempt has been made to bring the data and calculations up to date, as will be done in Sec. 10.13 in reviewing the history of gas injection in this field.

[3] Cf. Woods and Muskat, *loc. cit.*

[4] $\overline{L}$, $\overline{L}_o$, and S, β in $\overline{\lambda}$ still refer to units of stock-tank oil and are not to be confused with the bar notation of Sec. 9.5.

TABLE 1.—MATERIAL-BALANCE DATA FOR JONES SAND, SCHULER FIELD

Date	Reservoir pressure, psi (gauge)	Cumulative oil production, 10^6 bbl	Cumulative gas-oil ratio, ft³/bbl	Cumulative gas injection, 10^9 ft³	L_j, oil in place (10^6 bbl), calculated for $V_{gi} = 0$, $W = 0$	L_j, oil in place (10^6 bbl), calculated for $V_{gi} = 10^7$ bbl, $W = 0$	L_j, oil in place (10^6 bbl), calculated for $V_{gi} = 8.91 \times 10^6$ bbl, $c = 256.9$
1937:							
Sept. 17	3,520	0	760	0	—	—	—
1938:							
July...	3,125	2.818	853	0	111.10	83.63	74.25
Oct....	2,910	4.653	906	0	114.99	87.50	76.94
1939:							
Jan....	2,785	6.030	939	0	119.43	92.56	78.85
Apr....	2,650	7.360	960	0	118.96	90.92	75.65
July...	2,505	8.751	990	0	116.96	90.09	73.72
Aug....	2,501	9.011	997	0	120.71	93.75	75.52
Oct....	2,425	9.873	1,018	0	119.66	93.21	74.22
1940:							
Jan....	2,290	11.259	1,070	0	117.51	91.69	76.04
Apr....	2,125	12.619	1,200	0	117.29	91.99	74.23
July...	1,950	13.998	1,310	0	114.77	90.58	75.84
Aug....	1,878	14.462	1,350	0	112.11	88.35	72.30
Oct....	1,785	15.321	1,440	0	113.96	90.20	74.83
Dec....	1,670	16.132	1,500	0	111.88	88.45	73.93
1941:							
Jan....	1,630	16.552	1,520	0	112.25	88.93	74.58
Feb. 14	1,609	16.739	1,520	0	111.52	88.25	73.74
Feb. 28	1,608	16.929	1,519	0	112.56	89.26	74.39
Mar...	1,601	17.348	1,516	0	114.45	91.16	75.76
Apr....	1,575	17.757	1,515	0	114.34	91.17	75.63
May...	1,555	18.177	1,516	0	115.19	92.06	76.23
June...	1,537	18.583	1,520	0	116.19	93.14	77.04
July...	1,514	19.006	1,527	0.3724	115.77	92.78	76.68
Aug....	1,505	19.426	1,532	0.9543	115.69	92.73	75.92
Sept...	1,502	19.830	1,534	1.529	115.86	92.93	75.63
Oct....	1,506	20.249	1,537	2.136	117.07	94.13	76.65
Nov...	1,506	20.652	1,541	2.762	117.70	94.76	75.88
Dec....	1,500	21.068	1,545	3.402	117.82	94.89	75.70
1942:							
Jan....	1,498	21.474	1,549	4.006	118.31	95.39	75.50
Feb....	1,492	21.850	1,556	4.649	118.37	95.49	75.41
Mar...	1,493	22.261	1,565	5.392	119.23	96.34	75.33
Apr....	1,493	22.619	1,566	5.883	119.86	96.97	75.63
Average	.....	......	.....	.....	116.05	91.76	75.41

showing that the mean calculated value of the oil in place will decrease linearly with the volume assumed for the gas cap.

To evaluate the fluctuations in the individual L_j's by the method of least squares the individual deviations may be expressed as

$$\Delta_{oj} = \overline{L}_o - L_{oj}; \qquad \Delta_j = \overline{L} - L_j = \Delta_{oj} - B_j V_{gi} \quad : \quad B_j = \overline{\lambda} - \lambda_j. \qquad (7)$$

The mean-square deviations are then given by

$$\overline{\Delta^2} = \overline{\Delta_o^2} - 2V_{gi}\,\overline{B\Delta_o} + V_{gi}^2\,\overline{B^2}. \tag{8}$$

It thus appears that the mean-square deviation will be a quadratic function of V_{gi}, with a minimum at $V_{gi} = \overline{B\Delta_o}/\overline{B^2}$ and an absolute minimum

$$(\overline{\Delta^2})_m = \overline{\Delta_o^2} - \frac{\overline{B\Delta_o}^2}{\overline{B^2}}. \tag{9}$$

The percentage rms deviation is

$$\frac{100\sqrt{\overline{\Delta^2}}}{\overline{L}} = \frac{100\sqrt{\overline{\Delta_o^2} - 2V_{gi}\,\overline{B\Delta_o} + V_{gi}^2\overline{B^2}}}{\overline{L}_o - \overline{\lambda}V_{gi}}, \tag{10}$$

with a minimum at

$$V_{gi} = \frac{\overline{L}_o\,\overline{B\Delta_o} - \overline{\lambda}\,\overline{\Delta_o^2}}{\overline{L}_o\,\overline{B^2} - \overline{\lambda}\,\overline{B\Delta_o}}, \tag{11}$$

and a minimum percentage value of

$$\left(\frac{100\sqrt{\overline{\Delta^2}}}{\overline{L}}\right) = 100\sqrt{\frac{\overline{B^2}\,\overline{\Delta_o^2} - \overline{B\Delta_o}^2}{\overline{L}_o^2\,\overline{B^2} - 2\overline{\lambda}\,\overline{L}_o\,\overline{B\Delta_o} + \overline{\lambda}^2\,\overline{\Delta_o^2}}}. \tag{12}$$

In Fig. 9.6 are plotted the percentage rms deviations and the average calculated oil in place in the Schuler Jones Sand reservoir as a function of the assumed value of V_{gi}. While the deviations increase with increasing V_{gi}, the absolute minimum would be reached with a physically meaningless negative value of V_{gi}.* On the other hand there is evidence[1] that there was an initial gas cap in the field and that it occupied 3.7 per cent as much volume as the oil pay [$m = 0.037$ in Eq. (2)]. According to Eq. (6) and Fig. 9.6 this implies that $\overline{L} = 102.7 \times 10^6$ bbl, with a percentage rms deviation of 2.65 per cent and $V_{gi} = 5.5 \times 10^6$ bbl, on noting that $\beta_i = 1.45$ and $\overline{\lambda} = 2.429$.

These considerations clearly show that the criterion of constancy of the values of the oil in place, calculated by Eq. (1), in itself does not have sufficient resolving power to fix quantitatively the true magnitude of this and related reservoir parameters. In the above example a variation in the average oil in place of 20 million barrels would imply a change in the rms deviation only from 2.3 to 3.0 per cent. In the light of previously discussed approximations and uncertainties involved in the material-balance equation, deviations even of 3.0 per cent might be considered as quite

---

* The absolute value of the mean-square deviation has a minimum at a positive V_{gi}, as given by Eq. (9). The shift to a negative value of V_{gi} for a minimum percentage deviation results from the linear decrease of $\overline{L}$ with V_{gi} [cf. Eq. (6)].

[1] Old. *loc. cit.*

acceptable. While in some instances,[1] mean deviations less than 2 per cent have been obtained, it often happens that with the best data available these deviations still exceed 5 per cent.

Without entering into a detailed analysis of the errors inherent in the application of the material-balance equation, it may be noted that moderate uncertainties in the average reservoir pressure can alone cause spurious

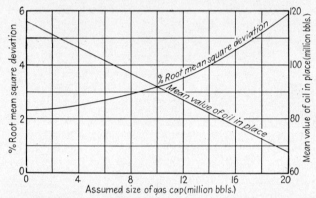

Fig. 9.6. The variation of the calculated mean values of the oil in place and the percentage rms deviation with the assumed size of the gas cap in the Jones Sand reservoir of the Schuler field, assuming no water intrusion.

variations in the calculated oil in place of the order of several per cent. Thus, upon expressing L_{oj} as

$$L_{oj} = aQ + bQ\overline{R}; \qquad a = \frac{\gamma\beta - S}{S_i - S + \gamma(\beta - \beta_i)}; \qquad b = \frac{1}{S_i - S + \gamma(\beta - \beta_i)}, \quad (13)$$

the fractional error caused in the evaluation of L_{oj}, due to an error Δp in the pressure, will be

$$\frac{\Delta L_{oj}}{L_{oj}} = \frac{Q}{L_{oj}}\left(\frac{da}{dp} + \overline{R}\frac{db}{dp}\right)\Delta p. \quad (14)$$

As both da/dp and db/dp are positive, the two terms will be additive. While the coefficient Q/L_{oj} will increase as the pressure declines, the parentheses will generally decrease. In order of magnitude, for the data pertinent to the Jones Sand at Schuler, it is found that

$$\frac{\Delta L_{oj}}{L_{oj}} \cong 0.002 \, \Delta p. \quad (15)$$

---

[1] Cf., for example, the Reed City field, Michigan (Woods and Muskat, *loc. cit.*), and Table 3. On the other hand, in one major gas-drive field in West Texas which has been thoroughly studied and for which the basic data have been gathered by the joint efforts of all the major operators, the rms deviation of the calculated oil in place for nine periods was 12.4 per cent. And six calculations for the oil in place in the Oklahoma City field (Katz, *loc. cit.*) gave a rms deviation of 10.3 per cent.

Thus errors of 10 psi in the reservoir pressure will lead to errors in L_{oj} of the order of 2 per cent. Because of the changing pressure distributions during the production history and natural uncertainties regarding the proper averaging procedure to be used, errors of 10 psi in the reservoir pressure cannot always be eliminated. Moreover the production data themselves, especially with regard to the gas production, are often subject to considerable uncertainty.[1] As a whole, therefore, fluctuations in the L_{oj}'s of the order of 2 per cent may well be expected even if the basic production mechanism is not in question.

While the determination of the absolute value of the oil in place, $\bar{L}$, merely by establishing the conditions of minimum deviation in the L_j's may thus be of questionable reliability, the situation in practice is generally subject to some reasonable control. The mere presence or absence of an initial gas cap is usually established with considerable certainty from the drilling, electrical-logging, or core-analysis records. And often the relative thickness of pay, compared with the total, occupied by the gas cap can also be determined from well and core data with fair precision. The latter permits the calculation of the quantity m of Eq. (2), and a reduction in the number of unknowns to one, on using Eq. (3). Under such circumstances or when it is known that there was no initial gas cap, the computation of the L_j's should be quite as trustworthy as the basic data entering in the material-balance equation, provided, of course, that it is also known that water intrusion is not a significant factor in the production mechanism. In any case, if the reservoir fluid and production data are available, it should be of value to make the material-balance calculations and to compare them with whatever independent estimates of the reservoir constants may have been indicated by geological and core information. However, if the latter are complete and well established, they should serve as better criteria in determining the production mechanism than the magnitude of the fluctuations in the oil in place as calculated by the material-balance equation.

Assuming that the gross reservoir parameters have been established, the material-balance equation can be used in predicting the future reservoir behavior and, in particular, the pressure vs. the cumulative recovery. In such calculations, however, the future gas-oil-ratio trend must be assumed, and the results will depend on the particular assumption made. On the

---

[1] An example of such uncertainties in basic reservoir data is provided by a comparison of three publications on the Jones Sand reservoir of the Schuler field, namely: W. B. Weeks and C. W. Alexander, *AAPG Bull.*, **26**, 1467 (1942); Old, *loc. cit.*; and Kaveler, *loc. cit.* The pressure data given in these papers differ on the average by 42 psi at corresponding survey dates and by a maximum of 91 psi. The reported gas-oil ratios differ by a maximum of 350 ft³/bbl and an average of 97 ft³/bbl, the latter being 7.5 per cent of the average ratio over the whole set of comparable data.

other hand, comparative calculations using different gas-oil-ratio histories will show the value of gas-oil-ratio control or of returning gas to the reservoir under assumed degrees of control over the rise in gas-oil ratio.

9.7. The Application of the Material-balance Equation—Partial Water Drives.—When water intrusion is suspected or known to occur, it is necessary to apply the more general form of Eq. 9.5(7) in making the material-balance analysis. This immediately brings to the fore a serious complication in addition to those discussed in the preceding section for simple gas-drive systems. This arises from the fact that the water-intrusion term is inherently a variable. Except in cases where the net water entry is kept from increasing by a deliberate pumping program at the edgewater boundary, W will be a continuously increasing function and not subject to approximation by an "average" constant value. In fact, W will continue to grow even if the field should be shut in.

To the extent that W is a totally unknown function and L and V_{gi} are also unknown, the value of the material-balance equation for determining these reservoir parameters collapses completely. Evidently both L and V_{gi} could then be chosen over a wide range of values, less than their true magnitudes, and W, together with its coefficients, could be calculated by inverting Eq. 9.5(7) so as to make the equation fit exactly for the observed production and pressure data. In fact, by using the notation of Eqs. 9.6(4) and rewriting Eq. 9.5(7) as

$$W = \frac{[S_i - S + \gamma(\beta - \beta_i)][L_{oj} - \lambda_j V_{gi} - L]}{\gamma}, \tag{1}$$

it will be seen that, even if the constancy of $L_{oj} - \lambda_j V_{gi}$ indicates the absence of significant water entry, a choice of L appreciably smaller than the mean value $\overline{L_o} - \overline{\lambda} V_{gi}$ will give a series of continuously increasing values of W as long as the reservoir pressures continue to decrease with increasing fluid withdrawals.[1] Such a variation of W may well appear in itself as a quite reasonable representation of a true water-encroachment history.

While a completely satisfactory solution of this problem when the initial oil in place is unknown has not been developed, an approach to a resolution of the difficulty can be based on the observation that the total water intrusion W_t should be expressible as a function of the pressure history of the field, i.e., by an equation of the form

$$W_t = cf(p,t), \tag{2}$$

---

[1] Of course, when the total production approaches the true ultimate recovery or the pressures approach atmospheric, any gross errors in assumed values of the reservoir contents will automatically reveal themselves by a breakdown in the fit of the later data with the material-balance equation or by other features of the general reservoir performance.

where the coefficient c is separated as a constant depending on fixed parameters of the water reservoir, such as its average permeability, thickness, water viscosity, and compressibility. For example, if the water entry is assumed to follow a steady-state behavior, W_t would have the explicit form[1]

$$W_t = c \int_0^t (p_i - p)dt, \qquad (3)$$

where p_i is the initial reservoir pressure and p the pressure at time t. If the water reservoir be considered as a compressible-homogeneous-fluid system, $f(p,t)$ will have a much more complicated form, such as is given in Eq. 11.5(15). In some applications that have been made, the latter has been approximated by[2]

$$W_t = c \int_0^t \frac{p_i - p}{\log t} dt. \qquad (4)$$

And even the further simplification of Eq. (3), namely,

$$W_t = c(p_i - p)t \qquad (5)$$

has been tried.

The important point about these expressions is that each reduces the unknown element in W_t to a constant c. This coefficient, in turn, is to be chosen, just as V_{gi}, so that the value of L will be substantially constant, as calculated at the various times at which the production and pressure data are available. Thus, by explicitly separating W into W_t* and the water produced W_p, Eq. 9.5(7) may be rewritten as

$$L = \frac{Q(\overline{R} - S + \gamma\beta) - V_{gi}(\gamma_i - \gamma) + \gamma W_p - \gamma W_t}{S_i - S + \gamma(\beta - \beta_i)}. \qquad (6)$$

Even on assuming that W_t is expressed by Eq. (2) or in any special form as Eqs. (3), (4), or (5), Eq. (6) will now contain three basic unknown constants if nothing else is assumed about the reservoir. For application to actual reservoirs, one may consider Eq. (6) simply as a linear equation in three unknowns, fill in the coefficients from the production and pressure data, and by a least-square analysis determine the best-fitting values of the constants L, V_{gi}, and c. If Eq. (6) be formally expressed as

$$\left.\begin{array}{l} L + \lambda_j V_{gi} + cA_j = C_j, \\[2mm] C_j = L_{oj} + \dfrac{\gamma W_p}{S_i - S + \gamma(\beta - \beta_i)} \\[3mm] A_j = \dfrac{\gamma f(p,t)}{S_i - S + \gamma(\beta - \beta_i)} \end{array}\right\} \qquad (7)$$

[1] Schilthuis, loc. cit.

[2] Woods and Muskat, loc. cit.

* As previously mentioned, any water deliberately returned to or injected into the oil reservoir may be considered as included in W_t or may be explicitly subtracted from W_p.

and λ_j, L_{oj} are given by Eq. 9.6(4), the best-fitting values of L, V_{gi}, and c will be the solutions of three linear equations with a symmetrical matrix, namely,

$$
\left.\begin{array}{l}
L + V_{gi}\overline{\lambda} + c\overline{A} = \overline{C}, \\
L\overline{\lambda} + V_{gi}\,\overline{\lambda^2} + c\overline{\lambda A} = \overline{\lambda C}, \\
L\overline{A} + V_{gi}\,\overline{\lambda A} + c\overline{A^2} = \overline{AC}_,
\end{array}\right\} \tag{8}
$$

where the bars indicate the mean value.

As most of the emphasis has been on the determination of L, the method heretofore used has been that of assuming V_{gi} and c and minimizing the fluctuations in L,* that is, by expressing the individual values of L as

$$
L_j = L_{oj} - \lambda_j V_{gi} - cA_j + D_j \tag{9}
$$

where D_j is the second term in C_j as given by Eq. (7), and applying a least-squares analysis similar to that outlined in Sec. 9.6. For practical purposes, however, this procedure should give essentially the same results as Eqs. (8)[1]. As may be anticipated from the simpler case of gas-drive systems discussed in the preceding section, here, too, the requirement of constancy of the individual values of L has very little discriminating power among the assumed values of V_{gi} and c. That is, relatively large variations in V_{gi} and c will result in but small changes in the fluctuations in L,† as illustrated by more generalized calculations for the Jones Sand reservoir of the Schuler field, summarized in Table 2, where in two cases the possibility of water intrusion was not initially excluded. In making the computations for the cases listed in Table 2 the individual constants that were not assumed to be 0 were determined to give the minimum rms deviation.

The function $f(p,t)$, which was used in calculating the A_j's [cf. Eq. (7)], was that listed by Old, which was computed using a compressible-fluid representation of the water reservoir. The values of W_t indicate the total water intrusions to April, 1942, corresponding to the values of c. The

---

* Old, *loc. cit.;* Woods and Muskat, *loc. cit.*

[1] Still another variation of the general material-balance method for fixing the reservoir characteristics by minimizing the spread between calculated and expected results is that of treating the reservoir oil content and the effective permeability of the water-drive aquifer as the adjustable constants, and choosing the latter so that the corresponding predicted pressure history deviates the least from that observed [cf. E. R. Brownscombe and F. Collins, *Jour. Petroleum Technology,* **1**, 92 (1949)]. But this, too, as well as further modifications of this procedure, suffers from the same low resolving power as the simpler analysis described here.

† The analogous situation with respect to Eqs. (8) is expressed by the fact that the total deviations will not vary rapidly with variations in the L, V_{gi}, and c from the values given by Eqs. (8). The calculations giving the results listed in Tables 2 and 3 were made by minimizing only the fluctuations in the L_j's defined by Eq. (9).

individual values of the L_j's for case 4 are listed in the last column of Table 1.

It will be noted that by proper choice of the water-intrusion factor c a smaller deviation can be achieved, even though V_{gi} be taken as 0, than

TABLE 2.—MATERIAL-BALANCE RESULTS FOR THE JONES SAND AT SCHULER UNDER DIFFERENT ASSUMPTIONS REGARDING THE RESERVOIR PARAMETERS

Case	c, bbl/day per psi	$10^{-6}W_t$ (total), bbl	$10^{-6}V_{gi}$, bbl	$10^{-6}\overline{L}$ bbl	Rms deviation in $\overline{L}$, %
1	0	0	0	116.05	2.33
2	194.3	10.6	0	101.68	1.85
3	0	0	2.95	108.88	2.44
4	256.9	14.0	8.91	75.41	1.64

by adjustment of V_{gi} with $c = 0$. Moreover, as is to be expected, the use of both constants c and V_{gi} permits a still greater reduction in the minimum rms deviation obtainable. Again, however, such improvements in the constancy of the L_j's are so small as hardly to provide a basis for dis-

TABLE 3.—COMPARISON OF MINIMUM-DEVIATION VALUES BY DIFFERENT METHODS OF EXPRESSING THE WATER INFLUX

Method of expressing water influx	Jones Sand, Schuler pool, Arkansas						Monroe reservoir, Reed City pool, Michigan				
	c	$10^{-6}a$, bbl	$10^{-6}W_t$ (total), bbl	$10^{-6}V_{gi}$ bbl	$10^{-6}\overline{L}$, bbl	Rms deviation in $\overline{L}$, %	c	$10^{-6}a$, bbl	$10^{-6}W_t$ (total), bbl	$10^{-6}\overline{L}$, bbl	Rms deviation in $\overline{L}$, %
Eq. (3)	5.6	0.109	11.3	7.85	82.5	1.74	59.3	1.348	6.61	63.2	1.06
Eq. (4)	43.0	0.130	12.6	4.34	88.5	1.60	356.0	1.329	6.61	62.8	1.09
Compressible-liquid theory	246.3	0.138	13.6	3.90	87.7	1.51	352.7	0.436	4.08	72.1	1.16

criminating among the corresponding values of $\overline{L}$. The economic significance of variations in the oil in place of 40 million barrels certainly seems too great to base an evaluation of an oil reservoir merely on the preference for a deviation of 1.6 per cent as compared with 2.3 per cent.

The exact formula, or function $f(p,t)$, used to calculate the variable part of the water-influx term also has little effect on the constancy of the calculated L_j's. In Table 3 are listed the minimum percentage rms deviations obtainable[1] for both the Schuler Jones Sand and the Monroe reservoir of the Reed City field, Osceola and Lake Counties, Michigan, using different functions $f(p,t)$. In addition, a constant a, also considered as an unknown, was added to the right-hand side of Eq. (2) as an attempt at

[1] Woods and Muskat, *loc. cit.*

an empirical correction to the early phase of the transient water influx represented by $f(p,t)$. In the Reed City field it was known that there was no free-gas cap initially, and V_{gi} was taken as 0 in all the calculations.

It will be evident from Table 3 that the minimum deviations are only slightly affected by the type of formula used for calculating the water influx, although here, too, the corresponding differences in the oil in place are quite appreciable. Moreover, while for Schuler the more complete compressible-fluid calculation of $f(p,t)$ gives a slightly lower mean per cent deviation, the least per cent deviation is obtained for the Reed City pool with the simple steady-state approximation for the water influx.

It may be noted that the total water intrusion of 10,600,000 to 14,000,000 bbl at Schuler corresponds to a cumulative oil production of 22,600,000 bbl, as of April, 1942. For the Reed City field the water influx of 4,000,000 to 6,600,000 bbl refers to Jan. 1, 1943, when the oil production totaled 11,460,000 bbl. Such substantial volumes of water entry should certainly be of significance with regard to the general reservoir performance and production mechanism. Yet the constancy of the calculated oil in place is, for practical purposes, virtually the same whether the water intrusion is considered as real or entirely ignored as being fictitious.[1]

It must therefore be concluded that material-balance calculations for partial-water-drive fields are beset with the same limitations as when applied to gas-drive reservoirs. On the other hand it should be observed that while the absolute magnitude of such deviations in the L_j's as may be considered simply as "fluctuations" cannot be used as an exact criterion for determining the reservoir constants, systematic trends in the calculated L_j's should be of significance. Thus upon replacing the incremental changes in the terms of Eq. (9) by differential equivalents, or time rates of change, as

$$\frac{dL_j}{dt} = \frac{d}{dt}(L_{oj} + D_j) + V_{gi}\left|\frac{d\lambda_j}{dt}\right| - c\frac{dA_j}{dt},$$

the types of changes in assumed values of V_{gi} and c that can correct various possible trends[2] in L_j and $L_{oj} + D_j$ can readily be listed. Such considerations show that there will generally be rather wide latitudes in the adjustments of both c and V_{gi} for stabilizing the values of L_j, although the relatively slow decrease of λ_j with decreasing pressure usually will mean that a small change in c may be equivalent to a rather large change, in the

[1] For the Reed City field the rms deviation, on assuming no water drive, is 1.58 per cent, with a mean value of oil in place of 88.6×10^6 bbl.

[2] If the basic reservoir fluid and production data are correct and if reservoir equilibrium conditions obtain, decreasing trends in the $L_{oj} + D_j$'s should not occur. However, such trends are sometimes observed for limited production periods (cf. Table 1).

opposite direction, of V_{gi}. Of course, if the $L_{oj} + D_j$'s are themselves substantially constant, the simplest procedure would be to keep V_{gi} and c zero. However, even then the constancy of L_j generally can still be maintained and often somewhat improved by varying both V_{gi} and c over appreciable ranges. As long as neither c nor V_{gi} is made negative, the material-balance method itself will impose no restrictions on their values except that of giving essentially constant values for the L_j's.

The distinction between a positive "trend" in the L_j's and erratic fluctuations is often subject to considerable uncertainty. The values listed in the last column of Table 1 certainly appear to be free of any definite trend.[1] In contrast the last half of the sets listed for $V_{gi} = 0$, 10^7, $W = 0$ would appear to show a real increasing trend. On the other hand the first half of each of these sets strongly suggests a declining trend in the L_j's. While such questions may be further clarified by more detailed statistical analyses, judgment based on experience in such studies will generally give the best guide for sound interpretation.

All these considerations point to the conclusion that the material-balance method should be used as a supplementary tool in interpreting reservoir performance rather than as the sole basis for reservoir-behavior analysis. And especially in partial-water-drive fields should it be applied only in conjunction with other reservoir data. In fact its most useful and appropriate application will lie in the calculation of the water intrusion when the initial reservoir fluid contents have been determined independently by geological, logging, and core data. The material-balance equation in the form of Eq. (1) will be suitable for such purposes. From such calculations, values of the encroachment coefficient c may be computed from Eq. (2) on choosing a suitable functional form for $f(p,t)$.

As in the case of gas-drive systems, once the value of the constants c, L, V_{gi} have been established, the material-balance equation, in the form of Eq. (6) or its equivalents, can be used to predict the future pressure behavior. Again the future gas-oil-ratio history will have to be assumed.[2] On the other hand in such calculations account may be taken of various degrees of gas return or water injection, and their effects on the maintenance of the reservoir pressure may be evaluated. While somewhat laborious,

---

[1] However, an extension of the calculations to include the subsequent production data leads to a marked decline in the L_j's for the last column of Table 1, to a value of 47.41×10^6 for August, 1946. On the other hand, the rising trends for the sixth and seventh columns are greatly accelerated, and the corresponding L_j's for August, 1946, are 132.4×10^6 and 109.4×10^6.

[2] In Chap. 10 a method of predicting reservoir behavior will be presented in which the gas-oil-ratio history will be determined automatically as a by-product of the pressure vs. recovery relation.

computations of this type involve nothing more than straightforward graphical or numerical trial-and-error procedures.[1]

As should be expected, Eq. (6) provides the condition that must obtain when the reservoir becomes subject to a complete water drive without further pressure decline. If γ, S, β refer to a stabilized pressure condition, the rate of fluid withdrawals must balance with the rate of water intrusion according to the relation

$$\frac{dW_t}{dt} = \frac{dW_p}{dt} + \frac{Q_g}{\gamma} + Q_o\left(\beta - \frac{S}{\gamma}\right), \tag{10}$$

where Q_o is the rate of oil production, Q_g the *net* rate of gas production (gas produced minus that returned), and dW_p/dt the rate of water production. The rate of water intrusion, dW_t/dt, should, of course, include any water injected into the producing formation, unless it has already been subtracted from dW_p/dt. Equation (10) gives the rate of water entry required for pressure stabilization at any fixed pressure and rate of fluid withdrawal or the rate of water injection required if the natural water drive is of insufficient intensity. Conversely, if the water-entry terms are known, the rate of fluid withdrawals that will permit pressure stabilization may be calculated by Eq. (10). These considerations provide the basis for interpreting the last two terms on the right-hand side of Eq. (10) as the rate of "space voidage" (cf. footnote, page 385).

Except for the requirement imposed by Eq. (10) the material-balance method breaks down in complete-water-drive systems if applied to an interpretation of the details of reservoir behavior. On the other hand, in the extreme case where the pressures are everywhere above the bubble point and the space voidage is necessarily provided entirely by the water entry and expansion of the oil, Eq. (6) still will formally retain its validity. For this case it will reduce to

$$L = \frac{\beta Q + W_p - W_t}{\beta - \beta_i}, \tag{11}$$

where now $\beta > \beta_i$. While, in principle, Eq. (11) will also permit a calculation of the oil in place, L, from observations of the pressure history, the accuracy of such determinations will depend almost entirely on that of the water-entry term W_t and of the formation-volume factor β. Serious uncertainties in these will vitiate any accuracy in the computed values of L. Again, however, if L has been established independently, Eq. (11) may be used to estimate the magnitude of the water intrusion.

[1] Examples of such calculations may be found in Carpenter, Schroeder, and Cook, *op. cit.* The results of illustrative calculations on the effects of gas return will also be given in Chap. 10, as applications of the more complete theory of gas-drive reservoirs to be presented there.

In view of the numerous limitations to the applicability of the material-balance equation, developed in the above discussions, it should be noted that the detailed and extensive treatment of this problem presented here and the method itself may still serve a number of useful purposes. It provides the medium for an instructive examination of the physical and thermodynamic processes involved in the liberation of natural gases from reservoir oils. It serves to emphasize the importance of the accurate determination and recording of reservoir-performance data, especially the gas-oil ratios and reservoir pressures. It makes clear that the gross pressure histories of oil reservoirs over limited time or cumulative production periods are not very sensitive to the details of the production mechanism or reservoir constants. It shows the need for serious effort in determining the gross volume and fluid contents of reservoirs by accurate core, logging, and geological data.

In spite of the fact that the material-balance equation should be applied only with care and full understanding of its limitations, it still represents a valuable tool for reservoir-engineering analysis. Its application involves only simple algebraic and arithmetic manipulation and requires neither specialized analytical training nor extensive computational facilities. If the volumetric reservoir constants are known, it affords the most direct means now available for determining the magnitude of the water intrusion from production-performance data. Under such circumstances it can be a valuable aid in establishing the production mechanism. Since it does not encompass dynamical reservoir characteristics, it does not suffice alone to predict future reservoir behavior. But if the reservoir parameters have been established on the basis of past performance or by independent means, then by assuming only the future gas-oil-ratio history the corresponding normal depletion behavior or those under gas- and/or water-injection pressure-maintenance operations can be readily calculated. Finally it may be observed that even when the electrical analyzer (to be discussed in Sec. 11.8) is available for making reservoir-performance predictions, the water-intrusion history, as calculated by inverting the material-balance equation [cf. Eq. (1)], provides the basis for determining the constants of the water reservoir.

9.8. Summary.—The producing mechanism controlling the performance of an oil reservoir depends on the nature of the energy available for oil expulsion and the manner in which it is used. The major types of reservoir energy associated with oil-producing reservoirs are (1) the energy of compression of the liquid phase within the oil-bearing rock, (2) gravitational energy, (3) energy of compression and solution of the gas content of the reservoir, and (4) the energy of compression of contiguous waters in mobile communication with the oil.

The energy of compression of the oil and water within the oil pay is of minor importance except during the initial phases of production from reservoirs containing highly undersaturated crudes and before water intrusion develops sufficiently to be an effective pressure-maintaining agent. Gravitational forces are of importance in the segregation of the fluid phases of different density, which thereby affect the efficiency of oil displacement by gas or water. As means for directly maintaining production rates, gravity effects do not play a major role until the reservoir pressures are substantially depleted and the oil seeps down toward the bottom of the pay and well bore by gravity drainage.

The gas content of oil reservoirs generally provides a major potential source of oil-expulsion energy. When not materially supplemented by the energy of invading waters, the use of the energy of the gas results in a "gas-drive" mechanism. If the gas is merely that originally in solution in the oil, the reservoir will produce by a "dissolved-gas" or "solution-gas" drive. If the oil pay is overlain by a free-gas zone and is produced so that there is a simultaneous expansion of the gas cap and gravity drainage of the oil down the flanks of the reservoir, the mechanism is termed a "gas-cap-expansion," or "gravity-drainage," drive.

If a large part or all of the volumetric withdrawals of a reservoir are replaced by the entry of water into the oil pay, it is being controlled by a "water drive." Depending on the reservoir structure the water intrusion may be predominantly lateral, as an "edgewater" encroachment, or the water may underlie a major part of the oil zone and invade the latter as a "bottom-water" drive.

In order to ensure efficient exploitation and to anticipate the future behavior of an oil reservoir one must first determine the type of drive controlling the production. This requires an analysis and interpretation of its early production performance. The latter is the composite record of observations on the reservoir pressure, gas-oil ratio, water production, motion of the water-oil and gas-oil contacts, and the variation of the individual well data throughout the field.

In water-drive fields the reservoir pressure may decline rapidly at first but will tend to stabilize as a pressure differential is established between the field and water reservoir so as to induce a rate of water intrusion sufficient to replace the fluid withdrawal (cf. Fig. 9.1). The quantitative aspects of this behavior will depend basically on the ratio of the fluid-withdrawal rate to the inflow capacity of the water reservoir. The pressure at which substantial stabilization may develop or the continued rate of decline will depend directly on the rate of fluid withdrawal. If the field should be shut in, the pressures will rise. The gas-oil ratio will not vary greatly during complete-water-drive production. The productive capacities of the wells will remain substantially constant if the water drive prevents growth of the free-gas phase in the oil zone. The pressures will be higher near the region of water entry and will fall off in the interior of the field if the reservoir is uniformly producing under water-drive action (cf. Figs. 9.2 and 9.3).

The reservoir pressure in gas-drive fields will depend mainly on the cumulative recovery. It will not be sensitive to the production rate, except as the latter may affect the gas-oil ratio. Even closing in the field will not result in a pressure rise. The pressure distribution within the field will reflect largely variations in local cumulative withdrawals relative to the local oil content of the pay (cf. Fig. 9.4). Except for an initial period when it may decline somewhat below its solution value, the gas-oil ratio will rise with increasing cumulative production to maximum values 5 to 10 times as great as the solution ratio. When the state of ultimate depletion is approached, the gas-oil ratio will begin to decline. The productive capacities of the wells will continually decline because of increasing oil viscosity and decreasing permeability to the oil phase.

The classification of producing mechanisms is basically arbitrary, except for the strict solution-gas drive or complete water drives. If the fundamental criterion were the magnitude of the actual ultimate recovery, many partial-water-drive reservoirs would be classed with the complete-water-drive systems, while others would more appropriately be grouped with gas-drive reservoirs. Another division on the basis of individual field recoveries would apply to gas-cap-expansion and gravity-drainage reservoirs. While the ultimate recovery is indeed of paramount importance from an economic standpoint, it represents a less fundamental basis. The general performance histories of complete-water-drive reservoirs are essentially independent of the magnitude of the residual oil, which determines the ultimate recovery. The same applies to partial-water-drive systems and also to those producing by gravity drainage. The quantitative values of the factors determining their ultimate recoveries affect only the details but not the gross performance features, which are largely controlled by the gas-drive mechanism. And even solution-gas-drive reservoirs would have substantially the same types of histories even if the permeability-ratio curves were such as to give recoveries comparable with those of water drives. For the purpose of reservoir-*performance* analysis, both gravity-drainage and partial-water-drive reservoirs will therefore be treated here as generalizations of solution-gas-drive systems, and only the complete water drive will be considered as a fundamentally different producing mechanism.

As all oil-producing reservoirs must obey the law of conservation of matter, an interrelation must exist between the total fluid withdrawals, the reservoir pressure, the initial oil and gas content of the reservoir, and such extraneous fluids as may enter the original producing section as the result of natural fluid invasion or artificial injection from the surface. This relationship [the material-balance equation; cf. Eq. 9.5(7)] involves as unknowns the original oil in place, the initial free-gas volume in the reservoir, and the volume of water intrusion, if any.

If it is known that there is no water entry, it would be possible in principle to determine the original oil in place and free-gas volumes, which are reservoir constants, by solving the equations obtained after introducing the pressure and production data at two or more time intervals. Or one of the constants could be so chosen that the calculated values of the other, for various time intervals, are substantially constant. However, such procedures often prove unsatisfactory, since rather large variations in the assumed values of the constants may have only small effects on the internal consistency of the equations (cf. Fig. 9.6).

If the possibility of water intrusion must be taken into account, the situation becomes even more ambiguous. For the water-intrusion term is inherently variable, increasing with time. The general material-balance equation then involves an unknown function as well as two unknown constants. To resolve this new difficulty the water-entry term can be expressed as a constant times a function of pressure and time representing the state of flow in the water reservoir [cf. Eq. 9.7(2)]. Several types of representation have been proposed and tried for describing the water influx. These include functions giving the outflow from a compressible-liquid system, that from a steady-state system, intermediate approximations, and even more drastic simplifications. In all cases these functions can be computed from the pressure vs. time history of the oil-producing system, the coefficient or scale factor being left as an unknown constant. The latter is then treated similarly to the two oil-reservoir constants, the initial oil in place and free-gas volume. That is, these three constants are then so determined as to achieve a maximum degree of internal consistency when the material-balance equation is applied to a series of time intervals during the course of the observed production history.

As is to be expected, with the additional degree of freedom provided by the water-intrusion coefficient, the uniqueness of the true solution for the reservoir constants becomes even less well defined than for gas-drive systems, even though the fluctuations among the calculated constants can be somewhat reduced. And if nothing be assumed as known about these constants, almost equally satisfactory approximate solutions of the equations will often be found that may differ greatly in their implications regarding the size of the reservoir and the nature of the basic mechanism controlling its performance (cf. Table 2).

While fundamentally the material-balance method of reservoir analysis is thus severely limited in its utility, it can still be of definite value if properly understood and applied. The observation of "trends" in the calculated constants, such as the initial oil in place, will suggest changes to be made in the values assumed for the other constants. Quite frequently the ratio of the initial free-gas volume to that of the oil volume can be determined from core and logging data. On introducing this information into the basic material-balance equation the number of unknowns can be reduced by one. And if it is known a priori that there was no initial gas cap or that there is no water intrusion, these facts should be applied directly to the equation and thus limit the uncertainty in the other unknowns. For if the reservoir has been established to be a pure gas-drive system, with no water entry and no initial gas cap (or one of known volume), the calculation of the oil in place should be quite trustworthy.

When water intrusion is known to be a factor, the most satisfactory application of the material-balance equation can be made when the oil in place and free-gas volume can be determined independently from core or geological data. The water entry can then be calculated by inverting the material-balance equation to express the water-intrusion term as a function of the oil-reservoir constants and the pressure and production history [cf. Eq. 9.7(1)]. In fact such computations will generally provide the most satisfactory method for determining the magnitude and history of the water entry.

If, by whatever means, the basic reservoir unknowns have been definitely determined, the material-balance equation can be used to predict the future behavior under assumed producing conditions. If the future gas-oil ratios be assumed, the nature of the pressure decline of gas-drive reservoirs can be calculated under direct depletion or gas-return operations. The effect of gas-oil-ratio control can be evaluated by comparing the calculated pressure vs. recovery relations as obtained for different assumptions regarding the future gas-oil-ratio history. The effect of water injection on the reservoir pressure behavior can also be computed both for gas and partial-water-drive systems. It is in applications of this type that the material-balance method can be of greatest value, although, as will be seen in the next chapter, still more powerful methods can be used when the permeability-saturation relationship of the producing formation is known.

CHAPTER 10

GAS-DRIVE RESERVOIRS

10.1. Introduction.—Most oil fields that are now depleted were produced under the gas-drive mechanism. The reason is that they were produced "wide open," with virtually no restrictions on the rates of fluid withdrawal. The entry of such boundary waters as may have been in contact with the oil pay was insufficient to keep pace with the production of oil and gas during the primary production phase. The production of wells and fields at their maximum capacity was encouraged by the close well spacings (less than 10 acres per well) common prior to 1930 and the necessity for the protection of each operator against drainage by his neighbors.

In the early thirties, "proration" controls for the limitation of production rates from fields and wells were instituted in most of the oil-producing states. Moreover the increased understanding of the physical mechanism of oil expulsion led to acceptance of the principle that in many reservoir rocks water will be a more efficient oil-expulsion agent than gas. Thus, as the combined result of state and voluntarily imposed controls of fluid-withdrawal rates, most oil fields in the States have been so produced in recent years that mobile bounding waters, if present, could contribute substantially to the replacement of the space voidage created by the oil and gas production.

Except where the oil reservoir is effectively sealed from communication with water-bearing formations or where the latter have inherently low water-supply capacities, rather few fields are currently being produced entirely by the solution-gas-drive mechanism. And even in such cases it is becoming common practice to begin the injection of gas or water during the primary production phase to retard the pressure decline or arrest it completely and increase the ultimate recovery beyond that which could be obtained merely by the expenditure of the gas energy originally available in the reservoir. On the other hand, except when the oil is highly undersaturated or the water reservoir has an exceptionally great intrusion capacity, the early life of practically all reservoirs will be controlled by the gas-drive mechanism. For even if the production rates are restricted, a pressure differential must be developed between the oil and water reservoirs to induce such rates of water entry as will arrest further pressure decline. Moreover pressure-maintenance operations are generally not undertaken until enough of a natural depletion history has developed to indicate the desirability or necessity of extraneous fluid injection.[1]

In any case a study of the performance of pure gas-drive fields provides the natural foundation for the consideration of partial-water-drive systems and the comparison between the basic gas-drive and water-drive oil-expulsion mechanisms. It will also serve to explain the broad features of the histories of the many fields, now depleted, that were exploited without controls and the action of water drives and to provide an understanding of the relatively low recoveries associated with these reservoirs.

Although an extended discussion of gas-drive reservoir performance will be given in the following sections, it should be emphasized that the theoretical material to be presented here refers to idealized systems. While the fundamental physical principles involved are quite well established, the analytical treatment of multiphase reservoirs is, at present, entirely impractical without the introduction of drastic simplifications. These include the assumption of complete reservoir uniformity, the neglect of the localization of actual fluid withdrawals through the medium of well bores, and the neglect of the effect of gravity in inducing fluid segregation and downstructure gravity drainage of the oil. Moreover, as a whole the analysis will serve only to relate average reservoir pressures to oil withdrawals and will not give the *time* histories of the reservoir or the effects of rates of withdrawal on the performance. These must be introduced as supplementary considerations or by additional approximation procedures, and means for doing so will be formulated for the case of partial-water-drive reservoirs. Accordingly no attempt will be made to analyze, interpret, or predict quantitatively the performance of actual gas-drive reservoirs. Nor are the discussions of gas-injection systems to be construed as accurately descriptive of the results to be expected from specific pressure-maintenance operations. On the other hand it is felt that the theoretically derived reservoir histories do reflect with substantial accuracy the behavior of such idealized systems as would satisfy the assumptions made in the analysis. These, together with the illustrative examples presented of observed oil-reservoir performance, should serve as guides in interpreting the production histories of actual gas-drive reservoirs. If the limitations of the theory and the nature of the modifying factors obtaining in practical systems are given proper consideration, the material presented in this chapter should provide a reasonable basis for planning the development and operation of gas-drive oil-producing reservoirs.

---

[1] Of course, in fields that are largely condensate producers, plans for gas return and pressure maintenance are usually made as soon as possible after development. However, the mechanism of production of condensate fields is inherently different from typical crude-oil-producing gas-drive fields and will be treated separately (cf. Chap. 13).

10.2. The Basic Equations for Solution-gas-drive Reservoir Histories.[1]
—In Sec. 7.7 were given the basic hydrodynamic equations governing the
flow of multiphase fluids through porous media. In principle these equations should and do govern the flow of the oil and gas in an oil-producing
reservoir when undergoing depletion by fluid withdrawal through wells
drilled into the producing formation. It would be a simple matter to
state the initial and boundary conditions that, together with the geometrical parameters and functions (empirical) characterizing the physical
properties of the petroleum fluids and rock, would analytically define any
particular oil- and gas-producing system. For example, in the ideally
simple case of a single-well system, closed physically by a termination
of the producing medium or by virtue of interaction with other wells,
beginning at uniform pressure p_i and oil saturation ρ_{oi}, and suddenly subjected to a preassigned bottom-hole pressure p_w, the initial and boundary
conditions would evidently be

$$p = \text{const} = p_i; \qquad \rho_o = \text{const} = \rho_{oi} \quad : \quad t = 0, \atop \frac{\partial p}{\partial r} = \frac{\partial \rho_o}{\partial r} = 0 \quad : \quad r = r_e; \qquad p = p_w \quad : \quad r = r_w,} \right\} \tag{1}$$

where r_e, r_w are the radius of closure and well radius, respectively, and it is
tacitly assumed that the water saturation is uniform and immobile.

Unfortunately, even for the simple initial and boundary conditions of
Eq. (1) the solution of Eqs. 7.7(1) constitutes an almost hopeless task.
At best, individual cases with numerically assigned parameters would have
to be treated by laborious numerical procedures.[2] And the effects of
changes in the many significant parameters could be determined only by
repeated calculations for each combination of individual assumptions.
Accordingly, from a practical point of view some type of major simplification or drastic assumption must be introduced if any quantitative treatment
is to be developed.

The gross approximation involved in the theoretical developments of the
following sections lies in the neglect of the isolated and concentrated foci
of fluid withdrawals—the wells—which are necessarily an integral part

---

[1] Cf. M. Muskat, *Jour. Applied Physics* **16**, 147 (1945), where the method of derivation used here was given, as well as generalizations to systems in which gas is being
returned. Previously an integrated representation of essentially the same physical
considerations and involving trial-and-error numerical or graphical procedures had been
described by J. Tarner, *Oil Weekly*, **114**, 32 (June 12, 1944) and referred to by E. C.
Babson, *AIME Trans.*, **155**, 121 (1944). An equivalent of these will be briefly discussed in Sec. 10.9.

[2] Such a numerical study has thus far been carried out only for a linear system with
initial and boundary conditions equivalent to Eq. (1) [cf. M. Muskat and M. W. Meres,
Physics, **7**, 346 (1936) and footnote p. 304].

of all actual oil-producing fields. The reservoir will be considered as equivalent to a uniform tank, subjected to a uniform distribution of fluid withdrawal, and entirely free of pressure gradients, just as in the case of the material-balance method (cf. Sec. 9.5). In fact the analysis will be based on the equation of continuity, which is essentially the material-balance equation in differential form but without an explicit introduction of the pressure-gradient terms, which induce the fluid flow. Thus, by recalling Eqs. 7.7(1), the left-hand sides of the equations will be replaced by their physical equivalents expressed as the actual rates of oil, gas, and water withdrawal, Q_o, Q_g, Q_w, per unit volume of the producing section, *i.e.*,

$$\left.\begin{array}{l} Q_g = -f\dfrac{\partial}{\partial t}\left(\dfrac{S_o\rho_o}{\beta_o} + \dfrac{S_w\rho_w}{\beta_w} + \gamma\rho_g\right), \\[2mm] Q_o = -f\dfrac{\partial}{\partial t}\left(\dfrac{\rho_o}{\beta_o}\right), \\[2mm] Q_w = -f\dfrac{\partial}{\partial t}\left(\dfrac{\rho_w}{\beta_w}\right), \end{array}\right\} \qquad (2)$$

where f is the porosity, ρ_o, ρ_w, ρ_g the oil, water, and gas saturations, β_o, β_w the formation-volume factors of the oil and water phases, S_o, S_w the gas solubility in the oil and water, and γ the relative density, compared with standard conditions, of the free-gas phase.

The fluid ratios will therefore be

$$\left.\begin{array}{l} R = \dfrac{Q_g}{Q_o} = \dfrac{\delta[(S_o\rho_o/\beta_o) + (S_w\rho_w/\beta_w) + \gamma\rho_g]}{\delta(\rho_o/\beta_o)}, \\[3mm] R_w = \dfrac{Q_w}{Q_o} = \dfrac{\delta(\rho_w/\beta_w)}{\delta(\rho_o/\beta_o)}, \end{array}\right\} \qquad (3)$$

where δ refers to increments in a unit time interval. Now in the integrated form of Eqs. (3), which is readily shown to be nothing more than the material-balance equation, such as Eq. 9.6(1) if $S_w = 0$, the gas-oil ratio R is retained in the equation as a measure of the gas production. Thus, in applying the material-balance equation to an interpretation of the past performance of a field, R simply represents the observed cumulative gas-oil ratio. But, in predicting the future behavior, R must be assumed as a function of the cumulative oil withdrawal. This, of course, is the basic weakness of the material-balance method, in its simple integrated form discussed in Chap. 9, when used as a tool to predict reservoir histories.[1]

The fundamental advance of the theory to be presented here is that the

---

[1] In Sec. 10.9, however, will be outlined a procedure for supplementing the material-balance equation with one taking into account the permeability-saturation relationship, so that the method becomes essentially equivalent to that developed in this section.

fluid ratios R and R_w will be forced to follow the implications of the permeability-saturation relationships for the particular rock in question. This is, of course, automatically provided by Eqs. 7.7(1), in which the detailed pressure distributions within the porous medium also are taken into account explicitly. On the other hand the regional pressure variations or those created locally by the wells—the foci of withdrawals—will be neglected here. The fluid ratios for the reservoir as a whole will be taken as those corresponding to the "typical," or "average," reservoir element, with which are associated the fluid saturations ρ_o, ρ_w, and ρ_g entering in Eqs. (3). Specifically they will be expressed by the relations developed in Sec. 8.1, namely,

$$\left.\begin{array}{l} R = S_o + \alpha(p)\Psi(\rho) + \xi(p)\Phi(\rho), \\[2mm] R_w = \dfrac{\xi(p)}{S_w}\,\Phi(\rho), \end{array}\right\} \qquad (4)$$

where

$$\alpha(p) = \frac{\gamma\beta_o\mu_o}{\mu_g}; \qquad \xi(p) = \frac{S_w\mu_o\beta_o}{\mu_w\beta_w}; \qquad \Psi = \frac{k_g}{k_o}; \qquad \Phi = \frac{k_w}{k_o}, \qquad (5)$$

and the argument ρ in $\Psi(\rho)$, $\Phi(\rho)$ indicates the composite fluid-saturation distribution.

Equations (4) were derived in Sec. 8.1 for linear steady-state flow. The derivation given there will lead to the same equations if applied to radial or spherical flow or to any type of curvilinear steady-state flow conditions. They will remain valid *locally* even in transient systems as long as differential driving agents, with respect to the different phases, as gravity or capillary forces, may be neglected. It is the assumption that Eqs. (4) apply throughout the reservoir as a whole, which constitutes the primary simplification of the present treatment. On the other hand this assumption involves simply the previously stated approximation that variations in p and ρ throughout the reservoir will be ignored and that the reservoir can be represented by an "average" differential element which can be considered as isolated and free from interactions with its surroundings. Subject to this approximation,[1] Eqs. (4) can be combined with Eqs. (3) to yield the following differential equations:

$$\left.\begin{array}{l} \left(1 + \dfrac{\mu_o}{\mu_g}\Psi\right)\dfrac{d\rho_o}{dp} + \dfrac{d\rho_w}{dp} = \rho_o\lambda(p) + (1 - \rho_o - \rho_w)\epsilon(p) + \rho_o\eta(p)\,\Psi + \rho_w\nu(p), \\[3mm] \dfrac{\mu_o}{\mu_w}\Phi\,\dfrac{d\rho_o}{dp} - \dfrac{d\rho_w}{dp} = \dfrac{\rho_o\mu_o}{\mu_w\beta_o}\Phi\dfrac{d\beta_o}{dp} - \dfrac{\rho_w}{\beta_w}\dfrac{d\beta_w}{dp}, \end{array}\right\} \qquad (6)$$

where:

$$\epsilon(p) = \frac{1}{\gamma}\frac{d\gamma}{dp}; \qquad \lambda(p) = \frac{1}{\gamma\beta_o}\frac{dS_o}{dp}; \qquad \eta(p) = \frac{\alpha}{\gamma\beta_o^2}\frac{d\beta_o}{dp}; \qquad \nu(p) = \frac{1}{\gamma\beta_w}\frac{dS_w}{dp}. \qquad (7)$$

[1] In fact, it can be shown that Eqs. (6) also follow directly from Eqs. 7.7(1), on dropping the gravity terms and those involving the gradients in the pressure.

These are the final basic equations for simple gas-drive reservoirs producing by depletion of the solution gas. Before discussing them in detail, it will be useful to simplify them somewhat to correspond to the situations most commonly obtaining in practice, namely, that in which the connate water, represented by ρ_w, is immobile.[1] $\Phi(\rho)$ will then be 0, and the second of Eqs. (6) will reduce to

$$\frac{d\rho_w}{dp} = \frac{\rho_w}{\beta_w}\frac{d\beta_w}{dp}, \tag{8}$$

with the solution

$$\rho_w = \frac{\rho_{wi}\beta_w}{\beta_{wi}}, \tag{9}$$

where ρ_{wi}, β_{wi} refer to the initial values of ρ_w, β_w. Equation (6) can then be rewritten as

$$\left(1 + \frac{\mu_o}{\mu_g}\Psi\right)\frac{d\rho_o}{dp} = \rho_o\lambda(p) + \left(1 - \rho_o - \frac{\rho_{wi}\beta_w}{\beta_{wi}}\right)\epsilon(p) + \rho_o\eta(p)\Psi + \frac{\rho_{wi}\beta_w\nu(p)}{\beta_{wi}}. \tag{10}$$

Equations (6) and (7) or (9) and (10) determine the variation of the oil and water saturations ρ_o and ρ_w as functions of the pressure. All the terms except ρ_o and ρ_w themselves and Ψ and Φ are to be considered as known functions of pressure, established empirically by measurements on the gas, oil, and water contained in and produced by the reservoir. Ψ and Φ are functions of the fluid saturations, to be derived from the permeability-saturation relationship of the particular producing formation constituting the reservoir.[2] Examples of these are included in the curves of Figs. 7.1 to 7.13. As both Eqs. (6) and Eq. (10) are of the first order, they can be integrated numerically quite readily by such standard procedures as the Runge-Kutta or Milne methods.[3]

The starting point for the integration of Eqs. (6) or (10) consists of the initial pressure and fluid distribution within the reservoir. If these refer to the virgin conditions, ρ_w will be the reservoir volume saturation of the connate water and the initial value of ρ_o will be $1 - \rho_w$. The end point for the integration is either atmospheric pressure or any higher pressure

---

[1] Although the strict immobility of the connate water, even above the water-oil transition zone, is a moot question, such mobility as it may have will generally still be of negligible importance from a practical standpoint.

[2] Of course, as in the case of the permeability itself—and porosity—the permeability-saturation relationship will usually be considered in gross practical applications as an average of those characterizing the individual strata comprising the whole producing section referred to as "the reservoir," although the actual permeability variations must be taken into account in attempting to analyze the quantitative details of reservoir performance.

[3] Cf., for example, J. B. Scarborough, "Numerical Mathematical Analysis," Johns Hopkins Press, 1930.

at which one may assume the reservoir will be economically depleted as a simple gas-drive system.

The integration of Eqs. (6) or (10) will give directly only the variations of the reservoir fluid saturations with the reservoir pressure. The fluid saturations may be translated into equivalent volumes of total stock-tank fluid withdrawals by the relations

$$\left.\begin{array}{l} \overline{Q_o} = 7{,}758.4f\left(\dfrac{\rho_{oi}}{\beta_{oi}} - \dfrac{\rho_o}{\beta}\right), \\[2mm] \overline{Q_w} = 7{,}758.4f\left(\dfrac{\rho_{wi}}{\beta_{wi}} - \dfrac{\rho_w}{\beta_w}\right), \end{array}\right\} \qquad (11)^*$$

where $\overline{Q_o}$, $\overline{Q_w}$ refer to the cumulative oil and water productions, in barrels per acre-foot of producing rock, of porosity f. In the case of Eq. (10), where the connate water is assumed to be immobile, $\overline{Q_w}$ will vanish, as it should, by virtue of Eq. (9). From the reservoir saturations the gas-oil ratio can be calculated by application of Eq. (4), and on multiplication by the incremental values of $\overline{Q_o}$ and summing the total gas withdrawal is obtained.[1]

The integration of Eqs. (6) or (10) automatically results in a prediction of the physical and economic ultimate recoveries to be obtained by the gas-drive depletion of the reservoir. The former will be given by the value of $\overline{Q_o}$ at atmospheric pressure. The latter will be the value of $\overline{Q_o}$ at the assumed abandonment pressure.[2]

Thus the equations developed above make it possible to predict the complete pressure and gas-oil-ratio histories, vs. the cumulative production, of solution-gas-drive reservoirs. But, as is to be expected from the approximations underlying the equations, they do not give directly information regarding the individual wells draining the reservoir. Within the accuracy of these approximations the production histories of pure gas-drive reservoirs as well as the ultimate recoveries are thus inherently determined independently of the well system, including their distribution and density. Moreover, as Eqs. (6) and (10) do not involve the time factor or the rates

---

* The parentheses of Eqs. (11) will give the cumulative withdrawals as fractions of the pore space.

[1] The total gas withdrawals can also be expressed as the difference between the initial and current gas contents of the reservoirs, *i.e.*, as

$$\overline{Q_g} = 43{,}560f\left[\frac{S_i\rho_{oi}}{\beta_i} - \frac{S\rho_o}{\beta} - \gamma(1 - \rho_o - \rho_w)\right]\text{ft}^3/\text{acre-ft},$$

if S and γ are expressed as cubic feet per cubic foot.

[2] While in practice well abandonment in gas-drive fields usually is determined by limiting rates of production, the pressure is used here to avoid the inclusion of the uncertain factors relating production rates to pressure.

of fluid withdrawal, the pressure-decline and gas-oil-ratio histories, as predicted by them as functions of the cumulative withdrawals, will also be independent of the production rates. Although these conclusions are inevitable consequences of the assumptions and method of derivation underlying Eqs. (6) and (10), there is no direct evidence that they are fundamentally in error. On the other hand it should be noted that such factors as gravity segregation between the oil and gas phases, the possible coning of overlying gas caps or underlying bottom waters, and capillary phenomena have been deliberately omitted from the considerations thus far. Under suitable circumstances these may materially modify the behavior of the completely idealized gas-drive system and must be taken into account in the investigation of the performance of actual oil-producing reservoirs and in planning their exploitation.

10.3. Theoretical Production Histories of Solution-gas-drive Reservoirs.—Although there is no basic difficulty in solving numerically the simultaneous Eqs. 10.2(6), the increased generality provided by the inclusion of the water-flow terms is of little practical interest in considering solution-gas-drive fields.[1] Moreover, because of the relatively low solubility of natural gases in water and correspondingly small variation of the formation-volume factor of the water phase with pressure,[2] it will suffice for illustrative purposes to neglect entirely the gas in solution and changes in the water phase. Accordingly the equations used as the basis for the following discussion will be

$$\left(1 + \frac{\mu_o}{\mu_g}\Psi\right)\frac{d\rho_o}{dp} = \rho_o\lambda(p) + (1 - \rho_o - \rho_w)\epsilon(p) + \rho_o\eta(p)\Psi, \\ R = S_o + \alpha\Psi, \quad \alpha = \frac{\gamma\beta_o\mu_o}{\mu_g}, \Bigg\}$$

(1)

where ρ_w will be taken as constant.

The particular gas-drive reservoir that will be considered here, to illustrate the implications of Eq. (1), will be comprised of a porous medium having a permeability-saturation relationship as shown by the solid curves of Fig. 10.1 and producing a gas and oil with physical properties as plotted

[1] The early or initial appearance of water in gas-drive fields is probably generally due either to water-saturated streaks lying between sections of the main oil-saturated formation or to bottom waters brought into the well bore by coning. While the exposure of sections of oil-water transition zones will also lead to initial-water production and such as would require treatment by the more general Eqs. 10.2(6), situations of this kind are usually avoided by modern completion practices, except for exceptional cases where the transition zone may comprise a major part of the whole oil-productive formation.

[2] The total volume expansion of gas-saturated formation waters under reservoir conditions, as compared with standard conditions, is only of the order of 3 per cent [cf. C. R. Dodson and M. B. Standing, *California Oil World*, **37**, 21 (Dec. 15, 1944)].

in Fig. 8.1, except that the oil viscosity will be assumed to be uniformly half of that indicated in Fig. 8.1. The initial reservoir pressure is assumed to be 2,500 psia. The connate-water saturation is taken as 30 per cent. As implied by the intercept of the solid k_g/k_o curve in Fig. 10.1 it is assumed

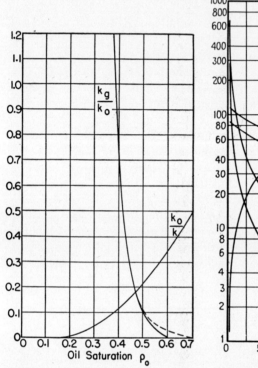

FIG. 10.1. The permeability-saturation relationships assumed in the calculation of solution-gas-drive histories. k_g, k_o are permeabilities to gas and oil. k = homogeneous-fluid permeability. Connate-water saturation assumed = 30%.

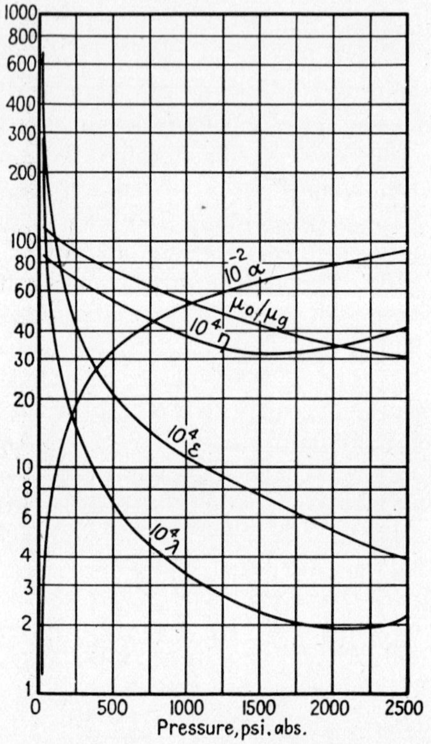

FIG. 10.2. The variation with pressure of the petroleum fluid functions α, λ, ϵ, η, and μ_o/μ_g, used in the calculation of solution-gas-drive histories.

that a free-gas equilibrium saturation of 10 per cent must be created before gas flow will develop.

It will be noted from Eq. (1) that before integration is attempted the functions λ, ϵ, and η must be determined. According to their definition by Eq. 10.2(7), they involve the derivatives of the basic functions γ, S_o, and β_o. The values for λ, ϵ, and η, obtained by carrying through the required procedures, are plotted in Fig. 10.2, together with α and μ_o/μ_g.

The results of the integration of Eq. (1), using these data, are shown in Fig. 10.3, plotted against the reservoir pressure p. The gas-oil ratios were computed from the second of Eqs. (1). The pressure and gas-oil-ratio data of Fig. 10.3 are replotted in Fig. 10.4 vs. the cumulative recoveries in units of the pore space, and as equivalent stock-tank oil recoveries for a 40-ft 25 per cent porosity stratum, the latter being obtained by application of Eq. 10.2(11).

While the reservoir pressure declines monotonically with the cumulative recovery, as might have been anticipated from general considerations, the

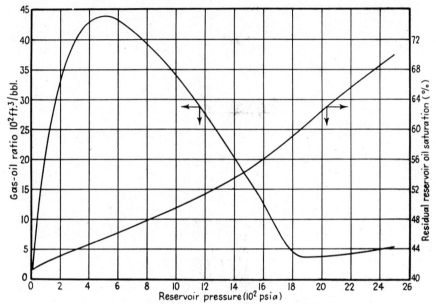

Fig. 10.3. The calculated variation of the gas-oil ratio and reservoir oil saturation with the pressure for a hypothetical solution-gas-drive reservoir. Connate-water saturation assumed = 30%. (*From Jour. Applied Physics, 1945.*)

gas-oil-ratio history, plotted in Figs. 10.3 and 10.4, shows several characteristic features. These are (1) an initial decline below the original solution value, (2) a subsequent sharp rise, and (3) an even more rapid decline after the maximum is reached.

The initial decline in gas-oil ratio, appearing in Figs. 10.3 and 10.4, is due entirely to the assumption of a nonvanishing equilibrium free-gas saturation. During the building up of the gas saturation to the equilibrium value for flow the producing gas-oil ratio must be the solution ratio. The latter, however, will decrease as the pressure falls, thus leading to the decline indicated in Figs. 10.3 and 10.4. The amount of the decline will clearly depend on the total pressure drop required before the equilibrium free-gas saturation is developed. Hence, as a fraction of the original solution ratio, the minimum gas-oil ratio will be smaller for lower initial gas solubilities. Evidently the total initial decline in gas-oil ratio will also increase as the equilibrium free-gas saturation increases. And as will be seen in Sec. 10.4, the gas-oil ratio will generally[1] immediately begin to rise from its solution value if there is no equilibrium free-gas saturation.

The rise in gas-oil ratio after the equilibrium free-gas saturation is reached is the result of the rapid rise in the gas permeability, or $\Psi(\rho)$, as

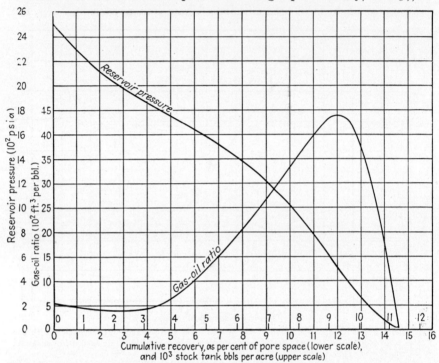

FIG. 10.4. The calculated pressure and gas-oil-ratio histories, for a hypothetical solution-gas-drive reservoir, as functions of the cumulative recovery, expressed both as a percentage of the pore space and as barrels per acre for a 40-ft 25 per cent porosity pay. Connate-water saturation assumed = 30%. (*From Jour. Applied Physics, 1945.*)

the oil saturation falls below the equilibrium value (cf. Fig. 10.1). Ultimately, however, the decrease in S and α as the pressure continues to decline counterbalances the rise in Ψ, and R reaches a maximum. As the reservoir pressure continues to fall, the atmospheric equivalent of the free-gas flow through the formation decreases, even though its reservoir volume rate of flow, compared with that of the oil, rises monotonically. As a result, after reaching the maximum the gas-oil ratio declines sharply until the abandonment or atmospheric pressure is reached.

The slope and curvature of the reservoir pressure curve reflect the current

[1] Cf., however, the footnote on p. 425.

gas-oil ratio. Thus during the initial decline of the gas-oil ratio the slope of the pressure curve (pressure drop per unit stock-tank recovery) decreases. When the gas-oil ratio begins to rise, the pressure curve changes in curvature and develops an increasingly steep decline. This is arrested, and another inflection occurs as the gas-oil ratio passes its maximum and begins its rapid decline. This behavior is, of course, to be expected on the basis of physical considerations.

It is also of interest to note that Eq. (1) can be rewritten as[1]

$$\frac{dp}{d(\rho_o/\beta)} = \frac{\beta + (R - S)/\gamma}{\rho_o\lambda + (1 - \rho_o - \rho_w)\epsilon - [(\rho_o/\beta)(d\beta/dp)]}. \qquad (2)$$

This shows that the pressure drop per unit stock-tank recovery is proportional to the current space voidage, except for the denominator, which gives the effect of changes in the residual-fluid properties associated with the variations in pressure.

The abscissa intercept (at atmospheric pressure) of the pressure curve in Fig. 10.4 gives the physical ultimate recovery. And the abscissa value at any chosen abandonment pressure will give the economic ultimate recovery. For the conditions underlying these illustrative calculations it will be seen that the physical ultimate recovery would occupy 14.5 per cent of the pore volume. And if the abandonment pressure be chosen as 100 psia, the economic ultimate recovery would be 13.8 per cent of the pore space. These recovery values represent 27.1 and 25.8 per cent of the original stock-tank oil content of the reservoir. On reference to Fig. 10.3 it will be found that these recoveries also imply ultimate free-gas saturations of 28.7 and 27.7 per cent, respectively. It is to be noted that these ultimate recovery values are provided automatically by the integration of Eq. (1) as the terminal points of the production history.

10.4. The Effect of the Reservoir Fluid and Rock Characteristics on the Production Histories of Solution-gas-drive Reservoirs.[2]—The example of theoretical gas-drive histories (Figs. 10.3 and 10.4) given in the preceding section is in itself only of illustrative significance. The numerical values are the composite result of all the specific numerical assumptions made in the use of the particular functions λ, ϵ, η, α, μ_o/μ_g, and Ψ, shown in Figs. 10.1 and 10.2. While the absolute magnitudes are thus of little interest, the

---

[1] By inverting Eq. (2) one obtains:

$$\left(1 + \frac{\mu_o}{\mu_g}\,\psi\right)\frac{d\bar{\rho}}{dp} = \bar{\rho}\lambda + (1 - \bar{\rho}\beta - \rho_w)\frac{\epsilon}{\beta} - \frac{d\beta}{dp}\frac{\bar{\rho}}{\beta},$$

where $\bar{\rho} = \rho_o/\beta$, which is basically equivalent to Eq.(1), but is often more convenient for numerical treatment.

[2] The results of this section are taken largely from M. Muskat and M. O. Taylor, *AIME Trans.*, **165**, 78 (1946), and *Petroleum Eng.*, **18**, 88 (December, 1946).

effect on the production histories and recoveries of changes in the assumed parameters and functions should be of practical importance.

Because of the nonlinear character of Eq. 10.3(1) it is not feasible directly to predict quantitatively from analytical considerations the effects of changes in the basic functions. On the other hand it will be obvious, for example, from inspection of Eq. 10.3(1), that only the ratio of viscosi-

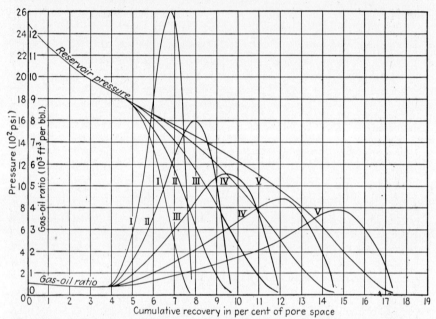

FIG. 10.5. The calculated pressure and gas-oil-ratio histories of hypothetical solution-gas drive reservoirs producing oils of different viscosities. For curves I, II, III, IV, and V the assumed gas-free oil viscosities are 11.04, 5.52, 2.76, 1.38, and 0.69 cp, respectively. Gas solubility at 2,500 psi assumed = 534 ft³/bbl in all cases. Connate-water saturation assumed = 30%. (*After Muskat and Taylor, AIME Trans., 1946.*)

ties, μ_o/μ_g, rather than the individual values will play a role in determining the production history. Moreover, if S and γ be changed by a common factor, the relation between ρ_o and p will not be affected, although R will be uniformly modified by that factor. Other qualitative inferences can also be drawn from the structure of Eq. 10.3(1), but these can be best demonstrated by numerical illustrations.

In studying the effects of the different physical parameters controlling the gas-drive performance it must be understood that there is generally a close interrelation between the various properties of petroleum-fluid systems. With a change in one, such as the solubility, will usually be associated a change in another, as the formation-volume factor of the oil. The assumption of specific changes in just one of the variables or of simultaneous changes in two may therefore be somewhat artificial from a practical point of view. On the other hand such an arbitrary procedure will better serve to show the sensitivity of the reservoir histories to the individual parameters than if all the factors were changed simultaneously.

As indicated above, only the ratio μ_o/μ_g enters in Eq. 10.3(1) to affect the relation between ρ_o and p. The magnitude of this effect is indicated by repeating the integrations leading to Figs. 10.3 and 10.4 with uniform changes (constant factors) in the curve for μ_o/μ_g in Fig. 10.2, throughout the whole pressure range, while retaining everything else fixed. The reservoir pressure and gas-oil-ratio vs. cumulative-recovery curves so found are plotted in Fig. 10.5, the different pairs of curves referring to a different atmospheric-pressure (reservoir-temperature) viscosity of the oil.[1]

Since Eq. 10.3(1) involves μ_o/μ_g only as a factor of Ψ, the curves of Fig. 10.5 coincide during the initial phases of the pressure decline, when $\Psi = 0$ and the equilibrium gas saturation is being built up. This part of the production history is independent of the fluid viscosities, within the basic approximations underlying Eq. 10.3(1).[2] However, as soon as the gas phase becomes mobile, the ratio μ_o/μ_g affects the relative loss in the gas and oil from the reservoir and the pressure-decline curves for the different-viscosity oils diverge. As is to be expected, the pressure decline, with cumulative recovery, becomes steeper with increasing oil viscosity, or μ_o/μ_g.

The gas-oil-ratio curves show even more strikingly the role played by μ_o/μ_g when the gas phase is mobile. From Eq. 10.3(1) it follows that, for a given pressure and oil saturation, R increases linearly with μ_o/μ_g. Hence, as the oil viscosity increases, the gas becomes dissipated ("bypasses") more rapidly, the pressure declines more sharply, and the ultimate recoveries become correspondingly reduced. Table 1 summarizes some of the numerical implications of Fig. 10.5 regarding the recovery and gas-oil ratio. In all cases the bubble-point (2,500 psi) solubility was kept fixed at 534 ft³/bbl, shrinkage[3] at 30.8 per cent, and connate water at 30 per cent.

---

[1] Since the viscosities of natural gases at reservoir temperatures do not change rapidly with the character (gravity) of the crude oil associated with them, the assumed changes in μ_o/μ_g have been attributed entirely to changes in the oil viscosity.

[2] In this region, where $\Psi = 0$, Eq. (1) can be formally integrated to give

$$\frac{\rho}{\rho_{oi}} = 1 - \frac{1}{\gamma} e^{-\int_p^{p_i} \lambda \, dp} \int_p^{p_i} \gamma\lambda \, dp \, e^{\int_p^{p_i} \lambda \, dp} .$$

[3] The shrinkage, in per cent, as used here is the formation-volume factor of the oil phase minus unity, multiplied by 100.

The data of Table 1 are plotted in Fig. 10.6 against the gas-free crude viscosity and its reciprocal. The latter permits extrapolation to viscosities

TABLE 1.—THE CALCULATED ULTIMATE RECOVERIES AND MAXIMUM GAS-OIL RATIOS FOR SOLUTION-GAS-DRIVE FIELDS PRODUCING CRUDES OF DIFFERENT VISCOSITIES

Gas-free crude viscosity, cp	Stock-tank recovery, % pore space		Stock-tank recovery, % initial oil		Free-gas saturation, % pore space		Max. gas-oil ratio, ft³/bbl
	Physical ultimate (atmospheric)	To 100 psia	Physical ultimate (atmospheric)	To 100 psia	Physical ultimate (atmospheric)	At 100 psia	
0.69	17.3	16.6	32.3	31.0	31.7	30.7	3,900
1.38	14.5	13.8	27.1	25.8	28.7	27.7	4,400
2.76	11.9	11.3	22.2	21.2	26.0	25.1	5,600
5.52	9.65	9.26	18.0	17.3	23.6	22.9	8,000
11.04	7.79	7.54	14.6	14.1	21.7	21.0	13,100

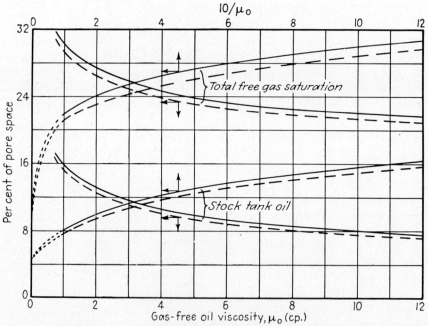

FIG. 10.6. The calculated recoveries, expressed as stock-tank oil and total gas space voided (free-gas saturation) of hypothetical solution-gas-drive reservoirs, as functions of the gas-free oil viscosity. Solid curves, recoveries to atmospheric pressure. Dashed curves, recoveries to 100 psia.

higher than the maximum listed in Table 1, as indicated by the dotted segments. These were drawn so that the free-gas space even at infinite viscosity would equal the equilibrium free-gas saturation of 10 per cent.

It will be seen that, while the decrease in recovery with increasing viscosity tapers off at high viscosity, the effect of the viscosity variation is substantial. The variation in the ultimate free-gas saturation roughly parallels that in the stock-tank recovery.

The effect of gas solubility can be treated similarly to that of the crude viscosity. However, to obtain a more realistic appraisal of the role played by the gas solubility, account must be taken of the fact that the expansion

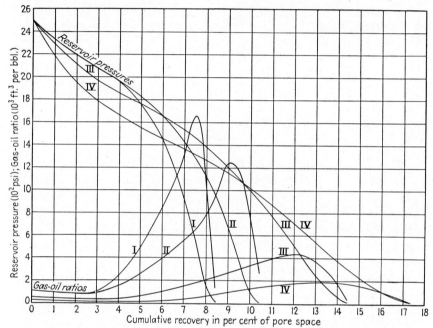

Fig. 10.7.   The calculated pressure and gas-oil-ratio histories of hypothetical solution-gas-drive reservoirs producing oils of different gas solubilities, shrinkages, and viscosities.   Curves I, II, III, and IV refer to solubilities and shrinkages indicated in the first four rows of Table 2, respectively.   (*After Muskat and Taylor, AIME Trans., 1946.*)

of the crude from atmospheric and the shrinkage factor will generally increase together with the amount of gas in solution (cf. Fig. 2.30). As a first approximation the shrinkage may be taken as proportional to the solubility. The results of a series of calculations in which the assumed solubility was varied,[1] with and without corresponding changes in the shrinkage, are plotted in Fig. 10.7.

The recoveries implied by the curves of Fig. 10.7 are summarized in Table 2, together with the results from two additional calculations given

---

[1] Here, as in all the calculations reported in this section, the physical parameters were changed uniformly over the whole pressure range by a constant factor corresponding to the terminal values indicated in the tables.

in the last two rows. Comparison of rows 1 and 2 shows again the direct effect of changes in the crude viscosity, similar to that indicated by Table 1, but for greater total solubility and shrinkage.

TABLE 2.—THE CALCULATED EFFECT OF GAS SOLUBILITY AND SHRINKAGE ON SOLU-TION-GAS-DRIVE RECOVERIES

Solubility at 2,500 psi, ft³/bbl	Gas-free crude viscosity, cp	Shrinkage from 2,500 psi, %	Stock-tank recovery, % pore space		Stock-tank recovery, % initial oil		Free-gas saturation, % pore space		Max. gas-oil ratio, ft³/bbl
			To atmospheric	To 100 psia	To atmospheric	To 100 psia	To atmospheric	To 100 psia	
1,068	2.76	61.6	8.41	8.17	19.4	18.9	31.1	30.3	16,600
1,068	1.38	61.6	10.4	10.1	24.1	23.4	33.4	32.5	12,500
534	1.38	30.8	14.5	13.8	27.1	25.8	28.7	27.7	4,400
267	1.38	15.4	17.3	16.2	28.6	26.7	25.4	24.1	1,900
1,068	1.38	30.8	17.2	16.7	32.1	31.1	31.6	30.7	9,400
534	1.38	0	26.9	25.8	38.4	36.8	26.9	25.8	3,150

The effect of changes in the solubility alone may be seen by comparison of rows 3 and 5. As would be expected, the recoveries are greater when the amount of gas available in solution is greater, though the increase due to doubling the solubility is equivalent to less than 3 per cent of the pore volume. On the other hand, if only the shrinkage is varied, the changes in recoveries are considerably more pronounced, as indicated by the data of rows 2 and 5.

The resultant effect of making simultaneous changes in solubility and shrinkage is indicated by comparison of rows 2, 3, and 4. These represent a balance between the tendency of the increased solubility to give greater recoveries and that of the increased shrinkage to reduce the recoveries, in which the latter definitely predominates. It is for this reason that rows 2, 3, and 4 show the apparently anomalous trend of decreasing re-coveries with increasing gas solubility. The controlling role played by the oil shrinkage is strikingly demonstrated by row 6, in which the purely arbitrary assumption of zero shrinkage was made. For this extreme case the recovery, expressed either as a fraction of the pore space or of the initial oil in place, is higher than for any of the others listed, although the free-gas saturation developed is actually smaller than for all others except that tabulated in row 4. If the oil were to suffer no shrinkage, all the free-gas saturation would of necessity have to be created by oil expulsion. But if the shrinkage is high, it will tend to create a space voidage and accelera-tion of the depletion of the gas energy in addition to that resulting from actual removal of the oil. In fact in the two cases listed in Table 2 for 61.6 per cent shrinkage (rows 1 and 2) a free-gas saturation of 26.7 per cent could be created by simply bleeding off the gas and cooling to standard temperature, without any oil expulsion whatever.[1]

The significance of the shrinkage in determining the actual oil recovery is further emphasized by the observation that, in spite of the fact that the free-gas space created by the production does increase with increased gas solubility, the stock-tank recoveries decrease (cf. rows 2, 3, and 4). In fact among these systems of 1.38 cp gas-free viscosity that with the greatest free-gas saturation (row 2) will give the least recovery. On the other hand it should be noted that, while the ultimate recoveries are greatest for the system with lowest solubility (row 4), the initial pressure decline is most rapid for this case (curve IV, Fig. 10.7). This, however, is quite reasonable, when it is observed that to create the free-gas saturation to replace a given volume of oil withdrawal the pressure drop required will increase as the solubility decreases, if the other significant factors are substantially the same.

The composite effect of such simultaneous variations of the physical properties of the petroleum fluids as may occur in practice may be derived by using the API gravity of the crude as the gross characterization parameter of the gas and oil system. While the crude-oil gravity by no means uniquely determines all the physical properties, it will be recalled from Chap. 2 that it does provide at least a semiquantitative means of correlating many of the observations which have been made on the gas solubility, shrinkage, and viscosity of individual oil and gas systems. In fact, by using the empirical correlation graphs presented in Chap. 2, it is possible to construct "typical," or average, curves for the significant fluid properties of gas-oil systems of different API crude gravities, as functions of the pressure, at reservoir temperature. The values so found for these functions, at 3,000 psi and the assumed reservoir temperature (190°F), are listed in Table 3.

TABLE 3.—VALUES OF FLUID PARAMETERS AT 190°F, ASSUMED FOR DIFFERENT CRUDE GRAVITIES

Crude gravity, °API	Gas solubility at 3,000 psi, ft³/bbl	Formation vol. of oil at 3,000 psi	μ_o, cp, 3,000 psi	μ_o, cp, 14.7 psi	Relative gas density at 3,000 psi	$10^2\mu_g$, cp, 3,000 psi
10	205	1.100	76	430	177	2.02
20	458	1.220	2.8	13.7	186	2.08
30	737	1.352	0.69	2.44	195	2.25
40	1,032	1.521	0.29	0.90	203	2.54
50	1,400	1.763	0.14	0.47	210	2.94

[1] While this would be impossible, except through the action of diffusion, in a porous medium, the technique of solubility and shrinkage determinations in bottom-hole sample analysis represents just this type of process (cf. Sec. 2.8).

Upon applying the complete curves for the fluid-property functions to the integration of Eq. 10.3(1) the gas-drive production histories can be calculated for the systems of different crude gravities. The results of such calculations are plotted in Fig. 10.8. The recoveries, referring to a

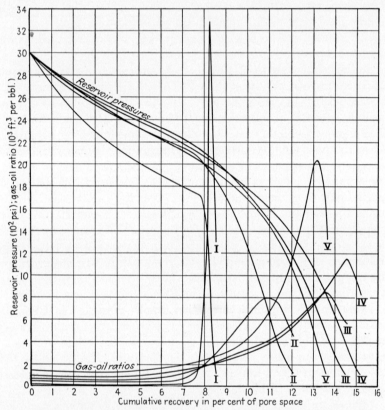

Fig. 10.8. The calculated pressure and gas-oil-ratio histories of hypothetical solution-gas-drive reservoirs producing oils of different API gravity. Curves I, II, III, IV, and V refer to crude oils of 10, 20, 30, 40, and 50° API, respectively. Connate-water saturation assumed = 25% in all cases. (*After Muskat and Taylor, Petroleum Eng., 1946.*)

state of practical ultimate depletion, taken as 100 psia, and maximum gas-oil ratios are plotted as a function of the crude gravity in Fig. 10.9. In all cases the connate-water saturation was taken as 25 per cent, the gas-oil permeability ratio as that shown in Fig. 10.1, and the initial reservoir pressure as 3,000 psia.

As would be expected from the previous results of this section, referring to calculations in which the individual fluid properties were varied separately, the absolute ultimate recovery increases at first with increasing crude gravity but reaches a maximum and then declines at gravities exceeding 40°API. This mainly reflects the effect of shrinkage in more than counterbalancing those of oil viscosity and gas solubility at gravities greater than 40°API. However, the recovery as a fraction of the initial stock-tank oil in place continues to increase with increasing gravity up to 50°API, although a decline would probably develop at still higher gravities. On the other hand the free-gas saturation developed at ultimate depletion (100 psia) increases with increasing crude gravity throughout the plotted gravity range and probably would continue to increase even up to gravities of condensate oils.

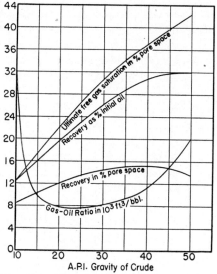

The rise in the maximum gas-oil ratio for gravities exceeding about 25°API is mainly due to the increasing gas solubility as the crude gravity increases (cf. Table 3). The extremely high value for the 10°API oil is largely a reflection of the very low recovery and high viscosity for this oil. In fact, as soon as the equilibrium gas saturation (10 per cent) is reached, in the 10°API system, the gas-oil ratio rises and [the pressure

Fig. 10.9. The calculated ultimate recoveries, free-gas saturations (at 100 psi), and maximum gas-oil ratios of hypothetical solution-gas-drive reservoirs, as functions of the crude gravity. (*After Muskat and Taylor, Petroleum Eng., 1946.*)

declines almost vertically (cf. Fig. 10.8), since the natural growth in the k_g/k_o ratio is tremendously magnified by the oil to gas viscosity ratio, μ_o/μ_g, to create a highly accelerated rise in the relative mobility of gas to oil. It is the associated rapid depletion of the gas that thus leads to the low ultimate recovery. Of course, as the oil viscosity decreases with increasing gravity, the tendency for the sharp rise in gas-oil ratio will be lessened and the maximum reached will decline until the effect of increased gas solubility becomes dominant.

As in the case of the other theoretical results presented in this section, the numerical values plotted in Figs. 10.8 and 10.9 should not be construed as quantitative predictions of actual reservoir behavior. The fixed assumptions regarding the connate-water saturation, the permeability-saturation relationship, the complete neglect of gravity effects, etc., severely restrict the absolute significance of calculations of this type. The physically significant content of such curves as shown in Figs. 10.8 and

10.9 lies in the trends of the results and their comparative values and in the physical interpretation of the reasons giving rise to the variations.

In principle the effect of the connate-water saturation on gas-drive performance and recovery can be evaluated in the same manner as the viscosity, solubility, and shrinkage parameters. It is clear, however, that to carry out such calculations one must know how the permeability-saturation relationship may vary with the connate-water saturation. From the general observation that the connate-water content of oil-producing sands usually increases with decreasing permeability (cf. Sec. 3.11) an associated change in the quantitative features of the permeability-saturation curves may also be expected. Unfortunately, however, available data on this point are too meager to warrant any specific assumptions. Accordingly it will be assumed here that the sand itself remains substantially the same and that, as indicated by Leverett and Lewis,[1] $\Psi(\rho)$ is a function only of the total liquid saturation, for moderate[2] values of the water saturation. Hence, for the extreme case where the connate-water saturation is arbitrarily taken as 0, $\Psi(\rho)$, as a function of the oil saturation, will still be given by the curve of Fig. 10.1 except for a shift of 0.30 in the values of the abscissa scale. Upon assuming further that the gas solubility at 2,500 psi is 534 ft^3/bbl, the shrinkage is 30.8 per cent, and the gas-free crude viscosity is 2.76 cp, the calculated pressure-decline and gas-oil-ratio histories are shown as curves III of Fig. 10.10. These curves indicate an ultimate recovery, in per cent of the pore space, of 10.2 per cent to atmospheric and 9.92 per cent to 100 psia, corresponding to free-gas saturations of 30.0 and 29.2 per cent, respectively, and recoveries of 13.3 and 13.0 per cent of the initial oil in place. Moreover the maximum gas-oil ratio is 12,900 ft^3/bbl.

These values are to be compared with those given in the third row of Table 1, which refers to the same system with a connate-water content of 30 per cent. It will be seen that while the free-gas saturation developed by the production will be greater if the connate-water saturation is zero, the oil recoveries both in absolute value as well as fractions of the oil in place will be decidedly smaller. This is again due to the greater space-voidage contribution of the oil shrinkage when its initial saturation is 100 per cent as compared with 70 per cent. On the other hand it is to be emphasized that it is only this factor that is evaluated by these calculations. It may well be that in comparing sands with inherently different connate-water saturations, the associated differences in their permeability-saturation

---

[1] M. C. Leverett and W. B. Lewis, *AIME Trans.*, **142**, 107 (1941).

[2] It is quite doubtful, however, that the effect of connate-water saturations as high as 30 per cent on the gas to oil permeability ratios will be negligible in actual consolidated sands.

characteristics may modify and possibly reverse the relative behaviors
indicated by the above comparison.

The sensitivity of gas-drive performance to the details of the permeability-saturation relationships of the producing rock may be estimated

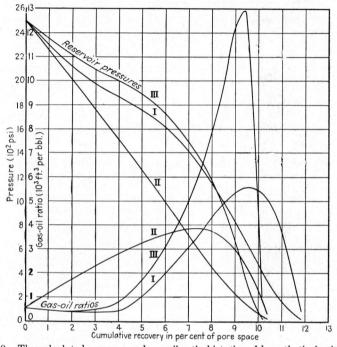

FIG. 10.10. The calculated pressure and gas-oil-ratio histories of hypothetical solution-gas-
drive reservoirs with different reservoir rock characteristics. Curve I, permeability-ratio
curve was solid curve of Fig. 10.1. Curve II, permeability-ratio curve was dashed curve of
Fig. 10.1. Curve III, connate-water saturation assumed = 0. Assumed gas solubility at
2,500 psi = 534 ft³/bbl, shrinkage from 2,500 psi = 30.8%, and gas-free oil viscosity = 2.76
cp in all cases. Connate-water saturation assumed = 30% for curves I and II. (*After
Muskat and Taylor, AIME Trans., 1946.*)

from the results of integrating the basic differential equation for different
types of $\Psi(\rho)$ curves. One basic feature of the curve used in all the above
calculations is the free-gas equilibrium saturation of 10 per cent. It is
this factor or assumption that has led to the initial declines in the gas-oil-
ratio curves and the concave-upward type of decline in pressure in all
the performance curves shown in previous figures.[1] This is confirmed by
curves II of Fig. 10.10, for which $\Psi(\rho)$ was taken as the dashed curve in

---

[1] While a nonvanishing equilibrium gas saturation will always lead to an initial
decline in the gas-oil ratio, such decline should also result if $\left|\dfrac{\partial \Psi}{\partial \rho}\right| < \mu_g/\mu_o\rho$ at the initial
pressure and saturation.

Fig. 10.1, with no equilibrium gas saturation, but with the other parameters the same as for curves I, which are reproduced from Fig. 10.5 (curves III). It will be seen that now the gas-oil ratio rises from the very beginning of production. And the pressure-decline curve does not become concave upward until the maximum in the gas-oil ratio is passed. As would be expected, the ultimate recoveries are lower if there is no equilibrium

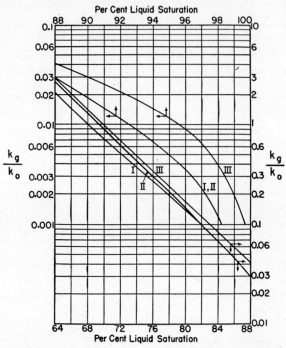

Fig. 10.11. Several types of permeability-ratio curves used in studying their effect on solution-gas-drive reservoir behavior. (*After Muskat and Taylor, AIME Trans., 1946.*)

free-gas saturation. At atmospheric pressure the recovery is 10.4 per cent of the pore space, and at 100 psia it is 9.77 per cent, as compared with 11.9 and 11.3 per cent for the case of a 10 per cent equilibrium gas saturation. On the other hand the maximum gas-oil ratio is lower than for curve I, since the total areas under the gas-oil-ratio curves, which are proportional to the initial solubility, must be the same for both.

The effect on the recoveries that may be caused by variations in the permeability-saturation characteristics at low liquid saturations or throughout the whole saturation range is illustrated by the results of comparative calculations using the curves of Fig. 10.11. These include three curves for $\Psi(\rho)$ that were considered as reasonable approximations to the true relationship in an actual oil field. In all cases they were extrapolated to zero at 100 per cent liquid saturation (no equilibrium gas saturation). The connate-water content was estimated as 20 per cent. The initial reservoir pressure was 1,727 psi, the gas solubility 758 ft³/bbl, and the shrinkage 36.0 per cent. The gas-free oil viscosity was 2.4 cp. The field had an initial gas cap occupying a volume 54 per cent of that occupied by the oil zone.[1] Curves I and II coincide in the region of high saturations and diverge when the total liquid saturation falls below 82 per cent, whereas curve III is higher than the others throughout the whole saturation range.

The ultimate recoveries (to 100 psi) and maximum gas-oil ratios found on using the different curves of Fig. 10.11 are summarized in Table 4.

TABLE 4.—THE CALCULATED EFFECT OF VARIATIONS IN THE PERMEABILITY-SATU-
RATION RELATIONSHIP ON RECOVERIES AND GAS-OIL RATIO

Permeability-ratio curve of Fig. 10.11	Stock-tank recovery to 100 psia, % pore space	Recovery, % initial oil	Free-gas saturation at 100 psia, % pore space	Max. gas-oil ratio, ft.³/bbl
I	16.7	28.4	33.3	14,800
II	16.2	27.5	32.7	16,500
III	15.5	26.3	31.9	15,500

As would be expected, the greatest values of $\Psi(\rho)$ (curve III) led to the smallest recoveries, and the highest recoveries would be obtained with the lowest $\Psi(\rho)$ curve (curve I). Moreover in the case of curve III the gas-oil ratio was found to rise most rapidly and resulted in a lower value of the maximum than for curve II. As a whole, however, Table 4 shows that the ultimate recoveries will not be very sensitive to the exact details of the permeability-saturation relationship, although in the absolute value of the oil recovery the small differences indicated in Table 4 might well be of commercial significance.

10.5. Gas-drive Reservoirs with Gas Caps but No Gravity Drainage.— The analysis of the previous sections can be readily generalized to cover gas-drive reservoirs initially overlain by gas caps, provided that the down-dip gravity drainage of the oil does not play a significant role in the production mechanism. This means that the gas cap will not expand appreciably into the oil zone. Rather its gas content will merely provide an additional supply of gas to be permeated and diffused[2] through the oil-saturated section. While in practice some degree of gravity drainage will of necessity be ever present, except when there is oil migration into

---

[1] The method of taking into account the presence of a gas cap, when there is no gravity drainage, will be outlined in Sec. 10.5.

[2] This refers to a mass movement rather than the molecular-diffusion process.

the gas cap due to excessive depletion of the gas-cap gas, it will be instructive to consider first the limiting case where only the effect of the
gas cap as a simple gas reservoir is evaluated. Moreover, as will be seen
in Sec. 10.15, there is still no really satisfactory method for treating
gravity-drainage problems. To take into account gravity drainage even
approximately it is necessary to introduce additional assumptions.

If the ratio of the gas-cap thickness to that of the oil-saturated zone be
denoted by H,* and the stock-tank equivalent of the oil in the gas cap
by $\overline{\rho_{oi}}$, the gas production from the system will be proportional to

$$Q_g = -\frac{H}{1+H}\delta[S\overline{\rho_{oi}}+\gamma(1-\rho_w-\beta\overline{\rho_{oi}})]-\frac{1}{1+H}\delta\left[\frac{S\rho_o}{\beta}+\gamma(1-\rho_w-\rho_o)\right]. \quad (1)$$

Similarly the oil production will be proportional to

$$Q_o = -\frac{1}{1+H}\delta\left(\frac{\rho_o}{\beta}\right). \quad (2)$$

Since it is assumed that only the oil zone is exposed to the well bore,
Eqs. (1) and (2) may be combined with the gas-oil-ratio equation

$$R = \frac{Q_g}{Q_o} = S + \alpha\Psi, \quad (3)$$

to give

$$\left(1+\frac{\mu_o}{\mu_g}\Psi\right)\frac{d\rho_o}{dp} = \rho_o\lambda + (1-\rho_w-\rho_o)\epsilon + \rho_o\eta\Psi$$
$$+ H\left[\beta\overline{\rho_{oi}}\lambda + (1-\rho_w-\beta\overline{\rho_{oi}})\epsilon - \overline{\rho_{oi}}\frac{d\beta}{dp}\right]. \quad (4)$$

By using the permeability-saturation curve of Fig. 10.1 and the fluid-
property functions of Fig. 10.2, the integration[1] of Eq. (4) for various
values of H leads to the pressure and gas-oil-ratio histories plotted in
Fig. 10.12. The ultimate recoveries and maximum gas-oil ratios are listed
in Table 5, and the former are plotted vs. H in Fig. 10.13. In all cases

---

* H is the same as the "m factor" sometimes used in the material-balance equation
[cf. Eqs. 9.6(2)]. It is also to be noted that throughout this chapter it is assumed
that the connate-water saturation in the gas cap is the same as in the oil zone, and that
the latter is initially fully saturated so that $\rho_{oi} = 1 - \rho_w$.

[1] For the range in ρ_o where $\Psi = 0$, Eq. (4) has the formal solution

$$\rho_o = (1+H)\rho_{oi} - \phi(p)\left[H\gamma_i\rho_{oi} + (1+H)\rho_{oi}\int_p^{p_i}\frac{\lambda}{\phi}dp + H\overline{\rho_{oi}}\int_p^{p_i}\frac{\left(\beta\lambda-\beta\epsilon-\dfrac{d\beta}{dp}\right)dp}{\phi}\right],$$

where $\quad \phi(p) = \dfrac{1}{\gamma}e^{-\displaystyle\int_p^{p_i}\lambda\,dp}$.

the gas solubility at 2,500 psi was taken as 534 ft³/bbl, the shrinkage from 2,500 psi as 30.8 per cent, and the gas-free oil viscosity as 2.76 cp.

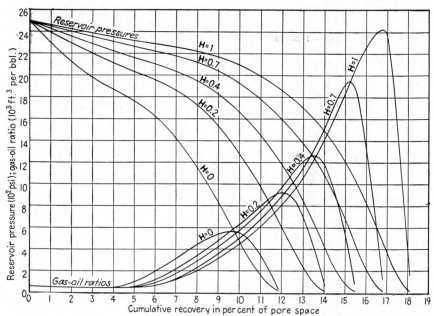

FIG. 10.12.   The calculated pressure and gas-oil-ratio histories of hypothetical solution-gas-drive reservoirs with various thicknesses of gas cap, but not subject to the action of gravity drainage. H = (thickness of gas cap)/(thickness of oil zone). Gas solubility at 2,500 psi assumed = 534 ft³/bbl; shrinkage from 2,500 psi = 30.8%; atmospheric pressure viscosity = 2.76 cp; connate-water saturation assumed = 30%. (*From Jour. Applied Physics, 1945.*)

TABLE 5.—THE CALCULATED ULTIMATE RECOVERIES AND MAXIMUM GAS-OIL RATIOS
FOR GAS-DRIVE FIELDS WITH VARIOUS THICKNESSES OF GAS CAP

H	Stock-tank recovery, % pore space		Stock-tank recovery, % initial oil		Free-gas saturation, % pore space		Max. gas-oil ratio, ft³/bbl	Total gas available, ft³/bbl of initial oil
	To atmospheric	To 100 psia	To atmospheric	To 100 psia	At atmospheric	To 100 psia		
0	11.9	11.3	22.2	21.2	26.0	25.1	5,600	534
0.20	14.1	13.6	26.3	25.3	28.3	27.4	9,300	872
0.40	15.5	15.0	28.9	28.1	29.8	29.0	12,700	1,209
0.70	16.8	16.4	31.4	30.7	31.2	30.5	19,500	1,715
1.00	18.1	17.7	33.8	33.1	32.5	31.9	24,300	2,222

The general trends shown by the curves in Figs. 10.12 and 10.13, and the data of Table 5, are all as would be expected from general considerations. The recoveries and free-gas saturations all increase with increasing thickness of gas cap, though the variation is slower than linear. The maximum gas-oil ratio does vary approximately linearly with the total amount of gas available. But the ultimate recovery increases over that

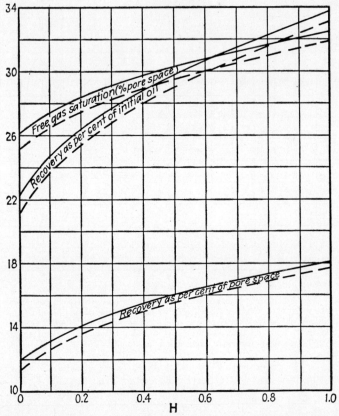

FIG. 10.13. The calculated recoveries and ultimate free-gas saturations of hypothetical solution-gas-drive reservoirs as functions of the gas-cap thickness. $H =$ (thickness of gas cap)/(thickness of oil zone). Solid curves, depletion to atmospheric pressure. Dashed curves, depletion to 100 psi. Assumed conditions were the same as for Fig. 10.12.

without a gas cap only by approximately 50 per cent, while the total gas content of the reservoir is increased by a factor of about 4.

It should be emphasized that these results are restricted by the assumptions that the gas cap itself does not bleed directly into the producing well bores and that it does not expand with the aid of gravity segregation and oil drainage. If the former condition does not obtain, the recoveries will be reduced and the reservoir pressures will decline more rapidly than indicated by Fig. 10.12. But if appreciable gas-cap expansion and gravity drainage develop, the recoveries will be increased and the pressure decline and gas-oil-ratio rise will be retarded. The numerical results of this section or their equivalents obtained by similar calculations based on Eq. (4) must therefore be applied with caution and due consideration of their limitations.

10.6. The Declines in Productivity Index and Production Rates in Solution-gas-drive Fields.—Although, as previously discussed, the method of treating solution-gas-drive reservoirs presented in the last several sections and expressed analytically by Eq. 10.3(1) ignores entirely the well system through which the reservoir is being depleted, it provides information closely related to the productivity of the wells. In Sec. 8.5 it was shown that for the limiting conditions of zero pressure differential the productivity index will be determined essentially by the ratio $k_o/\mu\beta$ pertaining to the interior regions of the reservoir. Because of the uncertainties associated with local well conditions and the numerical factors involved in the explicit formula for the productivity index, calculations of its absolute magnitude would be of questionable significance. However, the changes in the productivity index of a well as a reservoir is being depleted, as expressed by the ratio

$$\frac{PI}{(PI)_i} = \frac{k_o/\mu_o\beta}{(k_o/\mu_o\beta)_i},\tag{1}$$

where the subscript i refers to the initial conditions, should be subject to prediction with reasonable accuracy, provided that the well is representative of the field as a whole. If the reservoir pressure and oil saturation as a function of the cumulative recovery are known, Eq. (1) can also be evaluated as a function of the cumulative recovery. On carrying through such calculations the results plotted in Figs. 10.14 and 10.15 are obtained, for the systems whose pressures and gas-oil-ratio histories are given by Figs. 10.5 and 10.7, with ultimate recoveries as listed in Tables 1 and 2. In Fig. 10.16 is plotted the comparative behavior of fields producing different-gravity crudes.

The curves of Figs. 10.14 and 10.15 show the same types of coincidence as those of Figs. 10.5 and 10.7 during the initial phase of the production history while the equilibrium gas saturation is being built up and the saturation-decline history is independent of the oil viscosity. On the other hand, since at the terminal conditions, at a given abandonment or atmospheric pressure, the μ_o/μ_{oi}'s for the systems described by Figs. 10.5 and 10.7 are the same, the relative values of the productivity index will be determined—except for β—by the value of k_o, which, in turn, will depend on the ultimate oil saturations. It is for this reason that in Figs. 10.14 and 10.15 the minimum values to which the $PI/(PI)_i$'s ultimately fall, for the same β, decrease with decreasing residual-oil saturation, as

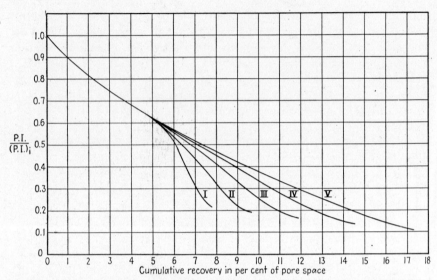

Fig. 10.14. The calculated decline histories of the productivity index of hypothetical solution-gas-drive reservoirs producing oils of different gas-free viscosities. For curves I, II, III, IV, and V, the gas-free oil viscosities are 11.04, 5.52, 2.76, 1.38, and 0.69 cp, respectively. Other assumed conditions were the same as for Fig. 10.5. (*After Muskat and Taylor, AIME Trans., 1946.*)

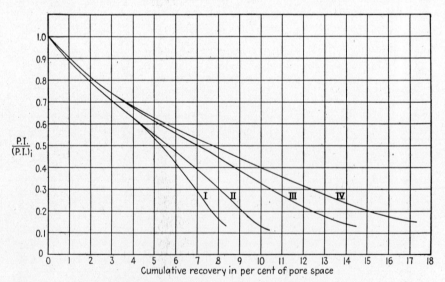

Fig. 10.15. The calculated decline histories of the productivity index of hypothetical solution-gas-drive reservoirs producing oils with different gas solubility and shrinkage. Curves I, II, III, and IV correspond to the pressure and gas-oil ratio histories of Fig. 10.7. (*After Muskat and Taylor, AIME Trans., 1946.*)

may be verified by reference to Tables 1 and 2. In the case of Fig. 10.16 the factors $(\mu_o\beta)_i/\mu_o\beta$ increase with increasing API gravity. However, this is more than counterbalanced by the effect of decreasing oil saturation, so that the resultant terminal values of $PI/(PI)_i$ decrease with increasing crude gravity. In all cases it appears that the productivity index should

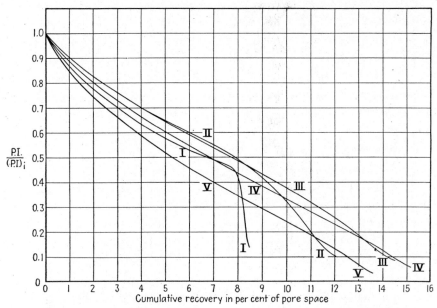

Fig. 10.16. The calculated decline histories of the productivity index of hypothetical solution-gas-drive reservoirs producing oils of different API gravity. Curves I, II, III, IV, and V refer to crude oils of 10, 20, 30, 40, and 50° API, respectively. (*After Muskat and Taylor, Petroleum Eng., 1946.*)

fall by a factor of 5 to 15 simply as the result of reservoir depletion and the associated changes in the factors determining the flow capacity of the wells. While satisfactory quantitative field data on the declines of the productivity index due to reservoir depletion are rather meager, they have been observed in several instances.[1] As pointed out in Sec. 8.7, data of this type should be of value in indicating the long-term changes in the nature of the reservoir fluids and their distribution during the course of the production history.

Superposed on these declines in the productivity index shown in Figs. 10.14 to 10.16 will be, of course, that in the reservoir pressure and in the total pressure differentials available for driving the oil to the well bores.

[1] R. V. Higgins, *U.S. Bur. Mines Rept. Inv.* 3657 (September, 1940), cf. also W. L. Horner, *Petroleum Eng.*, **17**, 133 (February, 1946), and E. W. McAllister, *AIME, Trans.*, **142**, 39 (1941).

Hence the total productive capacities, or potentials, of the wells will be reduced even more severely than would be indicated only by the curves of Figs. 10.14 to 10.16.

The variation of the productivity indexes as given by Figs. 10.14 to 10.16 also may be used to construct approximate production-decline curves for strict solution-gas-drive reservoirs. For example, upon assuming that all wells are similar and that $n(t)$ is the number drilled and producing at the time t, the production rate of the field as a whole may be expressed as

$$Q = n \, \Delta p \, \text{PI} = \frac{dP}{dt}, \tag{2}$$

where Δp is the common differential pressure under which the wells are producing, and P is the cumulative recovery at the time t. If the field is being produced without proration or other production controls, Δp may be taken as a fraction,[1] a, of the reservoir pressure. The latter in turn, as well as the PI, may be considered as functions of the cumulative recovery, as given by such curves as shown in Figs. 10.4, 10.5, 10.7, and 10.8. Hence Eq. (2) can be formally integrated as

$$\frac{1}{a(\text{PI})_i} \int \frac{dP}{p[\text{PI}/(\text{PI})_i]} = \int n \, dt, \tag{3}$$

implicitly giving the cumulative production as a function of time. As it stands, the right-hand side of Eq. (3) represents the cumulative producing time, as well-months or well-years, but may be explicitly expressed as a function of time if the well-development history $n(t)$ be considered as known. Functions by which drilling histories may often be approximated are

$$\left.\begin{aligned} n &= N(1 - e^{-bt}), \\ n &= N(1 - e^{-bt^2}), \\ n &= \frac{Nt}{b+t}, \end{aligned}\right\} \tag{4}$$

where N is the ultimate number of wells drilled.[2]

As previously indicated, to avoid the uncertain calculations of absolute values of the productivity index, $(\text{PI})_i$ may be taken as the initial PI of the earliest wells as actually observed, so that $a(\text{PI})_i p_i$ will represent the

---

[1] This, of course, is a simplifying assumption. If it is known that the wells will be operated under different conditions, as a fixed bottom-hole pressure or equivalent relation, this can be introduced into Eq. (2) and the subsequent integrations carried out in essentially the same manner as indicated here.

[2] In actual practice the number of producing wells is not a continuously increasing function of time. Because of the variation in the producing capacities of the individual wells they will be abandoned in sequence as they become unprofitable, rather than all remaining on production throughout the producing life of the field, as implied by Eqs. (4).

initial production rates Q_i of the first wells drilled.   Equation (3) can
then be rewritten as

$$\frac{1}{Q_i} \int \frac{dp}{(p/p_i)[\mathrm{PI}/(\mathrm{PI})_i]} = \int n(t)dt. \tag{5}$$

To illustrate the implications of this approximate treatment the integral
on the left-hand side of Eq. (5) has been evaluated for the hypothetical
reservoir producing a 30°API-gravity crude described by the data listed
in Table 3.   It was assumed that the reservoir area is 4,000 acres, of
40 ft thickness and 25 per cent porosity.   The ultimate well spacing was
taken as 40 acres per well.   The initial "potential," that is, $ap_i(\mathrm{PI})_i$, was
assumed to be 500 bbl/day.   Three different well-development programs
were assumed, as follows:

$$\begin{array}{lll}
(1) & : & n = 100(1 - e^{-0.7t}), \\
(2) & : & n = 20t \quad : \ 0 \leqslant t \leqslant 5, \\
    &   & = 100 \quad : \ t \geqslant 5, \\
(3) & : & n = 100.
\end{array} \tag{6}$$

The first is of the type of the first equation of Eqs. (4).   The second
implies a fixed annual drilling rate of 20 wells, until the 100 wells giving
the 40-acre spacing are completed in 5 years.   The last represents the
much more idealized case in which all the drilling is completed before
commencement of production.

The production-decline and cumulative-production vs. time curves, as
calculated with these assumptions, are plotted in Fig. 10.17.   Also plotted
in Fig. 10.17 are the pressure-decline curves, with the pressure expressed
as a fraction of its initial value.   The production-decline and cumulative-
production curves simulate qualitatively the corresponding actual curves
observed for the older uncontrolled gas-drive fields.   However, because
of the many simplifying assumptions underlying their construction, no
significance should be attached to the numerical values given in Fig. 10.17.
Moreover they are based on the assumption that no control is applied to
the field, a situation seldom obtaining under current operating practices.
On the other hand the method of calculation used in deriving the curves
of Fig. 10.17 can be applied also when a field is subjected to withdrawal
restrictions.   In such cases the production rate will evidently follow the
"allowable" rate until the latter represents the actual maximum capacity
of the field.   The corresponding value of the average production rate per
well will then give $\Delta p \cdot \mathrm{PI}$.   If Δp is again expressed as a fraction or function
of the current reservoir pressure, the value of the reservoir pressure or
cumulative production giving the required value of $\Delta p \cdot \mathrm{PI}$ can be deter-

mined from the calculated curves of pressure and PI vs. the cumulative recovery, such as those of Figs. 10.8 and 10.16. This reservoir pressure and cumulative production will represent the starting point for the unrestricted-production and declining-production-rate phase of the reservoir performance. Its time equivalent is readily computed from the production

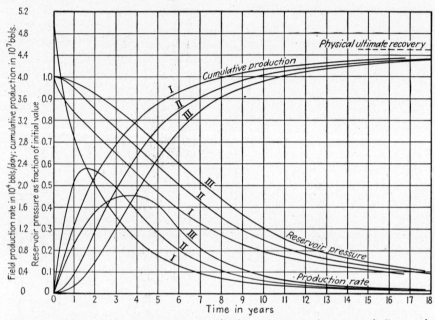

Fig. 10.17. The calculated approximate production rates, reservoir pressure declines, and cumulative production curves, vs. time, for hypothetical solution-gas-drive reservoirs with different well-development programs. Curves I, all wells (100) drilled before production is begun. Curves II, $n = 100(1 - e^{-0.7t})$ = number wells at time t (in years). Curves III, drilling is at a rate of 20 wells per year for 5 years. Total reservoir area = 4,000 acres; pay thickness = 40 ft; porosity = 25%. Maximum initial well capacity = 500 bbl/day. Basic reservoir-production history is that for a 30°API gravity crude plotted in Fig. 10.8.

rates during the period of withdrawal control. The subsequent time history of the field can then be calculated by exactly the same method as that outlined above for a reservoir producing "wide open" from the very beginning.

Finally it should be noted that the above-outlined method can also be used to predict the time variation of the pressure and production rate in reservoirs subjected to gas-injection operations. To the extent that the basic theory, to be developed and discussed in the next several sections, for determining the pressure vs. recovery history under pressure maintenance may be valid, the time can be introduced by Eq. (3) or (5) to give pressure- and production-rate-decline curves vs. time similar to those of

Fig. 10.17. This type of application may be of particular importance in evaluating and planning gas-injection programs.

10.7. Gas Injection into Gas-drive Reservoirs; Pressure Maintenance.— The method for treating gas-drive reservoirs presented in the last several sections can be generalized still further to include situations where gas is returned to the producing formation during the course of the primary-[1] production history. Again it will be limited by the assumption that gravity effects may be neglected. And it will be supposed that the injected gas is uniformly distributed and permeated throughout the reservoir, if there is no gas cap originally, and becomes ultimately diffused and dissipated through the oil zone even if injected directly into an overlying gas cap. The injected gas may be visualized as simply providing a partial or complete replacement for the loss of reservoir gas content associated with the oil production.

The operation of gas injection into an oil reservoir during the course of its primary-production history is generally termed "pressure maintenance by gas injection."[2] For the immediate effect of returning all or part of the produced gas from a reservoir must evidently be a retardation in the pressure decline. And if enough gas is injected, the decline will be completely arrested and the pressure will be maintained at its current value. In itself, such pressure maintenance would be of value only as means for prolonging the flowing lives of the producing wells and maintaining high productive well capacities. These benefits are generally of considerable importance, although when compared with the additional cost of the operations the net gain often may be left in doubt. However, when the maintenance of pressure represents an extension of the pressure vs. cumulative-recovery curve so as to imply an increased ultimate oil recovery, the operations may be more readily evaluated. It is this feature of the problem that will be considered here.

As is common practice, the "degree" of gas injection will be expressed here as the fraction r of the produced gas that is returned to the reservoir. The basic differential equation may then be derived exactly as in Sec. 10.5, with the single change that in Eq. 10.5(3) Q_g/Q_o is replaced by

$$\frac{Q_g}{Q_o} = (S + \alpha\Psi)(1 - r). \tag{1}$$

[1] The "primary"-production life refers here to that beginning with the initial field discovery and continuing until the original energy sources for oil expulsion are no longer alone able to sustain profitable producing rates.

[2] The term "repressuring" has also been used, although this is now usually applied to "secondary-recovery" gas-injection operations in reservoirs that have been already substantially depleted by the primary-producing mechanism. Moreover, in contrast to "pressure-maintenance" operations, those without fluid injection are often described as producing by "pressure depletion."

The equation for $d\rho_o/dp$ is then found to be

$$\frac{d\rho_o}{dp} = \frac{\rho_o\lambda+(1-\rho_w-\rho_o)\epsilon+\rho_o\eta\left(\Psi-\dfrac{rR}{\alpha}\right)+H\left[\beta\overline{\rho_{oi}}\lambda+(1-\rho_w-\beta\overline{\rho_{oi}})\epsilon-\overline{\rho_{oi}}\dfrac{d\beta}{dp}\right]}{1+\dfrac{\mu_o}{\mu_g}\left(\Psi-\dfrac{rR}{\alpha}\right)}, \quad (2)$$

where the notation is the same as that used in Eq. 10.5(4).

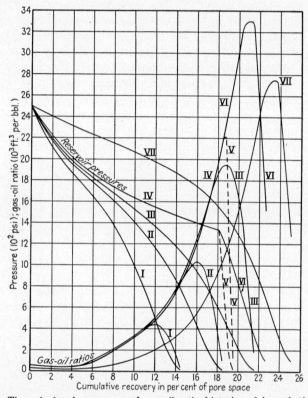

Fig. 10.18. The calculated pressure and gas-oil-ratio histories of hypothetical gas-drive reservoirs for various degrees of gas injection, with no gas segregation. Curves I, r (fraction of produced gas that is returned to the formation) = 0. Curves II, $r = 0.6$. Curves III, $r = 0.8$. Curves IV, $r = 1$ to a gas-oil ratio of 20,000 ft³/bbl. Curves V, continuations of curves IV with $r = 0$. Curves VI, continuations of curves IV with $r = 0.8$. Curves VII, $r = 0.8$, with an initial gas-cap thickness one-half the oil-zone thickness. Gas-oil ratios of curve VII are half of calculated values. Basic reservoir data are those used for Fig. 10.3.

By using again the solid $\Psi(\rho)$ curve of Fig. 10.1 and the curves of Figs. 10.2 and 8.1, and assuming that $H = 0$ (*i.e.*, no initial gas cap), the computed pressure and gas-oil-ratio histories for several values of r are plotted in Fig. 10.18. The curves for $r = 0$ are, of course, the same as those shown in Fig. 10.4. The produced gas-oil ratios were computed, as in previous cases, by Eq. 10.5(3).[1]

As is to be expected, the pressure decline is increasingly retarded and the oil recovery increased as a greater fraction of the gas produced is returned to the reservoir. At the same time the gas-oil ratios rise to increasing maximum values. Since in practice the operations could not be continued profitably at excessively high ratios, the value of r, or amount of gas returned, must be changed when the limit is exceeded. For a limit of 20,000 ft³/bbl, Fig. 10.18 implies that complete gas-return operations would have to be changed or terminated after a recovery of 18.2 per cent of the pore space and a pressure decline to 1,335 psi. If the gas injection were then stopped completely, the pressure and gas-oil ratio would follow the dashed curves. The pressure would undergo a precipitous decline, and after an initial continuation of the previous rise the gas-oil ratio would also drop very rapidly. At 100 psi the ultimate recovery would be equivalent to 19.3 per cent of the pore space, and at atmospheric pressure it would be 19.6 per cent. If, however, the injection operations are continued at the reduced rate of 80 per cent gas return,[2] the subsequent history will follow the solid segments of the curves (curves VI). The gas-oil ratio continues to rise—to 33,100 ft³/bbl—as it does even if gas injection is stopped completely, but the pressure decline is much more gradual. Thus the recovery at 100 psia will now be 22.7 per cent of the pore space. Table 6 gives a comparison of the more important features of the calculated gross production histories under various degrees of gas return.

It will be seen from Table 6 that both for the continuous gas-return ratio of 0.8 and that which ended up at 0.8, after the initial rate of 100 per cent return, the calculated increases in recovery over the natural depletion value ($r = 0$) will be at least 50 per cent. Of course, these increased recoveries can be gained only at the expense of handling large gas volumes. However, if such increases in recovery as are indicated by Table 6 and Fig. 10.18 should actually be obtainable under practical operating conditions, there is little doubt that the operations would be decidedly profitable. On the other hand it should be clearly understood that the above results in no sense warrant any universal implications regarding the additional recovery—if any—to be expected from gas injection in any particular reservoir. The only general significance of Table 6 and Fig. 10.18 lies in the *relative* results to be expected from various degrees of gas return under

---

[1] While the practical units of cubic feet per barrel have been used here for plotting the gas-oil-ratio curves, they must be expressed in common volume units as barrel per barrel when used in Eq. (2).

[2] If the gas is injected at a ratio r with respect to the oil produced, Eq. (2) will remain valid if rR is replaced by r.

TABLE 6.—THE CALCULATED EFFECTS OF GAS RETURN ON RECOVERIES AND GAS-OIL RATIOS IN GAS-DRIVE RESERVOIRS

Fraction of gas returned, r	Stock-tank recovery, % pore space		Stock-tank recovery, % initial oil		Free-gas saturation, %		Total gas produced, ft³/bbl, initial oil	Gas injected, ft³/bbl, recovery	Av. GOR, ft³/bbl, recovery	Max. GOR, ft³/bbl
	To atmospheric	To 100 psia	To atmospheric	To 100 psia	At atmospheric	At 100 psi				
0............	14.5	13.8	27.1	25.8	28.7	27.7	534 (to atmospheric)	0	1,970 (to atmospheric)	4,400
0.6............	18.5	17.8	34.6	33.3	33.0	32.0	1,335 (to atmospheric)	2,315 (to atmospheric)	3,858 (to atmospheric)	10,350
0.8............	21.6	20.7	40.4	38.7	36.3	35.2	2,670 (to atmospheric)	5,287 (to atmospheric)	6,609 (to atmospheric)	19,500
1.0 to a GOR of 20,000 ft³/bbl	18.2 (1,335 psi)		33.9 (1,335 psi)		28.7 (1,335 psi)		1,543 (to 1,335 psi)	4,553 (to 1,335 psi)	4,553 (to 1,335 psi)	20,000
0 to depletion..	19.6	19.3	36.6	36.1	34.1	33.6	1,994 (to atmospheric)	3,989 (to atmospheric)	5,447 (to atmospheric)	22,200
0.8 to 100 psi...	.....	22.7	.....	42.4	.....	37.2	3,897 (to 100 psi)	8,081 (to 100 psi)	9,191 (to 100 psi)	33,100
0.8 with H = 0.5	.....	25.1	.....	46.9	.....	39.8	6,446 (to 100 psi).	10,995 (to 100 psi)	13,744 (to 100 psi)	55,000

substantially ideal conditions where the reservoir is inherently suited to gas-injection operations. The absolute values are only of illustrative significance.

As an indication of the effect of an initial gas cap on the performance to be expected from gas-return operations, the special case where $H = 0.5$, with an 80 per cent gas return, is also plotted in Fig. 10.18. It will be seen that, even though no account has been taken of possible assistance from gravity drainage, the recovery, as a percentage of the initial oil in place, is increased by 8.2 per cent over that for $r = 0.8$ without an initial gas cap, or 21.2 per cent of the absolute value of the latter. On the other hand, the maximum gas-oil ratio will be almost three times as great as in the latter case, and the average gas-oil ratio will be more than twice as great.

The average gas-oil ratios listed in Table 6 were obtained by simply integrating the gas-oil-ratio curves of Fig. 10.18, except for the first three values listed. For the latter the averages were calculated by the formula

$$\overline{R} = \frac{S_i/(1 - r)}{\text{recovery}}, \tag{3}$$

where S_i is the solution ratio (534 ft³/bbl) and the recovery is the fraction of the initial oil, as listed in the fourth column. In this equation the gas left in the reservoir at atmospheric pressure is neglected. The values of total gas produced per barrel of initial oil are simply the average gas-oil ratios multiplied by the recoveries per barrel of initial oil. For the cases referring to a constant value of r the gas injected per barrel of recovery is given by r times the gas produced per barrel of recovery, *i.e.*, the average GOR. For those where r is not constant throughout the history the injected gas may be calculated as the sum of the produced gas and residual reservoir-gas content, minus the original solution gas.

The special case where all the produced gas is returned ($r = 1$) can be treated by formal integration of Eq. (2) or even more directly by applying the integrated form of the material-balance equation, to give the relation

$$Q(\gamma\beta - S) = [S_i - S - \gamma(\beta_i - \beta)]\left(\frac{\rho_o}{\beta}\right)_i + H[(\gamma_i - \gamma)(1 - \rho_w) - (\gamma_i\beta_i - \gamma\beta + S - S_i)\overline{\rho_{oi}}], \tag{4}$$

where Q is the recovery expressed as a fraction of the pore space. Thus the relation between the pressure and recovery will be independent of the permeability-saturation function $\Psi(\rho_o)$ and can be calculated by the algebraic equation (4). However, the producing gas-oil ratios will still be governed by Eq. 10.5(3). If there is no initial gas cap, Eq. (4) can be expressed as

$$\overline{Q} = 1 - \frac{\gamma\beta_i - S_i}{\gamma\beta - S}, \tag{5}$$

where $\overline{Q}$ is the recovery expressed as a fraction of the initial oil in place. Hence, if no regard is given to the gas-oil ratio, all the initial oil would be recovered by the time the pressure has fallen so that $\gamma = S_i/\beta_i$. However, since the oil permeability will become vanishingly small long before its saturation is reduced to zero, the gas-oil ratio will become infinitely large before the above limit of complete recovery could be reached.

None of the numerical examples discussed above, including that for $r = 1$, provided for complete pressure maintenance. In all the cases $(r > 0)$ the pressure decline was retarded rather than fully arrested. The injection rate required to effect a complete pressure stabilization, without the aid of water drive, is readily deduced from Eq. (2), namely,

$$1 + \frac{\mu_o}{\mu_g}\left(\Psi - \frac{rR}{\alpha}\right) = 0. \tag{6}$$

From this it follows that the rate of gas injection, Q_{gi}, must be

$$Q_{gi} = \gamma\left[\frac{(R - S)Q_o}{\gamma} + \beta Q_o\right] = Q_g + Q_o(\gamma\beta - S), \tag{7}$$

where Q_g is the rate of gas production.

The factor of γ is the current space voidage created by the oil and gas withdrawals. As is to be expected, Eq. (7) simply implies that the reservoir volume of the injected gas must equal the space voidage of the gas and oil production. In terms of the injection ratio $\bar{r}$, Eq. (7) leads to

$$\bar{r} = R + \gamma\beta - S. \tag{8}$$

In practice, however, complete pressure maintenance is seldom attempted, except when the operations are supported by a partial water drive. Plant processing of the produced gas and fuel requirements generally reduces the gas available for return to the reservoir to 70 to 85 per cent of that produced.

It is of interest to note from Eqs. (7) and (8) that, if $\gamma\beta < S$, complete pressure maintenance could be achieved even if less gas be injected than produced. This condition may be satisfied at low pressures. Its physical meaning is that the gas in solution in the oil occupies, through the medium of the oil phase, less volume than the same amount of gas would occupy as a free-gas phase at the same temperature and pressure.

10.8. The Effect of the Initial Conditions on the Effectiveness of Gas Injection.—The considerations of the preceding section all referred to situations in which the gas injection is undertaken at the very beginning of production. An important practical question pertains to the effect of delays in initiating the gas-return operations. For it is generally not feasible fully to evaluate the reservoir conditions and the production mechanism so as to establish the desirability or necessity of pressure maintenance until there has developed a substantial pressure decline and reservoir depletion. Theoretically this question can be answered by making comparative calculations using Eq. 10.3(1) or 10.5(4) to various pressures at which the gas injection may be undertaken and then applying Eq. 10.7(2), with the chosen r.* The initial conditions for the latter will be the final values of pressure and saturation at which the integration of Eq. 10.3(1) or 10.5(4) [or Eq. 10.7(2) with $r = 0$] is terminated. The

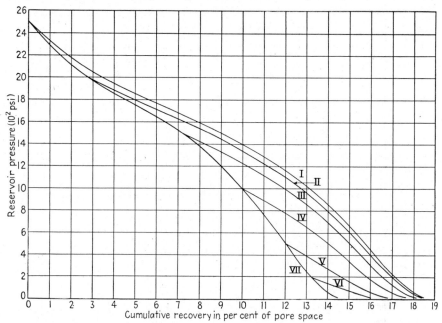

FIG. 10.19. The calculated pressure-decline histories of hypothetical gas-drive reservoirs as affected by the injection of 60 per cent of the produced gas at various stages of natural depletion. Reservoir pressure at start of injection = 2,500 (initial), 2,000, 1,500, 1,000, 500, and 200 psi for curves I, II, III, IV, V, and VI, respectively. Curve VII, no gas injection. Basic reservoir data are those used for Fig. 10.3.

results of such calculations are plotted in Figs. 10.19 to 10.22, for both 60 and 80 per cent gas-return programs. No limitation was placed on the maximum producing gas-oil ratios. It will be seen that theoretically there will always be a retardation in pressure decline as soon as gas injection is begun for any starting pressure. The initial gas-oil-ratio reaction will depend on the point in the production history when the gas injection is started, although up to a cumulative production of 11 per cent, for the systems considered, the gas-oil ratios differ so little for the various curves that they could not conveniently be shown on the scale used for the plots of Figs. 10.21 and 10.22. In all cases, however, if the gas injection is undertaken before the normal depletion gas-oil-ratio maximum has been

* A similar study based on the material-balance method (cf. Sec. 10.9), with equivalent results, has been reported by Tarner, *loc. cit.*

reached, the gas-oil ratios build up to still greater maxima, which increase as the starting pressure increases.

The total ultimate recoveries and free-gas saturations are plotted vs. the starting pressure for gas injection in Fig. 10.23. As expected, they all increase continuously as the gas injection is begun at increasing reservoir

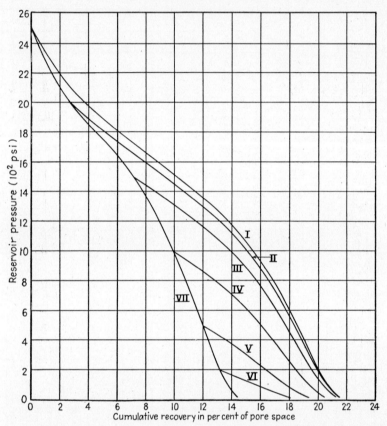

Fig. 10.20. The calculated pressure-decline histories completely analogous to those of Fig. 10.19, except that 80 per cent of the produced gas is returned to the producing formation.

pressures. It should be noted, however, that the curves rapidly flatten for starting pressures exceeding 500 psi. Thus by delaying an 80 per cent gas-return project, in the type of system under consideration, until the reservoir pressure has fallen to 1,000 psi, the loss in physical ultimate recovery will be, theoretically, only 1.1 per cent, in per cent of the pore space, and 0.9 per cent in the case of a 60 per cent gas-return program.

The implications of Figs. 10.19 to 10.23 are still further crystallized in Tables 7 and 8 listing both the recoveries and gas-injection volumes associated with the various gas-injection programs. Some of the data listed

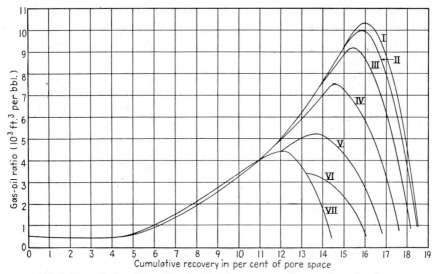

Fig. 10.21. The calculated gas-oil-ratio histories corresponding to those for the pressure decline of Fig. 10.19.

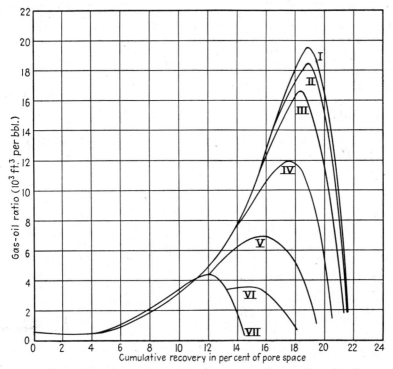

Fig. 10.22. The calculated gas-oil-ratio histories corresponding to those for the pressure decline of Fig. 10.20.

TABLE 7.—THE CALCULATED EFFECTS OF STARTING PRESSURE ON RECOVERIES AND GAS-OIL RATIOS FOR 60 PER CENT GAS-RETURN OPERATIONS

Gas injection started at psi	Stock-tank recovery, % pore space		Stock-tank recovery, % initial oil		Free-gas saturation		Total gas produced, ft³/bbl initial oil, to atmospheric	Gas injected, ft³/bbl recovery, to atmospheric	Gas injected, ft³/bbl initial oil, to atmospheric	Av. GOR, ft³/bbl recovery, to atmospheric	Max. GOR, ft³/bbl
	To atmospheric	To 100 psia	To atmospheric	To 100 psia	At atmospheric	At 100 psia					
2,500	18.5	17.8	34.6	33.3	33.0	32.0	1,335	2,315	801	3,858	10,350
2,000	18.4	17.6	34.4	33.0	32.9	31.8	1,238	2,047	704	3,599	10,000
1,500	18.2	17.3	33.9	32.3	32.6	31.5	1,106	1,687	572	3,263	9,200
1,000	17.6	16.7	32.9	31.2	32.1	30.8	966	1,313	432	2,935	7,500
500	16.8	15.7	31.4	29.3	31.2	29.7	739	653	205	2,353	5,200
200	16.0	14.6	29.9	27.3	30.4	28.5	590	187	56	1,974	4,400
None	14.5	13.8	27.1	25.8	28.7	27.7	534	0	0	1,970	4,400

TABLE 8.—THE CALCULATED EFFECTS OF STARTING PRESSURE ON RECOVERIES AND GAS-OIL RATIOS FOR 80 PER CENT GAS-RETURN OPERATIONS

Gas injection started at psi	Stock-tank recovery, % pore space		Stock-tank recovery, % initial oil		Free-gas saturation		Total gas produced, ft³/bbl initial oil, to atmospheric	Gas injected, ft³/bbl recovery, to atmospheric	Gas injected, ft³/bbl initial oil, to atmospheric	Av. GOR, ft³/bbl recovery, to atmospheric	Max. GOR, ft³/bbl
	To atmospheric	To 100 psia	To atmospheric	To 100 psia	At atmospheric	At 100 psia					
2,500	21.6	20.7	40.4	38.7	36.3	35.2	2,670	5,287	2,136	6,609	19,500
2,000	21.5	20.7	40.2	38.7	36.2	35.1	2,465	4,803	1,931	6,131	18,500
1,500	21.3	20.4	39.8	38.1	36.0	34.7	2,246	4,302	1,712	5,643	16,600
1,000	20.5	19.4	38.3	36.2	35.1	33.7	1,751	3,178	1,217	4,573	11,900
500	19.4	17.8	36.2	33.3	34.0	32.0	1,161	1,732	627	3,206	6,900
200	18.0	15.8	33.6	29.5	32.5	29.8	734	595	200	2,186	4,400
None	14.5	13.8	27.1	25.8	28.7	27.7	534	0	0	1,970	4,400

in these tables are plotted in Fig. 10.24, showing the increase in recovery, per unit volume of initial stock-tank oil, vs. the gas used in excess over the solution gas. In addition to the data taken directly from Tables 7 and 8, those of Table 5, in which the available gas-cap gas is represented as added gas, are also plotted in Fig. 10.24. As the viscosities of the oil phase assumed in the calculations giving the data of Table 5 were uniformly twice those used in the calculations underlying Tables 6 to 8, the numerical

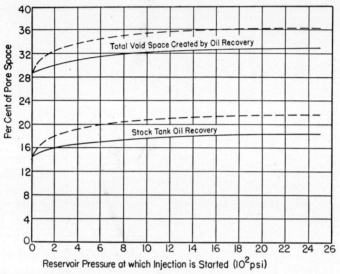

Fig. 10.23. The calculated variation of the ultimate recoveries and free-gas saturations (at atmospheric pressure) of hypothetical gas-drive reservoirs subjected to gas injection, as functions of the reservoir pressure at which the gas injection is started. Solid curves, 60 per cent gas return. Dashed curves, 80 per cent gas return.

results are not quite comparable. The somewhat lower efficiency of the additional gas available is therefore to be expected.

The difference between curves I and II indicates that for the same amount of additional gas supplied to the reservoir a greater recovery increase is obtained at the greater fractional return of the produced gas. As may be verified by reference to Tables 7 and 8, equal volumes of total added gas at the two different rates of return imply that the gas is injected at lower pressures for the greater rate of injection. The increased recovery under the latter conditions is evidently due to the greater reservoir volume throughput of the injected gas, in spite of the greater oil viscosity during the injection period. It thus appears that under the conditions assumed in the calculations the reservoir throughput volume is a more important factor in determining the gain in oil recovery than the average oil viscosity during the operations.

It is clear from Fig. 10.24, as well as Figs. 10.19 and 10.20, that the recovery increases most rapidly as the gas available for production is first increased beyond the solution value. The provision of additional gas will continue to increase the recovery, but at lesser efficiency. Hence, while commencement of gas return at the very beginning of production will give the maximum recovery, a delay in gas injection even until the pressure has suffered an appreciable decline need not be considered as disastrous.

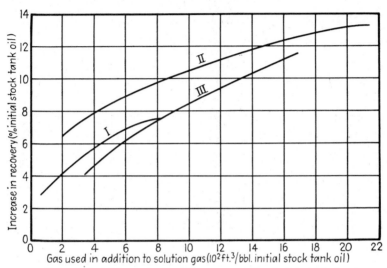

Fig. 10.24. The calculated increases in ultimate recoveries (to atmospheric pressure) of hypothetical gas-drive reservoirs as functions of the gas used in production in excess of the solution gas. Curve I, 60 per cent gas-return operations at different starting pressures (Table 7). Curve II, 80 per cent gas-return operations at different starting pressures (Table 8). Curve III, normal depletion of gas-drive reservoirs with different sizes of gas caps (Table 5).

By application of Fig. 10.24 or its equivalent an economic balance can readily be made between the increase in recovery and cost in handling the gas as the gas-injection operations are started at different pressures or at different rates of return. Without entering into such economic evaluations it may be anticipated that the economic optimum conditions for gas injection, without gravity drainage, will often correspond to less than 100 per cent gas return, and at a starting pressure lower than the initial reservoir pressure.

As in the case of the numerical data on gas injection presented in Sec. 10.7 it should be understood that the absolute values given above do not apply directly to specific reservoirs as they may occur in practice. Aside from the many assumptions regarding the physical data underlying these illustrative calculations, no account has been taken of the effect that gravity drainage may contribute to the producing mechanism. Moreover in all but one of the cases treated in this and the preceding section it was assumed that there was no initial gas cap overlying the oil zone, whereas most primary gas-injection operations are undertaken in reservoirs overlain by gas caps. Only the relative and comparative implications of the theoretical developments are of direct significance. On the other hand, in evaluating specific gas-injection programs, such factors as the effect of the operations in prolonging the flowing life of the producing wells and reducing operating costs, in the maintenance of producing well capacity, the availability of a market for the produced gas or legal restrictions on the production of gas without return must be taken into account.

While the probability of increasing oil recoveries by gas injection seems intuitively obvious, and the magnitude of such increases under idealized conditions can be formally computed by the method outlined in the preceding section, the modified reservoir performance under gas-injection operations is the resultant of distinct physical effects. The maintenance of pressure or retardation of the pressure decline will delay the evolution of the dissolved gas. This will tend to maintain a low reservoir oil viscosity, which will increase the sweep efficiency of the gas phase. Also of importance is the simultaneous effect of retarding the shrinkage of the reservoir oil. This will serve to minimize the decline in oil permeability and add to the gain owing to the lower oil viscosity. While the reduction in shrinkage implies also that the stock-tank equivalent of the residual reservoir oil will be lower at a given free-gas saturation, the gain due to this effect alone is difficult to evaluate. For if the economic limit on the continuance of gas injection should be determined by the gas-oil ratio, the higher pressures associated with an "inflated" state of the reservoir oil will involve greater gas-oil ratios, at a given free-gas saturation, and might lead to earlier abandonment than lower pressure operations where the oil phase is less expanded. Only if it is economically feasible to continue the operations to very high gas-oil ratios will it be possible to take full advantage of the oil expansion itself as a direct factor in materially reducing the unrecovered residual oil.

It is to be noted, too, that the very process of preventing the reservoir oil shrinkage by pressure maintenance implies that the energy of the solution gas is not used directly to move the oil to the well bore. Only that which escapes within the reservoir owing to the pressure decline which does occur contributes to the oil expulsion. The solution gas aids the lifting of the oil up the flow string and may subsequently do useful work or provide energy as a fuel after arriving at the wellhead. But the major burden of oil expulsion is carried by the injected gas. The apparent inefficiency in "conserving" and not using fully the solution-gas energy is more than counterbalanced by the increased oil expulsion and sweeping efficiency of the injected gas when applied to the low-viscosity "inflated" oil containing a maximum of solution gas. On the other hand the current local liquid-expulsion efficiency of the gas at a given gas-phase saturation is independent[1] of the manner in which this was developed. Hence the sweeping efficiency of the injected gas declines with increasing oil recovery in essentially the same manner as would that of the solution gas under the solution-gas drive.

10.9. The Material-balance Method of Calculating Gas-drive Reservoir Histories.[2]—The methods developed in the preceding sections for treating gas-drive reservoirs have all been based on the integration of first-order differential equations. These have referred to the behavior of a differential element of the reservoir, considered as isolated from but representative of the reservoir as a whole. While the equations are nonlinear and cannot be explicitly integrated in analytic form, except in very special cases [cf. Eq. 10.7(4)], they are subject to integration by well-established numerical procedures. Moreover the very process of numerical integration gives an illuminating and instructive insight into the role played by various fluid and rock properties in determining the pressure-production relationship. On the other hand the preparation of the functions α, λ, ϵ, and η, which appear in all the equations, is a rather laborious process. Since the integration itself is time consuming and requires extensive numerical manipulation, a brief outline will be given of an alternative procedure that involves only algebraic equations.

It will be recalled from the discussion of the material-balance equation in Sec. 9.5 that it is essentially a relationship between the oil production, gas production, and pressure, which may be expressed in the case of strict solution-gas-drive reservoirs as

$$Q_g = (\gamma_i - \gamma)V_{gi} + L[S_i - S - \gamma(\beta_i - \beta)] - Q_o(\gamma\beta - S), \qquad (1)$$

where Q_g is the cumulative gas production, Q_o the cumulative oil production, V_{gi} the initial free-gas volume, and L the original stock-tank-oil content. While in the applications of Eq. (1), discussed in Secs. 9.5 and 9.6,

---

[1] It is assumed that the permeability-saturation curves are functions only of the saturations and are free of hysteretic effects.

[2] The use of the material-balance equation in its integrated form was first reported by H. H. Kaveler, *AIME Trans.*, **155**, 58 (1944), as part of a study of pressure maintenance in the Schuler field. A qualitative outline of the method was also given by Babson, *op. cit.*, p. 120. And a detailed discussion of the numerical procedure with illustrative applications to gas-injection systems has been presented by Tarner, *loc. cit.* While the method is referred to in these publications as a trial-and-error procedure, the actual numerical steps involved are essentially the same as presented here.

Q_g, Q_o, V_{gi}, and L were considered as referring to the reservoir as a whole, there is no loss in generality in expressing them in units of a pore volume representing a typical, or average, sample of the reservoir. Thus,

$$Q_o = \left(\frac{\rho_o}{\beta}\right)_i - \frac{\rho_o}{\beta}; \qquad V_{gi} = H(1 - \rho_w - \beta_i\overline{\rho_{oi}}); \qquad L = \left(\frac{\rho_o}{\beta}\right)_i + H\overline{\rho_{oi}}, \qquad (2)$$

in the notation of Eq. 10.5(4). The pressure-recovery relationship can then be derived as follows. Upon assuming L and V_{gi} as known a pressure is chosen at a suitable increment below the initial value. If this pressure is considered as fixed, Eq. (1) becomes a linear relation between Q_g and Q_o or ρ, which is readily plotted. In terms of the saturation variables this equation may be written as

$$Q_g = H[(\gamma_i - \gamma)(1 - \rho_w) + (\zeta - \zeta_i)\overline{\rho_{oi}}] + (S_i - \gamma\beta_i)\left(\frac{\rho_o}{\beta}\right)_i + \frac{\zeta\rho_o}{\beta}, \qquad (3)$$

where $\zeta = \gamma\beta - S$. The permeability-saturation relation of the formation of interest is now applied in the form

$$Q_g = (S + \alpha\Psi)Q_o, \qquad (4)$$

where Ψ is the gas-to-oil-permeability ratio, considered as a function of ρ_o. This equation is plotted vs. ρ_o for the fixed pressure (fixed S and α),[*] taking into account the variation of Ψ with ρ_o. The intersections of the plots of Eqs. (3) and (4) will give the values of ρ_o and Q_g corresponding to the assumed pressure. The equivalent value of Q_o is calculated by Eq. (2).

The above procedure is then repeated after reducing the assumed pressure by an additional increment. While in Eq. (3) Q_g will refer to the cumulative gas withdrawals, only the increments from the previous values are to be used in Eq. (4). In this way a stepwise determination can be made of the relation between the gas and oil productions and the reservoir pressure. If gas is being injected into the reservoir, Q_g in Eqs. (1) and (3) will refer to the *net* gas withdrawal, that is, $(1 - r)Q_g$, where r is the injection ratio. However, in Eq. (4), Q_g will still represent the actual incremental gas production.

In this method, as well as that in which the differential equation is used, no account is taken of fluid segregation beyond that which may have given rise to the original gas cap. It is tacitly assumed that both the original gas in the gas cap and that injected are produced only by virtue of dissemination and movement through the oil zone. Moreover, to the extent that the differential equation used in this chapter must be solved numerically by stepwise procedures, it is basically equivalent to the above-outlined algebraic material-balance equation combined with Eq. (4). In fact

---

[*] Since the gas-oil ratio will vary during the interval, S and α, as well as Ψ, in Eq. (4) should refer to the mean values over the pressure and saturation interval.

the differential equations are nothing more than the result of formally applying incremental operations to Eq. (3) and eliminating the Q_g term by means of Eq. (4). As is to be expected, the treatment of specific systems by both the differential-equation and the material-balance method shows them to give the same results, within the errors inherent in the numerical processes involved.

10.10. Practical Aspects of Field Data Pertaining to Gas-drive Reservoir Performance.—Strictly speaking there is no conclusive evidence disproving or quantitatively confirming the theoretical developments of this chapter pertaining to gas-drive reservoirs. And it is unfortunately rather unlikely that such evidence will ever be fully established. Against examples of apparent failure of the theoretical predictions, claims could probably be made, justifiably, that the ideal conditions assumed in the theoretical analysis were not satisfied in the cited examples or that the physical reservoir data assumed to be known were not quantitatively trustworthy. On the other hand, "proof" of agreement between theory and field observation in particular instances might be questioned as being accidental, since it is extremely difficult to defend the absolute correctness of any quantitative evaluations of such reservoir data as effective pay thickness, true average permeability, the permeability-saturation relationship, the porosity, the connate-water saturation, etc.

The theory of gas-drive reservoir performance, as outlined in the preceding sections, was presented for two reasons. First it is not contradicted by field experience to the extent that its basic underlying assumptions are approximately satisfied in actual oil reservoirs. Second it represents a physically reasonable development from direct laboratory observations and general physical considerations. The implications of the assumptions and approximations made in the analytical treatment can be estimated, at least qualitatively. It provides a definite structure for the interpretation and quantitative evaluation—within its proper range of validity—of specific gas-drive reservoirs. Yet a test of its ultimate significance must be essentially of a statistical nature.

The hundreds of oil fields that were discovered before 1930 and that are now depleted are virtually worthless as sources of data on actual reservoir performance. Literally the only item of physical significance that can be derived from the available records of the great majority of currently depleted fields is the curve of field production rate vs. the time. Gas production was seldom measured, except when metered to gasoline plants or sold to transmission systems. Reservoir pressures were not a matter of interest and were not systematically recorded until after 1930. The measurement of the fluid properties did not become routine until about 1935. The term permeability was hardly more than a semiquantitative concept before 1930. General core analysis and well logging were simply beyond the horizon during the exploitation of most of the fields that are now substantially depleted.

This situation is especially unfortunate in that the older fields could have provided a type of data which is very difficult to obtain now. They were generally produced "wide open," at maximum capacity, with no controls. Hence in most cases their performance undoubtedly simulated well the simplest type of solution-gas-drive behavior. Gas-cap segregation and gravity drainage during the production stage were minimized, and it is very likely that water intrusion was as a rule too slow to keep pace with the fluid withdrawal and play a significant role in the production mechanism, except as a scavenger as the field neared abandonment. If satisfactory data were available regarding these older fields, they would probably constitute material of statistical significance in establishing at least trends and gross features of gas-drive reservoir performance.

Much more is known about fields of more recent development. It is common practice now to core the producing formation, to measure reservoir pressures, and to determine the thermodynamic properties of the reservoir fluids. There are, however, basic limitations to the value, at present, of the accumulated data on many of the fields still in their primary-production phase. One results from the fact that, since the early thirties, in most of the major producing states, regulatory bodies have imposed "proration" controls upon the production rates of individual wells and fields within their respective state boundaries. These function, in general, on the principle that the total production from the state is fixed in advance, at regular intervals, and this total is subdivided or prorated among the fields in the state. These prorated field "allowables" have generally been far below the current production capacities of the fields during their early producing lives, although the production rates of individual wells are usually allowed fixed preassigned minima, as 20 bbl/day in one state, except when the well is "penalized" for excessive gas production. The net result is that the annual rates of depletion of fields subjected to proration have been much slower than of those produced without proration controls. Accordingly many of these fields have thus far undergone relatively small percentages of their ultimate depletion, and their production histories are correspondingly far from completed. To estimate their actual ultimate recoveries and future histories thus often involves very questionable extrapolations.

Closely related to the prolongation of the producing lives of fields operated under proration controls is the direct effect on the reservoir behavior due to the limitation of the rates of withdrawals. This very process encourages the transformation into partial or complete water drives of what would have otherwise been simple solution-gas-drive production. Fluid segregation and downstructure oil gravity drainage also contribute more effectively to the reservoir performance. Generally this is to be regarded as a favorable situation and is not to be lamented, from the point of view of recovery efficiency. However, it has served drastically to reduce the number of solution-gas-drive fields on which data are available for analysis and comparison with theoretical predictions.

The process of elimination of material for a statistical analysis of actual gas-drive reservoir performance is carried still further by another development of recent years. This is the growing tendency to undertake pressure maintenance by fluid injection in the early life of a field. This is especially common in the case of fields which appear to be developing no substantial water drive even under restricted withdrawals, *i.e.*, those which are inherently of the solution-gas-drive type. In fact, under modern practices, fields are seldom "left alone," unless there is early and strong evidence of the action of a complete water drive. Their complete "natural" depletion performance becomes a matter of estimation and, often, purely theoretical prediction, rather than actual observation. This, too, is undoubtedly in most cases to be viewed as evidence of efficient engineering practice and is to be encouraged. In fact it represents a deep faith in the soundness of those physical principles that are used to guide such developments. But it also provides an explanation for the lack of systematic and statistical data on the gas-drive behavior of actual producing reservoirs.

Finally it should be noted that, even when a reservoir has been subjected to intensive specific study, the data that it is feasible to gather are never as complete as desired. The complete coring and logging of every well in a field would still provide only an infinitesimal sampling—one part in millions—of the reservoir rock. As pointed out in Chap. 3, the core-analysis data themselves are often subject to considerable uncertainty in interpretation. Only recently has the determination of permeability-saturation characteristics of rocks shown promise of being developed into a routine and well-established laboratory technique. Actual producing formations are virtually never strictly uniform, homogeneous, and free of stratification. The problems of averaging thus created are in no sense completely solved. And even the gross performance data are often subject to serious question. Reservoir pressures are sometimes in doubt because of insufficient build-up after shutting in the well, and they always involve the problem of averaging. The commonly observed vagaries of individual-well pressure fluctuations attest the fact that significant field pressures are basically a statistical inference. Water-production data are frequently subject to considerable question. And even more serious is the potential uncertainty generally inherent in estimates or records of actual gas production and gas-oil ratios. Except when all the produced gas is delivered to a processing plant or transmission line, it is simply not practical, in general, to measure directly and continuously the gas produced by individual wells or fields as a whole. Usually, therefore, the gas produced is computed by multiplying the incremental oil production from individual wells by values of gas-oil ratios measured during periodic gas-oil-ratio surveys, or the total field production by a field average gas-oil ratio. The uniqueness of such specific data and their applicability to the actual producing intervals between the surveys are assumptions that must at best be evaluated as necessary evils. And even if these be accepted, account must be taken of the fact that the reported gas-oil ratios usually refer to measurements at separators, at 30 to 75 psi, and must be corrected for the additional gas liberation on reduction to atmospheric stock-tank conditions.

These basic limitations to the analysis of actual reservoir performance must be recognized. Their discussion may be considered as an apology for the unhappy lack of conclusive field-observation evidence with respect to many aspects of reservoir behavior. For this situation is the result of practical difficulties, often beyond the control of the operator and engineer, rather than a consequence of a lack of appreciation of the importance of the problem. Nevertheless the circumstances underlying these limitations and their implications must be clearly understood if progress is to be made in establishing reservoir-engineering analysis on a sound basis.

10.11. Field Data on the Production Decline in Gas-drive Reservoirs.— In the light of the above considerations no attempt will be made to demonstrate the quantitative validity of the theoretical treatment of gas-drive reservoirs presented in the preceding sections by a detailed analysis of the actual performance of specific fields. Only the broad features of such performance will be illustrated by reference to observed field data. Thus several typical production-decline curves are plotted in Fig. 10.25.[1] As all these fields were largely depleted before reservoir-engineering analysis became an established practice, very few other pertinent data were gathered during their production histories. Nevertheless it may be noted that they are qualitatively similar to the theoretically computed production-decline curves (cf. Sec. 10.6) plotted in Fig. 10.17.

The peaks in the curves of Fig. 10.25 are evidently due to the gradual drilling developments of the fields, as discussed in Sec. 10.6. Several ex-

[1] Curves (1), (2), and (3) are plotted from data in papers of P. F. Martyn, G. D. Thomas, and A. F. Crider, respectively, pp. 1184, 1473, and 1658 of *AAPG Bull.* **22** (1938); curves (4), (5), (6), and (7) are plotted from data in papers of J. R. Reeves, W. C. Spooner, G. E. Burton, and J. J. Bartram, respectively, pp. 160, 196, 290, and 577 of "Structure of Typical American Oil Fields," Vol. II, AAPG, 1929.

amples of the behavior of individual wells or small groups completed at
essentially the same time are given in Fig. 10.26.[1] These show the broad
monotonic declines corresponding qualitatively to curve I of Fig. 10.17.

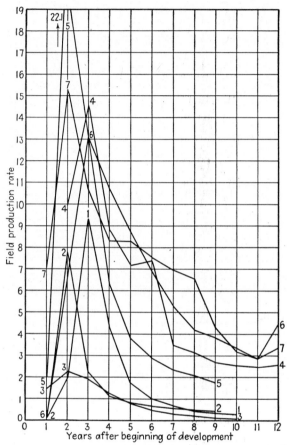

Fig. 10.25. Typical production histories of gas-drive reservoirs. (1) Refugio, Tex. (2) Carterville, La. (3) Bellevue, La. (4) El Dorado, Kan. (5) Homer, La. (6) Hewitt, Okla. (7) Elk
Basin, Wyo. Ordinate-scale units = 10^6bbl/year for (1), (3), (5), (6); 10^5bbl/year for (2),
(7); 2×10^6bbl/year for (4). Year of development = 1928, 1929, 1922, 1917, 1919, 1919,
and 1916 for (1), (2), (3), (4), (5), (6), and (7).

On the other hand, the irregularities in the curves of both Fig. 10.25 and
Fig. 10.26 should serve to emphasize that under practical operating conditions it is extremely seldom that individual wells or reservoirs as a whole
experience idealized and unperturbed histories. Changes in operating con-

[1] Curve (1) is plotted from data of J. L. Darnell "Manual for Oil and Gas Industry,"
1920; curves (2) and (3) are plotted from data of R. K. DeFord, p. 75, "Structure of
Typical American Oil Fields," Vol. II, AAPG, 1929.

ditions due to well repair, installation of artificial lift, variations in market demand, the development of parts of a common reservoir having different pay thickness or permeability, the redistribution of withdrawals over a

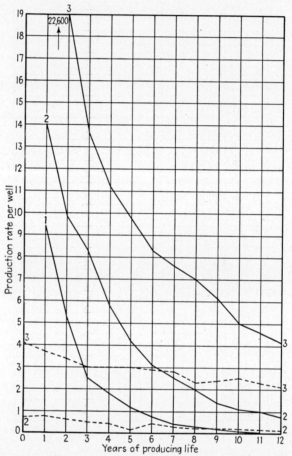

FIG. 10.26. Typical production histories of individual wells or small groups of wells drilled at the same time, producing from gas-drive reservoirs. (1) A property in Oklahoma having 5 wells the first year (1906) and 6 wells during remaining life. (2) Well #85, Florence field, Colorado. (3) Well #86, Florence field, Colorado. For both (2) and (3) initial year = 1896; lower dashed curves are continuations beyond 1907.

field, and well abandonment—all may lead to fluctuations from an "ideal trend," entirely aside from such gross operational factors as proration, gas-oil-ratio limitations, gas or water injection into the reservoir, or the institution of unitized field operation. Too great significance therefore should not be attached to individual curves or sets of data unless the role played by extraneous perturbing factors can be evaluated. In many cases only a statistical interpretation of the reservoir data will be warranted.

From a practical point of view it is often necessary to estimate the future performance and production histories of individual wells, groups of wells, or whole reservoirs, for long-term planning of the operations or for evaluation for sale, purchase, or tax purposes. This is generally done by extrapolation of past-performance observations. Of course, if the field should be subject to proration restrictions the future course of which is in itself uncertain, little can be done in predicting the future "decline" behavior. Moreover, if a field has been under proration control from the very beginning, its production while under such control does not represent a normal decline behavior and the whole extrapolation process becomes virtually meaningless. For completeness, however, the procedures that have been used in unprorated fields will be briefly outlined.

While production-decline curves are generally extrapolated by graphical manipulation, the more common methods may be related to analytical representations of the decline curves.[1] The simplest of these is the so-called "exponential," "geometric," "semilog," or "constant-percentage" decline, which implicitly or directly is based on the assumption that the production rate Q may be expressed by

$$Q = Q_o e^{-t/a}, \tag{1}$$

where t is the time (in days, months, or years), Q_o the initial rate, and a a constant. Equation (1) simply implies that a semilogarithmic plot of Q vs. t will be linear. In an equivalent form of Eq. (1), namely,

$$\frac{Q}{dQ/dt} = -a, \tag{2}$$

the analytical basis for the "loss-ratio" interpretation[2] of the exponential-decline data becomes evident. In this procedure the ratios of current production rates to the declines (ΔQ) over fixed time intervals are tabulated; if these are constant, the exponential decline is established as a basis for future extrapolation. Table 9 gives an example[3] of a case where this procedure evidently is quite satisfactory.

Another immediate consequence of Eq. (1) is

$$P = \int_o^t Q \, dt = a(Q_o - Q), \tag{3}$$

where P is the cumulative production. Hence a cartesian plot of cumulative production vs. the rate should be linear.

In view of Eq. (2) the constant a is evidently the reciprocal of the frac-

[1] A recent complete discussion of this problem is given by J. J. Arps, *AIME Trans.*, **160**, 228 (1945), who also includes an extended bibliography on the subject; cf. also S. J. Pirson, *Oil Weekly*, **122**, 45 (Sept. 9, 1946).

[2] Cf. R. H. Johnson and A. L. Bollens, *AIME Trans.*, **27**, 771 (1927).

[3] Arps, *loc. cit.*

tional decline in production rate per unit time. Thus in the case of the data of Table 9 the monthly decline would be, on the average, 1.15 per cent.

TABLE 9.—LOSS-RATIO TABULATION OF PRODUCTION-DECLINE DATA FOR A LEASE IN THE CUTBANK FIELD, MONTANA

Month	Year	Monthly production rate Q, bbl	Loss in production rate during 6 months' interval, ΔQ, bbl	Loss ratio (on monthly basis), $a = 6\dfrac{Q}{\Delta Q}$
July.......	1940	460		
January...	1941	431	-29	-89.2
July.......	1941	403	-28	-86.4
January...	1942	377	-26	-87.0
July.......	1942	352	-25	-84.5
January...	1943	330	-22	-90.0
July.......	1943	309	-21	-88.3
January...	1944	288	-21	-82.3

The "hyperbolic" or "log-log" production-decline curve, which has also been frequently used in numerical- or graphical-extrapolation procedures, may be defined by

$$\left.\begin{aligned}
\log \frac{Q}{Q_o} &= -\frac{1}{b}\log\left(1+\frac{bt}{a}\right), \\[2mm]
Q &= Q_o\left(1+\frac{bt}{a}\right)^{-1/b}.
\end{aligned}\right\} \tag{4}$$

or

By suitable adjustment of the constants a, b, a curve following Eq. (4) can be made linear on log-log paper.

The cumulative-production vs. production-rate curve defined by Eq. (4) can readily be shown to follow the equation

$$P = \frac{aQ_o^b}{1-b}(Q_o^{1-b} - Q^{1-b}), \tag{5}$$

which will give a linear log-log plot by suitable scale shifting. In the special case where $b = 1$, Eq. (5) breaks down, and its proper equivalent is

$$P = aQ_o\log\frac{Q_o}{Q}, \tag{6}$$

which will lead to a linear plot on semilog paper. The differential equation satisfied by Eq. (4), namely,

$$\frac{dQ}{dt} + \frac{Q}{a+bt} = 0, \tag{7}$$

implies that

$$\frac{d}{dt}\frac{Q}{dQ/dt} = -b, \tag{8}$$

which suggests that the first differences of the "loss ratios" should be constant, or that the loss ratio itself will vary linearly with the time. An example[1] where these conditions are approximately satisfied is given in Table 10, based on smoothed production data from a lease in Kansas

TABLE 10.—PRODUCTION-DECLINE LOSS RATIOS FOR A LEASE PRODUCING FROM THE ARBUCKLE LIMESTONE IN KANSAS

Month	Year	Monthly production rate Q, bbl	Loss in production rate during 6 months' interval, ΔQ, bbl	Loss ratio on monthly basis, $6\dfrac{Q}{\Delta Q}$	First Difference of Loss Ratio, $-b = \dfrac{1}{6}\Delta\dfrac{6Q}{\Delta Q}$
January..	1937	28,200			
July.....	1937	15,680	−12,520	− 7.51	
January..	1938	9,700	− 5,980	− 9.73	−0.37
July.....	1938	6,635	− 3,065	−12.99	−0.54
January..	1939	4,775	− 1,860	−15.40	−0.40
July.....	1939	3,628	− 1,147	−18.98	−0.60
January..	1940	2,850	− 778	−21.98	−0.50
July.....	1940	2,300	− 550	−25.09	−0.52
January..	1941	1,905	− 395	−28.94	−0.64
July.....	1941	1,610	− 295	−32.75	−0.64
January..	1942	1,365	− 245	−33.43	−0.11
July.....	1942	1,177	− 188	−37.56	−0.69
January..	1943	1,027	− 150	−41.08	−0.59

producing from the Arbuckle Limestone. The average value of b for these data is 0.51. A similar analysis[2] of 149 fields following this general type of decline behavior showed that in most cases b ranged between 0 and 0.4 and exceeded 0.7 in none of them.

It is evident that similar procedures can be devised for treating still more complicated analytical representations of the production-decline curves. And such have been developed for a variety of special cases corresponding to particular production-decline curves. Their general purpose is to provide nomographic charts or special forms of plotting the data so as to facilitate and remove the "personal equation" from the process of extrapolating past performance to give predictions of future behavior. However, as may be verified by a study of the decline curves plotted in Fig. 10.17, no simple analytical expression may be expected to describe the complete duration of the decline history. Such methods of treating production data as those discussed above must therefore be considered mainly as convenient empirical procedures, and no particular physical

[1] *Ibid.*

[2] W. W. Cutler, Jr., *U.S. Bur. Mines Bull.* 228 (1924).

significance should be attributed to the numerical values associated with
the various forms of the graphical or analytical representations.

**10.12. Field Observations on General Gas-drive Reservoir Perform-
ance.**—One of the few published complete sets of production records of
the performance of a gas-drive reservoir, unaffected by withdrawal restrictions, free gas cap, gas injection, or water-drive action, is that of the
upper Gloyd porous limestone reservoir of the Rodessa field extension in
Miller County, Ark. The pertinent data are plotted vs. time in Fig. 10.27[1]
and vs. the cumulative production in Fig. 10.28. The major part of the

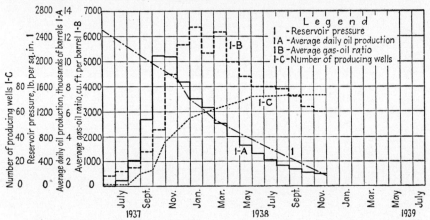

Fig. 10.27. The production and pressure history of the Upper Gloyd reservoir in the Miller
County, Ark., area of the Rodessa field. (*After Hill and Guthrie, U.S. Bur. Mines Rept.
Inv. 3715, 1943.*)

production history of this reservoir was completed in the 18 months' period
covered in Fig. 10.27, the pressure falling continuously from its initial
value of 2,500 psi in June, 1937, to 160 psi in December, 1938.

As the required data regarding the nature of the reservoir rock are not
available, no detailed analysis will be attempted for comparison with
theoretical predictions. It will be noted, however, that the gross features
of the field observations are at least qualitatively consistent with the
types of pressure, production-decline, and gas-oil-ratio curves calculated
theoretically in the earlier sections. Thus the gas-oil-ratio curves definitely
show a maximum falling in the latter half of the productive life of the
field, on a cumulative-recovery basis.[2] This is characteristic of all the

[1] This is reproduced from Fig. 43 of H. B. Hill and R. K. Guthrie, *U.S. Bur. Mines
Rept. Inv.* 3715 (August, 1943).

[2] Figure 10.44, giving the production history of the Wilcox Sand reservoir of the Oklahoma City field, also shows declining gas-oil ratios during the later stages of depletion.
This reservoir, however, has been subject to significant downflank gravity drainage,
as discussed in Sec. 10.17, which complicates the analysis of the details of its performance.

calculated gas-oil-ratio curves discussed in the preceding sections. The early rise in gas-oil ratio and approximately linear decline in reservoir pressure indicated in Fig. 10.28 are suggestive of curves II in Fig. 10.10, corresponding to a rock having no equilibrium gas saturation. The general form of the production-decline curve of Fig. 10.27 is qualitatively similar to curve II of Fig. 10.17 for an exponential type of drilling development. On

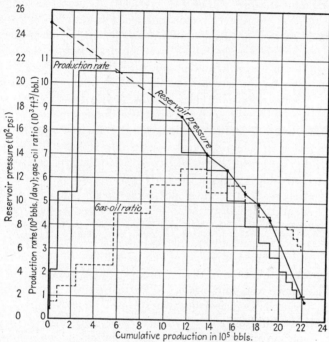

Fig. 10.28. The production and pressure history of the Upper Gloyd reservoir in the Miller County, Ark., area of the Rodessa field, plotted vs. the cumulative recovery.

the other hand such gross comparisons cannot be considered as quantitative "proof" of the theory. Such details as the dip in the gas-oil ratio during February, 1938, and the rather steep terminal decline in pressure indicated in Fig. 10.28, during a period when the gas-oil ratio was also falling, are not to be expected for an "ideal-behavior" reservoir. Very likely these apparently anomalous features can be explained in terms of certain changes or events in the operating history of the field that are not recorded graphically in Fig. 10.27. But they cannot be ignored in making quantitative interpretations of field data.

The examples to be given in the next section of the observed performance of pressure-maintenance operations will also serve to illustrate the early production histories of gas-drive reservoirs. These, as well as the case of the Rodessa field discussed above, are pertinent to the question of the occurrence of equilibrium gas saturations in actual oil-producing reservoirs. Virtually all published laboratory-determined permeability-saturation data, including the range of high liquid saturations, indicate the existence of an equilibrium free-gas saturation or at least a very slow build-up of the gas permeability (cf. Chap. 7). As seen in previous sections, a nonvanishing equilibrium gas saturation will lead theoretically to an initial decline in the gas-oil ratio. Yet no reservoir data published to date show initially declining gas-oil ratios that are free of ambiguity. It would appear, therefore, that either the samples which have been studied in the laboratory are not representative of actual producing reservoir rocks or that there is a serious discrepancy between the theoretical predictions and field observation. While many more data will be required before this situation can be clarified, it should be noted that the disagreement may be more apparent than real. Thus, as mentioned in the discussion of the material-balance equation in Chap. 9, gas-oil-ratio data, especially in the early producing life of a field, have been notoriously inaccurate or at least uncertain.[1] It is doubtful that the early gas-oil-ratio data on more than a small fraction of even recently developed fields can be considered as sufficiently accurate definitely to distinguish between a slight decline and slow rise in the gas-oil ratio. On the other hand, there are unpublished reports of several instances where declines in gas-oil ratio have been definitely observed, although these, too, may be subject to uncertainty.[2]

In interpreting this type of discrepancy it should be observed that the theory of gas-drive reservoirs presented in this chapter completely ignores the presence of wells as foci of fluid withdrawal. The region immediately about a well bore will evidently be further advanced in its local state of depletion than the reservoir as a whole. The actual producing gas-oil ratio should therefore reflect an oil saturation that is some type of average between the low values near the well bore and the higher saturations at remote points. It may well be that the slow building up of an equilibrium gas saturation in the interior of the producing formation may be masked by the rapid fall in saturation in the immediate vicinity of the well bore.

Aside from the general implication regarding the presence or absence of an equilibrium gas saturation, field data on gas-oil ratios can be interpreted

[1] For example, the initial decline in gas-oil ratio shown in Fig. 10.37 is probably in error, since the oil in the Canal field had initially a bubble point of only 2800 psi.

[2] A different type of evidence of the occurrence of nonwetting-phase equilibrium-saturation phenomena in actual reservoirs is provided by observations of the rise of the gas-oil ratios of condensate reservoirs being produced by pressure depletion (cf. Sec. 13.9). Such performance definitely implies that the condensate precipitated in the sand by retrograde condensation remains immobile at low saturations because of its discontinuous globular distribution.

to give the variation of the gas to oil permeability ratio as a function of the liquid saturation. For from Eq. 10.3(1) it follows that

$$\Psi(\rho_o) = \frac{k_g}{k_o} = \frac{R - S}{\alpha}. \tag{1}$$

Hence, if the pressure functions $S(p)$ and $\alpha(p)$ are known, k_g/k_o can be calculated from combined observations of the reservoir pressure and gas-oil

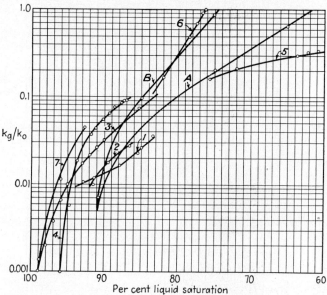

Fig. 10.29. The variation of the permeability ratio with the liquid saturation, as determined from field observations. Curve 1, West Texas San Andres Dolomite. Curve 2, Panhandle Dolomite. Curve 3, West Texas San Andres Dolomite. Curve 4, Louisiana Petit Lime. Curve 5, Oklahoma Wilcox Sand. Curve 6, Oklahoma Simpson Bromide Sand. Curve 7, Oklahoma Hunton Lime. Curve A, Unconsolidated sand. Curve B, Consolidated sand. Curves 1–7 represent field data, and A and B laboratory experiments. (*After Elkins, API Drilling and Production Practice, 1943.*)

ratio. If, in addition, the original pore volume and connate-water saturation have been determined, the residual average oil or total liquid saturation can be computed from the cumulative oil withdrawal to correspond to the value of k_g/k_o given by Eq. (1).

In Fig. 10.29[1] are plotted k_g/k_o vs. liquid-saturation curves for three dolomites, two limestones, and two sands, as determined in the above manner. Two curves for sands established by laboratory measurements are included for comparison. And in Fig. 10.30 are plotted similar composite and averaged data from West Texas dolomite fields,[2] together with

[1] This figure is taken from L. E. Elkins, *API Drilling and Production Practice*, 1946, p. 160.

[2] Figure 10.30 is taken from E. C. Patton, Jr., *AIME Trans.*, **170**, 112 (1947).

laboratory-determined curves on dolomite samples. In contrast to the curves obtained from laboratory experiments[1] those derived from field data all appear to leave no room for an equilibrium gas saturation that could exceed a few per cent at the most. This is to be expected, in view of the

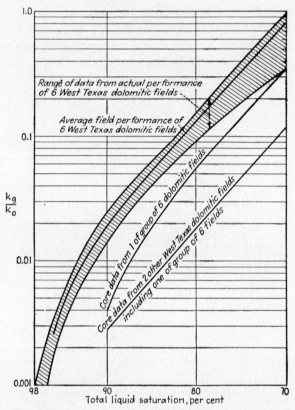

Fig. 10.30. A comparison of permeability-ratio curves as determined from dolomite reservoir performance and from laboratory experiments on core samples. (*After Patton, AIME Trans., 1947.*)

general field observation that the gas-oil ratios begin to rise as soon as or very shortly after production develops, as previously discussed. On the other hand it is to be noted again that the gas-oil-ratio data at high liquid saturations, even when reported and available, may be subject to considerable uncertainty, as is evidently the case with respect to curve 1 in Fig. 10.29.

Most of the data plotted in Figs. 10.29 and 10.30 are of too limited range definitely to establish analytic representations for the curves. However,

[1] Cf. Chap. 7.

they appear to approach approximately a semilogarithmic variation with the total liquid saturation, at values of the latter less than 80 per cent.[1] This relation can be expressed as

$$\log \frac{k_g}{k_o} = a - b\rho,$$ (2)

where ρ is the total liquid saturation. Except for the Wilcox Sand of the Oklahoma City field the values of a for the field curves lie in the range of 10 to 23, and b from 15 to 30. The laboratory data for the consolidated sand and the upper group of dolomites in Fig. 10.30 fall in these ranges, whereas those for the unconsolidated sands both fall below the indicated lower limits, and the lower laboratory-data line of Fig. 10.30 gives a value of a less than 10. For the Wilcox Sand $a \sim 1.2$ and $b \sim 3.6$.

While these empirically established permeability-ratio data are of interest in describing field observations, it should be understood that they represent the composite resultant of all those factors of actual reservoir behavior which do not obtain in the ideal systems on which theoretical calculations are generally based. Perhaps the most important of these is the fact that oil-producing reservoirs are never strictly homogeneous porous media. Even in massive limestone or dolomite formations the permeability variations often serve to create a substantial stratification, from the dynamical point of view, similar to that found in sandstone reservoirs. Accordingly the depletion will not be uniform throughout the producing section, and the oil saturations will be lower in the more permeable zones. As the basic dynamics of the solution-gas-drive oil-expulsion mechanism is nonlinear, the average gas-oil ratios for the composite system will be greater than would correspond to a liquid saturation averaged over the producing section as a whole. Hence when plotted against the average liquid saturation the field k_g/k_o data will lie higher than the actual curves for the individual strata, or as might be obtained by laboratory measurements on individual core samples. A similar effect is contributed by the inclusion in the total field gas-production rates those for wells which may be near an original or developed gas cap and are suffering from gas coning. As it is impractical to segregate the production data so that they will refer only to the true "average" saturation conditions, it is to be expected that the composite field permeability ratios will be higher than would be computed from laboratory determinations.

The implication of the above remarks is that actual solution-gas-drive reservoirs will generally appear to be operating less efficiently than cal-

---

[1] It should be noted that there is also a lower limit in ρ for the approximate validity of Eq. (2). Evidently the latter must break down when the oil saturation approaches its "equilibrium" value, and k_g/k_o becomes infinitely great as k_o becomes vanishingly small.

culated theoretically[1] by using k_g/k_o curves derived from laboratory core measurements. Nonuniform or stratified reservoirs will undergo an overlapping sequence of individual reservoir histories in the order of their permeabilities. The relatively rapid depletion of the most permeable strata will lead to apparently premature developments of high gas-oil ratios for the formation as a whole.[2] Even if the oil-expulsion processes be of substantially the same inherent efficiency for the strata of different permeability, the slower depletion rates of the tighter zones will lead to greater residual-oil contents in these parts of the pay when the production rate from the producing section as a whole falls to the economic limit for abandonment. Moreover the more rapid pressure decline in the more permeable zones will encourage an earlier intrusion of such mobile edgewaters as may bound the reservoir. A flooding out of these strata may force the abandonment of the well while the tighter strata still retain a large part of their original recoverable oil content.

A compensating factor in composite reservoirs suffering differential depletion is the cross flow from the tighter and higher pressure strata into the more depleted and lower pressure zones. Because of the large area available for this interzone migration and the short distances involved, the magnitude of this migration may be quite substantial. For example, if the average vertical oil permeability is 10 md and the viscosity is 1 cp, there will be a cross flow of 491 (bbl/day)/acre for a pressure gradient of 1 psi/ft. The expulsion efficiency of this migrating oil will be relatively low after reaching the more permeable strata, because of the higher gas

---

[1] A striking example is reported by G. H. Fisher (paper presented before the Petroleum Engineers Club, Dallas, Tex., Feb. 7, 1947), in which the laboratory-determined k_g/k_o curves for the Wasson field, Texas [A. C. Bulnes and R. U. Fitting, Jr., *AIME Trans.*, **160**, 179 (1945)], with an equilibrium free-gas saturation of 31 per cent gave a calculated ultimate gas-drive recovery of 40 per cent of the oil in place, whereas a recovery of only 17 per cent was obtained on using the k_g/k_o curve computed from the field performance, *i.e.*, Eq. (1).

[2] Conversely, abnormally rapid rises in the gas-oil ratio from strata that are not overlain by initial or developed gas caps may be considered as indicative of the presence of very high permeability streaks that are being depleted much more rapidly than the remainder of the pay. The presence of high-permeability streaks may also lead to an apparent sensitivity of the reservoir pressure to the production rate and abnormally sharp pressure declines following rapid increases in production rate. Such effects may be expected in fractured-limestone reservoirs. In a recently reported theoretical study of the simultaneous gas-drive depletion of noncommunicating stratified reservoirs (W. O. Keller, G. W. Tracy, and R. P. Roe, API meetings, Tulsa, Okla., March, 1949) it was confirmed that the differential depletion in the various strata should lead to an initially excessive rise in the apparent or field-calculated k_g/k_o curve. However, this initial trend would be followed by a marked flattening and crossing of the assumed laboratory-data curve, simulating that due to the development of gravity-drainage effects (cf. p. 470). The gas-oil ratio itself was found, in one case, to have an oscillatory character when plotted vs. the cumulative recovery, in contrast to the smooth uniform stratum behavior shown in Figs. 10.4, 10.5, 10.8, etc.

saturations and gas-oil ratios in the latter. The net effect, however, will be a greater economic recovery than if the different strata were completely isolated. The quantitative features of these interactions and the proper averaging procedure to be used in evaluating nonuniform reservoirs are still to be developed.

These latter considerations also show that the theoretically calculated physical ultimate recoveries of gas-drive reservoirs must be combined with economic factors in making estimates of actual recoverable reserves. Even if the basic permeability-saturation relationships were independent of the homogeneous-fluid permeability for the components of a single geological horizon, the economic ultimate recoveries will evidently decrease with decreasing permeability. Well abandonment in gas-drive reservoirs is generally determined by limiting rates of production required to pay for the cost of the operations. These limiting rates will be reached at higher average reservoir pressures and oil saturations in tighter pays than in higher permeability zones. The recoveries at the time of abandonment— the *economic* ultimate recoveries—will therefore be greater for the latter, even though the calculated physical recoveries will be substantially the same. No simple rule or formula will describe the variation of the economic ultimate recoveries with the permeability. Moreover the permeability itself will be but one of the physical factors involved. The oil viscosity may be expected to enter directly in a permeability-to-viscosity ratio. And the total recoverable reserves will be proportional to the pay thickness and porosity, except as the thickness will also affect the fractional recovery at the time of abandonment. On the other hand the variations with the permeability will generally be less for shallow producing formations than for deep reservoirs. The higher operating cost for the deeper fields will place the economic abandonment rate at a higher level where the cumulative recovery is still sensitive to the absolute value of the abandonment rate. Of course, in the limit of very tight and thin reservoirs the drilling and development costs themselves may make any exploitation and recovery whatever unprofitable, whereas a similar highly permeable formation may provide enough "flush" production to offer promise of pay-out.

Returning to the apparently anomalous permeability-ratio curve in Fig. 10.29 for the Wilcox Sand, it should be observed that it, too, reveals an important factor of actual reservoir behavior which is not yet within the scope of rigorous and quantitative reservoir analysis. Here again the actual producing conditions involve a variation in oil saturation not properly expressed by a gross average for the whole producing formation. In contrast, however, to the previously discussed situations where the zone of greatest depletion may appear to control the performance and lead to excessive gas-oil and permeability ratios, here the later stages of the production history have been apparently dominated by a zone of abnormally low depletion. Accordingly the gas-oil and permeability ratios fall below those to be expected of a uniform saturation system. The significant feature of this behavior is that the nonuniform depletion in this case is the result of a downstructure gravity drainage of the oil so as to replace the fluid withdrawals and maintain the oil saturations in the lower parts of the oil zone. The higher oil saturations tend to hold down the gas-oil ratios, which then appear to be abnormally low when correlated with the average oil saturation throughout the oil zone that would have obtained in the absence of gravity drainage.[1] The disclosure of phenomena of this type is in itself a valuable by-product of the field study of permeability ratios. The general problem of gravity drainage will be discussed more fully in Secs. 10.14 and 10.15.

10.13. Field Experience in Gas-injection Operations.—No attempt will be made here to prove quantitative agreement of field experience in gas-injection pressure-maintenance operations with the theoretical predictions of previous sections. While the field data to be presented do not contradict the theoretical developments or are subject to rationalization where the observed behavior appears to be anomalous, claims of quantitative agreement with calculated performance would be entirely unwarranted. It is seldom that enough detailed data are available on the nature of a reservoir to warrant the consideration of quantitative calculations of expected behavior as much more than a guide in evaluating plans for gas injection, since in practically all cases it is quite evident that the reservoirs are far from the uniform ideal systems which alone are as yet amenable to analytical treatment. Nor is the nature of the gas injection usually such as to simulate the process of direct diffusion and dissemination throughout the oil zone, as assumed in the theoretical considerations of Secs. 10.7 and 10.8. Such effects of gas-cap expansion and gravity drainage as may contribute to observed field performance are not as yet within the scope of quantitative evaluation by theory.

The following examples are presented to show the types of reservoir behavior under gas injection which may be expected under the practical conditions in which such operations are actually carried out. They may thus be considered as supplementary to or generalizations of the idealized performance predictions of Secs. 10.7 and 10.8. They represent the resultant effect of all the factors involved in real oil-producing reservoirs, with all their variations from the simple uniform prototypes underlying the theoretical treatment.

---

[1] It should be noted, however, that, while in the Oklahoma City field the gravity drainage has provided low gas-oil ratios in the downstructure wells, gas segregation within flat strata fully exposed to a well bore may lead to higher gas-oil ratios and less efficient recovery. The high values of the gas-to-oil-permeability ratios of Figs. 10.29 and 10.30 may also be due in part to such effects.

Figure 10.31[1] gives the reservoir pressure and gas-oil-ratio history of the
Lansing Lime formation of the Cunningham pool, Kingman and Pratt
Counties, Kan., discovered in 1932. The anticlinal oil reservoir, at a
depth of about 1,750 ft subsea, covers 1,400 acres, has a closure of about
75 ft, and contains 53 producing wells. The main pay is of variable character and is largely comprised of oölitic streaks, but the different zones
appear to be in communication through vertical fracturing. Its average

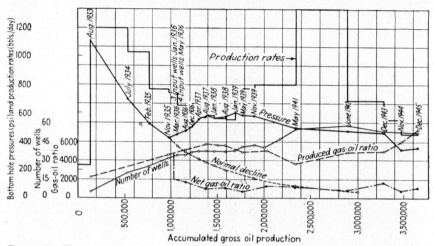

Fig. 10.31. The production and pressure history of the Lansing Lime reservoir of the Cunningham pool, Kansas. (*After Rae, Producers Monthly, 1946.*)

net thickness is 8 ft, and the average values of permeability and porosity
are 105 md and 11 per cent. The oil produced is of 31 to 36°API gravity.
The wells were acid-treated to stimulate production, the average increase
following the first treatment being about 300 per cent of the producing
capacity before acidization. The field had an initial gas cap. The original
reservoir pressure was 1,115 psi. Gas injection was undertaken after a
rapid pressure decline to 424 psi by January, 1936, and a cumulative production of 1,000,000 bbl. During the following 10 years 84 per cent of the
produced gas was returned to the formation through 3 to 5 injection wells,
with an additional production of 2,632,000 bbl but only 59 psi net pressure
decline. The initial rise in pressure and general pressure stabilization over
this long period indicates the supporting action of a partial water drive.
This is not surprising in view of the daily production of 500 bbl water,
although there is no evidence of a major water advance. The gas-oil ratios
remained substantially constant during the first 8 years of gas return

[1] The material pertaining to this field is taken from the paper of C. Rae, *Producers
Monthly*, **10**, 18 (October, 1946); cf. also R. B. Rutledge and H. S. Bryant, *AAPG Bull.*,
21, 500 (1937).

and then developed a marked rise following the addition of injection wells to the southwestern part of the field, where the pay is nonoölitic.

While the irregularity of the producing formation and high gas-oil ratios prior to gas injection might well have cast doubt on the success of the gas-return program, there can be little doubt that the performance shown in Fig. 10.31 and described above does represent satisfactory operations. The expected ultimate recovery of 4,500,000 bbl is 1,700,000 bbl greater than that indicated by an extrapolation of the behavior prior to gas injection. This corresponds to an increase in the recovery per acre-foot of pay from 250 to 402 bbl. It appears that in spite of the thin pay section there has been effective gas segregation, so that the injected gas largely remained trapped in the reservoir and helped to sustain the oil saturation within the oil zone. In any case, from a practical point of view, the Cunningham pool history must evidently be considered as an example of successful pressure-maintenance operations.

Pressure maintenance by gas injection in a field where complete reservoir data were available in planning the operations is illustrated by the Jones Sand reservoir of the Schuler field, Union County, Ark.[1] This reservoir, which was discovered in September, 1937, lies at a depth of 7,553 to 7,601 ft. By July 1, 1940, it was fully developed, with 146 producing wells on a 20-acre spacing. The entire Jones Sand section was cored in 136 wells, and very complete electrical and drilling-time logs were obtained on the field. The oil reservoir lies in an anticlinal trap of 90 to 135 ft closure. It had initially a small gas cap with a gas-oil contact at 7,270 ft subsea, and a water-oil contact lying at 7,370 to 7,380 ft subsea. The sand permeability is quite variable, ranging from 0 to 4,000 md and averaging 355 md. The average porosity is 17.6 per cent. The connate-water saturation is estimated as 15 per cent. The effective sand thickness varies from 0 to 70 ft and averages 37.5 ft. The initial formation-volume factor was 1.52. The stock-tank oil has a gravity of 34°API. The initial reservoir pressure was 3,520 psi gauge at 7,300 ft subsea. The reservoir originally contained 99 million barrels of residual oil according to the official records.

Although voluntary limitation of production was arranged in January, 1938, and state-regulated allocation was introduced in April, 1939, the early production history led to the development of large pressure gradients over the reservoir, as shown by the map of Fig. 10.32 for Feb. 15, 1941. These were due to relatively excessive oil withdrawals from the eastern part of

---

[1] The data presented here regarding this reservoir are taken from the records of the Arkansas Oil and Gas Commission, made available to the author through the courtesy of A. C. Godbold; cf. also Kaveler, *loc. cit.* While they differ in some details from those given in Secs. 9.6 and 9.7, these differences are not of great importance from the point of view of the pressure-maintenance history of the reservoir.

the field as compared with its local oil reserves, the higher gas-oil ratios in the eastern area, and limited migration from the west. The Jones Sand pool was unitized for common operation on Feb. 15, 1941, when the pressure had declined to about 1,550 psi and the average gas-oil ratio had

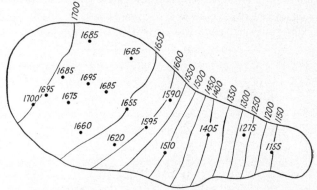

FIG. 10.32. The pressure distribution in the Jones Sand reservoir of the Schuler field, Arkansas, on Feb. 15, 1941, just prior to unitization. (*After Kaveler, AIME Trans., 1944.*)

risen to 2,700 ft³/bbl. Immediately the gas-oil ratios were reduced to an average of less than 1,500 ft³/bbl by simply limiting the withdrawals to less than 50 of the most efficiently producing wells and shutting in the remainder. A program for a return of 90 per cent of the produced gas

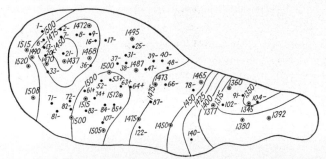

FIG. 10.33. The pressure distribution in the Jones Sand reservoir of the Schuler field, Arkansas, on Nov. 28, 1942, 21 months after unitization and 16 months after the start of gas injection. Plus signs indicate injection wells; minus signs indicate producing wells. (*After Kaveler, AIME Trans., 1944.*)

was begun in July, 1941, with an injection into 6 wells at the crest of the structure in the western part of the field. But before these operations in themselves could have had much effect, the pressure distribution had been so equalized that by July 25 the total gradient over the field was only about 100 psi. The pressure distribution as of Nov. 28, 1942, is shown in Fig. 10.33. The composite time history of the field to January, 1947, as taken from the records of the Arkansas Oil and Gas Commission, is plotted in Fig. 10.34.

While the initial arrestment of the pressure decline shown in Fig. 10.34 is due to the change of withdrawal and pressure distributions, following the inauguration of unitized operation, the strikingly persistent pressure stabilization during the next 5 years is evidently the result of the gas-injection program. The physical significance of the reservoir performance repre-

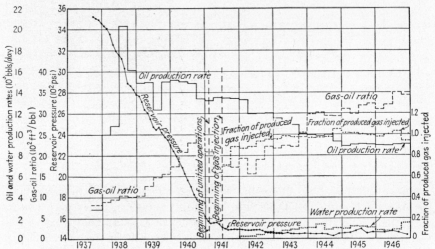

FIG. 10.34. The pressure and production history of the Jones Sand reservoir of the Schuler field, Arkansas to December 1946.

sented by Fig. 10.34 becomes apparent from Fig. 10.35, in which the pressure and gas-oil-ratio data are plotted vs. the cumulative production, together with the theoretically predicted behavior. The latter was computed[1] by using a material-balance method similar to that outlined in Sec. 10.9, with permeability-ratio data calculated from the actual performance history prior to unitization. In addition to predictions of the depletion behavior without gas injection, the calculated performance is indicated for a 90 per cent gas-return program.

It will be seen from Fig. 10.35 that as a whole the actual operational history has been even better, with respect to pressure decline and gas-oil-ratio rise, than predicted. The almost complete pressure stabilization, as compared with the predicted slow decline, is partly the result of the gas return at a rate higher than the 90 per cent used in the calculations, as is clear from Fig. 10.34. In addition, however, there has been a pressure-maintaining contribution due to water intrusion, since even 100 per cent gas return could not, in itself, provide replacement for both the oil and

[1] Kaveler, *loc. cit.*

gas space voidage.[1] The development of an appreciable water production in 1942 supports this interpretation.

The slower rise in gas-oil ratio than predicted is also probably the composite resultant of two factors. The permeability-ratio data, as determined from the early history and extrapolated to low liquid saturations, were undoubtedly too heavily weighted by the high gas-oil ratios prevailing in

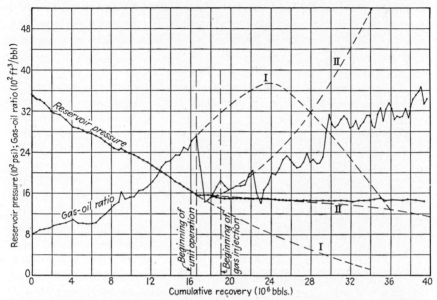

Fig. 10.35. The pressure and gas-oil histories of the Jones Sand reservoir of the Schuler field, Arkansas, vs. the cumulative recovery. Solid curves, observed. Dashed curves I, predicted for normal depletion. Dashed curves II, predicted, at the time of unitization, for 90 per cent gas return. (*After Kaveler, AIME Trans., 1944.*)

the low-pressure and more depleted eastern area of the field. The closing in of this area after unitization restricted the withdrawals to the less depleted part of the field, where the oil saturation was higher and gas-oil ratios accordingly lower. On the other hand, this very localization of the withdrawals to a limited part of the field would tend to accelerate the decline in oil saturation and rise in gas-oil ratios as the withdrawals were continued. It seems likely, therefore, that gravity drainage and gas segregation have been effective, to some extent at least, in sustaining the oil saturations in the downstructure withdrawal area. As gravity-drainage effects were neglected in deriving the theoretical performance, a maximum rise in gas-oil ratio would be predicted, if the permeability-ratio data were correct.

[1] The theoretical possibility of complete pressure maintenance with 100 per cent return [cf. Eq. 10.7(8)] did not apply to Schuler.

The operations as a whole to January, 1947, were evidently leading to very substantial benefits. Operating costs have been materially reduced by prolongation of the flowing lives and because of the small number of wells that have been used to produce the allowables from the field. Much of the produced gas that might otherwise have been wasted is being stored and saved in the reservoir for future sale and use. If the future operations should continue on the past favorable trend, the increase in ultimate oil recovery over the natural-depletion estimate of 34 million barrels may even exceed the 20 million barrels predicted at the time the gas-injection program was started.[1]

A gas-injection experiment[2] on a 750-acre section of a West Texas Grayburg Lime anticlinal reservoir serves to illustrate the complexities of operation as well as interpretation that may be sometimes encountered in pressure-maintenance programs. The producing section is a sandy dolomite with an average net pay thickness of 18 ft out of a gross thickness of 130 ft. The productive pay has a porosity ranging from 8 to 14 per cent and a permeability of 2 to 10 md. The oil gravity ranges from 33 to 37°API. The experiment consisted in converting one of the 26 producing wells to an injection well in October, 1942, after the reservoir pressure had declined to 1,275 psi from an initial value of 1,800 psi. During the following 22 months, 167 million cubic feet of gas was injected, after which the injection well was shut in for 6 months and then opened again as a producer. The history of these operations is recorded in Fig. 10.36.

A full explanation of all the details of the curves of Fig. 10.36 cannot be given. For example, since the reservoir had been producing entirely by the gas-drive mechanism, the pressure rise prior to the beginning of gas injection is quite anomalous, unless it be considered as only an apparent rise resulting from a pressure redistribution following the cut in oil production. On the other hand, for the first 11 months of injection, during which 52 million cubic feet of gas was returned to the reservoir—substantially all the gas produced—there developed no significant change or reaction. Yet almost suddenly in September, 1943, the gas-oil ratios began a steep rise and the pressure a precipitous decline. These developments are undoubtedly associated with the sharp increase in withdrawal rate made at the same time, although the exact reason is rather uncertain.

---

[1] The belief that water intrusion has materially contributed to the performance of the Schuler-Jones sand reservoir is now generally accepted, and the natural encroachment has even been supplemented by water injection since June, 1944 [cf. L. L. Jordan, *World Oil*, **128**, 121 (September, 1948)]. These factors, however, do not invalidate the conclusion that the gas-injection operations at Schuler have been successful from both recovery and economic standpoints, although the natural depletion recoveries might have been greater than originally estimated.

[2] Cf. Elkins, *loc. cit.*

It can, however, be rationalized if the reservoir rock be considered as a composite of a tight massive matrix through which is interspersed a substantially continuous and intercommunicating fracture system.[1] It would then be expected that sudden increases in withdrawal rate beyond the feeding capacity of the intergranular rock mass would lead to an abnormally

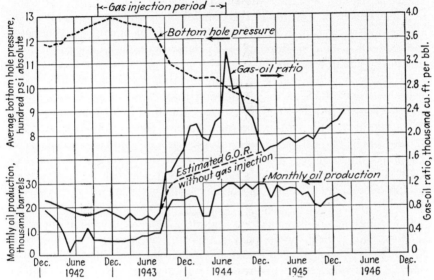

FIG. 10.36. The pressure and production history of a West Texas Grayburg Lime reservoir subjected to gas injection. (*After Elkins, API Drilling and Production Practice, 1946.*)

rapid depletion of the fracture system with its own inherent development of high gas-oil ratios. Aggravating this growth in gas-oil ratio would be the probable development of injection gas break-through via the excessively depleted fractures. The temporary decline in gas-oil ratio in March and April, 1944, following the reduction in withdrawal rate, and the continued abrupt fall after gas injection was terminated in June, 1944, are in qualitative accord with this interpretation.

The over-all operational history of this project shows that throughout the period of gas injection and until January, 1945, when the gas-oil ratio returned to the estimated normal trend, the gas actually produced was 305 million cubic feet greater than indicated by the trend prior to gas injection. This is 138 million cubic feet in excess of the total gas injected.

[1] A situation of this kind has been found to obtain in the West Edmond field, Oklahoma [cf. M. Littlefield, L. L. Gray, and A. C. Godbold, *AIME Trans.*, **174**, 131 (1948)]. An essentially equivalent interpretation has also been developed from a more detailed analysis of this project by L. F. Elkins and J. T. Cooke, AIME meetings, Dallas, Tex. (October, 1948).

This disturbing result can also be visualized as being due to the increased production rate during the second half of the injection period and to the slow bleeding of the intergranular rock into the excessively depleted fracture network. While an exact evaluation of this gas-injection project may be subject to doubt, it can hardly be considered as an outstanding success.

The last example of pressure-maintenance operations by gas injection to be considered here is that of the Canal field, Kern County, Calif.[1] This field was discovered in November, 1937, and development was completed on a 20-acre spacing in March, 1941. The structure is an elongated dome and lies at about 7,800 ft subsea, with a closure of about 150 ft. The main reservoir formation is a sand, covering 1,100 acres, of variable character interbedded with silt-shale streaks. The permeability shows poor well-to-well correlation. It ranges from 10 to 1,000 md and averages about 200 md. The porosity averages 22 per cent and ranges from 15 to 32 per cent. The connate-water saturation is estimated as 38 per cent. The initial pressure was 3,550 psi. The reservoir originally contained no gas cap. The oil had a bubble point of 2,800 psi, with a solution ratio of 750 ft^3/bbl and formation-volume factor of 1.29, at the reservoir temperature of 210°F.

Following the conclusion of an operating agreement among the three operators in the field, preliminary pilot plant gas injection was begun in June, 1941. Although there was some indication of gas break-through, a full-scale program with a compressor plant injection capacity of 12 million cubic feet per day was put into effect in August, 1942. The history of the reservoir is plotted in Fig. 10.37.

During the preliminary tests, gas was injected in a single well located at the crest of the structure. Another injection well, down the flanks along the major structural axis, was added when the full-scale program was started. Gas injection was started in May, 1943, in a third well on the structure flanks opposite the second well. Total gas injection by January, 1946, was 13,960,000,000 ft^3.

An interesting feature of these operations was the use of a tracer, ethyl mercaptan, with the injected gas to check on its motion. By January, 1942 (6 months later), the tracer was detected at a neighboring well 933 ft away. Within a year another well at the latter distance, and one twice as far away, showed the tracer. By February, 1943, 17 wells were producing gas containing mercaptan. In view of the limited total gas injection during these periods the appearance of the tracer at the producing wells definitely proves gas channeling through high-permeability streaks, rather than a uniform drive through the sand as a whole. The sharp rise

[1] The data on this project are taken from a paper by R. P. Mangold, API meetings, Los Angeles, Calif., June, 1946.

in gas-oil ratio since the latter part of 1942, as shown in Fig. 10.37, is also indicative of a nonuniform sweeping action.

In spite of the fact that the injected gas was not sufficient fully to replace the oil and gas withdrawals, the recorded pressure history indicates a substantial pressure-maintenance action. This might be due to water intrusion, although material-balance analyses have been inconclusive, or to

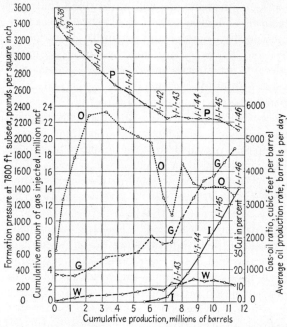

FIG. 10.37. The pressure and production history of the Canal field, California. Curves P, O, G, W, I refer to reservoir pressure, oil-production rate, gas-oil ratio, water cut, and cumulative gas injected. Plotted data for curves O, G, W are 6-month averages. (After Mangold, API meetings, 1946.)

variations in pressure and depletion in the different zones comprising the composite producing sand, so as to lead to spurious pressure observations. On the other hand, while the production capacity and rate of the field as a whole were satisfactorily maintained during the gas-injection operations, this was largely restricted to the upstructure wells. The wells downdip appeared to have been but little affected and declined in pressure and productivity as if there had been no gas injection. Gravity drainage did not seem to be a major factor in the operations. The effect of the gas injection on the ultimate recovery cannot be determined from the performance to January, 1946.

10.14. General Observations on Gas Injection.—The field examples discussed in the preceding section show that gas injection under some circumstances can be a means of achieving substantial increases in both oil recovery and profit, whereas under others its net beneficial effect may be at best subject to doubt. It is important to understand the factors determining the gross performance and success of the operations.

It will be clear from the theoretical treatments of Secs. 10.7 and 10.8 that if one calculates the recovery to be expected under gas injection, by Eq. 10.7(2) or its equivalent, one will always find that very appreciable increases in recovery should result. It would thus appear that gas injection should be a universally desirable and profitable operation. From a practical point of view, however, any such operations as gas injection that require considerable investment and operating costs should be evaluated on the basis of two questions, namely: Is it necessary? If so, will the particular program be economically profitable?

As to the necessity for gas injection the answer is negative if the reservoir is producing under a complete edgewater-drive[1] mechanism without an initial gas cap and provided that the inherent oil-expulsion efficiency by water flushing is greater than by gas drive, as is usually the case unless the connate-water saturations are high. However, if the field had an original gas cap and a pressure decline has developed so that the water intrusion exceeds the current withdrawals, there may be danger of the oil body being driven into the gas cap. The entry of oil into a gas cap will lead to a saturation of the invaded region with oil, to simulate the original oil zone. Even if this part of the formation should be subsequently flushed by the encroaching water, a residual-oil saturation of 20 to 30 per cent will still be left. If the gas cap were initially "dry," with zero or negligible oil saturation, the unrecovered oil in the gas cap after water flooding will represent a definite loss resulting from the water invasion. Hence, if the sweeping of the oil into the gas cap cannot be effectively prevented by controlling the magnitude and distribution of the oil production, gas injection into the gas cap may be warranted, even though the rate of water entry itself is sufficient to replace the fluid withdrawals.

The most common situation where the problem of gas injection arises is, of course, the reservoir operating under a gas-drive mechanism, with no evidence of substantial water-drive action. In such cases, as already mentioned, the idealized treatment of Secs. 10.7 and 10.8 will always appear to justify gas injection because of the prolonged flowing life, resulting from the retardation in pressure decline, and increased ultimate recovery. However, the assumed perfectly homogeneous and uniform reservoir rock

---

[1] If the producing formation is substantially isotropic and directly underlain by bottom waters, pressure maintenance by gas injection may offer a means to prevent the water invasion and premature drowning out of the producing wells (cf. Secs. 11.15 and 11.16).

underlying the analytical treatment will never obtain in practice. Indeed it is the uniformity of the producing formation which is the controlling factor in determining the degree to which the actual performance may approach the theoretical predictions.

In sandstone reservoirs the deviations from strict reservoir uniformity will generally be due either to lenticularity and lack of continuity in the producing formation or to permeability stratification. The former will lead to an irregular distribution of the injected gas within the reservoir and nonuniform and inefficient oil displacement. The latter will result in a channeling of the injected gas through the more permeable strata without an effective sweep of the tighter zones. Unless the high-permeability zones are separated from the less permeable strata by cross-flow barriers and can be sealed off, the gas injection may degenerate into a cycling of the gas through these strata with but minor gas invasion of the tighter parts of the rock. Where such conditions are suspected, gas injection should be undertaken at full scale only after a detailed and complete study of the reservoir formation and conservative estimates of the recoveries have been made.

In dolomitic or limestone reservoirs the type of nonuniformity that may seriously endanger the success of gas injection, in addition to gross stratification, is a fracturing of the producing rock.[1] Such fractures, in principle, are merely an extreme form of permeability inhomogeneity. If the fractures are widely distributed and in mutual intercommunication, they will provide low-resistance channels for the by-passing of the intergranular rock, since the latter is generally of very low permeability, unless it is oölitic. Accordingly it is unlikely that the injected gas will enter to an appreciable extent the tight component of the reservoir formation while it rapidly channels through the fractures between the injection and producing wells. Moreover the basic mechanism by which the oil in the intergranular rock may be produced will be the solution-gas-drive process or a bleeding into the fractures for subsequent flow into the wells. This will require the existence of a pressure differential between the fractures and the interior of the adjoining intergranular rock. To the extent that gas injection may serve to maintain the reservoir pressure, these gas-drive depletion and oil-expulsion processes will be retarded. And if the oil content of the fracture system is only a small part of the total oil reserves of the formation as a whole, gas injection may provide at best only a means for temporary gas storage and will contribute but little toward the achievement of increased oil recovery. It is possible that the Grayburg Lime

---

[1] Fracturing has also been observed in sandstone reservoirs. However, this is rather uncommon and generally does not appear to be a serious factor in gas-injection operations in sandstones.

gas-injection experiment, discussed in the preceding section, represents a situation of this type. In any case, calculated benefits of gas injection by the methods developed in Secs. 10.7 and 10.8, may well be meaningless and grossly misleading under such reservoir conditions.

Because of the above-discussed possibilities of failure of gas-injection operations it is important to establish first that artificial pressure maintenance will be necessary for efficient recovery. An initial rapid decline in reservoir pressure should not be hastily interpreted as proof of the absence of water drives.[1] These generally require the development of an appreciable pressure decline before sufficient water intrusion can be induced to provide substantial replacement of the fluid withdrawals (cf. Sec. 11.17). And even if this should not materialize, time should be taken to study thoroughly the nature of the reservoir rock and establish that the producing formation is inherently susceptible to successful gas-injection operations. Both the theoretical treatment of Sec. 10.8 and the actual field experience at Cunningham and Schuler, reviewed in the preceding section, show that if the reservoir conditions are actually favorable for gas injection, the operations will still be highly successful and profitable even if they are delayed until a substantial pressure decline has developed.

As previously indicated, in making calculations and predictions of the performance and recoveries under gas injection for specific reservoirs cognizance must be given to the nonuniformity of the producing formation. If the field has already undergone partial depletion by the solution-gas drive, extrapolations of the permeability-ratio curves derived from the past performance will provide a better basis for the predictions than scattered laboratory measurements, since the field data will include the effect of the nonhomogeneity of the reservoir. Or one may introduce a "conformance factor"[2] representing that fraction of the total productive formation that will be swept and affected by the injected gas, assuming that the remainder continues the simple solution-gas-drive performance. The resultant recovery is then simply the mean of those for solution-gas drive and under gas injection, weighted by the conformance factor. While, in principle, such a procedure does serve to correct approximately for the reservoir nonuniformity, the determination of the conformance factor is itself subject to much uncertainty. Comparison of field and laboratory permeability-ratio data, if available, may give limits to the conformance factor. But difficulties still remain as to the constancy of the factor and

[1] In fact, in complete-water-drive reservoirs, where the oil is undersaturated, the initial pressure declines will be even more rapid than if the pressures are at the bubble point and gas immediately starts to come out of solution (cf. Sec. 11.17).

[2] Cf. Patton, loc. cit., where a number of practical and economic factors related to gas injection are also discussed.

the interaction between the two hypothetical parts of the common reservoir. In any case, however, it is important to realize that except for the beneficial effects of gravity drainage, to be discussed in the next several sections, the recovery gains by gas injection calculated for uniform reservoirs will give only upper limits to those which may be expected to occur in practice.

To achieve maximum efficiency in gas-injection operations the reservoir should be produced as a composite unit through the medium of unitization or an equivalent operating agreement. The distribution of production and injection can then be so controlled as to ensure optimum reservoir reaction to the fluid injection. If the field has substantial closure, an original or developed gas cap,[1] and high rock permeability, the gas injection should be concentrated in the gas cap or structural crest and the production confined to the flanks. The producing wells should be completed so as to minimize gas coning and break-through from the gas cap. High-gas-oil-ratio wells should be shut in. Production rates should be restricted to encourage downflank gravity drainage of the oil and a uniform lowering of the gas-oil interface.

If the reservoir is flat and tight, an areal distribution of the injection wells should give more rapid and uniform returns. In its extreme form such a distribution may be developed as a regular network or pattern of injection wells interlacing a complementary network of producing wells, similar to those generally used in secondary-recovery operations, discussed in Chap. 12.[2] It is this type of operation to which the theoretical illustrative examples of Secs. 10.7 and 10.8 apply directly.

These considerations should not be interpreted as prescribing the manner of operation necessary for success in gas-injection programs. In fact, the previously discussed examples of Cunningham and Schuler may appear to contradict such a view. They are presented only as suggestions of the factors to be evaluated in planning full-scale pressure-maintenance programs.[3]

---

[1] If it is known that the reservoir will not be subject to water-drive action and that it is inherently susceptible to efficient gas-injection operations, these may be undertaken at the structural crests even prior to the development of a gas cap.

[2] However, the large number of injection wells required for such operations will generally make the use of network well distributions prohibitively expensive.

[3] In the last several years a number of reports have been published on the engineering analysis involved in planning pressure-maintenance operations as well as on the early histories of actual field operations with special emphasis on the reservoir performance [cf. in particular, Patton, *loc. cit.*; Elkins and Cooke, *loc. cit.*; J. R. Welsh, R. E. Simpson, J. W. Smith, and C. S. Yust, *Jour. Petroleum Technology*, **1**, 55(1949); J. A. Slicker, AIME meetings, Austin, Tex. (December, 1948); and W. O. Keller and R. A. Morse, AIME meetings, Dallas, Tex. (October, 1948)].

Finally it should be noted that even "full-scale" pressure-maintenance operations are seldom so in practice. Ten to fifteen per cent of the produced gas is generally used as fuel in the field operations. Moreover it is often inefficient to compress the gas produced below a minimum wellhead pressure. As a result, in the majority of cases only 60 to 80 per cent of the produced gas is actually returned to the producing formation. Under such conditions strict pressure maintenance is not to be expected, nor will it occur, unless a partial water drive is assisting in the fluid-withdrawal replacement. In fact to achieve complete pressure maintenance it will generally be necessary that the rate of gas injection even exceed the rate of gas production, as indicated by Eqs. 10.7(7) and 10.7(8). Most gas-injection operations provide only for a retardation in pressure decline rather than for strict pressure maintenance.

10.15. Gravity Drainage; General Considerations.—One of the outstanding problems in the evaluation of the behavior of oil-producing reservoirs is the quantitative treatment of the phenomenon of gravity drainage. There are three distinct aspects to the general problem of the role played by gravity in oil reservoirs. The first is that of the original segregation of the reservoir fluids prior to discovery and exploitation of the reservoir. This is expressed by the universally found sequence of the gas, oil, and water[1] with depth in accordance with their density, when they exist as distinct phases within a single reservoir zone. Such separation is evidently the result of the action of gravity, achieved through the medium of mass movement or molecular diffusion, and directed toward an ultimate state of equilibrium involving thermodynamic potentials, gravitational heads, and capillary forces (cf. Sec. 7.9). While there is reason to doubt that strict equilibrium is actually attained even over geologic time in all reservoirs, there can be little question that gravity has played the major role in creating whatever degree of equilibrium fluid segregation which ultimately obtains in oil-producing systems when first discovered. On the other hand, with respect to the fundamental problem of oil production, the action of gravity prior to field development serves mainly as an important factor in determining the "initial conditions" defining the individual reservoir under consideration.

By far the main reason for interest in the force of gravity lies in the role it may play in affecting the performance of the reservoir during the course of its primary-production history. Finally there is the period in the producing life of fields, at or near the state of complete pressure depletion,

---

[1] This refers to the gross segregation of the phases, rather than a mutually exclusive separation in which only a single phase occupies the whole of the local pore space above the water-oil contact.

when gravity may become the dominant agent for bringing oil into the well bore. Some aspects of this problem have been treated in Sec. 5.8 and will be further briefly considered below. However, the discussion to be given here will be restricted largely to the effects of gravity forces in controlling gas and oil segregation during the primary-producing phase of gas-drive reservoirs.[1]

While serious attention had not been devoted to the subject of gravity drainage until rather recently, one manifestation of its effect has been commonly observed ever since it became accepted practice to record producing gas-oil ratios. This is the formation of local gas caps during the course of production. Such gas caps are in themselves merely regions of high gas-oil-ratio production and may simply represent the result of high local oil withdrawal and depletion. However, the fact that in a statistical sense they generally are formed or first noted[2] in the structurally high parts of the reservoir is strongly suggestive of a significant contribution of upward gas migration into the gas cap and of oil drainage downflank, to give rise to a lowering of the oil saturation in the gas cap below that to be expected of the local solution-gas-drive depletion and withdrawals. Of course, gas-drive reservoirs by no means show this behavior universally. And a report of gas-cap formation in an individual instance does not in itself prove the existence of significant gravity drainage elsewhere. It is the general correlation of such gas caps with structure that constitutes a major item of evidence that gas and oil segregation often does occur during gas-drive production.

The frequently observed systematic downdip expansion of original gas caps or of those developed at the structural crest after production has begun is similar qualitative evidence of gravity drainage. However, here, too, the contribution of gravity segregation to such gas-cap expansion is to be measured only by the excess in reservoir gas saturation over that to be expected from the fluid withdrawals in the region invaded by the gas cap. Evidently, exact determinations of this contribution are difficult, and care must be taken in evaluating the field observations. An example in which such gas-cap expansion has been associated with substantial gravity drainage is illustrated by the gas-oil-contact contours shown in

---

[1] The problem of water coning in gas-drive reservoirs, which also directly involves gravity forces, has already been treated approximately in Sec. 5.9.

[2] There are cases, however, in which excessive local depletions have initially formed the equivalents of gas caps along a structural flank, which have subsequently migrated to the crest of the structure. And in some of the Near East fields producing from highly fractured limestones the gas evolved from solution apparently rises immediately to the structural crests, and the flank wells continue to produce almost indefinitely at their current solution ratios [cf. H. S. Gibson, *Oil and Gas Jour.* **46**, 48 (Apr. 8, 1948)].

Fig. 10.38 for the Parinas Sandstone reservoir of the Mile Six pool in Peru,[1] which has been subjected to complete pressure-maintenance operations throughout its history.

It should be noted that the material-balance equation or method, discussed in Chap. 9, does not, of itself, give any information regarding the action of gravity drainage. The material-balance method refers only to states of thermodynamic equilibrium and involves only total liquid- and gas-phase reservoir contents. Except as the phase distribution affects the gas-oil ratio, it does not enter the material-balance equation, which makes no distinction between gas-cap gas or free gas distributed within the oil zone.[2] The force of gravity, however, expresses itself as a dynamical phenomenon. It is because of this fact that the treatment of gravity drainage is such a difficult problem.[3]

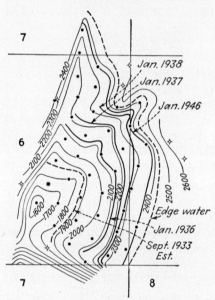

FIG. 10.38.   The structural contours and downdip migration of the gas-oil contact (heavy contours) in the Mile Six pool, Peru.

An estimate of the maximum possible rate of downdip oil drainage can be made as follows: Upon assuming the gas-oil contact to lie as indicated in Fig. 10.39, the downdip free-fall velocity of the oil will be

$$v_s = \frac{k_o \, \Delta\gamma g \sin\theta}{\mu_o},$$ (1)

---

[1] Fig. 10.38 is a composite replot of figures given by J. J. Mullane, *API Drilling and Production Practice*, 1944, p. 53, and R. E. Moyer, *Oil and Gas Jour.*, **46**, 251 (Dec. 27, 1947).

[2] In fields of high structural relief, however, the concentration by gravity drainage of the reservoir oil down the flanks of the structure, where the pressures are higher, might so change the effective average reservoir-fluid *p-v-T* characteristics that, if not corrected for, material-balance calculations of the initial-oil content could lead to decreasing trends [cf. E. P. Burtchaell, AIME meetings, Los Angeles, Calif. (October, 1948)].

[3] An approximate transient theory of gravity drainage has been recently reported [cf. W. T. Cardwell, Jr., and R. L. Parsons, *Petroleum Technology*, **11**, 1, (November, 1948)], in which the varying permeability with reduced saturation is taken into account. Capillary effects, however, are neglected in order to make the analysis tractable, and the quantitative significance of the theory is therefore somewhat uncertain.

where k_o is the permeability to oil, μ_o is the oil viscosity, $\Delta\gamma$ is the density difference between the oil and gas,[1] and θ is the dip angle. If h be the thickness of the oil zone normal to the direction of dip, the volume rate

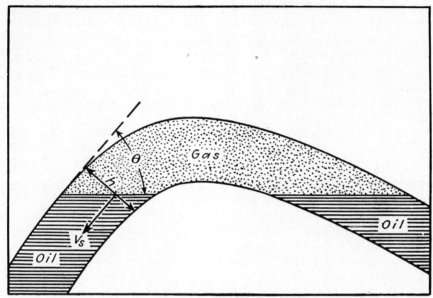

Fig. 10.39.

of oil downdip free-fall migration, in stock-tank measure, will be

$$Q_o = v_s h = \frac{k_o\,\Delta\gamma g h \sin\theta}{\mu_o\beta_o}, \tag{2}$$

per unit distance parallel to the strike, β_o being the formation-volume factor of the oil. The drainage per unit projected surface area of gas-oil contact is therefore

$$Q = \frac{k_o\,\Delta\gamma g \sin^2\theta}{\mu_o\beta_o} = 21.29\,\frac{k_o\,\Delta\gamma \sin^2\theta}{\mu_o\beta_o}\ (\text{bbl/day})/\text{acre}, \tag{3}$$

where k_o is expressed in millidarcys and $\Delta\gamma$ as a specific gravity. The corresponding rate of vertical fall of the gas-oil-contact plane will be[2]

$$v_z = 2.744 \times 10^{-3}\,\frac{k_o\,\Delta\gamma \sin^2\theta}{\mu_o\bar{f}}\ \text{ft/day}, \tag{4}$$

where $\bar{f}$ is the net porosity vacated by the oil drainage.

[1] If the gas phase is immobile and not continuous, there will be no buoyancy reaction on the oil due to the gas and $\Delta\gamma$ should be replaced by the oil density γ.

[2] In practice the gas-oil contact will not lie strictly in a plane because of permeability variations. Moreover, even if the permeability were uniform, the gas-oil contact would be a capillary transition zone rather than a sharp geometrical plane (cf. Sec. 7.9).

As an example of the order of magnitudes of the free-fall drainage and velocity implied by Eqs. (3) and (4), one may assume $k_o = 25$ md, $\Delta\gamma = 0.65$ gm/cc, $\theta = 20°,$* $\mu_o = 1$ cp, and $\beta_o = 1.3$. Then Eq. (3) gives an oil-drainage rate of 31.13 (bbl/day)/acre. This would suffice to replace an oil-withdrawal rate of 311 bbl per well per day on a 20-acre spacing pattern outside of the gas cap if the gas-oil-contact area were half of the outside productive area. The corresponding rate of fall of the gas-oil contact would be, by Eq. (4), 0.0417 ft/day, if the net voidage porosity is 0.125.

Under the conditions of restricted withdrawals now commonly practiced in the states, in accordance with state-imposed proration regulations or voluntary agreements, the above-indicated magnitudes of oil-drainage rates would evidently represent a significant degree of replacement of the oil production. On the other hand, since Eqs. (3) and (4) give only maximal effects of gravity drainage, they do not imply that the corresponding rates of downdip migration always actually occur in particular instances. The first criterion for their validity is that the gravity head is not opposed by pressure gradients. This requires a uniform reservoir pressure along the structure flanks so as to permit the free-fall gravity action. Increasing downdip pressures will oppose the gravity drainage. Rapidly decreasing downdip pressures will tend to mask the gravity-drainage effects and minimize the effectiveness of the gas-oil segregation. In extreme cases regional gas coning may develop. While the buoyant force on the gas phase will vanish under perfect free-fall gravity flow of the oil, upward countercurrent gas migration, when free fall is restricted, will tend to compensate for the reduced rate of oil drainage. In any case, Eqs. (3) and (4) should be considered as indicative of the maximum possibilities of gravity drainage rather than as formulas for quantitative evaluations of its magnitude.

Permeability stratification and nonuniformity of the producing formation will also minimize the effectiveness of gravity drainage. These will make it difficult to maintain a level gas-oil-contact plane. Excessive gravity drainage in a high-permeability stratum will lead to apparent gas channeling into downdip wells and bleeding of the gas-cap gas. Unless the stratum can be identified and sealed off, it will be impractical to take advantage of the slower gravity drainage in the less permeable parts of the pay without shutting in entirely the wells suffering from channeling.

While the above considerations are only of qualitative significance, Eq. (3) does indicate the relative effects of the physical reservoir param-

---

* While a dip of 20° is probably greater than the average among oil reservoirs, values as high as 70° have been observed (*e.g.*, the Santiago field, Kern County, Calif.) [cf. G. W. Ledingham, *AAPG Bull.*, **31**, 2063 (1947)].

eters. The role played by the structural slope is given by the term $\sin^2 \theta$. The relative magnitudes of this factor for $\theta = 5, 10, 15,$ and $20°$ are 1, 3.97, 8.82, and 15.4, respectively. Thus, if in the previous example the dip were $5°$ instead of $20°$, the maximum oil-drainage rate would be only 2.02 (bbl/day)/acre and the corresponding rate of fall of the gas-oil contact would be reduced to 0.0027 ft/day.

The direct driving force $\Delta\gamma$ is largely determined by the crude density and pressure. It decreases continuously with increasing API gravity but increases as the pressure is reduced. At 3,000 psi it will range from about 0.80 for a $10°$API oil to 0.35 for a $50°$API crude. At 100 psi the corresponding values of $\Delta\gamma$ will be approximately 0.95 and 0.70. Thus the total range of variation of $\Delta\gamma$ that may be encountered will be of the order of 3.

The composite factor $\Delta\gamma/\mu_o\beta$, which includes all the terms depending on the pressure, may vary over a range of the order of 500 among different reservoir-fluid systems, depending on the crude gravity and pressure. At all pressures it increases with increasing API gravity, from about 0.01 for a $10°$API crude to 1.4 for a $50°$API oil, at 3,000 psi. For the low-gravity oils it decreases appreciably as the pressure falls, having a value approximately 0.0025 for a $10°$API crude at 100 psi. For oils of $40°$API or greater there may be a slow decrease with declining pressure, or even a slight rise at intermediate pressures, as compared with 3,000 and 100 psi. Thus, as far as the reservoir fluids themselves are concerned, the maximum rates of gravity drainage will be much greater for higher than for lower API-gravity oils. They will not be sensitive to the reservoir pressure for high-gravity oils. While higher pressures will permit greater drainage rates of low-gravity crudes, they will still probably be too small in general to be of much practical significance.

The permeability to oil, k_o, also has a wide range of variation. Its value among known oil-producing reservoirs varies at least by a factor of the order of 1,000. Under uncontrolled production the composite mobility factor k_o/μ_o will affect similarly the withdrawal rate and gravity-drainage rate. As it is the relative magnitudes of these rates that determine the effect of gravity drainage on the gross reservoir performance, the permeability-to-viscosity ratio will therefore not be of major importance, for fixed structural conditions and well development, under "wide-open" flow. If, however, the withdrawal rates be severely restricted independently of the inherent productive capacity of the producing formation, the permeability-to-viscosity ratio may well be the controlling factor. For a fixed production rate in a given reservoir the extent to which the gravity drainage may replace the downdip oil withdrawal will then be directly proportional to k_o/μ_o. Conversely the same considerations indicate that unless k_o/μ_o

is inherently very small, it may be possible to achieve the beneficial effects of gravity drainage by limiting the withdrawal rates to an order of magnitude comparable with the gravity-drainage supply capacity of the reservoir.

10.16. Production Histories under Gravity Drainage and Gas-cap Expansion.—It is possible to construct a set of equations formally describing the production histories of gas-cap-expansion reservoirs analogous to the single differential equations considered in Secs. 10.3 to 10.8. Unfortunately, however, they are so complex that it would require a very extended program of numerical calculation to derive a complete picture of their quantitative implications. Moreover they involve uncertainties regarding the permeabilities to be used in computing the time rates of displacement of the gas-oil interface. The discussion to be given here will therefore be limited to a gross analytical description of the expansion history of the gas cap as a function only of the cumulative recovery.[1] No account will be taken of the time element. It will be assumed that the withdrawal rates and distribution are such as to permit a continuous and uniform fall in the gas-oil contact, with no loss of the gas-cap gas through the downstructure oil-producing zone. The residual oil in the expanded gas cap is to be considered as the result of the gravity-drainage mechanism rather than a sweeping action obtaining under solution-gas drive.

Upon representing the composite gas-cap and oil-zone reservoirs diagrammatically as indicated in Fig. 10.40 the gas content of the gas cap at any time can be expressed as

$$G = \overline{h_i}[\gamma(1 - \rho_w) - \overline{\rho_{oi}}\zeta] + (\overline{h} - \overline{h_i})[\gamma(1 - \rho_w) - \overline{\rho_{og}}\zeta]$$
$$= \overline{h_i}[\gamma_i(1 - \rho_w) - \overline{\rho_{oi}}\zeta_i] + rP_g; \qquad \zeta = \gamma\beta - S, \qquad (1)$$

where $\overline{h_i}$ is the initial average gas-cap thickness, expressed as a fraction of the total average oil and gas section; $\overline{h}$ is its fractional thickness at any stage of the production history; $\overline{\rho_{oi}}$ is the initial fractional oil saturation in the gas cap, in stock-tank measure; $\overline{\rho_{og}}$ is the residual-oil saturation, in stock-tank measure, in the oil-drained gas cap; P_g the total gas produced; r the fraction of the produced gas that is returned to the gas cap; and γ, β, S, ρ_w have their usual meaning. The subscript i generally indicates the initial values. It is assumed that the oil zone is initially saturated. Although it is to be expected that $\beta\overline{\rho_{og}}$ rather than $\overline{\rho_{og}}$ will be approximately constant, it will be assumed for simplicity that $\overline{\rho_{og}}$ itself is constant. Since it is further assumed that the produced gas is derived only from the oil zone, the total gas production P_g will be

$$P_g = (1 - \overline{h_i})S_i\frac{\rho_{oi}}{\beta_i} - (1 - \overline{h})\left[\gamma(1 - \rho_w) - \frac{\rho_o\zeta}{\beta}\right], \qquad (2)$$

[1] Cf. Muskat, *loc. cit.*

where ρ_o is the oil saturation in the oil zone and its water saturation ρ_w is taken to be the same as in the gas cap.[1] By combining Eqs. (1) and (2) it follows that

$$\bar{h}\left[\gamma(1-r)(1-\rho_w) + \zeta\left(\frac{r\rho_o}{\beta} - \overline{\rho_{og}}\right)\right] = \gamma_i(1-\rho_w)[r(1-\overline{h_i}) + \overline{h_i}] - \overline{\rho_{oi}}\zeta_i\overline{h_i}$$

$$- r(1-\overline{h_i})(1-\rho_w)\frac{\zeta_i}{\beta_i} - (\overline{\rho_{og}} - \overline{\rho_{oi}})h_i\zeta - r\gamma(1-\rho_w) + \frac{r\rho_o\zeta}{\beta}. \quad (3)$$

Equation (3) gives a relation between the gas-cap thickness $\bar{h}$, the pressure, through the functions γ, ζ, β, and the oil saturation ρ_o. Since the

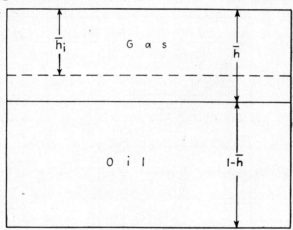

FIG. 10.40.

oil zone continues to produce by the solution-gas-drive mechanism, it will be assumed[2] that the relation between ρ_o and p is that given by Eq. 10.3(1) for the normal solution-gas-drive depletion history. The cumulative oil recovery as a fraction of the pore space is then computed by

$$P_o = (1-\overline{h_i})\frac{\rho_{oi}}{\beta_i} - (1-\bar{h})\frac{\rho_o}{\beta} - (\bar{h} - \overline{h_i})\overline{\rho_{og}}. \quad (4)$$

[1] Eqs. (1) and (2) do not provide directly for gas migration from the oil zone into the gas cap. Different degrees of migration represented by arbitrary fractions of the gas phase or gas content of the oil zone could be formally included in the analysis. For example, the special case of transfer to the gas cap of only the local free-gas phase and that dissolved in the residual oil ($\overline{\rho_{og}}$) immediately below the gas-oil contact could be taken into account by adding to the right-hand side of Eq. (3) the term $(1-r)$ $\int[\gamma(1-\rho_w) + S\overline{\rho_{og}} - \gamma\rho_o]d\bar{h}$. As this would considerably complicate the numerical treatment of Eq. (3) without substantially affecting the final results, it is not included in the following illustrative discussion.

[2] This assumption is not strictly valid. The accurate equation would involve dh/dp as well as $d\rho_o/dp$. A simpler rigorous procedure for determining the relation between ρ_o and p would be an application of the material-balance graphical or trial-and-error method, combining Eqs. (2) and (4).

The produced gas-oil ratio will still be given by

$$R = S + \alpha\Psi, \tag{5}$$

in the notation of Eq. 10.3(1).

For the special case where there is no initial free-gas cap $(\overline{h}_i = 0)$, Eq. (3) reduces to

$$h = \frac{r[(\gamma_i - \gamma)(1 - \rho_w) + (\rho_o\zeta/\beta) - (\rho_o\zeta/\beta)_i]}{\gamma(1 - r)(1 - \rho_w) + [(r\rho_o/\beta) - \overline{\rho_{og}}]\zeta}. \tag{6}*$$

By using the basic permeability and fluid data assumed in deriving the simple solution-gas-drive depletion histories of Figs. 10.3 and 10.4 the

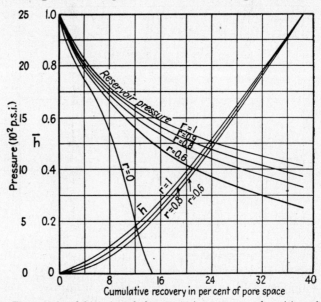

Fig. 10.41. The calculated histories of the reservoir pressure and position of the gas-oil contact, in hypothetical gas-drive reservoirs, with various degrees of gas injection and complete gas segregation. r = fraction of produced gas returned to the gas cap; $\overline{h}$ = fraction of effective formation thickness occupied by the gas cap; $r = 0$ represents normal solution-gas-drive depletion performance. (*From Jour. Applied Physics, 1945.*)

pressure and gas-cap-expansion histories, as calculated by Eq. (6), are plotted in Fig. 10.41 for various gas-return ratios r. The corresponding gas-oil ratios and oil-zone saturations are plotted in Fig. 10.42. The residual stock-tank-oil saturation in the developed gas cap, $\overline{\rho_{og}}$, was taken as 0.15.

It will be seen from Fig. 10.41 that the position of the gas-oil contact is insensitive to the gas-injection ratio and is mainly determined by the

* Equation (6) implies that there would be no gravity drainage for this case if there were no gas injection ($r = 0$). This, however, is due to the neglect of the upward gas migration and can be corrected for this effect by including in the numerator of Eq. (6) the term in the footnote of p. 492, or an equivalent assumed expression.

cumulative recovery. This, of course, is to be expected, since the growth or expansion of the gas cap is essentially a measure of the degree to which oil has been drained from the upper part of the formation and withdrawn through the oil zone. Because of the low residual-oil saturation assumed for the gas cap the depletion of oil from the gas cap by the gravity-drainage process gives the major contribution to the recovery, as compared with the decline in saturation in the oil zone.

It will be noted from Fig. 10.41 that for all cases of gas return the

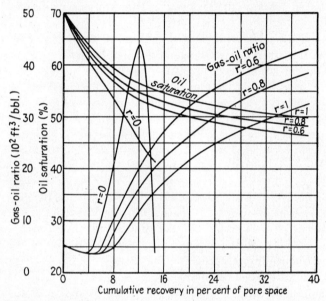

FIG. 10.42. The calculated histories of the producing gas-oil ratio and oil saturation in the oil zone corresponding to Fig. 10.41.

ultimate recovery has the same value, 38.5 per cent of the pore space or 72 per cent of the original oil in place. These are to be compared with the indicated ultimate recoveries, even to atmospheric pressure, without gas return or gravity drainage of 14.5 per cent of the pore space and 27.1 per cent of the original stock-tank-oil content. The high gravity-drainage recoveries are, of course, merely the direct result of assuming $\overline{\rho_{og}} = 0.15$. Hence, as implied by Eq. (4) and shown in Fig. 10.41, this recovery is independent of the amount of gas return. In fact, if the reservoir could be produced so slowly even without any gas return as to permit complete gravity drainage, with the same residual-oil content, the recovery would then, too, be 38.5 per cent of the pore volume. While the value $\overline{\rho_{og}} = 0.15$ has been chosen arbitrarily, these considerations show that it is the low residual-oil saturation resulting from the gravity-drainage process which makes this mechanism offer great promise for potential oil recovery. The gas injection serves mainly to maintain the reservoir pressure, prolong the flowing life, and provide high flow capacities. It is only to the extent that these conditions facilitate operations under which gravity drainage can be effective that pressure maintenance indirectly also represents an important contributing factor to high recovery by gravity drainage. As will be recalled from Secs. 10.7 to 10.8, distributed gas injection alone, without fluid segregation, would give recoveries that hardly even approach those indicated by Fig. 10.41 or Eq. (4) for the gravity-drainage mechanism.

The gas-oil-ratio histories plotted in Fig. 10.42 for the different gas-injection ratios are really nothing more than the single curve for $r = 0$ stretched out in accordance with the retarded-pressure-decline histories. Since the oil zone has been assumed to follow the normal depletion history, the relation between the gas-oil ratio and pressure will be independent of the gas-return ratio. However, when the different pressure vs. cumulative-recovery curves of Fig. 10.41 are applied to the same basic gas-oil-ratio vs. pressure history (cf. Fig. 10.3), the curves of Fig. 10.42 are obtained. The slower rise in gas-oil ratio for $r = 1$ thus simply reflects the slower pressure decline for $r = 1$ in Fig. 10.41. The absence of gas-oil-ratio maxima in the curves for gas return is due to the fact that even at the state of maximum recovery, the reservoir pressures have not fallen to 500 psi and the oil saturations to 45.5 per cent, as required for the development of a gas-oil-ratio maximum by the solution-gas-drive mechanism, for the particular system represented by $r = 0$.

The oil-saturation-decline curves are also plotted in Fig. 10.42. It will be seen that the average oil saturations in the shrinking oil zone remain substantially higher under gravity drainage than with normal depletion ($r = 0$), in spite of the much higher recoveries by the former. Thus, as previously indicated, the gravity-drainage mechanism not only inherently involves a high degree of depletion of the oil in the region of gas segregation but also provides at least a partial oil replacement for the downstructure withdrawals so as to retard the decline in oil saturation in the remaining oil zone. From a physical point of view it is this maintenance of high oil saturations in the oil zone by the downstructure drainage that limits the growth in the producing gas-oil ratios.

It should be understood that the numerical aspects of Figs. 10.41 and 10.42 have no quantitative significance with respect to practical applications. They have been presented only for illustrative purposes. They represent little more than direct implications of the various assumptions on which the analysis was based. Most serious of these is the value of 0.15 assumed for $\overline{\rho_{og}}$, the residual-oil saturation in the expanded gas cap.

It is only because this value is so much lower than that normally resulting from solution-gas-drive depletion that the calculated gravity-drainage recoveries were found to be more than twice those previously computed for the gas-drive mechanism. Unfortunately both field and laboratory data supporting the assumed value for $\overline{\rho_{og}}$, or any other, are extremely meager.[1] Admittedly the practical significance of the gravity-drainage or gas-cap-expansion oil-producing mechanism rests largely on the value of $\overline{\rho_{og}}$ that can be obtained under actual operating conditions. Much field and laboratory research is required to establish residual-oil data for the gravity-drainage mechanism. Undoubtedly it will be found that the value of $\overline{\rho_{og}}$ will be sensitive to the detailed rock structure and the magnitude of capillary forces, and quite possibly also to the extent to which pressure gradients superposed on the gravity forces may tend to modify the local fluid saturations near the gross gas-oil contact. Nevertheless, as will be seen in the following section, there is evidence that gravity drainage does contribute to the recovery process in actual oil reservoirs under favorable conditions, even though it is not yet feasible to evaluate its magnitude quantitatively.

10.17. Field Observations on Reservoir Performance under Gravity Drainage.—As previously noted, there are very few data pertaining to quantitative correlations between the basic characteristics of the gravity-drainage recovery mechanism and actual field performance. That gravity-drainage phenomena have been affecting the reservoir behavior in certain cases has been recognized. However it is only recently[2] that attempts have been made to segregate the contribution of the gravity-drainage process from those of other recovery mechanisms. In this section three such reservoir histories will be briefly reviewed, although no attempt will be made to provide quantitative interpretations of the observations.

The gas-oil-contact contours plotted in Fig. 10.38 for the Mile Six Pool, in Peru, show a gas-cap expansion over a vertical distance of more than 400 ft during the first 5 years of its productive life. Although the gas-cap expansion undoubtedly has been facilitated by the very effective gas-injection and pressure-maintenance operations instituted at the very beginning of production, it represents mainly an accompaniment of the downdip oil drainage. The fact that the downstructure oil saturation has been maintained at a high level, by virtue of the oil gravity drainage, is evidenced

[1] If the recently reported three-phase capillary-pressure displacement experiments [cf. H. J. Welge, *Petroleum Technology*, **11**, 1 (September, 1948)] be interpreted as approximating the potentialities of gravity-drainage oil-depletion processes, values of $\overline{\rho_{og}} = 0.15$ would appear to be a reasonable possibility in some cases.

[2] Cf., for example, L. F. Elkins, R. W. French, and W. E. Glenn [*Petroleum Technology*, **11**, (July, 1948)] on the pressure-maintenance operations in the Lance Creek field, Niobrara County, Wyo.

by the limited rise in gas-oil ratio during the producing life, as will be seen from Fig. 10.43.[1] The almost constant pressures maintained to 1946 also indicate that the injected gas largely has remained trapped in the gas cap and was not being used merely as a supplement to the dissolved gas for oil expulsion by the solution-gas-drive mechanism.[2] The high structural

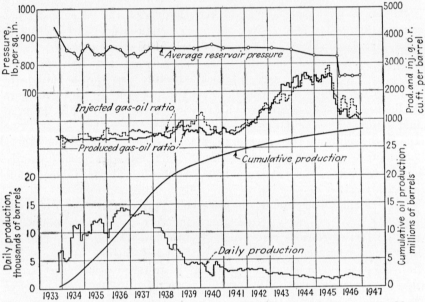

FIG. 10.43. The pressure and production history of the Mile Six pool, Peru. (*After Moyer, Oil and Gas Jour., 1947.*)

relief, good formation permeability—of the order of 1,000 md—and low-viscosity oil (40°API) have all been conducive to the development of significant gravity drainage and represent the conditions that are most favorable for achieving this type of reservoir performance. Aside from the high recovery to be derived from this field, estimated as 3 times the solution-gas-drive recovery, it is expected that 95 per cent of the ultimate recovery will be obtained by natural flow.

The Wilcox Sand reservoir of the Oklahoma City field has experienced a

[1] Fig. 10.43 is taken from Moyer, *loc. cit.*

[2] Although the production data of Fig. 10.43 do not of themselves eliminate the possibility of some water-drive action, no material change in water level has been observed in the field. The sudden drop in pressure in 1946 resulted from a well blowout in January, 1946, with a loss of 1,450 billion cubic feet of gas and 7,000 bbl of oil in 7 days. The marked decline in gas-oil ratios in 1945–1946 was due to the completion of new wells in the residual-oil-saturated section of the formation and the working over of old wells.

radically different type of history, although it, too, has enjoyed the benefits of gravity drainage. The pressure, production-rate, and gas-oil-ratio data for this reservoir are plotted in Fig. 10.44[1] vs. the cumulative production. Because of the extended period of development of the field (cf. Sec. 14.8) it is not feasible to make a detailed analysis of the reservoir behavior.

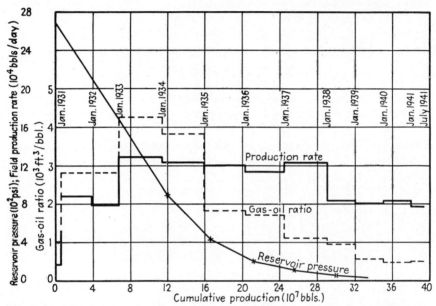

Fig. 10.44. The pressure and production history of the Wilcox Sand reservoir of the Oklahoma City field, Oklahoma.

Nevertheless the gross performance and data pertaining to the reservoir structure, rock, and fluids leave little doubt regarding the role played by gravity drainage.

The producing sand is comprised of well-sorted and -rounded sand grains with very little cementing material. The porosity averages 18 to 19 per cent. The permeability is very high, that of many samples exceeding 1,000 md. The sand lies on the flank of a general structure, with dips ranging to 15°. Its thickness varies from 0 at the upper unconformity closure by the Cherokee Shale to the east, to a maximum of 200 ft before being cut by the water table at the west. The initial stock-tank oil in

---

[1] These were taken from tabulations by D. L. Katz [*AIME Trans.*, **146**, 28 (1942)] but are subject to some uncertainty as the data in the paper do not seem to be entirely consistent. An earlier report on all the producing reservoirs of the Oklahoma City field was published by H. B. Hill, E. L. Rawlins, and C. R. Bopp, *U.S. Bur. Mines Rept. Inv.* 3330 (January, 1937).

place in the 6,900 productive acres has been calculated to be 1,070 million barrels. The oil gravity is 38 to 39°API.

Although subjected to proration throughout its life, the field was rapidly depleted by gas-drive production. Water invaded the lower part of the structure somewhat irregularly, and during the interval 1938 to 1941 the daily water production ranged from 10,900 to 17,700 bbl, although there has been no appreciable rise in water table since 1939.[1]

The most obvious indication of the action of gravity drainage in this reservoir is that, although by July, 1941, the pressures had become substantially atmospheric, the field was still producing at a rate exceeding 75,000 bbl/day from its 466 pumping wells. In view of the complete depletion of pressure, water invasion evidently was not supplying a substantial water-drive action. Further evidence is provided by the expansion of the gas caps. While the latter were originally of very limited area, with a gas-oil contact at − 5,052 ft in the northern part of the field and − 5,116 to − 5,164 ft in the southern extension, they had expanded by July, 1940, to depths ranging from − 5,196 to − 5,266 ft in the various strata and areas comprising the reservoir. Finally, oil-saturation determinations on cores and bailer samples, from wells drilled after the field had been substantially depleted, gave values ranging from 1.0 to 26 per cent above the gas-oil contacts and 52 to 93 per cent below the contact. The former averaged substantially less than those observed in typical sands depleted by the gas-drive mechanism. Moreover similar values were obtained in laboratory gravitational-drainage experiments on columns of consolidated Wilcox Sand, saturated with an oil of the same viscosity as the Wilcox crude at reservoir temperature (130°F).

As noted in Sec. 10.12, the existence of gravity drainage in the Wilcox Sand reservoir of the Oklahoma City field is also demonstrated by the anomalously low position of the permeability ratios vs. liquid-saturation curve (cf. Fig. 10.29). The low gas-oil ratios during the later stages of reservoir depletion (cf. Fig. 10.44), from which the permeability ratios are derived, reflect the high oil saturation in the zone from which the production is obtained. The cumulative withdrawals, if averaged uniformly over the whole producing section, would evidently imply an average residual-oil saturation that is lower than in the region replenished by gravity drainage. If the true permeability-ratio curve for the producing formation were known, the extent of the gravity drainage could be determined approximately from the shift in the liquid-saturation axis required to bring the field permeability ratios into agreement with the laboratory data, provided that the thickness of the zone of high oil saturation is known. For example, if the actual permeability-ratio curve for the Wilcox Sand were the same

---

[1] These observations refer to the date of the paper by Katz (*loc. cit.*).

as the unconsolidated-sand curve *A* of Fig. 10.29, then at the state where the observed permeability ratio was 0.3 the true average liquid saturation in the oil-producing part of the pay would be 71.4 per cent, in contrast to the indicated value of 63.5 per cent, which apparently represents an average liquid saturation as if it were uniformly distributed over the whole formation. The difference is a measure of the magnitude of the oil drainage expressed as a fraction of the pore space. When applied with caution and with full recognition of the limitations of the data, this type of interpretation of field permeability-ratio curves can serve to develop valuable information regarding the quantitative aspects of gravity-drainage phenomena.

Another example of the occurrence of gravity drainage under conditions of normal depletion without deliberate operating control is provided by the Lakeview field, Kern County, Calif.[1] This pool, which is a part of the Midway-Sunset field, was originally discovered in 1910 by the "Lakeview gusher," which produced 8,250,000 bbl, in about 18 months, from a depth of 2,225 ft. Failures in subsequent drilling led to abandonment of the field until 1935, when it was rediscovered in the Hallmark Sand at 2,545 ft. The sand is wedge-shaped and converges to a featheredge pinchout on the north flank of the Thirty-five Hill anticline. It has a 20° dip and thickness ranging from 0 to 200 ft. The oil gravity is 22°API. While there apparently was no free gas initially in this sand, the tremendous withdrawals from the Lakeview gusher led to the development of a gas cap at the crest of the structure. Detailed reservoir-performance records are not available. But the behavior of the 88 wells drilled after the rediscovery in 1935 has revealed a striking and progressive recession of the gas-oil contact. Thus whereas previous depletion had resulted in a gas-oil contact at 1,600 ft subsea on Jan. 1, 1935, the development after the rediscovery of the field led to a depression in the contact to − 1,700 ft by Oct. 1, 1936. It receded further to − 1,847 ft by Aug. 1, 1937, and on Feb. 1, 1938, it was found to lie at − 1,924 ft. While determinations of this type are not subject to high precision, there appears to be little question that a major downflank oil drainage has occurred in the Lakeview field. In fact this recession developed such regularity that the fall in the gas-oil contact appeared to be directly proportional to the oil withdrawals, at a rate of 1 ft/51,000 bbl. There has been some water intrusion, but this apparently has not played a major role in the performance. Aside from the initial uncontrolled gas-drive production of 8,250,000 bbl by the original discovery well, the subsequent pumping recovery of 12,000,000 bbl to June, 1938, appears to have been largely due to gravity drainage.

The examples of gravity drainage in the Oklahoma City field and the Mile Six field, discussed above, represent extremes in reservoir conditions and

[1] Cf. T. P. Sanders, *Oil and Gas Jour.*, **37**, 56 (June 23, 1938).

operation. While the latter produces from a normal sandstone reservoir, the Wilcox Sand at Oklahoma City is unique in that its connate-water content is the lowest thus far reported for any oil-producing reservoir, namely, 1 to 2 per cent. Moreover the crude produced from the field appears to wet the sand, preferentially with respect to water. In view of the abnormally low water saturations and oil-wetting preference it is very likely that a substantial part, if not all, of the sand surface is in direct contact with the oil phase. Under such conditions and in view of the uniform and favorable sand texture, very effective gravity drainage would be anticipated.[1] On the other hand, the displacement of the oil by a non-wetting phase, as water, would be expected to result in high residual-oil saturations. This also appears to be the conclusion derived from a study of the water-invaded region of the Wilcox Sand, which indicates the residual-oil saturations to be of the order of 50 per cent.

The Oklahoma City field observations also serve to show that under favorable conditions gravity drainage can contribute effectively to the oil recovery even after the reservoir pressure is substantially depleted. While comparable studies have not been reported for other reservoirs, it is not unlikely that in many of the older fields the long-persistent low rates of "settled production" represent, in part at least, the contributions of gravity drainage and oil-saturation redistribution in the producing formations. On recalling Sec. 5.8, the steady-state radial-flow gravity-drainage rate into a well is given by

$$Q = \frac{6.667 \times 10^{-4} \, k\gamma(h_e^2 - h_w^2)}{\mu\beta \log_{10} r_e/r_w} \text{ bbl/day,} \tag{1}$$

where k is the permeability in millidarcys; γ the specific gravity of the liquid; μ its viscosity; β its formation-volume factor; h_e the pay thickness or the fluid head, in feet, at r_e; and h_w the fluid head at the well radius r_w. Thus if $k = 100$ md, $\gamma = 0.75$, $h_e = 50$ ft, $h_w = 0$, $\mu = 3$ cp, $\beta = 1.1$, and $r_e/r_w = 2,640$, Eq. (1) gives $Q = 11.07$ bbl/day.[2] In order of magnitude, therefore, the simple radial-flow gravity drainage through a horizontal stratum might well suffice to account for some of the observations of continued low-rate or stripper production after the pressure has been substantially depleted. In fact it is difficult to find any other explanation for the prolonged oil bleeding from the reservoirs in many of the old fields that have been pumped to partial vacuum for a number of years.

The Mile Six field represents an extreme case of gravity drainage fol-

[1] The residual oil after drainage in an oil-wet sand may be considered as the equivalent of "connate oil." Because of the rounded sand character and slight cementation, "connate-oil" saturations as low as 10 per cent might well be expected.

[2] If h_e be taken as 100 ft and k as 500 md, as may obtain in the Oklahoma City reservoir, the resulting value of Q(221 bbl/day) would more than provide for the average production rate actually observed.

lowed by gas-cap expansion at substantially the initial reservoir pressure. While these operations undoubtedly have been successful, the contribution of the pressure maintenance to the gravity-drainage phase of the production mechanism cannot be quantitatively evaluated. As noted in Sec. 10.15, the factor $\Delta\gamma/\mu_o\beta_o$, in Eq. 10.15(3), for the maximum rate of gravity drainage decreases but slowly with decreasing pressure, for oils of 40°API gravity or higher. With regard to this factor, therefore, the pressure maintenance, in itself, could have had but little effect in the case of the Mile Six field.

The effect of the maintenance of pressure on the oil permeability k_o has probably been of greater significance. In so far as the decline in reservoir oil saturation due to fluid withdrawals is normally accelerated by the shrinkage associated with falling pressures, the reduction in the permeability to the oil is thus also acccentuated, and the rate of gravity drainage will be correspondingly reduced. It is likely, therefore, that the maintenance of pressure has retarded the decline in the rate of gravity drainage by minimizing the reduction in oil saturation. In addition the maintenance of pressure in the gas cap may have served to supplement the effect of the gravity forces in causing a downflank mass oil movement so as further to aid in keeping the oil saturation high in the oil-producing zone. Aside from these effects of uncertain magnitude the gas injection has undoubtedly been responsible for the sustained flowing life of the field. This, in itself, together with the potential gain in recovery by the sweeping action of the gas, such as is discussed in Secs. 10.7 and 10.13, would probably have been of sufficient value to justify the pressure-maintenance operations.

The most efficient type of gravity-drainage production would be that of the extreme idealized case where, by pressure maintenance, no gas is allowed to evolve in the oil zone. If the reservoir had an initial gas cap, such performance could be obtained, in principle, by building up the pressure in the gas cap above its initial value and producing the oil zone above its original bubble point. In an undersaturated reservoir it would be necessary to create a gas cap and gas-oil contact by gas injection and then develop a free-fall gradient down the structure flanks, while the minimum pressure was maintained above the bubble point. k_o would then be maintained at its maximal value, and the rates of withdrawals that could be sustained by gravity drainage would also be maximal. As these withdrawals would be produced with the solution-gas-oil ratio, the rate of gas injection required to maintain the pressure would remain substantially constant, and small as compared with that required in distributed gasinjection operations.

It is to be noted that, unless the reservoir is inherently susceptible to fluid segregation and gravity drainage and the downdip pressure gradients are not high compared with the gravity gradient, the gas injected for pressure maintenance will not remain trapped in the gas cap but will channel and diffuse through the oil zone in a manner assumed in the theoretical treatments of Secs. 10.7 and 10.8. When the injected gas does remain in the gas cap, the gravity drainage and replacement of the fluid withdrawals along the structural flanks must provide the direct mechanism of pressure maintenance in the oil-producing region. Otherwise the pressure in the latter would continue to decline, whereas that in the gas cap would be built up. Of course, as discussed in Secs. 10.7, 10.8, and 10.14, under favorable conditions gas injection without gravity drainage will in itself lead to substantial increases in recovery. And any regional downdip pressure gradient will also induce a migration of oil downstructure. On the other hand, if the reservoir structure and character of the rock and fluids are inherently favorable for substantial gravity drainage, as would be indicated by Eq. 10.15(3), pressure maintenance by gas injection will serve mainly as a means to prolong the flowing life and permit greater oil-withdrawal rates.

As previously pointed out, the basic merit of the gravity-drainage mechanism is that it provides a very efficient oil recovery within the expanded gas-cap zone. If the latter is created by true gravity drainage, rather than a general break-through and gas channeling near the gas-oil contact, its residual-oil saturation should be appreciably lower than can be achieved by the solution-gas-drive production mechanism. It should be observed, however, that the oil-drainage process at the gas-oil interface does not take place instantaneously. As the oil saturation is reduced, the permeability falls and further drainage is retarded.[1] The low residual-oil saturations associated with the gravity-drainage mechanism represent equilibrium values only. It will therefore not be feasible to maintain the maximum rate of gravity drainage given by Eq. 10.15(3) in the upper parts of the oil-saturated section even though the pressures are maintained above the bubble point. On the other hand the maintenance of pressure will keep the formation-volume factor of the oil high, so that whatever residual-oil saturation is left by gravity drainage will represent a lower equivalent of stock-tank oil than if it were trapped in the reservoir at low or atmospheric pressure.

10.18. Partial-water-drive Reservoirs.—In terms of frequency of occurrence the partial water drive is probably the most important oil-recovery mechanism.[2] For, as pointed out previously, most gas-drive reservoirs are

---

[1] Cf. Cardwell and Parsons, *op. cit.*

[2] In some producing districts, as Pennsylvania and California, even partial-water-drive reservoirs are rather uncommon. From a national standpoint, however, it is doubtful that strict solution or gas-cap drives represent the majority of oil reservoirs discovered to date.

bounded, in part at least, by mobile-water reservoirs, which supply water intrusion into the oil zone in varying degrees. On the other hand, except when greatly undersaturated, the great majority of actual reservoirs ultimately controlled by complete-water-drive action initially undergo a gas-drive history, while a sufficient pressure decline develops to induce enough water intrusion to replace the oil and gas withdrawals.

Partial-water-drive reservoirs will be considered here as those in which the water intrusion does not suffice to substantially replace all the reservoir space voidage caused by the oil and gas withdrawals. The partial-water-drive mechanism seldom controls the whole production history of a reservoir. At the beginning of the producing life it will generally be of little influence in the reservoir behavior, which will appear to be strictly gas drive. Ultimately, however, if the maximum supply capacity of the water reservoir is comparable with the rate of fluid withdrawals, the rate of water intrusion may become equal to the latter and the partial-water-drive performance will be transformed into that of a complete water drive.

As pointed out in Sec. 9.2, although the ultimate oil-displacement mechanisms of partial and complete water drives are the same and the ultimate recoveries are often comparable, the general performance of partial-water-drive reservoirs is more appropriately treated as a generalization of that of solution-gas-drive systems. The high recoveries associated with complete-water-drive performance are automatically provided in the generalized gas-drive analysis by simply choosing for the residual-oil saturation in the water-invaded area a value characteristic of the water-flushing displacement process. And while in the actual numerical illustrations a steady-state type of representation will be used for the water-influx history, this will be done only for convenience, the basic equations being applicable to compressible-liquid aquifers as well as steady-state water-supply reservoirs. At the same time the treatment will give an account of the fundamental gas-drive characteristics of the productive area. Since the physical structure of the theory will be largely based on concepts previously developed for gas-drive-reservoir analysis, the performance of partial-water-drive reservoirs will be presented here as a natural sequel to the previous discussions of solution-gas-drive and gravity-drainage producing mechanisms.

In many respects the behavior of partial-water-drive systems is similar to that of gravity-drainage- and gas-cap-expansion-drive reservoirs. In both, the oil-producing mechanism leading to the *local* oil expulsion from the producing wells is basically the gas-drive process. In both, the modifying external agent—gravity drainage or water intrusion—serves directly as means for moving oil from regions distant from the immediate locale of the withdrawals into that region, so as to retard the decline in oil saturation in the latter. In both, the external agents themselves leave lower residual-oil saturations and are more efficient in oil displacement than the solution-gas-drive mechanism and hence are the controlling factors in determining the ultimate recovery. In both, the area of production gradually shrinks as the producing wells near the current limits of the productive area are enveloped by the displacing fluid. And both types of producing mechanism are essentially rate-sensitive, in that the direct effect of the agent supplementing the gas-drive mechanism on the reservoir performance is dependent on the magnitude of the fluid-movement capacity associated with the gravity drainage or water intrusion as compared with the rate of reservoir fluid withdrawals.

In contrast it may be noted that whereas the magnitude of gravity drainage is, in a broad sense, substantially constant during the producing life, except for changes in the permeability, the rate of water intrusion depends directly on the past history. It always begins at zero and generally increases continuously with increasing reservoir withdrawals, unless the latter are subject to severe fluctuations or the water reservoir itself has a limited total fluid-expansion capacity. Moreover in gravity-drainage systems the main factor is the fluid redistribution within the original hydrocarbon reservoir volume. Hence, unless gas is injected for pressure maintenance, the pressures will continue to decline throughout the production history, even if the gravity drainage itself operates without restraint. In partial-water-drive reservoirs, however, the fluid redistribution is effected by the entry of extraneous fluid, and hence there is a continual reduction in the total void space constituting the productive oil and gas reservoir. This will lead to a corresponding retardation of pressure decline, which may even be completely arrested under favorable circumstances.

In spite of the common occurrence of partial-water-drive reservoirs their quantitative analysis and interpretation are even less advanced than those for the solution-gas drive and no more advanced than those for gas-cap expansion and gravity-drainage systems. The observed average pressure history of reservoirs of this type can be formally duplicated by suitable application of the electrical reservoir analyzer, to be discussed in the next chapter. However, in this method of analysis the inherent gas-drive characteristics of the oil-productive area do not enter as an integral part of the composite gas- and water-drive reservoir, and hence the method does not provide predictions of future behavior, except by supplementary trial-and-error procedures. While the theory to be presented in this section does take into account the basic gas-drive characteristics of the continually shrinking oil-productive area, the discussion will be limited largely to the formulation of the analytical method of treatment and an examination of several examples of hypothetical reservoir performance as calculated by the method.

As was assumed in the treatment of the simple gas-drive reservoirs the oil-producing formation will be taken here, too, as uniform in every respect. Its initial area will be denoted by A_o, and that at any time after water intrusion has begun by A. It will be assumed further that immediately behind the water-oil boundary the free-gas saturation is zero[1] and the residual reservoir oil saturation is ρ_{or}, a constant throughout the producing history. The gas and oil trapped in the water-invaded region will be considered as permanently lost[2] to the unflooded oil-producing part of the original reservoir. The latter, of area A, will be treated as producing by the solution-gas-drive process. It may then be shown that the rate of pressure change in the oil reservoir or at the initial oil-water boundary will be governed by the equation[3]

$$\frac{dp}{dt} = \frac{(dW/dt) - \beta Q_o[1 + (\mu_o/\mu_g)\Psi]}{\overline{Q_o}[\lambda\beta - \epsilon\beta - (d\beta/dp)] + \epsilon f A_o(1 - \rho_w) - \epsilon W}, \tag{1}$$

where Q_o is the oil-production rate, $\overline{Q_o}$ the total stock-tank-oil volume remaining within the original oil-productive area A_o, and W the cumulative net water intrusion within A_o, all these variables referring to unit formation thickness and the time t. The rest of the notation is that used in Sec. 10.3.

Equation (1) is only one of a set of three required completely to describe the pressure, oil-saturation, and water-invasion histories. The other two are

$$A\frac{d\rho_o}{dt} = \frac{A\rho_o}{\beta}\frac{d\beta}{dp}\frac{dp}{dt} - (\rho_o - \rho_{or})\frac{dA}{dt} - \frac{\beta Q_o}{f}, \tag{2}$$

$$(1 - \rho_w - \rho_{or})\frac{dA}{dt} = -A\left[\rho_o\lambda + (1 - \rho_w - \rho_o)\epsilon - \frac{\rho_o}{\beta}\frac{d\beta}{dp}\right]\frac{dp}{dt} - \frac{\beta Q_o}{f}\left(1 + \frac{\mu_o}{\mu_g}\Psi\right). \tag{3}$$

In principle Eqs. (1) to (3) suffice for a determination of the three basic unknown functions $p(t)$, $\rho_o(t)$, and $A(t)$. The production rate Q_o is considered as a known and preassigned function of time. $\overline{Q_o}$ is then given by

$$\overline{Q_o} = \frac{f\rho_{oi}A_o}{\beta_i} - \int_0^t Q_o(t)dt, \tag{4}$$

[1] It is unlikely that all the gas phase will be immediately displaced by the invading water. As this will not materially affect the theoretical behavior, the vanishing gas saturation is assumed here for the sake of simplicity.

[2] In accordance with the assumption of complete reservoir uniformity the producing formation is treated as being sharply divided between a clean oil-productive area A and an area $A_o - A$, which would produce only water. The corresponding theoretical recoveries and performance histories will therefore be more favorable than those to be expected in practice.

[3] It may be noted that if $W = 0$ and A_o be interpreted as the total thickness of oil zone and gas cap, Eq. (1) will also apply to gas-cap-expansion reservoirs and is one of the equations referred to on p. 490. And if $W = 0$ and A_o is taken as constant, Eq. (1) readily reduces to that for strict solution-gas-drive reservoirs (cf. footnote on p. 415).

where ρ_{oi} is the initial-oil saturation. $\overline{Q}_o$ is thus also a known function of t. The net water-intrusion terms W and dW/dt represent functions of the pressure and time, determined in detail by the nature of the water-bearing reservoir and operational history. If the water reservoir performs as a compressible-liquid system, W and dW/dt may be constructed from the analytical expressions given in the next chapter, corrected for the water production. If it can be approximated as a steady-state type of water intrusion and the water production is considered as shut in or completely returned to the formation, dW/dt and W will take the simple forms

$$\frac{dW}{dt} = c(p_i - p); \qquad W = c \int_0^t (p_i - p)dt, \tag{5}$$

where it is assumed that the pressure differential controlling the rate of water invasion is the same as the total average pressure drop within the oil-productive area, $p_i - p$, and that the coefficient c does not change during the production history.

From the nature of Eqs. (1) to (3) it is clear that the derivation of analytical solutions is literally impossible. But even their numerical treatment is so extremely laborious that an attempt to develop large classes of solutions is impractical. To obtain some idea as to their implications it will suffice to make the simplification that the production rate Q_o is constant and that the water intrusion follows the steady-state approximation of Eqs. (5). It is then convenient to introduce the basic dimensionless parameters

$$\overline{Q} = \frac{Q_o\beta_i}{fA_o(1 - \rho_w)}; \qquad w = \frac{cp_i}{Q_o}; \qquad \overline{A} = \frac{A}{A_o}; \qquad \bar{t} = \overline{Q}t. \tag{6}$$

$\overline{Q}$ is thus a measure of the production rate expressed as a fraction of the initial total reservoir oil content. And w is a measure of the flow capacity of the water reservoir, expressed as a ratio of the maximum steady-state intrusion rate to the production rate. $\overline{A}$ represents the residual-oil-productive area as a fraction of the original oil-reservoir area. $\bar{t}$ is a dimensionless time giving the cumulative recovery as a fraction of the initial reservoir oil content. In this notation Eqs. (1) to (3) may be written as

$$\left.\begin{array}{c} \dfrac{dp}{d\bar{t}} = \dfrac{w[1 - (p/p_i)] - \beta[1 + (\mu_o/\mu_g)\Psi]}{\theta(p) - \omega(p)\bar{t} - \epsilon w \int_0^{\bar{t}} [1 - (p/p_i)]d\bar{t}}, \\[4mm] \theta(p) = \lambda\beta + \epsilon(\beta_i - \beta) - \dfrac{d\beta}{dp}; \qquad \omega(p) = \theta(p) - \epsilon\beta_i; \end{array}\right\} \tag{7}$$

$$\overline{A}\frac{d\rho_o}{d\bar{t}} = \overline{A}\frac{\rho_o}{\beta}\frac{d\beta}{dp}\frac{dp}{d\bar{t}} - (\rho_o - \rho_{or})\frac{d\overline{A}}{d\bar{t}} - \frac{\beta(1 - \rho_w)}{\beta_i}; \tag{8}$$

$$(1 - \rho_w - \rho_{or})\frac{d\overline{A}}{d\overline{t}} = -\overline{A}\left[\rho_o\lambda + (1 - \rho_w - \rho_o)\epsilon - \frac{\rho_o}{\beta}\frac{d\beta}{dp}\right]\frac{dp}{d\overline{t}}$$

$$- \frac{\beta(1 - \rho_w)}{\beta_i}\left(1 + \frac{\mu_o}{\mu_g}\Psi\right). \quad (9)$$

While it is possible to solve these simultaneous equations numerically by standard procedures, it is convenient to use as checks the corresponding integrated material-balance equations. These may be expressed as

$$\rho_o = \frac{\beta}{\overline{A}}\left[\frac{1 - \rho_w}{\beta_i}(1 - \overline{t}) - \rho_{or}\int_{\overline{A}}^1 \frac{d\overline{A}}{\beta}\right]; \quad (10)$$

$$(1 - \overline{A})(1 - \rho_w) = \frac{(1 - \rho_w)w}{\beta_i}\int_0^{\overline{t}}\left(1 - \frac{p}{p_i}\right)d\overline{t}$$

$$+ \frac{\rho_{or}}{\gamma}\left[\varsigma\int_{\overline{A}}^1 \frac{dA}{\beta} + \int_{\overline{A}}^1 \frac{S\,d\overline{A}}{\beta}\right], \quad (11)$$

$$\frac{1 - \rho_w}{\beta_i}S_i = \frac{1 - \rho_w}{\beta_i}\overline{t}\overline{R} + \overline{A}\left[\frac{S\rho_o}{\beta} + \gamma(1 - \rho_w - \rho_o)\right] + \rho_{or}\int_{\overline{A}}^1 \frac{S}{\beta}d\overline{A}, \quad (12)$$

where $\overline{R}$ is the cumulative gas-oil ratio, and $\varsigma = \gamma\beta - S$. It will be noted that Eq. (10) expresses the requirement of reservoir oil-content conservation, Eq. (11) is the integrated equation of continuity with respect to gross reservoir volume, and Eq. (12) is the material-balance condition with respect to the reservoir gas content. In addition the incremental gas-oil ratios are required to satisfy the basic equation:

$$R = S + \alpha\Psi. \quad (13)$$

Illustrative calculations have been made using the above equations, assuming $\rho_w = 0.25$, $\rho_{or} = 0.20$, and values of $w = 0.5, 1, 3, 5$. The reservoir fluid and rock characteristics were assumed to be those previously used in Sec. 10.4 for the 30°API-gravity crude. The pressure and gas-oil-ratio histories corresponding to the above parameters are plotted in Fig. 10.45. The manner in which the productive area shrinks owing to water invasion is shown in Fig. 10.46. The variation of the reservoir oil saturation in the unflooded productive area is plotted in Fig. 10.47. In these figures the curves for $w = 0$, representing the strict solution-gas-drive performance, are also plotted for comparative purposes. As previously noted the basic independent variable $\overline{t}$ represents the cumulative recovery expressed as a fraction of the initial oil in place. And, as indicated by Eq. (6), the parameter w is the ratio of the maximum steady-state water-supply capacity of the aquifer to the withdrawal rate Q_o.

Considering first the curves for the pressure and gas-oil-ratio history, i.e., Fig. 10.45, it will be seen that the initial trends are quite similar to those for the strict solution-gas drive except that the pressure declines are increasingly retarded as w increases. The rise in the gas-oil ratio is also reduced, and the maximum becomes smaller and delayed to larger recovery values as w is made larger. These changes in the gas-oil ratio are, of course, to be expected, since the total areas under the curves should be independent of w except for the gas trapped in the flooded area. For $w = 0.5$ the gas-oil ratio finally declines sharply, and the pressure falls

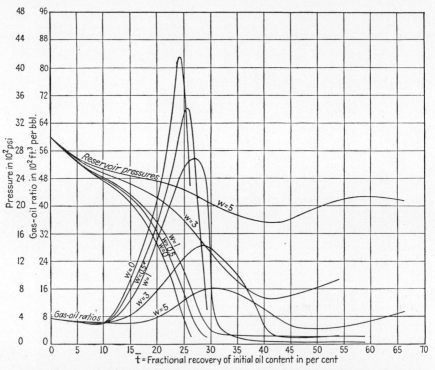

FIG. 10.45. The calculated pressure and gas-oil-ratio histories of partial-water-drive reservoirs. w = (maximum steady-state supply capacity of aquifer)/(oil-withdrawal rate). Residual-oil saturation in water-invaded zone assumed = 0.20.

to the assumed abandonment limit of 100 psi with a recovery of 29.3 per cent of the initial-oil content. While the latter is 3.1 per cent greater than that for $w = 0$, no radical change in the performance is evident. For $w = 1$, however, almost complete pressure stabilization develops at about 100 psi after a third of the initial-oil content is recovered. Shortly thereafter the gas-oil ratio becomes stabilized at the solution value. As will be noted by reference to Fig. 10.46, the curves for $w = 1$ have been terminated when the productive area has shrunk to 10 per cent of its initial value, which occurs after a recovery of 58.9 per cent of the initial oil in place. It is, of course, doubtful that the operating conditions after 90 per cent of the productive area has been flooded would still conform even approximately to the assumptions underlying the analysis.

A still more striking development in the production history is exhibited by the curves for $w = 3$ and 5. Here, instead of merely attaining a stabiliza-

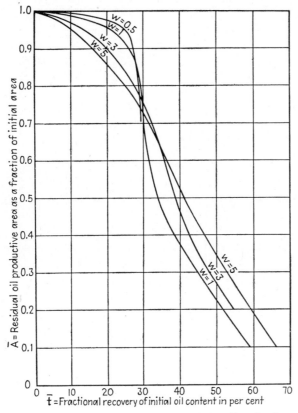

FIG. 10.46. The calculated histories of the shrinkage of the oil-productive area in partial-water-drive reservoirs. $\overline{A}$ = fraction of original reservoir area that is still uninvaded by water; w = (maximum steady-state supply capacity of aquifer)/(oil-withdrawal rate). Residual-oil saturation in water-invaded zone assumed = 0.20.

tion in pressure, the latter falls to a minimum and then rises. This, however, is not surprising when considered in connection with the gas-oil-ratio curves and Eq. (1). For, as is to be expected, Eq. (1) implies that dp/dt will become positive, *i.e.*, the pressure will begin to increase, when the numerator becomes positive. This will occur when the rate of water influx exceeds the rate of space voidage, which is the second term in the numerator of Eq. (1). On noting the marked declines in gas-oil ratio in the interval between 30 and 40 per cent recovery and taking into account the magnitude of w and the pressure decline, as indicated more explicitly in Eqs. (7), it is readily verified that the conditions for a rise in pressure should develop about the time shown by the curves of Fig. 10.45. As will be seen by reference to Fig. 10.47 the approach to a balance between the water-influx

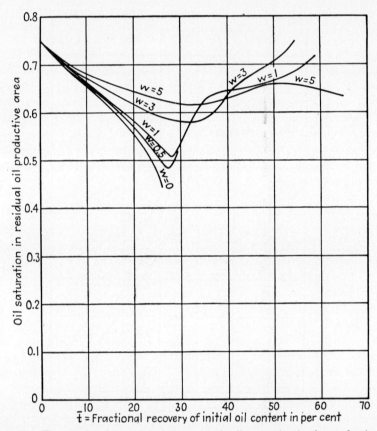

Fig. 10.47. The calculated histories of the reservoir oil saturation in the productive area of partial-water-drive reservoirs. $w =$ (maximum steady-state supply capacity of aquifer)/(oil-withdrawal rate). Residual-oil saturation in water-invaded zone assumed $= 0.20$.

and space-voidage rate is accelerated by the rise in reservoir oil saturation for $\bar{t} > 30$ per cent and by the fall in the value of Ψ. Moreover, as the latter trend continues until the oil saturation exceeds 65 per cent, the gas-oil ratio becomes the solution ratio[1] and it begins to rise again in accordance with the rise in pressure. The subsequent development of a maximum in the pressure for $w = 5$ reflects the reduction in the rate of water influx as

[1] It has been assumed in these calculations that the gas phase remains in equilibrium with the oil and will go back into solution if the pressure rises.

the pressure increases until it no longer balances the rate of space voidage owing to the fluid withdrawals.

As will be seen from Fig. 10.46 the shrinkage in productive area is rather slow at first for the lower values of w. By the time one-fourth the initial-oil content has been produced, only 5 per cent of the productive area will be flooded if $w = 0.5$. The areal invasion of the edgewater then becomes rapidly accelerated, and only 71 per cent of the initial area remains productive when the ultimate recovery of 29.3 per cent is produced. For $w = 1$ the sharp decline in productive area also develops after a recovery of about 25 per cent of the oil in place. While the rate of decline slows down after a 30 per cent recovery, it still continues at a rapid rate throughout the pressure-stabilization period until the assumed limit of 10 per cent of the original productive area is reached. For $w = 3$ and $w = 5$ the initial rates of shrinkage of the productive area are greater. These curves, too, show inflections but, as in the case of $w = 1$, ultimately assume an approximately linear decline with increasing recovery and would fall to a residual productive area of 10 per cent at 61 and 66 per cent recoveries.

The histories of the oil saturation in the uninvaded area, plotted in Fig. 10.47, show an interesting feature of the theory of partial water drives presented here. This is the development of minima and subsequent rises in the oil saturation, even for $w = 0.5$, whereas for the strict solution-gas-drive performance the oil saturation falls continuously as production proceeds. The formal analytical reason for this behavior can be derived from Eq. (8). In the latter the first and third terms are basically negative while the pressure is declining, whereas the second is positive. As was seen above, during the early producing history the rate of shrinkage of the productive area is rather slow. Hence the negative terms dominate Eq. (8) and the oil saturation declines. However, when the sharp rate of decline of the productive area develops, the second term in Eq. (8) becomes comparable with the sum of the other two. As the productive area continues to shrink and the rate of pressure decline also falls, the second term becomes equal to and then exceeds the negative terms. It is during this phase that the minimum in oil saturation and the subsequent rise take place.

From a physical point of view the rise in oil saturation, or resaturation of the productive area, simply reflects a greater volumetric rate of water intrusion than fluid withdrawals. That such a situation could develop for $w = 3$ and 5 is not surprising, since these values of w imply that the maximum potential water-supply capacity of the water reservoir is 3 and 5 times, respectively, the oil-withdrawal rate. Except for the voidage created by the free-gas withdrawals, pressure stabilization and resaturation should develop long before the pressures fall to atmospheric. However, in the case of $w = 1$, and especially for $w = 0.5$, the balance between the withdrawals and direct water intrusion would appear unattainable. Here the primary reason lies in the evolution and expansion of the gas dissolved in the oil trapped in the flooded area, as the pressure in this area declines together with that in the residual productive part of the reservoir. In fact a calculation of the rate of expansion of the trapped oil and gas at the minimum point for $w = 0.5$ in Fig. 10.47 shows it to exceed considerably the rate of water invasion. Thus the *net* water influx into the oil-productive area while the pressure is declining is actually greater than the rate of oil withdrawal, even though the rate of intrusion into the original oil-bearing reservoir is less than half the oil-withdrawal rate.[1]

Because of the complexity of the calculations involved in the solution of Eqs. (7) to (12) no simple physical explanations can be given for the rather peculiar shapes of the rising portions of the oil-saturation curves of Fig. 10.47 for $w = 1$, 3, and 5. They may partly reflect computational errors or approximations, although various criteria of consistency were continually applied. In the case of $w = 3$ the calculations were stopped when the oil saturation built up to its initial value. While it is possible to introduce appropriate modifications to derive the subsequent history, this latter undoubtedly would be of questionable significance. On the other hand there appears to be little reason to doubt the correctness of the broad features of the performance histories as implied by Figs. 10.45 to 10.47.

The ultimate recoveries implied by Figs. 10.45 to 10.47 are for $w = 0$, 0.5, 1, 3, and 5—26.2, 29.3, 58.9, 54.6, and 66.2 per cent, respectively, of the initial stock-tank-oil content of the reservoir. Those for $w = 0$ and 0.5 refer to an abandonment pressure of 100 psi. The values for $w = 1$ and 5 represent the recoveries by the time 90 per cent of the original productive area becomes invaded by water, although the pressures at such times would theoretically be 104 and 2,080 psi, respectively. The recovery of 54.6 per cent for $w = 3$ corresponds to the arbitrary termination of the calculations at the time of complete resaturation of the residual oil-productive area, which would develop, according to Fig. 10.45, when the pressure had risen again to 950 psi.

Because of these differences in the terminal states the indicated variations in ultimate recoveries do not quantitatively reflect those in the assumed values of w. However, it will be clear that the partial water drives are inherently capable of yielding considerably greater recoveries than the solution-gas-drive mechanism. Moreover these will increase with increasing water-supply capacity of the aquifer, as compared with the withdrawal rate, at comparable abandonment pressures or limits of total flooded area.

---

[1] If it had been assumed that free gas were also trapped immediately behind the water front, the gas-expansion effect would be increased and the resaturation of the residual productive area would occur even sooner than indicated in Fig. 10.47.

On the other hand it is to be observed that these higher recoveries associated with the partial water drives are directly the result of assuming the residual-oil saturation in the flooded area to be the low value of 20 per cent. This implies a maximum potential recovery by water displacement of 73.3 per cent, which is only 7.1 per cent greater than that found for $w = 5$ by the time 90 per cent of the productive area is flooded. If the residual-oil saturation immediately behind the water-oil front were 35 per cent, the maximum potential recovery would be reduced to 53.3 per cent. And if the connate-water saturation were 35 per cent rather than the assumed value of 25 per cent, the maximum possible water-displacement recovery would be still further reduced to 46.2 per cent. It will be clear, therefore, that in partial as well as complete water drives the total recoveries will be ultimately determined by the connate-water- and residual-oil-saturation parameters. There will be no fixed ratio of relative recoveries between partial water drives and solution-gas drives, and no generalizations are warranted regarding their relative merits.

Even if the basic parameters determining the recovery factors in such a theory as discussed here should be known, the actual recoveries may be considerably lower than calculated because of reservoir nonuniformity. While all producing mechanisms will be unfavorably affected by this factor, it will probably be more serious in the case of the partial-water-drive mechanism. The premature channeling of water in high-permeability streaks may result in "drowning out" the producing wells and forcing their abandonment while much of the reservoir is still unflooded. In fact in extreme cases it may be definitely beneficial to inject gas into the reservoir and maintain the pressure so as to prevent the entry of the water, although the stratification of the formation will also reduce the efficiency and aggravate the difficulties of such operations.

Under the assumption of steady-state water intrusion used in the illustrative discussion of Eqs. (1) to (3) and their reduction to Eqs. (7) to (12) the independent variable describing the performance history is the quantity $\bar{\iota}$ representing the cumulative fractional recovery. This, however, does not imply that the reservoir behavior will be independent of the production rate, as obtains in the case of the solution-gas-drive mechanism, according to the simplified theoretical treatment of this chapter. On the contrary the production rate is a primary factor in controlling the performance. For a fixed aquifer (that is, cp_i), w will be inversely proportional to the production rate. Thus the curves for $w = 0.5, 1, 3,$ and 5 in Figs. 10.45 to 10.47 are equivalent to operating conditions where the oil-withdrawal rates are in the ratio of $1 : 0.5 : 0.17 : 0.10$. On reference to Fig. 10.45 it will then be evident that even before the effects of pressure stabilization and rise develop, the normal decline trends are slower, as functions of the cumulative recovery, for the lower withdrawal rates.   This is in accord with the expectations of the general theory of complete-water-drive performance to be presented in the next chapter.

No attempt will be made here to correlate the above theoretical predictions of idealized reservoir behavior of partial water drives with actual field observations.   Published data on reservoirs that undoubtedly were controlled by partial water drives are not sufficiently complete to warrant any serious comparisons.   It is of interest to note, however, that there has recently appeared, in a purely incidental manner, rather striking evidence of the banking of oil ahead of a water front and a resaturation of a formation partly depleted by the solution-gas-drive mechanism, corresponding to the implications of Fig. 10.47.   This developed as the result of water-injection operations along the flanks of the structure of the Pettit Zone of the Haynesville field in Arkansas and Claiborne and Webster Parishes, La.   In spite of previous crestal gas injection, the flank wells had continued to produce by the solution-gas-drive mechanism, because of the low permeability of the limestone formation and the wide well spacing.   By the time water injection was started, many of these wells were producing with gas-oil ratios of 3,000 to 5,000 ft^3/bbl.   However, within $1\frac{1}{2}$ to 4 months after water injection was begun most of the immediate offsets of the injection wells began a systematic decline in gas-oil ratio.   A typical record is that shown in Fig. 10.48[1] for well 2–3, removed by 1,350 ft from injection well A–10–1.   In this case the reaction in gas-oil ratio became evident within 6 weeks after the water injection began.   Moreover the gas-oil ratio continued to decline even though the oil-production rate was increasing for 2 months.   Finally the solution ratio was reached, and the aeration of the oil column in the well bore became so poor that the well ceased flowing and had to be placed on the pump.

There can be no question that the history of well 2–3 and of a number of others like it reflects the resaturation of the oil-bearing formation about these wells by an oil bank formed by the injected water.   And the fact that in several of the wells the gas-oil ratios fell to the solution ratios indicates that the oil saturations were built up at least to the equilibrium values.   While it has often been assumed that oil banks are formed in water-flooding operations (cf. Sec. 12.13), there has been no evidence quite so conclusive as that demonstrated by Fig. 10.48.   Although the general features of the reservoir performance at Haynesville are quite different from those of the partial-water-drive mechanism discussed in this section, they do provide a confirmation of the possibility of reservoir resaturation implied by Fig. 10.47.

[1] Figure 10.48 is taken from P. S. Ervin, *API Drilling and Production Practice*, p. 80 (1947).

10.19. Summary.—The limitation of withdrawal rates by state or voluntarily imposed proration regulations, now practiced in most producing districts in the States, has greatly reduced the number of reservoirs producing throughout their lives by the pure solution-gas-drive mechanism. These restrictions have converted most potential gas-drive reservoirs into such as are controlled by partial or even complete water drives, except

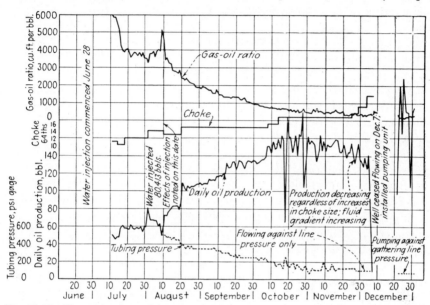

Fig. 10.48. The production and gas-oil-ratio histories of producing Well #2–3, 1,350 ft from water-injection Well #A–10–1, in the Haynesville field. (*After Ervin, API Drilling and Production Practice, 1947.*)

when the oil-bearing rock is completely sealed and closed. They have also tended to give the effect of gravity segregation between the oil and gas, combined with downstructure oil gravity drainage and gas-cap expansion, a greater chance to contribute to more efficient oil recovery. Moreover the increasing frequency of the injection of gas and/or water to maintain pressures and as supplements to whatever natural water-drive action may be present has further served to modify the normal gas-drive reservoir behavior that would obtain in many reservoirs if produced only through the medium of their original gas contents. There is nevertheless definite value in a thorough study of the simple solution-gas-drive reservoir performance. Such a study provides an understanding of the histories of the older fields, which were depleted by the gas-drive mechanism, and of those now being exploited which still must be produced in this manner. It may serve as a basis for comparison with the behavior of partial-water-drive, complete-water-drive, and gas-cap-expansion reservoirs. And it should give a description of the early history of most reservoirs, with initial reservoir pressures at the bubble point of the oil phase, before such other producing mechanisms as may ultimately control the performance become established.

While, in principle, the formulation of the equations of motion for heterogeneous-fluid systems, developed in Chap. 7, should suffice to describe the behavior of solution-gas-drive reservoirs, it is impracticable to apply them without major approximations. But if the presence of the producing wells be ignored and the reservoir is represented by a tank subjected to uniform withdrawals, a differential equation can be constructed relating the oil saturation to the reservoir pressure [cf. Eq. 10.3(1)]. On integration of this equation the pressure and gas-oil-ratio histories vs. the cumulative recoveries can be readily determined. This procedure also automatically gives the magnitude of the ultimate recovery, either at absolute pressure depletion or at any chosen abandonment pressure. The physical data entering this treatment are the gas solubility, formation-volume factor of the oil, gas density, and oil and gas viscosities, as functions of pressure at the reservoir temperature, and the gas-to-oil permeability-ratio vs. saturation curve for the producing rock.

Illustrative calculations by this method of the performance to be expected of idealized solution-gas-drive reservoirs show that the gas-oil ratio will decline initially from the original solution value, then rise rapidly to a maximum, and finally drop off sharply as ultimate depletion is approached (cf. Fig. 10.4). The initial decline is characteristic of permeability-saturation curves having a nonvanishing equilibrium free-gas saturation. The rapid rise results from the steep growth in the gas-to-oil permeability-ratio curve as soon as the equilibrium free-gas saturation is exceeded. The ultimate decline is largely due to the fall in gas density of the free-gas phase produced with the oil associated with the reduced reservoir pressures. The latter fall continuously, with slopes reflecting the trend in the gas-oil-ratio curve. The ultimate recoveries are in the range of those generally observed in practice and, for the particular permeability-saturation curve that was assumed, are of the order of 8 to 17 per cent of the total pore-space volume, or 14 to 32 per cent of the initial oil in place, depending on the physical properties of the petroleum fluids. The free-gas saturations developed at depletion range from about 21 to 32 per cent.

Comparative calculations of the effect of the oil viscosity on the recovery show that, as would be expected, the recovery will decrease with increasing viscosity. A factor of about 12 in the latter will change the recovery approximately twofold (cf. Table 1) for the range of viscosities investigated. Increasing amounts of gas in solution would, in themselves, lead to greater recoveries. But the associated increases in oil shrinkage may lead to resultant lower recovery values (cf. Table 2). In fact, by using the crude gravity as a composite index of the gas and oil characteristics and taking into account the interrelated variations of oil and gas viscosity, gas solubility, and oil shrinkage, it is found that the absolute ultimate recovery will be a maximum at about 40°API, and less at both higher and lower gravities (cf. Fig. 10.9). This is largely the result of the opposing effects of the changes in oil viscosity and shrinkage with the crude gravity. However, the percentage of initial oil recovered would increase monotonically with gravity from 10 to 50°API for the particular types of reservoir rocks and fluids assumed. The important role played by the shrinkage is also reflected in the theoretical result that the absolute recovery may decrease with decreasing connate-water content if the permeability-saturation relation be considered as fixed.

The production histories and ultimate recoveries will depend on the nature of the permeability-ratio curve for the producing rock as well as on the properties of the petroleum fluids. If there is no equilibrium free-gas saturation, the gas-oil ratio will generally begin to rise immediately on starting the withdrawals, but the maximum reached will be less than if there is an initial decline due to a nonvanishing equilibrium gas saturation (cf. Fig. 10.10). The ultimate recovery will be reduced. In general the recoveries will decrease as the gas-to-oil permeability ratios are increased (cf. Table 4).

If the effects of gas-cap expansion and gravity drainage be neglected, the theory for the simple solution-gas-drive reservoir can readily be extended to include those with initial gas caps [cf. Eq. 10.5(4)]. It is tacitly assumed in this treatment that the gas-cap gas is diffused through the oil zone and produced with the solution gas as the pressure declines. As would be expected, the recoveries increase with increasing gas-cap thickness (Table 5 and Fig. 10.13). An increase by a factor of 4 in the total gas initially available, including the gas-cap gas, as compared with the solution gas only, was found theoretically to increase the ultimate recovery by almost 50 per cent, and the free-gas saturation at depletion by somewhat less than 25 per cent, for the particular illustrative examples investigated.

Associated with the declines in reservoir pressure as a gas-drive reservoir is depleted is a continuous fall in the oil mobility. In particular the permeability-to-viscosity ratio will decrease, and hence the theoretical value of the productivity index (cf. Figs. 10.14 to 10.16). While calculations of the absolute magnitude of the productivity index are subject to question, their values during depletion, relative to those observed initially, may be approximated by ratios of the current permeability-to-viscosity ratio to that obtaining initially. On adding assumptions regarding the pressure drawdowns at the producing wells and the nature of the well-development history, an approximate theory can be derived to give the variation of the cumulative production with time. From these the production-rate and pressure-decline vs. time curves can be readily computed (cf. Fig. 10.17).

The basic differential equation relating the oil saturation to the reservoir pressure can be still further generalized to cover conditions where gas is injected to retard the pressure decline and increase the ultimate recovery. It is assumed here, too, that the gas injected in the pressure-maintenance operations is distributed and diffused uniformly throughout the oil-producing zone so as to be produced with the solution gas. The integration of this generalized equation shows that the injection into the reservoir of the produced gas will serve to retard the pressure decline and ultimately lead to higher recoveries, as expected on the basis of general considerations. The magnitudes of these effects will increase with the fraction of the produced gas that is returned. However, since the increased recoveries are generally associated with increases in the free-gas saturation, even though the retardation in oil shrinkage also is a contributing factor, the producing gas-oil ratios will tend to rise beyond the maximum values associated with the normal depletion mechanism. Thus, whereas the latter, for the hypothetical reservoir considered as an example, would be only 4,400 ft³/bbl, the maximum under 60 per cent gas-return operations would be 10,350 ft³/bbl and this would rise to 19,500 ft³/ bbl if 80 per cent of the produced gas were returned throughout the producing life (cf. Fig. 10.18). If all the produced gas were returned, the gas-oil ratio would rise to 20,000 ft³/bbl by the time the reservoir pressure had dropped from its initial value of 2,500 psi to only 1,335 psi. And if in the latter case the injection operations were discontinued at 1,335 psi, the pressure would drop sharply and would reach 100 psi with an additional recovery equal to only 1.1 per cent of the pore space. If the injection ratio were merely reduced to 80 per cent, the increased recovery to 100 psi would be 4.5 per cent of the pore space. The latter would represent an increase of only 2 per cent, in units of the pore space, as compared with an 80 per cent gas-return project throughout the producing life, although it would require about 50 per cent additional gas injection per barrel of oil recovered (cf. Table 6).

It is possible to calculate directly the pressure vs. cumulative-recovery relation without integration of the differential equation [cf. Eqs. 10.7(4) and 10.7(5)] if all the produced gas be returned. These relationships are independent of the gas-to-oil permeability-ratio curve of the reservoir rock.

As in most actual gas-injection operations only 60 to 80 per cent of the produced gas is returned, the effect will be only a retardation in pressure decline rather than strict pressure maintenance, unless there is a supporting water-drive action. To achieve complete pressure stabilization will in general require an injection ratio of more than 100 per cent [cf. Eq. 10.7(8)], although if the operations are conducted at low pressures, somewhat less than 100 per cent return may suffice in certain cases.

As would be expected, the total gain in oil recovery will be a maximum if the gas return is undertaken at the very beginning of the production history. However, the loss in recovery due to a delay in starting the gas injection is rather minor unless the starting pressure is appreciably less than half the initial pressure (cf. Fig. 10.23). The maximum gas-oil ratios, average gas-oil ratios, total gas injected, and gas injected per barrel of recovery will all decrease as the pressure at which the gas return is commenced decreases (cf. Tables 7 and 8). Conversely the oil recovery under similar conditions of operations will increase with increasing amounts of gas used in the operations (cf. Fig. 10.24).

While it appears that it is possible to predict theoretically rather completely the behavior of gas-drive reservoirs, the theory is actually severely restricted by the assumptions and approximations on which it is based. These include the assumption that the reservoir is strictly uniform throughout, that the withdrawals are distributed uniformly rather than through widely separated well bores, and that gravity does not contribute significantly to the performance. Unfortunately, however, it is difficult to evaluate quantitatively the effect of these approximations by comparison of the calculated behavior and observed field performance. For rather few of the newer fields are being produced under unrestricted withdrawals and by the simple solution-gas-drive mechanism. The limitation of withdrawals often permits the development of partial water drives or gravity drainage materially to modify the gas-drive performance. And when these do not materialize, fluid injection is often undertaken to retard the pressure decline artificially. On the other hand the older fields, now depleted, which were produced "wide open" and usually by the pure gas-drive mechanism, are of little value in that practically nothing is known about them except the productive area and production-decline curves. Means for taking bottom-hole-pressure data were simply not available, the techniques of core analysis had not been developed, and the significance of measuring accurately and recording gas production, except for sale, was not recognized. Since at best the problem of comparing theory and actual performance of gas-drive reservoirs is essentially statistical, it may be a long time before comparisons of real significance can be established.

From a practical point of view it is often necessary to estimate the future production decline of partly depleted fields that are no longer subject to withdrawal limitations. This can be done empirically on the basis of past performance by extrapolation procedures, even though they have no strict theoretical foundation. It is often found that the production decline follows an exponential law [cf. Eq. 10.11(1)]. This can be verified most simply by a linearity in a semilogarithmic plot of production rate vs. time or by a constancy of the ratio of current production rate to the decline in a previous fixed interval (cf. Table 9). A linear relation between the cumulative production and production rate is equivalent evidence [cf. Eq. 10.11(3)]. Once this type of variation is established, the future decline can be readily predicted.

A power-function relation between the production rate and time [cf. Eq. 10.11(4)] is also frequently observed. This can be linearized by suitable plotting on log-log scales. It also implies a constancy of the first differences of the ratios of production rate to previous incremental declines (cf. Table 10), and adjustable log-log linearity of cumulative production vs. the production rate [cf. Eq. 10.11(5)]. Here, too, extrapolation to give future behavior is a simple matter when the previous data are arranged to follow the prescribed relationship.

Many similar empirical expressions of production-decline data can and have been developed for the purpose of predicting future performance. This very range of representations in itself is evidence of the lack of uniqueness of any general formula that might be constructed for describing solution-gas-drive behavior. While the empirical extrapolation procedures are of value in fixing rules for estimating future decline data, they are, of course, all based on the assumption that the trend of the past history will continue to be followed for the duration of the range of extrapolation.

Very few complete solution-gas-drive reservoir-production histories have been reported thus far. Qualitatively those on which the records are available follow the theoretically expected behavior (cf. Figs. 10.27 and 10.28). The reservoir pressure declines continuously with the cumulative production and does not appear to be sensitive to the production rate. The production rate rises as drilling continues, reaches a maximum when development is completed, and then declines as the pressure is depleted. The gas-oil ratio rises with increasing cumulative recovery, reaches a maximum, and then also declines. However, in no case yet reported has the gas-oil ratio definitely shown an initial declining trend, as would be expected if the rock had a nonvanishing equilibrium free-gas saturation, although unpublished reports suggest that this may actually occur. On the other hand, failure to confirm this theoretical prediction does not in itself invalidate the basic foundations of the theory. Rather it may reflect the effect of factors, such as localized fluid withdrawals through well bores and nonuniformity of the pay, that were of necessity neglected in the theoretical calculations.

From simultaneous observations of over-all field gas-oil ratios and average reservoir pressures it is possible to calculate an effective value of the gas-to-liquid permeability ratio for the producing formation [cf. Eq. 10.12(1)]. When the corresponding cumulative withdrawals are translated into equivalent reductions in average oil saturations in the rock, curves can be constructed for the permeability ratio vs. the liquid saturation (cf. Figs. 10.29 and 10.30). At liquid saturations below 80 to 85 per cent these become approximately linear on semilogarithmic plotting. However, excluding the data for the Oklahoma City Wilcox reservoir, the field-performance curves generally lie higher than those determined by laboratory core measurements. The discrepancies are especially marked at high liquid saturations, where the field-determined data appear to indicate negligible or zero equilibrium gas saturations. This, of course, merely reflects the failure to observe initially declining gas-oil ratios. While gas-oil ratios, especially during the early development periods of fields, are notoriously inaccurate, the field data on gas-oil ratios or computed permeability ratios represent the composite effect of all those factors that are conveniently and necessarily ignored in the theoretical analysis of solution-gas-drive performance. Perhaps the most serious of these, which is present even if there are no complicating effects of water intrusion or gravity drainage, is the lack of uniformity of the producing section. This will lead to a non-uniform depletion in the various reservoir components comprising the formation as a whole. The resultant depletion history will be a superposition of those of the individual strata modified by their continuous mutual interaction and interzone fluid migration. As the permeability-saturation characteristics of reservoir rocks are basically nonlinear, the composite early history will reflect to an exaggerated extent the greater depletion in the more permeable and more highly depleted parts of the reservoir. If the whole section is open to production, the observed gross gas-oil ratios and gas-to-oil permeability ratios will appear to be abnormally high when plotted against liquid saturations, when the latter are averaged and assumed to be distributed uniformly throughout the formation.

The histories of gas-injection operations are even more difficult to interpret or predict quantitatively than those of uncontrolled solution-gas-drive reservoirs. In addition to all the complexities and uncertainties associated with the latter are those involved in the disposition and movement of the injected gas. While there is little reason to doubt the basic correctness of the theory of gas-injection systems, within its recognized limitations, and while it is desirable to calculate, when possible, the ideally expected performance history as a guide in planning gas-injection programs, the experience derived from actual pressure-maintenance operations provides a valuable supplement to the theoretical considerations. A study of examples of reported gas-injection operations definitely indicates that, where the reservoir conditions are inherently favorable, substantial increases in economic recovery and reduction in operating costs can be achieved. In the Lansing Lime reservoir of the Cunningham pool a return of 84 per cent of the produced gas, begun after the reservoir pressure had fallen from 1,115 to 424 psi, led to a marked arrestment of the pressure decline and an estimated increase in ultimate recovery, over the normal depletion recovery, of 61 per cent. The gas apparently largely remained segregated in the gas cap, and the gas-oil ratios showed virtually no rise during the first 8 years of the operations (cf. Fig. 10.31). The maintenance of pressure was probably materially assisted by partial-water-drive action.

A very extensive reservoir analysis and study was made of the Jones Sand reservoir in the Schuler field before it was unitized and subjected to gas injection. The early development of high gas-oil ratios was cut almost in half merely by unitizing the field and shutting in the wells having such ratios. This also led to a sharp reduction in the rate of pressure decline. Gas-cap injection undertaken shortly thereafter has been leading to the expected behavior. The latter was computed theoretically by using permeability-ratio data derived from the earlier depletion performance (cf. Fig. 10.35). Although the pressure maintenance was not undertaken until the pressures had dropped to 1,550 psi from an initial value of 3,520 psi, an increased recovery of 60 per cent of the primary is estimated.

A 22-month gas-injection experiment in West Texas in a Grayburg Lime reservoir, using a single injection well, indicated the difficulties that might be encountered in highly stratified or fractured formations. While no reaction was observed during the first 11 months of injection, a very sharp gas-oil-ratio rise and rapid pressure decline suddenly developed at the same time the oil-production rates were more than doubled. Six months after the injection was discontinued, following a threefold increase in gas-oil ratio, the latter returned to the normal trend anticipated without gas injection (cf. Fig. 10.36). The experiment appeared to indicate that no permanent gain would result from the operations, and the total gas production during the period of excessive gas-oil ratios was greater than expected from the normal trend by almost twice the gas actually injected into the reservoir.

The use of a tracer with the injected gas in the sand reservoir of the Canal field, California, proved its break-through to neighboring wells within 6 months after injection. A sharp rise in the gas-oil-ratio trend also developed shortly after the operations were begun (cf. Fig. 10.37). On the other hand the pressure decline was greatly retarded, although this effect was largely confined to the areas at the crest of the structure and near the injection wells. The gain in ultimate recovery is uncertain.

As these examples indicate, it is by no means axiomatic that gas injection per se will be an efficient, economically profitable, and desirable operation to be applied to all reservoirs. On the other hand, under favorable circumstances it may provide very substantial returns in oil recovery and savings in development and operating costs. The methods of theoretical treatment of gas-injection systems now available will always lead to predictions of large gains in recovery. However, these are based on the assumption that the reservoir is inherently susceptible to such operations. The main criterion to be satisfied is that the producing formation be substantially uniform and free of connected and extended high-permeability streaks and channels. If the latter are present, they must be subject to identification and sealing to prevent gas break-through. If the rock is permeated by an interconnected fracture system that must be kept open to provide a commercially profitable withdrawal capacity, gas injection may fail to achieve the major objective of a material increase in recovery. While downstructure gravity drainage and gas-cap expansion, ignored in the theoretical analyses, will aid the gas-drive recovery mechanism, reservoir inhomogeneities may completely nullify the effort and expense involved in gas-injection operations. Moreover there will be little point in undertaking a gas-injection project if the reservoir can be operated under the complete-water-drive mechanism without prohibitively restricting the withdrawal rates and if the oil-expulsion efficiency by water displacement is inherently greater than by gas sweeping.

Although the problem of gravity drainage is not yet amenable to quantitative theoretical treatment, it must be faced in practical oil-field operation. The tendency, from a statistical point of view, for gas-cap formation to be localized at the structural crests of oil reservoirs is evidence of the occurrence of upward gas migration. And numerous field histories definitely point to downdip drainage of the oil as the mechanism that serves to maintain the production long after normal gas-drive depletion would be expected to lead to abandonment.

It is possible to estimate the maximum downstructure supply capacity of a reservoir by virtue of gravity drainage [cf. Eq. 10.15(3)]. This is proportional to the oil permeability, oil density, and the square of the sine of the dip. It is inversely proportional to the oil viscosity and formation-volume factor. For an oil gravity of 30°API an oil-gas-density difference of 0.555, oil viscosity of 0.69 cp, formation-volume factor of 1.35, permeability to oil of 25 md, and dip of 20°, gravity drainage could provide an oil supply to the flank of the structure of 37.0 (bbl/day)/acre of gas-oil contact. To achieve this maximum rate of drainage it is necessary that there be no updip pressure gradient so that free-gravity-fall conditions can obtain without gas break-through and channeling. If the pressures increase downdip, the oil drainage will be inhibited, although this will be partly compensated by a buoyant force tending to cause an upstructure gas migration. If the pressures decrease downdip, there will be a mass down-structure movement of both the oil and the gas (if the latter has a non-vanishing permeability) superposed on the gravity drainage. This may lead to a break-through and local by-passing of the gas, as in ordinary gas-drive flow, and may destroy the uniform lowering of the gas-oil contact. This will be further aggravated by permeability stratification and non-uniformity in the producing formation.

From the factors entering the formula for maximum gravity drainage it follows that the conditions favoring such drainage are high formation dip, low oil viscosity, and high oil permeability. However, the important criterion for the effectiveness of gravity drainage is its rate as compared with the downstructure oil withdrawals. It is for this reason that the role played by gravity drainage and fluid segregation will be of increasing importance as the withdrawal rates are reduced. On the other hand the real significance of the gravity-drainage mechanism is that in the region invaded by the expanding gas cap, following the falling gas-oil contact, the oil saturation will be reduced below that which can economically be achieved by the solution-gas-drive mechanism. In fact there is evidence indicating that under proper circumstances the residual oil after gravity drainage may be as low as that left after water flooding. Hence, if full advantage could be taken of the gravity-drainage mechanism and if the reservoir structure and fluid properties were inherently favorable, the ultimate recovery might equal or exceed that which could be obtained by any other method.

Evidence of effective gravity drainage has been observed both in fields that have been subjected to complete pressure-maintenance operations and in fields that have been initially depleted by the gas-drive mechanism. An example of the former is the Mile Six field in Peru (cf. Figs. 10.38 and 10.43). Here, by injecting gas at an average rate exceeding the gas with-drawals from the very beginning of production, the pressures have been kept almost constant. The injected gas has been trapped in the gas cap, leading to a continuous expansion of the gas cap and fall in the gas-oil contact and very little rise in the gas-oil ratio. A recovery as high as three times that expected for solution-gas drive is estimated for this field.

The other extreme is represented by the Wilcox Sand reservoir of the Oklahoma City field, in which gravity drainage was masked by the normal solution-gas-drive mechanism until the reservoir pressures were almost entirely depleted. Nevertheless, oil has continued to drain down from the upper parts of the reservoir, leaving oil saturations of 1 to 26 per cent and supplying a production capacity for the field of 75,000 bbl/day at reservoir pressures of only 20 psi gauge. While the gravity-drainage effectiveness may have been favored by the apparent oil-wetting character of the Wilcox Sand, comparable results have been reported in other fields.

Optimum conditions for reservoir exploitation by gravity drainage would be achieved by gas injection at the very beginning of production and by creation of an artificial gas cap, if one is not found initially, so as to maintain the reservoir pressure everywhere above the bubble point. Gas evolution within the formation will then be completely prevented, and the permeability to the oil will be a maximum. The fluid-property coefficients involving the oil density, expansion factor, and oil viscosity will then also be a maximum. Basically, however, pressure maintenance by gas injection in gas-cap-expansion or gravity-drainage reservoirs serves mainly to facilitate the economic aspects of the operations, in making it possible for the rate of downstructure oil drainage to keep pace with practicable withdrawal rates. Moreover, even when the conditions within the oil body permit maximum oil mobility, the rates of withdrawal will have to be restricted to allow the drainage of the gas-oil interface to follow the mass downflank migration toward the area of production.

From the point of view of general reservoir performance, partial-water-drive systems may be appropriately considered as a generalized type of solution-gas-drive reservoir. The solution-gas-drive mechanism controls the oil-expulsion processes involved in the area of direct drainage by the producing wells. Many gross features of partial-water-drive performance are similar to those characterizing gas-cap-expansion and gravity-drainage reservoirs. Among these are the tendency for resaturation of the area of local oil depletion by mass movement of oil from the remote boundaries of the reservoir, the shrinkage of the productive area, the low residual-oil saturations immediately behind the boundary of the productive area, and the sensitivity of the pressure decline to the withdrawal rates. On the other hand, in contrast to gravity-drainage systems, in which the pressures will of necessity decline with increasing net withdrawals, the water intrusion in partial-water-drive reservoirs continually reduces the total productive volume, and its rate may build up to such values as will arrest completely the pressure decline.

The theory of partial-water-drive reservoirs may be formulated by a set of three simultaneous differential equations [cf. Eqs. 10.18(1) to 10.18(3)]. While these can be solved, in principle, for any type of aquifer supporting the solution-gas-drive system, the assumption of steady-state water influx considerably simplifies the equations and analysis [cf. Eqs. 10.18(7) to 10.18(12)]. It is then found that the primary variable defining the current state of the system is the cumulative recovery expressed as a fraction of the total initial stock-tank-oil content [cf. Eq. 10.18(6)]. The basic parameter reflecting both the water-supply capacity of the aquifer and the oil-withdrawal rates is the ratio w of the maximum possible steady-state water-influx rate to the oil-production rate. Upon choosing w and the various physical properties of the rock and petroleum fluids pertaining to a particular reservoir of interest, the solution of the governing equations will give the histories, vs. the cumulative recovery, of the reservoir pressure, the gas-oil ratio, the uninvaded oil-productive area, and the oil saturation in the latter.

A series of illustrative calculations of these histories for different values of w serves to show the changes in reservoir performance as the role played by the water drive increases. Thus for $w = 0.5$, in which the maximum possible rate of steady-state water influx is only half the oil-production rate, there is but little change in the pressure and gas-oil-ratio histories over that for the complete solution-gas drive, beyond a lowering in the maximum gas-oil ratio, a shift in the maximum to higher recoveries, and an increase in ultimate recovery at 100 psi from 26.2 to 29.3 per cent of the initial-oil content (cf. Fig. 10.45). For $w = 1$, however, these effects are markedly accentuated, and almost complete pressure stabilization develops at about 100 psi. In the case of $w = 3$ the pressure first falls to a minimum value of 650 psi and then rises as the rate of water influx begins to exceed the rate of space voidage associated with the oil withdrawal. And for $w = 5$, not only does a pressure minimum develop at 1,760 psi, but the rise is ultimately arrested at 2,140 psi when the reduced water-influx rate is again exceeded by the rate of space voidage. The gas-oil ratios also fall to minima, reach solution values, and subsequently rise as the pressures continue to increase.

The oil-productive area uninvaded by the encroaching edgewaters generally shrinks slowly at first vs. the cumulative recovery but falls sharply when the latter exceeds about 25 per cent of the initial-oil content (cf. Fig. 10.46). It is reduced to 71 per cent of its initial value by the time the pressure falls to 100 psi for the case of $w = 0.5$, with a recovery of 29.3 per cent. For $w = 1$ it shrinks to 10 per cent of its initial value when the reservoir pressure is still 104 psi, and the recovery is 58.9 per cent. Only 19.3 per cent of the initial-oil reservoir remains unflooded at a recovery of 54.6 per cent for $w = 3$. And the productive area is reduced to 10 per cent, for $w = 5$, at a recovery of 66.2 per cent and a reservoir pressure of 2,080 psi, the latter following previous minima and maxima.

The oil saturations in the uninvaded productive area show resaturation effects even for $w = 0.5$. They fall to minima of 0.485, 0.507, 0.580, and 0.615 for $w = 0.5$, 1, 3, and 5, respectively, and then rise. For $w = 1$, 3, and 5 they build up again to exceed the equilibrium saturation assumed for the rock, so as to destroy the gas mobility. And for $w = 3$ the calculations indicate that the previously depleted oil reservoir may become completely resaturated during the course of production.

The high ultimate recoveries, exceeding 50 per cent, for $w = 1$, 3, and 5, implied by the theoretical-performance curves, have no general significance.

They result from the assumption that the residual-oil saturation immediately behind the water front is 20 per cent and that the connate-water saturation is 25 per cent. It is these parameters that basically control the ultimate recoveries from partial as well as complete water drives. As either saturation is increased, the ultimate recoveries will be reduced. And if their sum should be of the order of 60 to 70 per cent, the water-drive recoveries will not greatly exceed that which would be obtained by the gas-drive mechanism. Moreover, reservoir nonuniformity may seriously reduce the water-drive recoveries, owing to channeling effects and the drowning out of the producing wells. Permeability variations will usually limit the actual recoveries from water-drive fields more than under gas-drive production.

Although no detailed comparisons have been made between the theoretically predicted performance of partial-water-drive reservoirs and that observed in practice, the possibility of resaturating a partly depleted oil-productive area by water displacement has been demonstrated in water-injection operations in the Haynesville field. In this field it has been found that within 6 weeks to 4 months after the start of water injection the gas-oil ratios of offsetting producing wells began to decline. From initial values of 3,000 to 5,000 ft³/bbl they fell in a number of cases to solution ratios (cf. Fig. 10.48) and even stopped the wells from flowing in several instances. The general occurrence of this behavior of the offset wells appears to prove definitely that the oil was being banked ahead of the water and was resaturating the partly depleted formation surrounding the producing wells. While the operations at Haynesville cannot be considered as equivalent in detail to that of natural partial-water-drive production, the physical mechanism of the water invasion and oil displacement is evidently similar to that assumed in the theoretical treatment of partial-water-drive reservoirs.

CHAPTER 11

COMPLETE-WATER-DRIVE RESERVOIRS

11.1. Introduction.—While complete-water-drive reservoirs represent a limiting type of oil-producing system, they are definitely of more than academic interest. Some of the most prolific oil fields in the world have been produced or are operating under complete water drives through all or substantial parts of their producing lives. The most celebrated example is, of course, the East Texas field. Others are the fault-line fields in Texas[1] —Powell, Wortham, Currie, Richland, and Mexia—the Frio reservoir at Thompson in Texas, many of the fields in Kansas, and some of the Arbuckle Limestone fields in Arkansas. Many of the more recently developed fields in Mississippi, as Pickens, Tinsley, Eucutta, and Heidelberg, produce such highly undersaturated crudes that they can develop no gas-drive components until the pressures have fallen to small fractions (of the order of $1/5$ to $1/2$) of their initial reservoir pressures. And since proration and the general limitation of fluid withdrawals have become common practice, many fields which may originally have had gas caps or in which free gas has developed following an initial gas-drive operation have become transformed into and produced at least temporarily as complete-water-drive fields. Yates and Conroe, in Texas, are examples of this type.

As pointed out in Chap. 9, complete-water-drive fields will be considered here as those in which substantially all the fluid withdrawals are replaced by the intrusion of water into the oil-bearing formation. This water intrusion may be all supplied directly from adjacent or underlying water-bearing strata. Or the latter may be supplemented by the return to the water or oil reservoir of produced water or of entirely extraneous water deliberately provided for injection. In a strict sense, unless the reservoir pressures are completely stabilized, without further pressure decline, the equivalence between the space voidage of the fluid withdrawals and the volume of water entry will not be exact. For during the course of the

---

[1] Since these fields were developed before the practice of gathering reservoir data, in their current sense, was well established and were moreover largely exploited by "wide-open," or uncontrolled, production, there is little evidence that complete-water-drive action actually persisted throughout their producing histories. From an oil-recovery standpoint, however, it is quite certain that water flushing must have been the ultimate oil-expulsion mechanism.

reservoir pressure decline above the bubble point, the liquid-phase expansion within the oil-bearing section will always provide some replacement for the space voidage. However, except during the initial phase of the production of highly undersaturated crudes, this will generally be so small that the contribution of the water entry may still be reasonably termed "substantially complete" if it differs from the fluid-withdrawal volume only by the reservoir liquid-expansion volume. For example, as noted in Chap. 9 (cf. also Sec. 11.9) during the course of production of 2,366,000,000 bbl from the East Texas field the pressure declined approximately 600 psi, and yet less than 2 per cent of the production was replaced by expansion of the liquids within the oil reservoir. While, as will be discussed in Sec. 11.17, even the small reservoir fluid expansion will control the initial pressure-decline history of undersaturated reservoirs, this phase of the producing life will generally be of negligible importance from the over-all oil-recovery standpoint. Since gas is the only other important fluid-replacement agent, complete-water-drive systems will be considered here as those in which the replacement of the fluid-withdrawal voidage by creation or growth of the gas phase is negligible compared with the total withdrawal volumes.

As implied by these remarks, complete pressure stabilization is a sufficient[1] but not necessary condition establishing the existence of a complete water drive. On the contrary, all water-drive reservoirs—except those producing from cavernous limestones—must initially suffer some degree of pressure decline in order to induce a sufficient rate of water entry to retard the decline and ultimately arrest it completely, even if this could occur. Moreover a permanent arrestment of the pressure decline, without the injection of extraneous fluid into the producing formation or continual reduction in net withdrawal rate, is not to be expected,[2] unless the water-supply reservoir performs as a steady-state incompressible-liquid system. Indeed one of the primary reservoir-engineering problems pertaining to water-drive reservoirs is the description and prediction of the pressure-decline transient. In fact much of the following discussion will deal with the nature of the fluid flow in the water reservoir—the aquifer—, the oil field itself representing essentially a sink or outflow boundary for the water-bearing formation. If the oil is undersaturated and the water entry is substantially equal to the reservoir volume of the oil production, the latter is used directly as the controlling boundary condition applied to the

[1] It is presupposed, of course, that the pressure stabilization is not the result of artificial pressure-maintenance operations.

[2] In certain fields, however, complete pressure stabilization has been observed to extend over long producing periods, suggesting that the water reservoir is being subjected to a constant pressure head, as if it outcropped in an ocean bottom.

water reservoir. On the other hand, even if the oil reservoir is producing by a partial gas drive, the amount of water entry can be computed by inverting the material-balance equation, as discussed in Sec. 9.7, and applying the calculated water-intrusion rates and observed boundary pressures to determine the characteristics of the water reservoir.

Since the general performance characteristics of water-drive reservoirs have already been discussed in Chap. 9, the following sections will deal mainly with the quantitative aspects of the pressure and flux histories of such reservoirs. In contrast to multiphase-flow systems, rigorous analytical or electrical-analogue treatments can be developed for all the basic types of complete-water-drive systems. On the other hand these theoretical investigations will involve simplifying assumptions regarding the uniformity and geometrical symmetry of the porous media. In certain instances their practical applicability to actual producing fields has been quantitatively confirmed by detailed analysis. For most reservoirs, however, complete analytical treatments are not feasible. Supplementary examples of observed water-drive performance histories will therefore be presented to illustrate some of their more complex features, as the development of water production and the areal advance of edgewaters, and which reflect such practical operating conditions and inhomogeneities in reservoir structure as are beyond the scope of the simplified theoretical analysis. Moreover the discussion of recovery efficiencies and factors associated with water-drive reservoir performance will be deferred to Chap. 14, since these are determined primarily by the details of the oil-displacement mechanism and rock structure and do not directly affect the pressure and fluid-withdrawal relationships.

11.2. Simplified Steady-state Treatment of Water Intrusion in Complete-water-drive Reservoirs.—It will be seen in the next section that if the linear dimension of an aquifer associated with a water-drive oil reservoir is of the order of 10 miles or greater the compressibility of the water must be taken into account in quantitatively describing its performance. It is the transient resulting from this compressibility that exerts a controlling influence on the long-term histories of most water-drive reservoirs. On the other hand the compressibility of the liquids within the oil reservoir gives rise to a transient behavior that is probably of even greater significance in determining the pressure reaction of the oil reservoir during its initial-production phase. In order to describe this early period of the production history it is convenient to neglect the compressibility of the water in the aquifer and approximate its flux capacity by a continuous succession of steady-state representations. While such a treatment will lead to erroneous predictions of virtually complete pressure stabilization very early in the producing life, when the withdrawal rate does not exceed the maximum influx capacity of the aquifer, it will nevertheless serve as an instructive introduction to the more involved analysis of reservoirs in which the aquifer itself is considered as a compressible-liquid system.

It will be assumed that the oil is saturated at a bubble point p_b, lower than the initial reservoir pressure p_i. Upon denoting the initial reservoir oil content and interstitial-water volume, as measured at p_b, by V_{oi} and V_{wi}, the cumulative oil withdrawal, measured at p_b, by P, and the volume of net water intrusion, at reservoir pressure p, by W, it readily follows that

$$V_{wi}e^{-\kappa_w(p-p_b)} + (V_{oi} - P)e^{-\kappa_o(p-p_b)} + W = V_{oi}e^{-\kappa_o(p_i-p_b)} + V_{wi}e^{-\kappa_w(p_i-p_b)}, \quad (1)$$

where κ_o, κ_w are the compressibilities of the oil and water. To first-order terms in κ,* Eq. (1) may be rewritten as

$$(V_{wi}\kappa_w + V_{oi}\kappa_o)(p_i - p) - P[1 - \kappa_o(p - p_b)] + W = 0. \quad (2)$$

For a steady-state supply of water from the water reservoir the net water intrusion may be expressed by[1]

$$W = \frac{c}{1 + \kappa_w p} \int_0^t (1 + \kappa_w p)(p_i - p)dt - W_p, \quad (3)$$

where c is the intrusion coefficient, corresponding to a productivity index for the water reservoir, the intrusion rate referring to the reservoir pressure, and W_p is the cumulative water production.

On differentiating Eq. (2) it follows that

$$(V_{wi}\kappa_w + V_{oi}\kappa_o - P\kappa_o)\frac{dp}{dt} + [1 - \kappa_o(p - p_b)]\frac{dP}{dt} - \frac{dW}{dt} = 0. \quad (4)$$

Upon retaining only the dominant term in dW/dt, as derived from Eq. (3), Eq. (4) becomes

$$(V_{wi}\kappa_w + V_{oi}\kappa_o - P\kappa_o)\frac{dp}{dt} + \left(c - \kappa_o\frac{dP}{dt}\right)p = cp_i - (1 + \kappa_o p_b)\frac{dP}{dt} - \frac{dW_p}{dt}. \quad (5)$$

For a constant production rate, defined by

$$P = Q_o t, \quad (6)$$

---

* Dropping higher order terms is equivalent to assuming a linear variation of the density with pressure. As the exponential relationship is also an approximation, the use of the linear form, when convenient, is entirely justified (cf. Sec. 4.4).

[1] Equation (3) is essentially equivalent to Eq. 9.7(3) except for the terms $\kappa_w p$. It implies that the pressure at the water-oil boundary equals that in the oil reservoir, p, which in turn is considered as constant over the reservoir area. The pressure at the distant boundary of the aquifer is also assumed to remain constant, p_i, in the steady-state representation of Eq. (3).

and neglecting the rate of water production, dW_p/dt, Eq. (5) has the solution

$$\frac{p}{p_i} = \frac{p_e}{p_i} + \left(1 - \frac{p_e}{p_i}\right)(1 - x)^\alpha, \tag{7}*$$

where

$$p_e = \frac{1 - (1 + \kappa_o p_b)r}{\alpha r \kappa_o}; \qquad r = \frac{Q_o}{c p_i} \quad : \quad \alpha = \frac{c}{\kappa_o Q_o} - 1 = \frac{1}{\kappa_o p_i r} - 1;$$

$$x = \frac{P}{V_{oi} + (\kappa_w V_{wi}/\kappa_o)}. \tag{8}$$

Thus x represents the cumulative production, expressed as a fraction of the total initial-fluid content of the oil reservoir, the connate water being lumped with the oil in proportion to its compressibility.[1] According to Eq. (7), p will approach a limiting value p_e as x approaches 1. The latter value would imply a displacement of all the oil, plus the connate water multiplied by κ_w/κ_o. Even if the connate water be considered as mobile when subject to water flushing,[2] the natural residual-oil content after water flushing will mean that in practice the value $x = 1$ will not be attained. On the other hand, as will be seen presently, this is of no consequence with respect to the description of the transient behavior, as Eq. (7) predicts such a rapid decline in p that virtually the whole of the ultimate pressure drop is developed by the time x reaches values of the order of 0.1.

Both the limiting pressure p_e and rate-of-decline parameter α depend on r. From its definition by Eq. (8), r is seen to be the ratio of the rate of oil production to the maximum possible rate of water intrusion. It therefore gives a measure of the rate of oil withdrawal in terms of the replacement capacity of the water reservoir.

As $\kappa_o p_i$ will generally be much smaller than 1, $\alpha r \kappa_o p_i$ can be approximated by 1. Upon dropping $\kappa_o p_b$ also,[†] Eq. (7) can be approximated by

$$\frac{p}{p_i} = 1 - r + r(1 - x)^\alpha. \tag{9}$$

Equation (9) is plotted in Fig. 11.1 for several values of r, assuming $\kappa_o p_i = 0.03$ (for the solid curves), corresponding to $p_i = 3,000$ psi and

---

* If dW_p/dt be taken as a constant w, Eq. (7) will still apply, provided that the p_i's in the expression for r in Eq. (8) be replaced by $p_i - w/c$, or if p_e is replaced by $p_e - (w/\alpha Q_o \kappa_o)$.

[1] The connate water need not be considered explicitly if the oil compressibility be given an effective value equal to $\kappa_o + \kappa_w V_{wi}/V_{oi}$.

[2] Cf. R. G. Russell, F. Morgan, and M. Muskat, *AIME Trans.*, **170**, 51 (1947).

† In fact there is no real loss in generality in omitting p_b entirely, although the existence of a bubble point must be taken into account in the physical interpretation of the results.

$\kappa_o = 10^{-5}$ per psi. The very rapid drop in pressure and approach to the asymptotic stabilization pressure $[1 - r \sim (p_e/p_i)]$ will be noted. Thus 99 per cent of the ultimate pressure drop will be attained after only 2.7 per cent of "ultimate depletion" $(x = 1)$ if $r = 0.2$; 5.4 per cent for $r = 0.4$;

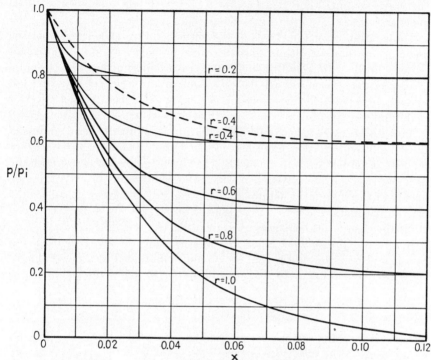

FIG. 11.1. The calculated variation of the reservoir pressure with the cumulative recovery in complete-water-drive systems, of the steady-state type, with uniform production rates. p/p_i = (reservoir pressure)/(initial pressure); x = (cumulative recovery)/(initial equivalent reservoir oil content); r = (reservoir liquid-withdrawal rate)/(maximum possible steady-state rate of water intrusion). For solid curves, $\kappa_o p_i = 0.03$. For dashed curve, $\kappa_o p_i = 0.06$. κ_o = compressibility of oil.

8.1 per cent for $r = 0.6$; 10.7 per cent for $r = 0.8$; and no more than 13.3 per cent[1] even if $r = 1$. Under such rapid pressure stabilization the assumption of uniform production rate and neglect of water production should not be especially serious approximations.

The above values for x indicate that the times required for an equivalent approach to the asymptotic final pressures will be essentially independent of the production rate. The effect of the oil compressibility is indicated by the dashed curve, for $r = 0.4$, for which $\kappa_o p_i$ was taken as 0.06, correspond-

[1] These values were calculated by using the values for α and p_e given by Eq. (8) rather than the approximation implied by Eq. (9).

ing to a compressibility twice as great as assumed for the solid curves. It will be seen that now 99 per cent of the ultimate pressure drop will develop by the time 10.7 per cent of the total reservoir content, as oil equivalent, has been displaced ($x = 0.1071$), whereas, as indicated above, the corresponding time required for $\kappa_o p_i = 0.03$ was just about half as great ($x = 0.054$).

The continuity of the slopes of the curves of Fig. 11.1 arises from the assumptions that the withdrawal rates are kept fixed and that no gas comes out of solution throughout the indicated pressure-decline history. In practical situations the curves will flatten sharply if and when the pressures fall to the bubble point and gas evolution begins. Hence, the curves of Fig. 11.1 could apply only to such values of p/p_i as exceed p_b/p_i.

As is to be expected, the pressure decline is a function not simply of the cumulative reservoir fluid displacement, represented by x, but also of the rate of oil withdrawal, defined by r. This is in direct contrast with the behavior of gas-drive systems, in which the pressure history is a function mainly of the cumulative oil production. Under gas-drive mechanisms the rate of production affects the pressure history only in so far as it may affect the gas-oil ratio and gravity segregation or drainage.

If, after the pressure has fallen to a value p_1 while the relative production rate has been r, the rate is changed to and maintained at a new value r_1, the subsequent pressure history may be shown to follow the equation

$$\frac{p}{p_i} = \frac{p_{e1}}{p_i} + \frac{p_1 - p_{e1}}{p_i(1 - x_1)^{\alpha_1}} (1 - x)^{\alpha_1}, \tag{10}$$

where x_1 is the value of the relative cumulative recovery [defined by Eq. (8)] at the time of the change, x is the total cumulative relative recovery including that at the withdrawal rate r, p_{e1} is the new ultimate depletion pressure, and α_1 is the new value of α corresponding to r_1. Upon making the same approximations as in Eq. (9), Eq. (10) may be rewritten as

$$\frac{p}{p_i} = 1 - r_1 + \frac{(p_1/p_i) - (1 - r_1)}{(1 - x_1)^{\alpha_1}} (1 - x)^{\alpha_1}; \qquad \alpha_1 = \frac{1}{\kappa_o p_i r_1} - 1. \tag{11}$$

To show the nature of the change in pressure history implied by Eqs. (10) and (11) it was assumed for several values of initial production rate r that after the pressure drop had reached 99 per cent of its ultimate value, r is changed by ± 0.2, that is, that $r_1 = r \pm 0.2$. The curves so calculated are plotted in Fig. 11.2. As expected, when $r_1 > r$, the pressure immediately starts to fall rapidly so as to approach its new asymptotic depletion value $1 - r_1$. Conversely, when $r_1 < r$, the pressure undergoes a rapid rise toward its new depletion value $1 - r_1$. It will be noted, however, that the initial rate of pressure rise, when $r_1 < r$, is steeper than the initial rate of fall when $r_1 > r$. The reason lies in the fact that, as shown in Fig. 11.1, inherently the transient period for pressure stabilization becomes smaller as r decreases. It arises analytically from the increase in the exponent α or α_1 as r decreases. As a whole the additional time required for approxi-

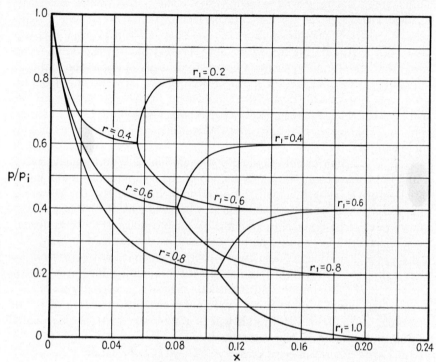

Fig. 11.2. The calculated effect of changes in the production rate on the reservoir pressure history in complete-water-drive systems of the steady-state type. $p/p_i =$ (reservoir pressure)/(initial pressure); $x =$ (cumulative recovery)/(initial equivalent reservoir oil content); $r =$ (reservoir liquid-withdrawal rate)/(maximum possible steady-state rate of water intrusion). $r = r$, initial value. $r = r_1$, changed value. $\kappa_o p_i = 0.03$. $\kappa_o =$ compressibility of oil.

mate stabilization at the new depletion pressure, in terms of the cumulative-production parameter x, is of the same order as if the field were initially produced at the new production rate.

In a similar manner the effects of still later changes in production rate could be determined. And by numerical integrations of Eq. (5) the pressure decline could be computed for any arbitrary and variable type of production-rate history, or function $P(t)$. Because of the basic limitations in the validity of Eq. (5) such additional calculations are hardly warranted. Those represented by Figs. 11.1 and 11.2, however, should serve to show that one of the fundamental variables controlling the pressure history in water-drive fields is the oil-withdrawal rate as compared with the water-supply capacity of the water reservoir. Moreover they emphasize the need for some degree of pressure decline if water entry is to develop at a sufficient rate appreciably to retard the pressure-decline rate. Finally, since the only withdrawal replacement means above the bubble point, except for the water entry, is the very small expansion of the reservoir liquids, associated with their compressibility, the pressure fall will be very rapid, in terms of the cumulative recovery. On the other hand it should be recognized that the simple concepts of complete pressure stabilization and fixed supply capacity of the water reservoir derive their meaning only from the steady-state representation of fluid movement in the water reservoir. As will be seen in the following sections, in compressible-liquid systems these terms no longer strictly apply, and the transient behavior is considerably more complex.

11.3. The Compressible-liquid Representation of Water-supply Reservoirs.—Before entering into a discussion of the compressible-liquid theory of water-drive systems, which will occupy much of the rest of the present chapter, it is well to understand clearly when and why the compressibility effects have to be taken into account. While no sharp rule can be formulated, it is possible to state the physical factors involved.

From the definition of complete-water-drive systems used here it follows that the oil reservoir itself does not directly provide a substantial part of the energy or means for the oil expulsion. These are derived from the adjoining and intercommunicating water reservoir. The reaction of the oil reservoir to the fluid withdrawals will therefore depend on the mechanics of the fluid motion in the water reservoir. The two will have to be in balance with a continuity of pressure and flux at their common boundary. Hence, as indicated in Sec. 11.1, except for details of pressure distributions within the oil reservoir, the latter can be represented simply as the cause or source of boundary conditions imposed on the water reservoir. Thus, the oil withdrawals will constitute a boundary condition expressed as a flux taken from the water reservoir. The latter will have to adjust itself to supply that flux. This will evidently require a lowering of pressure at the common oil-water boundary. It is this boundary pressure, corrected for and modified by the pressure distribution within the oil reservoir associated with the flow of the oil into the producing wells, that will represent the reservoir (oil) pressure. To determine the history of the latter it is therefore necessary to determine first the history of the oil-water-boundary pressure.

If the geometry of the water reservoir were known and it were assumed that at its most distant boundary the pressure remains fixed, it would be a relatively simple matter, by the methods of analysis discussed in Chap. 5, to calculate the extent to which the pressure at the water-oil boundary must fall to provide a steady-state water intrusion equal to any preassigned reservoir oil-withdrawal rate. That such intrusion would actually take place by steady-state flow is, however, an assumption to be investigated. Thus, supposing that immediately before beginning oil production the pressures throughout the water reservoir are constant, it will contain, per unit thickness, a mass of fluid equal to

$$M = \pi f \gamma_i (r_e^2 - r_f^2), \tag{1}$$

where it is assumed that the water reservoir is an annular system of external radius r_e and internal radius r_f and with a porosity f. γ_i is the water density corresponding to its initial uniform pressure. If a steady-state flow should be developed in the reservoir, with a density γ_f at r_f, its new mass content may be readily shown to be less than M by an amount

$$\Delta M = \pi f(\gamma_i - \gamma_f)\left(\frac{r_e^2 - r_f^2}{2 \log r_e/r_f} - r_f^2\right). \tag{2}$$

Now the steady-state mass flux capacity of the latter system is

$$Q = \frac{2\pi k \, \Delta\gamma}{\mu\kappa \log r_e/r_f}. \tag{3}$$

Hence, upon assuming $r_f \ll r_e$, the order of magnitude of the time required for the removal of the mass ΔM will be

$$t \sim \frac{\Delta M}{Q} \sim \frac{f\mu\kappa r_e^2}{4k} = \frac{\pi f\kappa r_e^2}{4\pi k/\mu}. \tag{4}$$

If $f = 0.25$, $\mu = 1$ cp, $\kappa = 3.10^{-6}$ per psi, $r_e = 10$ miles and $k = 100$ md, then $t \sim 7 \times 10^7$ sec. That is, it would take of the order of 2 years just to create the steady-state distribution. Under such conditions a steady-state approximation neglecting the compressibility of the system would evidently be of little significance in describing the immediate reaction of the reservoir to the change in density at r_f. On the other hand, if r_e were only 1 mile, $t \sim 8$ days and the approximation would not be so serious.

Equation (4) serves to exhibit the significant physical properties determining the duration of the transient behavior. Thus, as explicitly indicated in the last expression in Eq. (4), the transient time will be directly proportional to $\pi f\kappa r_e^2$, which represents the total mass change of the fluid content of the system per unit pressure drop. For this to be large and significant the system must be of large volumetric content ($\pi f r_e^2$), or the compressibility κ must be abnormally high, as might be the case if it contains a distributed gas phase. The transient duration will also be inversely proportional to k/μ, which is a measure of the flow capacity and the ease with which the change in mass content can be absorbed or removed from the system. Hence the transient period will be long in tight reservoirs and relatively short if the permeability is high.

An equivalent order-of-magnitude estimate of the times involved in the transient behavior of compressible-liquid systems may be made by assuming that the boundary pressures are being changed at a rate dp/dt. The corresponding rate of change of mass content, upon assuming $r_f^2 \ll r_e^2$, will then be

$$\frac{dM}{dt} = \pi f \bar{\gamma} \kappa r_e^2 \frac{dp}{dt}, \tag{5}$$

where $\bar{\gamma}$ is the average fluid density. This bears a ratio to the steady-state mass flux capacity [Eq. (3)] of

$$\frac{dM/dt}{Q} = \frac{f \mu \bar{\gamma} \kappa^2 r_e^2 \, dp/dt \log r_e/r_f}{2k \, \Delta\gamma} = \frac{f \mu \kappa r_e^2 \log r_e/r_f}{2k \, \Delta\gamma} \frac{d\gamma}{dt}, \tag{6}$$

where $d\gamma/dt$ is the rate of change of fluid density corresponding to dp/dt. It will be seen that the significant factors here are the same as in Eq. (4), that is, Eqs. (4) and (6) are related by

$$\frac{dM/dt}{Q} = \frac{2t \log r_e/r_f}{\Delta\gamma} \frac{d\gamma}{dt}, \tag{7}$$

where t is the value given by Eq. (4).

Equation (6) also gives a measure of the speed with which transient effects may be dissipated in flow systems. Thus for $r_e/r_f = 10$, $\Delta\gamma = 0.001$, $dp/dt = 1$ psi/day, and with the other constants the same as used in illustrating Eq. (4), Eq. (6) shows that $dM/dt \sim 11Q$. That is, the mass content of the system would change 11 times as fast as the steady-state capacity could provide for it. Evidently here, too, the transient effects arising from the fluid compressibility would have to be taken into account explicitly.

While these considerations have thus shown that in extended reservoirs of many square miles in area compressibility effects cannot be ignored, they also provide a justification for using steady-state approximations in treatments of individual-well homogeneous-liquid-flow systems, as discussed in Chap. 5. Thus on assuming that the radius of individual-well influence is 0.1 mile and that the oil compressibility is 1.5×10^{-5} per psi, Eq. (4) gives a time of the order of 10 hr for the absorption or removal of the change in mass content by steady-state flow, for $k = 100$ md, $\mu = 1$ cp, and $f = 0.25$. And Eq. (6) implies that, for a pressure change of 1 psi/day, dM/dt will be less than $0.1Q$ if $r_e/r_f \sim 2,600$. There should therefore be no great error involved in applying steady-state treatments to such systems. Moreover, if changes in boundary conditions should occur, it should be a satisfactory approximation to represent the associated time history as a continuous succession of steady states, each described by the corresponding instantaneous boundary values. Of course, if the porous medium contains a distributed gas phase, the compressibility may be enormously increased and the transients will be so stretched out that the steady-state treatment will become invalid. However, if the gas phase is mobile, the homogeneous-flow representation will be inherently inappropriate and the multiphase character of the flow will have to be taken into account explicitly.

The above discussion should not be considered as implying that all water reservoirs adjoining oil-producing formations are so large as necessarily to provide a sustained water-drive intrusion capacity to replace all the recoverable oil in the oil reservoirs. While in many instances the water reservoirs are indeed so extended that they may be treated as infinite during the producing life of the contiguous oil field, in others they may be so limited that their total volumetric expansion capacity will become evident long before all the oil will have been produced. This has been observed,[1] for example, at the Midway field, Arkansas, and it has been necessary to inject water into the reservoir to maintain an effective water-drive performance at the desired withdrawal rates. In the study of actual water-drive oil reservoirs one of the most important phases is an analysis of the geometry and flow capacity of the associated water reservoirs. It is for this reason that an extended treatment will be given here of the pressure and flux histories of various types of water reservoir, as if they were in themselves virtually models of the composite oil-producing and water-replacement systems.

As was seen in Sec. 4.4 the basic equation governing the homogeneous-fluid flow of compressible liquids is

$$a \, \nabla^2 \, \gamma = \frac{\partial \gamma}{\partial t}; \qquad a = \frac{k}{f \kappa \mu}, \tag{8}$$

it being assumed[2] that the density γ is related to the pressure[3] p by

$$\gamma = \gamma_o e^{\kappa p}, \tag{9}$$

---

[1] W. L. Horner, *API Drilling and Production Practice,* 1945, p. 27; cf. also Sec. 11.11.

[2] As noted in Sec. 4.4, Eq. (9) implies a compressibility independent of the pressure. Since this is not strictly true for actual liquids, κ is to be considered as the average value of the compressibility over the pressure range of interest. The approximation so introduced will not be serious in practical applications, as the variation in κ generally will be small and moreover will enter in most numerical equations in the combination a [cf. Eq. (8)], which involves reservoir-volume averages of k/f and pressure averages of μ. In fact the slow increase of μ with pressure (above the bubble point) may even more than compensate for the decrease in κ.

[3] While this definition implies that p is the gauge pressure, this is of no importance, since virtually all the derived quantities involving the pressure refer to pressure differences.

where γ_o is the atmospheric-pressure density. a is the analogue of the diffusivity in heat-conduction or diffusion systems, which are also governed by an equation similar to Eq. (8). The mass flux in compressible-liquid systems will be given by

$$\gamma\bar{v} = -\frac{k}{\mu}\gamma\,\nabla\,p = -\frac{k}{\mu\kappa}\nabla\,\gamma = -af\,\nabla\,\gamma. \tag{10}$$

While the analysis itself will be formally carried through in terms of the density function γ, the physical interpretation of the results will be usually expressed in terms of pressure differentials Δp, computed by

$$\Delta p = \frac{\Delta\gamma}{\gamma\kappa}, \tag{11}$$

derived from Eq. (9).

It may be observed from Eq. (8) that, if the independent variables of time and coordinates be expressed in dimensionless units, the time will take the form

$$\bar{t} = \frac{at}{r_o^2} = \frac{kt}{f\kappa\mu r_o^2}, \tag{12}$$

where r_o represents a linear dimension of the system. It is this composite value of $\bar{t}$ that will determine the time transients. For a fixed value of $\bar{t}$ the numerical value of t therefore will be affected by the physical and geometrical constants of the system exactly in the manner indicated by Eq. (4), which was derived independently to indicate the time scale of the transient phenomena. In fact Eq. (12) itself may be used directly to establish the conditions when compressibility and transient effects will be an essential part of the flow behavior.

For convenience and simplicity most of the analytical discussion of the following sections will be concerned with two-dimensional radial-flow systems. This does not imply that actually all water reservoirs are strictly circular or have radial symmetry. However, the cylindrical geometry provides a convenient approximation framework for many water reservoirs and also serves well to illustrate the basic physical features of the flow phenomena. Moreover a thorough understanding of the development of the analysis for radial systems will provide the background for constructing the corresponding solutions for reservoirs in which linear or other geometrical representations may be more appropriate. Accordingly the equation whose solutions will constitute the major content of the discussion will be the cylindrical coordinate form of Eq. (8), namely,

$$\frac{a}{r}\frac{\partial}{\partial r}\left(r\frac{\partial\gamma}{\partial r}\right) = \frac{\partial\gamma}{\partial t}. \tag{13}$$

Finally it should be noted that, while in the following analysis only the expansion due to the fluid compressibility is taken into account explicitly [cf. Eq. (9)], superposed effects due to the compressibility of the medium itself readily fall within the scope of the analysis. For it may be shown[1] that if the porous medium, as well as the entrained fluid, be considered as compressible, the governing differential equation is identical with Eq. (8), except that in the coefficient a the term κ is to be considered as the sum of the effective compressibilities of the fluid, the gross rock structure, and any equivalent expansion compressibility of the fluid content of interspersed shales and clays which are not otherwise included as part of the main flow system. The resultant expansibility of the composite rock and fluid system is usually expressed in hydrological applications as a "coefficient of storage," when multiplied by appropriate constants, and has served to explain the transient behavior of water-supply aquifers. Although studies of the latter have indicated that the fluid-expansion contributions of the rock structure and included clay beds may be even greater than that of the water itself,[2] it is doubtful that such conditions obtain in the majority of aquifers and oil-bearing formations involved in water-drive reservoirs. However, no such assumption is implied in the formal analytical discussions of this chapter. The numerical value of κ need be chosen only in making specific applications. In fact, since it is always associated with other physical parameters, which also are seldom known with high precision, in the coefficient a, the latter may be considered as a composite empirical constant reflecting the resultant expansibility due to whatever individual factors are actually of significance in the particular problem of interest. On the other hand, in the numerical discussion of hypothetical systems for illustrative purposes, values of κ corresponding only to the fluid phases will be used, since the contributions of the other sources of fluid expansibility will in general be totally unknown a priori.

11.4. The Pressure History of Complete-water-drive Systems Supplied by Water Reservoirs of Infinite Extent.—It is clear that if the water reservoir is so extended that its total fluid-expansion content is large compared with the oil reserves in the oil reservoir, the duration of the oil-production history may be terminated before an appreciable effect will have been exerted on the remote parts of the water-bearing formation. The interaction between the oil and water reservoirs will be then essentially the same as if the water reservoir were actually of infinite extent. Systems of this type will therefore be considered first as a preliminary to the treatment of the more realistic finite-reservoir representation.

---

[1] C. E. Jacob, *AGU Trans.*, 1940, pt. II, p. 574. While in Jacob's equation the dependent variable is the fluid head or pressure, it involves such approximations as would give equal validity to Eq. (8) with γ representing the fluid-mass content per unit volume of pore space (at the initial pressure).

[2] Jacob, *loc. cit.*, and *AGU Trans.*, 1941, pt. III, p. 783.

The specific system to be analyzed here will be defined by the initial and boundary conditions (cf. Fig. 11.3)

$$t = 0 \quad : \quad \gamma = \gamma_i(p = p_i),$$
$$r = r_f \quad : \quad 2\pi r \gamma v_r = q(t), \quad (1)$$

where the radius r_f is taken as the equivalent oil-field radius, assumed circular, and p_i is the initial pressure in the water reservoir. In actual

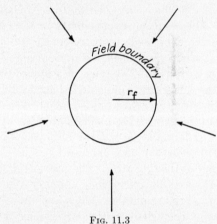

Fig. 11.3

complete-water-drive fields the mass flux $q(t)$ at the field boundary is to be considered as representing the fluid withdrawal within the field, except for that replaced by expansion of its own fluid content.

By means of the solution of Eq. 11.3(13), subject to the conditions of Eq. (1), the fluid density and pressure at $r = r_f$, or field boundary, can be determined as a function of time. This solution can be verified to be[1]

$$\gamma = \gamma_i + \frac{1}{\pi^2 f r_f} \int_0^\infty \frac{e^{-au^2t}[J_0(ur)Y_1(ur_f) - Y_0(ur)J_1(ur_f)] \, du}{J_1^2(ur_f) + Y_1^2(ur_f)} \int_0^t q(\lambda)e^{au^2\lambda} \, d\lambda. \quad (2)$$

where J_n, Y_n denote Bessel functions[2] of order n of the first and second kinds, respectively.

For the special case where $q(t)$ is a constant q_o, Eq. (2) reduces to

$$\gamma = \gamma_i + \frac{q_o}{\pi^2 a f r_f} \int_0^\infty \frac{(1 - e^{-au^2t})\,[J_0(ur)Y_1(ur_f) - Y_0(ur)J_1(ur_f)] \, du}{u^2[J_1^2(ur_f) + Y_1^2(ur_f)]}. \quad (3)$$

[1] The solution for this problem, applying to the special case of Eq. (3), has been derived by J. C. Jaeger in an unpublished manuscript.

[2] These functions are treated exhaustively by G. N. Watson, "The Theory of Bessel Functions," Cambridge University Press, 1922. Most of the properties required in developing the analysis given in this chapter are briefly outlined in M. Muskat, "The Flow of Homogeneous Fluids through Porous Media," Sec. 10.2, McGraw-Hill Book Company, Inc., 1937.

At the field boundary r_f, γ will therefore have the value

$$\gamma_f = \gamma_i - \frac{2q_o}{\pi^3 a f r_f^2} \int_0^\infty \frac{(1 - e^{-au^2 t})\,du}{u^3[J_1^2(ur_f) + Y_1^2(ur_f)]}. \tag{4}$$

On translating the decline in density $\gamma_i - \gamma_f$ to the corresponding pressure drop $\Delta p = p_i - p_f$ and introducing the dimensionless time variable $\bar{t}$, Eq. (4) becomes

$$\Delta p = \frac{2Q\mu}{\pi^3 k} \int_0^\infty \frac{(1 - e^{-z^2\bar{t}})\,dz}{z^3[J_1^2(z) + Y_1^2(z)]}; \qquad \bar{t} = \frac{at}{r_f^2}, \tag{5}$$

where Q represents the volumetric outflow per unit thickness at r_f, but measured at the surface,[1] that is, q_o/γ_o. In Fig. 11.4 a graphical evaluation

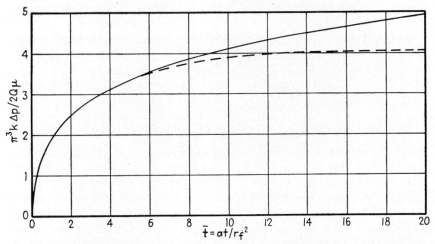

FIG. 11.4. The calculated pressure drop Δp vs. the time t plotted in dimensionless form, at the internal boundary of water reservoirs, with constant water-withdrawal rate Q per unit thickness. Internal-boundary radius $= r_f$; permeability of water reservoir $= k$; $\mu =$ viscosity of water; $\kappa =$ compressibility of water; $f =$ porosity; $a = k/f\kappa\mu$. Solid curve refers to an infinite water reservoir. Dashed curve applies to a finite water reservoir with the pressure kept fixed at an external radius that is 6.3 times r_f.

of Eq. (5) is plotted as the solid curve in dimensionless form, as $\pi^3 k\,\Delta p/2Q\mu$ vs. $\bar{t}$. As may be shown by an analysis of Eq. (5), Δp initially rises as $\sqrt{\bar{t}}$ and asymptotically assumes a logarithmic variation with $\bar{t}$. Thus in contrast to the steady-state approximation treated in the last section there

---

[1] To second-order terms in κ, the conversion between $\Delta\gamma$ and Δp can be made using any convenient value of γ [cf. Eq. 11.3(11)]. The values of flux represented by Q in this and the following sections refer to complete cylindrical systems. If the reservoirs are more appropriately described by sectors of angular width w, the equations relating pressure drops to flux will still remain valid provided that the values of Q used in the equations are the actual values multiplied by $2\pi/w$.

will be no strict stabilization pressure, although the rate of pressure decline will continually decrease. This, of course, is to be expected, since
the source of the fluid delivery by the water reservoir lies in the expansion
of its water content, which requires a continued decline in the pressure.
Associated with this decline in pressure is the fact that the infinite compressible-liquid aquifer has no sharply defined flow capacity to serve as a
unit for expressing the actual withdrawals, such as the ratio r of the preceding section [cf. Eq. 11.2(8)]. The efflux from such a system will vary
with the time even though the boundary outflow pressure is kept fixed
(cf. Sec. 11.5).[1] Moreover it will depend on the past history of the reservoir. On the other hand it is to be noted that in Eq. (5) Q enters essentially as a ratio to k/μ, which is a measure of the flow capacity of the water
reservoir to the extent that it is determined simply by the nature of the
rock and the fluid.

It should be observed that the physically impossible initial infinite rate
of pressure decline, implied by the proportionality of Δp to $\sqrt{t}$, and as
shown in Fig. 11.4, arises from the assumption that the water-influx rate
has a nonvanishing constant value from the very beginning. This in turn
is based on the tacit neglect of the fluid compressibility within the oil
reservoir. In practice the expansion of the oil-reservoir liquid content will
always provide a partial replacement for the fluid withdrawals, so that the
water-intrusion flux will build up gradually[2] even if the actual withdrawal
rate be kept strictly constant. While it would be difficult to take this
effect into account rigorously, this very early phase of the transient history
might be approximated by assuming an initial linear or parabolic rise of
the water flux across the field boundary, as indicated in Eq. (9) below.
From a practical standpoint, however, the analytical treatment given here
should suffice for illustrating the general features of water-drive-reservoir
behavior controlled by infinite aquifers. Additional refinements are hardly
warranted in view of the other gross idealizations underlying the analysis.

Under the basic assumptions and conditions defined by Eq. (1) the solid
curve in Fig. 11.4 is a universal curve and applicable to any infinite radial
reservoir of constant efflux rate, regardless of the actual numerical values
of the physical and geometrical parameters. To get an idea of its numerical
implications, specific values may be introduced in the dimensionless parameters. Thus if κ be taken as 4.5×10^{-5} per atmosphere, f as 0.25, μ

---

[1] In finite aquifers, however, in which the pressure is kept fixed at the external
boundary, steady-state conditions can be established that provide well-defined flux
capacities (cf. Sec. 11.6).

[2] It may be shown that for the steady-state type of water intrusion, discussed in Sec.
11.3, the ratio of the contribution to the withdrawal replacement by water influx to
that due to oil-reservoir liquid expansion is given, at any pressure p, by $(p_i - p)/(p - p_e)$.

as 0.5 cp, and k as 100 md, $a = 1.78 \times 10^4\,\text{cm}^2/\text{sec}$. If the equivalent field radius r_f is 2 miles, it follows that

$$\bar{t} = 0.01483t(\text{days}); \qquad t(\text{days}) = 67.45\bar{t}. \tag{6}$$

Then if Q be 1,000 (bbl/day)/ft,

$$\Delta p(\text{psi}) = 286.1 \left(\frac{\pi^3 k\,\Delta p}{2\mu Q} \right). \tag{7}$$

On applying these conversion factors to the solid curve of Fig. 11.4 it is found that after 1 month the pressure at the field boundary would decline by 400 psi, after 6 months the decline would be approximately 785 psi, and after 2 years the pressure would drop by about 1,200 psi. Now per foot of pay a 2-mile-radius, 25 per cent porosity reservoir will have a void space of 15,600,000 bbl, which may be assumed to contain of the order of 10,000,000 bbl of stock-tank oil. Thus a pressure drop of 1,200 psi will take place after a displacement of only 730,000 bbl of reservoir oil, which would be equivalent to only about 6 per cent of the original stock-tank-oil content of the field. And yet this would represent a complete-water-drive behavior[1] if the bubble point of the reservoir oil were as much as 1,200 psi below the initial reservoir pressure.

As is to be expected, Eq. (5) shows that at any particular time after production is started the pressure drop at the field boundary will be proportional to the water-influx rate Q. For a fixed withdrawal rate from the water reservoir, Q, it will ultimately assume a logarithmic increase with the cumulative withdrawals. For a given cumulative withdrawal the pressure drop will increase the greater the rate Q at which that withdrawal was maintained. For example, if for the water reservoir considered above the withdrawal rate were 500 (bbl/day)/ft of pay, it would take 4 years to develop a cumulative withdrawal of 730,000 bbl. Yet the pressure drop at that time would be only 724 psi. Similar examples may be drawn from Fig. 11.5, in which the pressure drop [calculated by Eq. (5)] has been plotted vs. the cumulative water influx for fixed values of the withdrawal rate. The withdrawal rate is expressed in terms of the quantity $\overline{Q} = Q\mu/k$, which has the dimensions of atmospheres^{-1}, and the cumulative influx as the dimensionless quantity $Qt/\pi r_f^2 f$, which is the cumulative water influx expressed as a fraction of the pore volume of the oil reservoir, of radius r_f. The water compressibility was taken as 4.5×10^{-5} per atmosphere.

A cross plot of Fig. 11.5, as shown in Fig. 11.6, gives directly the variation of the pressure decline, for fixed cumulative water influx, as a function of

---

[1] The term "complete water drive" is used here in a generalized sense, although in the numerical example more than half the total withdrawals would probably be supplied by reservoir liquid expansion for the first 6 months of production, and about 25 per cent of the withdrawals even during the first 2 years.

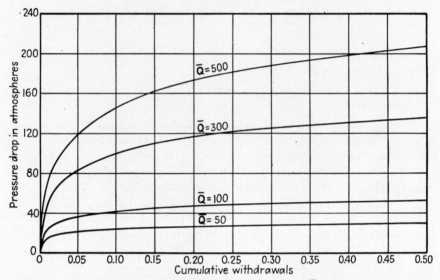

FIG. 11.5. The calculated pressure drop for fixed withdrawal rates $\overline{Q}$, at the internal boundary of an infinite aquifer, vs. the cumulative withdrawals, expressed as a fraction of the pore volume of the oil reservoir, of radius r_f. $\overline{Q} = Q\mu/k$; Q = withdrawal rate from the aquifer per unit thickness; μ = water viscosity; k = permeability of aquifer. Water compressibility assumed = 4.5×10^{-5} per atmosphere.

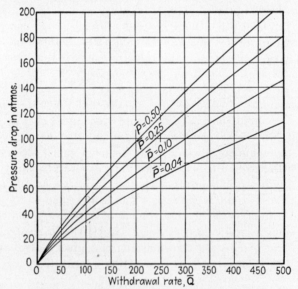

FIG. 11.6. The calculated pressure drop for fixed cumulative withdrawals $\overline{P}$, at the internal boundary of an infinite aquifer, vs. the withdrawal rate $\overline{Q}$. $\overline{P}$ = cumulative withdrawal from aquifer, expressed as a fraction of the pore volume of the oil reservoir, of radius r_f. $\overline{Q} = Q\mu/k$; Q = withdrawal rate from the aquifer per unit thickness; k = permeability of aquifer; μ = water viscosity. Water compressibility assumed = 4.5×10^{5} per atmosphere.

the rate of water withdrawal from the infinite aquifer. While the variation is slower than linear, it will be clear that the rate sensitivity of the pressure decline is an important feature of expansible water reservoirs.

An instructive but simple generalization of the above treatment pertains to the effect on the pressure-decline history, as plotted in Figs. 11.4 and 11.5, of a change in the influx rate. If the initial constant rate Q_0 is changed at dimensionless time $\bar{t}_0$ to Q_1, it may be readily shown from Eq. (2) that

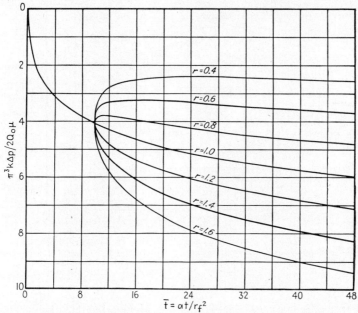

Fig. 11.7. The calculated pressure drop Δp vs. the time t, plotted in dimensionless form, at the internal boundary of an infinite water reservoir with an initial withdrawal rate Q_0 and a rate Q_1 after a time $\bar{t} = 10$; μ = water viscosity; κ = water compressibility; k = permeability of aquifer; f = porosity; r_f = internal radius of water reservoir; $r = Q_1/Q_0$; $a = k/f\kappa\mu$.

after the change is made, the pressure drop at the boundary r_f will follow the equation

$$\Delta p = \frac{2\mu}{\pi^3 k}[Q_0 I(\bar{t}) + (Q_1 - Q_0)I(\bar{t} - \bar{t}_0)], \qquad (8)$$

where $I(\bar{t})$ is the integral in Eq. (5). The first term represents the projected pressure-decline history if the rate had been maintained at Q_0. The second gives the effect of the change. As is to be expected, if Q_1 exceeds Q_0 the rate of decline will be accelerated, whereas if Q_1 is less than Q_0 the pressure drop will be reduced.

Equation (8) is plotted in dimensionless form, that is, $\pi^3 k \, \Delta p/2\mu Q_0$, vs. $\bar{t}$ in Fig. 11.7 for fixed values of the ratio Q_1/Q_0, denoted by r, it being assumed that the change from Q_0 to Q_1 takes place at $\bar{t} = 10 = \bar{t}_0$. In contrast to the similar dimensionless plot in Fig. 11.4 for constant Q the origin of coordinates is set in the upper left-hand corner so that the curves will directly parallel (except for a scale factor) the actual pressure-decline curves which might be computed for specific values of the physical constants. It will be seen that the immediate reaction to a change in water-influx rate is similar to that predicted by the steady-state theory (cf. Fig. 11.2). However, in contrast to the development of pressure stabilization in the latter

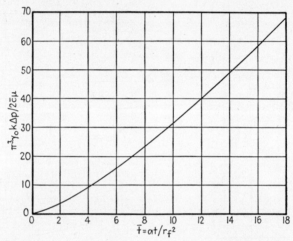

FIG. 11.8. The calculated pressure drop Δp vs. the time t, plotted in dimensionless form, at the internal boundary of an infinite water reservoir, of permeability k, with a linearly increasing water-withdrawal rate. r_f = internal radius of water reservoir; γ_o = water density; μ = water viscosity; $a = k/f\kappa\mu$; f = porosity; κ = compressibility of water; $\bar{c}$ = cr_f^2/a; c = constant rate of increase of mass withdrawal rate.

case the pressure decline here maintains an accelerated decline if the water-withdrawal rate is increased ($r > 1$). And if the withdrawal rate is reduced ($r < 1$), the pressure build-up reaches a maximum and then begins to decline again, though at a rate slower than it would have been at the same time if the withdrawal rate had not been changed. As is to be expected the magnitude and duration of the pressure increase is greater the greater the reduction in withdrawal rate.

The assumption of a constant withdrawal rate from the water reservoir for an extended period after production has begun is, of course, quite artificial from a practical standpoint. In the natural development of oil reservoirs the total field production rate will continually increase as additional wells are drilled, until an ultimate maximum field allowable is established, if proration is applied or other factors, as pipe-line outlets or market demands, impose a limit. And even if the field withdrawal rate itself were constant from the very beginning, the induced water inflow rate from the water reservoir still would rise gradually from zero, if there should be a component of gas drive in the replacement of the oil withdrawal or an expansion of the reservoir liquids if the latter are undersaturated. Although, as will be seen later, perfectly general types of water-flux histories can and must be considered in interpreting actual reservoir behavior, the

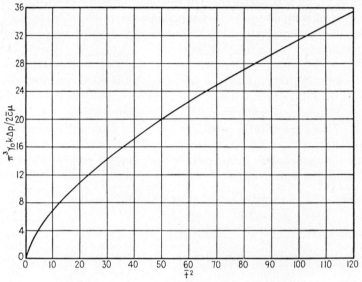

FIG. 11.9. The calculated pressure drop Δp at the internal boundary of an infinite water reservoir, with linearly increasing withdrawal rate, vs. $\bar{t}^2$. γ_o = water density; μ = water viscosity; k = permeability of water reservoir; $\bar{c}$ is proportional to the constant rate of increase of the mass withdrawal rate; $\bar{t}^2$ is proportional to the cumulative withdrawals.

simplest first approximation of a linear rate of increase is readily treated by application of Eq. (2). Thus if the mass-flux rate be expressed as

$$q = ct = \bar{c}\bar{t}; \qquad \bar{c} = \frac{cr_f^2}{a}, \tag{9}$$

it can be shown that the pressure drop at r_f is given by

$$\Delta p = \frac{2\bar{c}\mu}{\pi^3\gamma_0 k} \int_0^{\bar{t}} I(\bar{t})d\bar{t}, \tag{10}$$

where $I(\bar{t})$ is again the integral of Eq. (5).

The integral of Eq. (10), or the quantity $\pi^3\gamma_0 k \, \Delta p/2\bar{c}\mu$, is plotted vs. $\bar{t}$ in Fig. 11.8. As is to be expected the pressure drop here rises gradually with time. In fact, Fig. 11.8 implies that the pressure-decline curve would be convex upward, in contrast to the convex-downward character of the pressure-decline curves if the withdrawal rate starts out at a non-vanishing value (cf. Fig. 11.7). On the other hand, when plotted as a function of the cumulative withdrawal, the pressure drop is again convex upward, as shown in Fig. 11.9, where the abscissas are $\bar{t}^2$, which is proportional to the cumulative withdrawal.

11.5. Infinite Water Reservoirs with Radial Symmetry and Preassigned Pressures at the Circular Water-Oil Boundary.—While in most theoretical treatments of complete-water-drive systems the problem is largely one of relating the pressure history to a preassigned history of water withdrawal from the aquifer, it is nevertheless instructive to consider also the inverse problem. The latter involves an assignment of the pressure history at the original water-oil boundary, together with the initial pressure distribution, and the calculation of the flux across the boundary associated with the pressure history. If, as in Sec. 11.4, the aquifer is considered to be an infinite uniform permeable reservoir and it is assumed that the initial uniform density is $\gamma_i(p_i)$, the general density distribution at any time t, for a density history at r_f expressed by the function $f(t)$, can be shown[1] to be given by the equation

$$\gamma = \gamma_i[1 - v_0(r,t)] + \int_0^t f(\lambda) \frac{\partial v_0(t - \lambda, r)}{\partial t} \, d\lambda, \tag{1}$$

where

$$\left. \begin{array}{c} v_0(r,t) = 1 + \dfrac{2}{\pi} \displaystyle\int_0^\infty \dfrac{e^{-au^2t} U(u,r)\, du}{u[J_0^2(ur_f) + Y_0^2(ur_f)]}, \\[4mm] U(u,r) = J_0(ur) Y_0(ur_f) - J_0(ur_f) Y_0(ur). \end{array} \right\} \tag{2}$$

and

$v_0(r,t)$ is a solution of Eq. 11.3(13) and satisfies the conditions

$$v_0(r,0) = 0; \qquad v_0(r_f,t) = 1.$$

If $f(t)$ be approximated by a stepwise function, such as

$$f(t) = f_n \quad : \quad t_n \leqslant t \leqslant t_{n+1} \quad : \quad t_0 = 0, \tag{3}$$

Eq. (1) can be reduced to the form:

$$\left. \begin{array}{c} 0 \leqslant t \leqslant t_1: \\[2mm] \gamma = \gamma_i - (\gamma_i - f_0)v_0(r,t); \\[2mm] t_1 \leqslant t \leqslant t_2: \\[2mm] \gamma = \gamma_i - (\gamma_i - f_0)v_0(r,t) - (f_0 - f_1)v_0(r,t - t_1); \\[2mm] t_2 \leqslant t \leqslant t_3: \\[2mm] \gamma = \gamma_i - (\gamma_i - f_0)v_0(r,t) - (f_0 - f_1)v_0(r,t - t_1) - (f_1 - f_2)v_0(r,t - t_2), \\[2mm] \text{etc.} \end{array} \right\} \tag{4}$$

[1] Cf. H. S. Carslaw and J. C. Jaeger, *Philos. Mag.*, **26**, 473 (1938).

The volume[1] flux rate of water per unit thickness across r_f then has the value

$$0 \leqslant \bar{t} \leqslant \bar{t}_1:$$

$$Q = \frac{2\pi k}{\mu}(p_i - p_0)F(\bar{t});$$

$$\bar{t}_1 \leqslant \bar{t} \leqslant \bar{t}_2:$$

$$Q = \frac{2\pi k}{\mu}[(p_i - p_0)F(\bar{t}) + (p_0 - p_1)F(\bar{t} - \bar{t}_1)];$$

$$\bar{t}_2 \leqslant \bar{t} \leqslant \bar{t}_3:$$

$$Q = \frac{2\pi k}{\mu}[(p_i - p_0)F(\bar{t}) + (p_0 - p_1)F(\bar{t} - \bar{t}_1) + (p_1 - p_2)F(\bar{t} - \bar{t}_2)];$$

$$\text{etc.}$$

$$(5)$$

where

$$\bar{t} = \frac{at}{r_f^2}; \qquad \bar{t}_n = \frac{at_n}{r_f^2},$$

and

$$F(\bar{t}) = \frac{4}{\pi^2}\int_0^\infty \frac{e^{-\bar{t}z^2}\,dz}{z[J_0^2(z) + Y_0^2(z)]}. \tag{6}$$

The cumulative water influx into the oil reservoir, expressed as a fraction of the pore volume of the oil reservoir,[2] will therefore be given by

$$0 \leqslant \bar{t} \leqslant \bar{t}_1:$$

$$\frac{\int_0^t Q\,dt}{\pi f r_f^2} = \bar{P} = 2\kappa(p_i - p_0)G(\bar{t});$$

$$\bar{t}_1 \leqslant \bar{t} \leqslant \bar{t}_2:$$

$$\bar{P} = 2\kappa[(p_i - p_0)G(\bar{t}) + (p_0 - p_1)G(\bar{t} - \bar{t}_1)];$$

$$\bar{t}_2 \leqslant \bar{t} \leqslant \bar{t}_3:$$

$$\bar{P} = 2\kappa[(p_i - p_0)G(\bar{t}) + (p_0 - p_1)G(\bar{t} - \bar{t}_1) + (p_1 - p_2)G(\bar{t} - \bar{t}_2)],$$

$$\text{etc.,}$$

$$(7)$$

where

$$G(\bar{t}) = \int_0^{\bar{t}} F(\bar{t})d\bar{t}. \tag{8}$$

A plot of the function $F(\bar{t})^*$ is shown in Fig. 11.10. As implied by the first of Eqs. (5), this curve gives directly the decline in flux from the

[1] Again no distinction is made here in flux volumes at atmospheric pressure, as compared with the boundary pressure, since these are equivalent to terms in κ.

[2] It is assumed that the thickness of the oil reservoir equals that of the water reservoir. If this should be known to be incorrect, Eqs. (7) will still give the correct cumulative influx if multiplied by $\pi f r_f^2$.

* $F(\bar{t})$ is equivalent to the function G' plotted by W. Hurst, *AIME Trans.*, **151**,

water reservoir with time if the boundary pressure is kept fixed at p_0. In comparison with the steady-state flux capacity of a finite radial system, of boundary radii r_e, r_f, the function $F(\bar{t})$ is the analogue of the term $1/(\log r_e/r_f)$. It will be seen that, whereas initially the flux in the transient system will be very much greater than that in the finite steady-state analogue, it will decline continuously and ultimately fall below that in any fixed steady-state system, although the rate of decline steadily decreases. This decline is due to the continuous recession of the radius at which there is a significant pressure reaction from the lowering of pressure at r_f. In fact the transient-flux history may be considered as a sequence

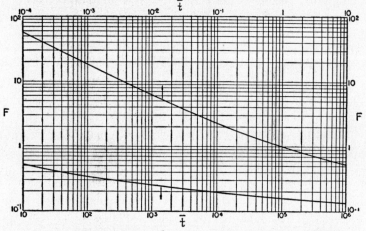

Fig. 11.10.   The function $F(\bar{t})$.   (*After Hurst, AIME Trans., 1943.*)

of steady-state values corresponding to increasing external-boundary radii given by $r_e = r_f e^{1/F}$.

By suitable superposition of elements of the curves of Fig. 11.10, according to Eqs. (5), the flux history for a stepwise changing boundary pressure can be readily constructed.

$G(\bar{t})$ is plotted[1] in Fig. 11.11. As indicated by the first of Eqs. (7) this curve shows the variation with time of the cumulative water influx if the boundary pressure is constant. It will be seen that as $\bar{t}$ increases and the rate of decline in the flux decreases, the cumulative influx gradually approaches an approximate linear increase with time. If the imposed pressure at r_f varies and can be represented by a stepwise approximation, as

---

57 (1943).   In the range of 10^{-2} to 10^3, $F(\bar{t})$ has also been tabulated by J. C. Jaeger and M. Clarke, *Royal Soc. Edinburgh Proc.*, (A) **61**, 229 (1942).

[1] $G(\bar{t})$ is equivalent to the function G of Hurst (*loc. cit.*), and Fig. 11.11 is taken from Hurst.

in Eq. (3), a superposition of elements of the curves of Fig. 11.11, according to Eqs. (7), will readily give the resultant cumulative-flux history.

The variation of the pressure distribution within the water reservoir, following a sudden lowering of pressure at the oil-water boundary to p_f,

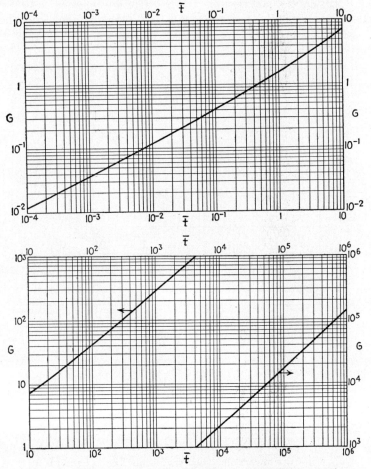

Fig. 11.11.   The function $G(\bar{t})$.   (After Hurst, AIME Trans., 1943.)

as given by the first of Eqs. (4), is plotted in Fig. 11.12.[1] The ordinates in Fig. 11.12 are ratios of the current pressure to the initial value,[2] if the

[1] The curves of Fig. 11.12 are replotted from those of A. Gemant [Jour. Applied Physics, 17, 1076 (1946)] derived in a study of a mathematically equivalent heat-conduction problem and plotted in a somewhat different form.

[2] The ordinates in Fig. 11.12 can also be considered as the ratios of the current and local pressure excess over the boundary pressure to the total pressure drop imposed at the boundary.

boundary pressure at r_f is assumed to be zero, and are plotted as functions of r/r_f for fixed values of the dimensionless time $\bar{t}$. The rapidly increasing values of $\bar{t}$ required to develop appreciable pressure declines as r/r_f is increased will be noted. Thus, whereas 20 per cent of the ultimate fall in pressure will develop at $r = 1.5r_f$ at $\bar{t} = 0.1$, it will require that $\bar{t} = 1,000$ before this drop will be observed at $r = 25r_f$. For the numerical example

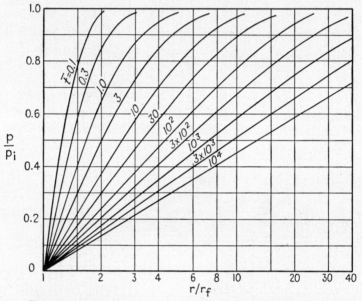

FIG. 11.12. The calculated pressure distribution in an infinite aquifer following a sudden lowering of the pressure (gauge) to zero at the internal boundary, of radius r_f; p = pressure at radius r; p_i = initial pressure; $\bar{t}$ = dimensionless time = $kt/f\kappa\mu r_f{}^2$; k = permeability of aquifer; f = porosity; μ = water viscosity; κ = compressibility.

considered in Sec. 11.4, $\bar{t} = 1,000$ would correspond to 185 years. And it will take 18 years for a 2 per cent drop in pressure to develop at $r = 25r_f$. It is clear, therefore, that from a practical point of view a reservoir for which the bounding radii have a ratio of 25 can be satisfactorily approximated by an infinite reservoir. On the other hand, at a radius $r = 10r_f$ about 3 per cent of the ultimate pressure drop would develop within about 22 months.

If $f(t)$ is approximated by a series of continuous linear segments, of slopes f_n', so that

$$\frac{df}{dt} = f_n' \quad : \quad t_n \leqslant t < t_{n+1}; \qquad t_0 = 0, \tag{9}$$

it can be shown that the density distributions may be expressed by the series of equations

$$0 \leqslant t \leqslant t_1:$$

$$\gamma = f(t) - f_0' \int_0^t [1 - v_0(r,\lambda)]d\lambda;$$

$$t_1 \leqslant t \leqslant t_2:$$

$$\gamma = f(t) - f_0' \int_{t-t_1}^t [1 - v_0(r,\lambda)]d\lambda - f_1' \int_0^{t-t_1} [1 - v_0(r,\lambda)]d\lambda;$$

$$t_2 \leqslant t \leqslant t_3:$$

$$\gamma = f(t) - f_0' \int_{t-t_1}^t [1 - v_0(r,\lambda)]d\lambda - f_1' \int_{t-t_2}^{t-t_1} [1 - v_0(r,\lambda)]d\lambda$$

$$- f_2' \int_0^{t-t_2} [1 - v_0(r,\lambda)]d\lambda;$$

etc.,

$$(10)$$

where v_0 is again the function defined by Eq. (2). The volume flux at r_f will then be given by

$$0 \leqslant \bar{t} \leqslant \bar{t}_1:$$

$$Q = -\frac{2\pi k}{\mu} p_0' G(\bar{t});$$

$$\bar{t}_1 \leqslant \bar{t} \leqslant \bar{t}_2:$$

$$Q = -\frac{2\pi k}{\mu} [p_0' G(\bar{t}) + (p_1' - p_0')G(\bar{t} - \bar{t}_1)];$$

$$\bar{t}_2 \leqslant \bar{t} \leqslant \bar{t}_3:$$

$$Q = -\frac{2\pi k}{\mu} [p_0' G(\bar{t}) + (p_1' - p_0')G(\bar{t} - \bar{t}_1) + (p_2' - p_1')G(\bar{t} - \bar{t}_2)];$$

etc.,

$$(11)$$

where the p_n''s are the time derivatives of the actual boundary pressures, corresponding to f_n', with respect to $\bar{t}$ as the time variable. Introducing now the new function $H(\bar{t})$ defined by

$$H(\bar{t}) = \int_0^{\bar{t}} G(\bar{t})d\bar{t}, \qquad (12)$$

the cumulative water influx, expressed as a fraction of the oil-reservoir pore volume, will be

$$0 \leqslant \bar{t} \leqslant \bar{t}_1:$$

$$\bar{P} = -2\kappa p_0' H(\bar{t});$$

$$\bar{t}_1 \leqslant \bar{t} \leqslant \bar{t}_2:$$

$$\bar{P} = -2\kappa [p_0' H(\bar{t}) + (p_1' - p_0')H(\bar{t} - \bar{t}_1)];$$

$$\bar{t}_2 \leqslant \bar{t} \leqslant \bar{t}_3:$$

$$\bar{P} = -2\kappa [p_0' H(\bar{t}) + (p_1' - p_0')H(\bar{t} - \bar{t}_1) + (p_2' - p_1')H(\bar{t} - \bar{t}_2)];$$

etc.

$$(13)$$

$H(\bar{t})$* is plotted in Fig. 11.13. This curve itself gives the variation of the cumulative water influx with time if the boundary pressure follows a linear decline from the outset. If the rate of pressure variation changes, a simple superposition, following the directions implied by Eqs. (13), will readily give the resultant cumulative-influx history.

While in practical applications the above stepwise approximations for the pressure variation at the original oil-water boundary will be most convenient for numerical computation, the volumetric water influx for

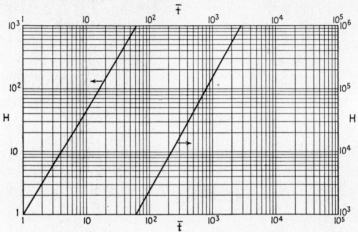

FIG. 11.13.   The function $H(\bar{t})$.   (*After Hurst, AIME Trans., 1943.*)

general continuous boundary-pressure variations, $p(t)$, can be expressed in a similar fashion, namely,

$$Q = -\frac{2\pi k}{\mu} \int_0^{\bar{t}} \frac{dp(\bar{\lambda})}{d\bar{\lambda}} F(\bar{t} - \bar{\lambda})d\bar{\lambda}. \tag{14}$$

Equations (11) are evidently an immediate consequence of Eq. (14). The cumulative influx is similarly given by

$$\int_0^t Q \, dt = -\frac{2\pi k r_f^2}{\mu a} \int_0^{\bar{t}} \frac{dp(\bar{\lambda})}{d\bar{\lambda}} G(\bar{t} - \bar{\lambda})d\bar{\lambda}. \tag{15}$$

Equations (13) readily follow from Eq. (15), when $dp/d\bar{\lambda}$ is a stepwise function.

The type of analysis given in this section has been applied[1] to compute the water influx in a Gulf Coast field, with a pressure history shown in Fig. 11.14. Assuming the water reservoir is of infinite extent and uniform throughout, the cumulative water intrusion was calculated by equa-

* This, too, is taken from Hurst (*loc. cit.*), whose function $\bar{G}$ is equivalent to $H(\bar{t})$.
[1] Hurst, *loc. cit.*

tions essentially equivalent to Eqs. (13). The constants used were k/μ = 0.2, $f = 0.28$, $\kappa = 3.6 \times 10^{-6}$ per psi, so that $a = 1.35 \times 10^4$ cm²/sec. = 1.255 $\times 10^6$ ft²/day. The oil-field radius was 7,850 ft, so that $\bar{t}$ = 2.037 $\times 10^{-2}t$ (days).

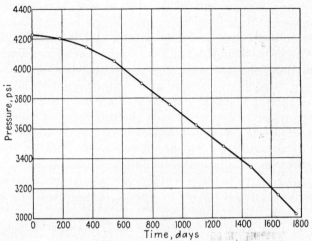

Fig. 11.14.   The observed reservoir-pressure history in a Gulf Coast field.   (*After Hurst, AIME Trans., 1943.*)

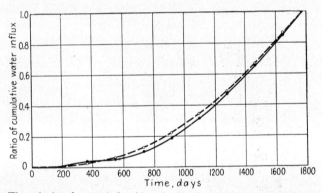

Fig. 11.15.   The calculated water-influx histories for the pressure curve of Fig. 11.14.   Solid curve calculated by material-balance equation.   Dashed curve calculated by the equivalent of Eq. 11.5(13).   Ordinates are ratios of cumulative influx to that at 1,766 days.   (*After Hurst, AIME Trans., 1943.*)

To eliminate the uncertainties associated with the absolute values of the physical constants the calculated magnitudes of the water influx were expressed as ratios to those computed for the date of the last pressure survey, that is, $t = 1,766$ days. The results are plotted as the dashed curve in Fig. 11.15. For comparison the water influx was also calculated by the material-balance equation [cf. Eq. 9.7(1)]. These values are plotted, also as ratios to that at $t = 1,766$ days, in Fig. 11.15 as the solid curve. In view of the inherent limitations of the material-balance method and the simplifications used in representing the water reservoir, the agreement between the two curves of Fig. 11.15 may be considered as entirely satisfactory and as a confirmation of the compressible-liquid analysis of the water-reservoir behavior.

11.6. Water Reservoirs of Finite Extent with Radial Symmetry and Circular Water-Oil Boundaries.—The assumption of an infinite water reservoir is obviously untenable from a strictly physical point of view. As previously noted, however, such an assumption should not be of serious consequence from a practical standpoint if the total expansion capacity of the water reservoir greatly exceeds the probable reservoir-volume recovery of the oil field. For under such circumstances it may be expected that even by the time of complete displacement of the oil from the oil reservoir the pressure decline in the aquifer will not have been transmitted to a substantial degree to the actual limits of the latter. In that case the water reservoir can be considered as infinite, at least during the producing life of the oil field.

An order-of-magnitude estimate of the conditions under which the water reservoir must be considered as finite may be derived[1] by equating a uniform expansion in the water reservoir to a displacement of three-fourths the pore volume of the oil reservoir. Assuming the pore volumes of the water and oil reservoirs to be V_w and V_o, respectively, the required equivalent uniform pressure drop in the water reservoir will then be

$$\Delta p(\text{psi}) = \frac{0.75 V_o}{\kappa V_w} = 2.5 \times 10^5 \frac{V_o}{V_w}. \tag{1}$$

Thus, even if the water-reservoir pore volume is 1,000 times that of the oil reservoir, an expansion such as would result from a uniform pressure drop of 250 psi would be required for a 75 per cent displacement. If it be assumed that the actual pressure distribution in the water reservoir is logarithmic and that the thickness is the same throughout the water- and oil-bearing strata, it may be shown that a mean pressure drop Δp implies a pressure differential $p_i - p_f$ between the water-reservoir boundaries, given approximately by

$$p_i - p_f = \frac{\Delta p (V_w/V_o) \log V_w/V_o}{(V_w/V_o) - \log V_w/V_o}, \tag{2}$$

so that

$$p_i - p_f = \frac{2.5 \times 10^5 \log V_w/V_o}{(V_w/V_o) - \log V_w/V_o} \sim \frac{2.5 \times 10^5 \log V_w/V_o}{V_w/V_o}. \tag{3}$$

---

[1] Essentially equivalent estimates to those presented here can be derived by a consideration of the transient-pressure-distribution curves of Fig. 11.12.

For a ratio $V_w/V_o = 1,000$, $p_i - p_f$ has the value 1,727 psi.

It thus appears that unless V_w/V_o is of the order of 1,000 or greater, the pressure decline during the whole production history will be so large that the finite character of the water reservoir will be very likely reflected in the details of the performance. In the early part of the productive life, however, corresponding to reductions in the factor 0.75 in Eq. (1), the pressure differential will be confined largely to the vicinity of the original water-oil boundary, and the finite extent of the water reservoir will not materially affect the pressure history.[1] On the other hand, if V_w/V_o is of the order of 100 or less, Eq. (3) indicates that it would be desirable to treat the water reservoir as a finite system from the very beginning, unless by virtue of a dispersion of free-gas phase[2] the effective compressibility of the water should be abnormally high. Such a treatment will now be given.

Considering again the water reservoir as a homogeneous radial system, the density distribution satisfying the boundary and initial conditions defined by[3]

$$\gamma = f_1(t) \quad : \quad r = r_1,$$
$$r\frac{\partial \gamma}{\partial r} = f_2(t) \quad : \quad r = r_2, \qquad (4)$$
$$\gamma = g(r) \quad : \quad t = 0,$$

may be shown[4] to be

$$\gamma = \pi \sum \frac{\alpha_n^2 J_0(\alpha_n r_1) U(\alpha_n r) e^{-a\alpha_n^2 t}}{J_0^2(\alpha_n r_1) - J_1^2(\alpha_n r_2)} \left[\frac{\pi}{2} J_0(\alpha_n r_1) \int_{r_1}^{r_2} rg(r) U(\alpha_n r) dr \right.$$
$$\left. - aJ_1(\alpha_n r_2) \int_0^t f_1(\lambda) e^{a\alpha_n^2 \lambda} d\lambda - \frac{aJ_0(\alpha_n r_1)}{\alpha_n r_2} \int_0^t f_2(\lambda) e^{a\alpha_n^2 \lambda} d\lambda \right], \quad (5)$$

where

$$U(\alpha_n r) = Y_1(\alpha_n r_2) J_0(\alpha_n r) - J_1(\alpha_n r_2) Y_0(\alpha_n r), \qquad (6)$$

and α_n is chosen so that

$$U(\alpha_n r_1) = 0. \qquad (7)$$

[1] At Magnolia, Ark., which initially contained a large gas cap and where the water intrusion has lagged somewhat behind the fluid withdrawals, the pressure history for the first 3 years could be fitted equally well by using an infinite-water-reservoir assumption and by one in which the water reservoir was considered as closed at a radius of 12 miles from the field [cf. W. A. Bruce, *AIME Trans.*, **151**, 110 (1943)].

[2] If the water is initially gas saturated, an evolution of gas and increase in the effective compressibility in the fluid content of the water reservoir should automatically develop as the pressure declines.

[3] For the general solution to problems in which the flux is specified over both boundaries, cf. Muskat, *op. cit.*, Sec. 10.11; and curves for the pressure decline at the internal boundary of a closed system, with a constant withdrawal rate, are given by T. V. Moore, R. J. Schilthuis, and W. Hurst, *Oil Weekly*, **69**, 19 (May 22, 1933), for different ratios of the external to the internal radii.

[4] Cf. M. Muskat, *Physics*, **5**, 71 (1934), and "The Flow of Homogeneous Fluids through Porous Media," Sec. 10.7, McGraw-Hill Book Company, Inc., 1937.

A special case of Eqs. (4) and (5) is that in which the pressure at the external water-reservoir boundary $(r_1 = r_e)$ is considered as fixed $(p_e = p_i \sim \gamma_i = f_1)$, and the flux $[f_2(t)]$ is specified at the original water-oil boundary $(r_2 = r_f)$. If the initial pressure or density distribution is uniform $[g(r) = \gamma_i]$, Eq. (5) reduces to

$$\gamma = \gamma_i - \frac{\pi a}{r_f} \sum \frac{\alpha_n J_0^2(\alpha_n r_e) U(\alpha_n r) e^{-a\alpha_n^2 t} \int_0^t f_2(\lambda) e^{a\alpha_n^2 \lambda}\, d\lambda}{J_0^2(\alpha_n r_e) - J_1^2(\alpha_n r_f)}. \tag{8}$$

If $f_2(t)$ is a constant q_o, the density decline at r_f will be given by

$$\gamma_i - \gamma_f = q_o \log \frac{r_e}{r_f} + \frac{2q_o}{r_f^2} \sum \frac{J_0^2(\alpha_n r_e) e^{-a\alpha_n^2 t}}{\alpha_n^2 [J_0^2(\alpha_n r_e) - J_1^2(\alpha_n r_f)]}. \tag{9}$$

The corresponding pressure decline at r_f will be

$$\Delta p = \frac{\mu Q_o}{2\pi k} \left\{ \log \rho + 2 \sum \frac{J_0^2(x_n \rho) e^{-x_n^2 \bar{t}}}{x_n^2 [J_0^2(x_n \rho) - J_1^2(x_n)]} \right\}; \quad \bar{t} = \frac{at}{r_f^2}; \quad \rho = \frac{r_e}{r_f}, \tag{10}$$

where Q_o is the volume flux rate per unit thickness corresponding to q_o[*] and $x_n = \alpha_n r_f$. Equation (10) is plotted in dimensionless form in Fig. 11.4 for $r_e/r_f = 6.3$, as the dashed curve.[1] It will be seen that, whereas the dashed and solid curves coincide to about $\bar{t} = 5$, the plot for Eq. (10) falls to approximately 82 per cent of that for the infinite water reservoir at $\bar{t} = 20$. This divergence reflects the assumption underlying Eq. (10) that the pressure is maintained at its initial value at r_e. In fact, as Eq. (10) shows directly, the pressure drop here will asymptotically approach a constant and steady-state value[2] given by

$$\Delta p_\infty = \frac{\mu Q_o}{2\pi k} \log \frac{r_e}{r_f}, \tag{11}$$

which, in the ordinate units of Fig. 11.4, has the dimensionless value 4.54 for $r_e/r_f = 6.3$. In the case of the infinite water reservoir the pressure drop at r_f will continue to increase until the pressure itself ultimately reaches atmospheric, regardless of Q_o, when it will no longer be possible to maintain the rate Q_o.

The manner in which the steady-state pressure distribution is built up for $r_e/r_f = 6.3$ is shown[3] in Fig. 11.16, in which the ratio of the excess in pressure at any point over the steady-state pressure at r_f to Δp_∞ is plotted

[*] Here, as in the similar problem for the infinite aquifer treated in Sec. 11.4, it will not be feasible in practice to impose suddenly on the aquifer a constant nonvanishing withdrawal rate, as Q_o, since the corresponding oil withdrawals will be initially provided mainly by the liquid-phase expansion within the oil reservoir.

[1] This curve is a replot of one computed by Hurst (*loc. cit.*), who derived Eq. (10) in somewhat different notation.

[2] It is, of course, presupposed that Q_o is limited to the maximum steady-state capacity of the system, so that $\Delta p_\infty < p_i$.

[3] W. Hurst, *Physics*, **5**, 20 (1934).

vs. r/r_f, for various values of $\bar{t}$.   This set of curves is, of course, independent of the absolute value of Q_o.

A formal solution of the basic equation 11.3(13) can also be derived if the pressure at r_f be specified, instead of the withdrawal rate.   For the

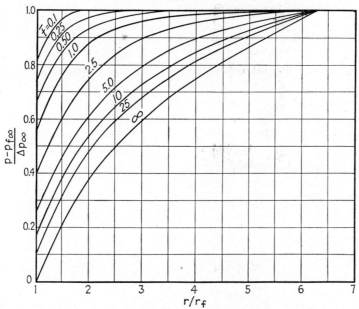

FIG. 11.16.   The calculated transient history of the establishment of a steady-state pressure distribution in a finite radial aquifer, at the internal boundary of which is suddenly imposed a constant withdrawal rate.

$$\frac{p - p_{f\infty}}{\Delta p} = \frac{\text{pressure excess over steady-state pressure at } r_f}{\text{total steady-state pressure differential}} ;$$

r_f = internal-boundary radius of aquifer; $r_e = 6.3r_f$ = external radius of aquifer, at which the pressure is maintained constant; r = radial distance; $\bar{t} = kt/f\kappa\mu r_f^2$; t = time; k = permeability of aquifer; f = porosity; μ = viscosity of water; κ = compressibility.   (After Hurst, Physics, 1934.)

special case where the density (pressure) at the external boundary r_e is maintained at the initial value for the reservoir as a whole, $\gamma_i (p_i)$, and the density at the radius r_f is a constant $\gamma_f(p_f)$, it may be shown[1] that

$$\gamma = \frac{\gamma_i \log r/r_f + \gamma_f \log r_e/r}{\log r_e/r_f}$$

$$- \pi(\gamma_i - \gamma_f) \sum \frac{J_0(\alpha_n r_e) J_0(\alpha_n r_f) U(\alpha_n r) e^{-\alpha \alpha_n^2 t}}{J_0^2(\alpha_n r_f) - J_0^2(\alpha_n r_e)}, \quad (12)$$

where

$$U(\alpha_n r) = Y_0(\alpha_n r_e) J_0(\alpha_n r) - J_0(\alpha_n r_e) Y_0(\alpha_n r), \quad (13)$$

and

$$U(\alpha_n r_f) = 0. \quad (14)$$

[1] M. Muskat, Physics, **5**, 71 (1934).

The volume flux per unit thickness across the original water-oil boundary, at r_f, is readily shown to be

$$Q = \frac{2\pi k(p_i - p_f)}{\mu}\left[\frac{1}{\log \rho} + 2\sum \frac{J_0^2(x_n\rho)e^{-x_n^2\bar{t}}}{J_0^2(x_n) - J_0^2(x_n\rho)}\right],$$

where

$$\bar{t} = \frac{at}{r_f^2} \quad : \quad \rho = \frac{r_e}{r_f}; \qquad x_n = \alpha_n r_f.$$

(15)

As is to be expected, Eqs. (12) and (15) indicate an asymptotic approach to the simple steady-state pressure distribution and flux rate as t becomes infinitely great. The transient decline in flux rate as represented by the bracket in Eq. (15) for $\rho = 5$ is plotted as curve I in Fig. 11.17. The excess over the asymptotic limit of $1/\log \rho = 0.62$ is the result of the fluid expansion between the boundaries r_e, r_f which is required for the establishment of the steady-state pressure distribution corresponding to the time-independent term in Eq. (12).

If the density at r_f should decrease linearly as

$$\gamma_f = \gamma_i - \epsilon t, \tag{16}$$

the generalization of Eq. (12) will be

$$\gamma = \frac{\gamma_i \log r/r_f + (\gamma_i - \epsilon t) \log r_e/r}{\log r_e/r_f}$$
$$- \frac{\pi\epsilon}{a}\sum \frac{J_0(\alpha_n r_e)J_0(\alpha_n r_f)U(\alpha_n r)(1 - e^{-a\alpha_n^2 t})}{\alpha_n^2[(J_0^2(\alpha_n r_f) - J_0^2(\alpha_n r_e)]}. \tag{17}$$

In the notation of Eq. (15) the mass-flux rate per unit thickness across the original water-oil boundary, at r_f, will be given by

$$Q = 2\pi f\epsilon r_f^2\left\{\frac{\bar{t}}{\log \rho} + \frac{1}{4(\log \rho)^2}[\rho^2 - 1 - 2\log \rho - 2(\log \rho)^2]\right.$$
$$\left. - 2\sum \frac{J_0^2(x_n\rho)e^{-x_n^2\bar{t}}}{x_n^2[J_0^2(x_n) - J_0^2(x_n\rho)]}\right\}. \tag{18}$$

The term in the braces, or $Q/2\pi f\epsilon r_f^2$, is plotted as curve II in Fig. 11.17, also for $\rho = 5$. Here the flux rate begins at 0 and after an initial rapid rise assumes an essentially linear increase with time, corresponding to the linearly decreasing boundary pressure at r_f.

For a physical system with constants used for the illustrations in Sec. 11.4, namely, $\kappa = 4.5 \times 10^{-5}$ per atmosphere, $f = 0.25$, $\mu = 0.5$, and $k = 100$ md, $\bar{t} = 1$ in Fig. 11.17 represents 67.45 days. And in the case of curve I the ordinate value 1 would correspond to 20.8 (bbl/day)/ft of sand per atmosphere drop in pressure differential. For curve II a value 1 for the braces would represent, with the above physical constants, a flux rate of 95.5 (bbl/day)/ft of sand for a rate of pressure decline at r_f

(= 2 miles) of 1 psi/day. For changes in the basic physical constants these scale factors will, of course, change in proportion.

Returning to Eq. (5) it may be noted that it will also apply if the flux $f_2(t)$ be specified at the external boundary r_e and the density, or pressure, $f_1(t)$, at the water-oil boundary r_f. For the special case where the water

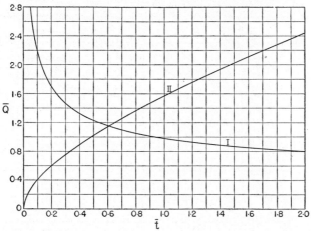

FIG. 11.17. The calculated transient-flux history from finite aquifers whose external boundaries are maintained at constant pressure. Curve I, pressure at internal boundary suddenly dropped and held at constant value. Curve II, internal-boundary pressure reduced continuously so that liquid density declines linearly with time. $\overline{Q}$ = dimensionless-flux rate, given by brackets of Eqs. 11.6(15) and 11.6(18); $\bar{t}$ = dimensionless time.

reservoir is of finite extent and closed, $f_2(t) = 0$,* and $f_1(t)$ is fixed at γ_f, Eq. (5) reduces to

$$\gamma = \gamma_f - \pi(\gamma_i - \gamma_f) \sum \frac{J_0(\alpha_n r_f) J_1(\alpha_n r_e) U(\alpha_n r) e^{-a\alpha_n^2 t}}{J_0^2(\alpha_n r_f) - J_1^2(\alpha_n r_e)}, \qquad (19)$$

where γ_i is the assumed initial uniform density distribution. The pressure decline at the closed boundary is then given by

$$p_e - p_f = \frac{2(p_i - p_f)}{\rho} \sum \frac{J_0(x_n) J_1(x_n\rho) e^{-x_n^2 \bar{t}}}{x_n[J_0^2(x_n) - J_1^2(x_n\rho)]}, \qquad (20)$$

in the notation of Eq. (15), and with x_n the root of

$$Y_1(x_n\rho) J_0(x_n) - J_1(x_n\rho) Y_0(x_n) = 0. \qquad (21)$$

The volume flux per unit thickness at r_f is also readily derived from Eq. (19) as

$$Q = \frac{4\pi k(p_i - p_f)}{\mu} \sum \frac{J_1^2(x_n\rho) e^{-x_n^2 \bar{t}}}{J_0^2(x_n) - J_1^2(x_n\rho)}. \qquad (22)$$

* If there should be an appreciable entry of surface waters into the reservoirs, the general equation (5) can be used, with $f_2(t)$ assigned a value corresponding to this extraneous influx.

The decline in pressure at the closed boundary, as well as that within the water reservoir, is shown in Fig. 11.18[1] for $r_e/r_f = 6.3$.

The ordinates in Fig. 11.18 are the ratios of the excess of pressure, over the fixed boundary (r_f) pressure, to the total initial pressure drop $p_i - p_f$.

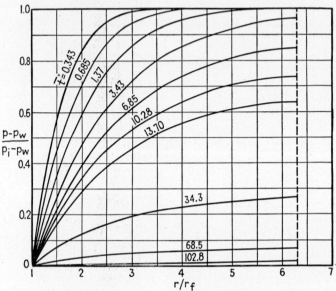

FIG. 11.18. The calculated transient history of the pressure decline in a finite closed aquifer at the internal boundary of which is suddenly imposed a constant pressure p_w.

$$\frac{p - p_w}{p_i - p_w} = \frac{\text{pressure excess over pressure at } r_f}{\text{total initial pressure differential}};$$

r_f = internal-boundary radius of water reservoir; r_e = external-boundary radius of water reservoir = $6.3r_f$; r = radial distance; $\bar{t}$ = dimensionless time. (*After Hurst, Physics, 1934.*)

It will be seen that an appreciable pressure drop at the closed boundary is indicated only for the curves with $\bar{t} \geqslant 3.43$.

The variation with time of the efflux rate is plotted in Fig. 11.19 for different values of r_e/r_f.* For convenience in plotting, the ordinates in Fig. 11.19 were chosen as proportional to the reciprocal of Q, that is, $k(p_i - p_f)/\mu Q$, or $1/4\pi$ times the reciprocal of the infinite series of Eq. (22). As is to be expected the curves for the different r_e/r_f's coincide initially but successively break away from the common envelope as the effect of the finite radius of the closed system comes into play and the flux rate begins to decline sharply. The envelope itself represents the decline in flux rate from an infinite water reservoir and is equivalent to $1/2\pi F(\bar{t})$, where F is the function plotted in Fig. 11.10.

[1] This is a replot of an equivalent set of curves of Hurst [*Physics*, **5**, 20 (1934)].
* *Ibid.*

The cumulative production P, as given by the integral of Eq. (22) and in the dimensionless unit of $P/fr_f^2\kappa(p_i - p_f)$, is plotted vs. the time in Fig. 11.20 for various values of r_e/r_f.* These, too, coincide at small $\bar{t}$ and diverge as the finite character of the closed system begins to limit the

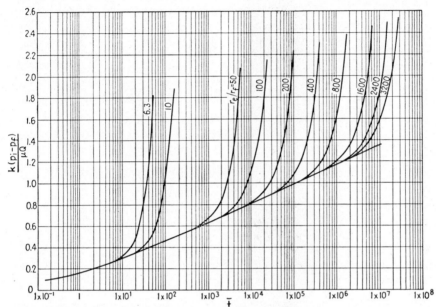

FIG. 11.19. The calculated variation, with the dimensionless time $\bar{t}$, of the efflux rate Q, from a finite closed water reservoir at the internal boundary of which is suddenly imposed a constant pressure p_f. p_i = initial pressure; k = permeability of aquifer; μ = water viscosity; r_f, r_e = internal- and external-boundary radii of aquifer. $\bar{t} = kt/\mu f\kappa r_f^2$; f = porosity; κ = water compressibility. (*After Hurst, Physics, 1934.*)

total expansion capacity. The asymptotic limits thus resulting are given by $\pi[(r_e^2/r_f^2) - 1]$ in the ordinate units of Fig. 11.20. The envelope, representing the limiting behavior of an infinite reservoir, is equivalent to the curve for $2\pi G(\bar{t})$, where G is the function plotted in Fig. 11.11.

11.7. Nonradial Water-drive Systems.—The analytical developments of the last several sections, which were based on the assumption of complete radial symmetry in the water reservoir, should serve to illustrate the broad features of water-drive pressure and flux performance. They will also often give semiquantitative estimates of the pressure and flux variations in compressible-water reservoirs even if the requirement of radial symmetry is not exactly satisfied. And in special cases, such as the East Texas field to be discussed in Sec. 11.9, they may be close enough approximations to provide the basis for a quantitative description of the observed

* *Ibid.*

pressure history. In general, however, it will be necessary to take into account the actual geometry of the water reservoir, which may deviate quite materially from the simple radial and symmetrical conditions treated previously. Moreover, in intrepreting or predicting the details of pressure

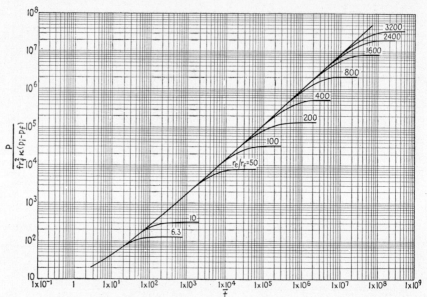

FIG. 11.20. The calculated variation, with the dimensionless time $\bar{t}$, of the cumulative efflux P, from a finite closed water reservoir at the internal boundary of which is suddenly imposed a constant pressure p_f. p_i = initial pressure; f = porosity; κ = water compressibility; r_f, r_e = internal- and external-boundary radii of aquifer. (*After Hurst, Physics, 1934.*)

distributions within the oil reservoir, above the bubble point of the oil, the assumption of general radial flow breaks down completely. In this section, therefore, a brief outline will be given of an analytical method that may be used to treat such general problems.

Perhaps the simplest procedure for constructing solutions in general compressible-liquid systems is that of making suitable syntheses of the elementary "instantaneous-sink" function,

$$\gamma = -\frac{q}{4\pi a f t} e^{-r^2/4at}, \tag{1}$$

which is a solution of Eq. 11.3(13), representing an instantaneous removal of q mass units of fluid at the origin ($r = 0$) at $t = 0$, but vanishing everywhere else at $t = 0$. If the sink is "permanent" and of strength $q(t)$ and the initial density is the constant γ_i, the corresponding solution is

$$\gamma = \gamma_i - \frac{1}{4\pi a f} \int_0^t \frac{q(\tau)}{t - \tau} e^{-r^2/4a(t-\tau)} \, d\tau. \tag{2}$$

If $q(\tau)$ is constant, q, Eq. (2) may be expressed as[1]

$$\gamma_i - \gamma = \frac{\mu \kappa q}{4\pi k} Ei\left(-\frac{r^2}{4at}\right), \tag{3}$$

where Ei denotes the "Ei function," which is tabulated in mathematical handbooks.

If γ has an initial distribution $g(x,y)$ over the infinite plane, the resulting solution may be expressed as

$$\gamma = \frac{1}{4\pi a} \left\{ \frac{1}{t} \int_{-\infty}^{+\infty} d\xi \int_{-\infty}^{+\infty} d\eta g(\xi,\eta) e^{-[(x-\xi)^2+(y-\eta)^2]/4at} \right.$$

$$\left. - \frac{1}{f} \int_0^t \frac{q(\tau)}{t-\tau} e^{-r^2/4a(t-\tau)} d\tau \right\}. \tag{4}$$

A generalization of Eq. (2) to a continuous linear source or sink distribution, with an initial uniform density γ_i, is given by:

$$\gamma = \gamma_i - \frac{1}{4\pi af} \int_0^t \frac{e^{-x^2/4a(t-\tau)} d\tau}{t-\tau} \int_{-\infty}^{+\infty} q(\eta,\tau) \ e^{-(y-\eta)^2/4a(t-\tau)} d\eta, \tag{5}$$

where the line source is assumed to lie along the y axis and to have a linear density $q(y,t)$. If the latter is taken as uniform, $q(t)$, along a length $2l$ from $-l$ to $+l$, Eq. (5) simplifies to[2]:

$$\gamma = \gamma_i - \frac{1}{4\pi af} \int_0^t \frac{q(t-\tau)e^{-x^2/4a\tau} d\tau}{\tau} \int_{-l}^{+l} d\eta e^{-(y-\eta)^2/4a\tau}. \tag{6}$$

For an analysis of the detailed effect of the actual well distribution upon the pressures within an oil reservoir, the contributions due to the individual wells can be formally summed to give a resultant expressed by:

$$\gamma = \gamma_i - \frac{1}{4\pi fa} \int_0^t \frac{d\tau}{t-\tau} \Sigma Q_i(\tau) e^{-[(x-x_i)^2+(y-y_i)^2]/4a(t-\tau)}, \tag{7}$$

where $Q_i(\tau)$ denote the flux rates for the various wells, located at (x_i,y_i). These may include virtual, or "image," wells, which are so distributed as to make the resultant density or pressure distribution satisfy boundary conditions further defining the flow system.

It should be noted that the above equations are based on the assumption that the reservoir in question is substantially of infinite extent and uniform

---

[1] Fluid-sink representations of the transient effects due to producing wells have been applied in hydrological investigations (cf. C. V. Theis, *AGU Trans.*, 1935, pt. II, p. 519; C. E. Jacob, *AGU Trans.*, 1940, pt. II, p. 574). The equivalent of Eq. (7) below has also been used in representing the fluid-head drawdowns due to multiple-well systems (cf. L. K. Wenzel and A. L. Greenlee, *AGU Trans.*, 1943, pt. III, p. 547).

[2] This special case has been used in a study of the East Texas field [cf. R. J. Schilthuis and W. Hurst, *AIME Trans.*, **114**, 164 (1935)].

throughout, as the fundamental solution [Eq. (1)] of which they are synthesized also involves that assumption. If the reservoir is bounded by a small number of linear segments, the effect of the boundary can be represented in certain cases by placing images across these boundaries of the actual flux distribution, as suggested in reference to Eq. (7). The use of these procedures to treat the composite oil- and water-reservoir systems also involves the assumption that the fluid compressibility, porosity, and thickness, or resultant expansion capacity per unit area, and flow resistance are essentially the same in the region of actual fluid withdrawal, *i.e.*, in the oil field, as in the water reservoir, which ultimately provides the major source of voidage displacement for the oil production. If there should be serious question about the validity of these assumptions, it will be necessary to treat the water reservoir and oil field separately and provide for their mutual interconnections by matching the separate pressure distributions through suitably formulated boundary conditions at the water-oil boundary. In the following section will be described an electrical device that can be used for the analysis of such composite systems as well as the water and oil reservoirs separately.

11.8. The Electrical Analyzer.—A powerful tool for the analysis of water-drive reservoirs when the water reservoir itself is not of simple geometry or uniform physical properties is the electrical analyzer.[1] This is based on the fundamental equation for flow of electric current in a dielectric medium, namely,

$$C \frac{\partial V}{\partial t} = \nabla \cdot (\sigma \nabla V), \tag{1}$$

where C is the local capacity per unit volume of the dielectric, σ is the specific conductivity, and V is the voltage. Now the general equation of homogeneous-fluid flow through porous media carrying compressible liquids is

$$f \frac{\partial \gamma}{\partial t} = \nabla \cdot \left(\frac{k}{\mu} \gamma \nabla p \right), \tag{2}$$

which may be rewritten as

$$f \frac{\partial \gamma}{\partial t} = \nabla \cdot \left(\frac{k}{\mu \kappa} \nabla \gamma \right), \tag{3}$$

on applying Eq. 11.3(9) for the relation between the density and pressure.

[1] This device was developed for application to oil-production problems by Bruce (*op. cit.*, p. 112), following earlier work by V. Paschkis and H. D. Baker [*AIME Trans.*, **64**, 105 (1942)], who showed how the basic heat-conduction equation could be solved by an electrical-analogy network and constructed an equivalent "analyzer" especially suited to the study of practical heat-conduction systems. The discussion here will be limited to an outline of the principles underlying the construction and operation of the analyzer. Some examples of its application will be presented in Sec. 11.10.

A comparison of Eqs. (1) and (3) thus immediately suggests the formal analogy given by

$$V \sim \gamma; \qquad \sigma \sim \frac{k}{\mu}; \qquad C \sim f\kappa; \qquad \bar{i} \sim \kappa \bar{v}_m, \tag{4}$$

where the compressibility κ, assumed constant, has been combined with f because it is physically representative of a capacitance rather than a resistance component. $\bar{i}$ in Eq. (4) denotes the current-density vector in the electrical system, and $\bar{v}_m$ the vector fluid mass flux.

Since changes in γ are generally very small, one may replace γ by the pressure as the basic dependent variable representing the state of the fluid. Upon introducing the linear approximation to Eq. 11.3(9), namely,[1]

$$\gamma = \gamma_o(1 + \kappa p), \tag{5}$$

and retaining only the dominant terms with respect to κ, Eq. (2) becomes

$$f\kappa \frac{\partial p}{\partial t} = \nabla \cdot \frac{k}{\mu} \nabla p, \tag{6}$$

which now is formally similar to Eq. (3) and hence permits the analogy with the corresponding electrical system if

$$V \sim p; \qquad \sigma \sim \frac{k}{\mu}; \qquad C \sim f\kappa; \qquad \bar{i} \sim \bar{v}, \tag{7}$$

where $\bar{v}$ represents the vector fluid volume flux.

The physical analogy implied by the above considerations is that the transient history in a compressible-liquid type of water reservoir is to be simulated by that in a dielectric medium of similar geometry, with distributed capacitance and conductivity related to the reservoir rock and fluid constants by Eq. (7), and subjected to initial and boundary conditions similar to those defining the reservoir. The pressure history and distribution in the latter will then be, except for a scale factor, identical with the voltage history and distribution of the electrical analogue. And the current densities in the latter will be identical, except for a scale factor, with the

---

[1] As discussed in Sec. 4.4 the decreasing compressibility with increasing pressure, as implied by Eq. (5), in a strict sense provides a better physical representation of the behavior of actual liquids than the constant compressibility implied by Eq. 11.3(9) and Eq. (3). The latter, however, has been used in the analytical discussions since it involves no further approximation as an equivalent of Eq. (2) and moreover serves to emphasize the physical role played by the fluid expansion in compressible-liquid systems. On the other hand, in operating the electrical-analyzer reservoir analogue and interpreting the results thus obtained, it is more convenient to use the fluid pressure as the primary physical variable, although it would be possible to retain Eq. (3) and apply a large scale factor to the density changes so as to make their product the analogue of the voltage and have a magnitude which is numerically comparable with that of the analyzer voltage.

fluid flux in the water reservoir. From a practical point of view this analogy in itself would be of little interest, since it would be seldom feasible to construct a continuous dielectric medium satisfying all the requirements of the analogy, especially if the water reservoir to be simulated were not strictly uniform throughout. Its significance arises from the fact that it is possible to approximate the continuous dielectric medium by a resistance-capacitance electrical network in which the individual resistance and capacity components can be independently adjusted and set to simulate the continuous distribution of these parameters. If the water reservoir be similarly visualized as an interconnected network of discrete rock units, with each of which may be associated definite values of the pertinent physical constants, a one-to-one correspondence between the electrical and flow systems can readily be established.

While, in principle, it is possible to make the electrical network simulate a three-dimensional system, it suffices for all practical purposes (since gravity effects are generally ignored) to consider the water reservoir as two-dimensional and treat the electrical analogue also as a two-dimensional network. The reservoir and fluid constants are accordingly to be considered as averaged over the vertical section of the water-bearing formation. On the other hand the thickness must be considered as explicitly absorbed in the term representing the local fluid expansion capacity. Moreover, since the reservoir is also to be simulated by a set of interconnected rock masses of finite volume and well-defined geometry, the effective fluid resistance of each must be introduced as the analogue of the corresponding resistance element. For practical application of the analogy, Eqs. (7) are therefore to be replaced by the system

$$V \sim p \quad : \quad R_e \sim R_f = \frac{\mu}{k} R_o; \qquad C \sim f_\kappa C_o; \qquad \bar{i} \sim \bar{v}. \qquad (8)$$

In Eq. (8) R_o is the purely geometrical resistance factor of the reservoir-volume element, and C_o is its actual bulk volume. For example, if the latter is essentially rectangular, of distance Δx in the direction of flow, Δy normal to this direction, and thickness h, R_o would be $\Delta x / h \, \Delta y$. If it is in the form of a sector subtending an angle $\Delta \theta$ at the center of polar coordinates and extending radially from r_1 to r_2, R_f would be $(\log r_2/r_1)/h \, \Delta \theta$, or $\Delta r / \bar{r} h \, \Delta \theta$, if the flow be approximated as linear, where $\Delta r = r_2 - r_1$ and $\bar{r} = (r_2 + r_1)/2$. For these two cases C_o would be $h \, \Delta x \, \Delta y$ and $h \bar{r} \, \Delta r \, \Delta \theta$, respectively. If the reservoir is inherently of an essentially linear or radial character or if a linear or radial component, bounded by streamlines, can be considered as representative of the whole reservoir, the corresponding electrical analogue would simply be the network shown in Fig. 11.21.

Once the geometrical and physical structure of the water reservoir is simulated, the initial and boundary conditions must be applied. The initial condition of uniform pressure, as usually obtains in the case of reservoir studies, is taken care of by charging all the condensers (the capacitance elements) to the same initial voltage. At the external boundary the initial voltage is permanently maintained fixed by a battery or equivalent voltage source, if the corresponding pressure is assumed constant. If the reservoir is considered as closed, the terminal resistance and capacitance elements are simply isolated from the external voltage supply.

In principle the boundary condition at the water-oil-boundary terminal of the electrical network can be applied either in the form of a preassigned

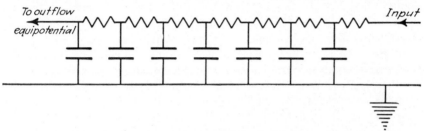

Fig. 11.21. A diagrammatic representation of the electrical analogue of a water reservoir.

pressure or voltage history or in that of a fluid-withdrawal rate or current history. It is the latter, however, that is usually used in actual reservoir studies. This is provided by a series of vacuum-tube circuits, each of which can be set to permit fixed current withdrawals from the network terminal and tied in to the latter by a controlling timing circuit in a sequence corresponding to the fluid-withdrawal history of interest.

To fix the numerical values of the electrical components it is convenient to introduce scale factors relating the electrical and fluid constants. These may be chosen as

$$\left.\begin{array}{l} V \text{ (volts)} = Lp \text{ (psi)}, \\ C \text{ (microfarads)} = Mf\kappa C_o(\text{bbl/psi}), \\ R_e \text{ (megohms)} = NR[\text{psi}/(\text{bbl/day})]. \end{array}\right\} \qquad (9)$$

From these it follows that the current and fluid-flux magnitudes will be related by

$$|\bar{i}| \text{ (microamp)} = \frac{L}{N} q \text{ (bbl/day)}, \qquad (10)$$

where q is the actual volumetric flux. Moreover the ratio of the time scales for the electrical and fluid systems, t_e and t_f, will be determined by

$$t_e \text{ (sec)} = MNt_f(\text{days}). \qquad (11)$$

If M is chosen to be of the order of 0.01 or 0.001 and N of the order of 10 or 100, production periods expressed in days will be covered in the electrical analyzer in a number of seconds of an order of magnitude lower. Thus a production interval of a year would require about a ½-min run on the analyzer. A scale factor of this order of magnitude is frequently used in practical applications.

If the water reservoir can be described completely in advance, the analyzer resistance and capacitance components can be set to correspond to those of the water reservoir, and then on imposing various voltage (pressure) or current (flux) initial and boundary conditions the voltage history at the oil-water-boundary terminal will parallel that of the pressure at the boundary. In general, however, only order-of-magnitude information will be available about the character of the water reservoir. In such cases the previous pressure history at the water-oil boundary can be used to determine the effective values of the geometrical and physical reservoir constants. This is accomplished by so adjusting and choosing the electric-circuit parameters that the voltage history recorded by the analyzer parallels the observed pressure history and substantially coincides with it on applying the scale-factor transformation of the first of Eqs. (9). The reservoir constants and their distribution, similarly translated by Eqs. (9), are then considered as representative of the actual reservoir parameters. While a choice of the constants to give a "best fit" with the observed pressure data could be based on a least-square-deviation analysis, attempts to make quantitative determinations of the reservoir constants are not warranted. In fact, experience and a thorough understanding of the geology of the water reservoir should provide an even more significant basis for evaluating the agreement between the analyzer and field-observed transients than mere numerical coincidences.

The water-reservoir properties as determined by the above-outlined procedure are, of course, of interest in themselves. For they may reveal features, such as barriers or faults, previously unsuspected and difficult to detect in general geological investigations. However, the immediate purpose of making such determinations is to provide a means for predicting future behavior. Once the water-reservoir characteristics have been established, predictions of future performance can readily be made by simply continuing the analyzer run under various assumed future withdrawal histories.

The water-withdrawal history which is imposed at the water-oil boundary to check the pressure history is not, of course, directly observable. If the reservoir oil is undersaturated and remains so throughout the pressure range of interest, it can be readily determined as the reservoir equivalent of the oil and water production minus the volumetric expansion of the residual oil-reservoir fluids associated with the observed pressure decline. Even if the oil-reservoir content is not known accurately, only a small error will be involved, since the total expansion of 10 million barrels of undersaturated reservoir fluid will be only of the order of 100 bbl/psi. However, if the oil is saturated and the oil reservoir contains a free-gas phase, the above simple procedure breaks down. The water influx can then be determined by application of the material-balance equation expressing the water influx in terms of the fluid withdrawals and initial reservoir contents, that is, Eq. 9.7(1). This requires that the initial-oil and free-gas volumes be known. Errors in the latter will lead to proportional errors in the calculated water-intrusion volumes. Unfortunately the uncertainty associated with the oil-reservoir volumetric constants is the weakest link in the whole process of analyzing the reservoir performance of systems containing a free-gas phase. On the other hand, as long as the past history of the reservoir can be simulated to a satisfactory approximation, the resultant representation developed for both the oil and water reservoirs should provide a reasonable basis for predictions of future behavior, provided that such extrapolations are not extended too far in the future.

It should be noted that in practice the water reservoir is generally represented by a linear-resistance-condenser network, as indicated in Fig. 11.21. Accordingly the actual water reservoir is approximated by an essentially cylindrical system, or sectors of cylinders, in which annular rings, bounded by approximately equipotential surfaces, are simulated by appropriate resistance-condenser network units.

While the electrical analyzer was originally developed as a tool for the study of the water reservoir itself, it has since been developed so as to permit direct application to combined oil- and water-reservoir systems. In fact the "pool unit" used to represent the oil reservoir comprises one of the major components of the composite instrument. Its scope also has been extended so as to include partial-water-drive reservoirs and such as contain a free-gas phase of varying compressibility or a segregated gas cap,[1] although the permeability variation within the oil reservoir due to changes in the fluid saturation cannot be taken into account by the linear circuits thus far used. Examples of its application to both complete- and partial-water-drive systems will be given in Sec. 11.10.

A photograph of the electrical analyzer is reproduced in Fig. 11.22.[2] The cabinet to the left contains the pool unit and its auxiliaries. At the top are switches for changing the pool resistors. In the center are the pool-unit jacks, representing the mesh points of the network, surrounded

[1] Cf. H. Schaeffer, *AIME Trans.*, **170**, 62(1947).

[2] This instrument, being used at the Gulf Research & Development Company, was built through the courtesy and cooperation of the Carter Oil Company and Standard Oil Development Company of New Jersey.

by terminal jacks for the pool condensers. Near the bottom are four rows of jacks for well resistor connections, and below these are multiple jack connections to the production- and injection-rate controls.

FIG. 11.22. An electrical analyzer.

At the top of the middle cabinet are a recorder for recording voltages and resistor shunts for both recorders. The second panel provides the controls for the two voltmeters and the switches for connecting them to different points of the network. The third panel holds the controls for the nine production-rate-control tubes with on-off switches at the top, rate-scale switches in the center, and manual rate controls at the bottom. The fourth panel contains water-drive resistor switches at the left and injection-rate controls at the right. The fifth panel has the main control switches and pilot lights for the analyzer. Behind the sixth panel (with a hinged door) are mounted the water-drive condensers. At the bottom of the cabinet is the filament and high-voltage power supply.

The right-hand cabinet has the recorder for measuring both currents and voltages at the top. The remaining panels are for the automatic controller, which provides the proper grid-bias voltage to both production- and injection-rate-control tubes at the proper time. There are nine groups of jacks for production rates and two for injection rates (lower right). In the center are terminal jacks to a voltage divider, which provides production rates from 0 to 100 per cent of a maximum value. A similar group of jacks to the right serves the same purpose for injection rates. At the lower left are the control switches and pilot lights for the automatic controller. Just below these are a group of jacks for making each of the 50 time intervals a multiple (2,3,4,6, or 10) of 2.5 sec, which is the basic time interval for the analyzer.

11.9. The East Texas Field.—While the electrical analyzer undoubtedly provides the most powerful means now available for the study of complete-water-drive systems, the basic principles can be illustrated by purely analytical methods. And where the water reservoir is of substantially uniform properties,[1] a treatment by the application of the analytical procedures discussed in the previous sections is virtually as simple as that using the electric analyzer, as far as the water reservoir itself is concerned. The East Texas field, in Rusk, Gregg, Upshur, and Smith Counties, Tex., discovered in 1930, provides a good example of the analytical method. In fact it was through such a study that the compressible-liquid theory of water-drive performance was first established.[2]

The East Texas field produces from the Woodbine Sandstone at a depth of 3,050 to 3,320 ft subsea. The general structure contours on top of this extended water reservoir in Northeast Texas are plotted in Fig. 11.23.[3] The contours in the field itself and an east-west vertical section are shown in Fig. 1.12. The reservoir is a stratigraphic trap lying on a monoclinal structure. It is about 42 miles long and 4 to 8 miles wide, covering about 134,500 productive acres. The average effective sand thickness is 35 ft. The produced oil has a gravity of 38 to 40°API. The reservoir temperature

---

[1] As noted in Sec. 11.8 the assumption of simple cylindrical geometry is usually made in electrical-analyzer applications, just as in the case of the analytical methods.

[2] M. Muskat, *Physics*, **5**, 71(1934); Schilthuis and Hurst, *loc. cit.* The treatment presented here follows that of the first reference. An essentially equivalent analysis was given in the latter reference, except that the field was represented by a finite line sink [cf. Eq. 11.7(6)], rather than a segment of a circular arc.

[3] Schilthuis and Hurst (*loc. cit.*).

is 146°F. The initial reservoir pressure was 1,620 psia, but the oil was saturated only to 755 psia, with a solution-gas–oil ratio of 365 ft³/bbl and formation-volume factor of 1.26. The average porosity is 25.2 per cent; permeability, 2,000 to 3,000 md; and connate-water saturation, 17 per cent. The original water-oil contact was at 3,325 ft subsea, and initially

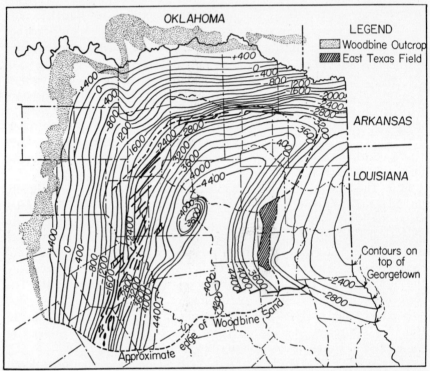

Fig. 11.23. The structure contours (subsea depths) on top of the Woodbine Sand in Northeast Texas. (*After Schilthuis and Hurst, AIME Trans., 1935.*)

71,620 productive acres were underlain by water. More than 25,000 wells have been drilled to the producing section.

As of Jan. 1, 1947, the average reservoir pressure was 1,023 psi. As a whole the oil reservoir at that time was thus still above the bubble point, although along the extreme southeastern edge of the field the pressures had fallen somewhat below the bubble point. The total initial stock-tank-oil content of the reservoir was of the order of 6,000,000,000 bbl. For a compressibility of 10⁻⁵ per psi, a pressure drop of 600 psi would result in an expansion of 36,000,000 bbl. Even if the contribution of the connate-water expansion and gas-evolution volume at the southeastern limits of the field were taken into account, approximately 98 per cent of the cumulative oil withdrawals of approximately 2,366,000,000 bbl by Jan. 1, 1947, must have been replaced by water intrusion. For all practical purposes the East Texas field can therefore be considered as producing under a complete water drive.

While the assumption of strict uniformity of the Woodbine water reservoir surrounding the East Texas field is certainly an extreme idealization, there is little direct evidence supporting any specific variation of the reservoir properties. Accordingly the Woodbine Sand outside the field will be taken as of uniform thickness, permeability, and porosity throughout. The field itself will be approximated by a concentrated withdrawal sink lying along a 20-mile-radius circle over an angular width of 120°. From the great areal extent of the Woodbine reservoir, as shown in Fig. 11.23, it is clear that it could be appropriately treated as an infinite reservoir over much of its producing life.[1] A generalization of Eq. 11.4(8) would then give the pressure history at the field boundary on applying the observed withdrawal history. However, since the reservoir is actually finite, the calculations will be based on the assumption that it is limited at a circular boundary 100 miles from the center of the radial sink representing the oil field. On the other hand it will also be assumed that throughout the production history to Jan. 1, 1947, the pressure at the external boundary remained fixed at its initial value of 1,620 psi.*

On the basis of these assumptions it is possible to calculate the pressure history to be expected at the water-oil reservoir boundary, $r = r_f = 20$ miles, from the observed or assumed water-withdrawal history, provided that the physical parameters of the water reservoir are known. The latter are the permeability k, porosity f, thickness h, water viscosity μ, and compressibility κ. Their values have been assumed to be as follows: $k/\mu = 2.65$ (darcys/cp), $f = 0.25$; $h = 123$ ft†; $\kappa = 5.3 \times 10^{-4}$ per atmosphere.

Neglecting entirely the above-mentioned fluid-replacement contributions, due to the expansion of the residual oil and water in the oil field, and the free gas in the southeastern part of the field, the water-withdrawal rate from the water reservoir has been taken as the sum of the reservoir volume equivalent of the oil-production rate plus the water-production rate, minus the rate of water reinjection into the Woodbine Sand. One of the outstanding features of the operation of the East Texas field is the organized

---

[1] The linear-sink representation of the oil field, used by Schilthuis and Hurst (*loc. cit.*), implies an effectively infinite extent of the water reservoir.

* While the assumption of closure at the external boundary would probably represent a better approximation, both assumptions, as well as that of infinite extension, should give essentially the same theoretical pressure-decline history during most of the producing life.

† The different thickness values given for the water and oil reservoirs represent gross averages and do not imply a discontinuity at their common boundary.

effort that has been made to return as much of the produced water as possible to the producing formation.[1] Since the start of these operations in 1938, a total of 644 million barrels of water produced from the field has been so returned, to July, 1947. During the first half of 1947 more than 90 per cent of the salt-water production, of the order of 480,000 bbl/day,

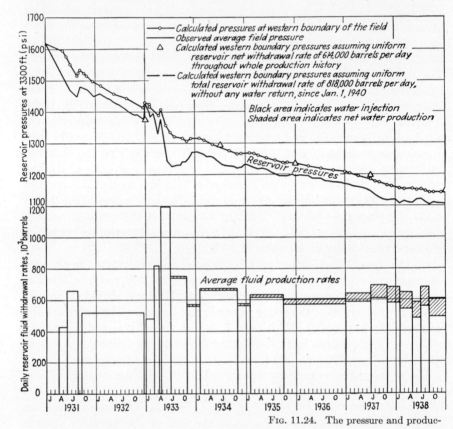

Fig. 11.24. The pressure and produc-

was injected into 75 wells along and beyond the western part of the field. While the water-return program was originally developed as a solution of the water-disposal problem created by the continually increasing water production, it was soon recognized to have a very important pressure-maintenance effect on the oil reservoir. In fact, as a first approximation the water return may be considered as simply reducing the net fluid withdrawal by an equal amount, as will be assumed here.

The production history of the field, averaged over intervals of approxi-

[1] A complete review of these operations and many of the production data used in the present discussion are given by W. S. Morris, *Oil and Gas Jour.*, **46**, 245 (Nov. 15, 1947); cf. also W. S. Morris, *Oil and Gas Jour.*, **45**, 92 (Aug. 31, 1946).

mately uniform net withdrawal rates, is plotted in the lower part of
Fig. 11.24.[1]   As the legend indicates, the ordinates are all expressed as
equivalent reservoir volumes.   The net withdrawal rates are represented
by the boundary between the black and hatched areas, or the total readings prior to the beginning of the water-return operations.   Upon denoting

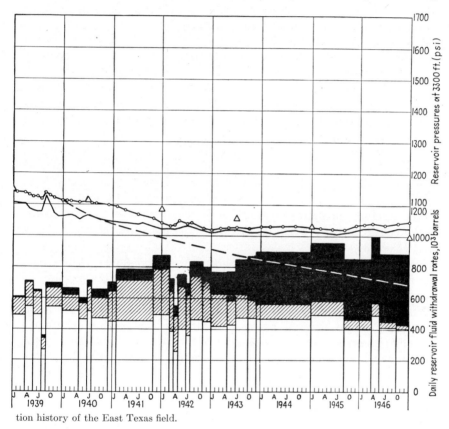

tion history of the East Texas field.

these net reservoir withdrawal rates by Q_n, the expected pressure decline
Δp at the original oil-water reservoir boundary will be given by

$$\Delta p = b[Q_n \log \rho + Q_1 J(\bar{t}) + (Q_2 - Q_1) J(\bar{t} - \bar{t_1}) + \cdots + Q(_n - Q_{n-1}) J(\bar{t} - \overline{t_{n-1}})], \quad (1)$$

where

$$J(\bar{t}) = 2 \sum \frac{J_0^2(x_n\rho) e^{-x_n^2 \bar{t}}}{x_n^2 [J_0^2(x_n\rho) - J_1^2(x_n)]}. \quad (2)$$

[1] The producing gas-oil ratios are not plotted in Fig. 11.24, for, within the limits
of accuracy of the measurements, they have remained at the solution ratio of 365
ft³/bbl throughout the production history to January, 1947.

ρ is the ratio of the boundary radii of the water reservoir and will have the value 5 for the present treatment of the East Texas field. $\bar{t}$ is the dimensionless time, at/r_j^2, where $a = k/f\mu\kappa$. Hence the numerical relation between $\bar{t}$ and t is

$$\bar{t} = 1.668 \times 10^{-4}t(\text{days}).\tag{3}$$

$\bar{t}_1, \bar{t}_2, \ldots, \bar{t}_n$ denote the values of $\bar{t}$, calculated by Eq. (3), corresponding to the times at which withdrawal rates change from Q_1 to Q_2, Q_2 to $Q_3, \ldots$, and Q_n to Q_{n+1}. b has the value $3\mu/2\pi kh$, the factor 3 correcting for the fact that the actual withdrawal rates Q refer to only $120°$ of a complete cylindrical system. For a net withdrawal rate of 10^5 bbl/day the coefficient of log ρ and the functions J has the numerical value 130, Δp being expressed as psi. Equations (1) and (2) are readily verified to comprise a direct generalization of Eq. 11.6(10) for the case of strictly constant withdrawal rates. The x_n's are roots of Eq. 11.6(7) for $\rho = 5$.

The theoretically predicted pressures at the water-oil reservoir boundary, calculated by Eqs. (1) and (2), taking into account the 49 intervals of approximately uniform withdrawals indicated at the bottom of Fig. 11.24, are plotted as the points on the uppermost curve in Fig. 11.24. The observed reservoir pressures, averaged over the entire oil field, are plotted as the continuous curve in Fig. 11.24. These pressures refer to a datum level of $- 3,300$ ft.

The absolute numerical values of the calculated pressures are, of course, the result of the particular assumptions used regarding the physical parameters describing the water reservoir. The significant feature of the calculated pressure curve is its general parallelism (to July, 1941) to the observed pressure history. Since it would be expected physically and is actually observed in the measurements of the pressure distribution within the field that the averages of the latter should be lower than those along the western boundary, the constants have been chosen to give calculated values higher than the observed averages. The flexibility available for adjusting the absolute level of the calculated pressure curve was largely limited to the water-reservoir thickness. Major changes in any of the other parameters would have altered the general transient pressure-decline trend, unless compensating changes were made in the remaining constants to leave the numerical coefficient of t in Eq. (3) substantially the same.

Upon accepting the constants leading to the calculated pressures as plotted, the differences from the measured pressures can be expressed as a correction factor proportional to the withdrawal rates, which would be a measure of the mean pressure gradient within the field. By applying such a correction factor to the calculated curve it is possible to obtain almost complete coincidence with the observed average pressures over most of the production history through 1940, except for the pressure peaks following sharp curtailments in the withdrawal rate. The calculated values of the latter are generally smaller than those observed. This apparent discrepancy, however, is to be expected as the result of the attenuation at the water-oil boundary of the large pressure reactions in the immediate vicinity of the actual fluid-withdrawal area.

While the persistent parallelism between the calculated and observed pressures for the first 10 years of the producing life shown in Fig. 11.24 undoubtedly represents a strong confirmation of the compressible-liquid-theory analysis of the field performance, the convergence of the curves during 1941 and 1942 might seem to nullify this interpretation. The immediate implication of the convergence is that the actual reservoir pressures apparently remained higher since 1941 than anticipated from the calculations. Since this tendency for pressure stabilization, as compared with the calculated decline, first developed during a period of marked increase in net withdrawal rates, it can hardly reflect a normal reaction of the water reservoir itself. Rather it represents a change in the quantitative physical meaning of the average field pressure, so as to give apparently higher average values than determined prior to 1941. In view of the rapidly increasing water production and the development of substantial water injection during 1940–1941, it is to be expected that the subsequent average pressure determinations would not be fully equivalent to those made when the field was operating simply under an essentially uniform withdrawal distribution. While the quantitative evaluation of the effect of the distribution of fluid withdrawal and injection on the average oil-field pressure would require an analysis of the composite water- and oil-reservoir systems, there is little doubt that qualitatively the above-indicated change in distribution would lead to apparently higher mean pressures. Accordingly the comparative trends of the calculated and observed pressures, even after 1941, as shown in Fig. 11.24, may still be considered as a satisfactory confirmation of the compressible-liquid treatment of the East Texas field water-drive behavior.

The triangles in Fig. 11.24 show the calculated pressures on the assumption that throughout its history the net withdrawal rate had been constant, at a value of 614,000 bbl/day, which is the gross average to Jan. 1, 1947, of the actual withdrawal rates plotted at the bottom of the graph. It is of interest that the general pressure-decline trend follows almost exactly the detailed pressure calculations, except for the minor fluctuations in the latter. The ultimate pressure drop as of Jan. 1, 1947, was about 50 psi smaller when the production-rate variations were taken into account, for the actual net withdrawal rates during the last several years of the calculated history were somewhat lower than the gross average.

A conservative estimate of the pressure-maintenance action of the water-injection program may be derived by treating the period since January, 1940, when water-return operations began to develop on a significant scale, as one of uniform withdrawal rate, averaged over the actual total withdrawals of oil and water,[1] without any water return. The pressure decline calculated on this assumption is that plotted as the dashed curve in Fig. 11.24. It will be seen that on the basis of this calculation the reservoir pressure drop would have been about 200 psi greater than that resulting from the actual operations. Because of the increasing amounts of water return and total withdrawals during 1945 and 1946, the use of average fixed withdrawals in computing the dashed curve would evidently lead to maximal values for the January, 1947, pressure. The estimate of a 200-psi pressure-maintenance effect, as of January, 1947, is therefore a minimal value.[2] The performance of the East Texas field thus gives an excellent illustration not only of the compressible-liquid water-drive mechanism but also of the pressure-maintenance effect by water injection as a supplement to the natural water intrusion.

The above-outlined calculations of the pressure history show the potentialities of simplified representations of water reservoirs. For all the physical and geometrical parameters, at least in the combinations μ/kh and $k/\mu f\kappa$, were assumed to be strictly uniform throughout the water reservoir. Certainly there would be only an infinitesimal probability that these quantities are actually constant throughout the Woodbine Sand west of the East Texas field. Nevertheless, whatever variations there may be apparently give resultant average properties that, when assumed strictly constant, suffice to represent the gross behavior of the reservoir in an entirely satisfactory manner.

While the values of k/μ, f, and h used in calculating the pressure-decline history of the East Texas field appear to be inherently reasonable in the light of general information available on the Woodbine Sand, the assumed compressibility of 5.3×10^{-4} per atmosphere is definitely higher than would be expected normally. In fact this is about 12 times as high as that given in standard tables. Of course, the fact that the observed pressure history has been closely duplicated by the calculated pressures, using this value of the compressibility, in no sense proves its quantitative validity. For a compressibility value half as great would have given identically the same results if the assumed boundary radii of the water reservoir

---

[1] The average total withdrawal rate between January, 1940, and January, 1947, was 818,000 bbl/day.

[2] Similar calculations of the effect of different net withdrawal rates on the pressure-decline history of the East Texas field, using the linear-sink representation of the field, are reported by S. E. Buckley, *API Drilling and Production Practice*, 1938, p. 140.

were both multiplied by $\sqrt{2}$. Moreover there is some flexibility in choosing the transient time coefficient indicated in Eq. (3). However, if the other physical constants be kept within reasonable limits, there is no doubt that an abnormally high compressibility is implied by the transient history of the field, even though its exact value may be somewhat uncertain.

As mentioned in Sec. 11.3, high effective compressibilities in water-drive transient systems may be due to changes in compression of the aquifer itself as the reservoir pressure changes. As the fluid pressure is reduced, the rock structure will absorb a greater part of the overburden loading. The resulting compaction will force an expulsion of the entrained fluids equivalent to a simple compressibility expansion. Included clays and shales may be even more susceptible to such compaction effects.

A different explanation for the high effective compressibility of the water can be based on the assumption that there is a dispersion of free gas throughout the water reservoir. Since the Woodbine water near the East Texas oil reservoir is undoubtedly undersaturated, the free-gas dispersion is to be visualized as comprised of localized gas "pockets" in the more distant parts of the sand and distributed with some degree of uniformity. They may constitute small gas fields or gas caps of oil fields. Such fields are known to be scattered throughout the Woodbine Sand west of the East Texas field. To have a resultant compressibility $\bar{\kappa}$ a gas-liquid mixture must contain a volume fraction x of gas given by

$$x = \frac{\bar{\kappa} - \kappa_o}{\kappa_g - \kappa_o}, \tag{4}$$

where κ_o, κ_g are the compressibilities of the separate liquid and gas phases and the solubility of the gas in the liquid is neglected. At a pressure of 100 atm a resultant compressibility 12 times a "normal" liquid compressibility of 4.4×10^{-5} per atmosphere could be obtained for a volumetric free-gas content of only 4.8 per cent. Hence a distribution of free-gas masses having a total volume 4.8 per cent of the gross pore volume of the Woodbine Sand would suffice to explain the high compressibility implied by the East Texas field performance. While the latter in itself cannot discriminate between the various possible physical mechanisms, general geological considerations appear to favor the interpretation based on the action of distributed free-gas masses.

11.10. The Smackover Limestone Fields.—The Smackover Limestone formation in Arkansas provides the basis for an interesting comparative study of oil fields subjected to the water-drive action of a common porous aquifer. Such a study has been made by the electrical analyzer.[1] As

[1] W. A. Bruce, *AIME Trans.*, **155**, 88 (1944). Except as indicated otherwise, the material of this section has been taken from the paper of Bruce.

shown in Fig. 11.25, the Smackover Limestone water reservoir covers an area of 10,000 to 20,000 sq miles. As of 1944, 12 oil- or condensate-producing fields, lying on anticlinal structures, had been discovered in the upper part of the Smackover formation, known as the Reynolds Lime.

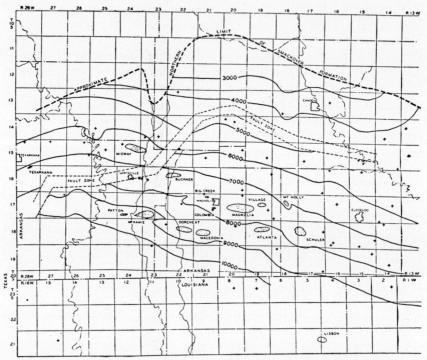

FIG. 11.25. A regional map of the Smackover area, showing the estimated areal extent of the pools and contours (in feet subsea) on top of the Smackover Limestone. (*After Bruce, AIME Trans., 1944.*)

All the pools show evidence of some degree of water-drive action. More than 75 dry holes penetrating the formation, together with the producing wells within the fields, serve to give a rather complete picture of the gross characteristics of the water reservoir. Except for the major fault indicated in Fig. 11.25 there appears to be substantial intercommunication throughout the aquifer. The porous section is 100 to 300 ft thick. The original reservoir pressures in the oil fields were directly proportional to their subsea depths, with a gradient of 51 psi/100 ft, which corresponds closely to the hydrostatic gradient of the formation waters.

The basic reservoir data characterizing the various fields producing from the Smackover formation are listed in Table 1.

TABLE 1.—RESERVOIR DATA ON FIELDS PRODUCING FROM THE SMACKOVER LIMESTONE

Pool	Producing wells (July 1, 1943)	Estimated no. of acres	Original reservoir vol., millions of bbl		Estimated av. laboratory permeability, md	Estimated av. laboratory porosity, %
			Gas zone	Oil zone		
Atlanta.....	22	1,000	2.4	23.8	1,280	15.0
Big Creek...	1	200	12.0	....	250	12.5
Buckner....	28	1,610	....	43.3	50	20.0
Dorcheat...	23	2,130	80.0	21.8	160	14.0
Macedonia..	17	2,250	100.0	24.0	230	14.0
McKamie...	17	2,800	288.6	.....	400	17.0
Magnolia...	116	5,600	65.0	360.0	1,500	17.0
Midway....	37	1,800	.....	180.0	140	21.3
Mt. Holly...	5	360	11.3	10.5	1,130	20.0
Schuler.....	15	1,200	4.7	34.4	1,500	18.5
Texarkana..	1	240	5.0	.....	170	14.0
Village.....	12	640	11.3	18.8	2,000	20.0

It will be noted that only two of the fields had no original gas caps. In fact, in these two, Buckner and Midway, the oil was initially undersaturated. Three of the reservoirs apparently are condensate producers only, with no black-oil zone of significance. The high permeabilities and moderate porosities reflect the oölitic character of the productive section of the Smackover Limestone.

TABLE 2.—PRODUCTION AND WATER-INFLUX DATA OF SMACKOVER LIMESTONE FIELDS

Pool	Oil, 10³ stock-tank bbl	Gas produced, millions of ft³	Water produced, 10³ bbl	Cumulative water influx, millions of bbl	Pressure change, psi (Δp)	Water influx, bbl)/ (day)/ (psi)	Cumulative water influx, 10³ bbl per Δp	Water influx, bbl)/ (day)/ (psi)/ (acre)
Atlanta....	3,322	4,346	229	5.49	425	11.9	12.9	0.012
Big Creek..	139	3,431	1	2.51	353	7.2	7.1	0.036
Buckner...	3,854	914	454	4.86	710	4.0	6.8	0.003
Dorcheat..	1,836	17,625	5	2.10	930	7.5	2.3	0.004
Macedonia.	580	7,017	3	.....	430			
McKamie..	2,747	13,148	....	3.34	245	17.2	13.6	0.006
Magnolia..	27,998	25,299	2,137	24.50	450	41.8	54.4	0.007
Midway...	2,225	581	52	1.84	240	18.6	7.7	0.008
Mt. Holly..	220	1,324	30	0.69	156	8.8	4.4	0.024
Schuler....	4,396	6,377	2,096	10.15	225	45.4	45.1	0.038
Texarkana.	22	650	....	0.55	89	17.4	6.2	0.073
Village....	1,802	7,930	518	8.15	170	25.2	47.8	0.038

In Table 2 are listed the fluid production and water-influx data, as of June 30, 1943. The latter were determined by application of the material-balance equation, assuming the volumes of oil and gas to be known, as given in Table 1. While the exact values of the calculated water intrusion depend on the reservoir-volume assumptions, independent evidence for the presence of substantial[1] water-drive action was available in most cases in observations of the early appearance of water in edge wells, increasing percentages of water production, or pressure rises following reductions in the withdrawal rates.

Electrical-analyzer studies have been made and reported on seven of the fields. The scale factors used and the average values of the water-reservoir constants required to reproduce the observed pressure histories are given in Table 3, where h represents the average thickness in feet and κ the compressibility in psi^{-1}. The numerical value for k, as used in Table 3, is 1.127 times its permeability in darcys. The factors L, N, M are those defined by Eq. 11.8(9).

Table 3.—Scale Factors and Aquifer Constants Used in Electrical-analyzer Studies of Smackover Limestone Fields

Pool	Arbitrarily assumed		Determined by analysis				
	$\dfrac{L}{N}$	MN	L	N	M	$k\dfrac{h}{\mu}$	κh
Atlanta.....	0.01	0.1	0.354	35.4	0.0028	2.55	0.00061
Buckner.....	0.01	0.1	0.1112	11.12	0.0090	1.28	0.00018
McKamie...	0.01	0.2	0.575	57.5	0.0035	4.14	0.00044
Magnolia....	0.001496	0.06	0.1367	91.3	0.0007	6.57	0.00231
Midway.....	0.00867	0.2775	0.1781	20.2	0.0137	3.45	0.00031
Mt. Holly...	0.02	0.3	0.493	24.65	0.0122	1.78	0.00021
Schuler.....	0.006	0.1	0.359	59.8	0.0017	4.32	0.00182
Average...	....	......	....	.....	......	3.44	0.00058

As indicated also by other geological evidence the values of the aquifer constants determined by the application of the electrical analyzer and listed in Table 3 imply that the water reservoir must have a substantially lower average permeability than the part of the Smackover occupied by the oil reservoirs. For if the permeabilities listed in Table 1 are combined with the water-reservoir constants of Table 3, effective compressibilities are obtained that, except for Buckner, are 50 to 100 times the normal value for water. Because of the rather extended drilling throughout the Smackover formation, it is doubtful that there is a distribution of an appreciable

---

[1] Except for the undersaturated reservoirs, Buckner and Midway, it is doubtful that any of the others have been producing strictly as "complete-"water-drive reservoirs.

number of undiscovered large gas masses, other than those associated with the known oil and gas fields, which could lead to such high effective compressibilities for the main body of the water reservoir. The electrical-analyzer treatment of these water-drive systems must therefore be considered as essentially empirical.

With respect to the detailed production histories of the Smackover Limestone fields only two examples, Buckner and Magnolia, will be discussed in this section. These will illustrate both the complete- and partial-water-

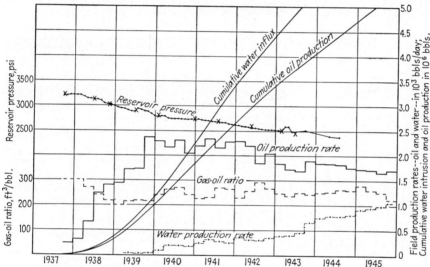

Fig. 11.26. The pressure and production history of the Buckner field, Arkansas. Crosses are reservoir pressures predicted with the electrical analyzer.

drive reservoirs. The history of the Midway field will be reviewed in Sec. 11.11 as an example of a pressure-maintenance operation by water injection.

Although water intrusion into the Buckner field, in Lafayette and Columbia Counties, Ark., is limited from the north by the fault zone, and there are other indications that it may be partly isolated, its production history through 1945 is rather typical of complete-water-drive performance. The pressure and production data and calculated cumulative water intrusion[1] to July, 1943, are plotted in Fig. 11.26.

Some of the pressure values obtained in the electrical-analyzer study of

---

[1] The production and pressure data plotted in both Fig. 11.26 and Fig. 11.28 are taken from the records of the Arkansas Oil and Gas Commission, made available to the author through the courtesy of G. R. Elliott. The water-influx curves were replotted from those of W. A. Bruce, *AIME Trans.*, **155**, 88 (1944).

the Buckner field are indicated as crosses in Fig. 11.26.[1] In this investigation the presence of the major fault as a barrier against free water influx from the north was taken into account, although the surrounding aquifer was otherwise treated as continuous and uniform and the field itself as a single intercommunicating reservoir.

The substantial constancy of the gas-oil ratios through 1945 reflects the undersaturated character of the Buckner crude. The fluctuations probably represent the errors often made in gas-oil-ratio measurements. The increasing rise in water production is the result of the gradual envelopment of producing wells by the advancing edgewaters.

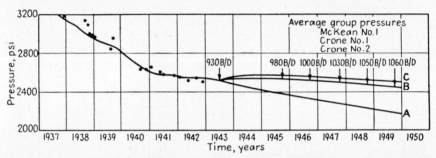

FIG. 11.27. The pressures observed and predicted to 1943, by means of the electrical analyzer, at a well group in the Buckner field, and predictions of subsequent pressures under various modes of operation. Curve A, no water injection, and field production rate = 2,475 bbl/day. Curve B, water injection as indicated, and field production rate = 2,475 bbl/day. Curve C, water-injection rate = 300 bbl/day, and field production rate = 1,700 bbl/day. Points are observed pressures. (*After Bruce, AIME Trans., 1944.*)

By definition the Buckner field must have been producing as a complete-water-drive reservoir, except for the reservoir liquid expansion, as long as the pressure remained above the bubble point of the oil. However, the increasing total withdrawals due to the rising water production may be expected ultimately to lead to gas evolution and partial-gas-drive performance unless the oil-production rate be cut drastically or a water-return program be instituted. One of the most useful applications of the electrical analyzer lies in the prediction of the effects of such changes in the operations. In Fig. 11.27 are shown the analyzer predictions regarding the average pressure history at a group of wells in the Buckner field for three modes of field operation after July, 1943. While the predicted reactions of the pressure to the indicated changes in the field operations are evidently of the type to be expected from general considerations, the quantitative

---

[1] The analyzer study was apparently based on slightly different production data than those plotted in Fig. 11.26 and gave a somewhat smoother pressure history than shown here. The differences, however, do not imply any serious uncertainty regarding the validity of the analysis.

magnitudes can be determined only by the application of the analyzer or equivalent analytical calculations.

The Magnolia field, Columbia County, Arkansas, is the largest of the Reynolds Lime reservoirs and was discovered in March, 1938. It provides an example of a partial-water-drive reservoir and its treatment by the electrical analyzer. That the water drive is not complete is immediately evident from the fact that the total water intrusion, to July, 1943, was only about 81 per cent of the cumulative oil and water production, as

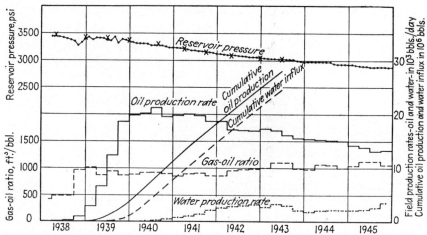

FIG. 11.28. The pressure and production history of the Magnolia field, Arkansas. Crosses are reservoir pressures predicted with the electrical analyzer.

indicated in Table 2. Moreover it originally contained a gas-cap volume one-sixth of the oil-zone volume. The oil was initially gas saturated at the reservoir pressure of 3,465 psi. The crude gravity is 38°API.*

The observed pressure and production history through 1945 and calculated cumulative water influx to July, 1943, are plotted in Fig. 11.28. In the electrical-analyzer study of this field the aquifer was assumed to be uniform, except for the barrier formed by the major fault zone. Some of the field pressures determined by the analyzer, using the water-influx data calculated by the material-balance equation and the constants given in Table 3, are shown as crosses in Fig. 11.28. It will be seen that the agreement with the observed pressures is quite as satisfactory as that obtained in the analysis of a complete-water-drive system, such as Buckner.

The slow rise in gas-oil ratios and limited decline in reservoir pressure indicate that the water drive has played a significant role in the early

---

* For a more detailed discussion of the geological aspects of the Magnolia reservoir and of its early history, cf. H. F. Winham, *AIME Trans.*, **151**, 15 (1943).

history of the reservoir performance by providing the replacement for a large part of the fluid-withdrawal voidage. It can be shown, however, that if the oil-withdrawal rate were increased to or allowed to exceed 25,000 bbl/day, without gas return, a controlling gas drive would quickly develop. If the gas-oil ratios be allowed to rise without restriction and the gas cap be depleted of its gas content, the oil would migrate into the gas cap, to be followed by an invasion from the encroaching edgewaters. Conversely the most efficient operating program from the point of view of oil recovery would result from the combined effect of the gas-cap expansion due to gas return and the flushing action of the natural water drive. If all produced gas were returned, approximately two-thirds of the field would be occupied by water and one-third by gas at the time of abandonment. The effects developed by 1960 of various possibilities of this type, as computed by the electrical analyzer, are listed in Table 4.

Table 4.—Electrical-analyzer Predictions of State of Magnolia Field in 1960 Following Various Methods of Operation

Operating conditions	Cumulative oil production (10^6 bbl)	Cumulative water influx (10^6 bbl)	Av. reservoir pressure, psi
100% gas return after 1945; liquid withdrawals of 20,000 bbl/day with 20% water production.......	146	120	2,650
100% gas return after 1945; withdrawal rate, with 20% water, reduced to hold the pressure at 2,900 psi..........................	116	110	2,900
Field gas-oil ratios limited to 2,000 ft³/bbl; oil-production rate continually reduced 3.2% per year, with 20% water	94	175	2,400
Uncontrolled gas and water production; oil-production rate held at 16,000 bbl/day	125	330	1,400

The loss of oil due to migration in the gas cap is a potential danger at Magnolia, as it is in other fields having both gas caps and active water drives. If the gas cap is initially "dry," an intrusion of oil will result in a loss of the residual oil left by whatever mechanism may ultimately control its recovery. Even if the gas cap should be subsequently invaded by water, the residual oil will still be of the order of 20 to 30 per cent of the pore space. It is to avoid such losses that gas injection into a gas cap may be warranted even when the reservoir is subject to strong water-drive action.

The close proximity of some of the Smackover Limestone fields, as shown in Fig. 11.25, raises the problem of interfield pressure interference. Within 10 miles of Magnolia are Atlanta, Big Creek, Dorcheat, Macedonia, Mt. Holly, and Village. It may well be expected that the cumulative water influx into these fields, as well as Magnolia, will lead to mutual pressure reactions superposed on their own fluid withdrawals. If by 1960 this influx, outside of the water invasion into Magnolia, totals 150 million barrels, a pressure drop of 75 to 250 psi may be induced at Magnolia.

More striking and serious is the effect of the production from Magnolia on the pressure at the nearby and much smaller Village pool. Since the latter has a strong water drive, the pressure decline due to the interference from Magnolia has been considerably greater than that due to its own production. Figure 11.29 shows the relative contributions of the Magnolia interference and the Village withdrawals, as determined by the electrical analyzer. It is evident that, if this interference were not recognized and taken into account

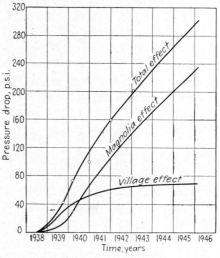

FIG. 11.29. The pressure-drop history at the Village pool, Arkansas, expressed as the resultant effect of the withdrawals at Village and Magnolia, and estimated by the electrical analyzer. Circles, observed pressure decline at Village. (*After Bruce, AIME Trans., 1944.*)

in the analysis of the performance of Village, the potential magnitude of its own water drive would be greatly underestimated.

11.11. Pressure Maintenance by Water Injection; The Midway Field.[1]—The Midway pool, Lafayette County, Ark., was discovered in January, 1942. It is the only major Reynolds Limestone reservoir lying north of the regional fault zone of the Smackover area (cf. Fig. 11.25). Its gross reservoir characteristics are indicated in Tables 1 and 2. It lies only 3½ miles north of the fault zone and 10 miles south of the northern limits of the Smackover formation. There is evidence that the water reservoir thins both to the east and west of the field. The structural closure within the

---

[1] The material of this section pertaining to the water-injection program at Midway is taken from Horner, *loc. cit.;* cf. also W. L. Horner and D. R. Snow, *API Drilling and Production Practice* (1943), p. 28. An extension of the production and pressure histories of the Midway field through 1947 is given by L. L. Jordan, *World Oil*, **128**, 121 (September, 1948).

field exceeds 200 ft. The original reservoir pressure was 2,920 psi at − 6,050 ft. The reservoir oil was initially saturated to a pressure of 2,528 psi, with a formation-volume factor of 1.24, and had a compressibility of 15 × 10⁻⁶ per psi, at the reservoir temperature of 182°F.

While the early history of the field indicated a rather strong water drive, the limitation of its surrounding aquifer made it obvious that it could not sustain a high rate of oil withdrawal throughout its producing life. With an electrical-analyzer study of the early history of the field as a guide,[1] a water-injection program was undertaken in April, 1943, to arrest the pressure decline by supplementing the natural water intrusion. The water was injected in four dry holes at the edges of the field, with casing set 115 ft or more below the water-oil contact, leaving approximately 130 ft of the formation exposed. By Sept. 1, 1945, more than 5 million barrels of water had been injected into four wells, including both the brine produced with the oil and fresh water from six shallow water wells within the field.

The production and pressure histories of the field to October, 1945, are plotted in Fig. 11.30. It will be seen that a definite pressure reaction was observed within 2 months after the water injection was started. After February, 1944, there was a consistent upward trend in the reservoir pressure, which was still continuing in September, 1945, although the total withdrawal rate had been increased more than 50 per cent above that before injection was commenced. Thus the field pressures as a whole were kept above the bubble point, and a change of the producing mechanism into a gas drive was prevented, although some rise in the gas-oil ratio developed in 1945.

It will be noted from Fig. 11.30 that the rates of water injection were almost equal to the rates of fluid withdrawal in 1945. It is clear, therefore, that if the natural water intrusion had continued undiminished with respect to its earlier rates, a much greater pressure rise would have developed than was actually observed. The latter evidently implies a declining contribution of the natural water influx, as anticipated from the limitations in the surrounding aquifer and the electrical-analyzer study. This is confirmed by computations of the actual rates of water influx from the pressure and production history, as shown in Fig. 11.31. It is obvious from Fig. 11.31 that the expansion capacity of the water reservoir surrounding the Midway field is of quite limited magnitude.

The water-injection operations at Midway are of historical interest in that they are the first which were undertaken in the primary-production period of a reservoir solely as a method of pressure maintenance. It will be recalled that the much more extensive water-injection program in the

[1] W. A. Bruce, *AIME Trans.*, **155**, 88 (1944).

East Texas field developed mainly as a means of disposal for the water production. While by Sept. 1, 1945, only 7,500,000 bbl of oil had been produced at Midway of an estimated ultimate recovery of 67,000,000 bbl, the history to that date certainly indicates successful operations and such as compare favorably with the more common programs of gas injection. Among the specific benefits derived from the water-injection project at Midway were (1) production rates in August, 1945, of 7,500 bbl/day without pressure decline, as compared with a maximum of 1,500 bbl/day if water injection were stopped; (2) reduction of lifting costs; (3) maintenance of high productivity indexes of the wells due to the prevention of gas evolution; and (4) an estimated increase in primary ultimate recovery of more than 35,000,000 bbl. Evidently the last in itself would completely compensate for the costs of the project, the operating expenses of which were averaging only $70 per day during June to August, 1945.

From a physical point of view the supplementing of a natural water drive by water injection involves no concepts or principles that are different from those entering the general problem of natural-water-drive reservoir performance. If the latter is inherently desirable and conducive to greater and more profitable oil recovery, there can be little danger of adverse effects resulting from water injection. The criteria on which the advisability of water-injection operations should be evaluated are largely economic. First, of course, it should be determined whether or not the natural water intrusion is itself sufficiently large to provide for complete fluid-withdrawal replacement. Such a determination should include an analysis of the future potential water-supply capacity of the aquifer. For, as seen in the case of the Midway field, if the supply capacity of the aquifer is of limited extent, it may provide only a temporary pressure-maintenance support for the fluid withdrawals, unless the latter be continually reduced. However, if the apparent failure of the natural water drive fully to replace the fluid withdrawals is due only to the lack of control over the latter and a balance could be established by nonprohibitive restriction and control of the field production rate and distribution, such measures may be economically preferable to water injection. The availability of water for injection, of suitable wells for injection, and the cost of the latter, if they have to be drilled for the purpose, are additional economic factors to be considered. On the other hand, if suitably located injection wells are available, the return of at least the produced water should be generally desirable.

The ultimate economic success of pressure maintenance by water injection presupposes, of course, the inherent susceptibility of the reservoir to maximum recovery by water drive. While this may be assumed for the majority of oil reservoirs, it is not a universal or axiomatic proposition.

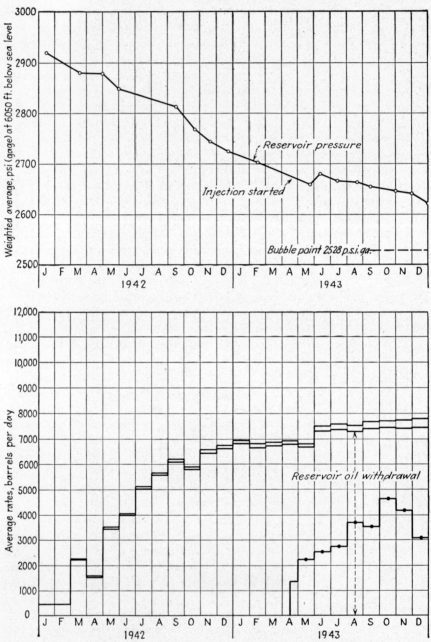

FIG. 11.30.   The pressure, production, and water-injection history for the Mid-

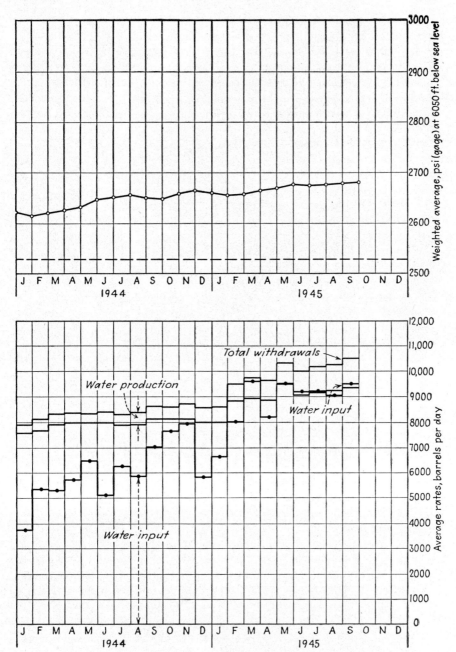

way field, Arkansas. (*After Horner, API Drilling and Production Practice, 1945.*)

If the reservoir initially had a gas cap and it is not feasible to control the gas and oil production to prevent pressure depletion in the gas cap, a strong water drive, whether natural or artificially supplemented by water injection, may force oil migration into the gas cap and loss of recoverable oil. Such loss may still be less than that to be gained by the water flushing of the original oil zone. But it is possible that the gross economic balance may favor the unsupported water drive without water injection, unless

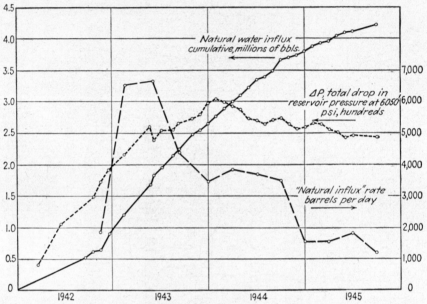

Fig. 11.31. The variation of the natural water influx and pressure drop in the Midway field, Arkansas. (*After Horner, API Drilling and Production Practice, 1945.*)

gas injection into the gas cap could be undertaken simultaneously. In fact, such gas injection to encourage gas-cap expansion and gravity drainage may in certain cases represent more profitable operations than a water-injection program.

If the average vertical permeability of the producing formation is comparable with that parallel to the bedding planes, so as to lead to a rapid conelike rise of bottom water under the producing wells near the edgewater boundary,[1] the difficulties of achieving high water-free and total oil recoveries along the field boundaries may be aggravated by injecting water into the aquifer. On the other hand, if the pay is highly stratified and the water tends prematurely to enter the wells through a continuous high-

---

[1] The general performance of bottom-water-drive reservoirs will be discussed in Sec. 11.13.

permeability stratum, which cannot be readily located and shut off in both producing and potential injection wells, water injection may be definitely inadvisable. Moreover, when the connate-water content of the oil zone is abnormally high, the inherent advantage of the water-drive mechanism, as compared with the gas drive, may be subject to question. Except for the pressure-maintenance effect itself the value of supplemented water intrusion as the result of water injection would, in such cases,

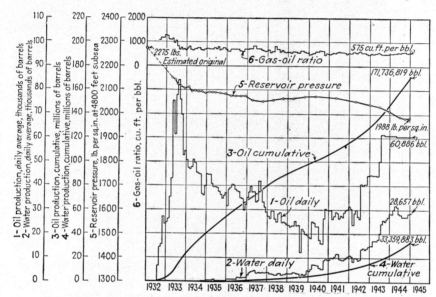

FIG. 11.32. The production and pressure history of the main Conroe Sand of the Conroe field, Texas. (*After C. J. Deegan, Oil and Gas Jour., 1945.*)

be of little value. And in extreme circumstances, such as may obtain in highly fractured limestones, accelerated water invasion may even be detrimental.

These observations should serve to emphasize that water injection, as well as all other operations to control reservoir performance, should be undertaken only after a study has been made of the particular reservoir of interest. No single method or program will have universal applicability. In many cases several possibilities may offer equal promise of success, within the accuracy of predictions of reservoir performance now possible. In others the only measures required to achieve efficient oil recovery may be those of controlling the total rate and distribution of fluid withdrawals. And in some it may be difficult to prove that virtually uncontrolled production will not give the maximum ultimate economic returns.

11.12. Further Examples of Gross Water-drive Performance.—In contrast to the situation regarding gas-drive reservoir performance, many rather complete reports are available on the producing histories of water-drive reservoirs. Supplementing those discussed in Secs. 11.9 to 11.11, several additional examples will be presented here, although no detailed analyses will be attempted.

The pressure and production history through 1944 of the main Conroe Sand of the Conroe field, Montgomery County, Tex., is plotted in

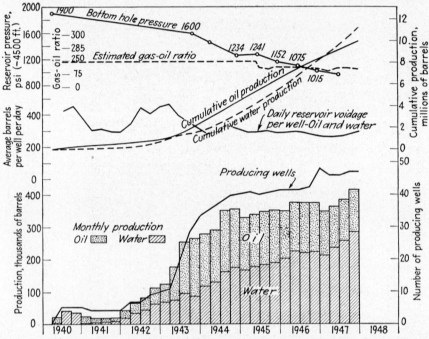

Fig. 11.33. The production and pressure history of the Eutaw reservoir in the Pickens field, Mississippi.

Fig. 11.32.[1] The oil zone of this field, discovered in 1932, is overlain in part by a gas cap. During its early development, with uncontrolled production rates, the gas-oil ratios were high and the pressure declined rapidly as if it were a gas-drive reservoir. The subsequent reduction of both oil and gas withdrawals quickly brought the field under the control of a natural water drive. The gas-oil ratio was reduced to approximate the solution ratio of the 38°API crude. The pressure was completely stabilized, and even began to rise, until the increased withdrawal rates due to war requirements and the continued growth in water production caused a resumption of the decline in 1941. A recovery of more than 60 per cent of the initial oil in place is estimated for this field.

[1] This is taken from C. J. Deegan, *Oil and Gas Jour.*, **44**, 100 (Oct. 6, 1945).

Figures 11.33 and 11.34, referring to the Pickens field, Madison and Yazoo Counties, and to Tinsley, Yazoo County, Miss., show the rapidly rising water production typical of the highly undersaturated reservoirs of Mississippi. The former produces a 40°API crude from the Eutaw-Wilburn Sand at about 4,500 ft, and the latter a crude of 35°API gravity from the Woodruff Sand at 4,250 ft. While the indicated gas-oil ratios are only estimates, there is no doubt that there has been no significant gas

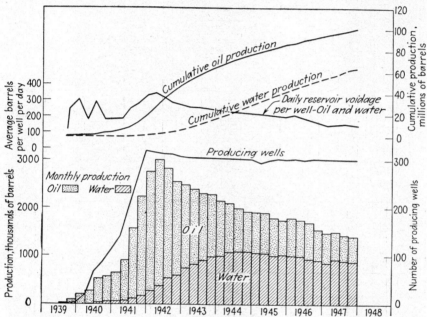

FIG. 11.34.   The production history of the Woodruff Sand reservoir of the Tinsley field, Mississippi.

evolution within the producing formation. The sustained total fluid production at Tinsley is to be noted, in view of the fact that by July, 1946, the reservoir pressure had declined to 210 psi from an initial value of 2,000 psi.

Another striking example of sustained total fluid production under complete water drive is shown in Fig. 11.35,[1] referring to the Hogg-Abrams producing unit, north flank, "A" sand reservoir, of the West Columbia field, Brazoria County, Tex. It will be seen that, except for fluctuations associated with individual well shutdowns or changes in local operating conditions, the total oil and water production rates remained substantially constant for 15 years, until proration forced a reduction in withdrawals.

[1] Cf. J. C. Miller, *AAPG Bull.*, **26**, 1441 (1942).

During this period of uncontrolled withdrawals—78 million barrels of total fluid—the average pressure declined only from 1,330 to 1,104 psi.   The

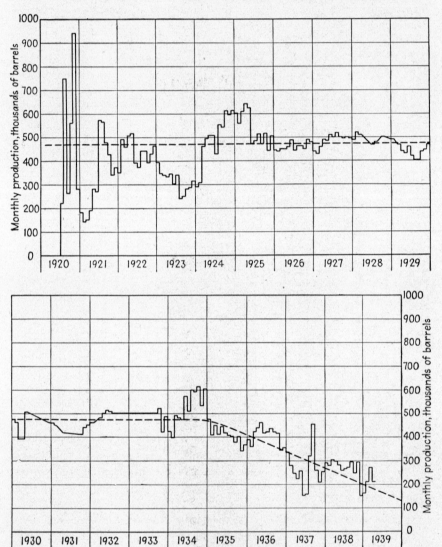

Fig. 11.35.   The total fluid-production history of the Hogg-Abrams producing unit, north flank, "A" sand reservoir of the West Columbia field, Texas.   (*After Miller, AAPG Bull., 1942.*)

observed gas-oil ratios, within the errors of measurements, did not exceed the solution ratio of 150 ft³/bbl.

It is to be expected that accompanying the increased water production of water-drive fields will be a general advance into the reservoir of the water-oil boundary. This is usually observed. Three of the boundary curves recorded over a 10-year interval in the Coalinga field, Fresno County,

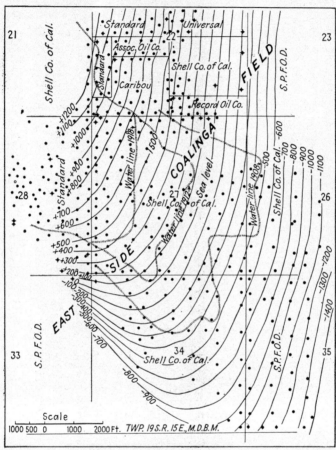

Fig. 11.36. A sketch showing the encroachment of edgewater in the Coalinga field, California. (*After Ambrose, U.S. Bur. Mines Bull., 1921.*)

Calif., are shown superposed on the structure contours in Fig. 11.36.[1] A plot of the water-encroachment history in the water-drive Ramsey pool, Payne County, Okla., is given in Fig. 11.37.[2] These water-oil contours evidently are not sharp boundaries of division between a region of clean oil production and one of complete water production. The latter are gen-

[1] This is taken from A. W. Ambrose, Underground Conditions in Oil Fields, *U.S. Bur. Mines Bull.* 195 (1921).

[2] Cf. M. B. Penn, API meetings, Oklahoma City, June, 1946.

erally separated by a zone of transition of varying content of water in the production, and the plotted boundaries of Figs. 11.36 and 11.37 represent the average areal position of the transition zone or the locations of a definite per cent of water-production contour. The type of transition zone that is often observed is illustrated by the water-production contours plotted in Fig. 11.38,[1] determined in a July, 1943, survey on the wells producing from the Waltersburg Sand of the Mt. Vernon pool, Posey County, Ind.

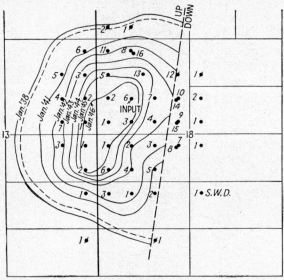

FIG. 11.37. The water-encroachment history in the Ramsey pool, Oklahoma. Solid contours indicate the position of the edgewater advance. (*After Penn, API meetings, 1946.*)

As may be readily inferred from the various examples of water-drive reservoir performance presented above, oil-producing wells may continue to produce oil for long periods after the first entry of water into the well bore. The time history of the rise in water content of the total fluid production follows no fixed pattern but will vary with the detailed mechanics of the production and character of the producing zone. In the case of extremely permeable cavernous types of reservoir, where there is only a negligible pressure drop in the formation and the oil virtually floats on the underlying water with a horizontal interface, wells have been observed to go to more than 95 per cent water within a few hours after its first appearance. Such situations, however, are exceptional. In most water-drive reservoirs it takes months or years for the percentage of water in the production to build up to 95 per cent. The exact value at abandon-

[1] Cf. K. B. Barnes, *Oil and Gas Jour.*, **43**, 42 (Feb. 3, 1945).

ment is also a variable depending on local economic factors, as the cost of lifting the water and of its disposal, its corrosive effect, the price of the oil, and general operating costs of the wells or lease. These may be such as to force abandonment shortly after water first appears, or the production may continue to be profitable until 99 per cent of the total is water.

In fact, gross average water-oil ratios throughout the whole producing life as high as 5 to 10 are not uncommon in some producing districts.

While the absolute value of water encroachment in individual fields is of significance only in relation to the fluid withdrawals from the fields, it is of interest to note the magnitudes that have been observed in various water-drive reservoirs. By calculating the water intrusion from the material-balance equation [cf. Eq. 9.7(1)] and representing the rate of water entry by the steady-state approximation, such as is indicated in Eq. 9.7(3), values of the encroachment factor c may be determined. Values so found for several fields that have complete or substantial water drives (cf. Fig. 9.1) are given in Table 5.[1] To the extent that the water entry follows

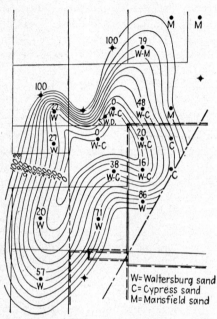

FIG. 11.38. The water percentage contours in the Waltersburg Sand reservoir of the Mt. Vernon pool, Indiana. (*After Barnes, Oil and Gas Jour., 1945.*)

a compressible-liquid expansion mechanism, these encroachment factors will not remain constant throughout the producing life. However, they

TABLE 5.—WATER ENCROACHMENT FACTORS

Field	Producing formation	Encroachment factor c	
		Bbl/(month)/(psi)	Bbl/(month)/(psi) /(acre-ft)
Schuler.........	Reynolds Lime	1,350	0.0860
Magnolia.........	Reynolds Lime	1,130	0.0021
Buckner.........	Reynolds Lime	126	0.0036
East Watchorn...	Wilcox Ordovician	364	0.0107
Ramsey..........	Wilcox Ordovician	396	0.0300
Turkey Creek.....	Frio	235	0.0426

[1] These data are taken from G. R. Elliott, *AIME Trans.*, **165**, 201 (1946).

do give an indication of the relative magnitudes of the water intrusion. Moreover, the values in the last column of Table 5, expressed per unit of total oil-reservoir volume, show a rough correlation with the steepness of the pressure-decline curves plotted in Fig. 9.1, although these, of course, reflect the rates of fluid withdrawals as well as the rates of water intrusion. In this respect it may be observed that whereas the equivalent steady-state encroachment factor in East Texas during 1943–1944 was approximately 23,000 bbl/(month)/(psi),* which is much larger than those given in Table 5, this corresponds to only 0.005 bbl/(month)/(psi)/(acre-ft). The latter is even smaller than most of those listed in Table 5 and is qualitatively consistent with the nature of the pressure-decline history as plotted in Figs. 9.1 and 11.24.

11.13. Bottom-water-drive Reservoirs[1]; The Physical Representation.— In the discussions of water-drive reservoirs presented thus far in this chapter only the gross performance features, from the point of view of the relation of pressure to withdrawals, were given consideration. The nature of the actual advance of water into the oil-producing formation was given cognizance only in the observation that in actual water-drive reservoirs the edge wells progressively go to water and a continually increasing component of water production generally accompanies the oil withdrawals. In deriving the pressure reaction of the reservoir, account must be taken of the total fluid withdrawals and also of such reduction in these as may correspond to a return of part or all of the water produced to the water-bearing stratum.

In considering, however, the details of the advance of the water-oil interface it is convenient to make a distinction between "edgewater encroachment" and "bottom-water" drives. In the first the motion of the water may be visualized as proceeding largely in a direction parallel to the planes of stratification. This will occur usually in relatively thin producing zones and in strata lying along structural flanks of appreciable dip. In the latter the water-oil interface will lie in a plane of zero or slight inclination, such as may obtain in thick strata or those having low relief. These idealized extremes will, of course, seldom occur in practice. Even when a field as a whole is controlled by an edgewater drive, the edge wells will generally show bottom-water-drive characteristics. Thus whereas, as seen in Sec. 11.9, the gross performance of the East Texas field can be fully described in terms of a general eastward advance of the Woodbine water, more than half the oil-productive area was originally underlain by water

---

* The value of c corresponding to the theoretical representation of its pressure-decline history, that is, $10^6/b \log \rho$ in Eq. 11.9(1), is 14,350 bbl/(month)/(psi).

[1] The material of this and the next three sections follows the treatment of M. Muskat, *AIME Trans.*, **170**, 81 (1947).

(cf. Fig. 1.12*b*). The wells in the western part of the field were thus potentially subject to a bottom-water drive, although the actual water intrusion in the field has been largely lateral because of widely disseminated shale breaks and marked stratification within the oil sand.

It should be understood that the distinction between edgewater and bottom-water drives made here refers to the individual well performance. Only the local details of the water-oil-interface motion are involved. The over-all behavior of both the oil and water reservoirs is still to be considered as being governed by the gross balance between fluid withdrawals and water-supply capacity, the latter, in turn, reflecting the physical properties of the aquifer and its fluid. Except for the details of individual well performance and the development of water production, the general interpretation and analysis of pressure vs. time or production histories, as developed in previous sections, will still apply whether the field be predominantly of the edgewater- or bottom-water-drive type.

In the edgewater-encroachment type of water-drive field the detailed production histories of the individual wells will be progressive, as the water moves in from the original boundaries and continually shrinks the productive area. The direct interplay between the advancing water and the oil-producing wells is not uniformly applied to all the wells in the field but continually shifts from the edgewater boundaries to the field interior, as production proceeds (cf. Figs. 11.36 and 11.37). The details of this advance will depend on how the wells directly affected by the edgewater or enveloped by it are produced. Physically the situation corresponds to what may be termed a multiple-well system produced by virtue of an advancing linear-drive flood. By using this representation it is possible to construct an approximation theory to describe the details of advance of the edgewater, by methods to be presented in the next chapter, although it is doubtful that such an analysis would be of great practical value. On the other hand, the gross effect of the shrinkage of the productive area, due to water invasion, on the nature of the fluid distribution in the unswept part of the oil zone, when the latter is producing by gas drive, can be calculated by the method outlined in Sec. 10.18.

In the case of the bottom-water-drive fields the situation is quite different. Here, at least in the limiting idealized case where the bottom water is spread uniformly under the whole productive area of the field, all the wells, if operated similarly, will experience similar or identical histories. From a qualitative point of view it is not difficult to predict the history of the production and the course of the water intrusion for such systems. Under the relatively high reservoir pressure obtaining within the water zone and at the base of the oil zone, and under the reduced bottom-hole well pressures, the water will be subjected to an appreciable pressure differential dis-

tributed over the oil-saturated zone.   The flow lines will be approximately
normal to the original water-oil interface and will be generally directed
upward until the part of the producing stratum penetrated by the well is
approached, when they will be deflected so as to proceed toward the well
bores.   A diagrammatic representation of this situation, though not to
scale, is shown in Fig. 11.39.

The mechanism of production and water entry implied by Fig. 11.39
should not be confused with the problem of "water coning" discussed in
Sec. 5.9.   It will be recalled that, while the approximate analytical theory
given of the latter in Sec. 5.9 was based on a homogeneous-fluid representation of the oil production, the es-
sentially horizontal character of the
fluid motion (cf. Fig. 5.19) was considered to be the result of a gas-
drive expulsion of the oil.   In this
treatment of water coning the bottom water represented merely a
mobile interface with the oil zone
rather than a withdrawal-replacement fluid.   Under approximately
stable and steady-state conditions,
therefore, the bottom water can lie

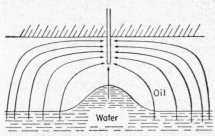

FIG. 11.39.   A diagrammatic representation
(not to scale) of the flow conditions about a
well producing by a complete bottom-water
drive.

statically in hydrostatic as well as in dynamical equilibrium beneath the oil
zone and exert no material influence on the oil production.   Although during the course of the gas-drive production the pressure will actually decline and the bottom water will rise into the oil zone, the local well behavior
can be treated as if at any state of the production history the contribution
of the slow transient water intrusion is of minor importance.

In the limiting and idealized analysis of bottom-water-drive reservoirs
to be presented here the water will be considered as the sole driving agent
for oil expulsion.   The flow of oil into the producing wells will be assumed
to take place only by virtue of its displacement by the rising water table.
Because the vertical velocity component at the water-oil interface will evidently be a maximum along the axes of the producing wells and will
increase as the bottoms of the wells are approached, the interface will develop a conelike form about each well.   However, it is only in the qualitative similarity in the shapes of the interface that the bottom-water-drive
system is analogous to that of water coning.

Largely as the result of analytical necessity a number of assumptions
have to be made in developing a tractable formulation of the bottom-water-
drive problem.   As previously suggested the analysis will be based on the

supposition that the rising water is the sole oil-displacing fluid, or that the reservoir pressures are above the bubble point throughout. Since the treatment will be mainly concerned with the local fluid movements about the well bores, steady-state conditions will be assumed. The general transient character of the water flow in the aquifer should not materially affect the problem. The permeability-to-viscosity ratio for the oil above the water-oil interface will be taken as the same as that for the water in the invaded part of the oil zone. While this will be strictly true in practice only by accident, available data do not support any other specific relative values for the ratios, and approximate equality may well obtain in many cases. If it be established that the permeability-to-viscosity ratio in the water-invaded region will exceed that in the oil-saturated zone, the entry of water into the producing wells will develop faster than calculated here and conversely if the latter is higher. However, the introducton of different values of the permeability-to-viscosity ratio in the oil- and water-invaded zones would so complicate the analysis as to make it almost intractable. For similar reasons the "stripping" of oil from the flooded parts of the oil zone will be neglected, and the residual oil immediately behind the water-oil interface will be considered as permanently trapped and locked in the pay.

The approximation that is potentially most serious will be the neglect of the density difference between the oil and water. With this assumption and the others already mentioned, all distinctions between the water and oil that may affect the fluid dynamics are ignored, and the composite oil and water system is replaced by a single homogeneous incompressible liquid. The water-oil interface will then be identified only as a geometrical-particle surface rather than a physical boundary between two fluids.

The neglect of the density difference between the water and oil implies that all the basic phenomena characterizing a bottom-water-drive system are independent of the *rates* of production. They will depend only on the geometrical and physical constants of the system and will not be changed by variations in the withdrawal rates. The latter will affect only the time scale. This, of course, cannot be strictly correct. The higher density of the water will tend to retard and flatten the conelike development of the water-oil interface. This effect will be greatest at low rates of production and where the producing pressure differential is very low. In extreme cases, where a well may be shut in for extended periods, the water elevation immediately about or below the well will decline, and an equalization of the interface level will tend to be established. However, it seems unlikely that where conditions of sustained production obtain the density difference between the oil and water will greatly modify the results to be developed here, except when the producing pressure differential is of the same order of magnitude as a head of the oil-zone thickness for a fluid with the differential density between the water and oil.

11.14. Bottom-water-drive Reservoirs; Analytical Treatment; Well Capacities.—In view of the physical representation to be used of the bottom-water-drive system, as discussed in the preceding section, the fluid motion will be governed by the Laplace equation. Considering at the outset the producing formation to be anisotropic, with vertical and horizontal permeabilities k_z and k_h, the potential equation in cylindrical coordinates (r,z) will be

$$\left.\begin{array}{c} \dfrac{k_h}{r}\dfrac{\partial}{\partial r}\left(r\dfrac{\partial \Phi}{\partial r}\right) + k_z\dfrac{\partial^2 \Phi}{\partial z^2} = 0, \\[2mm] \Phi = \dfrac{p - \gamma g z}{\mu}, \end{array}\right\} \tag{1}$$

where p is the pressure, γ the oil density, μ its viscosity, and g the acceleration of gravity.

If the original oil-zone thickness be denoted by h, the well radius by r_w, and the well penetration by b, and the origin of coordinates be set at the center of the well where it penetrates the producing formation, the boundary conditions will be

$$\left.\begin{array}{ll} \dfrac{\partial \Phi}{\partial z} = 0; & z = 0 \\[2mm] \Phi = \text{const} : & z = h \\[2mm] \Phi = \text{const} (\Phi_w) : & z \leqslant b; \quad r = r_w. \end{array}\right\} \tag{2}$$

The first condition expresses the closure of the oil zone by an impermeable stratum. The second implies that the pressure at the original horizontal water-oil interface remains uniform throughout the production history.[1] The third equation requires the potential over the well surface to be constant. Transforming the coordinates to provide an equivalent isotropic system and introducing dimensionless variables, as

$$\left.\begin{array}{ccc} z' = z\sqrt{\dfrac{k_h}{k_z}}; & h' = h\sqrt{\dfrac{k_h}{k_z}}; \\[3mm] \bar{z} = \dfrac{z'}{2h'} = \dfrac{z}{2h}; & \bar{x} = \dfrac{b}{2h}; & \rho = \dfrac{r}{2h'}; & \rho_w = \dfrac{r_w}{2h'}, \end{array}\right\} \tag{3}$$

Eq. (1) becomes

$$\frac{1}{\rho}\frac{\partial}{\partial \rho}\left(\rho\frac{\partial \Phi}{\partial \rho}\right) + \frac{\partial^2 \Phi}{\partial \bar{z}^2} = 0. \tag{4}$$

[1] This represents an approximation that will be considered separately in Sec. 11.16.

The boundary conditions of Eq. (2) take the form

$$\frac{\partial \Phi}{\partial \bar{z}} = 0 \; : \; \bar{z} = 0; \qquad \Phi = \text{const}(0) \; : \; \bar{z} = \frac{1}{2}; \\ \Phi = \text{const}(\Phi_w) \; : \; \bar{z} \leqslant \bar{x}; \qquad \rho = \rho_w. \Bigg\} \tag{5}$$

The well system in the $(\rho, \bar{z})$ coordinates is shown diagrammatically in Fig. 11.40.

Solutions of Eq. (4) to satisfy the boundary conditions of Eqs. (5) may be constructed by a synthesis of the contributions of flux elements along the well axis in a manner quite similar to that described in Sec. 5.4 for partly penetrating wells in strata sealed at both the top and the bottom. The only change required here is that at $\bar{z} = \frac{1}{2}$ the individual potential functions become constant (zero) rather than developing a zero normal gradient. If any particular well in question be considered as one in an infinite square network of spacing a, the potential

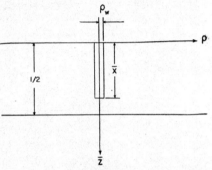

Fig. 11.40. A diagrammatic representation of a partially penetrating well producing by a bottom-water drive, in the dimensionless and isotropic coordinate system of Eq. 11.14(3).

function due to a flux element of strength $q\,d\alpha$ at a depth α, in units of $\bar{z}$, may be readily shown to be

$$d\Phi = 8q\,d\alpha \sum_{\text{odd}} \cos n\pi\bar{z} \cos n\pi\alpha \left[K_0(n\pi\rho) + \frac{I_0(n\pi\rho)K_1(n\pi\rho_e)}{I_1(n\pi\rho_e)} \right], \tag{6}$$

where K_0, I_0, K_1, I_1 are the independent Bessel functions of the third kind of zero and first order, respectively. ρ_e is defined by

$$\pi\rho_e^2 = \frac{a^2}{4h'^2} = \frac{\bar{a}^2}{4}; \qquad \bar{a} = \frac{a}{h}\sqrt{\frac{k_z}{k_h}}, \tag{7}$$

and it is assumed that the square-network elements isolating the individual wells are equivalent to closed circular disks of radius r_e and of the same area. The parameter $\bar{a}$ may be considered as the dimensionless well spacing. A constant linear flux density q_m, from $\bar{z} = 0$ to $\bar{z} = \bar{x}_m$, will therefore give a potential function

$$\Phi_m = \frac{8q_m}{\pi} \sum_{\text{odd}} \frac{\cos n\pi\bar{z} \sin n\pi\bar{x}_m}{n} \left[K_0(n\pi\rho) + \frac{I_0(n\pi\rho)K_1(n\pi\rho_e)}{I_1(n\pi\rho_e)} \right]. \tag{8}$$

On applying the limiting form of Eq. (8) suitable for small values of ρ and superposing the effects of different constant-flux segments of lengths

$\overline{x}_m$ and density q_m a resultant potential function Φ may be derived, of the form

$$\Phi = \Sigma q_m \Phi_m(\overline{z}, \rho, \overline{x}_m), \tag{9}$$

which will give an approximately constant potential over the well surface. Examples of the potential distributions over the well surface and below the well bottom along its axis, as determined in this manner, for infinite well spacing and $\rho_w = 0.001$, are plotted in Fig. 11.41. The indicated values of Φ_w are the potentials over the well surface, as averaged to the depths of penetration represented by $\overline{x}$.

The withdrawal rate from the well corresponding to a flux element of actual length b_m and density q_m is readily found to be:

$$Q_m = -2\pi k_h \int_0^{b_m} r\left(\frac{\partial \Phi_m}{\partial r}\right)_0 dz = -2\pi k_z \int_0^{r_e} r\left(\frac{\partial \Phi_m}{\partial z}\right)_h dr = 8\pi h q_m k_h \overline{x}_m. \tag{10}$$

Upon adding the similar contribution of the differential elements of strength q_p, represented by Eq. (6), which are generally required in developing a uniform well-surface potential by the above-mentioned synthesis procedure, the resultant production rate from each well in the network will be

$$Q = 8\pi h k_h (\Sigma q_m \overline{x}_m + q_p). \tag{11}$$

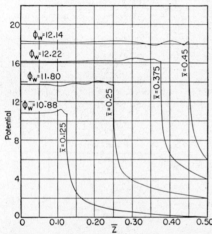

FIG. 11.41. The calculated adjusted potential distributions over the well surface and below the well bottom for partially penetrating wells producing by bottom-water drives. Φ_w = average adjusted well-surface potential; $\overline{x}$ = (well penetration)/(2 × oil-zone thickness); $\overline{z}$ = (depth)/(2 × oil-zone thickness). Values of 2, 4, and 6 have been added to potentials for $\overline{x}$ = 0.25, 0.375, and 0.45, respectively. (From AIME Trans., 1947.)

As the potential functions of Eqs. (6), (8), and (9) vanish at $\overline{z} = \frac{1}{2}$, the initial oil-water interface, the total potential drop corresponding to the withdrawal rate Q will be simply the approximate constant well-surface potential Φ_w given by Eq. (9). By defining a dimensionless production rate by

$$\overline{Q}_o = \frac{4(\Sigma q_m \overline{x}_m + q_p)}{\Phi_w} = \frac{Q}{2\pi h k_h \, \Delta \Phi}, \tag{12}$$

the actual rates in barrels per day for a pressure drawdown Δp will then be

$$Q = \frac{0.007082 h k_h \overline{Q}_o \, \Delta p}{\mu} \text{ bbl/day}, \tag{13}$$

where h is expressed in feet, k_h in millidarcys, Δp in psi, and μ in centipoises. Typical results of the numerical evaluation of Eq. (12) under different conditions as a function of the well penetration are shown graphically in Fig. 11.42.

The values of A in Fig. 11.42 represent the actual well spacing in acres per well corresponding to particular values of $\bar{a}$, ρ_w, and r_w, as given by

$$A = 5.74 \times 10^{-6} r_w^2 \frac{\overline{a}^2}{\rho_w^2}. \qquad (14)$$

To the extent that the actual well radius r_w is considered as fixed, varia-tions in ρ_w reflect those in h', de-fined by Eq. (3). Hence the three curves for $\overline{A} = \infty$ (infinite well spacing) correspond to producing-zone thicknesses varying by fac-tors of 5 or permeability ratios differing by factors of 25. The higher values of $\overline{Q_o}$ for those of greater ρ_w thus represent the in-creasing production capacities with decreasing pay thickness or in-creasing vertical permeability, as compared with the horizontal. The decreasing production capac-ities with decreasing well spac-ing, for fixed ρ_w, result from the decreasing driving area per well at the water-oil contact. The variation with the spacing, how-ever, is rather slow, as would be expected.

Curve 6 in Fig. 11.42 refers to a steady-state radial drive in an isotropic formation of 125-ft thick-ness[1] and with a ratio of external radius to well radius of 2,640, as taken from Fig. 5.11. The equiv-

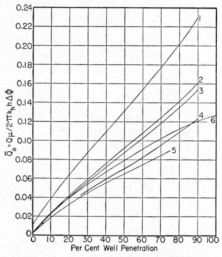

FIG. 11.42. The calculated production capaci-ties per unit formation thickness, in dimension-less units, of partially penetrating wells produc-ing by bottom-water drives. Curve 1, $\rho_w = 0.005$, $A = \infty$. Curve 2, $\rho_w = 0.001$, $A = \infty$. Curve 3, $\rho_w = 0.001$, $A = 1$. Curve 4, $\rho_w = 0.0002$, $A = \infty$. Curve 5, $\rho_w = 0.001$, $A = \frac{1}{5}$. Curve 6, partially penetrating wells in steady-state radial-drive systems. Q = pro-duction rate in reservoir measure; μ = oil viscosity; k_h = horizontal permeability; h = thickness of oil zone; $\Delta\Phi$ = potential or pressure drop; ρ_w = dimensionless well radius; A = spacing in acres per well. (*From AIME Trans.*, *1947.*)

alent spacing is 31.4 acres per well for a well radius of $\frac{1}{4}$ ft. It will be seen that the production capacities are of a comparable magnitude and vary with the well penetration in a manner similar to that for wells produced by bottom-water drives.

---

[1] The thickness is of lesser importance here, except at the low penetrations (cf. Fig. 5.13).

11.15. Bottom-water-drive Reservoirs; The Sweep Efficiency and Water-Oil Ratios.—Of greater interest than the production capacities of wells producing by bottom-water drives is the nature of the rise of the water-oil interface and the amount of clean oil that can be produced before water break-through. A measure of the latter can be conveniently expressed in terms of the "sweep efficiency," defined as the fraction of the volume of oil pay drained by the well that is swept out by the time the water first reaches the well. Its value will evidently be in the range of

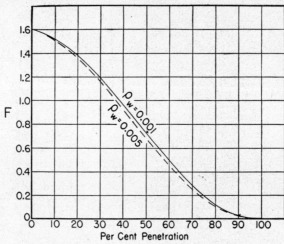

Fig. 11.43. The calculated variation of the sweep-efficiency function F, for bottom-water-drive systems, defined by Eq. 11.15(5), with the well penetration, for values of the dimensionless well spacing parameter $\bar{a} > 3.5$. For crosses, $\rho_w = 0.0002$. ρ_w = dimensionless well radius. (*From AIME Trans., 1947.*)

0 to 1. Low values will indicate that the water quickly enters the producing well while flushing out only a small part of the oil zone. On the other hand the limiting value 1 implies that all the producing section has been uniformly flooded by the time the water enters the producing wells. In general, if V be the volume of the oil zone, of thickness h, swept out by the rising water table at the time of the break-through, and a is the well spacing, the sweep efficiency E will be

$$E = \frac{V}{a^2 h}. \tag{1}$$

The value of V can be expressed as

$$V = \frac{Qt}{\bar{f}}, \tag{2}$$

where t is the time required for the water to break into the producing well, Q is the production rate (assumed uniform), and $\bar{f}$ is the microscopic displacement efficiency, *i.e.*, the porosity of the oil zone times the fraction of the pore space displaced by the invading water. As the water will evidently rise most rapidly along the well axis, t will be given by

$$t = \int_h^b \frac{dz}{v_z} = \frac{4h^2\bar{f}}{k_z} \int_{1/2}^{\bar{x}} \frac{d\bar{z}}{\frac{\partial\Phi}{\partial\bar{z}}}, \tag{3}$$

where the vertical velocity v_z and potential gradient $\frac{\partial\Phi}{\partial\bar{z}}$ refer to the well axis. On combining Eqs. (1), (2), and (3) with Eqs. 11.14(3), 11.14(7), and 11.14(11), it is found that E can be expressed as

$$E = \frac{F(x,\bar{a},\rho_w)}{\bar{a}^2}, \tag{4}$$

where

$$F = 32\pi(\Sigma q_m\overline{a_m} + q_p) \int_{1/2}^{\bar{x}} \frac{d\bar{z}}{\frac{\partial\Phi}{\partial\bar{z}}}. \tag{5}$$

Since the spacing parameter $\bar{a}$ enters the analysis directly only through the second terms in the brackets of Eq. 11.14(8), which vanish exponentially with increasing ρ_e or $\bar{a}$, F will be independent of $\bar{a}$ when the latter is sufficiently large. It may be shown that this will be the case when $\bar{a} > 3.5$. F will then become a function only of ρ_w and the fractional penetration $\bar{x}$. For this range of $\bar{a}$ the numerical values of F are plotted vs. the penetration in Fig. 11.43. It will be noted that these values of F are also rather insensitive to the magnitude of ρ_w.

Upon using the F given by Fig. 11.43 for the large values of $\bar{a}$ and separately determined values for $\bar{a} < 3.5$, the corresponding sweep efficiencies, expressed by Eq. (4), are plotted vs. $\bar{a}$ in Fig. 11.44 for fixed well penetrations and values of ρ_w. It will be seen that E decreases continuously with increasing $\bar{a}$. And as already indicated, for $\bar{a} > 3.5$, E varies inversely as $\bar{a}^2$.

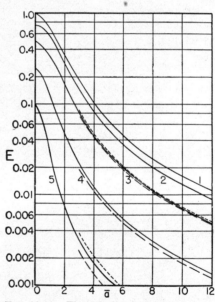

FIG. 11.44. The calculated variation of the sweep efficiency E with the well-spacing parameter a in bottom-water-drive reservoirs. For curves 1, 2, 3, 4, and 5, well penetrations are zero, 25, 50, 75, and 90 per cent. $\rho_w = 0.001$, 0.002, and 0.005 for solid curves, upper dashed curves, and lower dashed curves. ρ_w and $\bar{a}$ are defined by Eqs. 11.14(3) and 11.14(7). (*From AIME Trans., 1947.*)

For fixed pay thickness and permeability ratio the sweep efficiency will therefore be proportional to the well density, for $\bar{a} > 3.5$. The dashed curves in Fig. 11.44 show the effect of the dimensionless well radius ρ_w. Though the variation is small and almost vanishes completely at penetrations of 25 per cent or less, the sweep efficiency is seen to increase with decreasing ρ_w, or well radius, if h' be considered as fixed.

The very low values of the sweep efficiency indicated in Fig. 11.44 for the larger values of $\bar{a}$ give the justification for taking into account at the outset the possible anisotropy of the producing section. Thus, if the sand were strictly isotropic, $\bar{a}$ would be simply the ratio of the well spacing to the pay thickness. If the latter were 25 ft and the well separation were 660 ft (10 acres per well), $\bar{a}$ would equal 26.4. Hence even if the well just tapped the oil zone and $F = 1.6$, E would be only 0.0023. That is, less than $\frac{1}{4}$ per cent of the 250 acre-ft of the pay theoretically drained by each well would be flushed out by the time the rising water-oil interface reaches the wells. If the porosity were 25 per cent and the microscopic oil-displacement efficiency were 60 per cent, only 668 bbl of reservoir oil would be produced before the appearance of water. The quantitative significance of such a prediction is, of course, uncertain in view of the various assumptions underlying the analysis. However, there can be little doubt that the order of magnitude of the clean-oil production which would be recovered from such an isotropic formation is far lower than that actually produced in most cases from wells bottomed over a rising water table. To attempt to account for the recovery of thousands of barrels of clean oil from bottom-water-drive wells that are not located immediately over extended shale breaks, the anisotropy of the pay must evidently be considered.

The variation of the sweep efficiency with the penetration for fixed values of $\bar{a}$ is plotted in Fig. 11.45. Its variation with $1/\bar{a}^2$, which is proportional to the well density if the formation thickness and permeability ratio are considered as fixed, is shown in Fig. 11.46.

The numerical value of the total clean-oil production per well is given by

$$\overline{Q} = \frac{F\bar{f}h^3 k_h}{5.61\beta k_z}\, \text{bbl},\tag{6}$$

where β is the formation-volume factor of the oil and h is to be expressed in feet. If the value of $\bar{a}$ corresponding to h, k_h/k_z, and the well spacing exceeds 3.5, F can be read from the curves of Fig. 11.43. It will be seen that in this range the clean-oil recovery per well is independent of the absolute value of the spacing and is proportional to the horizontal to vertical permeability ratio and to the cube of the oil-pay thickness.

As is clear from Figs. 11.44 and 11.45 the absolute limit of the sweep

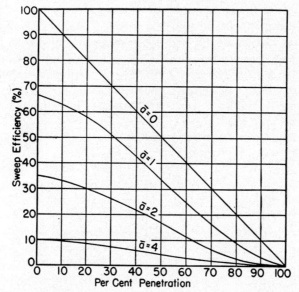

Fig. 11.45. The calculated variation of the sweep efficiency in bottom-water-drive systems as a function of the well penetration, for fixed values of the dimensionless well-spacing parameter $\bar{a}$. Dimensionless well radius = 0.001. (*From AIME Trans., 1947.*)

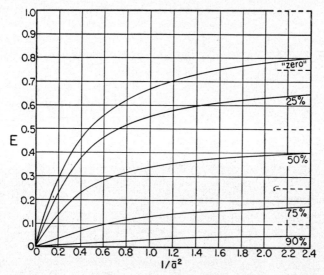

Fig. 11.46. The calculated variation of the sweep efficiency E with the reciprocal of the square of the dimensionless well-spacing parameter $\bar{a}$, for fixed well penetrations as given by notation on curves. Dashed segments indicate asymptotic limits of E at infinite well densities. (*From AIME Trans., 1947.*)

efficiency even if the water-oil interface rises strictly horizontally, as at zero well spacing ($\bar{a} = 0$), corresponds only to a flushing of the part of the pay below the bottom of the well. To compare the performance of wells of different penetration, before water entry, it would therefore be more appropriate to express the sweep efficiency in terms of the oil-saturated zone below the bottom of the well. Such an "effective" sweep efficiency $\overline{E}$ will be simply related to E as

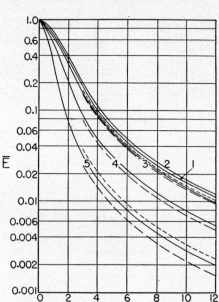

$$\overline{E} = \frac{E}{1 - 2\bar{x}}, \qquad (7)$$

so that complete flooding of the nonpenetrated pay by the time of water break-through will represent 100 per cent effective sweep efficiency in all cases. Figure 11.47 is a replot of Fig. 11.44 in terms of the effective sweep efficiency. The divergence of the curves of Fig. 11.47 shows that the effective sweep efficiency also decreases with well penetration at higher penetrations, although the trend is reversed at penetrations less than 40 per cent.

Fig. 11.47. The calculated variation of the effective sweep efficiency $\overline{E}$ [cf. Eq. 11.15(7)] with the dimensionless well-spacing parameter $\bar{a}$. Notation as in Fig. 11.44. (*From AIME Trans., 1947.*)

The low values of clean-oil recovery given by Eq. (6) and sweep efficiencies shown in Figs. 11.44 to 11.46, unless the producing formation is highly anisotropic, imply that an important part of the production history of the wells will involve the simultaneous production of oil and water. The treatment of this phase of the problem is much more difficult than that of computing the sweep efficiency and clean-oil recovery. However, it is possible to formulate a procedure for determining the rise in water percentage, although the numerical work involved is quite laborious.

By using the assumptions previously made in calculating the sweep efficiency and still ignoring the distinction between the oil-saturated and water-flooded parts of the pay, the streamline surface distribution in the flow system, both before and after water break-through, corresponding to the potential function of Eq. 11.14(9), can be shown to be given by

$$\overline{\Psi} = \frac{\Psi}{32hk_h} = \rho\Sigma q_m \sum_{\text{odd}} \frac{\sin n\pi\overline{z}\,\sin n\pi\overline{x}_m}{n}\left[K_1(n\pi\rho) - \frac{I_1(n\pi\rho)K_1(n\pi\rho_e)}{I_1(n\pi\rho_e)}\right]. \quad (8)^*$$

In Eq. (8), Ψ represents the actual stream function, and $\overline{\Psi}$ an equivalent form that is more convenient for numerical application. It may be readily verified that Eq. (8) satisfies the physical requirements for the stream function, namely, that it vanish at $\overline{z} = 0$ and $\rho = \rho_e$, and that at $\rho = 0$, $\overline{z} > \overline{x}$, Ψ assume the value for the total flux given by Eq. 11.14(11), except for the term q_p. A typical set of stream functions computed by Eq. (8)

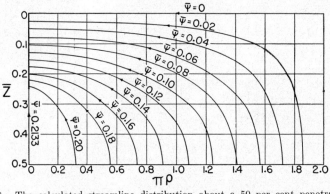

Fig. 11.48. The calculated streamline distribution about a 50 per cent penetration well producing by a bottom-water drive. Dimensionless well-spacing parameter $\overline{a}$ and well radius ρ_w equal $4/\sqrt{\pi}$ and 0.001. Coordinates $\overline{z}$, ρ, are defined by Eqs. 11.14(3). (*From AIME Trans., 1947.*)

is plotted in Fig. 11.48 for a well of 50 per cent penetration, with $\rho_w = 0.001$ and $\overline{a} = 4/\sqrt{\pi}$ (1.83 acres per well).

As the stream functions Ψ are related to the potential functions Φ in an anisotropic cylindrical coordinate system by the equations

$$\frac{\partial\Psi}{\partial z} = -2\pi rk_h\frac{\partial\Phi}{\partial r}; \qquad \frac{\partial\Psi}{\partial r} = 2\pi rk_z\frac{\partial\Phi}{\partial z}, \quad (9)$$

the time taken for a fluid particle beginning at the initial water-oil contact $\overline{z} = \frac{1}{2}$ to reach any arbitrary level within the oil pay will be

$$t = \frac{\pi\overline{f}}{16hk_h}\int\frac{r\,ds}{\dfrac{\partial\overline{\Psi}}{\partial n}} \simeq \frac{\pi\overline{f}}{16hk_h}\int\frac{r\,ds\,\Delta n}{\Delta\Psi}, \quad (10)$$

where ds is an element of the streamline surface followed by the particle;

---

* The contribution due to a differential element at $\overline{x}$ is given by the derivative with respect to $\overline{x}_m$ of the coefficient of q_m. In the calculations reported below the terms representing the differential-flux elements q_p were taken into account.

dn, Δn represent the element normal to ds; and $\Delta \overline{\Psi}$ is the change in $\overline{\Psi}$ over the distance Δn. Now if $\Delta \overline{\Psi}$ be considered as fixed, the integral in Eq. (10) will be $1/(2\pi\Delta\overline{\Psi})$ times the differential physical volume ΔV swept out by the fluid particles lying on the original water-oil interface between $\overline{\Psi}$ and $\overline{\Psi} + \Delta\overline{\Psi}$. This volume of revolution (about the $\bar{z}$ axis) may be expressed as an area ΔA measured between $\overline{\Psi}$ and $\overline{\Psi} + \Delta\overline{\Psi}$ when the stream functions are plotted in a $(\bar{z}, \pi^2\rho^2)$ coordinate system. Equation (10) then may be rewritten as

$$ t = \frac{\bar{f}}{32hk_h} \frac{\Delta V}{\Delta\overline{\Psi}} = \frac{\bar{f}h'^2}{4\pi k_h} \frac{\Delta A}{\Delta\overline{\Psi}}. \tag{11} $$

Now the rate of oil production can be expressed as

$$ Q_{\text{oil}} = \bar{f}\frac{\delta V}{\delta t}, \tag{12} $$

where δV is the increment of total oil-zone volume swept out between two neighboring water-oil interfaces in the time δt. If δV be expressed, similarly to ΔV, as an area δA in a $(\bar{z}, \pi^2\rho^2)$ plot of the interfaces, Eq. (12) will take the form

$$ Q_{\text{oil}} = \frac{32k_h h}{\delta} \frac{\Delta\overline{\Psi}\delta A}{\Delta A}. \tag{13} $$

Expressing $\Delta\overline{\Psi}$ as a fraction α of the total range $\overline{\Psi}_{\max}$, that is,

$$ \Delta\overline{\Psi} = \alpha\overline{\Psi}_{\max} = \frac{\alpha Q}{32hk_h}, \tag{14} $$

where Q is the total flux rate $(Q_{\text{oil}} + Q_w)$, Eq. (13) gives for the water-oil ratio R_w the expression

$$ R_w = \frac{Q_w}{Q_{\text{oil}}} = \frac{1}{\alpha}\frac{\delta\Delta A}{\delta A} - 1 = \frac{1}{\alpha}\frac{d\Delta A}{dA} - 1, \tag{15} $$

where the right-hand form of Eq. (15) represents the limiting equivalent when the time increments become infinitesimal. Thus the water-oil ratio plus 1 is given simply by the slope of the curve of incremental area between neighboring streamlines, ΔA, vs. the total area A under the interfaces, multiplied by $1/\alpha$. The state of the system associated with any particular value of the slope is determined by the corresponding value of A, which may be expressed in terms of the fraction of the total drainage volume swept out. This fraction is given by the ratio of A to the total area enclosed in the stream-function distribution $(\bar{z}, \pi^2\rho^2)$ plot.

While Eq. (15) could have been constructed directly from general considerations, the detailed derivation presented here serves to show the nature of the auxiliary numerical work involved in the evaluation of R_w. Moreover it will be seen that as a part of the whole procedure the shapes of the actual water-oil interfaces must be computed. On the other hand, if these interfaces and stream-function distributions could be determined independently, and especially the stream-function variation along the well surface, R_w could be simply computed from the relation

$$R_w = \frac{\overline{\Psi}_{\max} - \overline{\Psi}_i}{\overline{\Psi}_i}, \qquad (16)$$

where $\overline{\Psi}_i$ is the value of $\overline{\Psi}$ at which the interface intercepts the well surface. Again the water-oil ratio so found is to be associated with the area or volume under the interface having the well intercept $\overline{\Psi}_i$.

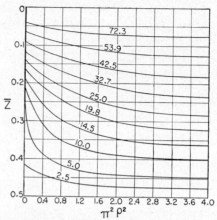

Fig. 11.49. The calculated water-oil interfaces about a well of 50 per cent penetration producing by a bottom-water drive. Dimensionless well-spacing parameter $\bar{a}$ and well radius ρ_w equal $4/\sqrt{\pi}$ and 0.001. $\Delta A =$ incremental area, proportional to time, in square centimeters, as measured on the original streamline plots. (*From AIME Trans., 1947.*)

The water-oil interfaces are evidently surfaces of constant time of particle travel from the original oil-water contact plane. They are formally defined by Eq. (10) but more conveniently determined by Eq. (11). From the latter it will be seen that the constant-time surfaces are also those of constant incremental area ΔA between streamline surfaces of separation $\Delta\overline{\Psi}$, measured in the $(\bar{z}, \pi^2\rho^2)$ coordinate system. Typical water-oil interfaces, defined by the constant values of ΔA, are plotted in Figs. 11.49 and 11.50 for well penetrations of 50 and 90 per cent, respectively, and an effective well spacing of 1.83 acres per well. The conelike character of the water-oil interfaces before break-through and their subsequent flattening are to be noted.

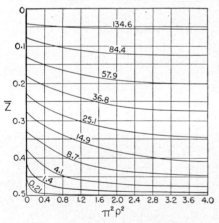

Fig. 11.50. The calculated water-oil interfaces about a well of 90 per cent penetration producing by a bottom-water drive. Conditions and notation are same as for Fig. 11.49. (*From AIME Trans., 1947.*)

Applying Eq. (15) to the water-oil interfaces, as plotted in Figs. 11.49 and 11.50 and the corresponding stream-function-distribution plots, the theoretical variation of the water-oil ratio with the cumulative production can be determined. Curves so obtained for "non-penetrating" wells and different well-spacing parameters are given in Fig. 11.51. Several similar curves for nonvanishing well penetration are plotted in Fig. 11.52. Since the cumulative production is expressed in Figs. 11.51 and 11.52 as the fraction of the total oil-zone volume per well flooded out, the abscissa intercepts evidently represent the sweep efficiencies plotted in Fig. 11.44. As the abscissa variable cannot physically exceed unity, the curves must of necessity converge and have a common vertical asymptote at this

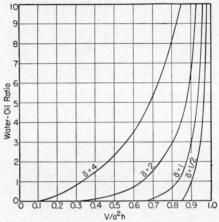

Fig. 11.51. The calculated variation of the water-oil ratio with the cumulative recovery in "nonpenetrating" wells producing by bottom-water drives. V = total volume of oil zone, per well, invaded by water; a = well separation; h = thickness of oil zone; $\bar{a}$ = dimensionless well-spacing parameter. (*From AIME Trans., 1947.*)

limit. It is of interest, however, that appreciable convergence of the curves develops by the time the water-oil ratios rise to the range of 5 to 10. Thus, whereas in the case of the nonpenetrating wells the clean-oil production would be 2.6 as great for $\bar{a} = \frac{1}{2}$ as for $\bar{a} = 2$ (a well density one-sixteenth as great), the total oil recovery at a water-oil ratio of 5 will be only 11 per cent greater for $\bar{a} = \frac{1}{2}$. Hence the gains in ultimate oil recovery, as limited by the rise in water-oil ratio to uneconomic values, will be much lower than those in clean-oil production due to the use of such well spacings or penetrations as may favor high sweep efficiencies.

11.16. The Role of Permeability Anisotropy and Well Spacing in Bottom-water-drive Reservoirs.—Perhaps the most important implication of the analysis given here of the behavior of wells producing by bottom-water drives concerns the nature of producing formations with respect to their isotropy. As noted in the preceding section, if the oil zone is assumed to be isotropic, the theoretical sweep efficiency will in general be so low that the period and amount of clean-oil production would be extremely limited.

Observations of appreciable clean-oil recovery from wells producing by water drive and completed over mobile bottom waters must therefore imply a high degree of effective permeability anisotropy. For example, to obtain an effective sweep efficiency in a 25-ft formation, with a 10 acre per well spacing, of 25 per cent, the ratios of horizontal to vertical permeability will have to be the values given in Table 6.

The values of $\bar{a}$ are the dimensionless well-spacing parameter, as given by Fig. 11.47 at $\overline{E} = 0.25$. It would have a value 26.4 if the pay were

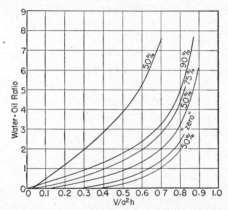

FIG. 11.52. The calculated variation of the water-oil ratio with the cumulative recovery in partially penetrating wells producing by bottom-water drives. V = total volume of oil zone, per well, invaded by water; a = well separation; h = thickness of oil zone; $\bar{a}$ = dimensionless well-spacing parameter. For uppermost curve, the spacing = 7.3 acres per well; for next four curves, spacing = 1.83 acres per well; for bottom curve, spacing = 0.20 acre per well. Notation on curves indicates well penetration. (From *AIME Trans.*, 1947.)

isotropic. The h''s represent the equivalent values of the oil-zone thickness to give $\overline{E} = 0.25$ in an isotropic medium.[1]

TABLE 6.—PERMEABILITY RATIOS REQUIRED TO GIVE 25 PER CENT EFFECTIVE SWEEP EFFICIENCIES IN A 25-FOOT OIL ZONE

Well penetration, %	$\dfrac{k_h}{k_z}$	$\bar{a}$	h'
"Nonpenetrating"	112	2.50	264
25	102	2.62	252
50	126	2.35	281
75	204	1.85	357
90	565	1.11	595

[1] Although the h''s listed in Table 6 correspond to values of ρ_w smaller than 0.001, which was assumed in applying Fig. 11.47, a correction for these differences would have but a very minor effect on the resultant values of k_h/k_z.

It is seen that in all cases the effective vertical permeability would have to be less than 1 per cent of the horizontal[1] and that if the formation were isotropic, its thickness would have to be 10 to 24 times as great as its actual thickness, to give $\overline{E} = 0.25$. This high degree of anisotropy would appear to be contrary to the results of core analysis, which generally show at the most only a single order-of-magnitude difference between the vertical and horizontal permeabilities and occasionally indicate a permeability perpendicular to the bedding planes that is even higher than that parallel to them. However, the anisotropy implied by Table 6 or equivalent calculations represents only a gross effective permeability contrast, rather than a strictly uniform dynamical characteristic similar to the optical or elastic anisotropy in crystalline materials. In most cases it may be the result of a dissemination through the rock of very thin or even microscopic shale or micaceous laminations. Such discontinuities in the rock structure would form parting planes for the cores. The layers adhering to the end faces would generally be removed in preparing the samples for measurement and thus would not be reflected in the permeability determinations. The permeability anisotropy resulting from such causes would be aggravated by the interference to vertical flow due to lenses or streaks of shale or clay, which are found in virtually all sandstone formations. In any case the vertical permeability of significance here is the resultant of whatever geometrical distribution of individual and localized strictly impermeable elements may be embedded in the gross porous rock matrix, which in itself may be fully isotropic. Indeed it is because such a resultant is almost impossible to derive from conventional core analyses that the interpretation of the performance of individual wells producing by bottom-water drives may be of particular value as a supplement to the analysis of the behavior of the reservoir as a whole.

There are no published reports of bottom-water-drive-reservoir analyses from the point of view presented here. However, long periods of clean-oil production, with recoveries of many thousands of barrels, have occasionally been observed in bottom-water-drive fields with the wells bottomed less than 5 ft above the original water-oil contact, with no evidence of intervening identifiable shale breaks or strictly impermeable barriers. On the other hand, in the case of a recently developed bottom-water-drive field in Mississippi producing a low-gravity oil, the oil-bearing sand is apparently so ideally uniform and isotropic that water has broken through in several wells after oil recoveries that agreed within 50 per cent of those predicted

---

[1] Because of the proximity to the well of the high-pressure driving source at the original water-oil contact the production capacities of the wells will still be comparable with those in radial-flow systems (cf. Fig. 11.42) in spite of the low vertical permeability, provided that the average horizontal permeability falls in the normally observed range.

by Eq. 11.15(6). While evidence of this type is as yet too meager to represent a real confirmation of the theoretical analysis, no contradictory observations have been reported for bottom-water-drive systems under conditions where the theory should be applicable.

Figure 11.46 implies that the oil recovery from bottom-water-drive fields will, in certain ranges of the physical parameters, increase with the well density. This, however, is purely a geometrical effect. It reflects in no way any influence of well spacing on the microscopic displacement efficiency of the local water-flushing mechanism. The latter is expressed by the factor $\bar{f}$ in the analytical equations and was explicitly assumed to be constant and uniform independently of the well spacing. If the well spacing should have any effect on the microscopic efficiency of oil displacement, it would have to be introduced as a separate factor determining the choice of $\bar{f}$ in Eq. 11.15(6).

The variation of the sweep efficiency and oil recovery with the well spacing is analogous to that which would obtain in areal water-drive or water-flooding systems if a linear array of wells were produced by the action of a parallel continuous line-drive source. In the latter case, certainly, if the wells were spaced widely as compared with their distance from the linear driving source, the local sweep patterns about each well would be completely separated, and the total oil recovery would be essentially proportional to the number of producing wells. Here, too, when the spacing is sufficiently wide there will be no overlapping between the local conelike water-oil interfaces coaxial with the individual wells, and the recoveries will be proportional to the well density. In both cases, too, the measure of close or wide spacing must be expressed in terms of the distance of separation between the driving source and well system. In the case of the bottom-water-drive reservoir this distance evidently should be the oil-zone thickness. Moreover the anisotropy of the formation will influence the effective well spacing. A low vertical permeability will tend to flatten the conelike apex of the water-oil interface to give the effect of a low well spacing, whereas a high vertical permeability will permit a rapid and concentrated cusping of the water-oil interface into the wells, with low sweep efficiencies. It is for these reasons that, as the analysis automatically provides, the real measure of the well spacing in bottom-water-drive systems is the dimensionless spacing parameter $\bar{a}$—the ratio of well separation to oil-zone thickness times the square root of the vertical to horizontal permeability ratio—rather than the well separation itself.

With regard to the true well-spacing parameter it may be noted that the ratios of well separation to oil-zone thickness will generally be such as to make the $\bar{a}$'s in isotropic formations fall in the range where the clean-oil recovery will be approximately proportional to the well density. However, in such instances the recovery from the wells will be so low as to make the production of questionable economic value in any case. On the other hand, where the sweep efficiencies for moderate spacings are inherently high, owing to the anisotropy of the pay, the corresponding values of the spacing parameter $\bar{a}$ will then automatically lie in the range where the clean-oil recovery is much less sensitive to the absolute well spacing. In any case the ultimate choice of the well spacing will be largely controlled by economic factors, as it must in all reservoir-development programs.

As pointed out in Sec. 11.13 the whole analytical theory of bottom-water-drive reservoirs presented here was based on the complete neglect of the density difference between the oil and water. All the conclusions derived therefrom hence appear to be entirely independent of the rates of well or field withdrawals. Unfortunately it is not feasible to evaluate accurately the seriousness of this assumption, although it is clear that its effect will become vanishingly small as the producing pressure differentials become very large as compared with the hydrostatic equivalent of a fluid column of thickness equal to that of the oil zone and density equal to the oil-water density difference. An upper-limit estimate of the role that could be played by the density difference can be made by comparing the differential gravity gradient to the calculated pressure gradient at the original water-oil interface along the well axis. For an oil-water density difference of 0.3 gm/cc the equivalent downward pressure gradient is 3×10^{-4} atm/cm. For the case of a 25-ft pay and "infinite" well spacing the pressure gradient at the original oil-water interface ($\bar{z} = 0.5$) is found to be (cf. Fig. 11.41) 1.56×10^{-3} atm/cm for a 50 per cent well penetration and 13.7×10^{-3} atm/cm for 90 per cent penetration, if the total pressure differential is 5 atm. Hence, even under the obviously extreme assumptions that the same relative magnitudes of gravity and pressure gradients apply over the whole water-oil interface and all along the well axis, the reduction in interface velocities and increases in clean-oil production would be of the order of only 20 and 3 per cent for the 50 and 90 per cent penetrations, respectively. Since the approximations made in deriving these estimates all tend to exaggerate the effect of gravity, it appears that under most conditions in actual practice, where the wells are produced continuously, the neglect of the density difference between the oil and water will not be of serious consequence.

Of course, as observed previously, if a well should be shut in there will always be a tendency for a recession in the water-oil interface. And in such cases where the permeability is so high that pressure differentials comparable with the differential fluid head of the oil zone suffice to give appreciable production rates, an improvement in sweep efficiency may well be achieved by restricting the withdrawal rate. Intermittent production will, in principle, also favor higher resultant sweep efficiencies. Nevertheless it seems unlikely that in most practical situations the value of such control of the production rates will be of major importance.

The assumption that the pressure remains uniform at the original oil-water interface is another factor leading to earlier break-through of water and lower sweep efficiencies, as calculated, than would occur in practice. For the fluid withdrawals will actually lower the pressure below the bottom of the well even at the original water-oil contact and will thus lessen the tendency for the rapid water rise along the well axis. This effect will be more serious for the higher well penetrations. It could be taken into account by setting the constant-pressure surface far below the initial water-oil interface[1] so as to reduce the effective well penetration. However, the choice of the exact position for this constant-pressure plane would be rather arbitrary. Moreover comparative calculations in cases with depressed constant-pressure boundaries show that while the corresponding sweep efficiencies may be appreciably increased, they usually will still be far too low to provide significant volumes of clean-oil production unless high degrees of anisotropy are assumed.

Finally it should be noted that the idealized performance described by the analytical theory may be complicated in practice by the effect of fluid motion along the planes of stratification, if the latter are not strictly horizontal. Even if the permeability normal to the bedding planes were zero, the water could still advance and enter the well from the distant and downdip parts of the pay by simply displacing the oil up the bedding planes. When the well penetration is high, such lateral migration may lead to an early appearance of water just as if the formation were isotropic. It is therefore necessary to analyze carefully all the available data regarding the nature of the reservoir rock and its structure in developing interpretations of bottom-water-drive well performance. The theory presented here should not be construed as implying a priori that wells producing under water drives and bottomed over a water-oil contact will necessarily lead to the calculated performance or that the producing strata will always be highly anisotropic. It is to be used only as an aid in the study of the well performance, if and when an investigation of the producing conditions indicates that they satisfy at least approximately the basic physical assumptions underlying the analysis.

11.17. Some Practical Aspects of Water-drive Performance.—The discussion in this chapter has emphasized the fluid dynamics within water reservoirs, which presumably provide the "drive" on their adjoining oil reservoirs. It has been seen that the pressure and flux histories of such

[1] Cf. L. F. Elkins, *AIME Trans.*, **170**, 109 (1947).

systems are subject to description and prediction when the reservoir characteristics are known. The oil reservoir itself may well have appeared to be playing a minor role, except as a means for fixing the boundary conditions for the water reservoir. Of course, only the gross features of the oil-reservoir behavior can be so determined.

The nature of the pressure distribution within the oil reservoir and the details of its production history require special consideration. If and while the pressures are above the bubble point of the oil, the fluid dynamics within the oil reservoirs still can be treated by the compressible- or even incompressible-homogeneous-fluid theory. The only generalization required over that used for describing the water reservoir is the introduction of a suitably distributed fluid-sink system corresponding to the actual well system. In treatments by the use of the electrical analyzer the well system, appropriately grouped, can be made an integral part of the water reservoir. Or it can be treated separately and "matched"[1] to the water-reservoir behavior. On the other hand, when the direct oil expulsion is caused by a solution-gas drive, the details of the pressure distribution within the oil reservoir cannot as yet be computed rigorously. However, the history of the average pressure and fluid-saturation conditions in the reservoir is subject to calculation, and this can again be matched to the water-reservoir behavior through the medium of the material-balance equation or its equivalent (cf. Sec. 9.7).

The details of the pressure distribution within a water-drive field reflect the withdrawal distribution and the geometry of the water-inflow boundary. In stratigraphic traps, with an essentially unidirectional drive, the pressure will decline continuously on receding from the water-oil contact toward the area of closure of the trap. The pressure contours in the East Texas field (cf. Fig. 9.2) illustrate this type of pressure distribution, although there, as in all reservoirs, the exact nature of the pressure variation is affected by the reservoir geometry and formation permeability, as well as the withdrawal distribution. For the simple case of a unidirectional linear drive into a reservoir of permeability k, uniform thickness h, and withdrawal per unit area Q and closed at the distant boundary, the pressure drop over the field will be

$$\Delta p = \frac{\mu Q L^2}{2kh} \tag{1}$$

where L is the width of the reservoir in the direction of the drive.

The average pressure $\bar{p}$ is given by the equation

$$\bar{p} = p_i - \tfrac{2}{3}\Delta p, \tag{2}$$

where p_i is the pressure at the water-oil boundary.

[1] This term is used here by analogy with the very similar situation common in electric- and electronic-circuit problems.

In anticlinal reservoirs, completely surrounded by invading edgewater and forming a radial system with a uniform areal withdrawal rate Q, the pressure drop between the center of the field and oil-water boundary will be given by

$$\Delta p = \frac{\mu Q R^2}{4kh},\qquad (3)$$

where R is the oil-reservoir radius. The weighted average pressure $\bar{p}$ will equal the mean of the pressures between the center of the field and that at the radius R and can also be expressed as

$$\bar{p} = p_i - \frac{\Delta p}{2}.\qquad (4)$$

These expressions should give the order of magnitude of the pressure gradients over complete-water-drive fields, although the exact values will depend on the detailed reservoir and producing conditions.

As has been previously emphasized the definition of the water-drive mechanism should be based, from a physical point of view, on the criterion of the displacement of the produced fluids by a water phase. Hence the water-drive action must be inferred from field performance. No a priori prediction can be made at the time of discovery that a given field will produce by the water-drive mechanism, although experience with other fields draining the same geologic horizon may be a valuable guide. If the initial reservoir pressure is abnormally high or low, as compared with the hydrostatic head at the reservoir depth, it may be inferred that the reservoir is sealed and is not in communication with a mobile water body, unless it is known that a pressure deficiency has been created by interference due to withdrawals from previously producing near-by fields. On the other hand the equivalence of the reservoir pressure to the hydrostatic head is no assurance of the development of effective water-drive action during the actual production history, although such equivalence may[1] indicate the possibility of water-drive performance.

Contrary to a rather widespread misconception, the existence of water-drive action does not require that the reservoir pressure show no substantial decline from its initial value. Actually such decline is necessary in order to develop rates of water intrusion as will replace the fluids withdrawn. The observation of a rapid initial pressure decline is in itself no evidence that an effective water drive will not develop. In fact, unless the formation has almost infinite permeability, by virtue of large fractures or cavernous channels of communication with the aquifer, as appears to

---

[1] When the edgewaters are immediately overlain by very low gravity crudes or tar zones, there may be a delay in entry and break-through of the water into the oil-productive section. In some reservoirs, however, such tar belts do not seem to have offered any abnormal resistance to the water invasion.

have been the case in some of the Mexican fields, the development of a substantial pressure differential between the reservoir and aquifer is a necessary preliminary to the creation of large rates of water intrusion. It will be recalled from Fig. 9.1 that, while a tendency for pressure stabilization became apparent later in the lifetimes of the various water-drive fields, they all showed rather rapid initial declines. This type of behavior is also evident in the theoretical predictions of water-drive pressure histories, as

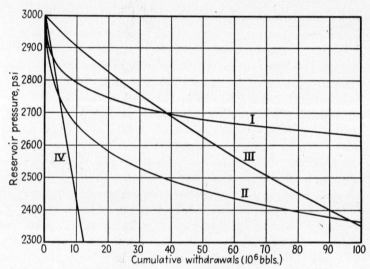

FIG. 11.53. The calculated pressure vs. cumulative-production histories of an idealized uniform reservoir if produced by different mechanisms. Curve I, complete water drive; reservoir fluid-withdrawal rate = 50,000 bbl/day. Curve II, complete water drive; reservoir fluid-withdrawal rate = 100,000 bbl/day. Curve III, solution-gas drive. Curve IV, reservoir fluid-expansion only.

may be noted for the steady-state treatment in Fig. 11.1 and for both finite and infinite compressible-liquid aquifers in Fig. 11.4.

Not only must complete-water-drive reservoirs initially suffer rather rapid decline rates, but if the oil is actually undersaturated the pressure will fall even faster during its early history than if it were saturated and producing by pure solution-gas drive. For under gas-drive production the gas evolution resulting from the pressure decline will provide a local oil-displacing medium and thus tend to sustain the pressure against immediate excessive pressure drop. Of course, when the gas phase develops sufficient mobility to by-pass the oil, the typical gas-drive depletion mechanism comes into play, with its characteristic rapid dissipation of the reservoir energy.

To illustrate these considerations, comparative calculations have been made on the pressure decline vs. the cumulative recovery to be expected in a hypothetical reservoir under various extreme mechanisms of oil expulsion. The results are plotted in Fig. 11.53. For the complete-water-drive mechanism the reservoir was assumed to have a radius of 3 miles, with no reservoir fluid expansion. The thickness of the aquifer was taken as 50 ft and its permeability as 500 md, and the water viscosity was assumed to be 0.5 cp. The initial reservoir pressure was taken as 3,000 psi. In Fig. 11.53 are plotted the theoretical pressure-decline (at the initial water-oil boundary) curves for water-influx rates and reservoir oil-withdrawal rates of 50,000 and 100,000 bbl/day. For the gas-drive mechanism the reservoir rock and fluid constants were taken as those used in the calculations of the gas-drive pressure-decline history of a reservoir producing a 30°API-gravity crude, discussed in Sec. 10.4. As the total pore space of the assumed oil reservoir of 18,095 acres and 25 per cent porosity was 1,755,000,000 bbl, a conversion of the cumulative-recovery scale of Fig. 10.8 gave the gas-drive curve of Fig. 11.53. As a final extreme producing mechanism it was assumed that there was neither a water nor a gas drive, so that the fluid withdrawals were replaced only by the residual reservoir fluid expansion. For a compressibility of 10^{-5} per psi the straight line of Fig. 11.53 represents the resultant pressure-decline history.

The curves of Fig. 11.53 show clearly that during the early production history the pressure decline will be the least rapid for the pure gas-drive mechanism. For a reservoir fluid-withdrawal rate of 100,000 bbl/day the water-drive mechanism without reservoir fluid expansion will show a more rapid pressure-decline rate, per unit of withdrawals, than the gas-drive reservoir until 15,000,000 bbl has been produced. And the actual pressure will remain lower than for the gas-drive reservoir until a cumulative recovery of 96,500,000 bbl. Even for the withdrawal rate of 50,000 bbl/day the water-drive reservoir will show lower pressures up to a cumulative recovery of 38,700,000 bbl and a more rapid rate of decline until 5,000,000 bbl has been produced.

These comparisons have inherently favored the gas-drive mechanism in so far as the pressure decline for the latter was calculated for an equilibrium gas saturation of 10 per cent. However, as the whole range of cumulative recovery plotted in Fig. 11.53 represents less than 6 per cent of the reservoir pore space, the relative positions of the various pressure-decline curves should not be substantially changed for any reasonable form of the permeability-saturation relationship of the producing rock. The neglect of the reservoir fluid expansion in the case of the water-drive reservoir tended to exaggerate the calculated pressure decline for the water-drive mechanism. The effect of the reservoir fluid expansion is shown by the straight line in Fig. 11.53. It will be seen that during the initial history the pressure decline will be retarded more by the reservoir fluid expansion than by the water-drive action. In fact without the aid of this expansion the water-drive pressures will begin to decline at an infinite rate if the withdrawals are started at a nonvanishing value. It will be clear, however, that even if the water-drive pressure-decline curves were corrected[1] for the reservoir fluid expansion, they would still remain below the gas-drive curve until substantial cumulative recoveries had developed.

While Fig. 11.53 illustrates the more favorable pressure performance of the gas-drive reservoir during the initial phases of the production history, it also shows the basic difference between the various producing mechanisms with respect to their long-term pressure history. That is, whereas in a pure gas-drive reservoir the pressure continues to decline with the cumulative recovery without any significant tendency for stabilization,[2] the rate of pressure decline in a complete-water-drive reservoir will continually decrease, unless the withdrawal rate be increased. Moreover, if the withdrawal rate should be decreased the pressure will be temporarily increased, whereas a pressure increase can never develop in a strict gas-drive reservoir even if the field be completely shut in. Hence, unless the producing rates are excessive as compared with the supply capacity of the aquifer, the average pressure during the producing life of a complete-water-drive reservoir will in general be higher than for a similar reservoir producing by gas drive. In fact, whereas economic oil depletion and abandonment in gas-drive fields generally imply an important degree of pressure depletion, the economic ultimate recoveries from water-drive reservoirs, which are usually limited by the rise in water production, may, if properly controlled, be obtained at abandonment pressures that are half or even greater fractions of the original pressure.

By far the most significant characteristic of water-drive reservoirs from the economic point of view is that their "yield" is generally materially greater than for gas-drive reservoirs. This is simply a consequence of the fact that in most reservoirs the local oil-displacement efficiency by water flushing is inherently greater than for the solution-gas drive. Whereas a

---

[1] As a first approximation this correction can be made by simply adding the cumulative recoveries by the reservoir fluid expansion to those for the pure water-drive mechanism, at the same values of the pressure. It may be readily shown that for both gas- and water-drive systems the initial slope of the pressure vs. withdrawals curve is equal to the reciprocal of the reservoir liquid volume times its compressibility (under pressure decline). It is because the latter is of the order of 10^{-4} per psi for saturated crude oils at high pressures, as compared with 10^{-5} per psi for undersaturated crudes, that the initial decline rates in gas-drive reservoirs will be less than in liquid-expansion or water-drive systems. In fact, under favorable conditions, the initial rate of pressure decline, vs. the cumulative recovery, can be used as an indication of the condition of saturation or undersaturation of the reservoir liquid. And if the reservoir fluid properties and initial-oil content are known, the size of the gas cap, if any, can also be determined, in principle, from the initial pressure-decline rate.

[2] The concavity of the gas-drive curve in Fig. 11.53 is the result of assuming an equilibrium free-gas saturation of 10 per cent and of the decreasing gas-oil ratio during the build-up of this saturation.

recovery by gas drive of one-third of the initial-oil content may be considered in many cases as quite satisfactory, a local displacement of 50 per cent by water flushing is usually about the minimum expectancy. And local recoveries of 75 per cent of the oil in place are not beyond reason. The interest in effective water-drive action thus has a very substantial economic basis. Hence, to the extent that water-drive performance represents the optimum recovery mechanism for a particular reservoir and that the development of such performance is subject to the control of the operator, it is evidently worth serious effort to achieve such a reservoir behavior. A more detailed discussion of actual recoveries to be expected in water-drive fields and data on such recoveries as already have been observed will be given in Sec. 14.13.

It should be noted, however, that water-drive performance per se does not always ensure a maximum oil recovery regardless of reservoir conditions. Reservoirs that are overlain by initial gas caps and supported by an active water drive may suffer in oil recovery if the water drive is not properly controlled. Since an effective water drive will tend to induce a mass movement of the oil body ahead of it, care must be taken that the oil will not be driven into the gas cap. The water drive must be so controlled and the pressure in the gas cap be maintained so as to prevent an upward movement of the initial gas-oil contact, even if this should require the injection of gas into the gas cap to ensure a downward pressure gradient toward the oil zone.

A factor that must be considered in predicting water-drive recoveries concerns the effect of permeability stratification in reducing the net gain to be derived from the water-drive mechanism, as compared with gas-drive oil expulsion. Even if the recovery efficiency from any particular stratum flooded out by water should be high, the rate of entry of the water will not be uniform in formations having a high degree of permeability stratification. It will evidently be greatest in zones of high permeability. And in extreme cases the water may channel through such zones before a substantial flushing has been developed in the tighter sections of the formation. Unless the watered-out stratum can be identified, located, and plugged off, the recovery of a major part of the oil from the lower permeability strata will be accompanied by water production. As zones of intermediate permeability become flooded out, the net water-oil ratio may become so high that it will no longer be profitable to continue production and drainage from the tightest zones of the formation. The gross recovery efficiency may then be considerably lower than anticipated for an ideal single zone[1] and homogeneous pay.

[1] Under homogeneous-fluid flow and steady-state conditions the effect of permeability stratification in water-drive reservoirs can be treated by the method developed for the cycling of condensate reservoirs (cf. Sec. 13.8), corrected for the different mobilities of

If the reservoir must undergo an initial gas-drive phase to induce the water entry, the high-permeability-zone channeling will be aggravated. For these zones will contribute most to the oil production and become depleted most rapidly. This will encourage the water entry into them to a degree that is greater than arises simply from the permeability differences. This will be partly counteracted by the fact that the permeability to water when flowing through a flooded zone is generally quite low compared with that to oil at high oil saturations. Nevertheless it is quite conceivable that in cases of extreme stratification the final oil recovery at the time of abandonment due to excessive water production may be even lower than if the same formation were completely sealed off from its bounding edgewaters and of necessity were produced by a solution-gas drive. In fact such a situation may well arise in highly fractured limestones where the major part of the oil content happens to lie in the intergranular tight rock and only a small fraction in the highly permeable fractures. Conditions of this kind apparently obtain in the West Edmond field, Canadian, Logan, Oklahoma, and Kingfisher Counties, Okla., where the water invasion in the western part of the field appears to be flushing out but little more than the fracture component of the producing formation.[1]

A water-drive reservoir may show even less recovery than would be expected from the solution-gas-drive mechanism also in substantially isotropic and unstratified formations produced by bottom-water-drive action. As shown in Sec. 11.15 the sweep efficiency of the rising water table in isotropic strata will generally be of the order of 1 per cent or less. Very little of the recoverable oil will therefore be produced water-free, and the continually increasing water cuts may force abandonment at total recoveries lower than could have been obtained by solution-gas drive.

The high recoveries generally associated with water-drive reservoirs are basically the consequence of the relatively low residual-oil saturations of 20 to 30 per cent left by the process of oil expulsion by water displacement. Since there is little evidence that this residual-oil saturation varies appreciably with the connate-water saturation, it is clear that the fraction of the original oil recovered will decrease as the connate-water saturation increases. In particular, if the connate-water content should be 45 per cent or more, the fractional recovery by water drive may be no greater than by

---

the oil and water. Such a treatment has been carried out recently (H. Dykstra and R. L. Parsons, API meetings, Los Angeles, Calif., May, 1948), assuming a probability distribution in permeability. The effect of the stratification alone, assuming equal mobilities for the water and oil, may be inferred from the results developed in Sec. 13.8.

[1] Cf. M. Littlefield, L. L. Gray, and A. C. Godbold, *AIME Trans.*, **174,** 131 (1948); cf. also H. S. Gibson, *World Oil*, **128,** 217 (June, 1948).

solution-gas drive. Such situations will not occur in the majority of producing reservoirs. However, these considerations should serve to show that the planning of the exploitation of oil reservoirs should not be based on general rules, even if these should have statistical significance. Interpretations and predictions of reservoir performance and recovery must be made on an individual basis in the light of the particular characteristics of the specific reservoir of interest.

Finally it should be observed that since virtually all actual reservoirs are stratified to some extent, their performance will represent the resultant of those of the individual strata comprising the whole of the producing section. If there were no cross flow or intercommunication between the different-permeability zones, the composite history would be a simple linear superposition, in parallel, of the transients of the individual homogeneous strata, such as are predicted by the theoretical analysis of this chapter, each governed by its own time scale. In such cases the separate zones would actually constitute distinct reservoirs having in common only the same producing well pressures. In many reservoirs, however, there will be relatively free intercommunication and cross flow between the layers of different permeability. These will modify the simple parallel behavior of the separated strata. Quantitative agreement of observed performance with predictions based on uniform stratum analysis is therefore not to be expected, except as the physical constants used in the latter are treated essentially as empirical parameters and interpreted as averages correcting both for the actual variations in the physical properties and for the interactions between the various parts of the composite formation.

11.18. Summary.—From a physical point of view and with respect to their gross pressure and production performance, complete-water-drive reservoirs may be defined as those in which the replacement of the reservoir fluid-withdrawal voidage by the creation or growth of a gas phase is negligible as compared with the total withdrawal volumes. Such conditions will generally obtain when the reservoir oil is highly undersaturated and is in communication with a mobile water body. During the early history of such reservoirs and before the pressure has declined sufficiently to induce an appreciable rate of water intrusion from the adjoining aquifer, fluid-withdrawal replacement will be provided largely by the expansion of the oil-reservoir liquid phase. While such conditions obtain and as long as the reservoir pressure remains above the bubble point, the producing mechanism will be, in a strict sense, neither of the gas-drive nor of the water-drive type. However, this phase of the production history will be of short duration if the aquifer is inherently capable of providing a significant water influx into the reservoir. Accordingly, from the point of view of over-all performance, a reservoir may be considered as operating by the water-drive mechanism as long as the rate of water intrusion becomes substantially equal to the rate of fluid withdrawals before the development of significant contributions from a free-gas phase, even though initially the reservoir liquid expansion may be the major withdrawal-replacement agent.

Many important oil fields were initially highly undersaturated and were automatically produced by the water-drive mechanism. Others, in which the reservoir oil was originally saturated, have developed complete-water-drive behavior after an initial gas-drive phase in which the pressure declined sufficiently to induce rates of water intrusion substantially equal to the reservoir fluid withdrawals. The latter situation has become increasingly common as oil- and gas-withdrawal rates have been restricted, since such restrictions facilitate the development of a balance between water influx and reservoir withdrawals. In still others the injection of water into the aquifer has supplemented the natural influx into the oil reservoir so as to achieve a resultant water intrusion sufficient to replace the fluid withdrawals.

If the water motion in the aquifer be represented as a sequence of steady states of homogeneous-liquid flow, the pressure history of the oil reservoir can be readily derived from the solution of an ordinary differential equation, taking into account the expansion of the oil-reservoir liquid content [cf. Eq. 11.2(4)]. The pressure is found to decline sharply at first and then rapidly to approach a stabilized value at which the water influx completely replaces the withdrawals (cf. Fig. 11.1). The ratio of the pressure drop at stabilization to the initial pressure equals the ratio of the reservoir oil-production rate (assumed constant) to the maximum water-supply capacity of the aquifer. If the production rate be changed, the pressure quickly restabilizes itself to a higher or lower value corresponding to the new rate (cf. Fig. 11.2).

While the steady-state representation of the water flow in the aquifer serves to show the general role played by the withdrawal rate in determining the pressure decline, the compressibility of the water must be taken into account in most specific applications. The duration of the transient associated with the water compressibility will be essentially proportional to the area of the aquifer, the water compressibility, and the water viscosity and inversely proportional to the permeability of the aquifer [cf. Eq. 11.3(4)]. It may be shown that for the areas and properties of aquifers which might support a water-drive oil-production mechanism throughout a substantial part of the producing life the transient effects and water compressibility should be taken into account in attempting a quantitative description of the reservoir behavior. On the other hand, similar considerations show that, for the more limited distances involved within the confines of oil reservoirs, the transient phenomena will be of very short duration and can generally be neglected, if the pore space does not contain a free-gas phase.

The formal treatment of compressible-liquid systems is simply comprised of the solution of a partial differential equation in the liquid density, which is identical in structure to the heat-conduction equation [cf. Eq. 11.3(8)]. For the special case where the aquifer is assumed to be of infinite extent and the water-oil boundary is taken as circular, with a constant water-withdrawal rate, a universal curve can be constructed for the pressure-decline history, using dimensionless pressure and time parameters (cf. Fig. 11.4). In contrast to the implications of the steady-state treatment of the aquifer flow, this curve shows that there will be no strict stabilization pressure in infinite aquifers. Rather the pressure drop will increase monotonically, though at a continually decreasing rate. In absolute value, however, the pressure decline will depend on the production rate (cf. Fig. 11.5) and will increase markedly, for fixed cumulative water withdrawals, with increasing rate of withdrawal (cf. Fig. 11.6). This sensitivity of the pressure history to the withdrawal rate is one of the most characteristic features of the water-drive production mechanism.

If an initial constant withdrawal rate is suddenly increased, the pressure will suffer a renewed sharp decline similar to that experienced originally. Conversely, if the rate be suddenly reduced, the pressure will rise sharply. However, the rise will not persist indefinitely but will be converted into a slow decline after reaching a maximum (cf. Fig. 11.7).

If the infinite aquifer be subjected to a sudden lowering of pressure at the water-oil boundary, the transient history of the water flux across the boundary is found to begin theoretically at an infinite value and to decline rapidly with increasing time, but with a continuously decreasing decline rate (cf. Fig. 11.10). The pressure decline within the aquifer gradually extends outward into the water reservoir (cf. Fig. 11.12), though the pressure reaction will, in general, not develop appreciably within the normal producing lives of water-drive fields at distances greater than 25 times the effective radius of the water-oil reservoir boundary. This type of analysis has been applied in computing the water-influx history into a Gulf Coast water-drive field and has been found to be in substantial agreement with that calculated by application of the material-balance equation (cf. Figs. 11.14 and 11.15).

While the finite limit of the aquifer will not come into play during the early transient history of most water-drive reservoirs, the later stages of the production history may ultimately be affected by such limits. This will usually occur if the volume of the water reservoir is appreciably less than 100 times that of the oil reservoir. Under such conditions it is best to treat the aquifer from the beginning as a finite system, with boundary conditions assigned for both the internal- and external-bounding surfaces. When the flux is assigned at one boundary and the pressure at the other, the density distribution in the aquifer can be expressed as an infinite series of products of Bessel and exponential functions [cf. Eq. 11.6(5)]. If the pressure (density) at the external boundary be kept fixed at its initial value and a constant withdrawal be taken from the internal-boundary surface, the pressure history at the latter may be expressed as the steady-state value plus a superposed exponentially declining transient series [cf. Eq. 11.6(10)]. The transient history can also be analytically determined if the pressures or densities are specified at both boundaries [cf. Eq. 11.6(12)]. And for finite closed aquifers, in which the pressure at the internal boundary is suddenly lowered, the pressure-decline history at the closed external boundary and flux transient at the constant-pressure surface, as well as the cumulative-production history, can readily be determined from general curves plotted in terms of dimensionless variables (cf. Figs. 11.18 to 11.20).

To study the pressure distribution within a water-drive reservoir due to localized fluid-withdrawal centers, as individual producing wells, it is convenient to use instantaneous or permanent fluid-sink functions [cf. Eqs. 11.7(1) and 11.7(2)]. In some cases these can be synthesized as integrals over continuous fluid-withdrawal distributions [Eq. 11.7(5)] or as sums of contributions due to individual wells [Eq. 11.7(7)]. Methods based on such functions may be of special value in treating analytically reservoirs of irregular geometry and in predicting the effects of localized water injection.

Virtually all analytical treatments must of necessity be limited to uniform reservoirs and those of limited geometrical complexity. It is possible, however, to generalize the scope of interpreting and studying water-drive reservoir performance by substituting for the actual system an equivalent electric circuit. The justification for this is based on the observation that the flow of current in a dielectric medium is formally analogous to that of a compressible homogeneous liquid through a porous medium [cf. Eq. 11.8(1)]. If the actual fluid density be expressed as a linear function of the pressure [cf. Eq. 11.8(5)], the electric voltage will correspond to the fluid pressure, the electrical conductivity to the ratio of the permeability to the viscosity, the local dielectric capacitance to the product of the porosity and compressibility, and the current-density vector to the vector volumetric flux [cf. Eq. 11.8(7)]. By constructing an "electrical analyzer" in which these quantities can be suitably assigned and adjusted by numerical scale factors [cf. Eqs. 11.8(9) and 11.8(10)], the theoretical pressure history in a reservoir can be determined by observing the voltage history in the corresponding electrical network. The latter may be conveniently recorded on a chart (cf. Fig. 11.22), with a time scale of $2\frac{1}{2}$ sec as the equivalent of a production period of the order of 1 month.

The first step in making an electrical-analyzer study of a water-drive reservoir is the determination of the physical and geometrical properties of the supporting aquifer. The best estimates possible for the aquifer constants, as derived from geological and related evidence, are introduced, after suitable transformation, into the electric circuit of the analyzer. The history of the water influx into the oil reservoir is then imposed at the circuit terminal representing the water-oil reservoir boundary. The voltage history then given by the analyzer is compared with the observed pressure history at the water-oil reservoir boundary or the equivalent average oil-field pressures. If the two agree, the assumed aquifer constants may be considered as representative of its actual physical and geometrical properties. If not, the electric-circuit constants are changed until satisfactory agreement is achieved. In this way an empirical representation of the aquifer may be developed and used for predicting the future pressure history of the oil reservoir under various assumed operating conditions.

If the reservoir oil is at a pressure above its bubble point, the pressure distribution and history within the field can be determined by means of the electrical-analyzer "pool unit" (cf. Fig. 11.22). The latter, constructed and treated in a manner similar to the aquifer circuit, permits the application of various withdrawal histories to individual wells or groups of wells within the field and the recording of the pressure (voltage) at points within the field. The effects of water injection within or external to the field can be similarly studied.

To the extent that the water-influx history into the oil reservoir can be established, the electrical analyzer can be applied to the analysis of partial- as well as complete-water-drive systems. And by further extensions of the basic electric circuits, even the effects of free-gas-phase expansion and changes in the resultant compressibility of the oil-reservoir fluids can be approximately treated.

While the East Texas field is of historical interest in that it gave rise to the compressible-liquid theory of water-drive performance, it also represents perhaps the outstanding example of the quantitative application of this theory. Aside from its tremendous economic importance—it having produced more than twice as much oil as any other field in this country—it is of particular interest from a physical point of view in that the supporting Woodbine aquifer permits a representation as an idealized system of uniform properties throughout. Accordingly, by using a radial-flow simplification with respect to the gross geometry of the aquifer, its pressure-decline history can be calculated theoretically by simply applying the known production history of the field [cf. Eq. 11.9(1)]. On introducing geometrical and physical constants in substantial agreement with those indicated by independently determined core, fluid-analysis, and geological data, except only the water compressibility, the computed pressure history at the water-oil boundary parallels almost exactly the observed pressures from the discovery of the field in 1930 to 1941 (cf. Fig. 11.24). The parallelism does not persist after 1941 because of the increasing amounts of water injection into the producing formation and the change in the quantitative significance of the measured average reservoir pressures. However, even by January, 1947, the difference between the calculated and observed pressure was only about 20 psi, as compared with that indicated by the trend before 1941.

The analysis of the East Texas field performance is of added interest in that account was taken of the continual growth in water production since 1935 and the large-scale water-return operations, to the Woodbine Sand, undertaken in 1940–1941. In fact, during 1947 the water produced appreciably exceeded the reservoir oil withdrawals, and more than 90 per cent of the former was returned through some 75 injection wells. If it were not for this water-return program, the pressure decline by Jan. 1, 1947, would have exceeded the actual decline by approximately 200 psi.

To obtain agreement between the observed and calculated pressure histories in the East Texas field it was necessary to assume that the effective compressibility of the water was about 12 times that normally measured for water. While this does not represent a quantitative empirical determination of the compressibility, it would be impossible[1] to obtain a corresponding degree of agreement over the whole producing life to January, 1947, if one assumed a normal value. Among the possible interpretations of this abnormally high compressibility a reasonable one appears to lie in the assumed presence of a distribution of large free-gas masses in the Woodbine reservoir occupying a volume of the order of 5 per cent of the total pore volume of the aquifer.

The Smackover Limestone formation in Arkansas represents another major aquifer supporting water-drive reservoirs by the compressible-liquid-expansion mechanism. Among the 12 oil or condensate fields producing from this formation (cf. Fig. 11.25), 7 have been successfully subjected to analysis by the electrical analyzer. These include both partial- and complete-water-drive reservoirs, with cumulative total influxes, as of June 30, 1943, ranging from 4,400 to 54,400 bbl/psi decline in pressure and rates of influx varying from 4.0 to 45.4 (bbl/day)/psi (cf. Table 2). As determined by the electrical analyzer the effective fluid conductivity of the aquifer supporting the various reservoirs ranged from 1,136 md-ft/cp to 5,830 md-ft/cp, with an average of 3,052 md-ft/cp. And the fluid-capacitance con-

---

[1] This has been verified by application of the electrical analyzer, which not only exactly duplicated the calculated curves of Fig. 11.24 for the same assumed constants but also showed that no reasonable changes in these constants or distributed variations in them could independently give an effect equivalent to the high compressibility over the 16-year producing history.

stants varied from 18×10^{-5} ft/psi to 231×10^{-5} ft/psi, with an average of 58×10^{-5} ft/psi (cf. Table 3).   As a whole, these empirically determined aquifer constants correspond to effective permeabilities that are considerably lower than those observed within the oil reservoirs.

In addition to reproducing the previously observed pressure histories (cf. Figs. 11.26 and 11.28) the electrical analyzer has been applied in these fields to predictions of their future performance under various assumed operating conditions (cf. Fig. 11.27 and Table 4).   Another interesting application has been made in the study of the interference effect that the withdrawals from the Magnolia field—the largest of the Smackover Limestone reservoirs—may be exerting on the near-by and much smaller Village pool.   In fact, it was so found that the Magnolia field was creating considerably greater pressure declines at Village than the production of the latter (cf. Fig. 11.29).

Another of the Smackover reservoirs, Midway, provides an example of pressure-maintenance operations by water injection.   Although the early field history of this undersaturated reservoir appeared to indicate a strong water-drive action, the supporting aquifer terminates only 10 miles north of the field and is cut by a fault $3\frac{1}{2}$ miles to the south (cf. Fig. 11.25), thus seriously limiting its total expansion capacity.   The early inauguration of water injection at the edges of the field quickly retarded the previous pressure-decline trend and ultimately resulted in a substantial rise in pressure (cf. Fig. 11.30).   Moreover an analysis of these pressure reactions shows that the natural water influx has actually been decreasing (cf. Fig. 11.31), as would be expected from the limited volume of the aquifer in communication with the reservoir.

The outstandingly successful water-injection program at Midway does not imply that such operations will be necessary or desirable in all reservoirs. While the return of the produced water to the reservoir or its supporting aquifer will usually be advantageous, there is little reason to provide extraneous waters for injection unless it has been established that the natural water influx will not suffice to sustain the desired withdrawal rates without excessive pressure decline.   Such operations should be evaluated primarily on the basis of economic factors.   Moreover, under certain conditions the supplementing of the water intrusion by water injection may even aggravate the operating problems.   This may occur when a gas cap overlies the oil zone and there is danger of driving the oil into the gas cap.   In formations of high effective vertical permeability the injection of water near producing wells may make it more difficult to achieve satisfactory sweep efficiencies from the rising water-oil interfaces.   And in extreme cases of permeability inhomogeneities, such as may obtain in highly fractured limestones where a major part of the oil reserves lies in the intergranular matrix, the water-drive mechanism may be inherently unsuited to efficient recovery. If the natural water intrusion in such reservoirs be supplemented by water injection, the difficulties arising from water by-passing and isolation of the intergranular rock may be unnecessarily aggravated.

Even without detailed analysis the broad characteristics of water-drive performance may usually be recognized from the general character of the production and pressure histories. Thus the tendency for pressure stabilization with decreasing withdrawal rates and approximate constancy of the gas-oil ratio, as shown by the Conroe field (cf. Fig. 11.32), is typical of water-drive behavior of fields producing gas-saturated oil with or without gas caps. Highly undersaturated water-drive reservoirs will naturally produce at constant gas-oil ratios until the pressures decline to the bubble point (cf. Figs. 11.33 and 11.34). They generally experience rather rapid growths in the water production, although the total fluid-production capacity may remain substantially constant over long periods—if unrestricted by proration (cf. Fig. 11.35). The water advances inward toward the areas most distant from the original oil-water boundary (cf. Figs. 11.36 and 11.37), the percentage of water production from the individual wells increasing as the aquifer is approached (cf. Fig. 11.38). The details of these developments will depend on the distribution of fluid withdrawals, the manner of well completion, reservoir structure, and the uniformity of the producing section.

By expressing the rate of water influx into a water-drive reservoir as an encroachment factor times the pressure decline, the former may be considered as a measure of the inherent supply capacity of the aquifer. While this factor will vary during the producing life of compressible-liquid water-drive reservoirs, values computed for periods of approximate pressure stabilization should be indicative of at least the order of magnitude of the fluid-withdrawal rates that can be supported by the aquifer. Among six fields, not including East Texas, whose water-influx histories have been interpreted in this manner, the encroachment factor ranged from 126 to 1,350 (bbl/month)/psi. When reduced to influx rate per unit reservoir volume, it varied from 0.002 to 0.086 (in barrels per month per psi per acre-foot) (cf. Table 5). The latter shows a general qualitative correlation with the steepness of the pressure-decline curves for the corresponding fields. Although the over-all encroachment factor for East Texas is approximately 23,000 (bbl/month)/psi, the reduced factor is only 0.005 (in barrels per month per psi per acre-foot).

While the gross performance histories of complete-water-drive reservoirs are determined by the volumetric and dynamical balance between the fluid withdrawals and the water intrusion from the aquifer, the details of the advance into the oil zone of the water-oil interface will, under certain conditions, have an important bearing on the actual oil recoveries. This will be of especial significance in the case of bottom-water-drive reservoirs or those areas in edgewater-drive fields where the wells are bottomed directly over the water-oil contact.   The close proximity to the wells of the high-pressure driving source will lead to an accelerated rise of the bottom waters along the well axes, thus forming conelike water-oil interfaces (cf. Fig. 11.39). Although these will be similar in shape to the elevated water-oil interfaces resulting from bottom-water coning in gas-drive reservoirs (cf. Sec. 5.9, Fig. 5.19), the physical mechanisms underlying the two phenomena are basically different.   Coning in gas-drive reservoirs represents an incidental effect of the upward pressure gradients associated with the essentially horizontal multiphase flow into the producing wells and may remain in substantial hydrostatic equilibrium with the oil-producing zone, whereas in the bottom-water-drive field the water-oil interface must of necessity continue to rise into the oil zone to replace the withdrawals from the latter.

Assuming complete reservoir uniformity and identity in physical properties, i.e., permeability, viscosity, porosity, and fluid density, of the composite fluid and porous medium above and below the water-oil interface, its shape and the nature of its advance as well as the production capacities of the wells can be computed theoretically. The latter are found to be of the same order of magnitude, and to vary with the well penetration in the same manner, as for wells treated as single-phase steady-state radial-flow systems (cf. Fig. 11.42). The resultant effect of the shape of the interface on the oil recovery may be expressed in terms of a "sweep efficiency," defined as the fraction of the gross volume of the oil zone drained per well that is flooded out at the time of first water production. This term is the three-dimensional analogue of the sweep efficiency used to describe the behavior of water-flooding and cycling operations (cf. Secs. 12.6 and 13.5). In the bottom-water-drive system the sweep efficiency is a direct measure of the volume of the clean-oil production.

As would be expected from physical considerations the analysis shows that for fixed well penetrations the sweep efficiency is mainly determined by a composite "dimensionless well-spacing parameter."   The latter is the ratio of the actual mean well separation to the oil-zone thickness, multiplied by the square root of the ratio of vertical to horizontal permeability.   The sweep efficiency decreases continuously with increasing well-spacing parameter (cf. Fig. 11.44).   Hence it decreases as the well separation is increased, as the pay thickness is decreased, and as the ratio of the vertical to horizontal permeability increases. At resultant values of the parameter exceeding 3.5 the sweep efficiency varies inversely as the square of the spacing parameter or in direct proportion to the well density. Under such conditions, which will generally obtain unless the oil zone is highly anisotropic, there is a fixed clean-oil production per well [cf. Eq. 11.15(6)], and the total from the reservoir will be simply proportional to the number of wells used. The amount of clean-oil production will increase with decreasing well penetration (cf. Fig. 11.45).

An especially interesting implication of the analysis is that, unless the permeability parallel to the bedding planes very greatly exceeds that normal to them, the absolute magnitudes of the sweep efficiencies will be so low—much less than 1 per cent—that, except for very thick oil zones, the clean-oil production will be but a small fraction of the normally expected total oil recovery. A high degree of effective permeability anisotropy must therefore be inferred when it is observed, in a bottom-water-drive field, that wells not completed immediately over extended shale breaks continue to produce clean oil for extended periods. The very low vertical permeabilities so implied may be due to microscopic shale laminations or equivalent localized barriers to vertical flow, rather than an inherent anisotropy in the basic rock matrix such as would be indicated by conventional core analysis.

Once the water breaks through to a well producing by a bottom-water drive, there will be a continual increase in the water-oil ratio. However, the rate of rise in the water-oil ratio will generally be slower for systems of low sweep efficiency. Hence the total oil recovery by the time a fixed water-oil ratio is reached will not differ so greatly among various systems as will their corresponding clean-oil recoveries (cf. Figs. 11.51 and 11.52).

One of the major assumptions underlying this bottom-water-drive theory is that the water and oil densities are equal. This implies that all geometrical aspects of the movement of the water-oil interface will be independent of the rates of fluid withdrawal. The latter will affect only the time scale. The basic question involved concerns the relative magnitudes of the upward pressure gradients driving the oil into the well bore and the downward gravity forces due to the actual density difference between the oil and water. If a well be shut in, the latter will undoubtedly lead to a fall in the apex of the previously developed conelike water-oil interface and an equalization of its elevation in the oil zone. Indeed, an intermittent producing program with long shut-in intervals between the producing periods may well lead to appreciably improved resultant sweep efficiencies. And even continued well production at such low rates that the pressure drawdowns do not greatly exceed the head of an oil-zone-thickness column with a fluid density equal to the difference between the water and oil will probably give greater volumes of clean-oil production than predicted by the simplified theory. However, in most cases where the pressure differentials must of necessity be very much larger than the maximum gravity heads, in order to obtain the desired producing rates, it is unlikely that the sweep efficiency will be appreciably affected by variations in the withdrawal rates.

Both theoretical analyses and field observations show that the average pressures during the producing life of a complete-water-drive field will usually be higher than in solution-gas-drive fields. Whereas abandonment of the latter generally is the result of a depletion of the pressure and gas energy to a state where the production rates of the wells are too low for profitable operation, the reason for abandonment of water-drive fields in most cases is the excessive production of water. The pressures and total fluid-production rates often show but little decline at abandonment as compared with those prior to the development of appreciable water production. During the initial phases of complete-water-drive reservoir production, however, the situation may be reversed. If the reservoir oil is highly undersaturated, the pressure will initially decline at a very rapid rate, except for the effect of the reservoir liquid expansion, until the induced water intrusion from the aquifer becomes comparable with the oil-withdrawal rate (cf. Fig. 11.53). And even when the effect of the reservoir fluid expansion is included, the pressure-decline rate will exceed that of a similar solution-gas-drive reservoir for an appreciable part of the early production history. In the latter the development of a free-gas phase, as soon as the pressure drops below the bubble point, immediately provides a replacement for the withdrawals and tends to minimize the decline. Moreover the decline rate in the complete-water-drive reservoir will be rate-sensitive. Hence, if the withdrawal rates are excessive, the pressures will quickly fall to the bubble point and lead to gas evolution and partial-gas-drive behavior. Such gas evolution, as has actually been observed, will also lead to sharp, though temporary, reductions in the pressure-decline rate, as if the water drive itself had suddenly become highly effective as a pressure-maintaining mechanism. Of course, as production continues, the pressures in pure solution-gas-drive fields will ultimately fall below those in the water-drive reservoir and the average over the whole producing life will be higher in the latter. Initially, however, the pressure-decline rate alone will usually not permit a unique interpretation in terms of the producing mechanism, which immediately or ultimately may control the performance. In fact, it may be expected that a reservoir containing a saturated crude will always start out as a gas-drive producing system whether or not it ultimately becomes subject to complete-water-drive control.

From an economic and practical standpoint the water-drive producing mechanism is of significance primarily because it generally provides the highest local efficiency of oil displacement and recovery. It is largely for this reason that the extended discussion in this chapter of the general performance characteristics of water-drive reservoirs, of means of identifying them, and of their reaction to various operating conditions and practices may be justified. In most porous media the oil left after it has been subjected to the expulsion forces of gas evolution and sweeping is considerably greater than that remaining after it has been flushed by water. Control of reservoir exploitation so that it will produce under a water-drive mechanism in contrast to that of solution-gas drive will usually be of great economic importance. On the other hand, this does not imply that the recoveries under complete-water-drive action will universally be greater in actual reservoirs than by any other producing method. Reservoirs initially overlain by gas caps that do not contain an appreciable oil content may suffer serious losses in recovery if water invasion is uncontrolled and permitted to drive the oil into the gas cap. Highly stratified producing formations may develop severe water channeling through the high-permeability zones, which, if not subject to control, may force abandonment of wells and productive acreage after total recoveries even less than might have been derived by gas-drive depletion. Since, in actual practice, all reservoir formations will have some stratification, the recoveries will always be correspondingly lower than would be expected of an ideal homogeneous stratum. Moreover their gross performance histories will represent the resultants of superpositions, with appropriate phase shifts, of the idealized behavior of the individual zones of different permeability, modified by cross-flow interactions as well as their own inherent areal inhomogeneities.

In extreme cases of permeability variations, such as may obtain in extensively fractured or intermediate limestones, water invasion will tend to by-pass and isolate the intergranular oil reserves. The recovery of the latter without excessive water production may require the free operation of the simple solution-gas-drive mechanism, while deliberately preventing, as far as possible, all water encroachment until the intergranular rock has been depleted. Substantially isotropic formations in communication with bottom waters may also give greater gas-drive than water-drive recoveries, owing to the very low sweep efficiencies of the bottom-water invasion pattern. And in formations of high connate-water saturations the residual oil after water flushing may represent as high a fraction of the initial-oil content as would be found under gas-drive expulsion. Finally, where effective gas segregation and gravity drainage can be achieved, the recoveries may compare quite favorably with or even exceed those which would result from complete-water-drive operation. In any case, while maximum oil recoveries by the water-drive mechanism may, in most reservoirs, be considered as an a priori reasonable working hypothesis, each reservoir must be carefully analyzed to make sure that it does not represent a situation where the reverse will actually be true.

CHAPTER 12

SECONDARY RECOVERY

12.1. Introduction.—Secondary-recovery operations are those of injection into a reservoir of gas, air, or water, after it has reached a state of substantially complete depletion of its initial content of energy available for oil expulsion or where the production rates have approached the limits of profitable operation. From a physical point of view, such operations may be considered simply as an extreme form of delayed pressure-maintenance operations. In fact, in contrast to the more common type of so-called "pressure-maintenance operations," which generally result only in pressure-decline retardation, with an incomplete replacement of the space voidage created by the fluid withdrawals, there is usually some build-up of reservoir pressure in secondary-recovery gas or water injection. Especially in the case of water injection ("water flooding") the rate of fluid injection, if conducted on a major scale, generally exceeds the volumetric withdrawals virtually throughout the whole course of the operations. This situation also often obtains in secondary recovery by gas injection, usually termed "gas repressuring."[1]

The basic physical difference between secondary-recovery and pressure-maintenance operations is, as already indicated, that the "initial condition" for the former is a state of virtually complete depletion of the reservoir pressure or the inherent natural oil-expulsion energy, whereas in the latter fluid injection is undertaken during the course of the primary oil-producing processes before such states are reached. Associated with this difference are others that are even more fundamental, though they, too, represent mainly differences in degree rather than kind. These are that (1) the oil saturation as a whole is lower, (2) the oil viscosity is higher, (3) the formation-volume factor of the oil is lower, (4) the surface tension of the oil is higher, (5) the interfacial tension between the oil and water is lower, and (6) initial differences in pressure or saturation distributions related to variations in the nature of the rock are generally accentuated in secondary-recovery as compared with pressure-maintenance operations. And these, too, have still further implications. The lower oil saturations imply that

---

[1] Although this term may not appear well chosen because of the tendency to confuse it with "pressure maintenance," it has had rather wide usage and will be considered here as synonymous with secondary-recovery operations by gas or air injection.

the free-gas saturations are higher.[1] The greater oil viscosity will increase
the loss in oil mobility resulting from the reduced oil saturation. The
lower formation-volume factor of the oil means greater stock-tank equivalent and economic value per unit of decrease in pore saturation by the
injection operations. And the accentuation of differences in saturation
and pressure within different parts or substrata of the reservoir as a whole
will aggravate the inherent tendencies for by-passing or channeling of the
injection fluids.[2]

Secondary-recovery operations offer the advantage that the pressures
required for fluid injection are often lower than those which would be required for injection under pressure maintenance. This is especially true
in the case of gas injection. While, in principle, it would appear that, in
water-flooding formations of moderate or high permeability, the gravity
head of the water column should suffice to give appreciable injection rates,
rather high surface pressures often are actually used in practice to accelerate
the oil production and shorten the life of the operations. Fluid injection
prior to reservoir depletion may not be feasible because the fluids do not
happen to be available at all or are in greater competitive demand for other
uses. The need to maintain all wells on production to meet market demands and the desire to avoid the cost of drilling special injection wells
may also provide a reason for delaying fluid injection until the primary-
production phase has been completed. On the other hand it should be
noted that, however urgent these reasons may be from the immediate
economic standpoint, there is practically no major physical factor making
secondary-recovery operations advantageous as compared with similar
fluid injection before substantially complete reservoir depletion has taken
place.

Purely secondary-recovery operations will undoubtedly become of decreasing practical importance after the currently depleted fields that were
produced by gas or partial water drives and are susceptible to such operations have been so treated. Nevertheless there are many secondary-recovery projects now in progress and many more possibilities for successful
application still to be developed. In fact the magnitude of these possibilities may be visualized on noting[3] that, of the 419,750 wells producing
oil in the United States in 1945, 299,146 were "stripper" wells—those
whose operating expense substantially equals the production income.[4]

---

[1] It is assumed here that there has been no natural water encroachment to replace
the oil voidage in the area subjected to secondary-recovery operations.

[2] With respect to changes in surface forces, hardly enough is known at present to
indicate definitely how significant the resultant effect will be.

[3] These statistics are taken from API "Petroleum Facts and Figures," 1947.

[4] While this definition is not precise, it roughly reflects the approach to well abandonment. Thus whereas the over-all average production rate per well in the United States
was 11.3 bbl/day, the average of the stripper-well production rate was 2.1 bbl/day.

There were more than 20,000 stripper wells in each of the states of Pennsylvania, Oklahoma, Texas, Ohio, and New York. Of the total productive acreage of all wells of 6,719,891, the productive area produced by stripper wells covered 3,104,410 acres. Pertinent, too, is the observation that the average daily production rate per well in the United States as a whole in 1945 was 11.3 bbl, and the average for the states of Pennsylvania and Ohio was 0.4 bbl/day each. While only a small part of the area now operated as stripper production may be susceptible to profitable secondary-recovery operations, that which is suitable undoubtedly contains many millions of potential oil recovery. It is evidently pertinent to review the reservoir-engineering principles involved in this phase of oil production.

Aside from the differences in physical conditions between secondary-recovery and pressure-maintenance operations previously mentioned, there is also a basic difference from a practical standpoint. This pertains to the nature of the well distribution commonly used. Fluid injection for pressure maintenance is generally concentrated beyond the boundaries of the oil-saturated zone so as to induce a gross movement of the gas-oil or water-oil boundary and shrink the volume of the oil-saturated section. Thus, gas-injection wells are usually located in a gas cap or structurally high regions of the reservoir so as to drive the oil downstructure en masse. And water injected during the primary-production phase is usually distributed near the water-oil boundary, or below it within the contiguous water reservoir, so as to achieve a general upstructure movement of the oil similar to that provided by natural water drives. In secondary-recovery operations, however, the injected gas or water is commonly distributed areally throughout the field, so that the individual injection wells are surrounded by producing oil wells as completely as possible. Structural considerations play only a minor role, if any, in determining the injection-well locations in most secondary-recovery projects.

There are a number of reasons for the use of the areal distributions of the fluid-injection wells in secondary-recovery operations. A very practical consideration often forcing such a distribution is the limitation of the part of the field being subjected to gas or water injection by the lease boundaries of the particular operator undertaking the program. In order to ensure maximum effect on the property of the operator and to avoid possible claims of damages from neighboring leaseholders, the injection wells must of necessity be located in the interior of such continuous area as is owned by the operator. Another major reason is that by a reduction in the average distance between the foci of fluid drive and withdrawal the rates of production developed and sustained by the operations will be increased

---

It is also of interest that the amount of oil left in the ground underlying the 9,103 wells abandoned in 1945 has been estimated as 18,627,000 bbl.

and the over-all life of the operations will be shortened. Finally it is usually possible to exercise closer control and to apply corrective measures more readily, if the need for these should develop, by an areal distribution of injection and producing wells than if they were respectively located along the boundaries of the gross oil-productive area.

When a sufficiently large continuous area is involved in the secondary-recovery program, it is now common practice, especially in water flooding,[1] to try to arrange the injection and producing wells in regular patterns, forming a composite interlaced network of the two types of wells. And often different lease owners will undertake cooperative fluid-injection pro-

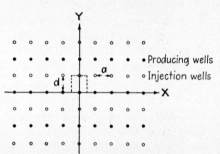

FIG. 12.1. A diagrammatic representation of a direct-line-drive well network. Dashed segment represents basic symmetry element.

grams so as to enlarge the continuous area of operations and to facilitate the development of regular-pattern well distributions. Injection wells may then be placed on the lease boundaries under arrangements for shared or cooperative operation. Historically the first type of regular network that was tried was the alternating-line-drive pattern, as illustrated in Fig. 12.1. Here the injection wells and producing wells are located with regular and equal spacing in parallel alternating lines, the wells of one type lying immediately opposite those of the other. Next followed the "five-spot" pattern, as indicated in Fig. 12.2. This may be visualized as a special type of "staggered"-line-drive network (cf. Fig. 12.7), in which the lines of one type are shifted parallel to themselves by half the spacing between the wells within the lines and the separation of the lines is also half the spacing within the lines. In the five-spot, each well of either type is uniformly surrounded by four of the other. This is the pattern most commonly used at present. Still another pattern that has been used is the "seven-spot," shown in Fig. 12.3. This is comprised of a hexagonal network in which either[2] the injection or the producing wells may be placed at the hexagon centers.

    [1] In gas injection, regular well patterns are used much less frequently because it is not economically feasible to provide as many injection wells as are required for the standard networks (cf. Sec. 12.18).
    [2] While no major distinction will be made here between these two distributions because they have the same basic steady-state network conductivity and sweep efficiency, the pattern with injection wells at the hexagon centers is sometimes referred to as a "four-spot." However, in the treatment of the transient history of water injection these two distributions will show different interference areas and correspondingly different sweep efficiencies (cf. Sec. 12.13).

While the physical principles underlying secondary-recovery operations
are quite well understood, the difficulties of quantitative description and

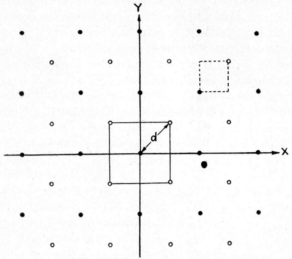

FIG. 12.2. The five-spot well network. Dashed segment represents basic symmetry element.

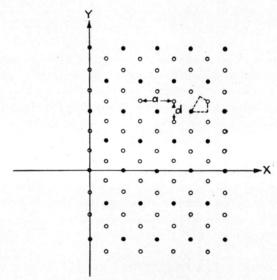

FIG. 12.3. The seven-spot well network. Dashed segment represents basic symmetry element.

prediction are fundamentally the same as in the analysis of the primary-
producing period of oil-bearing reservoirs and, in particular, those producing by the solution-gas-drive mechanism. Only special phases of the
general problem have been given quantitative treatments, and these have

involved a variety of simplifying assumptions. A completely satisfactory theory would evidently have to deal with heterogeneous-fluid systems. It would have to take into account the difference between the nature of the injection fluid and those displaced into the producing wells. It would have to be basically a nonsteady-state analysis, in view of the changing fluid distributions as the operations continue. Finally it would have to provide for the unbalance between the total fluid-injection and withdrawal rates, especially in water-flooding systems.

It is clear from these considerations that no simple homogeneous-fluid steady-state treatment will suffice to describe quantitatively the dynamics of actual reservoirs subjected to secondary-recovery operations. Nevertheless such an idealized theory of the well networks often used in secondary-recovery projects does serve to show their unique geometrical characteristics that may come into play in certain aspects of the operations. Even in water-flooding systems the injection-well transient, which dominates the period of fill-up of the void space (cf. Sec. 12.13), ultimately tends to approach the steady-state-flow conditions as the result of well interference. And in gas-injection operations, where the oil is essentially "swept" along with the gas, the resistance characteristics of the flow systems may be approximated by sequences of steady-state values, as the oil saturation is reduced. An outline of the steady-state network analysis will therefore be presented in the next several sections as a reference for such applications as may be made in the study of these special phases of secondary-recovery operations and as instructive illustrations of analytical procedures useful in the treatment of other reservoir flow problems.[1] However, since this theory is limited in its quantitative applicability, no distinction will be made between systems where the injection fluid is a gas and those where it is water, although it is recognized that the detailed mechanics of the oil-displacement processes are radically different in the two cases.

12.2. The Steady-state Flow Capacity of Direct-line-drive Networks.— The basic element for the homogeneous-fluid steady-state treatment[2] of well networks is the pressure distribution due to an infinite single-line array of wells, of spacing a, placed parallel to and at a distance b from the x axis, namely,

$$p(x,y) = \frac{Q\mu\beta}{4\pi kh} \log \left[\cosh \frac{2\pi(y-b)}{a} - \cos \frac{2\pi x}{a} \right], \qquad (1)$$

where Q is the production rate of each well, as measured at the surface, μ

---

[1] Cf., for example, condensate-reservoir cycling patterns (Sec. 13.5).

[2] This analysis, including the discussion through Sec. 12.9, follows that of M. Muskat and R. D. Wyckoff, *AIME Trans.*, **107**, 62 (1933).

the fluid[1] viscosity, β the formation-volume factor of the oil,[2] k the effective permeability of the fluid, and h the effective thickness of the permeable section. It will be readily verified that Eq. (1) is a solution of the potential equation for the steady-state homogeneous-fluid-flow pressure distribution [Eq. 4.4(4)]. As would be expected, it gives a pressure distribution periodic in x, with a periodicity a. It is symmetrical about $y = b$, the axis of the line array. But on receding from this line it rapidly approaches an asymptotic linear variation at distances equal to the well spacing, namely,

$$p = \frac{Q\mu\beta}{4\pi kh}\left(\frac{2\pi|y - b|}{a} - \log 2\right), \qquad (2)$$

thus smoothing out the "ripples" due to the individual wells and having a pressure distribution characteristic of a line drive of continuous and constant flux density. The isobaric contours [constant values c of the brackets in Eq. (1)] plotted in Fig. 12.4 illustrate these features, for the case $b = 0$, and in coordinate units x/a, y/a.

Since Eq. (1) does not involve directly the well radius or well pressure, it actually represents the distribution due to an array of mathematical sinks of infinitesimal radii and negative infinite pressures. However, for practical application the equivalent well pressure and radii may be introduced by simply noting that at a radial distance r_w, small compared with a, from the actual locations of the sinks the pressure p_w will be

$$p_w = \frac{Q\mu\beta}{4\pi kh}\log 2\sinh^2\frac{\pi r_w}{a}. \qquad (3)$$

To obtain the pressure distributions due to an areal line-drive well network as shown in Fig. 12.1, it need be noted only that the latter is comprised of a series of line arrays of producing wells (positive Q) with $b = 2nd$ and of a superposed but shifted series of arrays of injection wells (negative Q) with $b = (2n + 1)d$, n in each series representing any and all of the positive and negative integers. Adding the contributions of these individual arrays, and making provision for convergence requirements, one readily obtains

---

[1] It is tacitly assumed here that the injection fluid and the whole of the fluid system within the network is effectively incompressible, as water. However, under the other broad assumptions underlying the treatment, the formal analysis will also apply for gas injection provided that p is replaced by p^2 and Δp by Δp^2 and that Q is the mass flux per unit density.

[2] In most secondary-recovery operations β will equal simply the thermal expansion from ambient to reservoir temperature, if the initial solution gas has been substantially depleted.

$$p(x,y) = \frac{Q\mu\beta}{4\pi kh} \left\{ \log \left(\cosh \frac{2\pi y}{a} - \cos \frac{2\pi x}{a} \right) \right.$$

$$\left. + \sum_{1}^{\infty} (-1)^m \log \frac{4 \left[\cosh \frac{2\pi}{a}(y-md) - \cos \frac{2\pi x}{a} \right] \left[\cosh \frac{2\pi}{a}(y+md) - \cos \frac{2\pi x}{a} \right]}{e^{4\pi md/a}} \right\}, \quad (4)$$

where the injection rate of each individual injection well is taken the same

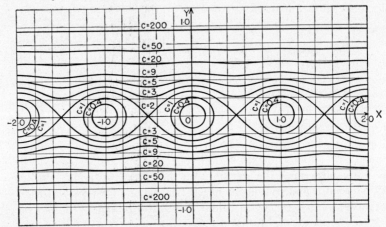

FIG. 12.4. The calculated steady-state homogeneous-fluid equipressure contours (constant values of c) about an infinite linear array of wells with unit spacing lying on the X axis; pressure $=$ const $\times$ log c.

as the production rate of each producing well, Q. In Eq. (4), Q has been taken as numerically positive for both producing and injection wells, and -1 coefficients have been applied to the pressure-distribution terms of the latter.

Equation (4) implies a common producing well pressure p_{w+} given by

$$p_{w+} = p(y^2 + x^2 = r_w^2) \cong p(x = 0, y = \pm r_w)$$

$$= \frac{Q\mu\beta}{4\pi kh} \left(\log 2 \sinh^2 \frac{\pi r_w}{a} + \sum_{1}^{m} (-1)^m \log \frac{16 \sinh^4 \pi m/a}{e^{4\pi md/a}} \right). \quad (5)$$

It will be noted that Eq. (5) is the same as Eq. (3), except for the series representing the contribution of the other line arrays of the composite network.

At the injection wells the well pressures p_{w-}, implied by Eq. (4), are

$$p_{w-} = p[(y-d)^2 + x^2 = r_w^2] \cong p(x = 0, y = d \pm r_w)$$

$$= \frac{Q\mu\beta}{4\pi kh} \left[\log 2 \sinh^2 \frac{\pi d}{a} - \frac{\log 16 \sinh^2 \pi r_w/a \ \sinh^2 2\pi d/a}{e^{4\pi d/a}} \right.$$

$$\left. + \sum_{2}^{\infty} (-1)^m \log \frac{16 \sinh^2 (m-1)\pi d/a \ \sinh^2 (m+1)\pi d/a}{e^{4\pi md/a}} \right]. \quad (6)$$

Denoting the net pressure differential $p_{w-} - p_{w+}$ by Δp, Eqs. (6) and (5) lead to

$$Q = \frac{2\pi k h\ \Delta p/\mu\beta}{\log\left[(\sinh^4 \pi d/a \sinh 3\pi d/a)/(\sinh^2 \pi r_w/a \sinh^3 2\pi d/a)\right]}, \quad (7)$$

where terms have been dropped that become negligible when $d/a > \frac{1}{4}$,

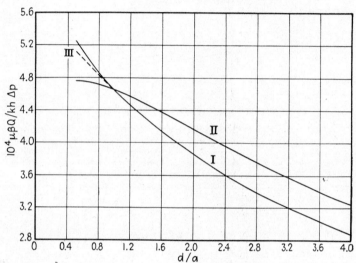

FIG. 12.5. The calculated steady-state homogeneous-fluid production capacities of line-drive well networks. μ = liquid viscosity (cp); β = formation-volume factor; k = permeability (md); h = formation thickness (ft); Δp = pressure differential (psi); Q = production or injection rate per well (bbl/day); d/a = (distance between injection and producing lines)/ (distance between wells within lines). Curve I, direct drive, a fixed at 660 ft. Curve II, direct drive, d fixed at 660 ft. Curve III, staggered drive, a fixed at 660 ft.

as will generally obtain in practice. On the other hand, when $d/a \geqslant 1$, the further simplification can be made to the form

$$Q = \frac{2\pi k h\ \Delta p/\mu\beta}{\pi d/a - 2\log 2 \sinh \pi r_w/a} = \frac{0.002254 k h\ \Delta p/\mu\beta}{\dfrac{d}{a} - 1.17 + \dfrac{2}{\pi}\log\dfrac{a}{r_w}}\ \text{bbl/day}, \quad (8)$$

where k is expressed in millidarcys, h in feet, Δp in psi, and μ in centipoises. This is seen to be the same as to be expected of a line drive between *continuous* line sources and sinks of separation $d + (2a/\pi)\log a/2\pi r_w$. The excess of the latter over d is a measure of the increased effective flow resistance between the actual well arrays, arising from the fact that the fluid must leave and enter the system from separated and individual well centers, rather than continuously distributed fluid line sinks and sources. Equations (7) and (8) are plotted as curves I and II in Fig. 12.5 as functions of d/a, with a fixed at 660 ft and d fixed at 660 ft, respectively. For

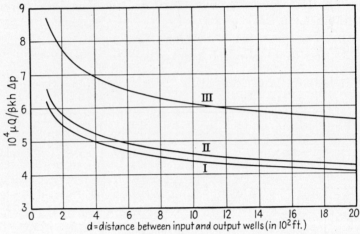

Fig. 12.6. The variation of the calculated steady-state homogeneous-fluid production capacities of various well networks with the distance d between the injection and producing wells. Curve I, direct line drive $(d = a)$. Curve II, five-spot network. Curve III, seven-spot network. Ordinates are relative production capacities of the producing wells, in the units of Fig. 12.5.

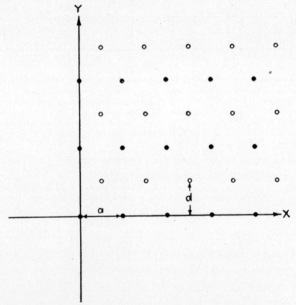

Fig. 12.7. A diagrammatic representation of the staggered-line-drive network.

the simple square network $(d = a)$, Eq. (8) is plotted vs. d as curve I in Fig. 12.6.

12.3. The Steady-state Flow Capacity of Staggered-line-drive Patterns.—The general staggered-line-drive network is shown diagrammatically in Fig. 12.7. By analogy with the method of analysis outlined in the preceding section, it may be verified that the pressure distribution in the staggered-line-drive pattern can be expressed by

$$p(x,y) = \frac{Q\mu\beta}{4\pi kh}\left\{\log\left(\cosh\frac{2\pi y}{a} - \cos\frac{2\pi x}{a}\right)\right.$$

$$+ \sum_1^\infty \log \frac{4\left[\cosh\frac{2\pi(y-2md)}{a} - \cos\frac{2\pi x}{a}\right]\left[\cosh\frac{2\pi(y+2md)}{a} - \cos\frac{2\pi x}{a}\right]}{e^{8\pi md/a}}$$

$$\left. - \sum_0^\infty \log \frac{4\left[\cosh\frac{2\pi(y-2md-d)}{a} - \cos\frac{2\pi x}{a}\right]\left[\cosh\frac{2\pi(y+2md+d)}{a} + \cos\frac{2\pi x}{a}\right]}{e^{4\pi d(2m+1)/a}}\right\} \cdot \quad (1)$$

Evaluating the pressure differential between the injection and producing wells, one finds[1]

$$\Delta p = p\left(\frac{a}{2}, d \pm r_w\right) - p(0, \pm r_w)$$

$$= \frac{Q\mu\beta}{2\pi kh}\left[\log\frac{\cosh^4 \pi d/a \cosh^3 3\pi d/a}{\sinh^2 \pi r_w/a \sinh^4 2\pi d/a \sinh 4\pi d/a}\right.$$

$$\left. + \sum_2^\infty \log\frac{\cosh (2m-1)\pi d/a \cosh^3 (2m+1)\pi d/a}{\sinh^3 2m\pi d/a \sinh 2(m+1)\pi d/a}\right]. \quad (2)$$

For $d/a > \frac{1}{2}$ the series term is negligible, and Eq. (2) may be rewritten as

$$Q = \frac{2\pi kh\ \Delta p/\mu\beta}{\log\dfrac{\cosh^4 \pi d/a \cosh^3 3\pi d/a}{\sinh^2 \pi r_w/a \sinh^4 2\pi d/a \sinh 4\pi d/a}}$$

$$= \frac{0.007082kh\ \Delta p/\mu\beta}{\log\dfrac{\cosh^4 \pi d/a \cosh^3 3\pi d/a}{\sinh^2 \pi r_w/a \sinh^4 2\pi d/a \sinh 4\pi d/a}}\ \text{bbl/day}, \quad (3)$$

in units of millidarcys, feet, psi, and centipoises for k, h, Δp, and μ.

It is of interest to note that, for $d/a \geqslant 1$, Eq. (3) reduces to Eq. 12.2(8). This shows that when the line spacing equals or exceeds the well spacing within the lines the staggering of the latter has no significant effect on the fluid resistance of the network. Equation (3) is plotted vs. d/a as curve III in Fig. 12.5.

12.4. The Steady-state Five-spot-pattern Flow Capacity.—While the five-spot network, shown in Fig. 12.2, can be treated directly as a suitable

---

[1] It is assumed throughout these analytical developments that $r_w \ll a$, and corresponding simplifications are made at various stages of the analysis.

superposition of individual line arrays, it is simpler to apply the observation that the five-spot is actually nothing more than a special case of the general staggered line drive, namely, that for $d/a = \frac{1}{2}$. Upon introducing this value into Eq. 12.3(3) and making the appropriate simplifications, it is readily found that for the five-spot

$$Q = \frac{\pi k h \, \Delta p/\mu\beta}{\log d/r_w - 0.6190} = \frac{0.003541 k h \, \Delta p/\mu\beta}{\log d/r_w - 0.6190} \text{ bbl/day,} \qquad (1)$$

in the units of Eq. 12.3(3). Equation (1) is plotted as a function of d as curve II in Fig. 12.6.

12.5. The Steady-state Flow Capacity of the Seven-spot Pattern.— Whereas in the previously discussed networks the number of injection wells equals the number of producing wells, the ratio may be 1:2 or 2:1 in the seven-spot pattern. However, the flow resistance is the same for the two types. And in both cases the complexity of the system of line arrays makes necessary a separate listing of the various groups of wells of which the composite network is comprised. For the pattern shown in Fig. 12.3 these are: producing wells at $(na, 3md)$, $[(n + \frac{1}{2})a, (3m + \frac{3}{2})d]$; injection wells at $[na, (3m + 1)d]$, $[na, (3m + 2)d]$, $[(n + \frac{1}{2})a, (3m + \frac{1}{2})d]$, $[(n + \frac{1}{2})a, (3m + \frac{5}{2})d]$.

Upon noting that the flux per well must be twice as great for the producing wells (in the case of Fig. 12.3) as for the injection wells, the pressure function resulting from the superposition of the above well groups may be shown to be

$$
\begin{aligned}
p = \frac{Q\mu\beta}{4\pi k h} \Bigg\{ &\log\left(\cosh\frac{2\pi}{a}y - \cos\frac{2\pi x}{a}\right) \\
&+ \sum_{1}^{\infty} \log \frac{4\left[\cosh\dfrac{2\pi}{a}(y - 3md) - \cos\dfrac{2\pi x}{a}\right]\left[\cosh\dfrac{2\pi}{a}(y + 3md) - \cos\dfrac{2\pi x}{a}\right]}{e^{12\pi md/a}} \\
&+ \sum_{0}^{\infty} \log \frac{4\left[\cosh\dfrac{2\pi}{a}\left(y - \dfrac{3d}{2} - 3md\right) + \cos\dfrac{2\pi x}{a}\right]\left[\cosh\dfrac{2\pi}{a}\left(y + \dfrac{3d}{2} + 3md\right) + \cos\dfrac{2\pi x}{a}\right]}{e^{4\pi d(3m + \frac{3}{2})/a}} \\
&- \frac{1}{2}\sum_{0}^{\infty} \log \frac{4\left[\cosh\dfrac{2\pi}{a}(y - d - 3md) - \cos\dfrac{2\pi x}{a}\right]\left[\cosh\dfrac{2\pi}{a}(y + d + 3md) - \cos\dfrac{2\pi x}{a}\right]}{e^{4\pi d(3m + 1)/a}} \\
&- \frac{1}{2}\sum_{0}^{\infty} \log \frac{4\left[\cosh\dfrac{2\pi}{a}(y - 2d - 3md) - \cos\dfrac{2\pi x}{a}\right]\left[\cosh\dfrac{2\pi}{a}(y + 2d + 3md) - \cos\dfrac{2\pi x}{a}\right]}{e^{4\pi d(3m + 2)/a}}
\end{aligned}
$$

$$-\frac{1}{2}\sum_{0}^{\infty}\log \frac{4\left[\cosh\frac{2\pi}{a}\left(y-\frac{d}{2}-3md\right)+\cos\frac{2\pi x}{a}\right]\left[\cosh\frac{2\pi}{a}\left(y+\frac{d}{2}+3md\right)+\cos\frac{2\pi x}{a}\right]}{e^{4\pi d(3m+\frac{1}{2})/a}}$$

$$\left.-\frac{1}{2}\sum_{0}^{\infty}\log \frac{4\left[\cosh\frac{2\pi}{a}\left(y-\frac{5d}{2}-3md\right)+\cos\frac{2\pi x}{a}\right]\left[\cosh\frac{2\pi}{a}\left(y+\frac{5d}{2}+3md\right)+\cos\frac{2\pi x}{a}\right]}{e^{4\pi d(3m+\frac{5}{2})/a}}\right\}\cdot(1)$$

By evaluating the pressure differential between the producing and injection wells, Δp, and making the appropriate reductions, as was done in the previous sections, it may be shown that

$$Q = \frac{4\pi kh\,\Delta p/\mu\beta}{3\log d/r_w - 1.7073} = \frac{0.004721kh\,\Delta p/\mu\beta}{\log d/r_w - 0.5691}\text{ bbl/day.}\qquad(2)$$

Equation (2) is plotted as a function of d as curve III in Fig. 12.6.

12.6. The Steady-state Sweep Efficiencies of Well Networks[1]; The Direct Line Drive.—In addition to their flow resistances, steady-state homogeneous-fluid well networks are characterized by "sweep efficiencies." These represent the fractions of the total area of the pattern or network element that are invaded by the injection fluid by the time it first reaches the producing wells. They are the two-dimensional analogues of the three-dimensional sweep efficiency described in Sec. 11.15 with respect to bottom-water-drive performance. They are always less than unity. The deficiency from unity arises from the differences in velocities within the network elements of the various paths that the injection fluid may take in traveling from the injection to producing wells. These differences, in turn, are related to the differences in pressure gradients along the various paths (streamlines) and their total lengths. In regular networks the shortest of these lies generally along the straight line directly connecting an injection well to its nearest neighboring producing well. The injection fluid thus reaches the producing wells most quickly along these paths. By the time these paths have been traversed the fluid following longer travel-time streamlines will have reached to various distances from the producing

---

[1] The discussion of Secs. 12.6 to 12.11 is presented mainly to complete the description of the purely geometrical aspects of regular well networks. While the steady-state theory gives an instructive picture of the tendencies for differential fluid movement over the different parts of a fluid-injection system and does have applicability to cycling operations in condensate-producing reservoirs (cf. Chap. 13), it will hardly represent even a semiquantitative treatment of the motion of fluids injected for secondary recovery. For this reason no detailed discussion will be given here of nine-spot floods, although these, too, have been given complete analytical treatments [cf. H. Krutter, *Oil and Gas Jour.*, **38**, 50 (Aug. 17, 1939), and M. Muskat, *Producers Monthly*, **12**, 14 (March, 1948)].

wells, leaving the intervening regions unswept. The area of this unswept part of the network or the fractional area actually swept out is measured by the sweep efficiency.[1]

When the pressure distribution within the well network is known, in analytical, graphical, or numerical form, and the position of the shortest travel-time streamline is also known, the sweep efficiency can be calculated without difficulty. Thus, if ds denotes an element of length along this streamline, the time required for the injection fluid to first reach the producing well will be

$$t = \frac{\mu \bar{f}}{k} \int_0^{s_t} \frac{ds}{\left|\frac{\partial p}{\partial s}\right|},$$ (1)

where s_t is the total path length between the injection and producing wells and $\bar{f}$ is the effective displacement porosity, denoting the fraction of the local bulk volume of the rock that is occupied by the injection fluid. The reservoir volume of the injected fluid at time t will be Qt per well, and it will occupy an area $Qt/h\bar{f}$. If the network-element area associated with each injection well be A, the sweep efficiency E will therefore be

$$E = \frac{\mu Q}{khA} \int_0^{s_t} \frac{ds}{\left|\frac{\partial p}{\partial s}\right|}.$$ (2)

On applying this procedure to the direct-line-drive pattern, it is noted first that the shortest travel-time streamline is that along the y axis from 0 to d (cf. Fig. 12.1). Hence

$$\frac{\partial p}{\partial s} = \left(\frac{\partial p}{\partial y}\right)_{x=0}$$

$$= \frac{Q\mu\beta}{2akh}\left\{\coth\frac{\pi y}{a} + \sum_1^\infty (-1)^m\left[\coth\frac{\pi(y-md)}{a} + \coth\frac{\pi(y+md)}{a}\right]\right\}.$$ (3)

By simplifying Eq. (3) by the observation that the neglect of the series terms beyond the first will lead to maximum errors of 0.4 per cent for

---

[1] This has often been referred to as "flooding" efficiency, as it has been given more widespread consideration in water-flooding operations. Although no explicit distinction will be made between water- and gas-injection systems in the analytical treatment of the sweep efficiencies of steady-state networks, the concept of a geometrical sweep efficiency has very little applicability to secondary-recovery gas-injection operations, where the displaced fluid—the oil—is "dragged" along by the parallel throughflow of the injected gas. In fact the steady-state sweep efficiencies are only of slight practical significance even for water-flooding operations because of the inherently transient character of the oil-displacement mechanism (cf. Sec. 12.12).

$d/a \geqslant \frac{1}{2}$, Eq. (2) can be integrated to give the result

$$E = \frac{1}{\pi(\cosh^2 \pi d/a - 2)(d/a)} \left(\cosh^2 \frac{\pi d}{a} \log \cosh \frac{\pi d}{a} - 0.6932 \sinh^2 \frac{\pi d}{a} \right). \quad (4)$$

For $d/a \geqslant 1.5$, this reduces to

$$E = 1 - 0.441 \frac{a}{d}. \quad (5)$$

The value of E given by Eq. (4) is plotted vs. d/a as curve I in Fig. 12.8.

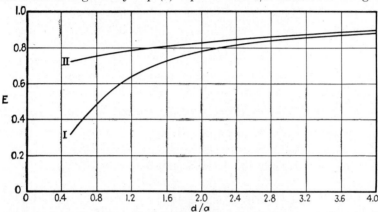

Fig. 12.8. The variation of the calculated steady-state homogeneous-fluid sweep efficiencies E of line-drive networks with $d/a =$ (distance between the injection and producing lines) /(well spacing within the lines). I, direct line drive. II, staggered line drive. (*After Muskat and Wyckoff, AIME Trans., 1934.*)

12.7. The Steady-state Sweep Efficiency of Staggered Line Drives.—

The treatment of general staggered-line-drive networks is complicated by the fact that the shortest travel-time streamline—the direct injection-producing-well connecting line—does not lie along one of the axes of the coordinate system which is most convenient for the construction of the pressure distribution (cf. Fig. 12.7). The pressure gradient along this streamline will therefore have the form

$$\frac{\partial p}{\partial s} = \frac{1}{\sqrt{1 + 4\bar{d}^2}} \left(\frac{\partial p}{\partial x} + 2\bar{d} \frac{\partial p}{\partial y} \right); \qquad \bar{d} = \frac{d}{a}. \quad (1)$$

The sweep efficiency may therefore be expressed by the integral

$$E = \frac{\mu\beta Q(1 + 4\bar{d}^2)}{2khad} \int_0^{a/2} \frac{dx}{\dfrac{\partial p}{\partial x} + 2\bar{d} \dfrac{\partial p}{\partial y}}. \quad (2)$$

A graphical evaluation of Eq. (2), on applying Eq. 12.3(1) to obtain the pressure gradients, for different values of $\bar{d}$, leads to the curve II plotted in Fig. 12.8.

The special case $(\bar{d} = \frac{1}{2})$ of the staggered line drive, which corresponds to the five-spot pattern, is given by Fig. 12.8 or by a direct analytical derivation[1] as

$$E(\bar{d} = \tfrac{1}{2}) = 0.715. \tag{3}$$

12.8. The Steady-state Sweep Efficiency of Seven-spot Networks.— For the seven-spot pattern a typical streamline of least travel time lies along the y axis, as may be seen from Fig. 12.3. On applying the pressure-distribution function of Eq. 12.5(1) and evaluating graphically the corresponding integral of Eq. 12.6(2), it is found that the efficiency has the value

$$E = 0.740. \tag{1}$$

Here, as in the other patterns, the minimum travel times, under steady-state conditions, or times for break-through of the injection fluid, can be calculated from the sweep efficiency by the equation

$$t = \frac{V\bar{f}E}{Q\beta}, \tag{2}$$

where V is the volume of the network unit associated with each injection well.

For direct line drives with $d/a \geqslant 1.5$ the time for injection-fluid break-through, corresponding to the steady-state sweep efficiency, is therefore

$$t_d = \frac{2da\bar{f}}{Q\beta}\left(1 - 0.441\,\frac{a}{d}\right). \tag{3}$$

For the five-spot pattern the injection-fluid break-through time is

$$t_5 = \frac{1.430d^2\bar{f}}{Q\beta}. \tag{4}$$

The seven-spot network time for injection-fluid break-through is

$$t_7 = \frac{1.922d^2\bar{f}}{Q\beta}, \tag{5}$$

where Q is the producing rate per well, in surface measure, and here, as well as in Eqs. (3) and (4), d is the distance between the injection and producing wells. It will be noted that in all cases the time of break-through is proportional to a square of the dimension of the network element. The volumes of fluid injection per network element required for developing break-through are evidently the products of the producing rates per well and the break-through times. On the other hand the steady-

[1] This is the value derived recently as a special case of a treatment of "nine-spot" flooding networks (cf. Muskat, *loc. cit.*), although earlier work indicated a value of 0.723.

state break-through times, as given by Eqs. (3) to (5), are of little significance in secondary-recovery water-injection operations because of the inherent transient character of a major part of their operating life (cf. Secs. 12.13 and 12.14).

12.9. The Steady-state Pressure Distributions in Well Networks.—
The numerical values of the steady-state flow capacities and sweep efficien-

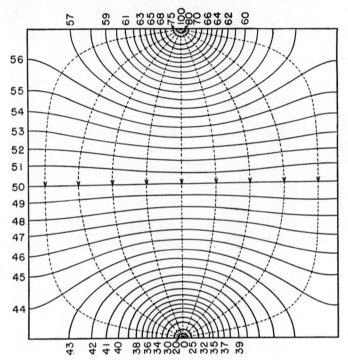

Fig. 12.9.   The steady-state homogeneous-fluid equipressure contours and streamlines in a two-well element of a direct-line-drive network.   Numbers represent percentages of the total pressure drop.   (*After Muskat and Wyckoff, AIME Trans., 1934.*)

cies given in the last several sections for the various well networks are evidently the results of the unique pressure distributions within these patterns.   These distributions could be calculated from the analytical expressions constructed for the evaluation of the flow capacities and sweep efficiencies.   However, it is simpler to determine the distributions empirically by measurements of the electrical equipotential curves on a sheet conduction model having the same geometry as the network element of the flow system.   Such models are most conveniently made of a uniform metal plate[1] or may be of an electrolytic nature.   In either case, since the flow

---

[1] For details of the measurements, see M. Muskat, "The Flow of Homogeneous Fluids through Porous Media," Sec. 9.21, McGraw-Hill Book Company, Inc., 1937.

of electric current and the voltage distribution are governed by Laplace's equation, as are the fluid velocities and pressures in steady-state homogene-

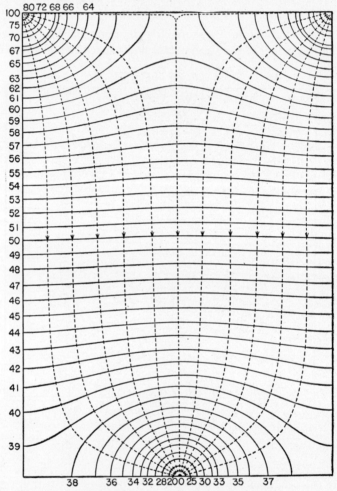

Fig. 12.10. The steady-state homogeneous-fluid equipressure contours and streamlines in a two-well element of a staggered-line-drive network. Numbers represent percentages of the total pressure drop. (*After Muskat and Wyckoff, AIME Trans., 1934.*)

ous flow in porous media, there will be a one-to-one correspondence between the two types of system.

The pressure distribution obtained by the sheet conduction model for a simple direct-line-drive network is plotted in Fig. 12.9.[1] For simplicity, only a single element of symmetry was used for the model, as this is repre-

[1] Figs. 12.9 to 12.12 are taken from Muskat and Wyckoff, *loc. cit.*

sentative of all similar two-well units in the complete direct-line-drive pattern. As the steady-state pressure distribution is independent of the absolute value of the pressure differential, the equipressure lines are numbered in percentages of the total pressure differential between the injection and producing wells. In Fig. 12.9 are also shown, as dashed curves, the positions of typical streamlines, giving the paths of fluid motion between

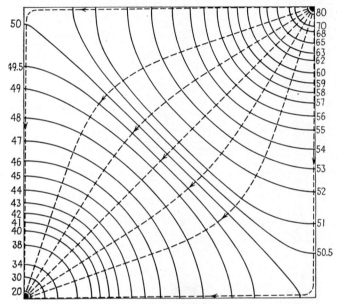

Fig. 12.11. The steady-state homogeneous-fluid equipressure contours and streamlines in a quadrant of a five-spot-network element. Numbers represent percentages of the total pressure drop. (*After Muskat and Wyckoff, AIME Trans., 1934.*)

the injection and producing wells. These were drawn in as orthogonal trajectories with respect to the equipressure curves.

For a staggered-line-drive network, with $d/a = 1.5$, the equipressure and streamline curves, as determined in the same way as those of Fig. 12.9, are shown in Fig. 12.10. The corresponding curves for a representative five-spot-pattern element are reproduced in Fig. 12.11. And those for a segment of symmetry in the seven-spot network are plotted in Fig. 12.12.

It will be noted in all the pressure-distribution plots of Figs. 12.9 to 12.12 that near both the injection and producing wells the equipressure curves are approximately circular. This implies that, as is to be expected, the flow is essentially radial near the wells. The pressure gradients are greatest near the wells in the regions of radial flow. They are least near parts of the boundaries of the symmetry elements where the streamlines change their direction sharply. If there is a one-to-one relation between the number of input and output wells, the nature of the pressure distribution is the same about each type. However, if there is less of one kind than the other, as in the case of the seven-spot (cf. Fig. 12.12), the pressure gradients are greater and more concentrated about the type of wells of smaller number.

While no further discussion of the pressure distribution in well networks

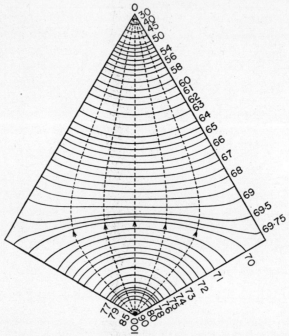

Fig. 12.12. The steady-state homogeneous-fluid equipressure contours and streamlines in a segment of a seven-spot-network element. Numbers represent percentages of the total pressure drop. (*After Muskat and Wyckoff, AIME Trans., 1934.*)

will be given here, it should be observed that the electrical models (sheet conduction or electrolytic) can be applied equally well for irregular well patterns for which analytical treatments would be impractical. Moreover they can be used for systems in which there may be differences in injection or producing rates among the various wells used for injection or production. The equivalent fluid capacity of the flow system can also be determined by suitable measurements of the electrical resistance of the model. On the other hand, when the well system is not of a regular geometry or the wells do not have equivalent rates of production or injection, it will in general be impossible to find in advance simple units of symmetry that are representative of the system as a whole. It will then be necessary to include all the wells in the same model or break it up into several models of smaller groups with overlapping elements for superposition into the composite pattern. Of course, due attention must then be given to the absolute voltages and current flows associated with the individual wells.

12.10. The Shape of the Injection-fluid Front: Steady-state Theory.— The sweep efficiencies discussed in Secs. 12.6 to 12.8 give the resultant areal coverage of the injection fluid by the time it reaches the producing wells, under steady-state conditions. To understand the physical basis of these sweep efficiencies it is instructive to examine the detailed shape of the advancing fluid front during its travel to the output wells. A knowledge of the shape of the injection-fluid front will also indicate the location of the unswept parts of the system, as these may be affected by the well pattern or the relative rates of injection and production among the wells in the pattern. While only the very simplest systems can be completely treated analytically, it is of interest to formulate the analytical problem as a preliminary to the discussion of the empirical procedures required for the treatment of the more complex well distributions of the types used in practice.

The tracing of the advancing injection-fluid front in steady-state flow may be considered as equivalent to that of determining the history of a line of fluid particles in a potential field. If the locus of fluid particles at any time t representing the front of the injection-fluid area[1] be denoted by $F(x,y,z,t)$ and the fluid velocity potential is $\Phi(x,y,z)$, then it can be shown[2] that F must satisfy the differential equation

$$\frac{\partial F}{\partial t} - \nabla \Phi \cdot \nabla F = 0. \tag{1}$$

If the boundary pressures are kept fixed and a set of orthogonal curvilinear coordinates $u(x,y,z)$, $v(x,y,z)$, $w(x,y,z)$ is introduced so that $u = $ const, for example, coincides with the equipotential surfaces, then

$$\nabla \Phi \cdot \nabla F = \frac{\partial F}{\partial u} |\nabla u|^2,$$

so that Eq. (1) becomes

$$\frac{\partial F}{\partial t} - |\nabla u|^2 \frac{\partial F}{\partial u} = 0, \tag{2}$$

where $|\nabla u|^2$ is to be expressed as a function of u, v, w. Equation (2) has the integral

$$F = t + \int \frac{du}{|\nabla u|^2} + g(v,w) = \text{const}, \tag{3}$$

[1] It is assumed here, as in the previous discussion, that there is no dynamical distinction between the region swept out by the injection fluid and that still unswept.

[2] M. Muskat, *Jour. Applied Physics*, **5**, 250 (1934).

where g is an arbitrary function to be chosen so that F assumes its initial preassigned form at $t = 0$. For two-dimensional systems, in which the stream functions are represented by $\Psi = $ const, Eq. (3) may be rewritten as

$$F(\Phi,\Psi) = t + \int \frac{d\Phi}{|\nabla \Phi|^2} + g(\Psi) = \text{const}, \tag{4}$$

in which the integral is to be evaluated along a streamline ($\Psi = $ const).

To illustrate the application of Eq. (4) it may be noted that for the flow into a single well at a distance d from a line drive lying on the x axis the potential and stream functions may be expressed as

$$\left.\begin{aligned} \Phi &= \frac{Q}{4\pi\bar{f}} \log \frac{x^2 + (y - d)^2}{x^2 + (y + d)^2} + \Phi_o, \\ \Psi &= \frac{Q}{2\pi\bar{f}} \tan^{-1} \frac{- 2dx}{x^2 + y^2 - d^2}, \end{aligned}\right\} \tag{5}$$

where Φ_o is the potential at the line drive, Q is the flux (in reservoir volume) per unit thickness into the well, and $\bar{f}$ is the net porosity occupied by the injection fluid. The denominator in the integral of Eq. (4) may then be shown to be

$$|\nabla \Phi|^2 = \frac{Q^2}{4\pi^2 \, d^2\bar{f}^2} (\cosh \eta + \cos \xi)^2,$$

where

$$\eta = \frac{2\pi\bar{f}(\Phi_o - \Phi)}{Q}; \qquad \xi = \frac{2\pi\bar{f}\Psi}{Q}. \tag{6}$$

The shape at any time t of the line of particles leaving the line drive at $t = 0$ is then given by

$$t = \frac{2\pi\bar{f}d^2}{Q \sin^2 \xi} \left[\frac{\sinh \eta}{\cos \xi + \cosh \eta} - 2 \cot \xi \tan^{-1} \tan \frac{\xi}{2} \tanh \frac{\eta}{2} \right]. \tag{7}$$

Equation (7) is plotted, together with the equipotential and streamline curves, in Fig. 12.13, where d is taken as the unit of length, $Q/\bar{f}$ as 2π, and Φ_o as 5. In these units the line of fluid particles will enter the well first at $t = \frac{1}{3}$ and in the original system at

$$t = \frac{2\pi\bar{f}d^2}{3Q}. \tag{8}$$

The total area swept out in this time t is therefore

$$A = \frac{Qt}{\bar{f}} = \frac{2\pi d^2}{3}, \tag{9}$$

or two-thirds the area of a circle of radius d.

In Fig. 12.13 will be seen the cusp formed as the injection fluid first enters the well. This is characteristic of all fluid-front contours as they approach producing wells and arises from the high pressure gradients near the wells, as shown in Figs. 12.9 to 12.12. The spreading of the cusp as production continues and its ultimate reversal are also to be noted.

As the case of an injection- and producing-well pair, lying along the

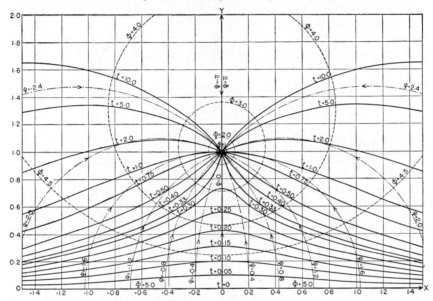

FIG. 12.13.   The calculated shapes of the injection-fluid fronts (t = const) in the steady-state homogeneous-fluid flow from a line drive ($t = 0$, $\Phi = 5$) to a producing well at a unit distance from the injection line. Φ = const: equipotential curves; Ψ = const: streamlines. (*From Physics, 1934.*)

y axis at $(0, \pm d)$, has potential and streamline distributions identical with Eqs. (5), the corresponding constant-time curves can be written down at once, by analogy with Eq. (7), as

$$ t = \frac{2\pi \bar{f} d^2}{Q \sin^2 \xi} \left(\frac{\sinh \eta}{\cosh \eta + \cos \xi} - \frac{\sinh \eta_o}{\cosh \eta_o + \cos \xi} - 2 \, \mathrm{ctn} \, \xi \, \tan^{-1} \tan \frac{\xi}{2} \tanh \frac{\eta}{2} \right. $$
$$ \left. + 2 \cot \xi \, \tan^{-1} \tan \frac{\xi}{2} \tanh \frac{\eta_o}{2} \right), \quad (10) $$

where $\eta = \eta_o$ is the equipotential defining the injection well at $(0, -d)$. The corresponding injection-fluid-front contours (t = const) and equipotential and streamline curves are plotted in Fig. 12.14, with the same units as for Fig. 12.13. Here the injection fluid does not reach the output well until

$$ t = \frac{4}{3} \frac{\pi \bar{f} d^2}{Q}, \quad (11) $$

which is twice the value of Eq. (8). And the total area swept out in that time is

$$A = \frac{Qt}{f} = \frac{4}{3}\pi d^2, \tag{12}$$

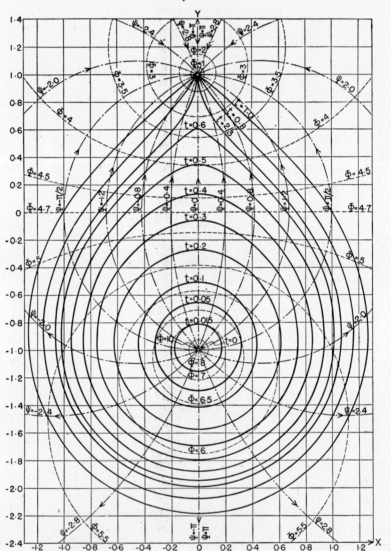

Fig. 12.14.   The calculated shapes of the injection-fluid fronts (t = const) in the steady-state homogeneous-fluid flow between a single injection- and producing-well pair. Φ = const: equipotential curves; Ψ = const: streamlines.   (*From Physics, 1934.*)

or only one-third the area of a circle of radius equal to the well separation. As in the case of Fig. 12.13 the injection-fluid front forms a sharp cusp as it enters the producing well. On the other hand, on leaving the injection well the fluid front remains approximately circular until it has advanced to about one-third the distance to the producing well. This is evidently due to the approximately circular potential and streamline distributions near the injection well, which also obtains in more complex well systems, as previously noted in the cases represented by Figs. 12.9 to 12.12.

When the well system is unsymmetrical or the distribution of injection or production rates is such as to make impractical an explicit solution of the basic potential equation for the pressure distribution, empirical methods must be used to determine the shape of the advancing injection-fluid front. This can be done entirely by means of electrolytic models, as will be discussed in the next section, or by a combination of electrical-model and graphical-integration analyses. In the latter method the potential distribution is found first by the use of an electrolytic or metallic sheet conduction model of the system. By graphical construction or by electrical determinations of the normals to the equipotential curves the streamlines between the input and output wells are then established. Computing from the potential distributions the corresponding pressure gradients, $\frac{\partial p}{\partial s}$, along the streamlines, the time taken for a fluid particle to travel a distance s from an input well or any previous location along a particular streamline can be graphically or numerically calculated by the equation

$$ t = \frac{\mu \bar{f}}{k} \int_0^s \frac{ds}{\left| \frac{\partial p}{\partial s} \right|}. \tag{13} $$

If the integral in Eq. (13) be evaluated along the various streamlines in the system, for different values of s, the loci of those values of s for which $t = $ const will represent the fluid fronts at the corresponding values of t.

These procedures may be considerably simplified by using a four-electrode probe comprised of two separately connected and crossed electrode pairs. By rotating the probe so that one electrode pair lies on an equipotential surface (zero drop between the two electrodes), the other, normal to it, automatically is then set parallel to the streamline at the center of the probe. The potential drop across the latter electrode pair gives a measure of the particle velocity or the pressure gradient required in Eq. (13). These can also be inverted by electronic circuits to permit a direct evaluation of the integral by stepwise summation, or the latter, too, can be effected electrically.[1] Particular applications of such "potentiometric-model" studies will be discussed in Sec. 13.7.

---

[1] Cf. B. D. Lee, *AIME Trans.*, **174**, 41 (1948). The theory of the potentiometric model will be outlined in Sec. 13.6.

12.11. The Shapes of Injection-fluid Fronts in Well Networks as Determined by Electrolytic-model Experiments.—The electrolytic model referred to in the preceding section provides an excellent method for giving directly a graphic representation of the steady-state history of advance of an injection fluid among groups of wells or regular networks that cannot be readily treated analytically. The principle of this model is based on the observation that the motion of ions in an electrolyte under an impressed potential gradient is exactly analogous to that of fluid particles in a porous medium of similar geometry and with a similar pressure-gradient distribution. More fundamental, in fact, is the exact equivalence of the steady-state pressure and electrical-potential distributions in uniform porous media carrying homogeneous fluids and electrolytes of similar geometry and boundary conditions. Both the pressure and potential distributions are governed by Laplace's equation [cf. Eq. 4.5(1)], and analogous to Darcy's law giving the velocity of fluid particles [cf. Eq. 4.3(5)] is Ohm's law expressing the vector current density as the specific conductivity times the gradient of the electrical potential.

If a layer of electrolyte is formed with boundaries geometrically similar to those of the porous medium of interest and in it are placed "injection" and "producing" electrodes with a space distribution and current fluxes corresponding to those of the injection and producing wells, the motion of the ions from the injection electrodes will simulate that of the injection fluid leaving the injection wells.[1] The envelope formed by the ions that have left the injection electrode at the same time will represent an interface between the injection and displaced fluids, and its motion will simulate the advance of the fluid-injection front as the injection is continued. The motion of the ions from the injection electrodes can be made visible by so choosing the constituents of the electrolyte and of the electrode fluids that a coloration develops in the electrolyte as it is invaded by the ions from the injection electrodes. In the initial development of this method[2] the injection electrodes were negative, and the injection fluid was simulated by the OH ions. The colorless electrolyte contained phenolphthalein, which turns red in the presence of OH ions. The holders of the electrolyte were simply pieces of blotting paper saturated with the conducting solution and cut to a shape similar to that of the part of the formation containing the wells of interest. By photographing the blotting paper at various times after beginning the experiment a permanent graphic history is obtained of the progress of the injection fluid toward the producing wells.

---

[1] The assumption is made here, as well as in the analytical treatments of the preceding section, of the complete identity between the injection and displaced fluid with respect to the permeability-to-viscosity ratio.

[2] R. D. Wyckoff, H. G. Botset, and M. Muskat, *AIME Trans.*, **103**, 219 (1932); R. D. Wyckoff and H. G. Botset, *Physics*, **5**, 265 (1934).

A modification[1] of this technique, which is more easily adapted to the study of irregular well patterns, makes use of a representation of the porous medium by an agar gelatin film, such as a 0.1 molar zinc-ammonium chloride solution containing 1 per cent agar, deposited as a thin layer ($\frac{1}{16}$ in.) in a mold forming the boundary of the area to be studied. The well

FIG. 12.15. An electrolytic gelatin model apparatus. (*After Botset, AIME Trans., 1946.*)

electrodes may be plastic tubes, $\frac{1}{2}$ in. in diameter and 1.5 in. in length, set in a bakelite cover plate and terminating in tips penetrating through the latter and into the gelatin. The injection (positive) electrodes are filled with 0.1 molar copper-ammonium chloride solution containing 1.5 per cent agar. The producing-well electrodes are filled with the same solution as in the gelatin field layer, but with the agar concentration increased to 1.5 per cent. The injection electrodes are connected through a bank of milliameters with rheostat controls to the positive side of a d-c voltage (to 1,000 volts) supply. To the negative side are similarly connected the

[1] Cf. J. S. Swearingen, *Oil Weekly*, **96**, 30 (Dec. 25, 1939); F. C. Kelton, *API Proc.*, **24** (IV), 199 (1942); H. G. Botset, *AIME Trans.*, **165**, 15 (1946).

producing-well electrodes. The blue coloration of the gelatin, developed by the invasion of the copper-ammonium ions, is photographed from the transparent underside of the model. A photograph of one form of the composite apparatus is reproduced in Fig. 12.15.

The simple blotting-paper model has been used effectively[1] in determining the idealized steady-state injection-fluid-advance histories in regular well patterns, where a single element of symmetry, containing but a single injection well and a single producing well, could represent the composite network. Thus for the direct-line-drive square network the various stages of injection-fluid advance, as determined on a network element defined by the dotted rectangle in Fig. 12.1, were found to be those shown in Fig. 12.16. The dark area in the last photograph represents a sweep efficiency of 57 ± 3 per cent, which compares favorably with that indicated by Fig. 12.8.

The outlines of the injection-fluid front at various times in a symmetry element (the dotted square in Fig. 12.2) of a five-spot network and the composite superposed history for a complete five-spot unit, as obtained with the blotting-paper model, are reproduced in Fig. 12.17. The sweep efficiency indicated by the most advanced interface corresponds to 75 ± 3 per cent, as measured in several tests. This is to be compared with the theoretically calculated value of 71.5 per cent [cf. Eq. 12.7(3)].

The seven-spot network affords two alternative methods of operation. The central wells of the hexagonal units could be used as injection wells and the peripheral wells as producers, making a ratio of 1:2 for the number of injection wells to the number of producing wells. Or the arrangement could be reversed, with a ratio of 2:1 between the number of injection wells to the number of producing wells. In either case the dotted area in Fig. 12.3 represents a symmetry element for the whole network and suffices for use in the electrolytic model. Again, using the blotting-paper model, the injection-front history for injection at the hexagon centers was found to be that shown in Fig. 12.18, which also includes the superposed history for the composite seven-spot element. The sweep efficiency derived from such model experiments was about 80 per cent, as compared with the calculated value of 74 per cent, [cf. Eq. 12.8(1)]. With the hexagon central well as a producer the injection-fluid fronts have the shapes shown in Fig. 12.19. The sweep efficiency derived from these models ranged from 77 to 80 per cent. As would be expected theoretically the steady-state sweep efficiency is independent of the direction of the fluid flow.

It will also be noted that no reference is made here to the exact values of the current or voltage used in the model experiments. As will be evident

[1] Preliminary tests showed the electrolytic model to reproduce satisfactorily the theoretically calculated injection-fluid fronts for the simple cases of a line drive into a single well and an isolated injection- and producing-well pair (cf. Figs. 12.13 and 12.14).

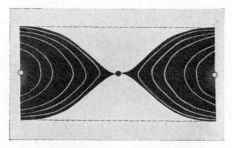

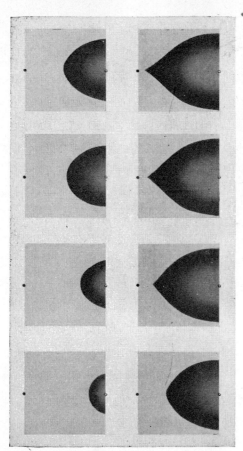

FIG. 12.16.   The photographic history of a direct-line-drive fluid-injection system, under steady-state homogeneous-fluid-flow conditions, as obtained with a blotting-paper electrolytic model.   (*After Wyckoff, Botset, and Muskat, AIME Trans., 1933.*)

Fig. 12.17. The photographic history of the injection-fluid fronts in a five-spot system, under steady-state homogeneous-fluid-flow conditions, as obtained with a blotting-paper electrolytic model. (*After Wyckoff, Botset, and Muskat, AIME Trans., 1933.*)

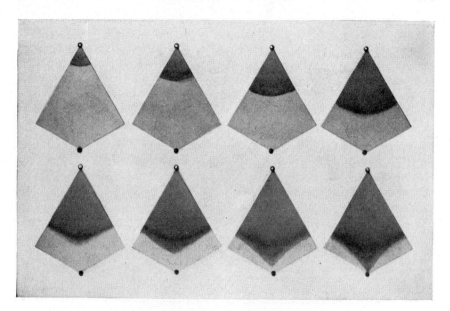

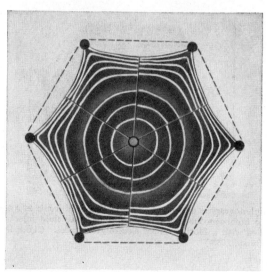

FIG. 12.18. The photographic history of the injection-fluid fronts in a seven-spot network, under steady-state homogeneous-fluid-flow conditions, with fluid injection in the hexagon centers, as obtained with a blotting-paper electrolytic model. (*After Wyckoff, Botset, and Muskat, AIME Trans., 1933.*)

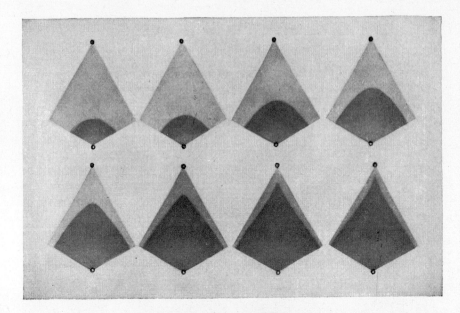

FIG. 12.19. The photographic history of the injection-fluid fronts in a seven-spot network, under steady-state homogeneous-fluid-flow conditions, with fluid injection at the hexagon corners, as obtained with a blotting-paper electrolytic model. (*After Wyckoff, Botset, and Muskat, AIME Trans., 1933.*)

from the theoretical considerations of the previous sections, the steady-state sweep efficiency is a property only of the geometry of the well distribution and is independent of the absolute values of the pressure, currents, nature of the fluids,[1] or permeability of the formation, as long as the latter is uniform and all the pressures and currents are kept the same. Moreover

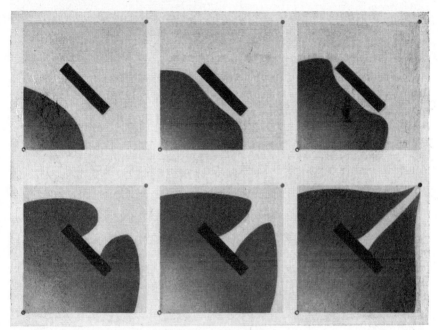

FIG. 12.20.   The photographic history of the injection-fluid fronts while moving around linear barriers in a five-spot network, under steady-state homogeneous-fluid-flow conditions, as obtained with a blotting-paper electrolytic model.   (*After Wyckoff, Botset, and Muskat, AIME Trans., 1933.*)

in the five-spot and seven-spot networks the geometry is by definition a fixed property of the system, and the steady-state sweep efficiencies are constants independent of the absolute well separations.   On the other hand in both the direct and staggered line drives the geometry can be changed by varying the separations between the lines as compared with the well spacings within the lines.   The sweep efficiencies then show corresponding variations, as plotted in Fig. 12.8.   Of course, if the individual-well injection or production rates should be deliberately varied among themselves, the

[1] As will be seen in Sec. 13.6 the travel times along the streamlines in gas-injection systems will depend on the gas density.   The steady-state sweep efficiencies therefore will not be entirely independent of the absolute pressures; nor will they be strictly identical with those for liquid flow, although these differences will not be of importance from a practical standpoint.

nature of the fluid-injection fronts and associated sweep efficiencies will be changed regardless of the basic geometry of the network.

Although the blotting-paper electrolytic model is rather unsuited to the study of irregular well patterns with varying fluid-injection and -producing rates, it can be used to show at least the qualitative effects of permeability barriers on the nature of the steady-state injection-fluid advance in regular

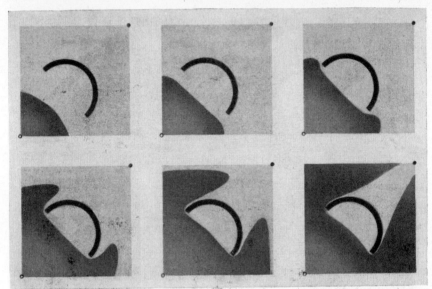

FIG. 12.21. The photographic history of the injection-fluid fronts while moving around curved obstructions in a five-spot network, under steady-state homogeneous-fluid-flow conditions, as obtained with a blotting-paper electrolytic model. (*After Wyckoff, Botset, and Muskat, AIME Trans.*, 1933.)

networks. The distortion in the injection-fluid front caused by a rectilinear obstruction lying normal to the injection-producing well diagonal in a five-spot unit[1] is illustrated by Fig. 12.20. While the flow resistance is undoubtedly increased by the barrier, it will be seen that it serves to push the injection fluid away from the central diagonal and increase the resultant sweep efficiency. A curved barrier, simulating an obstruction with a re-entrant angle toward the injection well, creates an extended "dead" zone and reduces the over-all sweep efficiency, as will be seen from Fig. 12.21.

Steady-state analogues of systems in which fluid injection is undertaken in a limited area with an irregular well distribution are more conveniently investigated by means of the gelatin model. Figure 12.22[2] illustrates the

[1] The use of a single network element in Figs. 12.20 and 12.21 to study the barrier effects implies the artificial assumption that such identical barriers occur in each five-spot unit.

[2] Figure 12.22 is taken from Botset, *loc cit.*

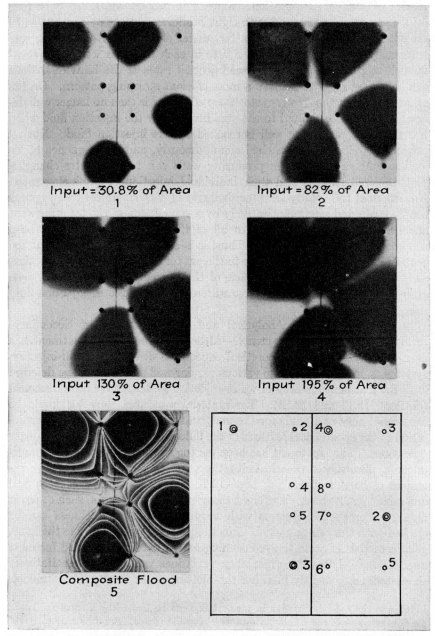

FIG. 12.22. The photographic history of the injection-fluid fronts in an injection project of limited area and with an irregular well distribution, under steady-state homogeneous-fluid-flow conditions, as obtained with a gelatin electrolytic model. Double circles indicate injection wells. (*After Botset, AIME Trans., 1946.*)

results of such an application. Since this project contained only four injection wells as compared with eight producing wells, the average withdrawal rate from each of the latter, under steady-state conditions, was necessarily only half the injection rate in each injection well. Moreover in such studies the producing and injection rates of the individual wells can be varied so as to achieve a most efficient sweeping pattern. On the other hand even the steady-state sweep efficiency is then no longer well defined, except when limited to the area invaded by the injection fluid when the very first producing well is reached by the injection fluid. Not all producing wells will be reached simultaneously, and the sequence of injection-fluid entry into the producing wells can be varied by changing both the well pattern and their individual injection or producing rates. Nevertheless the steady-state equivalents of various injection programs can be evaluated semiquantitatively by a comparison of the total unswept areas left after certain groups or all of the producing wells have been reached by the injection fluid. Thus, it is clear from Fig. 12.22 that the pattern illustrated would be quite inefficient for steady-state fluid injection, since a large area at the lower left of the lease remains uninvaded even after all the producing wells have suffered some degree of injection-fluid entry.

12.12. Limitations of Analytical and Model Studies of Secondary-recovery Fluid-injection Systems.—Although an extended treatment has been given here of the theoretical aspects of distributed fluid-injection systems, a comparison of the various regular well patterns will be deferred until the transient histories of water-injection systems have been discussed (cf. Secs. 12.13 and 12.14). The material presented thus far should be considered only as a guide for the understanding and qualitative interpretation of the geometrical features of the fluid motion in secondary-recovery operations. The treatment has been far too simplified to warrant quantitative application to practical situations. Except for the illustrative examples of barrier effects (cf. Figs. 12.20 and 12.21), both the analytical and model results have all referred to strictly uniform strata, such as never obtain in practice. Systems with varying formation thickness can be studied with the gelatin model. But the preparation of variable-thickness gelatin models is quite time-consuming and is hardly warranted for most applications. Lateral[1] permeability variations cannot be quantitatively simulated[2] conveniently in either the blotting-paper or the gelatin models

---

[1] Simple vertical stratification in permeability can be treated by a superposition of the flow behavior in the different strata with adjusted time scales (cf. Sec. 13.8).

[2] While it would be theoretically possible to adjust the gelatin-film thickness and conductivity to simulate areal variations in the reservoir thickness and permeability, the use of such refinements has not been reported as yet.

and are, of course, beyond the scope of tractable analysis without resorting to laborious numerical procedures. These effects can be taken into account in the potentiometric model and will be discussed in Sec. 13.6 in connection with their application to cycling operations.

The most serious limitation of both the analytical and electrical-model treatments of secondary-recovery fluid-injection systems arises from the basic assumption of steady-state flow. In the case of gas injection, where the oil displacement largely results from a sweeping action of the injected gas as it passes through the partly depleted oil-bearing formation, it may be possible to continue profitable operation and oil recovery for some time after the injected gas has broken through and an approximate steady state has been established. The steady-state considerations of the last several sections will then give a qualitative indication of the directions of motion of the injected gas, although the concept of a sweep efficiency will lose its precise meaning. However, in water flooding by water injection the steady-state conditions will usually not develop until the operations are virtually completed, as far as the recovery from the individual zones is concerned. A major part of the operating history will be controlled by the transient period during which the fluid-injection rates greatly exceed the withdrawal rates and the water fills out the depleted formation while banking up the oil ahead of it. After water break-through to a producing well has developed in a particular stratum, following the displacement of the oil bank, the water-oil ratios for that stratum may be expected to rise rapidly and the oil recovery after a steady-state liquid flow is established will probably be only a minor part of the total from the operations.

The physical implication of this situation, in the case of water flooding, is that during the transient period of fill-up of the depleted oil pay the pressure distribution and fluid motion will be determined largely by the injection wells themselves. Because of the continued low withdrawal rates during this period the producing wells will exert but little influence on the behavior of the injected water. The producing-well pattern will play only a minor role as compared with that indicated by steady-state treatments. The radial spread of the injected water will persist considerably longer than would be inferred from the electrical-model experiments (Figs. 12.16 to 12.19) and the steady-state pressure distributions (Figs. 12.9 to 12.12). Unless the injection-well separations exceed the injection-producing-well distances by a factor of the order of 2, the distortion of the radial injection-fluid front will first develop as the result of interference and close approach of the oil banks of the neighboring injection wells. The fluid-injection front will not begin to cusp toward the producing wells until it has approached much more closely to the producing wells than is suggested by the steady-state model experiments. For the injected fluid will not "know" about the presence of the producing wells until it comes within the short range of the weak pressure sinks of the latter.

On the basis of these considerations it is to be expected that the actual sweep efficiencies in the water flooding of uniform strata will be greater than the steady-state values in which the total fluid-withdrawal rates are assumed to equal the total injection rates. The steady-state network conductivities or production capacities, however, will bear no direct relation to those observed during the transient period. The injection-well capacities will appear to be higher and those of the producing wells will be lower than the common values predicted by the steady-state theory. While the latter gives an instructive picture of the purely geometrical flow properties of well networks or distributions of the type commonly used in secondary-recovery operations, they should not be applied quantitatively except under special circumstances.

12.13. The Transient History of Water-injection Wells.[1]—While it is not feasible to treat the transient phase of fluid-injection systems in as complete a manner as the steady-state analysis of the preceding sections, it is possible to construct an approximate theory for the decline in intake rate of water-injection wells during the "fill-up" period until the development of well interference. It will be assumed that the injected water spreads radially outward, reducing the residual-oil saturation to ρ_{or}, and banking the oil ahead into an annular ring, as shown diagrammatically in Fig. 12.23. A free-gas saturation ρ_g will be assumed[2] left in both the oil bank and watered area. If the initial-oil saturation, assumed uniform, is ρ_{oi} and the water saturation is ρ_w, then

$$(r_e^2 - r_o^2)(1 - \rho_w - \rho_g - \rho_{oi}) = (r_o^2 - r_w^2)(\rho_{oi} - \rho_{or}), \tag{1}$$

where r_w, r_o, r_e are the well radius, radius of the watered area, and external radius of the oil bank. By neglecting r_w as compared with r_o, it follows from Eq. (1) that

$$\frac{r_e^2}{r_o^2} = \frac{1 - \rho_w - \rho_g - \rho_{or}}{1 - \rho_w - \rho_g - \rho_{oi}} \equiv \frac{f_w}{f_o}, \tag{2}$$

where f_w, f_o are the effective fill-up porosities in the watered area and oil bank, i.e., the actual porosity times the numerator and denominator, re-

---

[1] The material of this and the next sections is based on the work of S. T. Yuster, *Pennsylvania State College Bull.*, **40** (No. 3), 43 (Jan. 18, 1946), and S. T. Yuster and J. C. Calhoun, Jr., *Producers Monthly*, **9**, 40 (November, 1944), although several modifications and generalizations have been introduced in the treatment given here. A discussion of still other aspects of injection-well behavior is given by P. A. Dickey and K. H. Andresen, *API Drilling and Production Practice*, 1945, p. 34.

[2] It is usually assumed that $\rho_g = 0$ (cf. also Sec. 10.18). Since there is no direct field evidence on this point, the value of ρ_g has been left arbitrary for the sake of generality.

spectively, of the second of Eqs. (2). The cumulative volume of water injection will be

$$V = \pi h f_w (r_o^2 - r_w^2),\tag{3)*}$$

where h is the effective formation thickness.

Assuming that the instantaneous rate of water injection, Q, is determined

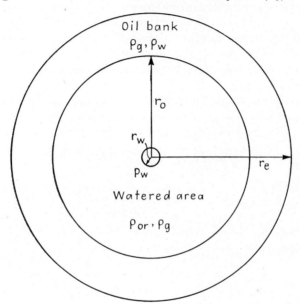

FIG. 12.23. A diagrammatic representation of the radial spread of water from an injection well.

by the steady-state radial-flow equation [cf. Eq. 6.2(5)], it may be expressed as

$$Q = \frac{2\pi k_w h (p_w - p_f)/\mu_w}{\log r_o/r_w + \alpha \log r_e/r_o}; \qquad \alpha = \frac{k_w \mu_o}{k_o \mu_w},\tag{4}$$

where p_w is the injection pressure (at the formation), k_w, k_o are the effective permeabilities to water and oil, μ_w, μ_o are the water and oil viscosities, and p_f represents the residual formation pressure at the time of injection. As the ratio r_e/r_o is constant, by Eqs. (2), the combination of Eqs. (2), (3), and (4) leads to a relation between the cumulative injection and injection rate as

$$\log \frac{V}{\pi h f_w r_w^2} = \frac{4\pi k_w h (p_w - p_f)}{Q\mu_w} + \alpha \log \frac{f_o}{f_w}.\tag{5}$$

Thus the logarithm of the cumulative-injection volume should vary linearly with the reciprocal of the injection rate. A plot of $\log V$ vs. $1/Q$,

* While Eq. (3) appears to imply that the connate water is not moved along with the injected water, it remains valid even if all the connate water is displaced ahead of the injected water, provided that $r_w \ll r_o$.

if linear and satisfying Eq. (5), will have a slope proportional to k_w/μ_w, from which the latter can be determined, assuming the injection pressure p_w is kept fixed. The intercept for $1/Q = 0$ should give the effective well radius r_w, if α and the fill-up porosities are known.

It may be noted that the linear relation between log V and $1/Q$ and the same slope theoretically should remain valid even if the oil banking be neglected ($f_o = f_w$). In the latter case only the common value of the fill-up porosity and pay thickness will be required to determine r_w from the log V intercept. It follows from Eq. (5) that if a constant-injection rate is maintained, the injection pressure will increase as the logarithm of the cumulative injection (or time).

The time history of the injection rate or cumulative-injection volume, under constant injection pressure, can readily be determined by integrating Eq. (5). Upon noting that Q is the time derivative of V, it follows that

$$1 + e^{1/\overline{Q}}\left(\frac{1}{\overline{Q}} - 1\right) = \frac{4k_w t\,\Delta p}{\mu_w f_w r_w^2}\left(\frac{f_w}{f_o}\right)^\alpha \equiv \bar{t}, \tag{6}$$

$$1 + \overline{V}(\log \overline{V} - 1) = \frac{4k_w t\,\Delta p}{\mu_w f_w r_w^2}\left(\frac{f_w}{f_o}\right)^\alpha \equiv \bar{t}, \tag{7}*$$

where $\overline{Q}$, $\overline{V}$ are the dimensionless-injection rates and cumulative-injection volumes, defined by

$$\overline{Q} = \frac{Q\mu_w}{4\pi k_w h\,\Delta p}; \qquad \overline{V} = \frac{V}{\pi h f_w r_w^2\left(\dfrac{f_o}{f_w}\right)^\alpha}. \tag{8}$$

In the notation of Eq. (8), Eq. (5) becomes simply

$$\log \overline{V} = \frac{1}{\overline{Q}}. \tag{9}$$

In these units, Eqs. (6) to (9) represent universal relations applicable to any injection well penetrating a uniform stratum.

Equations (6) and (7) are plotted in Fig. 12.24. It will be seen that after the initial sharp fall in $\overline{Q}$ it assumes a very slow decline, given asymptotically by $\overline{Q} \sim 1/\log \bar{t}$. The asymptotic rise in cumulative injection is slower than linear and corresponds to $\overline{V} \sim \bar{t}/\log \overline{V}$. For conversion of these dimensionless units to days, barrels per day, and barrels the factors are

$$t\,(\text{days}) = \frac{39.51 r_w^2 \mu_w f_w (f_o/f_w)^\alpha}{k_w\,\Delta p}\,\bar{t} \quad : \quad Q\,(\text{bbl/day}) = \frac{0.01416 k_w h\,\Delta p\overline{Q}}{\mu_w};$$

$$V\,(\text{bbl}) = 0.5595 h f_w r_w^2\left(\frac{f_o}{f_w}\right)^\alpha \overline{V}, \tag{10}$$

* The implication of Eq. (7) that $\overline{V} = 1$ at $\bar{t} = 0$ is of no practical significance, and arises from the neglect of r_w as compared to r_o and r_e in deriving Eqs. (2) and (5).

where k_w is expressed in millidarcys, r_w and h in feet, Δp in psi, and μ_w in centipoises. Thus, by taking $(f_o/f_w)^\alpha = 1$, as has been assumed in most applications that have thus far been made of these equations, and an effective well radius of 1 ft, as may have been developed by shooting, Eqs. (10) imply that, for an injection rate of 0.0035 (bbl/day)/(psi)(md-ft) and 1 cp viscosity, $\bar{Q}$ will be 0.247. From Fig. 12.24, $\bar{t}$ would then be 175, and $\bar{V}$ will be 57. The latter, by Eqs. (10), implies that the cumulative injection, for $f_w = 0.1$, will be 3.19 bbl/ft of pay, and the elapsed time of

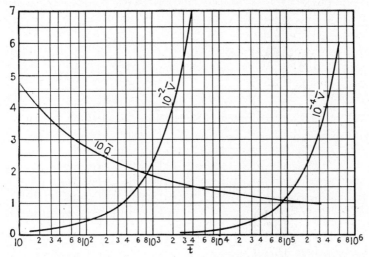

Fig. 12.24. The theoretical variation of the dimensionless injection rates $\bar{Q}$ and cumulative-injection volumes $\bar{V}$, in water-injection wells, with the dimensionless time $\bar{t}$, during the radial-expansion phase. $\bar{Q}$, $\bar{V}$, and $\bar{t}$ are defined by Eqs. 12.13(6) to 12.13(8).

injection would have been 691.3 days if the injection pressure differential were actually only 1 psi or 0.691 day if it were 1,000 psi. If two of the basic field data are known, as the injection rate or cumulative injection at a certain time, one of the physical constants, such as the effective permeability to the water, or fill-up porosity, can be determined by application of Eqs. (10) and Fig. 12.24, assuming, of course, the basic validity of these relationships.

An example of field data showing the observed relationship between the water-injection rate and cumulative injection is given in Fig. 12.25.[1] The ordinates are logarithms, to the base 10, of the cumulative injection in barrels, and the abscissas are the reciprocals of the injection rate in barrels per day. Each point represents 1 month of elapsed time. It will be seen that the theoretically predicted linear variation of log V with $1/Q$ [cf. Eq. (5)] is followed for a period of more than 1 year. The formation thick-

<hr />

[1] This is taken from Yuster, *loc. cit.*

ness was 10 ft, and the injection pressure was 2,000 psi. On combining these data with the slope of the curve an effective permeability to water of 0.082 md was computed. As the air permeability of the formation into which the water was being injected was 1.9 md, the equivalent relative permeability is 0.043. By taking f_w and f_o as 0.0625, the extrapolated log V intercept gave an effective well radius of 14 ft. This large radius was apparently the result of shooting the well, with 3.1 qt of explosive per foot.

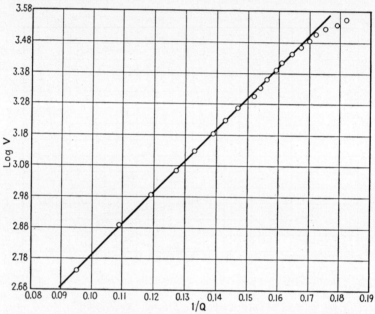

Fig. 12.25. A plot of field data from a well in the Bradford area on the variation of the cumulative water injection V in barrels vs. the injection rate Q in barrels per day. (*After Yuster, Producers Monthly, 1945.*)

12.14. The Interference between Water-injection Wells.—The radial spread of the injected water and the oil bank evidently cannot continue indefinitely even if the formation is strictly uniform. Interference and distortion will certainly develop as soon as the oil banks from two neighboring injection wells come in contact. This will give an upper limit to the time and cumulative-injection volumes for which Eqs. 12.13(6) to 12.13(9) will retain their validity. Expressed as a per cent areal sweep per network element the range of validity will be $100\pi a/8d$ for the direct or staggered line drive (with $a < 2d$), or 39.3 per cent for the square network, 78.5 per cent for the five-spot, 90.7 per cent for the hexagonal network with the central injection well (the four-spot), and 60.5 per cent for peripheral injection in the hexagonal array. These represent minimal sweep efficiencies,

and they are seen to be different for the regular seven-spot and the four-spot. Moreover the latter and the five-spot interference sweep area, even though they are minimum values for the sweep efficiency, exceed those for the steady-state-flow conditions. Of interest, too, is that, whereas the steady-state sweep efficiency for the direct line drive increases as d/a increases (cf. Fig. 12.8), the reverse is true for the interference sweep area. Thus when $d/a = 1$, the sweep efficiencies are, respectively, 57 per cent and 39.3 per cent. However, for $d/a = \frac{1}{2}$, they are 31 per cent and 78.5 per cent.

The values of the dimensionless cumulative injection, $\overline{V}_i$, at which injection-well interference develops, under the assumption[1] $(f_o/f_w)^\alpha = 1$, are readily shown to be

Direct[2] line drives (cf. Fig. 12.1),

$$\overline{V}_i = \frac{1}{4}\left(\frac{a}{r_w}\right)^2 \quad : \quad a \leqslant 2d,$$

$$= \left(\frac{d}{r_w}\right)^2 \quad : \quad a \geqslant 2d. \tag{1}$$

Five-spot (cf. Fig. 12.2),

$$\overline{V}_i = \frac{1}{2}\left(\frac{d}{r_w}\right)^2. \tag{2}$$

Seven-spot [peripheral injection (cf. Fig. 12.3)],

$$\overline{V}_i = \frac{1}{4}\left(\frac{d}{r_w}\right)^2. \tag{3}$$

Seven-spot [central injection (four-spot)],

$$\overline{V}_i = \frac{3}{4}\left(\frac{d}{r_w}\right)^2. \tag{4}$$

It will be noted that in the five- and seven-spots d represents the separation between the injection and producing wells. In the line drives, d denotes the distance between the injection and producing lines. The values of $\bar{t}$ for interference development are readily obtained from those for $\overline{V}$ by applying Eq. 12.13(7) or by reference to Fig. 12.24.

After interference has developed between the injection wells the radial expansion of the watered area and oil bank will be distorted and the equations of Sec. 12.13 will no longer be valid. Unfortunately the motion during the subsequent period until the oil bank reaches the producing wells is too complex for quantitative treatment. An approximation, how-

[1] If the oil bank is taken into account, Eqs. (1) to (4) would remain valid if $(f_o/f_w)^{\alpha-1}$ were equal to unity, that is, $f_o = f_w$, or $\alpha = 1$.

[2] For staggered line drives the first of Eqs. (1) applies in all cases.

ever, can be made by assuming that during this interval the injection rate declines linearly with time until it falls to the steady-state value, such as is given by the analysis of Secs. 12.2 to 12.5. In dimensionless units the latter are

$$\overline{Q}_s \text{ (direct line drive)} = \frac{1}{2\pi d/a + 4 \log a/2\pi r_w}, \tag{5}$$

$$\overline{Q}_s \text{ (staggered line drive)} = \frac{1}{2 \log \dfrac{\cosh^4 \pi d/a \cosh^3 3\pi d/a}{\sinh^2 \pi r_w/a \sinh^4 2\pi d/a \sinh 4\pi d/a}}, \tag{6}$$

$$\overline{Q}_s \text{ (five-spot)} = \frac{1}{4 \log d/r_w - 2.476}, \tag{7}$$

$$\overline{Q}_s \text{ (seven-spot)} = \frac{1}{3 \log d/r_w - 1.7073}. \tag{8}$$

The cumulative-injection curve is extended in accordance with the linearly decreasing injection rate, until the whole network element is filled up. Thereafter it will continue with a constant slope equal to the steady-state injection rate after the latter has been reached.[1] The interference dimensionless time interval is thus

$$\Delta \bar{t} = \frac{\overline{V}_f - \overline{V}_i}{(\overline{Q}_i + \overline{Q}_s)/2}, \tag{9}$$

where $\overline{V}_f$ is the cumulative dimensionless total fill-up volume, $\overline{V}_i$ that when interference develops, as given by Eqs. (1) to (4), $\overline{Q}_i$ the injection rates corresponding to $\overline{V}_i$, and $\overline{Q}_s$ the steady-state rates, as given by Eqs. (5) to (8).

To illustrate these procedures it will be assumed again that $r_w = 1$ ft. Then, for a 300-ft injection-well separation in a five-spot pattern, $d = 212.1$ ft, and $\overline{V}_i = 22,500$. The corresponding values of $\overline{Q}_i$ and $\bar{t}$ are 0.09979 and 2.029×10^5. These give the terminal values for the radial-expansion phase. From Eq. (7), $\overline{Q}_s = 0.05276$. As $\overline{V}_f = 28,648$, $\Delta\bar{t} = 80,600$, so that the steady-state conditions will begin at $\bar{t} = 2.835 \times 10^5$. The histories of the dimensionless-injection rate and cumulative injection for this five-spot system will therefore be as shown in Fig. 12.26. The numerical equivalents of the time scale in days and of $\overline{Q}$ and $\overline{V}$ in barrels per day and barrels can readily be computed by means of Eq. 12.13(10) after fixing the values of k_w, μ_w, h, Δp, and f_w.

In practice the transition between the radial-expansion transient and

---

[1] In the steady-state region there is no need to take into account the fill-up porosities. However, it is convenient to use the same definitions of the variables and conversion factors [Eqs. 12.13(10)] throughout the calculations in order to preserve dimensionless scale continuity in such plots as Fig. 12.26.

steady-state condition will undoubtedly occur more smoothly than indicated in Fig. 12.26, even in a single stratum of uniform permeability. However, the uncertainty in the approximation made in constructing the transition region is probably no greater than that in the physical data involved in making a numerical application of the theory.

Figure 12.26 and the theoretical considerations on which it is based refer

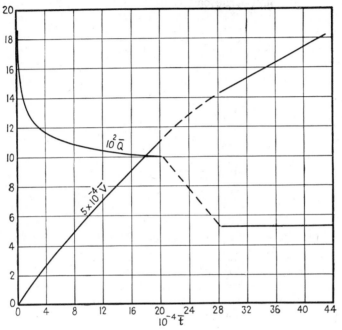

Fig. 12.26. The calculated time variation of the water-injection rates $\bar{Q}$ and cumulative-injection volumes $\bar{V}$, in dimensionless units, for a five-spot pattern with an injection-well separation of 300 ft and effective well radius of 1 ft. Dashed segments represent linearized approximations for the interference period between the radial expansion and steady-state conditions.

only to a single uniform zone. If the formation is stratified and cross flow is not an important factor, a parallel superposition of the transient histories can be applied to describe the behavior of the composite system. All that is necessary is a change in the time, injection-rate, and cumulative-injection scales for the individual strata in accordance with their permeabilities and thicknesses, as required by Eqs. 12.13(10). The time scales and the duration of the noninterference transient will be reduced in inverse proportion to the permeabilities. The injection rates will be proportional to the millidarcy-foot values for the separate zones, and the cumulative injections will be proportional to their respective thicknesses. Figure 12.27 gives the results of such a calculation for a five-spot pattern, assuming the formation to be comprised of three layers, each 10 ft thick, of effective water permeabilities 5, 10, and 15 md, an injection pressure of 1,000 psi,* $\mu_w = 1$ cp, $f_w(f_o/f_w)^\alpha = 0.1$, $r_w = 1$ ft, and a 300-ft injection-well separation. It will be seen that the radial-expansion transients terminate after 160.4, 80.2, and 53.5 days, for the 5-, 10-, and 15-md strata, respectively, and complete fill-up or steady-state flow begins after 224.1, 112.0, and 74.7 days. The resultant history of the composite system is plotted in Fig. 12.28. It will be observed that the transition intervals of the injection histories

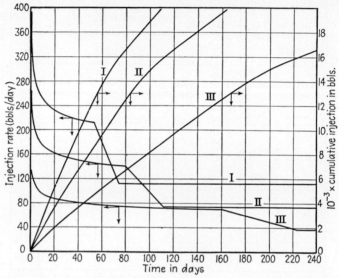

FIG. 12.27. The calculated histories of the injection rates and cumulative-injection volumes for a water-injection well in a five-spot pattern in three 10-ft strata of effective permeabilities to water of 15(I), 10(II), and 5(III) md. Injection-well separation = 300 ft; effective well radius = 1 ft; water viscosity = 1 cp; $f_w(f_o/f_w)^\alpha$ assumed = 0.1.

of the individual strata, shown in Fig. 12.27, are not so pronounced in the composite curves of Fig. 12.28.

An application of this type of calculation to an actual field problem is illustrated by Fig. 12.29,[1] which gives the observed and computed transient histories for a five-spot water-injection well in the Bradford, Pa., area. The formation taking the water was considered as divided into four zones of average air permeabilities of 0.5, 1.62, 4.01, and 9.85 md, with thicknesses of 10.7, 11.8, 8.9, and 1.7 ft, respectively. The water-injection-well

---

* It is assumed that all the zones are uniformly and completely depleted. In practice, however, where the formation may be highly stratified the differences in degrees of oil and pressure depletion in the various zones may appreciably modify the simplified theoretical predictions.

[1] This is taken from Yuster and Calhoun, *loc. cit.*

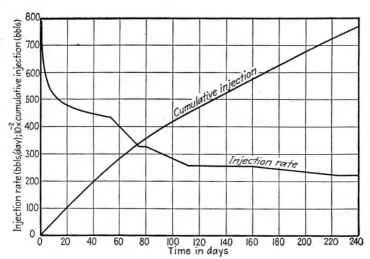

Fig. 12.28. The calculated water-injection histories of the composite three-layer system with individual histories as plotted in Fig. 12.27.

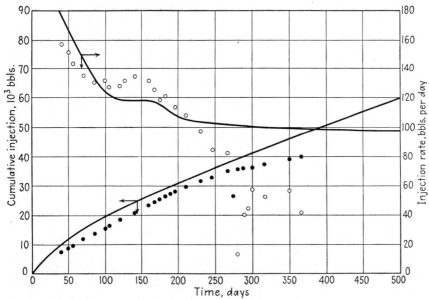

Fig. 12.29. A comparison of calculated (continuous curves) and observed (circles and points) water-injection data in a five-spot injection well. (*After Yuster and Calhoun, Producers Monthly, 1944.*)

separation was 300 ft, the average sand porosity was 14 per cent, the connate-water saturation was 36 per cent, the initial-oil saturation was 30 per cent, and the residual-oil saturation after flooding was taken as 25 per cent. The injection pressure differential was 1,835 psi. From plots of log V vs. $1/Q$ the relative permeability to water was determined to be 38.2 per cent, and the effective well radius was 18.6 ft. While the actual injection rates become much lower than those calculated after about 200 days,* the agreement between the observed and computed data seems close enough to represent at least a gross confirmation of the basic physical structure of the theory presented here.

The oil production will be derived mainly during the transition intervals between the radial-expansion transient and the complete-fill-up steady-state flow. In multilayer systems, such as may be inferred from typical permeability logs, it is to be expected that the oil production will show a continuous build-up and decline history until steady states are developed in the tightest strata, even though in the individual zones the period of oil expulsion may be of only limited duration. Because of the complex dynamics of the flow of oil and water in the transition interval, resulting from injection-well interference, and the final cusping of the oil banks into the producing wells, the detailed history of the latter cannot be derived rigorously. While some predictions could be developed by making assumptions regarding the manner in which the oil bank is squeezed into the producing wells during the transition period, no theory of this type has been reported as yet.[1]

Although the well patterns actually used in water-flooding programs are often determined by the distribution of wells already drilled for the primary-production phase, it is of interest to note the comparative features of the various networks as indicated by the theoretical considerations. For such purposes the well separations in the different well distributions are to be chosen so that they have a common total well density. In both the direct and staggered line drives the area per well is da, or $da/2$ per injection well and per producing well. The well density in the five-spot network is 1 well per d^2 ft^2 or $d^2/2$ ft^2 per injection well and per producing well, where d is the injection-producing-well distance. In the normal seven-spot, with peripheral injection, the area per well is $\sqrt{3}d^2/2$, where d is the injection-producing-well distance. The area per producing well is $3\sqrt{3}d^2/2$, and per

---

* This abnormal fall in injection rates was apparently due to plugging effects.

[1] The assumption that the oil is produced from each stratum during the time between the fill-up of the original gas space and the total steady-state fill-up has been recently applied by F. E. Suder and J. C. Calhoun, Jr., (API meetings, Tulsa, Okla., March, 1949). Empirically, it has been found that the oil-production-rate histories often follow approximately an equation as: $Q = ate^{-bt}$.

injection well it is $3\sqrt{3}d^2/4$. In the seven-spot with central injection (the four-spot) the total well density is the same as in the regular seven-spot. However, the injection-well area is now $3\sqrt{3}d^2/2$ ft² per well, and that of the producing wells is $\sqrt{3}d^2/4$ ft² per well. Thus for common well densities

$$da = d_5^2 = \frac{\sqrt{3}d_7^2}{2}, \tag{10}$$

where the subscripts 5, 7 refer to the five- and seven-spot patterns.

Assuming $d_5/r_w = 1,000$, and $d = a$ for the line drive, the dimensionless characteristics of the four basic patterns are listed in Table 1. It will be

TABLE 1.—COMPARISON OF VARIOUS FLOODING NETWORKS

	Radial-expansion injection volume, $10^{-5}\,\overline{V_i}$	Injection rate per injection well at $\overline{V_i}$, $10^2\,\overline{Q_i}$	Time at $\overline{V_i}$, $10^{-6}\,\overline{t_i}$	Fill-up injection volume, $10^{-5}\overline{V_f}$	Injection rate per injection well at $\overline{V_f}$, $10^2\,\overline{Q_s}$	Time for fill-up, $10^{-6}\,\overline{t_f}$
Line drive $(d = a)$...........	2.5	8.046	2.857	6.366	3.501	9.553
Five-spot	5.0	7.621	6.061	6.366	3.975	8.417
Seven-spot (peripheral injection)	2.887	7.954	3.341	4.775	2.60	6.919
Seven-spot [central injection (four-spot)]	8.660	7.315	10.974	9.549	5.20	12.395

seen that the radial-expansion injection volume is least for the square line drive and greatest for the four-spot. The times required for the development of interference are in the same order as the injection volumes, and the corresponding injection rates are in the reverse order. The fill-up volumes are proportional to the areas to be flooded per injection well. The steady-state injection rates, after fill-up, are least for the seven-spot and maximal for the four-spot. The same applies with respect to the time required for fill-up.

It should be noted that for equal total well densities an area which would contain 1 injection well in a line drive or five-spot would include the equivalent of ⅓ of an injection well for a seven-spot pattern and ⅔ of an injection well for the four-spot network. Thus the total throughflow rates in both the latter would be equivalent to $10^2\overline{Q_i} = 3.467$, as compared with 3.501 and 3.975 for the line drive and five-spot. This lower value of $\overline{Q_i}$ for the seven-spot as compared with the five-spot will be beneficial in retarding the rise in water cuts as the most permeable zones become filled up and their oil has been flushed out.[1] However, the main advantage of the seven-spot is its relatively short fill-up time. This is evidently largely due to the higher injection-well density in the seven-spot pattern in which peripheral injection is used. Of course, in choosing a pattern the relative costs of injection and producing wells will also be an important economic factor.

12.15. Field Experience in Water Flooding.—From both historical and technical points of view the Bradford field, McKean County, Pa., and the

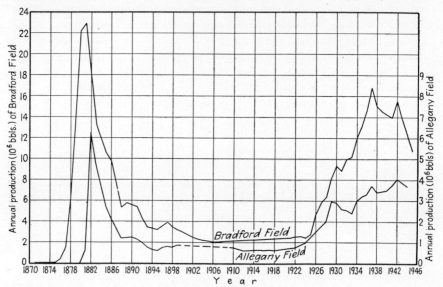

Fig. 12.30. The production histories of the Bradford field, Pennsylvania, and Allegany field, New York.

Allegany field in New York represent undoubtedly the outstanding examples of water-flooding operations developed to date. The first planned water-flooding projects were undertaken in these fields. Much of the technical understanding and operating experience underlying this phase of reservoir engineering was the outgrowth of the multitude of individual projects undertaken in the Bradford and Allegany fields.

The production histories of these two reservoirs are plotted in Fig. 12.30.[2]

[1] The fact that the relative permeabilities to water in flooded zones are quite low will also tend to lessen the differential rates of flooding, which would otherwise be in proportion to their air permeabilities.

[2] The curve for Bradford was plotted from tabulations of G. G. Bauer, *Producers Monthly*, **11**, 29 (November, 1946); that for the Allegany field and the general discussion given here pertaining to it are taken from G. W. Holbrook and W. H. Young, Jr., *API Drilling and Production Practice*, 1945, p. 65. For a detailed discussion of the Bradford field history and practices, cf. C. R. Fettke, Chap. 8, "Secondary Recovery of Oil in the United States," API, 1942.

Sec. 12.15]   *SECONDARY RECOVERY*   695

The Bradford field was discovered in 1871. It covers an area of 85,000 acres, and the depths of the producing sands (members of the Third Bradford Sand) range from 1,100 to 2,100 ft. While virtually no reservoir data were gathered during its primary solution-gas-drive producing life, subsequent core studies have shown the productive zones to have an average porosity of 14.5 per cent. Their permeabilities range from 2 to 600 md, with an average lying in the interval of 7 to 10 md. The initial-oil saturations at the time of flooding averaged about 40 per cent. Natural water leakage through defective wells from nonproductive sands developed on an appreciable scale in 1907 and led to the gradual rise in production rate beginning at that time (cf. Fig. 12.30). Deliberate water injection began in 1921, after being legalized by statute. By June, 1941, 41,000 acres were covered in water-flooding projects, at an average total well spacing of approximately 1 acre per well. Of the 480 million barrels of oil produced in this field by January, 1946, 235 million barrels has been obtained by secondary-recovery methods, practically all of which resulted from water flooding.[1]

The Allegany field, which includes one large and several smaller pools in Allegany and Steuben Counties, N.Y., was discovered in 1879. Its composite area totals 58,400 acres. While the general structure is anticlinal, oil production is largely controlled by lenticularity. The main producing horizon among several pays is the Richburg Sand, at an average depth of about 1,250 ft. The average pay thickness is 18 ft. The general rock characteristics are similar to those at Bradford. The crude gravity, as at Bradford, is approximately 42°API.

The production history of the Allegany field is also quite similar to that of Bradford. After the passing of the "flush" dissolved-gas-drive production phase a slow rise in production rate developed about 1912, due to accidental water flooding resulting from improperly plugged abandoned wells. Deliberate water flooding began about 1920, and for the next 22 years the field production rates increased by a factor of about 7. On Jan. 1, 1944, the field contained 16,650 producing wells, of which about 6,500 were natural producers and the remainder were drilled in the course of water-flooding development. The former produced less than 500 bbl/day during the second quarter of 1944, out of a total of 10,225 bbl/day. The number of water-input wells in Jan. 1, 1944, totaled about 6,500, covering an area of 16,000 to 18,000 acres. Water-injection rates per well in the Allegany and Bradford fields, and in the eastern district generally, have been of the order of 1 to 5 (bbl/day)/ft of pay. The cumulative production as of Jan. 1, 1945, was 113,250,000 bbl, of which approximately half has been obtained by water flooding. The latter has been obtained in

[1] The long producing lives of the Bradford and Allegany water-flooding operations are due to the gradual extension of the projects to new parts of these fields.

about 20 years, whereas the equivalent primary recovery required 45 years. The average spacing corresponds to 2 to 3 acres per five-spot unit.

Since the proving of the practical effectiveness of water flooding in Pennsylvania and New York, numerous smaller scale projects have been undertaken in other parts of the country,[1] especially in the Mid-Continent district and Texas. Systematic surveys have been made of the water-flooding possibilities in new areas.[2] For illustrative purposes several such projects will be briefly reviewed here.

A highly successful water-flooding project is that conducted on the Davis lease of the Woodsen shallow field, Throckmorton County, Tex.[3] The pro-

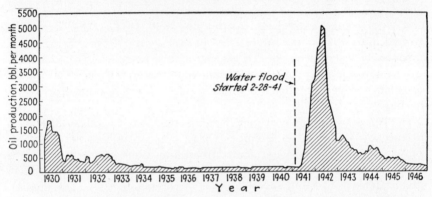

Fig. 12.31. The oil-production history of the Davis lease in the Woodsen field, Texas, during the primary-production and subsequent water-flooding operations. (*After Dean, Oil and Gas Jour., 1947.*)

ducing formation is an upper Tannehill Sand at 323 to 333 ft. The field was discovered in December, 1929, and 184 oil wells were drilled on its 400 acres. The solution-gas-drive primary-production phase was virtually completed by 1941, when the 82 remaining producers were yielding less than $\frac{1}{3}$ bbl/day per well. The average sand thickness is 15.6 ft, average porosity is 26.5 per cent, and the permeability ranges from 137 to 620 md, with an average of 413 md. The residual-oil saturation in cores averaged 20.5 per cent,[4] and the average core-water saturation was 54 per cent. The oil gravity is 30°API.

---

[1] A general survey of secondary-recovery operations in the United States is given by P. D. Torrey, Chap. 1, "Secondary Recovery of Oil in the United States," API, 1942; other chapters of this book give more detailed discussions of practices and experiences in individual states; the results of special state surveys are also given in a series of reports issued by the U.S. Bureau of Mines, *e.g.*, D. B. Taliaferro and D. M. Logan, "History of Water Flooding of Oil Sands in Oklahoma," *U.S. Bur. Mines Rept. Inv.* 3728 (November, 1943).

[2] Cf., for example, G. H. Fancher and D. K. Mackay, "Secondary Recovery of Petroleum in Arkansas—A Survey," Arkansas Oil and Gas Commission, (1947).

[3] Cf. P. C. Dean, *Oil and Gas Jour.*, **49**, 78 (Apr. 12, 1947).

[4] The low value is presumably due to core flushing.

Water injection was started Feb. 28, 1941, with treated lake water, later mixed with the produced water. The Davis lease, of 32 acres, was developed on a five-spot pattern, using 21 oil wells and 19 injection wells. Injection surface pressures have been 325 to 350 psi. Injection rates were of the order of 5 (bbl/day)/ft of sand. The lease production during February, 1941 was 111 bbl.

The oil-production history during the primary-production period and under water flooding is plotted in Fig. 12.31. From 3 bbl/day of oil pro-

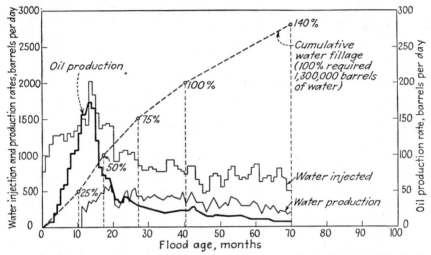

Fig. 12.32. The oil-production, water-injection, and water-production histories of the Davis lease in the Woodsen field, Texas, during water-flooding operations. (*After Dean, Oil and Gas Jour., 1947.*)

duction at the beginning of the flood the production rate quickly rose to a peak of 170 bbl/day within 13 months. The cumulative water-flood–oil production to Jan. 1, 1947, was 78,807 bbl, or 2,396 bbl/acre, almost twice the previous cumulative recovery by gas drive of 39,795 bbl, or 1,210 bbl/acre. This recovery exceeds by about 70,000 bbl that which would have been expected from a continuation of the previous operations without water flooding. The history of the water injection and water production is plotted in Fig. 12.32.

The injection-water–oil ratio sharply dropped from the initial values, of the order of 100, to a minimum of about 10 shortly before the peak in oil production was reached. It then followed a rising trend, which resulted in ratios exceeding 50 after the fifth year of the operations. The water-oil ratios in the production remained zero until the first water break-through after 11 months. It reached a value of 10 after 20 months and varied between 20 and 30 in the sixth year. By Jan. 1, 1947, the water injected totaled 1,926,000 bbl, or 24.4 times the oil produced and 1.4 times the water required to fill up the total pore volume of the sand. In December, 1946, the lease production averaged 7 bbl/day of oil and 160 bbl/day of water.

Not quite so striking, but still entirely satisfactory, were the results of water flooding of a 40-acre lease in the Prideaux field, Archer County, Tex., as shown in Fig. 12.33.[1] The formation involved is the Gunsight Sand at 685 ft. The average porosity is estimated as 25 per cent and the

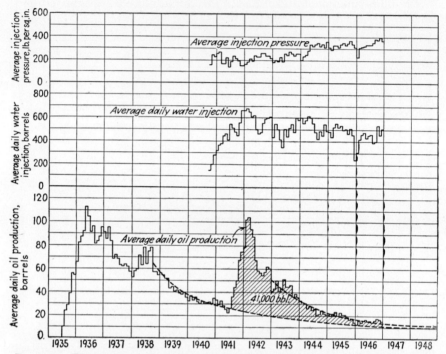

Fig. 12.33. The production history of a lease in the Prideaux field, Texas, subjected to water flooding. (*After Taliaferro and Guthrie, Producers Monthly, 1947.*)

permeability as 500 to 600 md. The initial development of the lease occurred in 1935, and gas injection was undertaken in the latter part of 1936. The primary peak oil production was 113 bbl/day, in May, 1936. In October, 1940, when water flooding was started and shortly before gas injection was discontinued, the 11 producing wells had a daily rate of about 30 bbl of oil and 10 bbl of water. The total oil production to that date was 113,340 bbl, which was equivalent to 270 bbl/acre-ft. For the water-flooding operation 1 oil producer and 1 gas-injection well were converted to water injection, and 4 new wells were drilled for the purpose. The total oil recovery to January, 1947, since the beginning of water flooding was

[1] Cf. D. B. Taliaferro and R. K. Guthrie, *Producers Monthly*, **11**, 15 (March, 1947).

84,080 bbl, of which approximately 41,000 bbl was in excess of that normally expected from the previous decline curve (cf. Fig. 12.33). The water injected totaled 1,100,000 bbl, or 13.1 times the oil produced during the period of flooding operations.

The oil-production curves for two flooding projects in Nowata County, Okla., are plotted in Fig. 12.34.[1] The Alluwe Oil Corp. operations (curve 1)

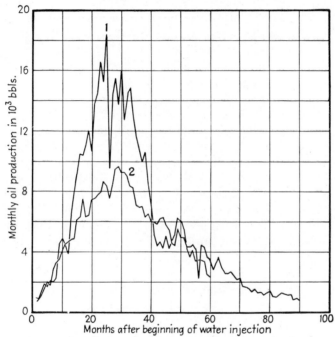

Fig. 12.34.   The oil-production histories during water injection of the Alluwe Oil Corp. (1) and H. J. Walter Co. (2) leases in Nowata County, Okla.

were conducted on an area of 165 acres with 77 water-injection and 73 oil-producing wells (330 ft between like wells) arranged in a five-spot pattern. The flooded formation is a 55-ft zone of Bartlesville Sand at 500 ft, which had been subjected to vacuum[2] and gas injection before water injection was started. The producing rate of the area at the beginning of water injection, July, 1937, was 15 bbl/day. A rise in production rate began

[1] The plots of Fig. 12.34 were made from tabulations of Taliaferro and Logan, *loc. cit.* The material given here pertaining to these projects was taken from the same source, where many other Oklahoma water-flooding operations are discussed in detail.

[2] "Vacuum" refers to operations in which the wellhead pressures in the producing wells are maintained below atmospheric, to increase the pressure differential over the formation and stimulate the production rates in depleted fields. Such practices used to be common before fluid-injection operations became well established.

in September, 1937, and reached a peak of 18,248 bbl during July, 1939. The injection water was variously comprised of creek water, Verdigris River water, and produced water. It was aerated, treated, and filtered before injection at about 600 psi. By July 1, 1942, 5,641,500 bbl of water had been injected, with an oil production of 456,350 bbl since the start of water flooding, or an over-all water-oil ratio of 12.4. The oil recovery is equivalent to 2,766 bbl/acre.

The H. J. Walter Co. operations, with an oil-production history shown as curve 2 in Fig. 12.34, covered 208 acres developed on a five-spot pattern. The oil-producing formation is a 22-ft Bartlesville Sand at a depth of 475 ft. Prior to flooding it had produced 2,000 bbl/acre by primary methods (gas drive) and 500 bbl/acre by gas and air injection. Its total production rate was about 20 bbl/day at the beginning of the flood. The operations were started as a "delayed" flood; and while water injection began on July 24, 1934, the producing wells were not drilled until January, 1935. By July 1, 1942, there were 65 new oil-producing wells and 73 new water-injection wells in the project. The producing wells were originally completed as pumping wells, but about 75 per cent were subsequently converted to flowing wells. The injection water was first taken from a shallow fresh-water bed and later brought in by pipe line from the Verdigris River. Injection pressures have been 450 to 600 psi. To July, 1942, the oil recovery by flooding had been 387,500 bbl, or 1,863 bbl/acre. The metered injection water totaled 3,970,600 bbl for a gross water-oil ratio of 10.2.

The water-flooding history of an intensively developed 25-acre lease in the Bradford field is plotted in Fig. 12.35.[1] The net pay thickness of the Third Sand in this area was 33 ft. Its average porosity was 14 per cent, and the average permeability of one core was 6.45 md. The injection-producing-well separation in the five-spot pattern was 212 ft. The injection pressures were 1,500 to 1,600 psi. The total oil recovery after 114 months of operation was equivalent to 5,755 bbl/acre, or 169 bbl/acre-ft. The gross water-oil input ratio was 4.3, and the average produced water-oil ratio was 1.5. The total water injection per acre after 90 months was 20,135 bbl. It will be noted from Fig. 12.35 that within $5\frac{1}{2}$ years the total fluid-producing rate became approximately equal to the water-injection rate. The oil recovered in this period was 92 per cent of the total after $7\frac{1}{2}$ years. The latter represented 15 per cent of the total pore volume of the part of the reservoir being flooded.

12.16. Practical Aspects and Requirements for Water Flooding.—The illustrative examples of water-flooding operations presented in the preceding section have been deliberately chosen to show the types of results obtained in successful projects. Not all those which have been tried have

---

[1] Figure 12.35 was plotted from a tabulation of Fettke, *loc. cit.*

been profitable.   The criterion of success is, of course, the comparative value of the additional oil produced and the cost of the operations.   No simple rules can be given to prescribe the amount of additional recovery required for a pay-out and profit.   The gross returns will depend on the price of the oil as well as its volume.   The price of the oil in various water-flooding districts has varied by more than a factor of 2.   The cost of the operations will depend on the cost of drilling new wells or reworking old

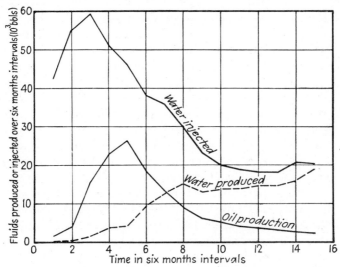

Fig. 12.35.   The water-flooding history of a lease in the Bradford field, Pennsylvania.

wells, the number of new and reworked wells needed, the cost of lifting produced water, the availability and cost of the water supply and of its treatment, and the water-oil ratio in the production.[1]   Some of these factors will vary with the depth of the formation to be flooded.   Evidently no single formula will automatically evaluate these factors, except as it is merely a symbolic summation of all the items comprising the operating costs.

The primary criterion for successful water-flooding operations is the availability of a sufficient supply of recoverable oil.   This implies that the average residual-oil content at the time of beginning the flooding, left by previous recovery methods, is appreciably greater than that which may be

[1] A general discussion of the economic aspects of secondary-recovery operations is given by P. D. Torrey, Chap. 2, "Secondary Recovery of Oil in the United States," API, 1942; cf. also W. B. Berwald, *Oil and Gas Jour.*, **46**, 317 (Nov. 15, 1947).   A discussion of the economics of the water-flooding operations in both the Prideaux and Woodsen fields described in the preceding section is given by W. Krog, *Producers Monthly*, **12**, 14 (March, 1948).

expected to result from the flooding operations. As there is no reason to anticipate the artificial water-flushing process to be significantly different in oil-displacement efficiency[1] from that due to a natural water drive, reservoirs that have been produced by water drive will, in general, not be suitable for water flooding. Likewise, reservoirs in which gravity drainage has been the controlling recovery mechanism will usually contain too little "floodable" oil for successful water flooding. It is the depleted gas-drive reservoirs with only moderate connate-water saturations that, as a class, may be expected a priori to have enough residual-oil saturation to warrant water flooding. Even these, however, will be automatically eliminated if they have suffered accidental water flooding over the reservoir as a whole because of leaking casing and intrusion of stray waters. Moreover, when the connate-water saturation is so high (of the order of 50 per cent) that the residual-oil saturation left by gas drive is comparable with that resulting from water flushing, water flooding will again offer little promise of success. Water flooding of formations with residual-oil saturations as low as 25 per cent will always be a hazardous undertaking, and in most[2] districts profitable operations will be unlikely unless the residual-oil saturation exceeds 35 per cent.

Substantial reservoir uniformity is the other major physical criterion for successful water flooding. Unfortunately it is even more difficult to fix the limits of uniformity than that of the required residual-oil saturation. It is clear, however, that, if, for example, the formation contains a stratum of limited thickness with a permeability 25 times that of the average of the remainder of the section, rapid channeling and by-passing will develop. Unless this zone can be located and shut off, the average water-oil ratios will soon become too high for profitable operation in most fields. The channeling tendency due to the permeability contrasts will be aggravated by the lower depletion pressure that may obtain in the highly permeable zones. Moreover these strata will contain less residual oil than the tighter layers, and their flooding will lead to relatively lower oil recoveries than the flushing of the less depleted members of the pay.

[1] There is some evidence, however, that the residual-oil saturations left by water flushing may be lower when the water invades a sand with a partial gas saturation than when it is displacing oil from a fully saturated formation; cf. H. Dykstra and R. L. Parsons, API meetings, Los Angeles, Calif., May, 1948.

[2] For California projects it has been estimated that 55 per cent residual-oil saturation is desirable to ensure successful water flooding [cf. N. Van Wingen and N. Johnston, *Oil and Gas Jour.*, **45** (July 13, 1946)]. It may be noted, too, that whereas initial-oil contents of 350 to 400 bbl/acre-ft have given profitable flooding operations in the Bradford area in Pennsylvania, successful flooding of the Bartlesville Sand, Oklahoma, has generally required initial-oil contents of 600 to 700 bbl/acre-ft. These differences are, of course, the consequence of both economic and physical factors, as the price of the oil, reservoir uniformity, etc.

Areal continuity and uniformity of the flooded reservoir are also a prerequisite for profitable operation.   Lenticular pays will lead to an irregular geometry of the injection front with low sweep efficiencies.   Formation lenticularity will make it only a matter of chance if the well pattern is suited to an efficient recovery of the oil being displaced by the injected water.

In the close-spacing water-flooding projects used in shallow fields an increased oil-production rate due to water injection usually develops within 2 to 6 months after the beginning of injection.   The time in specific projects will evidently depend on the absolute well spacing, rate of injection, fill-up porosities, and formation uniformity.   It is generally planned so that the operating life does not exceed 10 years.   The cumulative water-oil ratio for most successful floods lies in the range of 8 to 20.   The major part of the oil production is obtained after an input of about 3 to 5 reservoir fill-up volumes.

A five-spot pattern is most often used in actual flooding operations.   The theoretical differences between the various ideal patterns generally do not warrant changing radically the pattern that will fit most conveniently into the well distribution available from the primary-production development. Some operators consider it undesirable to convert old producers into injection wells, whereas others have not found such conversion disadvantageous. There is no evidence that the well spacing is in itself an important factor in determining the oil-recovery efficiency.   Its choice is controlled by the economic balance between the investment cost and the effect of the spacing on the operating life.[1]   For a given injection rate the time scale for the recovery will be proportional to the square of the spacing distance.   The average absolute permeability[2] is likewise a factor entering mainly in the economic evaluation of a project and the choice of spacing rather than the estimation of ultimate recoveries or efficiency.   There is no evidence that a particular permeability range is especially advantageous for water flooding, although very low permeabilities will necessitate high injection pressures and close spacings to complete the operations in a reasonable time, and they will be correspondingly less profitable than projects in higher permeability strata, if other factors are the same.

It is generally assumed, on the basis of operating experience, that profitable water flooding is not feasible if the reservoir oil viscosity exceeds 20 to 25 cp.   The completion of operations when the reservoir crude is of high

---

[1] A detailed discussion of the economic aspects of water-flooding operations, especially with respect to the role played by the well spacing, has been given by S. T. Yuster, *Producers Monthly*, **12**, 18 (November, 1947); cf. also J. K. Barton and R. G. Prentice, *Producers Monthly*, **12**, 27 (February, 1948).

[2] The permeabilities of significance are the effective permeabilities to the water and oil, rather than the air permeabilities.

viscosity requires a longer period of high water-oil-ratio production. Once water break-through has developed in the most permeable zones, the rate of flushing of the tighter strata will be relatively low if the oil viscosity is high, and the water-oil ratios will be correspondingly greater. In cases of extremely high oil viscosity the microscopic displacement efficiency by the water may also be greatly reduced, and the water may channel through as if it were a nonwetting phase, leaving high residual-oil saturations. For moderate-viscosity oils, however, it is doubtful that the flushing efficiency is greatly affected by the oil viscosity. "Flood-pot" tests[1] on cores from Mid-Continent Sands showed the same residual-oil saturation, statistically, for a viscosity range of 6 to 20 cp and definitely higher residual-oil saturations only for samples in which the oil viscosity was 49.3 cp.

In the eastern water-flooding operations emphasis is placed upon achieving maximum water-injection rates. Injection pressures are maintained as high as possible without leading to lifting of the overburden (0.8 to 1.4 psi/ft depth) or equivalent development of water break-through when the bottom-hole pressures become of the order of magnitude of the pressure head of the overburden.[2] While the primary purpose of high injection pressures and rates is the reduction in operating life, there is no agreement regarding the effect of injection rates on the efficiency of the recovery on the basis of either field or laboratory investigations. Experiences in the Mid-Continent district have been interpreted[3] as implying that low injection rates are more conducive to high recovery efficiency, whereas the reverse has been claimed[4] as the implication of flooding experience in Pennsylvania. Greater recoveries have also been reported[5] in using low velocities of water advance in laboratory flooding experiments on cores from Mid-Continent reservoirs. On the other hand, tests[6] on long Second Venango cores from Pennsylvania

[1] Cf. W. A. Heath, *API Proc.*, **24** (IV), 171 (1943). Cf. also H. Dykstra and R. L. Parsons, API meetings, Los Angeles, Calif., May, 1948, who found the variation in flooding recovery with the oil viscosity to be rather slow for oil-water viscosity ratios less than 10.

[2] Cf. S. T. Yuster and J. C. Calhoun, Jr., *Producers Monthly*, **9**, 16 (February, 1945), and Dickey and Andresen, *loc. cit.*, p. 34.

[3] R. C. Earlougher, *AIME Trans.*, **151**, 125 (1943).

[4] H. M. Ryder, *Petroleum Eng.*, **62**, 119 (January, 1945); cf. also *Producers Monthly*, **11**, 18 (May, 1947).

[5] Earlougher, *loc. cit.*

[6] R. A. Morse and S. T. Yuster, *Producers Monthly*, **11**, 19 (December, 1946). Unfortunately the cores used in these experiments were inherently unsuited for water flooding and gave residual-oil saturations of 35 to 50 per cent. Earlier work on Bradford cores [J. C. Calhoun, Jr., R. L. McCormick, and S. T. Yuster, *Petroleum Eng.*, **16**, 82 (January, 1945)] gave increasing recoveries with increasing pressure gradients, but these experiments appear to have been seriously affected by end effects. Flooding experiments on unconsolidated sand columns have also been reported as showing greater

showed no definite effect of the pressure gradient on the residual-oil saturation, and still others[1] on cores from the First Venango Sand did indicate increased recoveries with increasing pressure gradient. While the reported discrepancies possibly may be due to differences in the nature of the formations and fluids involved, it is not clear, from a physical point of view, why a significant effect of the velocity on the recovery should be expected, unless the order of magnitude of the gradients is equivalent to the threshold pressure for displacing such segregated oil masses as may be trapped in localized parts of the rock. In any case, until the matter is clarified it would appear that for practical purposes the simple economic factors related to the injection rates may serve as the primary control in choosing the injection pressures, up to the break-through value. Similar economic considerations, rather than differences in the physical processes, are the determining factors in planning the operations and the injection- and producing-well pressures so that the producing wells are to be operated either by pumping, as is the most common practice, or by flowing.[2]

Delayed flooding, in which the producing wells are not placed on production or drilled at all until the injection wells have taken enough water to substantially fill up the void space, is based on the sound principle of attempting to equalize the water entry into the tight and permeable strata before water break-through. In some cases, as in Allegany field projects, it has been reported as achieving higher recoveries than conventional flooding practices. For successful application of this method it is necessary

---

recoveries when high flooding velocities were used [F. G. Miller, *U.S. Bur. Mines Rept. Inv.* 3505 (October, 1941)]. Still other flooding experiments on a long consolidated sandstone core (1,660 md) containing both interstitial water and oil showed no effect of pressure gradient on the oil recovery over a range of 0.70 to 3.42 psi/ft [cf. C. R. Holmgren, *Petroleum Technology*, **11**, 1 (July, 1948)].

[1] R. V. Hughes, *Producers Monthly*, **11**, 8 (July, 1947); cf. also J. N. Breston and R. V. Hughes, *Journal of Petroleum Technology*, **1**, 100 (1949). In the latter paper are reported the results of a number of flooding experiments with a First Venango consolidated sand core and a core from Rushford, N.Y., similar to the Third Bradford Sand. For both cores the oil recoveries decreased with decreasing pressure gradients, the curves dropping sharply at gradients of 0.5 psi/ft and about 7 psi/ft for the First Venango and Rushford cores, respectively. While there appears to be no reason to doubt the immediate validity of these experiments, they make it difficult to understand the generally accepted high recoveries from complete-water-drive fields (cf. Sec. 14.13), if these or similar data should be considered as applicable to other sand formations and to natural water-flooding processes.

[2] Cf. R. C. Earlougher, *Oil Weekly*, **122**, 38 (Aug. 26, 1946); T. F. Lawry, *Oil Weekly*, **123**, 54 (Nov. 18, 1946); S. T. Yuster, *Producers Monthly*, **12**, 18 (November, 1947). In additional experiments on the effect of the pressure gradient on water-flooding recovery, H. Dykstra and R. L. Parsons, API meetings, Los Angeles, Calif., May, 1948, found no variation in recovery below interface advance velocities of 2 ft/day, with only a slight effect, of doubtful reality, at velocities as high as 40 ft/day.

that the various strata be separated by shale breaks or equivalent barriers so as to be free of vertical intercommunication. Otherwise the more rapid initial water intrusion into the more permeable zones may lead to cross flow into the tighter layers ahead of the water advance in the latter, thus trapping the intervening oil masses, so that their displacement to the well will be associated with excessive water-oil ratios.

In actual water-flooding operations, attention must be given to the treatment and nature of the injection water to prevent corrosion and plugging of the formation and to achieve an over-all maximum injection capacity. Corrosion is usually prevented by maintaining a high pH in the water by the addition of lime, sodium bisulfite, or similar chemicals.[1] Aeration, the addition of bactericides, and filtration[2] are measures commonly used to minimize plugging of the formation by the injection water. And acidization has had successful field trials in the Bradford area,[3] as a means of increasing the water-injection capacity.

As a result of studies of the interaction of intergranular clays with water and its effect on the water permeability (cf. Sec. 3.7), it appears that the use of brines or low pH waters should be definitely advantageous,[4] as compared with fresh water, in achieving high injection rates when the formation is a "dirty" sand.[5] The laboratory experiments have been confirmed by field tests in the Bradford field, which show the injection capacities to be markedly increased on changing from fresh water to brine. However, in clean sands or limestones fresh waters have been entirely satisfactory.[6]

The problem of minimizing the unfavorable effects of permeability stratification has been given much study in the last several years. While no universal remedy has been developed, various methods have been tried with successful results under favorable circumstances. Selective shooting,[7] in which the explosive is mainly concentrated opposite the tighter strata, so as to equalize their local injection capacity as compared with the more permeable zones, is coming into widespread use in water-flooding opera-

---

[1] J. DePetro, *Producers Monthly*, **9**, 16 (June, 1945); R. J. Pfister, *Producers Monthly*, **9**, 24 (June, 1945).

[2] F. B. Plummer, *Producers Monthly*, **9**, 20 (June, 1945).

[3] R. J. Pfister, *Producers Monthly*, **10**, 24 (July, 1945).

[4] R. V. Hughes, *Producers Monthly*, **11**, 13, 12, 10 (February, March, April, 1947); R. V. Hughes and R. J. Pfister, *Petroleum Technology*, **10**, 1 (March, 1947); L. E. Miller, *Producers Monthly*, **11**, 35 (November, 1946).

[5] If high connate-water saturations are associated with the clay content, the inherent susceptibility of the formation to successful flooding will be low because of the low oil saturations that will probably be left by the primary gas-drive depletion.

[6] In all cases the injected water should be chemically "compatible" with the connate water so as to avoid internal precipitation and plugging.

[7] H. M. Ryder, *Producers Monthly*, **6**, 16 (April, 1942).

tions.　Chemical plugging of undesirable exposed strata[1] has been used successfully under suitable conditions.　Very favorable results have been reported[2] in the selective plugging, or shutoff, of high-permeability strata in water-injection wells by treatment with stabilized resin emulsions.[3]

Even during the early development of water-flooding techniques, means were sought for increasing the microscopic displacement efficiency of the water-flushing mechanism.　It had been assumed that by adding suitable agents to the water it would "wash" the sand more effectively than would untreated water.　Sodium carbonate was used in early experiments, with negative results and some evidence that plugging resulted from its use. More recently, surface-active or neutral agents of various kinds have been tried in laboratory experiments.[4]　Among the various types of additives studied, oil-soluble wetting agents[5] and water-soluble surface-inactive compounds gave no increases in water-flooding recoveries.　Reactive gases injected into cores ahead of the water gave small reductions in residual-oil saturations, but not enough to be promising for practical application.　Only water-soluble wetting agents appreciably increased the oil recoveries.　Unfortunately, however, as might have been anticipated, such surface-active and polar compounds were highly adsorbed by the rock.　While there were definite increases in oil recovery, the cost of providing for enough wetting agent to replace the adsorption losses was such as far to exceed the value of the increased recovery.

Even if the problem of the adsorption of the surface-active compounds were overcome, it is doubtful that the use of addition agents will be of value in increasing the primary displacement efficiency, since it is the connate water rather than the injected water that appears to follow directly behind the oil bank, as is indicated both by laboratory[6] and field studies.[7] The action of injection-water additives should therefore be limited to the stripping phase of the oil-displacement process, assuming that the residual-

[1] Cf. H. T. Kennedy, *Oil Weekly*, **123**, 61 (Nov. 18, 1946).

[2] D. Martin, K. H. Andresen, and F. W. Ellenberger, *Producers Monthly*, **11**, 18 (April, 1947).

[3] The complete success of these procedures depends on the separation of the different strata by shale breaks or the absence of appreciable cross flow within the formation.

[4] P. L. Terwilliger and S. T. Yuster, *Producers Monthly*, **11**, 42 (November, 1946).

[5] Oil-soluble wetting agents added to oil have been found to lower the residual-oil saturations [cf. J. N. Breston and W. E. Johnson, *Producers Monthly*, **10**, 28 (May, 1946)]. It is doubtful, however, that such agents could have practical applicability in water-flooding operations.

[6] R. G. Russell, F. Morgan, and M. Muskat, *AIME Trans.*, **170**, 51 (1947); cf. also A. P. Clark, Jr., *Producers Monthly*, **11**, 11 (July, 1947).

[7] It is a common observation that the first water produced in water-flooding operations is saline, even though the injected water is fresh; cf., for example, C. H. Riggs, *Producers Monthly*, **9**, 12 (May, 1945).

oil mobility is not completely destroyed by the primary flushing mechanism. While the threshold pressures for displacing a discontinuous oil phase should be reduced on lowering the water-oil interfacial tension, such effects, if they occur, will be associated with high water-oil ratios. In any case, there is no evidence from field experience that the use of addition agents is of practical importance.

In evaluating the probability of successful water-flooding operations, appeal must be made to the basic criteria previously mentioned, namely, reservoir continuity and uniformity and the presence of relatively high residual-oil saturations.[1] The determination of the former will involve the application of core analysis, logging data, the geological interpretation of the reservoir structure, and a study of the primary-production performance of the various wells in the field. The latter is a more difficult problem. Even if new wells are drilled and cored for the purpose of obtaining oil-saturation data, drilling-fluid invasion and flushing of the residual oil will make most such determinations highly questionable, although it is apparently felt in some districts that, with cable-tool coring, water invasion is not a serious factor.[2] Chip coring has been reportedly[3] used in Pennsylvania quite successfully, but the samples available for analysis are quite small. An alternative method is to compute the residual-oil saturation from the past cumulative recovery. If the latter, expressed in barrels, be denoted by P and the residual-oil saturation by ρ_{or}, the two will be related as

$$\rho_{or} = \beta_f\left(\frac{1 - \rho_w}{\beta_i} - \frac{P}{7{,}758.4Ahf}\right), \tag{1}$$

where β_f is the formation-volume factor of the oil at depletion, $i.e.$, at the beginning of the operations, β_i the initial formation-volume factor, ρ_w the connate-water saturation, A the productive area, h the net pay thickness, and f the porosity. The connate-water saturation can be determined by coring with oil or the capillary-pressure method, as discussed in Chap. 3. While Eq. (1) will give only the average residual-oil saturation, it will at the same time provide a direct measure of the gross oil content remaining in the formation. Of course in most of the fields that are now depleted the various data involved in Eq. (1), except for the cumulative recovery and productive area, will not be available except by the drilling and coring of new wells and the analysis of the reservoir fluids.

Even if the residual-oil saturation has been established or estimated,

---

[1] It is assumed that the porosities and pay thickness are sufficiently high for the residual-oil saturations to represent a satisfactory total volume of recoverable oil.

[2] There have been reports, however, that cable-tool cores show even more flushing than cores taken with rotary tools under similar conditions [cf. H. M. McClain, *Producers Monthly*, **11**, 31 (April, 1947)].

[3] Cf. H. M. Ryder and D. T. May, *Producers Monthly*, **2**, 16 (March, 1938).

the final saturations after flooding, or the water-flooding recovery, are still subject to considerable uncertainty. These are often determined experimentally by "flood-pot" tests, in which partly or completely oil-saturated cores, with their connate water, are flooded in the laboratory and subsequently analyzed for their residual-oil content. If the samples are representative and the experiments are not seriously affected by end effects, the results should be significant, although there may be some uncertainty that these requirements have been satisfied. Often a residual-oil saturation, as 20 to 25 per cent, is simply "estimated" and used as a basis for the recovery evaluation. Unfortunately, in view of the many other uncertainties associated with the final prediction of the recovery to be expected, it is doubtful that higher precision than such "estimates" is warranted,[1] once it has been determined that the formation will take water and reduce the oil saturations to a value in the expected range. In any case the evaluation should be made conservatively. Moreover sweep-efficiency factors, perhaps of the order of 50 per cent, should be applied to correct not only for areal incompleteness of the flooding but also for the effects of stratification. And unless successful experience is available on other parts of the same reservoir or producing formation, experimental pilot floods should always precede the development of large-scale programs. Moreover effort should be made to unitize the operations over as large a continuous area as possible, so as to achieve increased operating efficiency and the more complete accumulation of reservoir information, provide greater flexibility in pattern development, and permit easier reservoir control.

It will be evident from the examples of water-flooding operations cited in Sec. 12.15 and many others that have been analyzed that additional recoveries comparable with those obtained by the primary solution-gas-drive recovery mechanism (of the order of 150 to 200 bbl/acre-ft) are not impossible. In extreme cases the water-flooding recovery may be even twice that prior to water flooding. On the other hand the operations may be profitable with considerably lower recoveries than those derived from the primary production, if the development and operating costs are correspondingly low. The ranges in recovery actually observed are fully consistent with the differences in residual-oil saturation that may be expected between the water-flushing and solution-gas-drive oil-producing mechanisms (cf. Chap. 14).

12.17. Secondary Recovery by Gas Injection; Theoretical Considerations.—Although the deliberate injection of gas into a reservoir to stimulate oil recovery was first tried in 1903, in the Macksburg field, Washington

---

[1] However, the recent studies of H. Dykstra and R. L. Parsons, API meetings, Los Angeles, Calif., May, 1948, in which the effects of permeability stratification were taken into account, suggest a procedure for calculating water-flood recoveries that seemed to check quite well with those observed in several actual water-flooding projects.

County, Ohio, long before even experimental water flooding was attempted, gas injection as a secondary-recovery operation has not been so systematically and intensively studied as water flooding. Perhaps this has been due to the less severe operating problems and lower investment costs required to undertake gas injection, so that potential failures did not represent such great risks. In any case it is only in the last few years that any investigations, either experimental or theoretical, have been reported on attempts to develop even a semiquantitative description of the response of depleted reservoirs to gas injection. While the theory to be presented here has been compared with field observations only in isolated instances, it constitutes the only formal procedure thus far developed for correlating the behavior of depleted reservoirs, subjected to gas injection, with the basic physical principles of fluid flow through porous media. On the other hand it is to be recognized that it is an approximate treatment and refers only to the declining-production phase, after the initial rise and peak in production rate has been passed.

If the producing formation is substantially depleted (by solution-gas drive) prior to gas injection and the free-gas-phase saturation has become continuous, it is unlikely that there will develop an oil bank ahead of injected gas[1] comparable with that occurring in water flooding. The oil saturation will undoubtedly be reduced more rapidly near the injection wells. However, the expansion of the free-gas phase on approaching the producing wells will tend to limit the build-up of the oil saturation as the producing well is approached. While there will be a transient period of rising production rate associated with the first passage of the injected gas to the producing wells, a quantitative description of this phase of the problem and of the variable saturation distribution would be extremely difficult. For the purpose of predicting the gross behavior of gas-injection systems after their production peaks have passed it will suffice to assume that the oil saturation is uniform throughout. Moreover, since the formation will be considered as completely depleted with respect to its solution-gas content, its contribution to the gas production will be neglected,[2] and the production of the oil will be assumed to result only from the sweeping or stripping by the injected gas.

On the basis of these and other assumptions, to be indicated below, it is possible to relate the decline in oil saturation to the total gas-injection volume, by applying the permeability-saturation characteristics of the

---

[1] Such banking of oil by gas drive is also unlikely in normal pressure-maintenance operations, although such effects are suggested by observations that have been made in a gas-injection project in Louisiana, and is exhibited in laboratory experimentation when the oil displacement is controlled completely by capillary-pressure forces.

[2] This assumption will, of course, be invalid if the rate of gas production appreciably exceeds the rate of injection.

porous medium.[1]   Most data on the latter (cf. Sec. 10.12) indicate that
after a free-gas saturation of the order of 20 to 25 per cent has been built up,
the gas-to-oil permeability ratio k_g/k_o varies approximately exponentially
with the oil saturation, ρ_o [cf. Eq. 10.12(2)], *i.e.*,

$$\frac{k_g}{k_o} = e^{a-b\rho_o}; \qquad b\rho_o = a - \log \frac{k_g}{k_o}. \tag{1}$$

If the solution gas be neglected,

$$\frac{k_g}{k_o} = \frac{\mu_g}{\gamma\mu_o\beta_o} \frac{Q_g}{Q_o}, \tag{2}$$

where μ_g, γ are the viscosity and relative density of the injected gas, μ_o, β_o
are the viscosity and formation-volume factor of the oil, and Q_o, Q_g are
the local volume fluxes of oil and gas.   If now the pressure variation through
the system be also neglected, the factor $\mu_g/\gamma\mu_o\beta_o$ can be taken as constant.
Moreover, under the assumption of instantaneous steady-state flow, which
basically underlies this treatment, Q_o and Q_g may be considered as proportional or equal to the total throughflow of oil and gas, respectively.   It
follows, then, that

$$b\frac{d\rho_o}{dt} = -\frac{b\beta_o}{Ahf}Q_o = \frac{1}{Q_o}\frac{dQ_o}{dt} - \frac{1}{Q_g}\frac{dQ_g}{dt}, \tag{3}$$

where A is the area involved in the operations, h the pay thickness, and
f the porosity.

The integral of Eq. (3) is readily shown to be

$$\left. \begin{aligned} Q_o &= \frac{Ahf}{b\beta_o}\frac{d\log V}{dt} = \frac{AhfQ_g}{b\beta_o V}, \\[2mm] V &= \int_0^t Q_g\,dt + V_i, \end{aligned} \right\} \tag{4}$$

where

and represents the cumulative gas injection, or throughput, plus an equivalent amount V_i corresponding to the initial oil-production rate ($t = 0$).
The cumulative oil recovery $\overline{Q}$ between cumulative gas-injection volumes
V_1 and V_2, at t_1 and t_2, is

$$\overline{Q} = \frac{Ahf}{b\beta_o}\log\frac{V_2}{V_1}. \tag{5}$$

If the gas or injection throughput rate Q_g is constant, Eq. (4) gives for
the time variation of the oil-production rate

$$Q_o = \frac{Ahf}{b\beta_o(t + V_i/Q_g)}. \tag{6}$$

---

[1] The treatment given here follows that of M. Muskat, *Producers Monthly,* **10,** 23
(February, 1946); cf. also R. J. Day and S. T. Yuster, *Producers Monthly,* **10,** 27 (November, 1945).

Thus the oil-production rate varies inversely as the time or the cumulative gas-injection volume.

Although Eq. (6) cannot be expected to be of universal validity because of the many assumptions underlying its derivation, it is of interest that the oil-production histories from some gas-repressuring projects have been

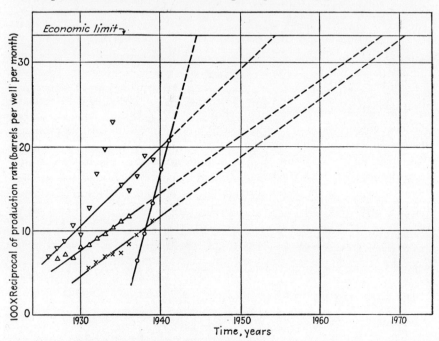

FIG. 12.36. A plot of reciprocals of the oil-production rate vs. the time for several gas-injection leases in the Titusville–Oil City area of Pennsylvania. o, lease I. ∇, lease V. Δ, lease VII. x, lease IX. (*After Day and Yuster, Producers Monthly, 1945.*)

found to follow Eq. (6) with a good approximation. Figure 12.36 gives several examples of linear relations between the time and the reciprocal of the production rate, as required by Eq. (6), for several projects in the Titusville–Oil City district of Pennsylvania.[1] It will be seen that the data show satisfactory linearity except for lease V, where a reduction in injection rate during 1931–1934 may have been the cause of the deviations of the corresponding production-rate points. The upper horizontal line of Fig. 12.36,

[1] Figure 12.36 is taken from Day and Yuster, *loc. cit.*, who in turn replotted the basic data of P. A. Dickey and R. B. Bossler (Chap. 9, "Secondary Recovery of Oil in the United States," API, 1942). The latter authors suggest that an empirical equation governing the oil-production-decline histories has the form $Q_o = A + B(t - C)^{-n}$, which is evidently a generalization of Eq. (6). Cf. also J. C. Calhoun, Jr., *Oil and Gas Jour.*, **47**, 95 (Jan. 6, 1949) and 95 (Jan. 20, 1949) for plots of field data confirming Eq. (5).

corresponding to a production rate of 3 bbl/month per well, indicates the simple procedure for predicting the length of the operations if a lower limit of production rate be set for profitable operation. Of course, Eq. (6) and such plots as those of Fig. 12.36 refer only to the period of declining oil production, after the transient-period rise that always follows the starting of gas injection when the latter is effective at all.

If the production-decline data do give straight lines when plotted as in Fig. 12.36, their slopes should represent the value of $b\beta_o/Ahf$. As all factors except b will be determinable from the general reservoir data pertaining to the formation being repressured, b can be calculated from the slope.[1] The values so found for the leases referred to in Fig. 12.36 and another lease M are I, 166.7; V, 90.1; VII, 42.6; IX, 40.8; M, 55.2. Increasing values of b imply more rapid rises in the gas-to-oil permeability ratio, [cf. Eq. (1)] and less efficient gas-drive oil expulsion. It will be recalled from Sec. 10.12 that the values of b derived from both field data on the primary performance of solution-gas-drive reservoirs and laboratory permeability-saturation studies on consolidated porous media had a range of 15 to 30. While the larger values for b derived from Fig. 12.36 may represent in part differences in the basic character of the producing formations involved in the gas-injection projects in Pennsylvania and those of the solution-gas-drive reservoirs referred to in Sec. 10.12, it is likely that they largely reflect the inherently lower efficiency of the gas-injection operations. For it would be expected that permeability stratification and inhomogeneities would more seriously impair the efficiency of oil recovery by fluid injection than the solution-gas-drive mechanism governing the natural depletion period. Moreover, direct-gas-drive experiments[2] on long cores of Pennsylvania sands, in which permeability variations would be less serious than in actual formations, gave a lower range of values of b, corresponding to the desaturation of their oil content with continual gas throughflow, namely, 21.9 to 39.7, as the pressure gradient was varied.

The change in oil saturation, $\Delta\rho_o$, corresponding to any time interval during which the oil-production rate declines from Q_1 to Q_2, for constant injection rate, can readily be computed once the coefficient b has been determined. By combining Eqs. (1) and (2) it follows that

$$\Delta\rho_o = \frac{1}{b} \log \frac{Q_1}{Q_2}. \tag{7}$$

The corresponding cumulative recoveries are evidently $Ahf\,\Delta\rho_o/\beta_o$. Calculations of the saturation reductions and total recoveries, as well as the

---

[1] If the gas-injection or throughput rate Q_g is variable, an application of Eq. (5) will be more convenient for the determination of b.

[2] Day and Yuster, *loc. cit.*

observed recoveries, for the previously mentioned leases are given in Table 2.[1] It will be noted that the declines in oil saturation corresponding

TABLE 2.—RECOVERIES IN GAS-INJECTION PROJECTS

Lease	Calculated cumulative recoveries (10^3 bbl)	Observed cumulative recoveries (10^3 bbl)	Reduction in oil saturation, %	Time period
I	11.3	11.5	0.81	1936–1940
V	—	146.2	1.64	1915–1940
VII	117.4	114.2	2.43	1926–1940
IX	53.2	53.0	2.33	1930–1940
M	48.8	47.7	1.68	1933–1945

to the cumulative recoveries do not exceed 2.5 per cent,[2] even when the operations have been continued for more than 10 years.

The throughput gas-oil ratios are also readily obtained from the simplified theory outlined above. From Eq. (4) it follows that the gas-oil ratio R is

$$R = \frac{Q_g}{Q_o} = \frac{b\beta_o}{Ahf} V = \frac{R_i}{V_i} V, \tag{8}$$

where R_i is the gas-oil ratio corresponding to V_i. Thus the gas-oil ratio will increase in proportion to the cumulative gas-injection volume.[3] It will therefore also increase linearly with the time if the injection rate is constant. On applying Eq. (5), R can be expressed in terms of the cumulative oil recovery $\overline{Q}$ as

$$R = R_i e^{R_i \overline{Q}/V_i}, \tag{9}$$

indicating that R will increase exponentially with the cumulative recovery. These relations show at once the increasing inefficiency of the gas injection as the operations continue and the oil saturations decline.

To the extent that steady-state gas throughflow is actually established, the steady-state homogeneous-fluid network theory developed in previous sections can be applied to calculate the relation between the gas throughput and the pressure differential. If $\overline{Q_s}$ denotes the dimensionless steady-state flow capacity of the network, as given by Eqs. 12.14(5) to 12.14(8), the value of Q_g will be given by

[1] This tabulation is taken from Day and Yuster, loc. cit.

[2] In view of the fact that in the lease VII project the total gas-drive depletion recovery was only 150,000 bbl (cf. footnote, p. 727), it would appear that the oil-saturation reductions listed in Table 2 may be considerably lower than actually achieved.

[3] The linear relation between the gas-oil ratio and the cumulative gas volume implied by Eq. (8) has also been confirmed by field observations in several instances [cf. J. C. Calhoun, Jr., Oil and Gas Jour., 47, 103 (Jan. 13, 1949)].

$$Q_g = \frac{2\pi k_g h \, \Delta p^2}{\mu_g} \overline{Q}_s, \tag{10}$$

where Δp^2 is the difference in squares of the pressure between the injection and producing wells, h the pay thickness, μ_g the gas viscosity, and k_g the current value of the gas permeability. As injection is continued, k_g will increase, and the pressure drop required to maintain the gas throughput should decrease, if plugging does not develop.

It was noted that the laboratory-determined values of b for Pennsylvania cores varied for different pressure gradients. These values were obtained from semilogarithmic plots of the residual-oil saturation vs. the total gas throughflow,[1] which were linear below oil saturations of about 85 per cent. The variation in b from 21.9 to 39.7 corresponded to those in the pressure gradient of 3.9 to 0.13 psi/ft. The higher gradients thus gave higher oil-recovery efficiencies. Apparently the higher gradients even overcame the effect of the reduced-volume throughputs within the cores associated with the higher mean pressures. Aside from the practical importance of these observations, if established as generally valid,[2] is their significance in implying that the permeability-saturation relationship depends on the pressure gradient. This, in turn, implies that either the gas or the liquid relative permeabilities or both are not determined merely by the rock and fluid saturations but are also sensitive to the pressure gradient.

12.18. Field Experience in Gas Injection.—In most states other than Pennsylvania and New York many more gas-injection projects have been undertaken than water-flooding projects. However, except for those referred to in Sec. 12.17, very few attempts have been made to interpret the observations in terms of basic reservoir properties. Nevertheless to illustrate the general character of the operating histories of gas-injection projects several examples will be briefly reviewed in this section.[3]

Figure 12.37 gives the oil-production history of the eastern part of the Delaware-Childers field,[4] Nowata County, Okla. This field was discovered in 1906, on completion of a well at 622 to 638 ft in the Bartlesville Sand. After reaching a peak of 4,800,000 bbl in 1909 the production rate declined rapidly. Vacuum was applied in 1913 with no perceptible effect.

[1] R. J. Day and S. T. Yuster, *Pennsylvania State Coll. Bull.* 39 (Jan. 19, 1945).

[2] More recent experiments on a long consolidated core (1,660 md) containing both interstitial water and oil gave more efficient gas-drive recoveries at low pressure gradients in the range of 0.25 to 1.04 psi/ft (cf. Holmgren, *loc. cit.*), although the ultimate recoveries appeared to be greater at the higher pressure gradients.

[3] Still other examples of gas-injection projects, in North Texas, are discussed by R. Gouldy and G. Stine, *Producers Monthly*, **12,** 21 (March, 1948).

[4] This figure and the related discussion are taken from K. H. Johnston and C. H. Riggs, *U.S. Bur. Mines Rept. Inv.* 4019 (December, 1946). The gas-injection curve in Fig. 12.37 was added from another plot in the report.

By 1925, when air injection was started, 23,000,000 bbl of oil had been produced, from the 6,549 developed acres of the eastern part of the field. The permeability of the producing formation averages 60 to 90 md, and its average porosity is approximately 20 per cent. The gross oil-productive volume of the eastern part of the field was 128,000 acre-ft.

The initial gas-injection operations were considerably expanded after 1932, and by July, 1945, there were 482 injection wells in the field, with a compressor system capable of supplying 4,550,000 ft^3 of air and 13,400,000 ft^3

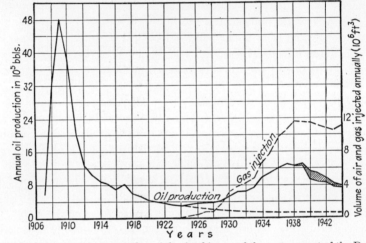

Fig. 12.37. The oil-production and gas-injection history of the eastern part of the Delaware-Childers field, Oklahoma. Shaded area represents contribution due to water flooding.

of gas per day. Injection pressures ranged from 20 to 160 psi. The production rate of the area at the time injection was started was 26,000 bbl/month. Both staggered-line-drive and five-spot patterns were used over the injection area. By July, 1945, 51,000,000,000 ft^3 of air and gas had been injected. During the injection of this volume of air and gas approximately 17,000,000 bbl of oil was produced. It is estimated that 13,000,000 bbl of this total was in excess of that which would have been produced without fluid injection. The gross injection-gas–oil ratio of this additional oil was thus 3,920 ft^3/bbl. Of the 6,448 acres developed for secondary-recovery operations, 5,570 acres were being repressured on July 1, 1945. In 10 per cent of the area the recovery exceeded 5,000 bbl/acre. It is estimated that the oil saturation had been reduced to 47.6 per cent by primary-production methods from its original value of 65 per cent, at initial reservoir conditions. By July, 1945, it had been reduced further to an average value of 39.8 per cent, as the result of air and gas injection.[1]

---

[1] Evidently this rather large decrease in oil saturation, as compared with the decreases observed in the Titusville–Oil City area of Pennsylvania (cf. Table 2), implies considerably lower values of the constant b [cf. Eq. 12.17(7)].

The crosshatched area in Fig. 12.37 represents the additional recovery due to water-flooding part of the field. Of the four flooding projects that have been tried three were failures and were abandoned. Of the total indicated gain in recovery of 680,000 bbl, 615,000 bbl was obtained from the fourth flood on a 96-acre area, started in 1937. The latter recovery, to July, 1945, was the result of injecting 9,800,000 bbl of water.

A gas repressuring operation that has been economically successful, al-

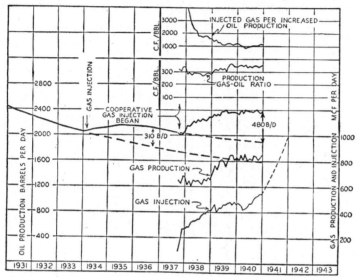

FIG. 12.38. The production history during the gas-injection operations in the north end of the Homer field, Louisiana. (*After Vitter, "Secondary Recovery of Oil in the United States."*)

though the increased recovery has been hardly even comparable with that in the Delaware-Childers field, is the project in the Homer field, Claiborne Parish, La., started in 1934. The formation involved is the Nacatoch Sand, at 675 to 1,150 ft below sea level, to the north of the major fault in the field, and covers 1,347 acres. The Homer field was discovered in January, 1919, and reached a peak production of 22 million barrels in 1920. The producing zone of the Nacatoch Sand averages 50 ft in thickness. It is calcareous and poorly cemented, with an average porosity of 31 per cent and permeability of 300 md. The reservoir was rapidly depleted by a solution-gas drive. The average recovery to January, 1939, was 25,000 bbl/acre. Vacuum was applied to the field in 1921.

Gas injection was first tried on a 180-acre lease in February, 1934. The success achieved in these operations (cf. Fig. 12.38[1]) led to the start of a cooperative project over the whole of the north end of the field in Septem-

---

[1] Figure 12.38, as well as the discussion of the Homer operations, is taken from A. L. Vitter, Jr., Chap. 21, "Secondary Recovery of Oil in the United States." API, 1942.

ber, 1937. By the end of 1940 the operations included 255 producing wells and 23 injection wells, with an injection rate of 540,000 ft³/day. It will be noted from Fig. 12.38 that in January, 1941, the production rate was approximately 800 bbl/day greater than the extrapolated value indicated by the previous production history. The increased recovery by virtue of the cooperative injection, from September, 1937, through December, 1940,

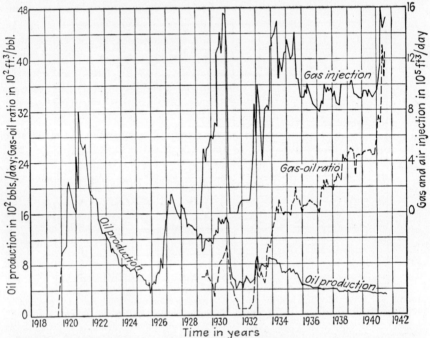

Fig. 12.39. The production history of the Red River field, Oklahoma, before and during the gas-injection operations.

was 368,000 bbl, and the total since injection was first tried in 1934 was 879,000 bbl. The volume of injected gas required per barrel increase in recovery is also plotted in Fig. 12.38. In December, 1940, it was approximately 1,000 ft³/bbl. This would be reduced to 700 ft³/bbl if the gain due to the initial injection operations were included.

An example of a gas-injection project showing the reaction of the reservoir to changes in the operating conditions is provided by the history of the Red River field, Tillman County, Okla., as shown in Fig. 12.39.[1] The field was discovered in October, 1919. The Cisco producing section lies at 1,500 to 1,625 ft. Its net thickness over the 270-acre block in the western part of the field, which had produced about 90 per cent of the total recovery and in which the gas-injection operations were conducted, averages

[1] The discussion related to Fig. 12.39 is taken from D. H. Stormont, *Oil and Gas Jour.*, **40**, 32 (Dec. 18, 1941). Figure 12.39 itself is a replot of graphs from the same source.

about 48 ft. It is lenticular and is limited by pinch-out of the sand. Its primary production was by gas drive, and there has been no evidence of significant water intrusion. The production rate, which had been declining since 1921, rose sharply following a well-deepening and new drilling program carried out in 1926.

By May, 1929, when gas injection was started, 4,375,300 bbl of oil had been produced, or 16,200 bbl/acre. The production rate was approximately 1,000 bbl/day. Gas was initially injected in 10 wells, chosen from among the 76 producers and distributed along the edges of the field. Vacuum, which had been previously applied to the field, was removed gradually to minimize the loss in production rate. During this period the API gravity of the oil rose from 39 to 41°, and the gasoline content of the casing-head gas dropped noticeably. While the injection rate into the 10 edge wells was being increased from an initial value of 60,000 to 500,000 ft³/day, evidence of channeling developed with but little response in the oil-production rate. Fifteen new wells were then drilled over the crest of the structure and in the interior of the field. By July, 1930, after this new program was completed, the production rate had risen to a peak of 1,570 bbl/day, or about 400 bbl/day more than when vacuum was being applied. The gas-injection rate was raised in the meantime to 1,000,000 ft³/day. In November, 1930, the gas injection was supplemented by air injection, which was maintained at a rate of about 375,000 ft³/day to March, 1931. At that time, gas injection was suspended because of a doubling of the price of the gas, and the air injection was also stopped because of breakage and corrosion difficulties that had been attributed to the use of air. As is evident from Fig. 12.39 the oil-production rate dropped sharply. Gas injection was started again in 4 wells in November, 1931, and full-scale injection was resumed in September, 1932. The production rate then rose to only about half of that prior to discontinuance of the injection. A flood in the latter part of 1935 caused another temporary shutdown and a drop of about 100 bbl/day production when operations were started again.

The gas-oil-ratio history of this project is also plotted in Fig. 12.39. Except for variations associated with the detailed operating conditions the gas-oil ratio rose gradually to about 2,600 ft³/bbl in January, 1941. After January, 1941, however, it showed a rapid increase and reached the value of 4,250 ft³/bbl by June, 1941, indicating the development of serious channeling. In September, 1941, gas was being injected in the 15 newly drilled wells, and a total of 275 bbl/day was being taken from 70 producing wells.

It is estimated that without gas injection the total ultimate recovery, to an abandonment rate of 100 bbl/day for the field, would have been 5,739,000 bbl. By September, 1941, the actual recovery had been 7,180,400 bbl. This represents an increased recovery of almost 5,400 bbl/acre. Continuation of the gas-injection operations to abandonment was estimated to lead to a total recovery of 7,492,000 bbl, which would give a gain due to gas injection of 31 per cent of the primary-production recovery.

12.19. Practical Aspects of Secondary-recovery Operations by Gas Injection.—It is almost axiomatic that the physical requirements for success in secondary-recovery gas-injection operations are basically the same as for water flooding, namely, continuity and reasonable uniformity of the producing formation and relatively high residual-oil saturations. Again, however, it is not feasible to define these requirements in a quantitative manner. But a recognition of these criteria even in general terms should serve to emphasize that predictions of success should be made conservatively if there is any doubt about their being satisfied in the particular project under consideration.

From a purely economic standpoint, reservoir uniformity is not so exacting a requirement for successful gas injection as it is for water flooding. Gas by-passing is expensive and wasteful because of the cost of supplying and compressing the gas, which is providing little oil expulsion. The same factors lead to increased operating expenses when channeling develops in water-flooding operations. In the latter, however, there is the additional lifting cost of handling excessive water production, whereas the operating cost of producing wells is not so sensitive to the gas-oil ratio. This difference is one of the reasons why water flooding has been rather unsuccessful in the Venango Sands of Pennsylvania, whereas air and gas injection has been quite profitable in spite of the variability of the formations.[1]

The significance of the residual-oil saturation in gas injection is somewhat different from its significance in water flooding. In the latter it is the difference between the residual-oil saturation at the beginning of flooding and the value, such as 25 to 30 per cent, that may be left by the water-flushing process that, together with the sweep-efficiency factor, is a measure of the expected recovery. In gas injection, however, it is the increase in gas saturation before the gas-oil ratios become excessive that often limits the recovery.[2] Thus an increase in gas saturation from 30 to 40 per cent will represent the same recovery when the initial-oil saturation is 35 per cent and the connate-water saturation is 35 per cent as when the oil saturation is 50 per cent and the connate-water saturation is only 20 per cent, if the porosity is the same in both cases. However, whereas gas injection might be quite successful under the first set of conditions, it would be most likely to be a total failure if applied to a formation having 35 per cent oil

---

[1] The high connate-water saturations and low oil saturations may have also been of importance in favoring gas injection as compared with water flooding.

[2] In laboratory experiments it is always possible to reduce the oil saturation to that left by water flooding by simply continuing the throughflow regardless of the gas-oil ratio. Direct evaporation probably contributes appreciably to the recovery if the experiments are continued to the range of high gas saturations.

saturation but only a 20 per cent connate-water content.   As implied by
the discussion of Sec. 12.17 the steady-state throughput gas-oil ratios will
increase exponentially as the free-gas saturation increases.   On the other
hand, when the connate-water saturations are high (45 per cent or greater),
the low oil saturations left even by the solution-gas-drive mechanism may
make additional recovery by the injection of gas also uneconomical, owing
to the low oil mobility.

When the connate-water saturations are low, the potential oil recovery
by water flooding will usually exceed that by gas injection.   From a practical standpoint, however, gas injection may still offer economic advantages,
if a gas supply is available, for there is usually less risk involved in utilizing
old wells with moderate reworking expense, thus minimizing the investment
cost.   Gas-injection operations are frequently developed without regard
to pattern regularity and with many more producing than injection wells,
as indicated in Table 3.[1]

TABLE 3.—WELL SPACING OF INJECTION AND PRODUCING WELLS IN REPRESENTATIVE
REPRESSURING PROJECTS

State	Average depth, ft.	Average pay thickness, ft.	Acres per injection well	Acres per producing well
Pennsylvania...	800	35	10.20	2.14
	1,000	40	5.40	1.30
	500	22	4.60	1.37
	2,200	27	41.50	11.30
West Virginia..	900	12	16.00	4.00
	1,700	15	423.00	29.10
	2,250	10	600.00	70.00
	2,300	25	150.00	17.50
Ohio..........	3,000	27	187.50	18.75
	300	25	19.50	3.12
Kentucky......	950	40	20.00	4.80
	1,000	40	22.40	5.42
	325	25	21.50	4.85
Illinois........	1,500	20	41.50	15.20
Kansas........	—	47	41.60	13.00
	—	40	53.00	9.60
Texas.........	1,150	45	7.40	3.70
	2,600	23	80.00	11.20
	1,900	35	16.00	3.00
	1,850	17	30.00	15.00

[1] Table 3, as well as Tables 4 and 5, are taken from P. D. Torrey, Chap. 7, "Secondary
Recovery of Oil in the United States," API, 1942.

As is evident from the examples of the last section the oil-recovery gains from gas injection, even when successful, are generally decidedly lower than those derived from successful water-flooding projects. The initial reaction to the gas injection is usually noted within 1 to 6 months, and unless there is a continual expansion in the operations, the peak in oil production is often observed in the second year. This peak usually does not exceed twice the rate before gas injection. Some typical data on injection and produced gas-oil ratios, as well as the injection pressures used in various oil-producing states, are listed in Table 4.

TABLE 4.—DATA RELATING TO THE GAS USED IN VARIOUS REPRESSURING OPERATIONS

State	Average intake pressure, psi	Average intake vol., ft³/bbl of oil recovered	Average producing gas-oil ratio, ft³/bbl
Pennsylvania....	—	11,000	
	—	23,600	
	—	15,600	
West Virginia...	87	10,250	
	31	3,670	
	40	11,500	5,690
	165	10,600	
	25	3,500	
	130	10,000	
	122	10,300	
	35	13,000	
	150	10,000	
	190	10,500	
Ohio..........	80	2,890	
	200	3,100	
	48	11,360	
Kansas........	160	800	2,700
	240	1,100	600
	135	1,450	2,200
Louisiana......	15	1,500	300
Texas..........	150	1,200	300
	230	2,600	4,300
	400	1,700	3,600
	75	150	130
	440	630	1,100

Ranges or average recoveries from repressuring projects in various formations are tabulated in Table 5.

TABLE 5.—ESTIMATED OIL RECOVERIES OBTAINED FROM AIR AND GAS REPRESSURING
OPERATIONS

State	Producing formation	Estimated ultimate recovery, bbl/acre
Pennsylvania..........	Venango Sands	220–2,570
	Bradford	150
	Gordon	1,000–1,400
West Virginia........	Big Injun	1,000–1,700
	Keener	1,500
Kentucky............	Wier	3,050
	Corniferous	2,700–3,300
Kansas..............	Bartlesville	1,250–2,000
Oklahoma...........	Bartlesville	1,300
Louisiana...........	Nacatoch	4,600
	Blossom	1,300

The low recovery indicated in Table 5 for the Bradford Sand is not typical of all the repressuring projects tried at Bradford. On the other hand, gas injection has not been widely used at Bradford simply because even the successful operations did not yield recoveries comparable with those obtained by water flooding.

The initial rise in production rate following the beginning of gas injection is largely due to the increase in pressure drop above the residual differential remaining at the termination of the normal depletion gas-drive phase.[1] If gravity drainage is the source of the "settled" production, the injection pressures will be superposed on the fluid head providing the gravity seepage. In addition the injected gas may serve as a supplementary driving pressure to the slow bleeding by solution-gas drive from the relatively undepleted tighter strata. While the lower permeability of these zones and their higher residual pressures will both tend to retard their invasion by the injected gas, whatever gas does enter these parts of the formation will give a higher displacement efficiency than the gas channeling through the depleted high-permeability strata.

The problem of permeability stratification and by-passing in gas injection is often remedied by the use of packers to segregate the zones of channeling when the latter can be identified and located and when the wells have not been shot. Selective plugging has also been tried. Successful experimentation on the injection of water, or "slugging," to reduce the permeability

[1] The increase in production rate after the normal depletion rates have become unprofitably low may itself lead to a significant part of the additional recovery by gas injection, especially in deeper and tighter formations.

to gas in the looser strata, which would take most of the water, has also been reported.[1] Here, too, the absence of appreciable cross-bedding permeability, due to shale breaks or equivalent lithologic conditions, is a prerequisite for the complete success of these remedial measures.

Although laboratory experiments gave negative results,[2] intermittent gas injection has apparently been found profitable in counteracting the effect of channeling in some nonuniform formations. This practice is based on the principle that the more rapid depletion of the gas content and pressure in the highly permeable zones, on continuing operation of the producing wells after shutting in the injection wells, will induce a crossflow bleeding into them from the tighter strata. On resumption of injection the oil will be swept out from the partly resaturated permeable layers more rapidly and efficiently than if it had to be driven out directly by gas invasion into the tight strata while by-passing continues in the permeable zones. If the parts of the formation of different permeability are in communication so as to permit cross flow and the shutdown periods of the injection wells are made sufficiently long (of the order of months),[3] it is reasonable to expect beneficial results from this method. The negative character of the laboratory experiments was probably due to the absence of such degrees of stratification in the test cores as obtain in actual extended producing formations.

The alternation of injection wells, to direct the injected gas to regions not rapidly swept by the gas from previous groups of injection wells, has been tried as a means of increasing the sweep efficiency. This procedure may be considered as essentially equivalent to a rotation of the sweep patterns for the individual groups of injection wells and should theoretically reduce or completely eliminate the "dead" zones left by the separate patterns (cf. Figs. 12.16 to 12.19). It is doubtful, however, that the gain in sweep efficiency so achieved will more than compensate for the economic loss due to reduced injection rates and lengthened operating life when the injection wells are used only on a "part-time" basis. Moreover, the composite sweep pattern of all the injection wells, if used simultaneously, will also be of higher efficiency than that of the individual groups, if their distribution is chosen properly.

Some of the laboratory data referred to in Sec. 12.17 indicated that higher gas-injection rates will increase the recovery efficiency. As in the case of

---

[1] F. Squires, *Oil and Gas Jour.*, **43**, 67 (Mar. 10, 1945); cf. also D. E. Menzie, *Producers Monthly*, **12**, 28 (November, 1947).

[2] R. F. Nielsen and S. T. Yuster, *Petroleum Eng.*, **62**, 200 (January, 1945).

[3] Beneficial results have been reported [H. D. Brown, Jr., *Producers Monthly*, **6**, 7 (June, 1941)] from intermittent gas injection using periods of the order of hours, although it is difficult to understand how the cyclic character of such operations could affect anything but the area immediately surrounding the injection wells.

water flooding the purely geometrical features of the steady-state sweep patterns and their efficiencies are theoretically independent of the absolute pressures and rates. Moreover, as would be anticipated on the basis of physical considerations, it has been verified by laboratory experiments[1] that it is the gas throughflow volume at the *mean* pressure in the sand, rather than the volume at atmospheric pressure, which determines the oil recovery. While high injection pressures would give accelerated rates of oil production, it would be expected that the oil displacement per unit surface volume of gas would be lower than in using low pressures, although the latter disadvantage might possibly be more than counterbalanced by the value of the higher recovery rate. However, if the local oil-displacement efficiency is inherently greater at the higher pressure gradients, such as is implied by the previously mentioned laboratory studies,[2] the gain from high injection pressures will be still greater. While there has been a report[3] of increased recovery efficiency in field tests at high pressure gradients, it would be desirable to obtain many more field data to establish the range of validity of this effect.

The problem of evaluating the susceptibility of a particular reservoir to successful secondary-recovery operations by gas injection is similar to that involved when water flooding is under consideration. Aside from establishing the continuity and reasonable uniformity of the productive section the amount of residual oil must be estimated. Of course, if the formation has already been invaded by water, from whatever source, gas injection should be given little consideration.[4] Likewise, areas that have been effectively depleted by gravity drainage may be eliminated as favorable prospects. On the other hand, if the pay has been produced only by the solution-gas drive and it satisfies the requirements of continuity and uniformity and has a moderate or low connate-water saturation, the success of gas injection will then depend largely on such economic factors as the availability of a gas supply, the amount and cost of new drilling or reworking of old wells that will be required, and the permeability of the formation. The latter will determine the injection capacities of the wells and the relation between the operating life and the well spacing.

The absolute value of the oil saturation often will not be so critical a factor as in the case of water flooding. It is the free-gas saturation that

---

[1] D. E. Menzie, R. F. Nielsen, and S. T. Yuster, *Producers Monthly*, **11**, 14 (December, 1946).

[2] Cf., however, the work of Holmgren, *loc. cit.*

[3] Cf. *Producers Monthly*, **10**, 14 (August, 1946).

[4] It has been reported that accidentally flooded sands have been dewatered and repressured successfully [cf. P. A. Dickey, *API Proc.*, **24** (IV), 158 (1943)]. These are undoubtedly exceptional situations and do not indicate a high probability of success in other flooded formations.

will largely control the efficiency of oil displacement by the injected gas, if the oil saturation appreciably exceeds the lower limit for mobility. Whereas a determination of the residual-oil saturation alone would suffice for the prediction of the probable success of a water-flooding project, if it is known to be of sufficient thickness, porosity, continuity, and uniformity, a knowledge of both the water and oil saturations is required for gas-injection evaluation. On the other hand, since the ultimate free-gas saturation created by the solution-gas-drive mechanism is definitely limited (cf. Chap. 10 and Sec. 14.12), the possibility of further reducing the oil saturation appreciably before developing excessive gas-oil ratios may be reasonably assumed even though the exact value of the oil saturation is uncertain, provided, however, that it is known that the interstitial-water saturation does not exceed 35 per cent. Except for the effects of imperfections of areal sweep efficiencies and channeling due to permeability inhomogeneities, a reduction in the oil saturation of 4 to 8 per cent should then be economically feasible for moderate- or high-gravity crudes, if the basic cost factors are in the range encountered thus far in successful operations.

The oil viscosity is a more important factor in gas repressuring than in water flooding. The tendency for gas by-passing and the value of the gas-oil ratio, at any total liquid saturation, will be proportional to the oil viscosity. The oil recovery and free-gas saturations developed by the solution-gas-drive depletion process will be quite low for crudes lower than 20°API (cf. Fig. 10.9), which, in itself, would tend to favor additional recovery by gas injection. However, these low recoveries are mainly due to the high crude viscosities, and the viscosity effect is greatly accentuated after the dissolved gas has been liberated and the pressures are reduced to near atmospheric. Low-pressure gas-injection recoveries will therefore also be highly inefficient in spite of the lower initial-gas saturations. It is doubtful that gas injection for secondary recovery can generally be successful economically if the API gravity of the oil is lower than 20°. In principle the economic limit of gas-injection operations can be determined from the permeability-saturation relationship by the equation

$$\frac{k_g}{k_o} = \frac{\mu_g R}{\gamma \mu_o \beta_o},$$
(1)

where R is the limiting gas-oil ratio for profitable operation, γ the relative gas density at the mean pressure in the formation, μ_o, μ_g the oil and gas viscosities, β_o the formation-volume factor of the oil, and k_g/k_o the gas-to-oil permeability ratio. By solving for the latter the average residual-oil saturation at abandonment could be calculated if its variation with the oil saturation is known.

The well spacing in gas-injection operations has usually been determined by the economic balance between the investment and operating costs and the economic value of the operating life, as affected by the well spacing, taking into account the existing well distribution, formation permeability, and the available gas supply. As is clear from Table 3, wide ranges of spacings have been used throughout the country. The operating life may be expected to vary approximately as the inverse of the injection-well density or in proportion to the square of the average separation between the injection and producing wells. However, it has been reported[1] that gas-injection operations in a group of leases in the Pennsylvania Venango district show the actual recoveries to decrease rapidly with increasing well spacing, when the latter is expressed as the geometric mean spacing[2] of the producing and injection wells. Thus, whereas at a mean spacing of 2 acres per well the recovery is indicated to be about 90 bbl/acre-ft, that at a mean spacing of 4 acres per well is only approximately 35 bbl/acre-ft. It is possible that this high sensitivity of the ultimate recovery to the spacing may be due to the lenticular and irregular character of the Venango Sands. However, if this relationship or an equivalent form should be established for other formations, it would evidently be an important factor in the choice of the optimum spacing.

No fixed rules can be given for the recoveries to be expected as the result of gas injection. In exceptional cases, recoveries as high as the primary production have apparently been obtained.[3] It would appear, however, that 50 per cent of the primary recovery should be considered as an upper limit for the great majority of successful operations and 20 to 30 per cent an average expectancy. While some gas-injection operations have been continued for more than 20 years, more than half the recovery is usually obtained in the first 3 to 5 years, unless the development is progressively extended.

The comparative values of natural gas and air as repressuring media also depend largely on economic factors. The higher solubility of natural gas in the reservoir oil makes it theoretically preferable to air, although it is doubtful that this is of major practical significance. However, the use of air often involves more serious corrosion problems, the danger of developing explosive air-gas mixtures, the formation of oxidation products, which are potential plugging agents, increased paraffin formation, and the dilution of the produced gas so that it is unsuitable for fuel. Nevertheless air has

[1] Cf. Dickey and Bossler, *loc. cit.*

[2] If the ratio of producing to injection wells is B and the gross average spacing is S, the geometric mean spacing is readily shown to be $(B + 1)S/\sqrt{B}$.

[3] In the lease VII project of Table 2 the gas-drive depletion recovery from 1898 to 1925 was 150,000 bbl. It is expected, however, that before abandonment this lease will produce 206,000 bbl by gas injection. While this will appreciably exceed the primary recovery, it still would represent only 44 bbl/acre-ft (cf. Dickey and Bossler, *loc. cit.*).

been used successfully, and there is little evidence that it has been basically inefficient.

12.20. Summary.—The injection of water or gas into a reservoir for purposes of oil displacement after the completion of the primary-production phase involves the same physical processes as when such injection is undertaken during the latter period. The basic differences lie in the "initial conditions." As compared with these conditions when water or gas is injected for "pressure maintenance," the oil saturation is lower, the oil viscosity is higher, the formation-volume factor of the oil is lower, and variations in fluid distributions within the reservoir are greater under secondary-recovery operations. The composite effect of these factors is to decrease the over-all efficiency of the fluid injection when applied after completion of the normal depletion processes. While it would therefore be advantageous to plan fluid injection during the primary-production phase and thus make unnecessary the secondary-recovery operations, this is not always possible. Moreover most of the older fields were produced to depletion before the value of efficient primary-production control was recognized. Of necessity, in these reservoirs fluid injection can be applied only as a secondary-recovery method. Because of the enormous potentially recoverable reserves involved in such operations a clear understanding of the pertinent reservoir-engineering principles is essential to avoid losses of comparable magnitude.

One of the unique features of secondary-recovery operations is the interlaced network distribution of injection and producing wells commonly used. These include the line-drive (cf. Fig. 12.1), five-spot (cf. Fig. 12.2), and hexagonal networks (cf. Fig. 12.3). If in the latter the peripheral wells of the hexagons are used for injection, it is commonly termed a "seven-spot" pattern. If the central wells are the injection wells, it is often called a "four-spot." Although severely limited in quantitative applicability to actual secondary-recovery operations, the purely geometrical dynamical characteristics of these well patterns may be expressed in terms of their behavior under steady-state homogeneous-fluid flow.

It is convenient to synthesize the steady-state pressure distributions in the various regular networks by superpositions of appropriately spaced and distributed linear arrays of wells. Each of these gives a periodic type of pressure distribution with radial equipressure contours close to the individual wells, but merging into that resulting from a line of continuous and constant flux density on receding from the line by a distance equal to the well spacing (cf. Fig. 12.4). On summing the effects due to alternating lines of injection and producing wells the resultant injection or producing capacity of the line-drive network is readily obtained [cf. Eq. 12.2(7)]. As is to be expected, this is found to increase as the spacing within the lines is increased compared with the line separation (cf. Fig. 12.5). Staggered-line-drive networks (cf. Fig. 12.7), in which the injection and producing lines are mutually shifted parallel to themselves, are treated similarly. But such staggering has no appreciable effect on the well capacities if the line spacing exceeds the well spacing within the lines. On the other hand, in the special case when the line separation is half the spacing within the lines the network becomes identical with the five-spot pattern, and the well capacities are increased [cf. Eq. 12.4(1)]. The same analytical procedure, though somewhat more laborious, leads to an expression for the production and injection capacities of wells in hexagonal networks.

The analytical expressions for the pressure distributions in the various well networks can be also applied to compute their steady-state sweep efficiencies. These are defined as the fractions or percentage of the total network area swept out by the injection fluid by the time it first enters the producing wells. The actual area swept out is proportional to the injection rate multiplied by the shortest travel time along a streamline between the injection and producing well. The ratio of this area to the total associated with each injection well is the sweep efficiency. On carrying through calculations of this type for the direct-line-drive network it is found that the sweep efficiency will increase continuously from 31.4 to 88 per cent, as the ratio of the injection and producing line separation to the well spacing within the lines varies from 0.5 to 4.0 (cf. Fig. 12.8). For the same range of variation of the latter the sweep efficiency of the staggered line drive is found to increase from 71.5 to 89.5 per cent. Since the five-spot network is basically of fixed geometry, it has a single value for its sweep efficiency, which calculation shows to be 71.5 per cent [cf. Eq. 12.7(3)]. Likewise the hexagonal network has the unique value of 74.0 per cent [cf. Eq. 12.8(1)], independently of its absolute dimensions and whether the fluid is injected in the central or peripheral wells. Moreover all the steady-state calculated sweep efficiencies are independent of the pressure differentials or injection and production rates, as long as all the injection wells have the same injection pressure and all the producing wells are at the same pressure. The time taken for the injection fluid to reach the producing well is proportional to the square of the well spacing and varies inversely as the injection rate.

The pressure distributions in regular well networks, which control the fluid motion and sweep efficiencies of steady-state homogeneous-fluid systems, are conveniently determined by means of electrical models. The use of such models is based on the fact that the voltage distribution in electrical conductors is governed by Laplace's equation, which also determines the pressure distribution in steady-state homogeneous-fluid flow in porous media. Such models, made of metallic sheets, show that the pressure distribution is essentially radial near the injection and producing wells. In the region between the wells, however, the equipressure contours change their character markedly, depending on the geometry of the network (cf. Figs. 12.9 to 12.12).

The shape of the injection-fluid front as it expands from the injection well and progresses toward the producing wells can be calculated, in principle, if the pressure distribution has been determined [cf. Eq. 12.10(4)]. The analysis, however, is quite laborious and has been carried through only for a few cases of very simple geometry, including the line drive into a single well (cf. Fig. 12.13) and the single well pair of one injection and one producing well (cf. Fig. 12.14). If the pressure distribution is known graphically in the form of equipressure contours, the motion of the injection-fluid front can be calculated by numerical or graphical integration procedures [cf. Eq. 12.10(13)].

It is much easier, however, to obtain a graphic picture and permanent photographic record of the shape of the injection-fluid fronts by using electrolytic models. The potential distributions in these, too, are governed by Laplace's equation. Moreover the ionic motion is controlled by Ohm's law, in complete analogy to Darcy's law for fluid motion. By simply saturating a piece of blotting paper, cut in the shape of a symmetry element of a regular well network, with an electrolyte containing phenolphthalein as an indicator, it will become colored red progressively as the OH ions enter the paper from the negative (injection) electrodes. The colored area is geometrically similar to that which would be swept out by the injection fluid in a porous medium with the same well distribution. By photographing the blotting paper periodically the history of advance of the injection fluid can be permanently recorded. Such procedures have been carried out for the line-drive network (cf. Fig. 12.16), the five-spot (cf. Fig. 12.17), and the hexagonal pattern, both with central (cf. Fig. 12.18) and peripheral injection (cf. Fig. 12.19). These models show that in all cases the injection fluid will first spread out radially and the front will remain approximately circular for a third or more of the distance to the nearest producing well. It then becomes distorted to conform to the geometry of the network. After advancing for three-fourths or more of this distance the front begins to sharpen and finally forms a cusp as it breaks into the producing well. The sweep efficiencies determined from these models agree satisfactorily with those calculated theoretically.

If the saturated blotting paper is replaced by a gelatin film, which develops a coloration on invasion by the ions from the injection electrodes, the electrolytic model becomes much more flexible and can be used in the study of irregular well distributions. The gelatin model also permits changing the injection and producing wells when swept out by the injection fluid or converting them to injection wells. The term "sweep efficiency" loses its precise quantitative significance when not all the producing wells are reached by the injection fluid simultaneously. However, the character of the injection-fluid front still permits a semiquantitative evaluation of various well distributions and their operating conditions and graphically reveals the location and shape of such "dead" areas as may be left by the sweep of the injection fluid (cf. Fig. 12.22).

While the steady-state analytical and model studies of injection- and producing-well networks used for secondary recovery provide an instructive description of the strictly geometrical features of the multiple-well systems, they are directly applicable only to very limited aspects of actual secondary-recovery operations. It is only after the assumed steady-state conditions, underlying the theoretical or model treatments, have been developed that their quantitative implications will have any validity in practice. In the case of water flooding a major part of the productive life in the individual strata will have passed by the time steady-state conditions will be established. During a large part of the operating history prior to that time the water injection will generally exceed greatly the total fluid withdrawals from the injection wells, and the steady-state theory will represent a very poor approximation of the transient behavior. Until interference develops between the injection wells, the water will expand radially about each injection well, and largely independently of the producing-well pattern. The sweep efficiency will no longer be defined by an invariable pressure distribution. In the case of gas injection into depleted formations the steady-state representation may serve as a semiquantitative approximation during an appreciable part of the operating life, as the rates of fluid injection and withdrawals usually will not be so completely out of balance as in the earlier phases of water-flooding operations. However, since the oil expulsion by the gas is largely due to a continued sweeping action as the gas flows through the partly depleted formation, rather than to a banking of the oil ahead of the gas, the sweep-efficiency concept loses most of its significance.

A theory of the transient history of water-injection wells can be developed by assuming a radial expansion of the injected water and a banking up of the oil ahead of it. As the radius of the watered area increases, the flow resistance to further water entry rises and the injection capacity declines. It is thus found that the injection rate, for a fixed injection pressure, will decrease as the reciprocal of the logarithm of the cumulative water injection [cf. Eq. 12.13(9)]. The time history of the injection rates and cumulative injection can also be calculated analytically [cf. Eqs. 12.13(6) and 12.13(7) and Fig. 12.24]. While the applicability of the theoretical considerations has not been "proved" in a general sense, field data in certain cases appear to conform quite satisfactorily with the analytical relationships (cf. Fig. 12.25).

The radial-expansion phase of the water-injection transient must evidently break down when the watered-out areas and the oil banks of neighboring injection wells come in contact. Interference develops then, and the radial fluid-injection front becomes distorted in accordance with the geometry of the well distribution. Ultimately, when the volume to be swept is completely filled up, steady-state conditions will be established. The steady-state-flow rates [cf. Eqs. 12.14(5) to 12.14(8)] are in general lower than those for the transient phase when interference first develops. By a linear extrapolation between these two rates the injection history can be approximately calculated during the transition interval (cf. Fig. 12.26).

The single-zone theory of radial injection-fluid expansion and the subsequent interference period can be generalized to stratified systems by a simple superposition procedure. It is only necessary to change the time scales, injection rates, and cumulative injections in accordance with the physical properties of the individual strata and make appropriate additions (cf. Figs. 12.27 and 12.28). An application of this type to an actual water-injection project (cf. Fig. 12.29) gave sufficiently close agreement to confirm the general correctness of the theoretical treatment.

A comparative study of the various regular well networks with the same total well density shows that the radial-expansion fill-up volumes will be in the ratio of 1.00:2.00:1.15:3.46 for the square-line-drive, five-spot, seven-spot, and four-spot patterns, respectively. The injection rates when interference develops will be in the ratio of 1.00:0.95:0.99:0.91 for the same sequence of patterns. The relative times for development of interference will have the ratios 1.00:2.12:1.17:3.84. The relative total fill-up injection volumes are as 1.00:1.00:0.75:1.50. The steady-state injection rate per injection well will vary in the ratio 1.00:1.13:0.74:1.49. And the total times for fill-up to the steady-state conditions will be in the ratios 1.00:0.88:0.72:1.30. For these calculations the equivalent five-spot injection-producing-well separation has been taken as 1,000 times the effective well radius. While other economic factors will usually control the choice of the pattern, the seven-spot has the advantages of relatively low fill-up times and steady-state throughflow capacity.

Water flooding, as a deliberately planned operation for secondary recovery, had its origin in the Bradford and Allegany fields, in Pennsylvania and New York, about 1920–1921. The shallow depths of the producing formations in these fields, their local uniformity and relatively high residual-oil saturations, coupled with the high market value of the oil, all led to outstandingly successful operations. At Bradford the recovery during the 25 years of water flooding has been almost equal to that obtained during the 50 years of primary solution-gas-drive production, although only about half the field has been subjected to flooding. Similar results have been obtained at Allegany (cf. Fig. 12.30).

The extension of the flooding experience developed in the Bradford and Allegany fields to other areas, especially the Mid-Continent area and North Texas, has shown that equally successful results can be obtained in other formations under favorable conditions.   In the Woodsen field, Texas, the recovery under water flooding has been twice that previously recovered by primary-production methods (cf. Fig. 12.31).

Although quantitative precision cannot be attributed to predictions of the increased recovery to be gained by water flooding, the general criteria of success and the principles for estimating such recoveries are quite well established.   To minimize channeling and achieve good sweep efficiency the oil-bearing formation should be continuous and substantially uniform. In addition, there is the obvious requirement that the total additional recoverable oil suffice to more than pay for the operations.   This places minimum limits on the effective pay thickness, porosity, and residual-oil saturation.   The latter is the most important, since if it is too low the operations will fail regardless of the porosity and thickness.   Hence reservoirs that have produced by water drive or have otherwise been invaded by water are inherently unsuitable.   Likewise formations that have been effectively depleted by gravity drainage will have little chance of success. The minimum oil saturation for which profitable operations can be anticipated even in shallow formations is of the order of 35 to 40 per cent.

The determination of the oil saturation is itself a difficult problem.   Core-analysis methods are beset with the complications of drilling-fluid flushing, although this is not considered to be serious in some districts.   An alternative is the calculation of the residual-oil saturation as the difference between the original-oil content and the cumulative production expressed as equivalent saturations.   The recovery by water flooding is then usually estimated as that corresponding to a reduction in oil saturation from its value at the beginning of flooding to that left after flooding.   The latter is often assumed to be 20 to 25 per cent or is taken as the value indicated by laboratory flooding tests.

Five-spot patterns are most commonly used in actual flooding operations. The well spacing and injection rates are usually chosen so that the operating life will not exceed 10 years, although the cost of drilling new wells or reworking and refitting old wells must be given consideration.   The average permeability is a significant factor mainly through its influence on the injection rates and well spacing.

Both low flooding velocities and rapid flooding have been claimed as necessary for efficient oil recovery on the basis of both field experience and laboratory data.   In any case high injection rates have the certain advantage of giving a reduced operating life, and in Pennsylvania injection pressures are usually kept at a maximum up to the point where there is danger of formation parting, which appears to develop when the bottom-hole pressure is approximately equivalent to the overburden pressure head.

Various methods have been tried to minimize the effect of permeability stratification in leading to channeling. One is "delayed" flooding, in which the producing wells are either not drilled at all or not put on production until enough water has been injected to provide for a substantial fill-up of the free pore volume. For this method to be successful it is necessary that the strata of different permeability be so separated as to be free of cross flow and intercommunication within the reservoir or that the vertical permeability be negligible. Selective shooting to equalize the local injection capacity of the various permeability zones is in widespread use in some districts. And selective plugging has been tried in a variety of techniques.

Special methods of water treatment have been developed to minimize corrosion and formation plugging. Studies of the effect of different waters on the permeability of dirty sands have led to the conclusion that low-pH waters or brines will give greater injection capacities. On the other hand no practical success has resulted from attempts to increase the displacement efficiency of the flood waters by adding to them special active agents. Laboratory investigations have shown that even where such improvements can be achieved, the loss of the additive due to adsorption by the rock will make their use economically prohibitive at present prices.

In well-planned and successful floods the first reaction at the producing wells to the water injection is usually observed within 2 to 6 months. The major part of the additional oil recovery is obtained after an injection of 3 to 5 reservoir fill-up volumes. The cumulative injection-water–oil ratio generally lies in the range of 8 to 20. If there is any doubt about the probable success of full-scale flooding operations, an experimental pilot flood should be tried first. And to achieve maximum sweep efficiencies, operating economies, and the most flexible control over the operations, as large a contiguous area as possible should be flooded under unitization or cooperative arrangements.

Although gas injection was first tried some 17 years before experimental water flooding, only the phase of declining production after the initial rise has been given a theoretical treatment. By assuming that the oil production is due to a uniform desaturation of the formation with a corresponding increase in gas-oil ratio, as determined by the permeability-saturation relationship, it is found that the rate of oil production will be proportional to the logarithmic derivative of the cumulative gas injection or throughflow [cf. Eqs. 12.17(4)]. For a constant gas-injection rate the reciprocal of the oil-production rate should increase linearly with the time [cf. Eq. 12.17(6)]. The cumulative oil production between two time periods should be proportional to the logarithm of the ratio of the cumulative gas injection at the end and beginning of the time interval. And the gas-oil ratio will theoretically increase linearly with the cumulative gas-injection volume and exponentially as the cumulative oil recovery.

A study of the oil-production-decline histories of several gas-injection projects in the Titusville–Oil City districts of Pennsylvania has shown them to follow the theoretically predicted behavior (cf. Fig. 12.36). From the slopes of the linear plots of the reciprocals of the production rate vs. the time a basic constant of the permeability-saturation relationship can be determined. The constants so calculated were of the same order of magnitude as, but definitely higher than, those established directly by gas-injection laboratory experiments on long cores of Pennsylvania sandstones and also those implied by the primary gas-drive production histories of solution-gas-drive reservoirs. This suggests that the effect of permeability inhomogeneities and stratification is more serious in affecting recovery efficiencies by fluid injection than during the primary-production phase of gas-drive reservoirs. On the other hand the laboratory experiments showed this constant to vary with the pressure gradient, indicating that either the gas or the liquid relative permeabilities or both were sensitive to the pressure gradient in these particular studies.

One of the notably successful gas-injection projects of the Mid-Continent district is that started in 1925 in the Delaware-Childers field, Nowata County, Okla. After a primary-gas-drive depletion recovery of 23 million barrels an additional recovery of 17 million barrels was obtained by air and gas injection of 51 billion cubic feet to July, 1945. Of this recovery, 13 million barrels, or more than 2,000 bbl/acre, is attributed directly to the air and gas injection. In the Homer field, Claiborne Parish, La., the increased recovery by gas injection from 1934 to 1941 totaled 879,000 bbl from 1,347 acres. In December, 1940, this recovery was being obtained at a rate of 1 bbl/1,000 ft^3 of injected gas. And in the Red River field, Oklahoma, the gain in ultimate recovery due to gas and air injection was estimated as about 31 per cent of the primary recovery.

The basic requirements for success in gas repressuring are the same as for water flooding. However, reservoir uniformity is not so critical a criterion as in the case of water flooding. Formations of variable permeability and lenticular strata have been profitably subjected to gas injection, although the inherent probability of success will always be greater for reservoirs of substantial uniformity and continuity.

With respect to the fluid content of the formation the free-gas and total liquid saturations may be of equal significance to the oil saturation. The economic limit for profitable gas injection is controlled by both the gas and oil saturations, which in turn determine the gas-oil ratio. For the

same oil saturation, success is more likely when the connate-water saturation is high than when it is low.   Conversely, however, for a given free-gas
saturation the probability of successful gas injection will increase with decreasing interstitial-water content.   Fields depleted by solution-gas drive
with intermediate connate-water saturations of the order of 35 per cent
may be even more profitable if subjected to gas injection than to water
flooding, if other factors are also favorable.   On the other hand, if the
connate-water saturation exceeds 45 per cent, it is doubtful that either gas
or water injection, following a solution-gas drive, will be economically
successful.

Gas-injection operations are less frequently developed in regular patterns
than water-flooding projects.   Many more producing wells are ordinarily
used than injection wells (cf. Table 3).   The average injection-well spacing
in some districts has been of the order of several hundred acres per well.
In successful projects some response in the producing wells is generally
noted within 1 to 6 months after injection is started.   The oil recoveries
from profitable operations usually exceed 1,000 bbl/acre and have been
as high as 5,000 bbl/acre in some projects.

In input wells that have not been shot, packers have been used to segregate zones where channeling is taking place.   Selective plugging and water
"slugging" have also been tried to minimize by-passing.   Intermittent
injection and shifting of the injection-well patterns have been used to
achieve improved recovery efficiencies.   None of these procedures is a
universal remedy for channeling difficulties, nor do they always ensure
profitable returns.   Their success is dependent on the local conditions, the
absence of formation cross flow, and the manner of application.

Laboratory experiments indicating that the recovery efficiency by gas
injection will increase with increasing injection pressures and pressure gradients have apparently been confirmed in one field project.   There is no
evidence, however, as to the general validity of these observations, and con-
flicting results have been reported on the basis of other experimental investigations.

Low reservoir oil viscosities will be conducive to lower gas-oil ratios in
gas-injection operations.   Hence, lower residual-oil saturations can be obtained for fixed gas-oil-ratio abandonment limits [cf. Eq. 12.19(1)] with
higher gravity crudes.   The well spacing is usually determined in gas-
injection operations by the simple economic balance between investment
cost and the effect of the spacing on the operating life, although some data
have been reported indicating increased recoveries at low well spacings.
There is no evidence that air and gas have inherent significant differences
as oil-expulsion media.   When gas is available at a reasonable cost, it has
usually been favored.   Air often leads to corrosion difficulties; it involves

explosion hazards; and the probability of forming with the oil oxidation products that may plug the sand and of increased paraffin formation is greater when air is used.

Recoveries by gas injection vary over an even greater range than the primary solution-gas-drive recoveries. In exceptional cases the latter have been equaled by those due to repressuring. Unsuccessful projects have, of course, usually resulted in very minor increased recoveries. When the reservoir conditions are inherently susceptible to gas injection, 20 to 30 per cent of the primary recovery may be reasonably expected.

CHAPTER 13

CONDENSATE RESERVOIRS

13.1. Introduction.—The primary problem involved in the study of the various type of oil-producing reservoirs discussed in previous chapters has been the dynamical interaction between the fluids and their porous-media carriers. The physical and thermodynamic properties of the fluids have mainly played the role of parameters affecting only the details of the performance. Condensate-producing reservoirs are unique in that it is the thermodynamic behavior of the petroleum fluids that is the controlling factor in their performance and economic evaluation. It is for this reason that they will be given here a separate treatment, although their dynamical aspects are controlled by the same basic laws of fluid flow through porous media[1] as govern the production of crude oil and natural gas.

13.2. General Considerations Regarding Condensate-reservoir Fluids.—Condensate reservoirs produce a liquid phase commonly called "condensate" or "distillate." In contrast to crude oil it is usually a colorless or straw-colored[2] liquid and generally has an API gravity of 48° or higher. As compared with crude-oil production, it is associated with high gas-oil ratios, of the order of 10,000 ft^3/bbl or greater. From a physical point of view the most important characteristic of the condensate is that it is not necessarily a liquid in the reservoir from which it is produced. In most cases it is a liquid phase developed from a hydrocarbon mixture that is in a single or dew-point gas phase under the reservoir conditions. Such a phase transformation may take place within the reservoir as the result of pressure decline, by a process of isothermal retrograde condensation (cf. Sec. 2.5), although the liquid so formed in the producing formation will generally remain trapped and provide only a negligible part of the liquid product recovered at the surface. The major part of the condensate actually produced at the surface is that condensing from the gas by more

---

[1] Except for the correction of the wet-gas permeability due to the presence of the connate water the homogeneous-fluid theory will govern the dynamics of condensate reservoirs when the pressure is maintained by "cycling" (cf. Sec. 13.4). And even when produced by pressure depletion the homogeneous-fluid representation should still provide a satisfactory approximation from a practical standpoint.

[2] The dark coloration sometimes observed in condensate liquids is probably due in most cases to contamination with slight amounts of crude oil or fine dispersions of dark bituminous material picked up by the petroleum fluids from the reservoir rock on their passage toward the producing wells.

general retrograde-phase transformations during the simultaneous decline in pressure and temperature as the reservoir fluid rises up the flow string. This liquid-phase recovery is often supplemented to an important extent by various methods of processing the gas arriving at the wellhead or passing through the separators so as to extract additional liquefiable hydrocarbons that are still in the gas phase on reaching the surface.

Although condensate-producing fields are usually considered as being

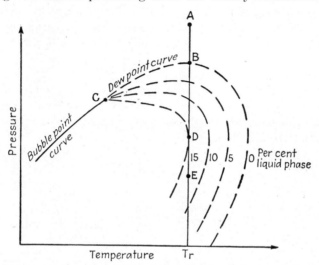

Fig. 13.1.   A diagrammatic representation of the gross phase behavior of the reservoir fluid from a condensate-producing formation. T_r denotes the reservoir temperature and C the critical point.

comprised simply of gas-phase reservoirs, the universality of such conditions is neither observed nor to be expected.   If the condensate-bearing reservoir fluid is an undersaturated gas, *i.e.*, a single phase above the dew point, as indicated by A in Fig. 13.1, it cannot be in equilibrium with a liquid phase and no liquid phase will be present if equilibrium obtains.   If, however, it is a saturated vapor,[1] at its dew-point pressure, as at B, it *may* coexist with a liquid phase and the composite liquid and gas will then be essentially equivalent to a normal segregated two-phase system.   If, as is usually the case, the gas phase may be visualized as simply the excess beyond that which will go into solution in the liquid, the latter will be a "crude" oil with its characteristic dark color and relatively low API gravity.   Under equilibrium conditions the composition of this "black" oil will be the same as that of the first liquid phase condensing from the gas if its temperature

---

[1] In most condensate-producing reservoirs the dew-point pressure, at the reservoir temperature, of the initially produced fluids has been found to be the same as the reservoir pressure, within the uncertainties of the experimental determinations.

or pressure were lowered. Except for the special retrograde-condensation property of the gas the reservoir as a whole could be considered as a normal crude-oil reservoir overlain by a gas cap.

As would be expected from these considerations, there is no thermo-dynamic limitation to the relative amounts of the liquid (crude-oil) phase and the condensate-bearing gas phase that may initially comprise the composite reservoir. Crude-oil rims have been found underlying most condensate-bearing gas reservoirs. In some, such as the *D* Sand of the Benton field, Bossier Parish, La., oil zones definitely appear to be absent. On the other hand, where they are present their size and content may be so small as to be of entirely negligible significance, or they may be large enough far to exceed in value the gas-cap contents.

An understanding of the unique properties of condensate-producing gases is facilitated by reference to their composition, as compared with crude-oil and natural-gas mixtures. Such comparative analyses (in mole per cent) for typical hydrocarbon systems of both types are given in Table 1.

TABLE 1.—TYPICAL COMPOSITIONS OF A CONDENSATE-PRODUCING GAS AND A CRUDE-OIL–GAS MIXTURE

	Saturated vapor			Crude-oil–gas mixture		
	Gas	Conden-sate	Reservoir fluid	Gas	Oil	Reservoir fluid
Methane.......	85.69	—	82.38	80.53	0.31	45.26
Ethane........	4.45	—	4.28	5.37	0.14	3.07
Propane.......	3.64	0.19	3.51	3.85	0.33	2.30
Iso-butane.....	1.57	2.53	1.61 }	3.70	0.97	2.50
n-Butane.......	3.06	2.22	3.03 }			
Iso-pentane....	0.35	6.77	0.60 }	2.09	1.97	2.04
n-Pentane......	0.45	6.37	0.68 }			
Hexanes	0.34	17.36	0.99	1.17	2.49	1.75
Heptanes and heavier......	0.45	64.56	2.92	3.29	93.79	43.08
Total......	100.00	100.00	100.00	100.00	100.00	100.00
Mol. wt. of heptanes and heavier......	.....	.....	133	.....	200	

It will be seen that, while the separated gas phases in each case are not greatly different in composition,[1] the liquid condensate has a decidedly

[1] The gas phase of the crude-oil–gas mixture of Table 1 is considerably richer in the heavier liquefiable components than the reported compositions of natural gases usually indicate. The latter, however, generally refer to separator-gas samples, whereas that of Table 1 refers to the stock-tank gas obtained by direct flashing of the bubble-point reservoir liquid.

lower content of heptanes and heavier than the oil, and moreover the average molecular weight of these components is considerably lower for the condensate. But more important still is the fact that, whereas in the crude-oil–natural-gas mixture there are 1.27 moles of gas per mole of liquid, the corresponding ratio for the condensate system is 25. It is these composition characteristics that give the condensate-bearing reservoir fluid its unique properties.

Condensate reservoirs were not discovered, or at least recognized as such, until the early thirties. They have since been found with increasing frequency, especially in the Gulf Coast area. This is undoubtedly to be associated with the increasingly deep drilling of the last 10 years.[1] While this is usually attributed simply to the higher pressures and temperatures prevailing at the greater depths, it is the pressure and the composition of the hydrocarbon mixtures, rather than the temperature, that are the controlling factors. As indicated in Fig. 13.1 and noted in Sec. 2.5 isothermal retrograde condensation on pressure decline from the dew point will occur only at temperatures above the critical and at pressures near[2] the critical pressure. Since the critical temperatures of condensate-producing fluids correspond to some type of average[3] of the individual components, they will be exceeded by the reservoir temperatures even at very shallow depths, and hence the factor of temperature would not alone limit the occurrence of condensate fields to the greater depths. On the other hand their critical pressures will generally be considerably greater than those of the separate constituents and will be approached or exceeded by the reservoir pressure only in the deeper fields. Of course the probability of occurrence of condensate *types* of hydrocarbon mixtures may also be inherently greater at the higher pressures and temperatures found in the deeper horizons. However, the relationship between the nature of the petroleum fluids and depth, and general reservoir conditions and sedimentary history, is only one of the

[1] A rough statistical analysis of condensate fields discovered to 1945, based on a survey of 224 fields by J. O. Sue and J. Miller, *API Drilling and Production Practice*, 1945, p. 117, shows that about 88 per cent of these were found at depths exceeding 5,000 ft and that the average depths of about 60 per cent were greater than 7,000 ft.

[2] The range of pressures over which isothermal retrograde condensation may occur, in relation to the critical, depends on whether the pressure-temperature phase diagram is of the type of Fig. 2.8a or 2.8c. The pressures must be less than the critical for Fig. 2.8a, and they may be either greater or less for Fig. 2.8c. Most condensate-fluid systems that have been studied appear to follow the latter, as indicated also by Fig. 13.1, so that the pressures must be only in the "range" of the critical.

[3] While such averaging appears to obtain in the case of binary hydrocarbon mixtures (cf. Fig. 2.16), the critical temperatures of more complex systems often deviate sharply from molar averages and may even be completely outside the range of the critical temperatures of the individual components [cf. C. K. Eilerts, V. L. Barr, N. B. Mullens, and B. Hanna, *Petroleum Eng.*, **19**, 154 (February, 1948)].

problems of the origin of oil whose solutions are still hardly beyond the state of nebulous speculation.

As the dew-point pressure is the natural starting point for the consideration of the phase behavior of condensate-reservoir fluids, it is instructive to see how the dew-point pressure[1] may vary with the temperature and gross composition of the hydrocarbon mixture. Expressing the latter in terms of the API gravity of the stock-tank oil (condensate) and the gas-oil ratio of the composite system, a correlation of data obtained in a study of fluids from five San Joaquin Valley (California) fields is reproduced in Table 2.[2]

TABLE 2.—DEW-POINT PRESSURES OF VARIOUS HYDROCARBON MIXTURES AT THREE
TEMPERATURES

(In psia)

Temperature	Oil gravity, °API	Gas-oil ratio, ft³/bbl					
		15,000	20,000	25,000	30,000	35,000	40,000
100°F	52	4,440	4,140	3,880	3,680	3,530	3,420
	54	4,190	3,920	3,710	3,540	3,410	3,310
	56	3,970	3,730	3,540	3,390	3,280	3,180
	58	3,720	3,540	3,380	3,250	3,140	3,060
	60	3,460	3,340	3,220	3,100	3,010	2,930
	62	3,290	3,190	3,070	2,970	2,880	2,800
	64	3,080	3,010	2,920	2,840	2,770	2,700
160°F	52	4,760	4,530	4,270	4,060	3,890	3,650
	54	4,400	4,170	3,950	3,760	3,610	3,490
	56	4,090	3,890	3,690	3,520	3,380	3,270
	58	3,840	3,650	3,470	3,320	3,200	3,110
	60	3,610	3,430	3,280	3,150	3,040	2,960
	62	3,390	3,240	3,100	2,990	2,890	2,810
	64	3,190	3,060	2,930	2,820	2,740	2,670
220°F	54	4,410	4,230	4,050	3,890	3,750	3,620
	56	3,990	3,780	3,600	3,440	3,300	3,180
	58	3,700	3,480	3,280	3,110	2,970	2,850
	60	3,430	3,210	3,030	2,880	2,760	2,660
	62	3,150	2,970	2,800	2,670	2,570	2,480
	64	2,900	2,740	2,590	2,470	2,380	2,300

[1] Here, as well as in all the discussion in this chapter with respect to condensate reservoirs and cycling, the dew-point pressure refers only to the upper branch of the dew-point curve between the critical and cricondentherm temperatures; these are sometimes termed "retrograde" dew points.

[2] This is taken from B. H. Sage and R. H. Olds, *AIME Trans.*, **170**, 156 (1947). The gases used in the experiments were separator samples. Their methane content ranged from 82.5 to 89.5 per cent for the mixtures from the five fields studied. The heptanes and heavier content of the separator-liquid samples varied from 48.97 to

It will be seen that in all cases, for fixed temperature and gas-oil ratio, the dew-point pressure decreases with increasing oil gravity, the rate of variation being greatest at low gas-oil ratios and the higher temperatures. In the range of the high gas-oil ratios listed in the table[1] the dew-point pressure increases monotonically with decreasing gas-oil ratio, for fixed oil gravity and temperature. Its sensitivity to the gas-oil ratio is greatest for the lower gravity oils. Its variation with the temperature is not monotonic, and for the systems to which Table 2 refers it is greater in most cases at 160°F than at either 100 or 220°F.

13.3. The Depletion History of Condensate-producing Reservoirs.—If it were not for the retrograde phenomenon, a condensate-bearing single-phase reservoir would perform simply as a gas field. The recovery of condensate would be proportional to the amount of gas produced. And except for possible reservoir shrinkage by water intrusion and the slow variation of the deviation factor of the gas with pressure, the reservoir pressure would decrease linearly with increasing cumulative recovery. It is the potential loss of the liquid content of the gas phase due to retrograde condensation as the pressure declines that is the crux of the problem of evaluating and predicting the performance of a condensate-producing field.

As in the case of crude-oil–natural-gas mixtures the phase relations can be determined for either of two types of process, flash and differential. The former, in which the composition and total mass of the mixture are held fixed while the pressure and volume are varied, corresponds to a simple path as *BDE* in Fig. 13.1, if the process is isothermal. The variation in the amount of liquid phase forming during such paths is illustrated by the curves of Fig. 13.2[2] for mixtures of condensate and gas from the Paloma field, Kern County, Calif. For three of the curves the temperature was 250°F and the gas-oil ratios were 5,361, 7,393, and 14,439 ft³/bbl. For the fourth the temperature was 190°F, and the gas-oil ratio was 7,393 ft³/bbl. It will be seen that the volume of liquid has a maximum in all cases, except for curve III, for which the maximum apparently lies below 1,000 psi.

73.11 per cent. The precise determination of dew-point pressures is a rather difficult problem. It is often chosen simply as the pressure-axis intercept of liquid-condensation curves such as those of Figs. 13.2 and 13.3 but can also be established by visual observations on the vanishing or first appearance of the liquid phase in glass-capillary or variable-volume cells fitted with suitable windows [cf. H. T. Kennedy, *Petroleum Eng.*, **11**, 77 (July, 1940); W. F. Fulton, *API Drilling and Production Practice*, 1939, p. 354; and J. P. Sloan, API meetings, Shreveport, La., May, 1946, and *Oil and Gas Jour.*, **46**, 158 (Mar. 25, 1948)].

[1] At lower gas-oil ratios the dew-point pressures may develop a maximum and then decline (cf. Fig. 13.2 below).

[2] These are plotted from tabulations of R. H. Olds, B. H. Sage, and W. N. Lacey, *AIME Trans.*, **160**, 77 (1945).

These maxima of retrograde condensation correspond to the point D in Fig. 13.1. The decline in liquid content at lower pressures simply represents the normal vaporization process. This dividing point recedes to higher pressures at the lower gas-oil ratios,[1] *i.e.*, for the fluids richer in condensate content. And as would be expected, the total liquid condensation increases with decreasing gas-oil ratio. Moreover a comparison of curves II

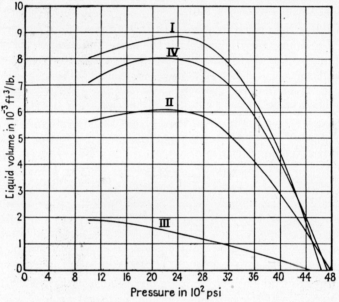

Fig. 13.2. The liquid-condensation curves for combined gas and condensate samples from the Paloma field, for fixed temperatures T and gas-oil ratios R. For curves I, II, and III, $T = 250°F$, $R = 5,361$, $7,393$, and $14,439$ ft³/bbl. For curve IV, $T = 190°F$, $R = 7,393$ ft³/bbl. The volume units refer to 1 lb of total mixture.

and IV, as well as reference to Fig. 13.1, shows that the retrograde liquid accumulation decreases with increasing reservoir temperature.

Although the fixed-composition pressure decline does not correspond to the process occurring in practice, it serves to illustrate one of the basic problems of operating a condensate reservoir by pressure depletion. Thus, on noting that the specific volumes of the single-phase dew-point fluids for curves I, II, III, and IV are 0.04909, 0.05450, 0.06901, and 0.04888 ft³/lb, it follows that the maximum condensation volumes will represent, respectively, 18.1, 11.2, 2.75, and 16.4 per cent of the original hydrocarbon

---

[1] The maximum condensation pressures are rather high in Fig. 13.2 because of the relatively low gas-oil ratios of the mixtures. They decrease with increasing gas-oil ratios or with decreasing composite reservoir fluid density and often lie in the range of 1,000 to 1,500 psi for actual systems.

volume or pore space. If ρ_l are these values expressed as fractions and the connate-water saturation is ρ_w, the maximum possible total average liquid saturation that could develop by this process would be $\rho_w + \rho_l(1 - \rho_w)$. Even if the condensate is added to the water as a continuous liquid phase, its permeability would evidently be extremely low. It is more probable, however, that the condensate would be distributed as a dispersed phase and have no permeability whatever except possibly under the very high pressure gradients near a well bore.[1] The condensate will therefore remain trapped and lost until partial revaporization sets in after the pressures fall below the point of maximum condensation. It is this potential loss of the liquid content of the reservoir fluids that plays a major role in evaluating the reservoir and in determining the method of development and operation.

While the curves of Fig. 13.2 demonstrate the basic retrograde characteristics of condensate-reservoir fluids and the resulting danger of loss in condensate recovery, they are not of quantitative significance for two reasons. Even if the pressure decline in practice followed the flash process, the total liquid-phase separation plotted in Fig. 13.2 would not represent actual volumes of stock-tank liquid product. For this liquid phase will contain appreciable concentrations of the lighter hydrocarbons, which would separate as a gas at atmospheric conditions. More fundamental is the fact that if the pressure declines at all, it is the result of a removal of part of the reservoir fluids. As only the gas phase will be produced, because of the lack of mobility of the condensed-liquid phase, the composition of the system will constantly change. The reservoir will therefore undergo a differential process of pressure decline and depletion. The amount of liquid condensation under such conditions will evidently be lower than when all the gas phase is maintained in contact with the liquid. It is from experiments in which the pressure decline in the sample container is caused by gas-phase withdrawal that the phase and volumetric data simulating the depletion performance of an actual condensate reservoir can be obtained.

The gross phase and composition characteristics of a condensate-producing reservoir undergoing pressure depletion are illustrated in Fig. 13.3,[2] obtained by a combination of experimental data and calculated analyses for a gas-cap fluid having a dew point of 2,960 psi at a reservoir temperature of 195°F. It will be noted that here the maximum liquid condensation represents only 8.2 per cent of the hydrocarbon pore space. Moreover the volume that the C_4+ fraction would occupy at 60°F is only about 70 per cent of the total at the maximum retrograde point. The C_4+ content of

---

[1] This situation is entirely analogous to the immobility of a low-saturation distributed gas phase until it builds up to the "equilibrium saturation" (cf. Sec. 7.6).

[2] Figure 13.3, as well as Figs. 13.4 and 13.5, are taken from M. B. Standing, E. W. Lindblad, and R. L. Parsons, *AIME Trans.*, **174**, 165 (1948).

the produced gas, *i.e.*, the gas phase in equilibrium with the reservoir liquid, declines from its initial value of 5.4 gal/10^3 ft^3 to a value of 3.2 gal/10^3 ft^3 as the liquid condensation increases in the reservoir. Shortly after normal vaporization sets in, the C_4+ content of the gas rises sharply, while the

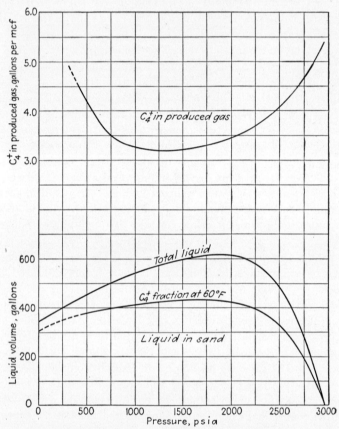

Fig. 13.3. Experimental curves for the liquid condensation and C_4+ content of the produced gas, as a function of pressure, resulting from the depletion of a condensate-bearing reservoir. Liquid volumes refer to an initial hydrocarbon space volume of 10^3 ft^3. (*After Standing, Lindblad, and Parsons, AIME Trans., 1948.*)

reservoir liquid volume declines on further pressure reduction. If all the C_4+ in the produced gas were converted to a liquid phase, the upper curve of Fig. 13.3 would also give the variation in the effective gas-liquid ratio of the composite well stream during the producing life.

The detailed composition history of the condensed-liquid phase in the reservoir is shown in Fig. 13.4. It will be seen that the concentrations of the lighter and more volatile components, methane, ethane, and propane, decrease continuously with declining pressure.[1] At the same time the heaviest component, C_7+, increases monotonically in concentration. The intermediate constituents tend to fall somewhat in concentration at first,

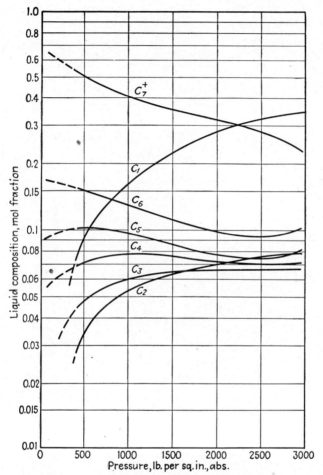

FIG. 13.4. Experimental curves for the composition history of the reservoir liquid phase for the system described by Fig. 13.3. (*After Standing, Lindblad, and Parsons, AIME Trans., 1948.*)

then rise for varying pressure intervals and, in the case of C_4 and C_5, begin another decline in the lower pressure range. The API gravity of the reservoir liquid phase will evidently show a continual decrease as the pressure declines.

[1] The composition of the first liquid phase formed on pressure decline should be identical with that of any reservoir liquid or gas-saturated crude oil lying below and in equilibrium with the condensate-reservoir dew-point gas.

The composition of the gas phase or produced gas corresponding to Figs. 13.3 and 13.4 is plotted vs. the pressure in Fig. 13.5. The general behavior of the liquefiable components, C₄ and heavier, simulates that of

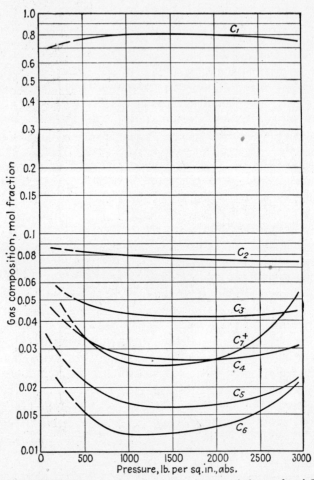

FIG. 13.5. Experimental curves for the composition history of the produced fluids for the system described by Fig. 13.3. (*After Standing, Lindblad, and Parsons, AIME Trans., 1948.*)

the composite C_4+ curve of Fig. 13.3. The concentrations of the volatile constituents, C_1 to C_3, do not change greatly from their initial values in the dew-point fluid. Such curves are of particular value in predicting the nature of the product to be recovered from the reservoir at any state of depletion, as well as the cumulative recoveries. Thus, the cumulative number of moles of the ith component, N_i, per unit volume of hydrocarbon reservoir space, recovered by the time the pressure has declined from its initial value p_i to p is

$$N_i = \frac{1}{RT} \int_p^{p_i} \frac{n_i \, dp}{Z},\tag{1}$$

where n_i is the current mole fraction concentration, as given by Fig. 13.5, R is the gas constant per mole, T the reservoir temperature, and Z is the deviation factor of the gas, which usually can be calculated with satisfactory accuracy from the composition (cf. Sec. 2.7). The assumption of a slow variation of Z and neglect of the liquid-phase volume, implied by the form of Eq. (1), should not involve serious errors from a practical standpoint. From the cumulative molar recovery, as calculated by Eq. (1), the total liquid-phase recoveries of various heavy-component groupings can be computed. Moreover by application of the equilibrium ratios (cf. Sec. 2.10) the separation of all the components at the surface between the gas and liquid phases can be calculated, and their individual contributions to the cumulative recoveries can be determined as a function of the pressure. The cumulative recoveries computed in this manner for another condensate-producing reservoir are plotted in Fig. 13.6.[1] The curve indicated as "stable condensate" represents the total liquid phase produced in the stock tank through the separators.

The original hydrocarbon contents of the reservoir fluids to which Fig. 13.6 refer were, per 10^6 ft^3 of pore space, 17,490, 18,710, 21,220, and 23,650 bbl of n-C$_5$+, i-C$_5$+, n-C$_4$+, and i-C$_4$+, respectively. From Fig. 13.6 it is seen that, if the reservoir were produced by simple depletion to a pressure of 500 psi, the cumulative production will contain 8,600 bbl of n-C$_5$+, 9,770 bbl of i-C$_5$+, 11,380* bbl of n-C$_4$+, and 13,680 bbl of i-C$_4$+, per 10^6 ft^3 of pore space. These represent 49.2, 52.2, 53.6, and 57.8 per cent of the original content, respectively. Conversely, 50.8, 47.8, 46.4, and 42.2 per cent of these components, respectively, would be lost in the reservoir if it were produced to 500 psi by pressure depletion.

It should be noted that the curves of Fig. 13.6, as derived by an integration of composition curves such as those of Fig. 13.5, on applying Eq. (1),† refer to the total recoverable liquid products from the well stream. This implies that the well fluids are processed by a hydrocarbon-extraction plant, so as to remove the condensable components from the separator gases. If only the stable liquid components of the stock-tank condensate were recovered without processing of the gas, a considerably lower part of the

---

[1] This is taken from E. W. McAllister, *California Oil World*, **38** (No. 20), 19 (1945).

* This is not indicated on Fig. 13.6 but is tabulated separately by McAllister (*ibid.*).

† Usually, however, the variation of Z is neglected, and the areas under the curves of Fig. 13.5 are taken as equivalents of the integral of Eq. (1)

original condensable hydrocarbon content would be recovered. The difference will generally be of increasing importance as the gas-oil ratio increases, since the greater volumes of gas phase can carry off correspondingly higher fractions of the total condensable product. In some cases as much

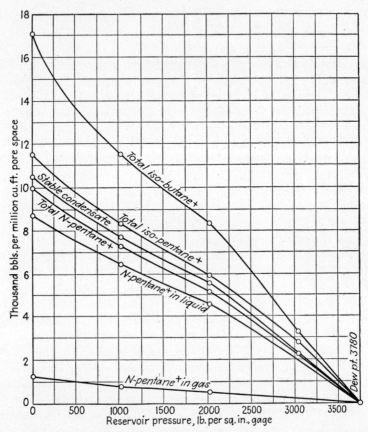

FIG. 13.6. Experimental curves for the cumulative recoveries of the heavier components of a condensate-bearing fluid obtained by reservoir pressure depletion. (*After McAllister, California Oil World, 1945.*)

condensable liquid product will be lost in the gas phase as is recovered directly, so that the recoveries of the C_4+ components without plant extraction or multiple stage separation might be half of that indicated by such curves as Fig. 13.6. The magnitude of the recovery without gas processing can be determined by calculating the current gas–liquid-phase separation, by whatever separator system is used, from the composition of the produced fluid, as given by curves such as those of Fig. 13.5, and integrating over the pressure decline by means of Eq. (1).

Additional numerical illustrations of the gross performance of actual condensate reservoirs under normal depletion, as predicted by experimental studies on a set of six condensate-bearing fluids, are reproduced in Table 3.[1] While no systematic correlations can be derived from the six examples of Table 3, since they vary so widely in the controlling conditions, the order of magnitude of the general numerical features of the performance is clearly indicated. Thus the stable-condensate recoveries by pressure depletion for the six samples fall in the range of 44.5 to 65.8 per cent of the initial contents. The corresponding reservoir losses range from 55.5 to 34.2 per cent. It will be noted, too, that, whereas all the produced stable condensate will be recovered by three-stage separation, the pentanes plus recoveries from the produced wet gas will be of the order of 70 to 80 per cent. And from 75 to 90 per cent of the produced butanes will be lost in the separator gas if it is not further processed.

In contrast to crude-oil-producing reservoirs the gross depletion history of condensate systems is substantially independent of the dynamical characteristics of the reservoir, it being assumed that it is not subject to significant water intrusion. The relation between the cumulative recovery and the average pressure may be constructed similarly to Eq. (1) and may be verified to be

$$\overline{P} = \int_p^{p_i} \frac{dp/p_i}{Z/Z_i},\qquad (2)*$$

where $\overline{P}$ is the cumulative recovery expressed as a fraction of the total initial molar content and p_i, Z_i are the initial pressure and deviation factor of the reservoir gas. The liquid condensation has been neglected, in constructing Eq. (2), both with respect to its volumetric displacement of reservoir gas-phase volume and its hydrocarbon content. It has already been noted that the volume occupied by the reservoir condensate will generally be small. And while a substantial part of the total condensable component of the hydrocarbon mixture may be retained in the reservoir liquid phase, this will still represent but a small part of the total molar content of the reservoir.

As Z is a slowly varying function of the pressure, Eq. (2) implies an approximately linear decline of the pressure with the cumulative molar recovery. In terms of the condensate recovery the pressure will decline more rapidly than linearly,[2] up to the pressure of maximum retrograde

---

[1] This table is a composite of those given by B. W. Whiteley, AIME meetings, College Station, Tex., December, 1947.

* Equation (2) will be quantitatively accurate in "dry-" gas fields, where there is no reservoir condensation, either because the state of the reservoir gas lies on the lower dew-point-curve segment, or because its cricondentherm temperature is lower than the reservoir temperature (cf. Sec. 13.10).

[2] This will be partly compensated by the decrease in Z as the pressure declines.

TABLE 3.—THE PRESSURE-DEPLETION PERFORMANCE OF CONDENSATE RESERVOIRS AS DETERMINED BY THE ANALYSIS OF VARIOUS RESERVOIR-FLUID SAMPLES

Sample	(1) Bbl/10⁶ ft³	(1) %	(2) Bbl/10⁶ ft³	(2) %	(3) Bbl/10⁶ ft³	(3) %	(4) Bbl/10⁶ ft³	(4) %	(5) Bbl/10⁶ ft³	(5) %	(6) Bbl/10⁶ ft³	(6) %
Original content:												
Butanes	33.8	100	7.9	100	17.2	100	24.0	100	14.1	100	12.6	100
Pentanes plus	64.5	100	26.6	100	53.0	100	85.3	100	55.9	100	94.0	100
Hexanes plus	—	—	23.5	100	42.5	100	—	—	46.3	100	83.3	100
Stable condensate	60.1	100	22.5	100	45.4	100	85.1	100	44.0	100	93.5	100
Produced in wet gas:												
Butanes	33.4	98.8	6.8	86.0	15.0	87.2	23.6	98.3	13.3	94.3	12.4	98.4
Pentanes plus	32.4	50.2	17.0	63.9	32.7	61.6	55.9	65.5	29.5	52.8	51.7	55.1
Hexanes plus	—	—	14.0	59.5	24.8	58.4	—	—	23.5	50.7	42.9	51.5
Stable condensate	26.7	44.5	13.0	57.8	29.9	65.8	51.1	60.1	21.8	49.5	44.7	47.8
Retrograde loss in reservoir:												
Butanes	0.4	1.2	1.1	14.0	2.2	12.8	0.4	1.7	0.8	5.7	0.2	1.6
Pentanes plus	32.1	49.8	9.6	36.1	20.3	38.4	29.4	34.5	26.4	47.2	42.3	44.9
Hexanes plus	—	—	9.5	40.5	17.7	41.6	—	—	22.8	49.3	40.4	48.5
Stable condensate	33.4	55.5	9.5	42.2	15.5	34.2	34.0	39.9	22.2	50.5	48.8	52.2
Stock-tank liquid recovery by three-stage separation:												
Butanes	3.2	9.5	—	—	—	—	4.2	17.5	1.6	11.3	3.1	24.6
Pentanes plus	23.8	36.9	—	—	—	—	45.4	53.2	21.4	38.3	41.6	44.4
Hexanes plus	—	—	—	—	—	—	—	—	20.6	44.5	37.2	44.6
Stable condensate	26.7	44.5	13.0	57.8	29.9	65.8	51.1	60.1	21.8	49.5	44.7	47.8
Loss in separator gas:												
Butanes	30.2	89.3	—	—	—	—	19.4	80.8	11.7	83.0	9.3	73.8
Pentanes plus	8.6	13.3	—	—	—	—	10.5	12.3	8.1	14.5	10.1	10.7
Hexanes plus	—	—	—	—	—	—	—	—	2.9	6.2	5.7	6.9
Stable condensate	0	0	0	0	0	0	0	0	0	0	0	0
Reservoir pressure, psia	5,670		4,965		3,765		3,780		4,267		4,595	
Reservoir temp., °F	124		207		228		247		197		265	

condensation, because of the increasing gas-oil ratios (cf. Fig. 13.3). At still lower pressures the slope of the pressure vs. oil-recovery curve should decrease somewhat as the gas-oil ratios decrease.

As a gross index of the "richness" of condensate-reservoir gases and of their changes during the producing life it is convenient to use the terms "gas-liquid," "gas-oil," or "gas-condensate ratios,"[1] although values of these ratios are, of course, of significance only if the separator conditions are fixed. For the initial characterization of a condensate reservoir the gas-condensate ratio may serve to indicate whether it is "rich" or "lean." While these terms do not have precise definitions, it is generally agreed that a gas-condensate ratio of 15,000 ft³/bbl or less implies a rich gas and that if the ratio exceeds 40,000 ft³/bbl the reservoir gas is lean.

It should be noted, however, that for the strict evaluation of condensate-producing reservoirs the terms "gas" and "liquid" no longer have quantitative significance. As shown by Fig. 13.5 the "gas" entering the well bores will vary considerably in composition during the producing life. It is, indeed, just this variation that reflects the basic phenomenon of retrograde condensation in the reservoir, and the trapping of the condensed-liquid phase in the producing formation, if the latter is operated by simple pressure depletion. And as indicated in Fig. 13.4 the composition of the reservoir liquid phase may vary during the producing life even more than the composite produced fluids.

The alternative to the fluid phase as a basis for the identification of the hydrocarbons produced from a condensate reservoir is evidently a composition grouping. A procedure sometimes used is that of classifying the total C_4+ content of the reservoir or produced fluids as "plant product" and the remainder as gas. The plant product, so defined, corresponds to the total stable liquid phase that is usually extracted by gas-processing plants. In making detailed economic appraisals of condensate-reservoir-development programs it may be necessary further to subdivide the produced fluids into components such as "stable condensate," gasoline fractions, "liquefied gases," etc., which are individually specified by composition. However, an accounting in terms of the i-C_4+ and C_7+ constituents will often suffice for preliminary economic-evaluation purposes.

It is possible to develop a material-balance equation for condensate reservoirs generalizing Eq. 9.5(6) for crude-oil fields, and also free of the assumptions underlying Eq. (2). However, in place of the usual p-V-T characteristics of the reservoir fluids it involves such empirical properties of the gas and liquid phases that their determination would be almost as

---

[1] An equivalent representation often used is the C_5+ or C_4+ content of the gas in gallons per 10^3 ft³, since 1 gal/10^3 ft³ corresponds to a gas-liquid ratio of 42,000 ft³/bbl if all the heavier components were transformed to the liquid phase.

involved as the experimentation required to establish Figs. 13.3 to 13.6. Because of this severe limitation to its applicability it will not be considered further here. Equation (2) and simple modifications for the cases of gas injection or water intrusion should suffice for most practical purposes as far as the condensate reservoir itself is concerned.

13.4. Cycling—General Considerations.—As implied by Eq. 13.3(2) the fractional recovery of the molar hydrocarbon content of a condensate-producing reservoir will be approximately equal to the fractional decline in pressure. Hence, if the initial pressure is 4,000 psi, approximately $87\frac{1}{2}$ per cent of the original hydrocarbon content will be recovered by the time the pressure declines to 500 psi. On the other hand, as was noted in Sec. 13.3, in some cases about half of the heavy components (C_5+) may still be left in the reservoir owing to retrograde condensation. Because of the considerably greater market value of the liquid hydrocarbon products the loss of the heavy fractions will represent a substantial part of the total value of the initial hydrocarbon mixture. Thus, if the latter corresponds to a gas-liquid ratio of 10,000 ft^3/bbl, if the market price of the dry gas is \$0.10 per thousand cubic feet, and if the average of that of the liquid products is \$2.50 per barrel, a recovery of $87\frac{1}{2}$ per cent of the gas and 60 per cent of the liquid products would be equivalent to only 67.9 per cent of the initial value of the hydrocarbon reserves. And the 40 per cent of the condensable hydrocarbons left in the reservoir would represent 28.6 per cent of the gross initial value. The economic significance of the liquid products lost by condensation within the reservoir will, of course, depend on the inherent richness of the reservoir fluid, or the initial gas-oil ratio, and the actual magnitudes of the retrograde losses. In principle, however, it is generally possible to prevent at least a large part of this loss by operations termed "cycling," which will now be considered.

Cycling is simply the process of injecting "dry" gas into a condensate-producing reservoir to replace the reservoir fluid withdrawals—the "wet" gas—so as to maintain the reservoir pressure[1] and prevent retrograde condensation of a liquid phase within the porous medium. Cycling is sound from a physical point of view. Its practical value in specific cases is determined entirely by the economic balance[2] between the cost of the operations and the gain in recovery as compared with pressure depletion. The former depends mainly on the additional well-development cost required

[1] This could also be accomplished by water injection, although the economics of water injection except as a supplementary measure for pressure maintenance will usually compare unfavorably with gas injection.

[2] For a general discussion of the economic aspects of cycling cf. W. H. Woods, *Oil and Gas Jour.*, **46**, 89, 99 (Aug. 16, 23, 1947); cf. also E. Kaye, *AIME Trans.*, **146**, 22 (1942); E. O. Bennett, R. C. Williams, and G. O. Kimmell, *Petroleum Eng.*, **13** (No. 10), 99 (1942).

for the cycling program, the amount of compression needed to bring the processed[1] gas to the injection-wellhead pressures, and the volume of gas to be handled to ensure a reasonable operating life. The latter is given essentially[2] by the product of the fractional initial-oil content, per unit hydrocarbon pore volume, which would be left in the reservoir by pressure depletion, times the total hydrocarbon pore volume swept out by the injected gas during the life of the operations, if equivalent gas-processing plants be assumed available with and without cycling. The retrograde-condensation losses, which are the basic reason for considering cycling at all, have been discussed in the preceding section and, as noted there (cf. Table 3), can be determined by suitable experimentation with and analysis of the original reservoir fluids. The reservoir volume swept out by the injected dry gas is largely controlled by the geometry of the injection- and producing-well pattern and the uniformity of the reservoir formation. These will be discussed in the next several sections.

13.5. Analytical Determinations of the Sweep Efficiencies of Cycling Patterns.—The basic method of calculation of sweep efficiencies for cycling patterns is the same as that outlined in Sec. 12.6 for the steady-state homogeneous-fluid treatment of secondary-recovery operations. Since the dry-gas-injection rates in cycling are frequently substantially equal to those of the wet-gas withdrawals, the steady-state representation should provide a practical and reasonably accurate approximation. The assumption of equal viscosities and deviation factors for the dry and wet gases will also involve errors of negligible significance, compared with the basic idealization of reservoir uniformity that underlies virtually all analytical treatments.[3] And for purposes of simplicity the sweep-efficiency analysis will be carried through as if the fluid were an incompressible liquid. Although the sweep efficiencies, flow pattern, and injection fronts for gas flow will not be strictly identical to those for liquids, the differences will not be of importance unless the total pressure differential between the injection and producing wells is very large (cf. Sec. 13.6).

In contrast to the areal interlaced distribution of injection and producing wells commonly used in secondary-recovery operations, cycling patterns are generally developed by segregating the injection and pro-

---

[1] In all cycling operations the separator gas is passed through extraction plants to remove the condensable hydrocarbons before returning the stripped gas to the formation. In fact this additional liquid-product recovery itself often represents a large part of the total gain from the cycling operations.

[2] A more detailed discussion of the comparative recoveries and economics of cycling and pressure-depletion operations will be given in Sec. 13.10.

[3] In Sec. 13.8 the theory will be given of the effect of permeability stratification. However, lateral and areal variations in permeability and thickness are, for practical purposes, beyond the scope of tractable analysis.

ducing wells. Because of the smaller potential value of the recoverable reserves in condensate-bearing reservoirs[1] and their greater average depths it is necessary to minimize the well investment. In fact, average well spacings of 320 acres per well are often used in condensate fields. While higher well densities would shorten the operating life, the increased cost of the wells and of the larger capacity gas-processing and compression plants severely limits the total well density. Moreover the number of injection wells often is considerably smaller than the number of producing wells.

To make most efficient use of the rather small number of wells drilled for cycling programs the injection and producing wells are generally located along the reservoir boundaries. If the reservoir area is approximately rectangular, the injection wells may be placed along one side and the producing wells along the opposite side, so as to give an "end-to-end" sweep. The injection wells may also be distributed at the center of the reservoir and the producing wells along the whole peripheral productive boundary. Or they can be inverted, with the injection wells placed at the boundary. The particular pattern to be used must, of course, be fitted to the gross geometry of the reservoir in question.

While an interior well distribution for the same total well density would permit more flexible control over the operations, a more complete accumulation of reservoir information, and the earlier detection of the development of channeling or the effects of reservoir inhomogeneities, the over-all sweep efficiencies in uniform formations will be greater for greater average separations between the injection and producing wells. The "dead" areas in sweep patterns are often concentrated about the producing wells, owing to the cusping of the injection fluid as it enters the region of the local pressure distributions created by the producing wells. These dead areas will be rather insensitive to the nature of the injection-well distribution and their location, provided that the relative production and injection rates are kept fixed, and as long as the average separation between wells of the same kind is appreciably less than the average separation of wells of different kind. Hence the fractional loss in sweep area represented by these unswept regions will decrease as the total area to be swept, or the distance between the injection and producing wells, increases.

The simple case of an end-to-end sweep between parallel single lines of injection and producing wells can be easily treated analytically.[2] Upon

---

[1] Even a "rich" reservoir gas, which would produce at a ratio of 10,000 ft³/bbl of condensate, would have a total condensate content of only 215 bbl/acre-ft of reservoir volume, if the pressure is 300 atm., temperature is 200°F, deviation factor is 0.9, and the porosity and connate-water saturations are each 25 per cent.

[2] Since this treatment is given only to illustrate the basic features of cycling patterns in which the injection and producing wells lie oppositely along the actual reservoir

placing for convenience the x axis parallel to and midway between the injection and producing wells (cf. Fig. 13.7), the pressure distribution will be

$$p(x,y) = \frac{Q\mu}{4\pi kh} \log \frac{\cosh 2\pi(y-d)/a - \cos 2\pi x/a}{\cosh 2\pi(y+d)/a - \cos 2\pi x/a}, \tag{1}$$

where Q is the common injection and producing rate, in reservoir measure, of the wells in the two lines, μ the gas viscosity, k the permeability, h the pay thickness, a the well separation within the lines, and $2d$ the distance between the lines. Equation (1) evidently is a simple superposition of the pressure distributions for the individual lines as given by Eq. 12.2(1).

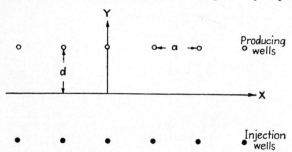

FIG. 13.7. A diagrammatic representation of a parallel-line cycling pattern.

The fluid velocity between the injection and producing wells along the y axis is therefore

$$v_y = -\frac{k}{\mu\bar{f}} \frac{\partial p}{\partial y}\Big|_{x=0} = -\frac{Q}{2a\bar{f}h} \frac{\sinh 2\pi d/a}{\sinh \pi(y-d)/a \, \sinh \pi(y+d)/a}, \tag{2}$$

where $\bar{f}$ is the displacement porosity, *i.e.*, the porosity times the fraction of the pore space occupied by the dry gas.

The shortest time of travel between the injection and producing wells will therefore be

$$t = 2\int_0^d \frac{dy}{v_y} = \frac{2\bar{f}a^2h}{Q}\left(\frac{d}{a}\coth\frac{2\pi d}{a} - \frac{1}{2\pi}\right), \tag{3}$$

implying a sweep efficiency E given by

$$E = \frac{Qt}{2adh\bar{f}} = \coth\frac{2\pi d}{a} - \frac{a}{2\pi d}. \tag{4}$$

As is to be expected, E increases uniformly from 0 at $d/a \ll 1$, to 1 at $d/a \gg 1$. Thus for $d/a = 0.1$, 1, and 5, E will be 0.204, 0.841, and 0.968,

boundaries, the analysis is based on the simplifying assumption of an infinite-reservoir area. The infinite-medium assumption will also be made in the discussion of the circular cycling pattern, although it is possible to treat in both cases also the finite-reservoir systems.

respectively. The unswept area per unit injection- and producing-well pair
is readily seen to be

$$A = 2ad\left(1 - \coth\frac{2\pi d}{a}\right) + \frac{a^2}{\pi}. \tag{5}$$

Hence for $d/a \geqslant 1$ the unswept area at the time of first dry-gas entry into
the producing wells has the constant value a^2/π, independent of the exact
value of d. Equations (4) and (5) verify the previously outlined qualitative

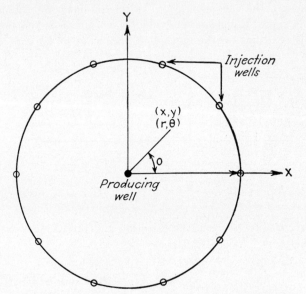

Fig. 13.8. A diagrammatic representation of a circular cycling pattern.

considerations indicating the increasing sweep efficiency with increasing
separations between the injection and producing wells.[1]

Equation (1) also describes the pressure distribution between a continuous line drive at $y = 0$ and the line of producing wells along $y = d$. The
sweep efficiency will be given by Eq. (4) also for this case, although the
shapes of the injection-fluid fronts will be quite different. The dead areas
per producing well will be half of those of Eq. (5) for the end-to-end sweep
pattern.

Peripheral injection into a circular ring of wells, in an infinite area, and
production from the center (cf. Fig. 13.8), or the same basic pattern with
the injection and producing wells reversed can also be treated analytically.
Thus it may be verified that the pressure distribution and stream functions,

---

[1] This principle is also well illustrated and demonstrated by the curves of Fig. 12.8,
which show that in infinite line-drive networks the sweep efficiency also increases as
the separation between the injection and producing lines increases.

p and Ψ, are given by the real and imaginary parts, respectively, of the complex potential function

$$p + i\Psi = \frac{Q\mu}{2\pi kh}\left[\log z - \frac{1}{n}\log(z^n - R^n)\right],\tag{6}$$

where R is the radius of the circular ring, Q is the injection or producing rate of the central well, and z is the complex coordinate variable $x + iy$, or $re^{i\theta}$ (cf. Fig. 13.8). It follows from Eq. (6) that

$$p(r,\theta) = \frac{Q\mu}{2\pi kh}\left[\log r - \frac{1}{2n}\log(r^{2n} + R^{2n} - 2r^nR^n\cos n\theta)\right] + \text{const.}\tag{7}$$

The fluid velocity between the injection and producing well along $\theta = 0$ is, then,

$$v_r = -\frac{k}{\mu\bar{f}}\left(\frac{\partial p}{\partial r}\right)_{\theta=0} = -\frac{Q}{2\pi h\bar{f}r}\cdot\frac{R^n}{R^n - r^n}.\tag{8}$$

The time of travel between the injection and producing wells is therefore

$$t = \int_R^0 \frac{dr}{v_r} = \frac{\pi h\bar{f}nR^2}{Q(n+2)},\tag{9}$$

and the sweep efficiency for the area enclosed by the circular ring of wells will be

$$E = \frac{n}{n+2}.\tag{10}*$$

For a single injection- and producing-well pair ($n = 1$), Eq. (10) gives $E = \frac{1}{3}$, which agrees, as it should, with that found in Sec. 12.10 [cf. Eq. 12.10(12)]. Equation (10) also shows that the sweep efficiency rapidly approaches unity as n increases and that the gain resulting from additional wells decreases as $2/(2 + n)^2$.

By exactly the same method it may be shown that if the injection and producing wells are both equally spaced at angles $2\pi m/n$, $m < n$, on concentric rings of radii with a ratio R, the sweep efficiency will be

$$E = \frac{n}{R^2(R^n - 1)}\left(\frac{R^{n+2} - 1}{n + 2} - \frac{R^n - R^2}{n - 2}\right), \quad n \neq 2.\tag{11}†$$

---

* Equation (10) has also been derived by a somewhat different procedure by B. D. Lee' *AIME Trans.*, **174**, 41(1948).

† For $n = 2$ the equation involves a logarithmic term. For $n = 1$, however, it becomes equivalent to $E = \frac{1}{3}$, as it should when properly interpreted. For the similar problem of a continuous circular line drive of radius r_e into a concentric circular ring of radius R, of n producing wells, which is of interest in the development of complete-water-drive reservoirs, the sweep efficiency may be shown to be

$$\cdot\ 1 - \frac{2R^2}{n - 2}\left\{\frac{1 - (R/r_e)^{n-2}}{r_e^2 - R^2}\right\}, \quad \text{for } r_e \gg R.$$

It will be clear that, when $R \gg 1$, Eq. (11) reduces to Eq. (10). But at moderate values of R and n there will be an appreciable difference between the sweep efficiency for the concentric rings and a single central well. Thus if $R = 5$ and $n = 3$, Eq. (11) gives $E = 0.508$, whereas Eq. (10) gives $E = 0.60$. And for $R = 10$, $n = 3$, Eq. (11) gives $E = 0.574$. This, too, implies greater sweep efficiencies as the injection- and producing-well separations are increased.

By using potential-theory methods, such as applied in Sec. 12.6 for the calculation of sweep efficiencies in regular well networks, the characteristics

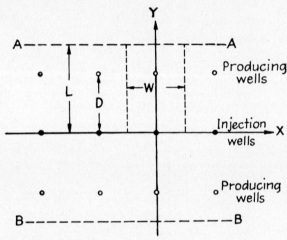

FIG. 13.9. A diagrammatic representation of a bilateral cycling pattern.

of bilateral cycling patterns have also been determined[1] analytically. Here, as seen in Fig. 13.9, the injection wells are placed along the central axis of the reservoir and the producing wells on either side—or conversely—the field boundaries being represented by AA, BB. The geometry of this system is determined by the length-to-width ratio of the basic rectangle, L/W, and the ratio of the separation between the injection and producing lines to the spacing within the lines, D/W (cf. Fig. 13.9). Typical results[2] of the analysis are illustrated by Figs. 13.10 and 13.11 for $D/W = 1.25$ and 1.75, respectively, L/W being 1.75 in each case. To the left of Figs. 13.10 and 13.11 are the equipressure contours ($p = $ const) and streamline ($\Psi = $ const) distributions. To the right are injection-fluid fronts, on each of which are indicated the fraction of wet gas in the production, the fraction

[1] Cf. W. Hurst and A. F. Van Everdingen, *AIME Trans.*, **165**, 36 (1946).

[2] Only the upper halves of the pressure and streamline distributions and fluid fronts in a bilateral system are plotted in Figs. 13.10 and 13.11, since by symmetry those in the lower halves will simply be their reflections in the X axis.

of total wet gas displaced, and the total gas processed, or throughflow, divided by the gas originally in place.

Figures 13.10 and 13.11, as well as calculations for other cases, show again the increasing sweep efficiency as the distance between the injection and producing wells is increased. Thus, for Fig. 13.10, $E = 0.492$, whereas

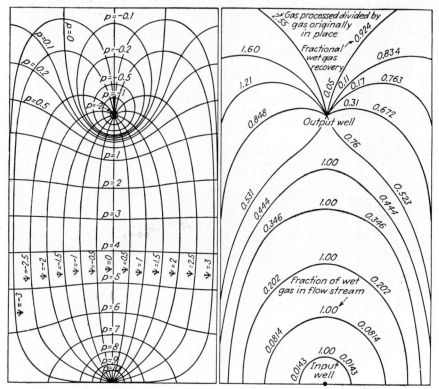

FIG. 13.10.   The calculated pressure (p) and streamline (Ψ) distributions and injection-fluid fronts in a bilateral cycling pattern in which $L/W = 1.75$ and $D/W = 1.25$. $L =$ half width of field; $W =$ well spacing within the injection and producing lines; $D =$ separation between the injection and producing lines.   (*After Hurst and Van Everdingen, AIME Trans., 1946.*)

for Fig. 13.11, in which the producing wells are placed at the very limits of the reservoir,[1] $E = 0.741$. If D/W were 1.00, with $L/W = 1.75$, E would fall to 0.369, or only half of that for $D/W = 1.75$. It should be noted, however, that here by far the greater part of the gain due to the increased separation between the injection and producing lines results from the improved sweep behind the producing wells, in the area between them and the assumed reservoir boundaries AA, BB. Whereas as fractions of the

[1] This special case also gives the sweep efficiency in an end-to-end sweep in a finite reservoir with the wells located along the actual boundaries, with $d/a = 0.875$.

total reservoir area the sweep efficiencies vary as 0.369, 0.492, 0.633, and 0.741 as D/W is changed from 1.00, 1.25, 1.50 to 1.75, in terms of the areas only between the injection and producing lines they are 0.646, 0.689, 0.738, and 0.741, respectively.

The composition of the produced gas, expressed as the fractional wet-gas

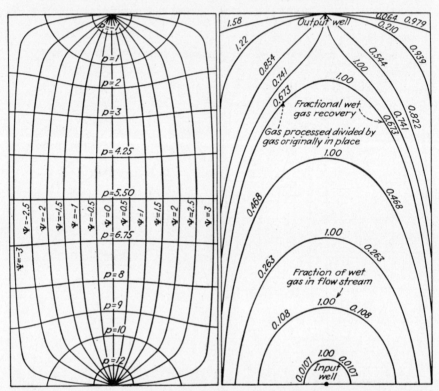

FIG. 13.11. The calculated pressure (p) and streamline (Ψ) distributions and injection-fluid fronts in a bilateral cycling pattern in which $L/W = 1.75 = D/W$. L = half width of field; W = well spacing within the injection and producing lines; D = separation between the injection and producing lines. (*After Hurst and Van Everdingen, AIME Trans., 1946.*)

content, for a bilateral drive is plotted in Fig. 13.12, for the cases shown in Figs. 13.10 and 13.11 and also for $D/W = 1.00$ and $D/W = 1.50$. Thus after a throughflow equal to the original gas volume the wet-gas contents of the produced gas for the four systems will range from 19 to 36 per cent, the latter referring to the case of maximum separation between the injection and producing wells. The total gas processed, or throughflow, by the time the wet-gas content falls to 15 per cent will be, for the four cases, 1.26, 1.33, 1.46, and 1.35, respectively, times the original reservoir gas content. These are evidently proportional to the total operating lives, for equal throughflow rates.   While the differences are rather small and would appear to favor the pattern with the shortest distance between the producing and injection lines and the lowest sweep efficiency, the total wet-gas recoveries are quite appreciably affected by the line separations and sweep efficiencies.   The total wet-gas recoveries vs. the gas throughflow, corresponding to Fig. 13.12, are plotted in Fig. 13.13.   From this figure and reference to Fig. 13.12, it will be seen that at the 15 per cent limit of wet-gas content the total wet-gas recoveries will be 64, 78, 90, and 96 per cent, respectively, of the original wet-gas content of the reservoir.

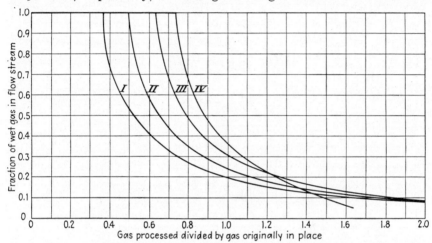

Fig. 13.12.   The calculated composition of the gas produced from bilateral cycling patterns vs. the gas processed for different separations between the injection and producing wells, D.   I, $D/W = 1.00$.   II, $D/W = 1.25$.   III, $D/W = 1.50$.   IV, $D/W = 1.75$.   In all cases $L/W = 1.75$. L = half width of field; W = well spacing within the injection and producing lines.   (*After Hurst and Van Everdingen, AIME Trans., 1946.*)

These differences, of course, will be of greater economic significance than those in the total gas processed and demonstrate the basic advantage of the higher sweep efficiency of the bilateral pattern with greater separation between the injection and producing lines.

It should be emphasized that all the considerations of this section refer to reservoirs of uniform permeability and thickness.   The study of cycling systems by electrical models, as will be discussed in the following two sections, are also primarily limited to single-zone reservoirs, although the variations of permeability and thickness within a single stratum can be treated by the potentiometric model.   On the other hand, when the formation is known to be comprised of substantially distinct layers of different permeability, the study of the composite motion of the injection fluid will require a supplementary analysis of the superposed histories of the individual zones.   This will generally imply an appreciable lowering of the resultant sweep efficiency. The theory of multilayer systems will be given in Sec. 13.8.

It should be noted that theoretically the times of injection-fluid break-through, areas swept out, and sweep efficiency will be the same if the injection and producing wells are interchanged, provided that the pressures at all injection wells are the same and all the producing well pressures are equal. The choice among the two possibilities will depend on practical

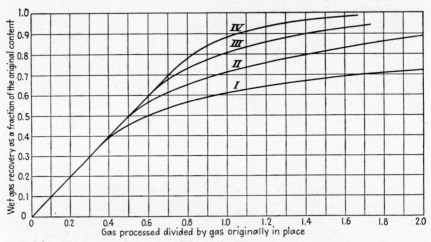

FIG. 13.13. The calculated variation of the total wet-gas recovery in bilateral cycling systems vs. the volume of gas processed for different separations between the injection and producing wells, D. I, $D/W = 1.00$. II, $D/W = 1.25$. III, $D/W = 1.50$. IV, $D/W = 1.75$. In all cases $L/W = 1.75$. L = half width of field; W = well spacing within the injection and producing lines. (*After Hurst and Van Everdingen, AIME Trans., 1946.*)

considerations, such as the relative cost of injection and producing wells, the presence of bounding oil rims, the mobility of underlying waters, etc.

13.6. The Theory of Potentiometric Models.—Although the sweep efficiencies of general well patterns in two-dimensional uniform media can be determined by use of the electrolytic gelatin model described in Sec. 12.11, the potentiometric model is basically more accurate. Moreover it is more flexible in making it possible to treat systems of variable permeability and porosity, which, for practical purposes, are beyond the scope of the gelatin model, and variable-thickness formations can be studied with it much more conveniently than with gelatin models. The general principle of operation of the potentiometric model, when the four-electrode probe is used, has been outlined in Sec. 12.10.[1] For homogeneous two-

---

[1] The technique of operating electrolytic potentiometric models in which the fluid motion is calculated directly from the pressure or potential distributions is described by W. Hurst and S. N. McCarty, *API Drilling and Production Practice*, 1941, p. 228. The construction and operation of the four-electrode-probe models, which are much

dimensional systems, of constant thickness, the basic analogy between the electrical model and the flow system simply rests on the observation that the electrical potential corresponds to the pressure and the current density to the fluid flux. While this analogy still obtains in more complex systems of variable thickness and permeability, the construction of the electrical model to give an accurate analogue requires more detailed consideration.[1]

The equations of continuity for steady-state current flow in an electrolytic medium and fluid flow in a porous body are

$$\nabla \cdot \bar{i} = \nabla \cdot \bar{v} = 0, \tag{1}$$

where $\bar{i}, \bar{v}$ are the vector current density and mass flux, respectively. Since all model studies of the type considered here are based on a two-dimensional idealization in both the electrical- and porous-media-flow systems, Eqs. (1) may be expressed, by virtue of Ohm's and Darcy's law, as

$$\nabla \cdot \sigma \nabla V = 0 = \nabla \cdot \frac{\gamma k h}{\mu} \nabla p, \tag{2}$$

where σ is the equivalent electrical conductivity, V the voltage, γ the gas density, k the permeability[2] to the gas, μ its viscosity, h the local effective pay thickness, and p the pressure. As γ, μ are, in principle, functions only of the pressure, the second half of Eqs. (2) can be formally simplified to

$$\nabla \cdot k h \nabla \Phi = 0, \tag{3}$$

where:

$$\Phi = \int \frac{\gamma}{\mu} dp. \tag{4}$$

If there is areal geometrical similarity between the reservoir in question and the model and if both have the same source and sink distributions, corresponding to the injection and producing wells, the electrical model will have a voltage distribution identical, except for scale, with that for Φ,

---

more convenient and determine simultaneously the streamline distributions and the potential gradients along the streamlines, have been reported by Lee, *loc. cit.* It may be noted also that the basic principle of determining the shapes of fluid-injection fronts from the potential distributions had been stated and applied to a five-spot well network by R. D. Wyckoff, H. G. Botset, and M. Muskat, *AIME Trans.*, **103**, 219 (1932), using a conducting metallic sheet to establish the equipotential contours.

[1] The discussions of the analogy in the literature are restricted to systems of uniform thickness and permeability, except for the treatment of transient-water-drive reservoir histories by the electrical analyzer (cf. Sec. 11.8). The detailed analysis given here follows that of M. Muskat, *Petroleum Technology*, **11**, 1 (November, 1948).

[2] This should be the *effective* permeability in the presence of the connate water.

provided σ is made everywhere proportional to kh. The variability of σ is obtained by varying the depth h_e of the layer of electrolyte, so that

$$\sigma = \sigma_o h_e = akh \qquad (5)$$

where σ_o is the specific conductivity of the electrolyte and a is a scale factor. Thus gross geometrical similarity and a variation of the electrolyte thickness in proportion to the millidarcy-feet of the formation will ensure formal equivalence between the voltage and Φ distributions. If the reservoir formation is to be approximated as one of uniform permeability, the electrolytic bath is to be made geometrically similar to an isopach map of the section.

It should be noted that the porosity does not enter in the construction of the electrolytic-model analogue,[1] although, as will be seen presently, it is involved in the derivation of the fluid-front contours from the pressure distributions. The primary criterion for equivalence between the model and the actual reservoir is the creation of geometrically similar *potential* fields, which are determined only by the thickness, permeability, and boundary conditions. The basic function of the model is to give an empirically measurable solution of Eq. (2) for the pressure distribution. The determinations of injection-fluid fronts is essentially nothing more than an interpretation of the pressure distribution, as may be obtained by suitable numerical, graphical, or electrical manipulation of its characteristics.

To derive the character of the fluid motion it is noted that the rate of local fluid advance along the streamlines will be given by

$$\frac{ds}{dt} = \frac{v}{\bar{f}} = \frac{k}{\mu \bar{f}} |\nabla p| = \frac{k}{\bar{f}\gamma} |\nabla \Phi|, \qquad (6)$$

where v is the local volumetric flux along the streamline and $\bar{f}$ is the displacement porosity, *i.e.*, the actual porosity times the fraction of the pore space displaced by the invading fluid.[2] The time of travel over an element of length ds along a streamline will therefore be

$$dt = \frac{\bar{f}\gamma \, ds}{k |\nabla \Phi|}. \qquad (7)$$

Hence, if the potential[3] distribution represented by Φ is known, Eq. (7) will permit the stepwise integration of the time of advance of the fluid

---

[1] "Iso-vol" or constant-pore-volume analogues, which have been used in some model studies, will not give correct potential distributions, except when both the net hydrocarbon porosity and effective permeabilities are constant.

[2] It is intuitively probable, and it has been essentially confirmed experimentally, that in cycling operations there is no appreciable mixing between the injected dry and displaced wet gas. Under such conditions $\bar{f}$ is the total hydrocarbon porosity.

[3] Φ may be conveniently considered here as a potential function, although it does not satisfy the simple Laplace equation.

front. To carry through this procedure with the aid of the voltage distribution in the potentiometric model, the scale factors L, M may be introduced as

$$ds_M = L\, ds_R$$
$$V = M\Phi \tag{8}$$

where ds_M is a linear distance in the model and ds_R the corresponding distance in the reservoir and M is, in effect, the ratio of the total voltage between two points in the model to the corresponding difference in Φ in the reservoir. It then follows that Eq. (7) can be rewritten as

$$dt = \frac{Ma}{\sigma_o L^2}\, \frac{\gamma h \bar{f}}{h_e}\, \frac{ds_M}{|\nabla V|}. \tag{9}$$

If, as in the use of the four-probe electrodes, the potential drop ΔV is measured along the streamlines over the fixed electrode separation Δs_m, the corresponding fluid-travel-time increments will be

$$\Delta t = \frac{Ma}{\sigma_o L^2}\, \frac{\gamma h \bar{f}}{h_e}\, \frac{\Delta s_m^2}{\Delta V}. \tag{10}$$

By summing such increments along the individual streamlines the constant-time surfaces can be plotted. These evidently correspond to the various fluid-injection fronts, or interfaces between the injection and displaced fluids.

It is to be noted that, if the coefficient $\gamma h \bar{f}/h_e$ is variable, the sum of the reciprocals of ΔV will not alone suffice to determine quantitatively the shapes of the injection-fluid fronts. On the other hand, in most practical applications it will be necessary to make such approximations as will permit simplifications of Eq. (10). Thus, if permeability variations are neglected, Eq. (10) reduces to

$$\Delta t = \frac{M}{kL^2}\, \gamma \bar{f} \frac{\Delta s_m^2}{\Delta V}, \tag{11}$$

where k is the assumed uniform permeability to the fluids involved. If $\bar{f}$ is also considered as constant, the only remaining variable in Eq. (11), except for ΔV, will be γ. Since γ does not vary rapidly in cycling systems except near the injection and producing wells, it should suffice to neglect its variation outside of these regions, if average values are used, in actual field studies. In fact such approximations would appear to be inherently reasonable if the variations in $\bar{f}$ and k are also neglected.

As indicated by Eqs. (6) the velocity of advance of the fluid front will be proportional to the pressure gradient whether the fluids involved are gases or liquids. The pressure distributions, however, will be quite different in the two cases. On the other hand, since the distributions in the function

Φ will be the same for gases and liquids for a given reservoir formation, the shapes of the injection-fluid fronts will be different for gases only because of the factor γ in Eq. (7). The assumed equivalence in these fronts between liquids and gas thus implies the neglect of the variation of the gas density γ.

Although the shapes of injection-fluid fronts will be independent of the fluid viscosities, the absolute times of travel will be proportional to the viscosity. Moreover, even aside from the effect of the viscosity, the sweep rates will be different for gases and liquids, for the same terminal pressures.

While the density factor in Eq. (7) makes almost impossible a strict analytical treatment of the motion of gas injection-fluid fronts, it does not present unsurmountable difficulties in using the potentiometric model if it is felt desirable to take it into account. For it is only necessary that the density distribution be calculated from the potential distribution, and its local value be multiplied into the reciprocals of the gradients, according to Eq. (7) or (10), to obtain the time increments. Such a stepwise evaluation of the latter would in any case be required if the permeability or displacement porosity is variable. Actually, however, even the effects of the latter are usually neglected in practical applications, and the travel times are determined simply by summing the reciprocals of the potential increments, ΔV. From a practical standpoint, attempts to take into account the refinements associated with the variations in permeability and displacement porosity usually will not be warranted, since it will be very seldom that their variation will be known with any certainty. When these effects must be neglected of necessity, it is doubtful that the corrections due to the density variations would be justified except in the immediate vicinities of the individual wells. It is for this reason that no attempt has been made in the theoretical analyses of Sec. 13.5 to treat the problem of sweep efficiency as if the fluids involved actually had variable density.[1] Nor will consideration be given to this effect in the next section, where an illustrative example of cycling studies by use of the potentiometric model will be discussed, although the model experiments were made without correcting for the gas-density variation.

13.7. An Illustrative Application of the Potentiometric Model to Cycling Systems.—An instructive example of the way in which the potentiometric model has been applied in planning cycling operations is shown in Fig. 13.14.[2] The reservoir in question had a continuous water-gas contact around the boundary of the structure, except at the north, where it was

[1] A calculation carried through for a circular injection ring with a central producer, taking into account the variable gas density, gave sweep efficiencies that were higher than Eq. 13.5(10), corrected for an average density, for $n < 5$, and lower for $n > 5$.

[2] This is taken from D. L. Marshall and L. R. Oliver, *AIME Trans.*, **174**, 67 (1948).

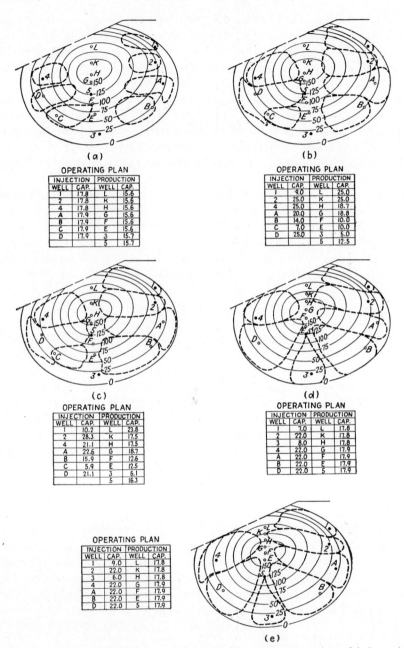

(a)

OPERATING PLAN

INJECTION		PRODUCTION	
WELL	CAP.	WELL	CAP.
1	17.8	L	15.6
2	17.8	K	15.6
4	17.8	H	15.6
A	17.9	G	15.6
B	17.9	F	15.6
C	17.9	E	15.6
D	17.9	3	15.7
		5	15.7

(b)

OPERATING PLAN

INJECTION		PRODUCTION	
WELL	CAP.	WELL	CAP.
1	9.0	L	25.0
2	25.0	K	25.0
4	25.0	H	18.7
A	20.0	G	18.8
B	14.0	F	10.0
C	7.0	E	10.0
D	25.0	3	5.0
		5	12.5

(c)

OPERATING PLAN

INJECTION		PRODUCTION	
WELL	CAP.	WELL	CAP.
1	10.2	L	23.8
2	28.3	K	17.5
4	21.1	H	17.5
A	22.6	G	18.7
B	15.9	F	12.6
C	5.9	E	12.5
D	21.1	3	6.1
		5	16.3

(d)

OPERATING PLAN

INJECTION		PRODUCTION	
WELL	CAP.	WELL	CAP.
1	7.0	L	17.8
2	22.0	K	17.8
3	8.0	H	17.8
4	22.0	G	17.9
A	22.0	F	17.9
B	22.0	E	17.9
D	22.0	5	17.9

OPERATING PLAN

INJECTION		PRODUCTION	
WELL	CAP.	WELL	CAP.
1	9.0	L	17.8
2	22.0	K	17.8
3	6.0	H	17.8
4	22.0	G	17.9
A	22.0	F	17.9
B	22.0	E	17.9
D	22.0	5	17.9

(e)

FIG. 13.14. Dry-gas invasion fronts, as determined by a potentiometric model, for various plans for cycling a condensate reservoir. Contours indicate sand thickness. Operating plans indicate assumed injection and production rates of individual wells in 10^6 ft^3 day. Measured sweep efficiencies for plans a, b, c, d, and e are 27, 63, 72, 67, and 76 per cent respectively. (*After Marshall and Oliver, AIME Trans., 1948.*)

limited by a fault. Five wells had already been completed in the reservoir, and the cycling patterns studied were chosen so as to use as many of these wells as possible. The net thickness variation of the condensate-bearing sand is indicated by the isopach contours in Fig. 13.14.

In order to prevent water entry into the reservoir and producing wells the latter were located on the crest of the structure, and the injection wells were distributed along the flanks on both sides of the line of producing wells. The wells already drilled are designated by numbers and the suggested new locations by letters.

In the first pattern tried the injection rates were made the same for all the injection wells, and the producing wells were produced at equal rates, with a total withdrawal rate equal to the total injection rate of 125 million cubic feet per day. The dry-gas invasion pattern at the time of first break-through into producing well F is shown in Fig. 13.14a. It will be seen that at the time of break-through the dry gas injected to the east of the producing wells still had far to go before reaching the latter. The total invaded area covered by the dry-gas contours represented only 27 per cent of the reservoir volume.

By using the same well locations but changing the individual injection and producing rates as indicated in the operating plan, the predicted dry-gas boundary at the time of first break-through, in well #5, as determined by the potentiometric model, is plotted in Fig. 13.14b. The increased injection rates in wells #2, #4, A, and D, heavier withdrawals from wells L and K, and restricted injection and producing rates from wells C and #3 have evidently led to a greatly improved sweep efficiency, namely, 63 per cent of the reservoir volume at the time of break-through. This pattern, however, was nevertheless unsatisfactory, because of the large unswept area east of the producing wells, and some of the required producing rates exceeded the actual capacities of the wells. A further readjustment of the injection and producing rates improved the sweep efficiency to 72 per cent (cf. Fig. 13.14c) but still left a considerable unswept area in the central thicker part of the reservoir.

A basic rearrangement of the well distribution was then tried, as shown in Fig. 13.14d. Well C was omitted, and D was moved farther south. E and F were shifted north of #5, and #3 was converted to an injection well. The individual injection and producing rates are indicated in the operating plan in the figure. The resulting sweep efficiency was found to be 67 per cent. A final modification, in which the proposed producing-well locations F, G, H, K, and L were moved still farther north and undrilled injection wells A, B, D were shifted to the south, gave a dry-gas-invasion pattern at first break-through as shown in Fig. 13.14e and a sweep efficiency of 76 per cent.

It will be evident from these progressive changes in plans that without further drilling than that involved in the program of Fig. 13.14e the latter probably represents a close approximation to the optimum cycling pattern, on the assumption of reservoir permeability uniformity made in the study. In actually carrying out this program additional information was obtained that indicated the reservoir conditions to be somewhat more complex than anticipated, and appropriate changes in the cycling operations were made after further model investigations. However, the preliminary phase of the investigation, as represented by Figs. 13.14a to e, should suffice to show the flexibility and power of the potentiometric model in studying the effects of both well location and the distribution of the relative injection and production rates on dry-gas-invasion fronts.[1]

13.8. The Effect of Permeability Stratification in Cycling[2] Operations.— One of the basic problems involved in the successful operation of cycling programs is that arising from dry-gas by-passing due to permeability stratification. It has been seen in previous sections that means are available for determining satisfactorily the areal sweep efficiencies and injection-fluid fronts for arbitrary distributions of well locations and injection or producing rates. These are applicable, however, only to individual strata. If the reservoir is comprised of a series of individual layers of substantially different permeabilities, the sweep processes will proceed in each one at rates approximately in proportion to their permeabilities. Hence, if some of the strata have much higher permeabilities than the remainder, the wet-gas displacement and dry-gas break-through will develop in them much sooner than in the remainder, and while an appreciable part of the reservoir as a whole is still unswept. The resultant sweep efficiency, *i.e.*, the total fractional wet-gas displacement at the time of first dry-gas break-through, will thus be reduced in proportion.

---

[1] Additional examples of applications are given in the paper of Marshall and Oliver (*ibid.*); cf. also the results of a study of the cycling operations in the Grapeland field, Houston County, Tex., by means of an electrolytic gelatin model, reported by F. C. Kelton, *API Proc.*, **24** (IV), 199 (1943). It should be noted, however, that from a practical standpoint the predictions of reservoir performance given by the electrical models cannot be more accurate than the assumed reservoir data. It is the uncertainty in the latter that ultimately limits the quantitative significance of model-study predictions.

[2] The general theory given here is also applicable to the study of stratification effects in water-drive reservoirs or primary-water-injection operations. The numerical results, however, will be modified by the additional factor of the ratio of the mobility (permeability-to-viscosity ratio) of the water to that of the oil. If this ratio exceeds 1, the stratification effects will be accentuated, whereas they will be lessened, as compared to cycling operations, if the water mobility is lower than that of the oil. This type of treatment of water-drive reservoirs with probability distribution of permeability has been developed by H. Dykstra and R. L. Parsons, API meetings, Los Angeles, Calif., May, 1948.

From a physical point of view the problem of permeability stratification can be readily treated without difficulty. If the different-permeability strata are mutually separated by shale breaks or are otherwise free of cross flow, they can be considered simply as a system of parallel reservoirs. But even if there is potential intercommunication normal to the bedding planes, this will not be of significance if the pressure distributions in the individual zones are substantially the same. The latter condition will obtain if the permeabilities and thicknesses of the separate members are uniform over the reservoir area or if the product of the permeability and thickness varies in a parallel manner for the different strata. In general, therefore, unless it is definitely known that appreciable cross flow is taking place or that the various-permeability layers are not continuous over the reservoir, the composite flow history can be approximated by a simple parallel superposition of those of its components.

While the superposition history of multilayer systems can be constructed by obvious graphical procedures, it can be formulated analytically as follows[1]: It will be assumed that the permeability k and displacement porosity $\bar{f}$ are continuous functions of a depth coordinate z along the well bore. The rate of dry-gas inflow[2] per unit thickness in the lamina at depth z may evidently be expressed as

$$Q(z) = ck(z), \tag{1}$$

where c is a constant determined by the areal geometry of the reservoir,[3] the well distribution, and their relative injection and producing rates. For a fixed cycling pattern and operating plan the composition of the produced gas in a uniform zone will be a function only of the total gas throughflow, expressed as a fraction of the hydrocarbon pore volume. The rate of wet-gas production from a unit-thickness lamina at z at the time t will therefore be

$$Q_w(z) = ck(z)F\left[\frac{ctk(z)}{A\bar{f}(z)}\right], \tag{2}$$

where F denotes the functional variation of the wet-gas fraction in the produced gas with the total gas throughflow, as determined by the well distribution and their relative fluxes, and the argument of F represents the

[1] The general theory of cycling with continuous permeability stratification presented here, including applications to probability and linear permeability distributions in addition to the results for the exponential distribution, is taken from M. Muskat, *Petroleum Technology*, **11**, 1 (November, 1948). Studies of discontinuous permeability distributions have been reported by Hurst and Van Everdingen, *op. cit.*, and Standing, Lindblad, and Parsons, *loc cit.*

[2] The rates of flow and volumes of throughflow discussed in this treatment refer to reservoir rather than surface measure.

[3] In addition to lateral uniformity and continuity of all the productive strata it is assumed here that these are all penetrated by each producing and injection well.

cumulative gas throughflow divided by the hydrocarbon volume available at z, A being the reservoir area. The fraction of wet gas in the total effluent from the stratified formation, at the time t, will then be

$$R_w(t) = \frac{\int_0^H k(z)F[ctk(z)/A\bar{f}(z)]dz}{\int_0^H k(z)dz}, \qquad (3)$$

where H is the total thickness of permeable pay. Equation (3) defines the composition history as a function of time. This can be related to the total fractional reservoir sweep by noting that the total wet gas produced at time t is

$$\overline{Q_w}(t) = \int_0^t dt \int_0^H Q_w(z)dz = c\int_0^t dt \int_0^H k(z)F\left[\frac{ctk(z)}{A\bar{f}(z)}\right]dz. \qquad (4)$$

The fractional reservoir sweep is then simply

$$\overline{V} = \frac{\overline{Q_w}(t)}{A\int_0^H \bar{f}\, dz}. \qquad (5)$$

In applying these equations it is convenient to consider the multilayer formation rearranged so that the permeability-to-displacement-porosity ratio $k/\bar{f}$ increases with z. Upon denoting the argument of the function F, $ctk/A\bar{f}$, by u, it follows from the definition of F that

$$\int_0^\infty F(u)du = 1; \qquad \int_S^\infty F(u)du = 1 - S; \qquad F(u) = 1 \;:\; u \leqslant S, \quad (6)$$

where S is the geometrical sweep efficiency in a uniform stratum.

Now at such values of t before any break-through has developed, *i.e.*, for

$$t \leqslant \frac{AS}{c}\left(\frac{\bar{f}}{k}\right)_{z=H} \equiv t_b, \quad F = 1,$$

and by Eq. (3),

$$R_w(t) = 1; \qquad \overline{Q_w}(t) = Qt; \qquad \overline{V} = \frac{Qt}{A\int_0^H \bar{f}\, dz}, \qquad (7)$$

where

$$Q = c\int_0^H k(z)dz, \qquad (8)$$

and is simply the total injection rate.

At times t between t_b and the time for break-through in the tightest zone, *i.e.*, for $t_b \leqslant t \leqslant t_m \equiv (AS/c)(\bar{f}/k)_{z=0}$,

$$R_w(t) = \frac{c\int_0^{z_0} k(z)dz + c\int_{z_0}^H k(z)F(u)dz}{Q}; \qquad u = \frac{ctk(z)}{A\bar{f}(z)}, \qquad (9)$$

where z_0 is such that

$$\frac{k(z_0)}{\bar{f}(z_0)} = \frac{AS}{ct}.$$

The cumulative wet-gas recovery will be

$$\overline{Q}_w(t) = ct \int_0^{z_0} k(z)dz + SA \int_{z_0}^{H} \bar{f}\, dz + c \int_{z_0}^{H} k(z)dz \int_{A\bar{f}S/ck(z)}^{t} F\left[\frac{c\tau k(z)}{A\bar{f}(z)}\right] d\tau. \quad (10)$$

After break-through has developed in the tightest zone, *i.e.*, for $t > t_m$,

$$R_w(t) = \frac{c \int_0^{H} k(z)F(u)dz}{Q}, \quad (11)$$

where now $F(u) < 1$. And the cumulative wet-gas recovery will be given by the general expression of Eq. (4).

To illustrate these relationships it will be assumed that

$$F(u) = 1 \quad : \quad u \leqslant S; \qquad F(u) = e^{(S-u)/(1-S)} \quad : \quad u \geqslant S. \quad (12)$$

This form satisfies Eq. (6) and roughly approximates the calculated variation of F in special cases (cf. Fig. 13.12). It will be further assumed that the permeability distribution is exponential,[1] as defined by

$$k(z) = ae^{bz/H}, \quad (13)$$

and that the displacement porosity $\bar{f}(z)$ is a constant, f.

Upon introducing the notation

$$\bar{t} = \frac{t}{t_b}; \qquad r = \frac{t_m}{t_b}; \qquad b = \log r, \quad (14)$$

where $t_b = AS\bar{f}e^{-b}/ac$, an evaluation of the above general equations, using Eqs. (12) and (13), gives

$$\bar{t} \leqslant 1 \quad : \quad R_w(\bar{t}) = 1; \qquad \overline{Q}_w(\bar{t}) = Qt; \qquad \overline{V}(\bar{t}) = \frac{S(r-1)\bar{t}}{rb}; \quad (15)$$

$$1 \leqslant \bar{t} \leqslant r \quad : \quad R_w(\bar{t}) = \frac{1}{r-1}\left[\frac{r}{\bar{t}} - 1 + \frac{(1-S)r}{S\bar{t}}\left(1 - e^{[S/(1-S)](1-\bar{t})}\right)\right], \quad (16)$$

$$\overline{V}(\bar{t}) = \frac{1}{b}\log \bar{t} + \frac{S}{b}\left(1 - \frac{\bar{t}}{r}\right)$$
$$- \frac{(1-S)}{b}e^{S/(1-S)}\left[Ei\left(-\frac{S\bar{t}}{1-S}\right) - Ei\left(-\frac{S}{1-S}\right)\right]; \quad (17)$$

[1] This type of permeability distribution implies a constant percentage change in permeability per unit of depth, and greater thicknesses of the low-permeability layers when the absolute permeability range is fixed. The ratio e^b of the maximum permeability ae^b to the minimum value a provides a convenient index of the exponential distribution and will be termed the "stratification constant" r [cf. Eq. (14)].

$$\bar{t} \geqslant r \quad : \quad R_w(\bar{t}) = \frac{r(1-S)e^{S/(1-S)}}{S\bar{t}(r-1)}\left[e^{-S\bar{t}/(1-S)r} - e^{-S\bar{t}/(1-S)}\right], \quad (18)$$

$$\overline{V}(\bar{t}) = 1 - \frac{1-S}{b}e^{S/(1-S)}\left[Ei\left(-\frac{S\bar{t}}{1-S}\right) - Ei\left(\frac{-S\bar{t}}{(1-S)r}\right)\right]. \quad (19)$$

In all cases the total throughput at the time $\bar{t}$, as a fraction of the net reservoir pore volume, is

$$\overline{Q}(\bar{t}) = \frac{S(r-1)\bar{t}}{rb}. \quad (20)$$

It will be readily verified that these expressions are continuous at their mutual contact points. At the time of break-through in the tightest zone ($\bar{t} = r$) these equations imply that

$$R_w(r) = \frac{(1-S)/S}{r-1}\left\{1 - e^{[S/(1-S)](1-r)}\right\}, \quad (21)$$

which reduces to the coefficient for $r \gg 1$. And

$$\overline{V}(r) = 1 - \frac{1-S}{b}e^{S/(1-S)}\left[Ei\left(\frac{-Sr}{1-S}\right) - Ei\left(\frac{-S}{1-S}\right)\right], \quad (22)$$

which has the asymptotic value for $r \gg 1$

$$V(r) \sim 1 - \frac{(1-S)^2}{Sb}. \quad (23)$$

In the limiting case of a uniform reservoir, $b \to 0$ and $r \to 1$. Equations (15) to (19) then reduce to the following forms:

$$\left.\begin{aligned} \bar{t} \leqslant 1 \quad &: \quad R_w(\bar{t}) = 1; \qquad \overline{Q}_w(\bar{t}) = Qt; \qquad \overline{V}(\bar{t}) = S\bar{t}; \\ \bar{t} \geqslant 1 \quad &: \quad R_w(\bar{t}) = e^{(S-S\bar{t})/(1-S)} = F(S\bar{t}); \\ V(\bar{t}) &= 1 - (1-S)e^{(S-S\bar{t})/(1-S)} = 1 - (1-S)F(S\bar{t}). \end{aligned}\right\} \quad (24)$$

In Eqs. (24), $\bar{t} = t/t_b = Qt/AH\bar{f}S$. Equations (24) can, of course, also be derived from first principles.

In the limit of 100 per cent areal sweep efficiency, that is, $S = 1$, Eqs. (15) to (19) reduce to

$$\left.\begin{aligned} \bar{t} \leqslant 1 \quad &: \quad R_w(\bar{t}) = 1; \qquad \overline{V}(\bar{t}) = \frac{(r-1)\bar{t}}{rb}; \\ 1 \leqslant \bar{t} \leqslant r \quad &: \quad R_w(\bar{t}) = \frac{1}{r-1}\left(\frac{r}{\bar{t}} - 1\right); \\ \overline{V}(\bar{t}) &= \frac{1}{b}\left(1 - \frac{\bar{t}}{r} + \log \bar{t}\right); \\ \bar{t} \geqslant r \quad &: \quad R_w(\bar{t}) = 0; \qquad \overline{V}(\bar{t}) = 1. \end{aligned}\right\} \quad (25)$$

For the intermediate time interval, $\overline{V}$ and the total throughflow, $\overline{Q}$, expressed as fractions of the reservoir hydrocarbon pore volume, can be related directly to R_w as

$$\left.\begin{aligned}\overline{V} &= 1 - \frac{1}{b}\left\{\log\left[1 + (r-1)R_w\right] - \frac{(r-1)R_w}{1 + (r-1)R_w}\right\}, \\ \overline{Q} &= \frac{r-1}{b[1 + (r-1)R_w]}.\end{aligned}\right\} \tag{26}$$

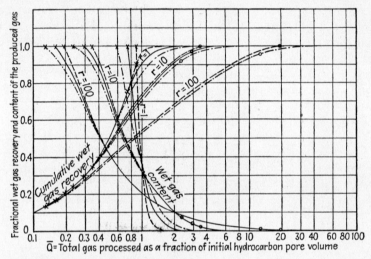

Fig. 13.15. The calculated variation of the fractional wet-gas content of the produced gas and of the total fractional wet-gas recovery vs. the total gas processed from cycling operations in exponentially stratified formations, for various areal sweep efficiencies S and ratios r of the maximum to minimum permeability. ----, $S = 0.90$ ———, $S = 0.75$. ·—·—·, $S = 0.60$. Crosses denote states of first dry-gas break-through; circles represent states of break-through in the tightest zones. (*From Petroleum Technology, 1948.*)

To illustrate these general relationships, the wet-gas content and cumulative wet-gas recovery have been plotted in Fig. 13.15 vs. the total gas throughflow for $S = 0.60, 0.75, 0.90$ and for $r = 1, 10,$ and 100. $r = 10$ corresponds to a ratio of the maximum to minimum permeability equal to 10, and this ratio is 100 for $r = 100$. $r = 1$ represents the strictly uniform reservoir. The abscissa values $\overline{Q}$ represent the total gas injection or production divided by the total hydrocarbon pore volume. $\overline{Q}$ is related to the argument $\bar{t}$ of Eqs. (14) to (19) as $\overline{Q} = (r-1)S\bar{t}/br$. The crosses in Fig. 13.15 denote the states of first dry-gas break-through in the most permeable zone, and the circles indicate break-through in the tightest layer. The curves for $r = 1$ simply reflect the functional form assumed for F, as required by Eqs. (24).

It will be noted from Fig. 13.15 that, whereas in a uniform formation the dry-gas break-through will develop after a total throughflow equal to the sweep efficiency S, dry gas will first appear in the producing wells for $r = 10$ after a throughflow of only 23.4, 29.3, and 35.2 per cent of the total hydrocarbon pore volume, for $S = 60, 75,$ and 90 per cent, respectively. The former will also represent the fraction of total wet-gas content produced by the time of first dry-gas break-through. And for $r = 100$ the corresponding break-through periods will represent recoveries of 12.9, 16.1, and 19.4 per cent of the original wet-gas content.

For dry-gas break-through in the tightest layers, with $r = 10$, the total gas processed will correspond to 2.34, 2.93, and 3.52 times the reservoir hydrocarbon pore volume,[1] for $S = 0.60, 0.75,$ and 0.90. By that time the wet-gas content of the produced gas will be 7.41, 3.70, and 1.23 per cent, respectively. And the total wet-gas recovery will be 92.2, 97.2, and 99.56 per cent of the initial reservoir content. For $r = 100$ the volumes of gas processed before break-through in the tightest layer will be 12.90, 16.12, and 19.35 times the reservoir hydrocarbon pore volume, for $S = 0.60, 0.75,$ and 0.90. The produced gas will then have wet-gas contents equal to 0.68, 0.33, and 0.11 per cent, respectively. And the total wet-gas recoveries will be 96.1, 98.6, and 99.99 per cent of the initial wet-gas content of the reservoir.

As shown in Fig. 13.15 and implied by Eq. (15) the total gas processed and wet-gas recovery by the time of first dry-gas break-through are directly proportional to the areal sweep efficiency S. And the cumulative wet-gas-recovery curves remain somewhat higher for all values of $\overline{Q}$ after break-through for the higher values of S. The wet-gas-content curves, however, first tend to merge and ultimately cross, although the latter divergence is so slight for $r = 10$ and 100 that it could not be shown on the scale of Fig. 13.15. For $r = 1$ the crossing point lies at $\overline{Q} = 1$, by virtue of the functional form assumed for $F(u)$.

The variation of the total wet-gas recovery vs. r by the time the wet-gas content falls to fixed limits, at which further processing may become unprofitable, is plotted in Fig. 13.16. As is to be expected the recovery curves decrease continually with increasing values of r or degree of stratification. For high values of r the recovery assumes an approximately logarithmic decline with increasing r. Figure 13.16 shows that the effects of stratification may be far more serious in limiting the total condensate recoveries than the areal sweep efficiency. Thus for $r = 100$, which does not represent an abnormally high degree of stratification as compared with

---

[1] The reservoir hydrocarbon volume used as a base for expressing the abscissa variable in Fig. 13.15 and in these comparisons is the actual net pore volume, or the initial total wet-gas content in reservoir measure.

those values commonly observed, the total recovery at an abandonment limit of 15 per cent wet-gas content will be only 61 per cent even if the areal sweep efficiency is 90 per cent.

The curves in Fig. 13.16 for $R_w = 1$ represent the fractional recoveries

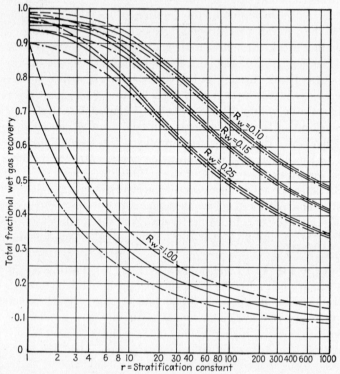

Fig. 13.16. The calculated variation of the total fractional wet-gas recovery in cycling operations vs. the stratification constant, r = ratio of maximum to minimum permeability, in exponentially stratified formations, for fixed fractional wet-gas-content abandonment limits, R_w. ——, $S = 0.90$. ——, $S = 0.75$. ·—·—, $S = 0.60$. S = areal sweep efficiency. (*From Petroleum Technology, 1948.*)

at the time of first dry-gas break-through. They are given by

$$\overline{V}(R_w = 1) = \frac{S(r - 1)}{r \log r},$$ (27)

and represent the composite sweep efficiency resulting both from the well pattern and permeability stratification. It will be seen that even for $r = 100$ the permeability stratification will reduce the over-all sweep efficiency almost by a factor of 5 compared with the areal sweep efficiency S. It is evidently because of the continued cycling operation to rather low wet-gas contents (after the initial dry-gas break-through) that the total wet-gas recoveries in practice will represent significant fractions of the original reservoir contents, as indicated by the upper curves of Fig. 13.16.

The total volumes of gas throughflow or processed, in reservoir measure, and as fractions of the total reservoir hydrocarbon volume, are plotted vs. r in Fig. 13.17, for various abandonment limits for the wet-gas content.[1]

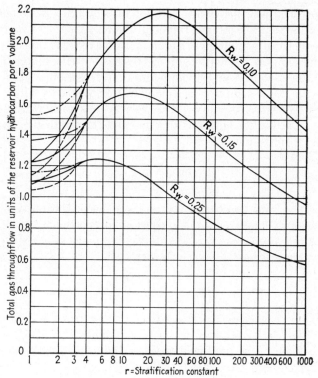

FIG. 13.17.   The calculated variation of the total gas throughflow, in units of the reservoir hydrocarbon pore volume, in cycling operations vs. the stratification constant, r = ratio of maximum to minimum permeability, in exponentially stratified formations, for fixed wet-gas-content abandonment limits, R_w. - - - -, $S = 0.90$. ——, $S = 0.75$. ·—·—, $S = 0.60$. S = areal sweep efficiency. (*From Petroleum Technology, 1948.*)

It will be observed that these are affected by the areal sweep efficiency only at the lower values of r.   In fact, for $r > 5$, the total throughflow to abandonment is, for practical purposes, independent of S.   Moreover the curves all show maxima in the range of r of 5 to 30 and then decline as r is still further increased.   The initial rise in the curves of Fig. 13.17 is due to the increasing throughflow required to give the approximately constant total wet-gas displacements indicated by Fig. 13.16 for low values of r.   The ultimate declines reflect the corresponding reductions in

---

[1] The total gas throughflows for $R_w = 1.00$ are evidently equal to the total wet-gas recoveries for $R_w = 1.00$ plotted in Fig. 13.16.

780 PHYSICAL PRINCIPLES OF OIL PRODUCTION [CHAP. 13

total wet-gas recovery, shown in Fig. 13.16 at high stratification ratios, which can be swept out by relatively small volumes of injected gas. It will also be noted that the volumes of gas processed will vary more rapidly with the abandonment limit of wet-gas content than the total wet-gas recovery.

If the permeability distribution is not satisfactorily represented by a continuous function, the integral representation used in the above discussion can be transformed into discrete summations by obvious procedures. For a rapid approximate evaluation of the by-passing effects due to permeability stratification a discontinuous permeability distribution may be used, together with the assumption of complete sweeping $(S = 1)$ at the time of first dry-gas break-through. This corresponds to a continuous-line-drive representation for both the injection and the producing wells.

The total throughflow rate will then be

$$Q = c \sum_1^N k_n h_n, \tag{28}$$

where N is the total number of layers, k_n is the permeability of the nth layer, h_n is its thickness, and c is a constant proportional to the pressure differential and also gives expression to the geometry of the system. The sweep-out time for the nth layer will then be

$$t_n = \frac{A\overline{f_n}}{ck_n} = \frac{\overline{f_n}T\Sigma k_n h_n}{k_n\Sigma h_n \overline{f_n}}, \tag{29}$$

where $\overline{f_n}$ is the displacement porosity for the nth zone, A is the total sweep area, and T is the time for a complete throughflow of the hydrocarbon reservoir volume. The fractional wet-gas content of the gas at any time t is then

$$R_w(t) = \frac{\displaystyle\sum_1^j k_n h_n}{\displaystyle\sum_1^N k_n h_n}, \tag{30}$$

where j is such that $t_j > t > t_{j+1}$, and it is assumed that the various layers are numbered in a sequence of decreasing t_n's or increasing $k_n/\overline{f_n}$'s. The fractional sweep of the total reservoir wet-gas content at the time t is

$$\overline{V}(t) = \frac{ct\displaystyle\sum_1^j k_n h_n + A\displaystyle\sum_{j+1}^N \overline{f_n} h_n}{A\displaystyle\sum_1^N \overline{f_n} h_n} = \frac{t\displaystyle\sum_1^j k_n h_n}{T\displaystyle\sum_1^N k_n h_n} + \frac{\displaystyle\sum_{j+1}^N \overline{f_n} h_n}{\displaystyle\sum_1^N \overline{f_n} h_n}. \tag{31}$$

When $t = t_j$, the time for break-through in the jth layer,

$$\overline{V}_j(t) = \frac{\dfrac{\overline{f}_j}{k_j} \sum_1^j k_n h_n + \sum_{j+1}^N \overline{f}_n h_n}{\displaystyle\sum_1^N \overline{f}_n h_n}. \tag{32}$$

Equation (32) is evidently the summation equivalent of the integral representation obtained by appropriately simplifying Eq. (10), namely,

$$\overline{V}(t) = \frac{\dfrac{\overline{f}(z_0)}{k(z_0)} \displaystyle\int_0^{z_0} k(z)dz + \int_{z_0}^H \overline{f}\, dz}{\displaystyle\int_0^H \overline{f}\, dz} \tag{33}$$

where z_0 is the depth at which break-through first develops at the time t.

As an example of the application of Eqs. (30) to (32) it may be noted that for a simple hypothetical case of four layers of equal thickness and displacement porosity, but permeabilities in the ratio of $1:5:10:25$, the sequence of break-throughs will come at values of 100, 80, 65, and 41 per cent of the total wet-gas content. The wet-gas contents of the produced gas after these break-throughs in the three most permeable strata will be 2.44, 14.63, and 39.02 per cent, respectively.

It should be noted that even the generalized treatment giving rise to Figs. 13.16 and 13.17 is based on the assumptions of an exponential permeability distribution [cf. Eq. (13)], the neglect of the variation of the connate-water saturation with the permeability, and the functional form of the function F represented by Eq. (12). Recent statistical studies of permeability distributions[1] of formation samples taken from well bores indicate that Gaussian distributions often represent a closer approximation to those occurring in actual reservoirs than exponential distributions. While the analysis similar to that given here for the exponential distribution has also been carried through[2] for the probability distribution, it is somewhat more complex and has been limited to 100 per cent areal sweep efficiencies $(S = 1)$. However, the general results are quite similar to those of Figs. 13.16 and 13.17, although the stratification index is different from that defining the exponential distribution. Since the actual permeability variation will seldom be known with precision, the exponential approximation may suffice for most practical applications.[3] Similar con-

[1] Cf. J. Law, *AIME Trans.*, **155**, 202 (1944).

[2] Muskat, *loc. cit.* The corresponding results for linear permeability distributions are also given in this work.

[3] If the actual distribution cannot be approximated satisfactorily by a single exponential function, it may be possible to resolve the composite section into segments

siderations apply with respect to the function F. While Figs. 13.16 and
13.17 will not be quantitatively valid in practice, they should provide
reasonable estimates of the effect of permeability inhomogeneities even
when applied to specific reservoirs.[1] In any case they serve to show that an
efficient areal well pattern will not alone ensure high cycling recoveries
and that the variations in zonal permeability and the abandonment limit
of wet-gas content of the produced gas will ultimately control the effectiveness of cycling operations.

13.9. Field Observations on Condensate-producing Reservoirs.—Although more than 200 condensate reservoirs had been discovered by 1945
and 37 cycling plants had been built by November, 1944, the published
data on their performance are extremely meager. And even these are
limited mainly to observations that the gas-oil ratios increased during the
producing life of reservoirs in which cycling was not undertaken at all
or where the pressure was not fully maintained by cycling. No comparisons
have been reported of the complete and detailed composition histories of
the produced fluids and the reservoir pressure or cumulative production
with the predictions from laboratory studies of the original reservoir fluids,
such as discussed in Sec. 13.3. On the other hand, as will be seen below,
there is good field evidence that the basic features of condensate-reservoir
performance are governed by the retrograde-condensation phenomena.
Moreover, when substantially complete cycling has been tried, the prevention of retrograde losses has been generally confirmed by the constancy of
the composition of the produced gas until the development of dry-gas
break-through.

One of the first reports of rising gas-liquid ratios in condensate reservoirs
producing by pressure depletion, without cycling, was that on the La
Blanca field, Hidalgo County, Tex.[2] The condensate-bearing reservoir in
this field was discovered in 1937 at a depth of 7,500 ft in the Frio Sand.

---

with individual approximating exponential representations. If, however, the probability
or linear distributions definitely appear to be indicated by the core-analysis data, the
corresponding cycling efficiency curves (Muskat, *loc. cit.*) should, of course, be used.
The argument of the Gaussian or probability distribution referred to here and elsewhere in this section is the logarithm of the ratio of the permeability to the median
permeability.

[1] It should be noted, however, that the idealized assumptions of perfect stratification
and areal continuity underlying the theoretical analysis will tend to accentuate the dry-
gas break-through by-passing effects as compared to those which may occur in practice.
And the use of a constant porosity $\bar{f}$ also implies an underestimation of the fractional
cycling recoveries, since in actual reservoirs the unswept tighter parts of the reservoir
may be expected to contain higher connate-water saturations and lower initial hydrocarbon reserves.

[2] F. V. L. Patten and D. C. Ivey, *Oil Weekly*, **92**, 20 (Dec. 12, 1938). Figure 13.18
is taken from this source.

The oil (condensate) gravity was 55°API. The original gas-condensate ratio was approximately 55,500 ft³/bbl, at the reservoir pressure of 4,200 psi. No attempt was made to cycle the field. By the time the pressure declined to 3,800 psi, the gas-condensate ratio had doubled to 111,000 ft³/bbl. The subsequent rise to 384,000 ft³/bbl at 2,180 psi is plotted in Fig. 13.18 as the solid curve. The dashed part of the curve is an extrapolation to indicate the predicted future behavior during the decline in pressure below 2,180 psi. It was estimated that 65 per cent of the liquid content of the

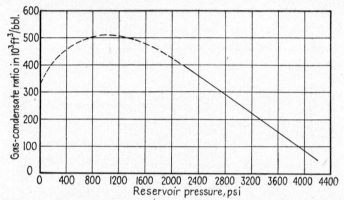

Fig. 13.18.   The variation of the gas-condensate ratio observed for the production from the La Blanca field with declining reservoir pressure.   Dashed segment represents extrapolation. (*After Patten and Ivey, Oil Weekly, 1938.*)

reservoir gas at 3,800 psi would be lost at ultimate pressure depletion and 82 per cent of the original condensate content at 4,200 psi. While the retrograde characteristics of the La Blanca reservoir fluids are not available for comparison with Fig. 13.18, there can be little doubt that the latter reflects the continued condensation and trapping of the liquid phase in the reservoir formation.

Similar, though not as pronounced, declines in condensate recovery per unit gas volume have been observed[1] in the 8,200-ft horizon of the Big Lake field, Reagan County, Tex. When discovered in 1929 the reservoir pressure was 2,270 psi, and the separator-fluid recoveries were in the ratio of 36 to 38 bbl/10⁶ ft³ of gas. When the last gas well was completed in March, 1933, the reservoir pressure had fallen to 1,225 psi and the rate of condensate recovery to 28 to 30 bbl/10⁶ ft³ of gas. The API gravity of the separator liquid rose from an initial range of 61 to 63° to 73 to 75° in 1938. The latter change also reflects the dropping out and trapping of the heavier wet-gas components in the reservoir, so that the residual-liquid content of the produced gas had an average lower molecular weight.

[1] Cf. E. V. Foran, *AIME Trans.*, **132**, 22 (1939).

The variation of the condensate recovery from 19 wells in the La Gloria field, Jim Wells County, Tex., over a 3-year period is plotted[1] in Fig. 13.19. Although this field was being cycled, the reservoir pressure apparently was not fully maintained. None of the wells in the group had yet been invaded by the dry gas, and in the unswept area the decline in pressure led to liquid condensation in the same manner as in uncycled reservoirs.

Perhaps the most complete reported study of the behavior of an actual condensate reservoir is that of the Bodcaw Sand of the Cotton Valley field,

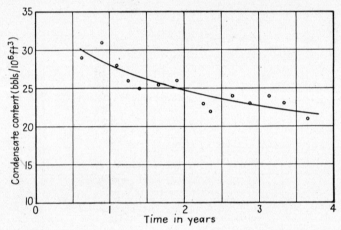

Fig. 13.19. The history of the average condensate content (C_5+) of 19 producing wells in the La Gloria field.

Webster Parish, La. The Bodcaw Sand is the most important member of five condensate-producing strata lying at depths of 8,100 to 8,600 ft and discovered in 1937. A small oil rim underlay the 18,295 acres of the anticlinal gas-cap reservoir. The reservoir temperature was 238°F, and the initial pressure was 4,000 psi gauge. An analysis of recombined separator-gas and -liquid samples showed the fluid to be at the dew-point pressure within the reservoir and to have a condensable liquid (C_4+) content of 113.98 bbl/10^6 ft^3, corresponding to a gas-liquid ratio of about 8,770 ft^3/bbl.

Prior to unitization for cycling and the beginning of gas injection in May, 1941, 81.3×10^9 ft^3 of gas, 5,438,500 bbl of condensate, and 944,800 bbl of oil had been produced. As the water intrusion from the bounding aquifer did not suffice to replace these withdrawals, the pressure declined continuously to approximately 3,220 psi. The volumetric calculation of the original gas content of the condensate reservoir, based on an average porosity of 16.2 per cent, connate-water saturation of 25.4 per cent, and average effective thickness of 23.8 ft, gave 510×10^9 ft^3.

[1] The original data were given by J. O. Lewis, *AIME Trans.*, **170**, 202 (1947).

Although 86 wells had been originally completed in the Bodcaw Sand, many were shut in when cycling was undertaken.   In March, 1946, 6 crestal wells were being used for injection, and 30 wells along the flanks and near the lower boundary were maintained on production.   Some of the shut-in wells were used for test purposes to study the sweep of the dry-- gas downstructure.

The gross compositon history of the produced gas is plotted in Fig. 13.20.[1] It will be seen that the data based on the recoveries from the cycling plant

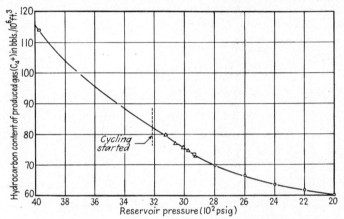

Fig. 13.20.   The variation of the hydrocarbon content of the gas produced from the Bodcaw Sand of the Cotton Valley field with the reservoir pressure. $\square$, recombined reservoir-fluid analyses. $\triangle$, plant-yield data. $\bigcirc$, extrapolated predictions based on laboratory analyses. (*After Miller and Lents, API Drilling and Production Practice, 1946.*)

agree closely with those predicted from the laboratory analyses.   The continued pressure decline and fall in the condensate content of the produced gas during the cycling operations do not imply a failure of the latter, since less than 80 per cent of the produced gas was returned to the formation.   Moreover; whereas the pressure drop of 700 psi to May, 1941, when cycling was begun, resulted from a withdrawal of only 81,300,000,000 ft^3 of gas, the subsequent withdrawal of 206,700,000,000 ft^3 of gas to March, 1946, led to an additional pressure drop of only 360 psi.   And the decline in content of condensable product of the produced gas in the latter period was 8.7 per cent, as compared with 28.3 per cent before cycling was started. That the continued decline in the curve after cycling was undertaken was due to retrograde liquid accumulation in the reservoir rather than dilution by dry gas was established by a detailed comparative study of the producing

[1] Figure 13.20, as well as the discussion given here of the Bodcaw Sand cycling operations, is taken from M. G. Miller and M. R. Lents, *API Drilling and Production Practice*, 1946, p. 128: cf. also R. L. Hock, *Oil and Gas Jour.*, **47**, 63 (Nov. 4, 1948).

characteristics of individual wells distributed over the whole producing area.

An interesting feature of the operation of the Bodcaw cycling plan was the periodic testing of the wet-gas content of wells that had been shut in prior to envelopment by dry gas, so as to determine the nature of the wet-gas displacement. The results of these tests are plotted in Fig. 13.21. The cumulative productions for these wells to January, 1946, ranged from 715×10^6 ft^3 for well E to 5.4×10^9 ft^3 for well J. It will be noted that

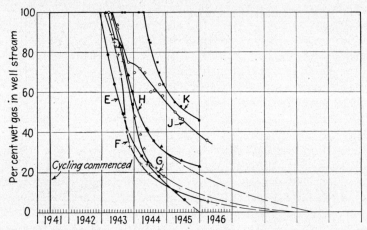

Fig. 13.21. The history of the approach of injection gas at various wells in the Bodcaw Sand of the Cotton Valley field, as obtained by periodic well tests on the wet-gas content of the well fluid. (*After Miller and Lents, API Drilling and Production Practice, 1946.*)

the wet-gas content dropped quite sharply, indicating a rather rapid and uniform sweeping of the dry gas past the test wells. An electrolytic-model study, in which the thickness variations were neglected, indicated that the dry gas should have appeared in wells H, J, and K, within 768, 830, and 1,347 days after cycling was started. The break-through times actually observed were 667, 720, and 1,056 days, respectively, which may be considered as agreeing quite satisfactorily with the predictions, in view of the simplifications made in the model study.

Using an average permeability distribution for the Bodcaw Sand, as based on core-analysis data from a number of wells, predictions were made of the expected dilution history of the producing wells, applying a simplified theory corresponding to Eqs. 13.8(28) to 13.8(32). These showed satisfactory agreement with the observed data plotted in Fig. 13.21. In fact the steep declines in wet-gas content implied by the curves of Fig. 13.21 were thus shown to reflect simply the rather uniform character of the formation and the absence of high degrees of stratification in the productive section. Related calculations of the reservoir volumes swept out by the injected gas, using the field average-permeability profile, also were in reasonable agreement with the implications of individual well tests on the spread of the dry gas.

Because a number of producing wells are spaced closely along the lower boundary of the gas cap, the very high areal sweep efficiency of 95 per cent was indicated by the electrolytic-model investigation. In fact the well distribution roughly simulates the circular cycling patterns discussed in Sec. 13.5 and is inherently susceptible to the achievement of high sweep efficiencies. Moreover there has been some intrusion of edgewaters, which has facilitated the sweep into the producing wells of the gas between the latter and the limits of the gas cap. From the actual reservoir performance to March, 1946, when 49 per cent of the reservoir content had been produced, it was estimated that the ultimate recovery will be 85 per cent of the wet-gas reserves, with a processing of a gas volume equivalent to 115 per cent of the initial-gas content.

13.10. Practical Aspects of Condensate-reservoir Development.—The basic practical problem arising in the exploitation of a condensate-producing reservoir evidently lies in the decision as to whether or not it is to be cycled. This is, of course, little else than an economic problem. Because of the many factors involved no simple formula or rule can express the economic balance between the operating profit under cycling and that under pressure-depletion operations.

It is instructive to illustrate the economic factors by reference to a hypothetical situation. Thus, if the gross condensate-reservoir volume is 200,000 acre-ft and the porosity and connate-water saturations are each 25 per cent, the total hydrocarbon pore volume will be 16.34×10^8 ft³. If the reservoir pressure is 300 atmo ($\sim 4,500$ psi), the reservoir temperature is 200°F, and the gas deviation factor is 0.9, the wet-gas content, in surface measure, will be equivalent to 4.29×10^{11} ft³. If the initial gas-liquid ratio is 40,000 ft³/bbl, the liquid content of the reservoir gas will be 10,725,000 bbl. Assuming that 80 per cent of this is recoverable by cycling, including subsequent blowdown, and 45 per cent by pressure depletion alone, without a plant, the corresponding recoveries will be 8,580,000 and 4,826,000 bbl, respectively. A plant of 125,000,000 ft³/day would be required for a cycling life of about 12 years with a throughput of 1.25 reservoir volumes. At a plant cost of $65,000 per 10^6 ft³/day capacity this would require an initial plant investment of $8,125,000. For an average injection capacity of 9,000,000 ft³/day per well, 14 injection wells would be needed. If their cost is $125,000 per well, the injection-well investment would be $1,750,000. When the plant operating expenses and flow-line costs are added and account is taken of the discounted present worth of the deferred income under cycling, it is clear that the value of the increased liquid recovery of 3,754,000 bbl will hardly compensate for the increased cost of the cycling operations over that involved in simple pressure depletion.[1]

If, however, the gas-liquid ratio of the same reservoir were 10,000 ft³/bbl, the value of the increased recovery would be multiplied approximately[2] by a factor of 4 and the balance in favor of cycling would evidently be substantial. These numerical values are, of course, only of illustrative significance and are not applicable to specific reservoirs. However, they should serve to show the order of magnitude of some of the major economic factors. Much more detailed accounting of all the cost items and a more precise evaluation of the recovered products in terms of the individual major components[3] will be necessary in order to derive a complete economic appraisal of the cycling possibilities in actual reservoirs.

It should be observed that the difference between the initial condensate content times the fractional sweep volume under cycling and the initial content times the pressure-depletion recovery factor is not a complete measure of the potential gain by cycling. For after the cycling itself must be terminated, because of the fall in wet-gas content of the produced gas until it is no longer profitable to reinject it, the reservoir can then be produced by simple pressure depletion. The unswept part of the formation will then provide an additional condensate recovery similar to that which would be obtained by pressure depletion if there had been no cycling. If $\overline{V}$ is the estimated fractional wet-gas recovery during the cycling phase with complete pressure maintenance, such as is indicated by Fig. 13.16, R_d the fractional condensate-recovery factor under pressure depletion, and $\overline{R}$ the total fractional condensate recovery, the latter would thus be

$$\overline{R} = \overline{V} + (1 - \overline{V})R_d. \tag{1}$$

Equation (1) is commonly used in predicting condensate recoveries. If $\overline{V}$ is taken as the resultant sweep efficiency to the cycling abandonment limit, the first term will need correction when applied to practical systems. Because of the shrinkage losses due to liquid extraction, the sale of low-pressure vapors, the use of some of the produced gas for fuel, although compensated in part by the fact that the returned dry gas generally[4] has a

[1] While these comparisons suggest that cycling would not be warranted economically under the assumed conditions if there were a current market for the dry gas, the operation of a gasoline- or liquid-extraction plant with pressure depletion might well be, in this case, the optimum economic-development method.

[2] With the increase in richness of the gas by a factor of 4 there would probably be associated an increased reservoir loss due to retrograde condensation. Moreover the details of the plant design and costs would be different.

[3] Cf., for example, W. H. Woods, *Oil and Gas Jour.*, **46**, 94 (Aug. 23, 1947).

[4] In fact, under certain conditions the deviation factors may increase to such an extent, on passing from the wet to the dry gas, as more than to compensate for the liquid-

higher deviation factor than the reservoir wet gas, the withdrawals will not be fully replaced and there will be some decline in pressure and retrograde loss even during the cycling operations, unless "make-up" gas is purchased. The first term on the right-hand side of Eq. (1) will therefore usually be reduced by a factor of 0.85 to 0.95.

Equation (1) must be supplemented by similar relations applying to the liquefiable products other than the stable condensate. For these the depletion recoveries R_d will generally be higher than for the stable condensate. The corresponding gains over pressure depletion will be lower in proportion. The value of R_d in Eq. (1) for these components will also depend on the manner of treating the wet gas after cycling is discontinued. Moreover, in comparing the relative recoveries of these components by cycling and pressure depletion, consideration should be given to the possibility of operating a gasoline or low-pressure hydrocarbon extraction plant without returning the stripped gas to the formation. On the other hand the dry gas following up the wet gas in the strata that were not completely swept during cycling will partly vaporize the retrograde liquid accumulation and make it available for recovery at the surface. Although it is difficult to evaluate all these factors accurately,[1] they should not be ignored arbitrarily in attempting to make quantitative comparative analyses of different methods of producing condensate reservoirs.

While the immediate purpose of cycling is the displacement of the wet gas at its dew point so as to prevent retrograde losses in the reservoir, the processing of the produced gas itself will generally provide a substantial part of the over-all gain from the cycling method of exploitation. Processing plants of modern design extract 50 to 75 per cent of the propane, 80 to 98 per cent of the butanes, and virtually all the pentanes plus content of the wet gas. As indicated by Table 3 the recovery of the butanes, even by three-stage separation, may be only 10 to 25 per cent of that in the produced gas, if not processed in an extraction plant.

The recovery factors will largely control the gross economics of various development methods. But they do not alone determine the advisability of cycling. If there is no satisfactory market for the gas, cycling may serve as a means of gas storage as well as a method for efficient recovery

---

extraction shrinkage losses [cf. D. L. Katz and C. M. Sliepcevich, *Oil Weekly*, **116**, 30 (Feb. 26, 1945)].

[1] Still another factor that should be taken into account when possible, although no evaluation of it has been reported, is the assumption that the uniform pressure-depletion and blowdown recovery factors as determined by calculation or laboratory experimentation are applicable to the actual stratified or nonuniform reservoirs. In the latter the depletion recovery factors should be averaged and weighted according to the permeability distribution and corrected, if possible, for cross flow. And the blowdown recovery factors will reflect primarily that associated with the tight parts of the pay and hence will be somewhat lower than that for a uniform depletion throughout the reservoir.

of the condensable-liquid products of the wet-gas reservoir content. For example, the gas content alone of the hypothetical reservoir considered above would be worth roughly 50 million dollars if it were stored at the original reservoir pressure until it could be sold at a price of $0.12 per thousand cubic feet. This is evidently much greater than the value of the liquid content, if the gas-liquid ratio were 40,000 ft^3/bbl. Similar consideration should be given to the gas-storage aspect of cycling if there is a probability that the market value of the gas will rise, even if it could be sold immediately. On the other hand, if the sale price of the gas were to remain constant, its storage until completion of the cycling would represent a loss in present worth due to the interest discount. In addition the slower recovery of the condensate under cycling than would be possible under unrestricted pressure depletion will lead to a reduced present worth of the liquid recovery. This time factor will also be of importance in determining the size or capacity of the cycling plant. Still other factors are the tightness of the formation, which will determine the number of injection wells required, the well costs, which will be controlled largely by the depth, the reservoir uniformity, possibilities of faulting, and the areal sweep efficiency. Each reservoir must be evaluated separately in the light of its unique characteristics.[1]

Although the primary emphasis of the discussion of this chapter has been on the exploitation of the condensate reservoir itself, such reservoirs often are underlain by crude-oil zones of sufficient size so that the development of the latter must be made an integral part of the program for producing the gas cap. Among the various possible methods of operation the most efficient, from the point of view of recovery, will be that in which the fluid withdrawals are limited to the crude-oil zone while pressure is maintained in the condensate reservoir by gas injection. The injection wells are then to be so located as to sweep the wet gas into the oil zone. In this manner the retrograde losses in the gas cap will be prevented, and the oil-recovery efficiency will be enhanced by virtue of the general pressure-maintenance effect and the contribution of the gravity-drainage mechanism. Moreover the condensate accumulation that may develop in the swept part of the black-oil zone, owing to the lower pressures in the withdrawal area, will be recoverable because it will add to the residual-oil saturation and have a nonvanishing permeability. In fact the mixing of the condensate with the crude oil and the lowering of its viscosity will tend to increase the black-oil recovery factor. The sweep efficiency of the gas

[1] Examples of studies made in planning cycling operations in actual fields are the reports of N. Williams, *Oil and Gas Jour.*, **46,** 99 (Dec. 6, 1947), on the Lake St. John field, Concordia and Tensas Parishes, La., and of W. L. Horner and E. G. Trostel, *Oil and Gas Jour.*, **46,** 77 (Mar. 4, 1948), on the Benton field, Bossier Parish, La.

cap will also be high, since the "dead" areas usually left behind the peripheral producing wells in the simple gas-cap cycling will be eliminated. If it is not feasible to confine the withdrawals entirely to black-oil wells and shut them in progressively as they become enveloped by the displaced wet gas until the oil zone is depleted, the recovery efficiency will not be seriously impaired as long as the production in the original oil reservoir is so controlled that there is a net downflank migration of the oil.

Simultaneous black-oil production and cycling of the gas cap, as is common practice, should lead to high recovery efficiency provided that a net pressure gradient is maintained from the gas cap toward the oil zone. If the gradients are reversed, there may be serious losses in crude-oil recovery due to oil migration into the gas cap. And if the wet-gas-oil contact is held fixed, the black-oil zone will become depleted as if there were no gas cap and no pressure maintenance, although it may still be subject to the favorable effects of gravity drainage. If a potentially active water drive is available, the necessity for maintaining pressure in the gas cap and downdip gradients near the gas-oil contact will be even more imperative.

Production of only the crude-oil zone, with no gas return to the gas cap, will lead to pressure decline and retrograde liquid accumulation in the latter. Cycling of the gas cap following the depletion of the oil reservoir could theoretically lead to complete condensate recovery as the combined result of residual-wet-gas displacement and revaporization of the liquid phase. However, the crude-oil recovery may be appreciably lower than under simultaneous cycling and oil production because of the reduced effectiveness of the gravity-drainage recovery mechanism and the lack of pressure maintenance. Moreover, if a gas-processing plant is not installed until cycling is undertaken, much of the intermediate condensable-liquid content of the separator gases produced during the oil-zone depletion will be lost. Such operations will also defer the income from the gas-cap production and reduce its present worth.

Shutting in the oil zone until the gas cap is completely cycled will evidently ensure high condensate recovery. Subsequent production of the oil reservoir before blowdown of the gas cap will lead to substantially the same recovery as when the oil-zone depletion precedes the gas-cap cycling. Analogous to the latter, there will be involved here a deferment of the income from the black-oil reservoir if its depletion is delayed until after the cycling is completed.

These general observations do not represent universal comparative evaluations of the various possible methods of exploiting composite crude-oil and condensate reservoirs. In considering a specific reservoir all its individual characteristics and the practical and economic aspects of its development must be taken into account. What may appear to be the most efficient operating plan theoretically may be entirely impractical in a particular situation because of unique practical conditions involved in actual operation. The market demand for the gas, condensate, or oil, state regulations, the richness of the wet gas, the cost of the wells, the presence or absence of water drives, the relative size of the oil zone, the faulting of the reservoir, the reservoir pressure, and other related factors may singly or in combination virtually force the adoption of one method of operation in one field, whereas a different method may be necessary in another reservoir that is apparently similar in some respects.

When the condensate reservoir is bounded directly by edgewater rather than an oil rim, the operating problems are simplified. If the edgewaters can provide a sufficiently active water drive to maintain the pressure near the dew point, without prohibitively restricting the withdrawal rates, gas return will evidently be unnecessary.[1] However, such active water drives are very unlikely except in highly fractured limestones, although it is conceivable that by supplementary water injection a practical degree of pressure maintenance could be achieved in some cases. Moreover the development of appreciable rates of natural water intrusion will require that some pressure decline take place in the condensate reservoir. On the other hand, if the edgewaters have any mobility whatever, the location of the producing wells near the water-gas contact will induce the entry of water, due to the superposition of the local pressure drawdowns and the slow pressure decline in the gas cap associated with the incompleteness of the gas-withdrawal replacement. By placing the producing wells at the crest of the structure not only will the production be water-free,[2] but it will be possible to achieve high areal sweep efficiencies by completing the injection wells below the edge of the gas cap and into the water-bearing formation. In this way a driving pressure toward the producing wells will be exerted on the gas even at the very limits of the condensate reservoir, and the development of dead spaces that would be left behind the injection wells if completed within the gas-bearing formation will be prevented. This increase in areal sweep efficiency should more than counterbalance the higher flow resistance in the water zone above the injection wells. While some of the water will be displaced into the gas cap, it will be gradually dispersed and should become trapped before reaching the producing wells. Aside from the achievement of increased sweep efficiency

[1] While an equilibrium gas saturation, of the order of 5 to 15 per cent, might be trapped in the water-invaded zone, the loss of this gas would probably not be serious as compared with the savings in the cost of the gas-injection facilities of a cycling program.

[2] This refers to the entry of edgewater into the producing wells. The production of condensate is always accompanied by fresh water, which had been held in the vapor phase in the reservoir and drops out, together with the condensate, as the temperature and pressure are reduced from those in the reservoir.

the location of the injection wells below the water-gas contact affords the possibility of using dry holes that may have been drilled in delineating the reservoir boundaries and that would otherwise be abandoned as useless. Although this procedure seems rather unorthodox, it has been successfully applied in several fields in the Gulf Coast area.[1]

It has already been noted that in cycling operations there is usually an incomplete replacement of the total fluid withdrawals. This results in a slow pressure decline and some liquid condensation. There is another cause of retrograde liquid accumulation during cycling operations. This is the pressure drawdown about the producing wells. Because all the production must pass through the annular ring immediately bounding the well bore, the retrograde condensation will quickly build up the pore saturation near the well bore until it becomes mobile and is swept into the well with the gas phase. The area in which the limiting saturation has been built up will gradually expand outward from the producing well as production continues. The dew-point pressure of the well effluent will correspond to that at the limit of the area of condensate mobility and should theoretically increase as the region of mobility expands, provided that the reservoir pressure as a whole remains fixed.

An estimate of the rate of build-up of the liquid-phase saturation can be made by evaluating the equation

$$\frac{d\rho}{dt} = \frac{Q}{2\pi r h f} \frac{dp}{dr} \frac{dC}{dp}, \tag{2}$$

where Q is the production rate, h the formation thickness, f its porosity, dp/dr the pressure gradient at the radius r, and C the liquid content of the wet gas per unit volume, in surface measure. By assuming a steady-state radial pressure distribution, a reservoir pressure of 4,000 psi, and a drawdown of 500 psi, the pressure gradient at a radius of 1 ft will be approximately 50 psi/ft. For $h = 50$ ft, $f = 0.25$, and $Q = 5 \times 10^6$ ft³/day, the factor of dC/dp becomes 3.18×10^6 psi/day. The value of dC/dp near the dew point of some of the condensate-reservoir fluids for which data have been obtained is of the order of magnitude of $10^{-6} - 10^{-7}$ cc/cc/(psi). It thus follows that the area within a radius of 1 ft about the well bore will be filled up with liquid phase to the point of mobility within a few hours or several days at the most. As the rate of fill-up will vary inversely as the square of the radius, it will take about 1.7 years for a condensate saturation of 20 per cent to develop at 100 ft from the well bore even if dC/dp is 10^{-6} cc/cc/(psi).* The rate of condensate-saturation build-up is also

----

[1] H. L. Hensley, *Oil and Gas Jour.*, **45**, 84 (May 3, 1947).

* If the effect of the liquid accumulation on the pressure distribution is neglected, the history of the liquid-saturation build-up as a function of both time and the radial distance can be readily calculated by applying Eq. (2). Results of this type have been

essentially proportional to the square of the drawdown or the square of the production rate. While this continued accumulation may become of serious magnitude in tight formations producing a very rich gas, it will probably not represent a major loss factor in most cycling operations. Moreover by the time the cycling is completed at least a part of the liquid phase will be revaporized as the dry gas sweeps through the area about the well bore.

Although the basic premise underlying the previous discussion of the operation of condensate reservoirs has been that the liquid-phase accumulation in the formation resulting from retrograde condensation is essentially irrecoverable, this is not strictly correct from a physical standpoint. It is true that except near the producing well bores the pore saturation of condensate liquid will generally be too low to have any mobility and will remain trapped as the residual wet gas is produced. This is confirmed by such observations as are recorded in Figs. 13.18 to 13.20, which show the condensate content of the well fluid to decrease with declining pressure in the same manner as the content of the reservoir gas phase alone. However, it is, in principle, possible to vaporize or "dry up" the condensed-liquid phase on exposing it to a sweep of dry gas. It may readily be shown that under equilibrium conditions the number of moles of dry gas, N, of mole fraction composition represented by n_{id}, required completely to vaporize a mole of liquid of molar composition n_{io} is given by

$$N = \frac{\Sigma(n_{io}/K_i) - 1}{1 - \Sigma(n_{id}/K_i)},\tag{3}$$

where the K_i's are the equilibrium ratios at the reservoir temperature and pressure. Thus, for example, on assuming the reservoir liquid-phase composition to be that plotted in Fig. 13.4 and the dry gas to be comprised of 88.14 per cent methane, 8.46 per cent ethane, 2.94 per cent propane, 0.18 per cent butanes, 0.08 per cent pentanes, 0.09 per cent hexanes, and 0.11 per cent heptanes +, it is found that it[1] would take 20.8, 46.8, 56.9, and 55.4 moles of dry gas to vaporize one mole of the retrograde liquid formed by pressure depletion to 2,500, 1,500, 1,000, and 500 psi, respectively.

This possibility of revaporizing the condensed liquid phase by contact with dry gas raises the question whether dew-point cycling is inherently necessary and whether dry-gas sweeping at lower pressures may not suffice

---

reported by M. G. Arthur, API meetings, Los Angeles, Calif., May, 1948. In this work it is also shown that the condensate accumulation near the well bore should have only a negligible effect on the well productivity.

[1] A reservoir temperature of 200°F was assumed, and the equilibrium ratios were taken from C. H. Roland, D. E. Smith, and H. H. Kaveler, *Oil and Gas Jour.*, **39**, 128 (Mar. 27, 1941).

to recover the original reservoir condensate content. As a strictly physical problem it is evident that such low-pressure cycling can lead to recoveries as high as by dew-point cycling. In fact, a detailed analysis[1] for a hypothetical condensate reservoir containing a wet gas whose depletion history is described by Figs. 13.3 to 13.5 shows this to be the case. Stepwise calculations were made of the pickup of the condensed liquid by dry gas as it passed through a linear column of sand that had been previously produced

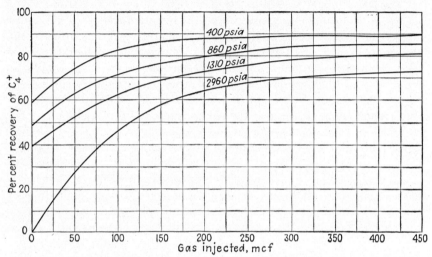

Fig. 13.22. The calculated variation of the total C_4+ recoveries from a hypothetical condensate reservoir vs. the cumulative gas injection, per 1,000 ft³ of hydrocarbon pore space, during cycling operations at various reservoir pressures. (*After Standing, Lindblad, and Parsons, Petroleum Technology, 1947.*)

by gas withdrawal to various limiting pressures. After so determining the vaporization and sweep history in a uniform-permeability zone the composite behavior was computed for a multilayer system with a Gaussian permeability distribution, with a standard deviation of 0.7985 in the log of the permeability, and a 75 per cent areal sweep efficiency. The per cent C_4+ recoveries as a function of the total gas injected at various pressures, derived in this manner, are plotted in Fig. 13.22.

It will be seen from Fig. 13.22 that whereas the recoveries per unit volume of injected gas, *i.e.*, the slopes of the curves, are greatest for the dew-point cycling (2,960 psia), the total recoveries for given volumes of injected gas increase with decreasing cycling pressure. This, in itself, is not surprising and simply reflects the fact that the multiple throughflows required for effective sweeping through the tighter zones and revaporization of the condensed-liquid phase when injected at low pressure may represent a gas

[1] Cf. Standing, Lindblad, and Parsons, *op. cit.*

volume, in surface- or atmospheric-pressure measure, even lower than the small number of throughflows which would be used at the dew point or high pressure. Thus the pore volumes of injected gas, for 100 per cent sweep efficiency, required to obtain a total recovery of 1,059 gal of C_4+ per 1,000 ft^3 of hydrocarbon pore space were found to be 2.0, 5.1, 5.9, and 7.8 for cycling pressures of 2,960, 1,310, 860, and 400 psi, respectively. But these represented volumes of injected gas, in surface measure, of 364,000, 396,000, 295,000, and 171,000 ft^3, respectively.

These numerical values apply only to the hypothetical system described by Fig. 13.22. For different reservoir fluid and dry-gas compositions the relative volumes involved will be different. If the volume of liquid accumulation on pressure decline is greater or if its heavy-component content is greater and of higher molecular weight than represented by the system of Figs. 13.3 to 13.5, the volume of dry gas required for revaporization will be correspondingly greater and the low-pressure cycling may require more gas injection than dew-point cycling for equivalent recoveries. The degree of permeability stratification and total areal sweep efficiency will also affect the comparative effectiveness of low-pressure and dew-point cycling.

All the calculations for low-pressure cycling on which Fig. 13.22 is based assume complete equilibrium between the injected dry gas and the local liquid phase. This, however, does not seem to be unreasonable, in view of the highly dispersed character of the liquid phase and the long distance of travel and time of exposure of the dry gas before leaving the formation. Moreover, laboratory experiments, in which the dry-gas velocity was considerably greater than that obtaining in actual reservoirs, showed the liquid-phase vaporization to follow the equilibrium predictions.[1]

Even if the actual reservoir conditions should simulate those assumed for the illustrative example discussed above, the curves of Fig. 13.22 in themselves do not prove that low-pressure cycling is desirable from a practical and economic standpoint. An important factor not taken into account at all in Fig. 13.22 is that the completion of the cycling phase does not represent the state of abandonment. As noted previously the dry-gas content alone in a reservoir that has been cycled at the dew point would represent a very substantial economic asset. Moreover the unswept part of the reservoir will still contain all its original condensate content. Evidently in actual operations a reservoir that has been cycled at the dew-point pressure or at any pressure above the final abandonment pressure would be allowed to bleed off its residual dry and wet gas by pressure depletion to abandonment. The recovery of condensate from the previously

---

[1] Experiments on condensate-fluid depletion in sand-filled vessels also show equilibrium behavior during both the retrograde and normal vaporization phases [cf. C. F. Weinaug and J. C. Cordell, *Petroleum Technology*, **11**, 1 (November, 1948)].

unswept volume should therefore be added to that obtained during cycling to give a proper comparison of the total recovery under dew-point and low-pressure cycling. Since only a small number of pore volumes will be passed through the formation under high-pressure cycling, the unswept volume will be higher than when the cycling is conducted at low pressure. The additional depletion recovery will be correspondingly greater for the former. Supplementary calculations of the amount of this recovery for the reservoir conditions underlying Fig. 13.22 show that the resultant recoveries for dew-point cycling will be substantially the same as for low-pressure cycling, for the same final abandonment pressure. In fact this conclusion appears to be generally applicable to relatively rich reservoir gases.

There are factors other than the total condensate recovery that are of importance in the comparative values of cycling at different pressures.[1] For the same plant capacity the total operating life would be shorter under low-pressure cycling. However, to maintain such plant capacities during a low-pressure cycling the number of wells would be considerably greater than during cycling at the dew point, since for a fixed pressure differential the injection- and producing-well capacities will be proportional to the mean reservoir pressure. This can be partly compensated by using higher pressure differentials. But this will require increased compression costs. The cost of the flow lines and much of the plant equipment also will be greater in handling, at low pressure, gas volumes comparable with those processed at high pressure. There is in addition the psychological factor that an operator may be reluctant sharply to cut his current income from gas sales during a preliminary depletion phase and at the same time make the large investments required in starting a gas-return program. Failure for such reasons to undertake the cycling operations would then certainly lead to large losses in liquid recoveries, if the reservoir were inherently susceptible to cycling. It should also be noted that, whereas the total C_4+ recovery from the reservoir may theoretically be as high with low-pressure as dew-point cycling, much of liquefiable material other than the stable condensate recovered may be lost under the former type of operations. Unless an extraction plant were built for operation during the initial pressure-depletion period, a large part of the intermediate hydrocarbons will be lost in the separator gases.

[1] For a more detailed discussion of the practical aspects of the problem cf. E. O. Bennett, *California Oil World*, **40**, 25 [October (2d issue), 1947]. A quantitative analysis by M. G. Arthur (API meetings, Los Angeles, Calif., May, 1948) of the economics of cycling a hypothetical reservoir at various pressures shows that the interest-discount factor and the future prices of the products may ultimately control the present worth of the net profits. For constant prices he finds that cycling at intermediate pressures will give maximum profits, while high-pressure cycling will be somewhat more profitable for increasing price trends, for the particular reservoir and economic conditions investigated.

There is much less question about the relative merits of low-pressure and dew-point cycling when the condensate reservoir is underlain by an oil zone of substantial size. A preliminary depletion phase will then lead to reduced oil recoveries, as well as a loss in the intermediate liquefiable hydrocarbons of both the condensate and solution gases. Moreover, to take advantage of the possibilities of a reduced operating life for the field, assuming there is a market for all available gas, the gas withdrawals during the depletion period will have to be considerably more rapid than would correspond to the normal depletion of the oil zone. It is likely that at best the oil-reservoir depletion could only keep pace with that in the gas cap; but this would nullify any pressure-maintenance and gas-sweep effects on the oil recovery. However, if there were a lag in the depletion of the oil zone, there would be serious danger of its encroachment into the gas cap, which would lead to still greater losses in oil recovery. In such composite reservoirs, high-pressure cycling, while involving longer operating lives, would undoubtedly lead to greater over-all efficiency in recovery, if cycling is basically desirable for the development of the gas cap. Of course, if there is an active water drive in the field, the difficulties of preventing migration of the oil into the gas cap during the depletion phase become greatly aggravated, and high-pressure cycling will be the only safe way to operate the field.

Although not concerned directly with condensate reservoirs, related to them are two other types worthy of brief mention. It will be recalled from Sec. 13.2 that the occurrence of retrograde isothermal condensation is limited by the requirement that the reservoir temperature lie between the critical and cricondentherm temperatures of the hydrocarbon mixture and that the reservoir pressure lie at least in the range of the critical pressure. It follows that if the reservoir temperature exceeds the cricondentherm temperature, the hydrocarbon mixture will be in a single phase regardless of the pressure. Thus there will be no retrograde condensation within the formation,[1] and the reservoir will perform as a gas field even though both condensate and gas will be produced at the surface. There will be no need for cycling or pressure maintenance for the purpose of ensuring the recovery of the condensable hydrocarbon content. Fields of this type have been discovered and developed as gas fields, although the composition of the initial production alone would suggest that they are lean condensate reservoirs.

If a condensate-producing wet gas were carried along its dew-point phase-diagram boundary curve with a continual reduction in temperature past the critical temperature, the saturated vapor would change into a bubble-point liquid (cf. Fig. 13.1). If the latter were now produced through

---

[1] The reservoir fluid will follow a pressure-temperature phase-diagram path represented by a straight line parallel to *ABDE* in Fig. 13.1 and lying entirely to the right of the dew-point curve.

a well bore, the surface fluids recovered initially would be identical with that which would have been obtained if the production had its origin in a dew-point gas reservoir, since the phase separation at any terminal state, .such as atmospheric conditions, is independent of the initial state. Within the bubble-point liquid reservoir, however, a gas phase would develop immediately on pressure decline.[1] Moreover, if considered as the equivalent of a crude-oil system, the reservoir liquid would have an abnormally high formation-volume factor. From the considerations of Sec. 10.4 it is clear that the solution-gas-drive recovery of the heavy-liquid components would be extremely low in spite of the low viscosity of the liquid phase. The produced gas will be rich in condensable-liquid components, and if processed by an extraction plant it will contribute appreciably to the total liquid-phase recovery. Within the reservoir the gas-phase saturations will quickly build up until the gas permeability is very high compared with the liquid, after which the liquid will simply shrink in place by continued gas evolution as the pressure declines owing to flow of the gas phase into the well bore. While a large part of the lighter and intermediate hydrocarbon content of the reservoir will thus be recovered, most of the C_7+'s will probably remain trapped in the formation.

The loss in recovery due to the rapid liquid-phase shrinkage can be prevented in such reservoirs by pressure maintenance. If the reservoir can be produced by gravity drainage, with gas injection into the structural crest, the direct recovery will be given by the difference between the average initial- and residual-liquid-phase saturations. The subsequent pressure depletion should lead to some additional recovery of the intermediate hydrocarbon components in the residual oil. Continued gas injection or cycling may also result in at least partial vaporization of the residual-liquid phase. Natural water drives or water injection will likewise serve to prevent the liquid-phase shrinkage and to reduce the residual-liquid saturation. However the latter will remain essentially unrecoverable even if water injection be discontinued and the pressure be allowed to decline. The methods of field operation must be evaluated separately for each specific reservoir in the light of the economic factors pertinent to the particular reservoir of interest.

While the hypothetical situation considered above would lead to a production with gas-liquid ratios characteristic of condensate reservoirs, fields of this type have been found[2] with substantially lower gas-oil ratios. The

---

[1] The reservoir fluid will follow a pressure-temperature phase-diagram path represented by a straight line parallel to *ABDE* in Fig. 13.1 and lying to the left of the point *C*.

[2] The Rattlesnake field, San Juan County, N. Mex., appears to have been this type of reservoir, although it was discovered (1924) before means for analyzing the reservoir conditions were developed [cf. A. H. Hinson, *AAPG Bull.*, **31**, 731 (1947)]. A com-

liquid phase itself in such cases is definitely of the condensate type. But the gas content is relatively low, and the critical temperature of the mixture apparently exceeds the reservoir temperature. Instead of leading to complete dew-point vaporization the reservoir temperature and pressure suffice only for the creation of a bubble-point liquid phase.

It will be obvious without detailed discussion that condensate-reservoir cycling cannot be effectively conducted unless the operations are unitized. Competitive pressure depletion in one part of the field and cycling in an adjoining and intercommunicating part of the reservoir will evidently defeat the purposes of the gas injection. Moreover pressure gradients will be developed leading to migration losses of the reservoir fluids from the cycled area.[1] The flexibility of well location for efficient dry-gas sweeping and all the operating economies that would be possible through unitized development will be seriously reduced. In fact the unitized operation may be considered as an axiomatic *sine qua non* for all cycling programs.

13.11. Summary.—Condensate reservoirs are unique because of the special thermodynamic properties of the reservoir fluid. The latter generally constitutes a saturated vapor within the reservoir and suffers retrograde condensation of a liquid phase when the pressure is reduced. Its over-all composition is largely comprised of methane and the intermediate hydrocarbons. The liquid phase formed from the vapor is usually straw-colored and of high API gravity. The critical temperature of the mixture is lower than the reservoir temperature, and the critical pressure must be of the same order of magnitude as the reservoir pressure. The average molecular weight of the heavy components is generally considerably lower than that of crude oils. The gas-liquid ratio of the production from condensate reservoirs is decidedly higher than that commonly associated with crude-oil–natural-gas systems. Reservoir "wet" gases are considered as "rich" if the gas-condensate ratio is of the order of 10,000 ft³/bbl, but many condensate fields produce with ratios of 50,000 ft³/bbl or greater.

---

plete *p-v-T* analysis of a particular reservoir fluid of this type has been reported by J. P. Sloan (AIME meetings, Dallas, Tex., October, 1948). An approximate treatment of the expected solution-gas-drive performance of a hypothetical reservoir containing this fluid indicates a recovery of 16.0 per cent of the initial "residual" oil content, corresponding to only 2.5 per cent of the pore space, and an almost vertical rise of the gas-oil ratio to 3.4×10^6 ft³/bbl after the equilibrium gas saturation has been developed (cf. M. Muskat, *Journal Petroleum Technology*, in press). Still another recent study of bubble-point-liquid reservoirs producing condensate-type fluids is that on the North Lindsay field, McClain County, Okla., reported by A. B. Cook, G. B. Spencer, F. P. Bobrawski, and E. J. Dewees, *Petroleum Eng.*, **19**, 158 (September, 1948).

[1] While regional fluid migration, in itself, is a virtually necessary and desirable feature of cycling operations, especially in composite condensate and oil reservoirs, such migration will lead to an inequitable distribution of the recoveries unless the field is unitized and property lines are ignored in the location of both the injection and producing wells.

As soon as the pressure is reduced below the dew point in a condensate-bearing reservoir, by fluid withdrawal, liquid will condense in the formation. This condensation process (retrograde) will continue until the pressure falls to a value of 1,000 to 2,000 psi, depending on the initial wet-gas composition and reservoir temperature. Beyond this point normal vaporization will develop, and the reservoir liquid volume will decrease (cf. Fig. 13.3). As the total liquefiable components comprise only a small part of the wet gas, even a flash condensation of the whole reservoir wet-gas content would lead to a build-up of the liquid saturation to only 5 to 18 per cent of the pore volume (cf. Fig. 13.2). The accumulation will therefore remain trapped in the formation, and only the gas phase will be produced. Because of this separation of the liquid phase the composition of the produced-gas phase will continually change (cf. Fig. 13.5), and the gas-condensate ratio will rise until the pressure of maximum retrograde condensation is reached. The reservoir liquid-phase composition will also change continuously as increments are added to it from the partly denuded residual gas (cf. Fig. 13.4). The liquid trapped in the reservoir by pressure depletion may amount to 30 to 60 per cent of the original reservoir content. A larger fraction of the heavy components will be lost in this manner than the intermediate constituents, although a large part of the latter that are produced may be carried away by the separator gases if the well fluids are not processed by extraction plants (cf. Table 3). The losses within the reservoir will evidently be greater as the fractional content of the heavy components in the wet gas increases.

While it is convenient to use the terms "gas-oil" and "gas-condensate ratios" in describing the production from condensate reservoirs, by analogy with crude-oil–natural-gas systems, they are not sufficiently precise for purposes of detailed economic evaluation. The compositions of the produced-gas and liquid phases will vary during the producing life and will be sensitive to the separator conditions. A more satisfactory description of the reservoir content and production is one based on the composition of the fluids. The production history under pressure depletion can then be expressed in terms of the cumulative recovery of the individual components or appropriate groupings of them vs. the reservoir pressure (cf. Fig. 13.6).

For practical purposes the dynamical behavior of a condensate reservoir may be considered as identical with that of a normal gas reservoir. The cumulative molar production will therefore decrease approximately linearly with the reservoir pressure [cf. Eq. 13.3(2)].

The obvious solution to the problem of preventing the retrograde losses of the heavier components resulting from reservoir pressure decline is pressure maintenance by fluid injection. When the fluid injected is the produced gas stripped of its liquefiable content, the process is called "cycling."

One of the first steps in planning cycling operations is the choice of the well distribution that will lead to an efficient sweep of the wet reservoir gas by the injected gas. While the well pattern must be adjusted to the basic geometry of the reservoir, the theoretical analysis of the simpler systems suffices to indicate the order of magnitude of the sweep efficiencies that can be achieved and the important factors controlling the sweep efficiency. By analogy to the similar problem in secondary-recovery operations the sweep efficiency is conveniently defined as the fraction of the cycled area swept out by the time of first dry-gas break-through.

When the reservoir can be approximated by a rectangular area, a convenient cycling pattern is the "end-to-end" sweep, in which the injection wells are distributed on one side and the producing wells on the other. For a uniform infinite porous medium the fraction of the area between the wells swept out by the time of first gas entry into the producing wells may be derived analytically in explicit form [cf. Eq. 13.5(4)]. It is found that the unswept area is independent of the separation between the injection and producing wells as long as the separation equals or exceeds half the spacing between wells of the same kind [cf. Eq. 13.5(5)]. It follows that the sweep efficiency increases with the distance between the injection and producing lines, for fixed separations of wells of the same kind. This result has considerable generality and is also verified in the case of the infinite-line-drive networks.

The cycling pattern in which the injection wells are equally spaced over a circular ring and a single producing well is placed at the center, or conversely, has a sweep efficiency given simply by the ratio of the number of wells in the ring to this number plus 2 [cf. Eq. 13.5(10)]. The efficiency thus rapidly approaches unity as the number of wells in the ring increases. Similar though somewhat more complicated results may be derived if both the injection and producing wells are distributed on concentric rings with equal and uniform angular spacing [cf. Eq. 13.5(11)].

Finite rectangular areas cycled with a bilateral pattern, with injection wells along a central line and producing wells along the opposite parallel sides (cf. Fig. 13.9), or conversely, also can be treated analytically. Here, on taking into account the finite-reservoir area the effect of locating the producing wells (or injection wells) at various distances from the boundary is clearly demonstrated. Thus, the over-all sweep efficiency is increased from 36.9 per cent, when the line of producing wells is located at a distance from the central injection line that is 57 per cent of the reservoir half width, to 74.1 per cent, when the producing line is placed along the actual limits of the reservoir. A calculation of the produced-fluid composition after dry-gas break-through shows the wet-gas content to fall sharply and reach a value of 15 per cent of the composite produced fluid by the time the total gas processed is of the order of 1.3 times the reservoir hydrocarbon pore volume (cf. Fig. 13.12). The exact value of the total gas processed by the time an abandonment limit such as 15 per cent wet-gas content is reached does not vary rapidly with the producing-well location. However, the total wet-gas recoveries increase from 64 to 96 per cent as the producing wells are moved to the reservoir boundary from an initial distance from the injection wells that is 57 per cent of the reservoir half width (cf. Fig. 13.13).

Cycling patterns of more complex geometry can be effectively studied only by means of electrolytic models. While gelatin models provide a quick visual record of the history of the injection-fluid motion, they do not have the precision and flexibility of the potentiometric models. With the latter the injection-fluid fronts can be determined for a system of variable thickness and permeability almost as easily as for one of strictly uniform characteristics. Such potentiometric models may be made by adjusting the depth of the electrolyte so as to be proportional to the product of the permeability and effective pay thickness at the corresponding point in the actual reservoir [cf. Eq. 13.6(5)]. On locating in the electrolytic bath a distribution of input- and output-current electrodes which is geometrically similar to that of the injection and producing wells, with respective currents proportional to the injection and producing rates, the voltage distribution in the model will be proportional to an effective potential function given by the pressure integral of the fluid density to viscosity ratio [cf. Eq. 13.6(4)]. The local fluid velocities will be proportional to the voltage gradients along the streamlines. The latter are conveniently measured by a four-probe electrode, two of which are set to lie along an equipotential, so that the other pair, normal to the first, lies along the streamline. The time increments for fluid movement along the streamlines are directly proportional to the product of the gas density and displacement porosity and inversely proportional to the permeability and voltage gradient.

In practical applications, where the permeability variations are neglected, the electrolytic-bath thickness is made geometrically similar to the isopach map of the formation. While in principle the gas-density variation will affect the details of the injection-fluid motion, it will be an important factor mainly near the injection and producing wells and is usually neglected. Models of this type have been successfully applied in the determination of optimum cycling well patterns (cf. Fig. 13.14) as well as in the interpretation of actual field observations.

If there is any flexibility whatever in the choice of the well locations, it is generally possible to find a distribution that will lead to a sweep efficiency of 60 to 80 per cent of the area to be swept, without requiring a prohibitively large number of wells. This, however, does not in itself ensure an effective sweeping of the wet reservoir gas. The areal sweep efficiency of a uniform stratum will generally be reduced materially by the permeability stratification of the formation. Such stratification will lead to a superposition history of the sweep patterns in the individual layers in which dry-gas break-through will develop in sequence according to their permeability. The resultant behavior of the composite system can be readily formulated in general analytical expressions for any type of permeability variation.

A convenient though approximate representation of the permeability stratification in producing formations is given by an exponential variation of the permeability with depth, on assuming that the different strata are arranged in a sequence of increasing permeability with increasing depth [cf. Eq. 13.8(13)]. The ratio of the maximum to minimum effective permeability then provides a simple stratification parameter controlling the behavior of the composite formation. On assuming also the wet-gas content of the effluent from the individual zones after dry-gas break-through to decrease exponentially with the total gas throughflow [cf. Eq. 13.8(12)], the time history of the composition of the production and the total wet-gas recovery can be readily computed for different stratification parameters (cf. Fig. 13.15). The effective resultant sweep efficiency at the time of first dry-gas break-through is then found to decrease approximately as the inverse of the logarithm of the stratification parameter [cf. Eq. 13.8(15) and Fig. 13.16]. Thus, when the latter is 10, the first dry-gas break-through will develop after only 35.2 per cent of the reservoir has been swept, even if the areal sweep efficiency is 90 per cent. And for stratification ratios of 100 the composite sweep efficiency will be only 19.35 and 12.9 per cent for areal sweep efficiencies of 90 and 60 per cent.

The severe reduction in sweep efficiency due to permeability stratification is compensated in practice by the fact that cycling operations are usually continued after the initial dry-gas break-through until the well effluent contains only 10 to 25 per cent of wet gas. The total fractional recoveries of wet gas to such cycling abandonment limits will evidently be much greater than the composite sweep efficiencies (the recoveries at first dry-gas break-through) and will increase as the abandonment limit of wet-gas content is reduced. While these, too, will decline as the stratification parameter increases, this reduction is not serious until the stratification ratio exceeds 10 (cf. Fig. 13.16). For greater values the total cycling wet-gas recovery decreases in an approximately logarithmic manner with the stratification ratio, and the latter will ultimately control the over-all efficiency of the cycling operations. The areal sweep efficiency itself will be of secondary importance except in highly uniform formations.

The total gas throughflow during cycling operations increases at first with increasing stratification parameter, reaches a maximum, and then declines (cf. Fig. 13.17). Over the range of areal sweep efficiencies and cycling abandonment limits usually occurring in practice the total gas throughflow will not exceed about 2.2 times the initial reservoir hydrocarbon volume and may be considerably lower when the small recoveries from highly stratified formations are swept out. The areal sweep efficiency again will affect the volume of total gas throughflow only in the range of low stratification parameters.

While many condensate reservoirs in which the original dew-point pressures have been maintained by cycling have been produced with substantially constant effluent composition until dry-gas break-through, but few detailed reservoir studies have been reported for reservoirs that have been produced by pressure depletion. However, those which have been so produced, such as the La Blanca field in Texas, have shown continually increasing gas-condensate ratios as the pressures have declined (cf. Fig. 13.18). Such observations confirm not only the basic thermodynamic phenomenon of retrograde condensation but also the immobility of the condensed-liquid phase formed in the reservoir. In the case of cycled reservoirs in which the reservoir pressure has not been completely maintained, such as the La Gloria field in Texas and the Cotton Valley field in Louisiana, similar though less rapid rises in the gas-condensate ratio have been observed (cf. Figs. 13.19 and 13.20). Moreover an analysis of the performance of the Bodcaw Sand reservoir of the Cotton Valley field showed the composition history of the produced fluid actually to follow that predicted from laboratory experiments on the phase behavior of the reservoir hydrocarbon mixture. Special tests on shut-in wells in this field gave data on the sweep of the dry gas through the Bodcaw Sand (cf. Fig. 13.21) and indicated close conformance with theoretical predictions based on the measured permeability variation in the formation. As the latter is relatively uniform and the well distribution is favorable for high sweep efficiency, it is anticipated that 85 per cent of the wet-gas reserves will be recovered, with a processing volume equal to 115 per cent of the initial-gas content.

The fact that in principle the hydrocarbon content of condensate reservoirs can be recovered almost completely by cycling does not imply that such cycling is universally desirable. As in the case of all reservoir exploitation the method used must be chosen on the basis of economic considerations. Because of the complexity of the latter the criterion of maximum profit cannot be formulated as simple and universal rules. It is evident, however, that the controlling factors will be the richness of the wet gas, the size of the reservoir, and its uniformity. Lean gases inherently suffer lower retrograde losses on pressure decline, and the total value of such losses will be correspondingly lower. Hence, except when there is no market for the gas, cycling of reservoirs producing at ratios of 50,000 ft³ of gas per barrel of condensate will usually be undesirable from an economic standpoint. Reservoirs of small volume will also be unfavorable for cycling because of the limited value of the retrograde losses resulting from pressure depletion. The low sweep efficiencies of highly stratified formations will likewise make questionable the economic success of cycling operations.

In estimating the recovery to be expected under cycling one must add to that obtained during the cycling phase the pressure-depletion recovery of the unswept reservoir volume which will be produced during the blow-down of the cycled reservoir. It is the resulting increase in recovery by the composite program of cycling and pressure depletion, as compared with simple pressure depletion, that must be balanced against the investment costs of the gas-processing and compressor plant, the wells required for injection, the gas-injection lines, and the associated operating costs. The deferred income from the dry gas, in case there is an immediate market for it, as well as the reduced present worth of the liquid products, because the operating life of the reservoir will usually be longer under cycling than under pressure depletion, must also be considered.

The retrograde losses of liquefiable hydrocarbons within a reservoir by complete pressure depletion will usually range from 30 to 60 per cent of the initial reservoir content. However, an appreciable part of the corresponding potential hydrocarbon recoveries of 70 to 40 per cent will be carried away by the separator gases, if the latter are not further processed. Successful cycling itself should lead to at least 50 per cent over-all recovery, and the subsequent pressure depletion should yield additional liquefiable products to make the total equal to 60 to 80 per cent of the original reservoir contents of condensable products.

When the condensate reservoir is bounded by a crude-oil zone of appreciable size, the development program must be so chosen as to achieve maximum recovery from both. The most efficient method would be a limitation of withdrawals to the oil zone with sufficient gas return to the gas cap fully to maintain the pressures while sweeping the wet gas into the oil wells. If it is not feasible to defer the recovery of condensate, the gas cap may be cycled simultaneously with the oil withdrawals, provided that a regional pressure gradient is maintained from the gas cap to the oil section so as to prevent migration of oil into the condensate reservoir and induce a pressure-maintenance action on the oil zone. Postponement of oil withdrawals until the gas cap is completely cycled will provide efficient condensate recovery, although the deferment of the oil recovery may not be economically feasible. Moreover the oil recovery, obtained without pressure maintenance, will not be quite so high as under simultaneous complete cycling and oil production. On the other hand a delay in cycling or gas return to the gas cap until the black-oil zone is depleted will lead to pressure decline and retrograde losses in the gas cap, which could be largely prevented by the other development methods. Of course, in practice all the numerous economic and physical factors pertaining to the particular reservoir of interest will have to be evaluated in choosing the operating program.

When edgewaters bound the gas cap, it will be undesirable to place the producing wells near the water-gas contact. The injection wells, however, may be advantageously completed within the water-saturated section, or "dry holes" near the water-gas contact can be used as injection wells. In this way the whole of the gas cap will be susceptible to dry-gas sweeping, and unswept "dead" areas will be kept to a minimum. Dry-gas injection below the gas-water contact has been applied successfully, and no evidence has developed that the water-saturated section above the injection wells offers permanent and serious resistance to the gas flow.

Even under complete pressure maintenance by cycling the drawdown about the producing wells will create local condensation of liquid. As all the produced fluid must pass through the annular zones immediately surrounding the well bore, the liquid will rapidly accumulate there until a saturation for mobility is built up. Thereafter the additional liquid condensation will be swept into the well bore. As production continues, the zone of saturation to the limit of mobility will expand away from the well bore. The rate of growth of the liquid saturation at any point will vary inversely as the square of the distance from the producing well and directly as the square of the producing rate or pressure drawdown. The resulting loss in liquid recovery should not be of serious magnitude except in very tight formations producing exceptionally rich gases.

While the retrograde liquid accumulation is immobile with respect to displacement by the gas flow, except near the well bore where the saturation has been built up to the point of mobility, it is subject to revaporization on exposure to dry gas. This observation raises the question whether dew-point cycling and the complete prevention of retrograde condensation is inherently necessary and whether the liquid phase that would be formed by pressure depletion could not be effectively recovered by low-pressure dry-gas cycling. Here, too, the obvious fact that it is *physically* possible to recover all the hydrocarbons by low-pressure cycling is of little significance except as it may be proved that such recovery is *economically* feasible. And even if the latter be confirmed, its practical value depends on the costs of and profits from such operations as compared with those of high-pressure or dew-point cycling.

A detailed analysis of the recoveries which can be obtained by cycling at various pressures following preliminary pressure depletion, for a particular set of reservoir conditions, showed that for these the total condensable-liquid recoveries to a given uniform abandonment pressure are essentially independent of the sequence of cycling and depletion and will require substantially the same volume of gas processing. The economic factors, however, require special consideration. If the plant capacities for dew-point and low-pressure cycling be the same, the latter will lead to shorter operating lives and increased present-worth values of the recovery. But to operate a low-pressure cycling system at the same throughput rate as at high pressures will require more wells and greater investments in flow lines and plant facilities. Failure to process the produced gas during the initial depletion phase will also lead to substantial losses in the intermediate liquefiable hydrocarbons. No simple rule will indicate the optimum cycling pressure. However, if there should be an increasing price trend for the gas and plant products during the operations, cycling at or near the dew-point pressure should generally lead to maximum present-worth profits. And when the condensate gas cap is underlain by a crude-oil belt of substantial size, the requirement of maximum recovery from the latter will usually leave little question as to the economic advantage of high-pressure cycling over low-pressure gas-return operations. Such advantage will become still more decisive if there should be a strong water-drive action in the field.

Since isothermal retrograde condensation will take place only at temperatures between the critical and cricondentherm temperatures of the hydrocarbon mixture, the same system would behave simply as a normal gas if it should occur in a reservoir with a temperature exceeding the cricondentherm. At the surface temperature and pressures the separation into condensate and gas would be exactly the same as if it were produced from a typical condensate reservoir, but there would be no phase change within the reservoir. Cycling would be entirely unnecessary, and the reservoir could be produced by pressure depletion as a gas field.

If, conversely, the reservoir temperature were below the critical, the reservoir fluid, if in a single saturated phase, will be a bubble-point liquid. Again the initial surface-production stream will appear identical to that which would be produced if the reservoir fluid were a saturated vapor. Within the reservoir, however, rapid gas evolution and liquid-phase shrinkage will develop and will result in a small recovery owing to direct liquid-phase expulsion. Pressure maintenance by gas or water injection and mass displacement of the reservoir liquid will be necessary in order to achieve high recoveries of the heavy hydrocarbon components.

Once it has been determined that cycling is the optimum method for field exploitation, it is necessary to unitize the operations to achieve maximum efficiency. Competitive development will make impossible the control over the well pattern and withdrawals necessary for effective dry-gas sweeping, will induce fluid migration and inequitable distribution of the recovered products between the competitive areas, and may make it impractical to maintain the pressures at the dew point to prevent retrograde losses.

CHAPTER 14

WELL SPACING, RECOVERY FACTORS, AND RECOVERABLE RESERVES

14.1. Introduction.—It will be evident to the reader that in the previous chapters no attempt has been made to provide specific instructions for the exploitation, development, or operation of actual oil-producing reservoirs. Nor has it been the purpose of the discussion presented thus far to provide explicit formulas for predicting quantitatively the recoveries from specific reservoirs or for evaluating them as items for sale or purchase. It has been an aim of this work to provide an exposition of the physical principles underlying the behavior of oil reservoirs so as to permit an understanding of their performance when observed in practice and an anticipation of the broad features of their performance from the consideration of basic data gathered during their development. Although these physical principles have been considered as well established and there is no definite evidence to the contrary, there are still many detailed aspects of reservoir phenomena in need of clarification and much work is yet to be done in correlating idealized theoretical predictions and actual field observations before quantitatively significant formulas directly applicable to the complex systems comprising real reservoirs can be constructed. A tabulation of formulas presuming to give quantitatively accurate values of optimum development conditions or recoveries would, at the present time, be both premature and misleading.

While it has served the purpose of this work to limit the discussion to methods and principles of reservoir behavior, it must be recognized that the *practice* of reservoir engineering demands explicit commitments in numerical form regarding general field development and evaluation. The location of wells and their spacing must be chosen. And the recoverable reserves must be estimated before plans can be made for large-scale expenditures on development drilling, the construction of pipe lines, gasoline plants, etc. In fact, from a practical standpoint, the ultimate goal of the science of reservoir engineering is simply the teaching of methods of development and operation of actual oil reservoirs to achieve the maximum efficiency in oil recovery. In this chapter, therefore, will be given a survey of the present status of the problems of well spacing and recovery estimation, although, in spite of their importance, the solution of these problems is perhaps in a less satisfactory state than most other phases of reservoir engineering.

14.2. Well Spacing—The Negative Side.—There is no conclusive evidence that by increasing the density of wells draining an oil reservoir by small increments, over such values as are commonly used in practice, the physical ultimate oil recovery[1] will be increased appreciably or that the economic ultimate recovery will be either increased or decreased appreciably. There is no conclusive evidence that by decreasing the density of wells draining an oil reservoir by small increments, below such values as are commonly used in practice, the physical ultimate recovery will be decreased appreciably or that the economic ultimate recoveries will be either increased or decreased appreciably. The exact *shapes* of the curves of physical or economic ultimate recovery vs. the well density are not known with certainty. In a broad sense it is the determination of such curves that constitutes the "well-spacing problem."

Limiting points on the recovery vs. the spacing curves are established axiomatically. At zero well density the recoveries, both physical and economic, are zero. In the limit of infinite well density the ultimate physical recovery, if variable at all, is maximal. The economic recovery is, however, probably zero, as a development with unlimited well density would immediately imply[2] an indefinitely large "negative profit." Qualitatively, by assuming for the present[3] that the physical ultimate recovery may depend on the well spacing, the relationships of the latter and the net profit to the well density would follow curves of the type shown in Fig. 14.1. But the all-important factor deliberately not specified in Fig. 14.1 is the well-density *scale*. This is the crux and the basic unknown of the well-spacing problem. Although the relative position of the maximum in the economic return curve and the asymptotic approximation to the maximum recovery will have no fixed relationship nor will correspond necessarily to that indicated in Fig. 14.1, the two are interrelated through economic factors and the density scale, though not specified numerically, is shown as common for both curves only in a symbolic sense.

[1] The term "physical ultimate recovery" is used to denote the maximum possible recovery, under a given production mechanism, regardless of the time required and of the cost of maintaining the operations. The "economic ultimate recovery" refers to that which can be recovered at a profit to the time of actual abandonment.

[2] In fact any well density with an investment cost exceeding the net value of the total recoverable oil would, if predictable in advance, imply a zero economic ultimate recovery. In practice, however, the economic ultimate recovery is usually limited by the vanishing of the *current* operating profit rather than the total net income, although the failure to develop at all "uncommercial" pays represents the zero economic ultimate recoveries resulting from predictions of net loss from any development.

[3] As will be seen in Secs. 14.4 and 14.5 there is actually no significant evidence that the physical ultimate recoveries do vary with the well spacing.

It is important to understand the reasons underlying the negative char-acter of the above statements. Evidence that presumably could be con-clusive must of necessity be one of three kinds, namely, (1) analytical, (2) laboratory scale experimentation, or (3) field observations. As yet none of these has led to conclusions entirely free of doubt and qualifications.

There simply have been no successful attempts to derive an answer to the well-spacing problem by strictly analytical means. The reason is obviously

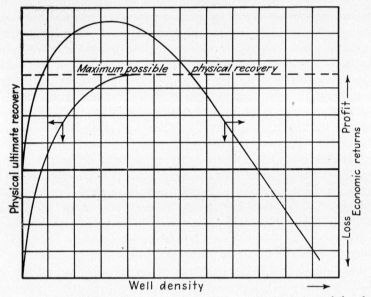

FIG. 14.1. A gross diagrammatic representation of the possible variations of the physical ultimate oil recovery and profit vs. the well density.

the practical impossibility of solving the fundamental governing dynamical equations [Eqs. 7.7(1)], even for two-dimensional gravity-free systems, without making such approximations as may inherently distort or mask completely whatever role, if any, the well spacing may play. But one transient solution of the basic equation has been reported, and that[1] was strictly numerical, limited to a one-dimensional pure gas-drive system, and was not sufficiently accurate to provide a "proof" regarding well-spacing propositions. A worthy challenge to computers and analysts would be a systematic program of developing solutions—numerical or graphical—for Eqs. 7.7(1).[2]

[1] M. Muskat and M. W. Meres, *Physics*, **7**, 346 (1936).

[2] With the advent of powerful electronic computing-machinery installations, de-veloped during and since the last war, the possibilities of studying the effect of well spacing on solution-gas-drive recovery, as implied by Eqs. 7.7(1), have been greatly

Evidence from laboratory experimentation on the well-spacing problem itself is also virtually nonexistent. Aside from the inherent difficulties of techniques involved in the experimental work the basic problem is that of scale. Several linear solution-gas depletion experiments have been reported.[1] But small-scale systems of a few feet may well fail to reveal how the details of oil expulsion may be affected by varying the distances of drainage by hundreds of feet, if there should be such effects. In principle it might be anticipated that an over-all dimensional scaling could be introduced to overcome this difficulty, and such dimensionally scaled-down model experimentation has been proposed.[2] However, this involved other complications, and while it may merit further investigation it has not been applied as yet to the well-spacing problem.

Of field data there is no such paucity as in analytical treatments or laboratory tests. But when examined from the point of view of giving "proof" of conclusions or implications, most of the data reported until recently are of little value. As noted in Sec. 10.10 the many older fields, which are now depleted, were developed and produced without gathering the data required to describe the physical nature of the reservoir. In fact, about all that is known of these fields is their total ultimate recoveries or the recoveries per acre. Even the conversion to recoveries per acre-foot of pay is seldom feasible with any degree of certainty.

The fields of more recent development are, as a whole, of limited significance at present with respect to the well-spacing problem, simply because their ultimate recoveries can be determined only by extrapolation of their past performance. Such estimates are necessarily uncertain if the extrapolation must be made over an extended range. The severe limitations of withdrawals, which have been common practice during the last 18 years, have served greatly to prolong the producing lives, so that the recoveries to date of many of the more recently discovered fields are only small fractions of the ultimate values. Moreover the increasing tendency to undertake fluid injection for pressure maintenance and increased oil recovery during the primary-production phase makes it difficult to establish a common basis for even attempting to establish a relationship between recovery and well spacing. The automatic conversion of potential gas drive to partial- or complete-water-drive performance has further served to complicate the interpretation of ultimate recoveries in terms of effects of well spacing, if any.

---

enhanced. In fact, such an investigation is currently under way by the author in collaboration with the International Business Machines Corporation, using the Selective Sequence Electronic Calculator developed by I.B.M.

[1] H. G. Botset, *AIME Trans.*, **136**, 91 (1940).

[2] M. C. Leverett, W. B. Lewis, and M. E. True, *AIME Trans.*, **146**, 175 (1942).

Even if all the pertinent reservoir data were known and the complete production history of a particular field were available, the ultimate recovery, in itself, would not directly indicate the role that may have been played by the well spacing. This can be inferred only from the comparative recoveries among different fields with different well spacings, but with all other significant features of the reservoirs and operating history substantially the same. As it is extremely improbable that all the requirements for strict comparisons of this type will ever be satisfied among actual reservoirs, the problem becomes essentially statistical. This further aggravates the problem of gathering the composite mass of reservoir and recovery data required to obtain results of significance.

To minimize the uncertainties regarding the similarity in reservoir properties among different fields, attempts have been made to compare the recoveries of various tracts in the same field developed with different well spacings. However, the very proximity of such tracts and their intercommunication through the common producing formation may be the source of serious uncertainty as to the significance of data of this type. As long as withdrawals were largely uncontrolled, the densely drilled areas were undoubtedly being depleted more rapidly than those of wider spacing. The more rapid decline in pressure that would necessarily develop in the former would lead to pressure differentials between the various well-density areas. Unless there are barriers to flow between the tracts, the pressure differentials will induce migration of oil toward the more densely drilled areas. This will provide more oil for recovery from the latter, while at the same time reducing the reserves of recoverable oil in the wider spacing regions. The resulting ultimate recoveries will thus tend to show greater yields from the closely spaced tracts, even if the well spacing in itself had no effect on the recovery. Unless such contributions of fluid migration within a field can be evaluated and corrected for, the well-spacing data will be of little significance, except as they may be indicative of maximal effects.

It is because of the above considerations that in a strictly scientific sense the relationship between well spacing and ultimate recovery still represents an unsolved problem. This, however, does not imply that nothing whatever is known about the problem. Some general positive aspects will be considered in the next section, and the available statistical data on actual recoveries in fields of different well spacings will be reviewed in Sec. 14.7.

14.3. Well Spacing—The Positive Side.—Reduced to its essentials the problem of well spacing is simply that of the relation between the magnitude and efficiency of oil drainage and the distance from the foci of withdrawals —the producing wells. Obviously such a relationship, if it could be found, would involve a great variety of variables, including the time, the rock properties, and the characteristics of the petroleum fluids. Moreover it will depend on the nature of the producing mechanism. A single universal curve or formula giving the variation of oil drainage with distance from a well is thus hardly to be expected.

The matter of drainage is largely one of degree. Certainly recoveries of hundreds of thousands of barrels of oil from a single well producing from a formation of moderate thickness imply that in such cases the oil has moved great distances in reaching the well. Except for those who still cling to the theory of the fixed radius of drainage, derived from a misinterpretation of the Jamin effect (cf. Sec. 7.6), it is now generally agreed that there is no limit to the range of ultimate *pressure* reaction in a porous medium of continuous fluid transmissibility. The uncertainty is associated with the magnitude of the reaction, the time required for its transmission to distant points, and the relationship between the oil displacement or depletion and the pressure reaction.

It is generally considered as intuitively certain that there should be and is a decreasing efficiency of oil drainage with distance from a well bore. The only element of uncertainty appears to be the rate of the variation and the range of distances at which it becomes of significance. Yet no proof has been given of this apparently obvious proposition. The most common argument is the claim that, since there is only a finite amount of energy associated with each unit volume of oil, it will suffice to move the oil only a limited distance to the well bore. This, in itself, however, implies that the frictional energy dissipation per unit distance of travel is essentially a constant, independent of the velocity or local reaction of the fluid element. But this is not the case. In fact, as a first approximation, it may be anticipated that the factor of distance of travel to the well bore will generally be compensated by an equivalent change in velocity, so that the resultant total energy dissipation will be substantially independent of the path length.

That fluid drainage through porous media over long distances has occurred and does occur can hardly be considered debatable. While the detailed mechanism is still not fully clarified, it is generally agreed that in many cases the accumulation of oil and gas in reservoir traps has been the result of a migration to the trap from the original source beds, which may have been many miles distant. The long-continued supply of water from aquifers supporting water-drive oil-reservoir production, as in the case of the East Texas field, implies mass movements of water over distances of many miles. In fact, in the case of the Woodbine aquifer, it appears that at least the pressure reaction due to the production from the East Texas and Van fields has definitely been exerted at the Hawkins field, Wood County, Tex., some 14 miles northwest of East Texas, by lowering the pressure and gas-oil and water-oil contacts in the eastern side of the fault at Hawkins.[1] Similar pressure reaction obtains at the Village pool, Columbia County, Ark., owing to interference through the Smackover Limestone aquifer by withdrawals from the Magnolia pool, also in Columbia County, Ark. (cf. Sec. 11.10).

Interlease migration within fields and pressure depletion in undeveloped areas by withdrawals in other intercommunicating parts of a common reservoir reflect the existence of fluid movement over many hundreds of feet. From the fact that under favorable circumstances the application of the material-balance equation leads to reservoir contents in agreement with volumetric estimates it can be inferred that there is substantial dynamical interaction throughout the reservoir, including the areas between producing wells. The actual areal advances of edgewaters over the whole of producing reservoirs represent large-scale fluid movements over distances comparable with the gross reservoir dimensions (cf. Figs. 11.36 and 11.37). The fundamental principle underlying the gravity-drainage and gas-cap-expansion mechanism is based on the concept of a regional and extended fluid movement down the reservoir flanks. And the generally accepted practice of shutting in wells of high gas-oil ratio, by independent action or under unitized operation, as effective reservoir-energy-conservation measures, rests on the tacit assumption that the gas so conserved will be used to displace oil from the area of the shut-in wells to the more distant wells producing at low gas-oil ratios.

Direct well interference observations, under favorable circumstances, as will be discussed later, definitely prove interwell communication. In fact, when due account is taken of the time-scale factor, it is difficult to find a single major type of evidence proving the absence of dynamical interaction throughout continuous reservoir strata, except as obvious geologic discontinuities or local modifications in rock character create definite barriers to fluid movement. References to "wide" or "close" well spacing or effective or deficient drainage must therefore basically reflect opinions only regarding the *completeness* or effectiveness of pressure and oil-content depletion equilibrium, in relation to the time scales pertinent to actual oil-producing operations. While it would be entirely unjustified to claim that uniform and complete oil depletion can be obtained in a fixed time in all reservoirs at any arbitrarily "wide" spacing, it would be equally presumptuous to advocate a universal requirement of a fixed "close" spacing as necessary for efficient recovery under all conditions.

14.4. Physical Considerations Regarding Well Spacing; Water-drive Fields.—It will be clear that from a physical point of view the well-spacing

---

[1] E. A. Wentland, T. H. Shelby, Jr., and J. S. Bell, *AAPG Bull.*, **30**, 1830 (1946); cf. also K. M. Fagin, *Petroleum Eng.*, **17**, 92 (August, 1946).

problem can have significance only under conditions where it is assumed that the reservoir is substantially uniform throughout and is subject to development by groups of uniformly spaced wells which are individually identical or equivalent. If the producing formation is highly variable, lenticular, or faulted, the well spacing must evidently be sufficiently close to ensure that each major lens or separate unit is penetrated by at least one well. And if the productive characteristics of the individual wells depend on their exact location, as may well obtain in practice, the physical aspects of the well-spacing problem will become submerged in the statistics of the relationship of well density and well position to the inherent productivity of the wells. Hence in theoretical considerations of well spacing it is necessary to idealize the problem from the outset, so that the well spacing will be the only major factor that may affect the performance and recovery. Moreover, as in all phases of reservoir engineering, it is also necessary in the case of the well-spacing problem to define first the nature of the producing mechanism under discussion before attempting predictions of well or reservoir behavior.

Considering first reservoirs producing by complete water drives, of the edgewater type, there appears to be no reason for expecting an effect of well spacing on the physical ultimate recovery, except with respect to the purely geometrical details of the water-invasion pattern at the water-oil boundary. By definition the complete-water-drive mechanism implies that there is a sufficient supply of water and water-expansion energy in the aquifer, or available through water injection, to invade and flush out the whole of the oil reservoir. The well density will, of course, affect the production capacity of the oil reservoir as a whole. High well densities will provide greater total withdrawal rates with lower pressure differentials within the producing formation. But the gross pressure and water-intrusion history will depend mainly on the total field withdrawals rather than the number of wells used. Since the energy for the oil expulsion, in reservoirs in which there is complete-water-drive action throughout their producing lives, is provided by an effectively infinite source and the average distance of travel of the invading water in covering the oil reservoir is essentially independent of the well density, the over-all physical ultimate recovery should not be materially affected by the well spacing.

There is a possibility, however, that the geometrical sweep efficiency of the advancing edgewater may depend somewhat on the well density and in this manner affect the economic ultimate recovery. This factor refers to the area flooded by the invading water when the latter first breaks into the nearest line of producing wells. As may be shown by applying the theory outlined in Sec. 12.6, if the advancing edgewater be considered as a continuous line-drive flood, the fraction of the area between the original water-oil boundary and the nearest line of producing wells flooded out when the water reaches the latter will be greater the closer the spacing is made between the wells in the producing line. If the line-drive representation be assumed to remain valid as successive lines of producing wells are flooded out and the edgewater advances across the field, a net gain in sweep efficiency would result from the higher well-density developments.[1] However, it is doubtful that, under practical conditions, where the producing wells can often be operated profitably while still producing 98 per cent water, the theoretical differences in sweep efficiency with respect to the clean-oil production would lead to material differences in over-all recovery as the average well spacing is varied within practical limits.

To the extent that the geometrical sweep efficiency may be of importance in complete-water-drive reservoirs a well network that is uniformly distributed over the area of the reservoir is not the most efficient method of development. As may be inferred from the discussion of cycling patterns in condensate-producing fields, in Sec. 13.5, complete sweeping could be achieved in a circular reservoir, with edgewater drive, by the idealized extreme of having simply a single producing well at the center of the reservoir. As a single well seldom provides sufficient withdrawal capacity for a whole reservoir, even if produced wide open, a ring of wells about the center of the field, of relatively small radius, with a central well drilled later after the ring has been flooded out, should still give greater geometrical sweep efficiencies than the uniform well distribution. If the reservoir is monoclinal and the edgewater advance is largely unidirectional, a line of producing wells along the most distant boundary of the field will give the maximum sweep efficiency. Well distributions of this type, of course, cannot be used unless the field as a whole is operated as a unit, as is common practice in cycling condensate reservoirs. However, these considerations show that, when such operation is feasible, the problem of well spacing in edgewater-drive fields becomes of minor importance as compared with that of the regional grouping and location of the wells.

The geometrical-sweep-efficiency factor may play a more important role in bottom-water-drive than in edgewater-drive reservoirs. This will be especially true in the case of producing formations that are not highly anisotropic. As was seen in Sec. 11.15 the sweep efficiency for wells of fixed penetration in bottom-water-drive systems is determined by the dimensionless well-spacing parameter $\bar{a} = (a/h)\sqrt{k_z/k_h}$, where a is the actual well separation, h is the original oil pay thickness, and k_h, k_z are the effective horizontal and vertical permeabilities. If $\bar{a} > 3.5$, the local conelike water-

[1] Even this gain will be affected as much by the geometry of the well network as by its average density.

oil interfaces below each well will merge with the original oil-water contact plane without overlapping those of its neighboring wells. The clean-oil recovery will simply be a constant for each well [cf. Eq. 11.15(6)]. The total clean-oil recovery from a given area will therefore vary inversely as the square of the well separation or in proportion to the well density. However, this effect of the well spacing is the result of the geometry of the rise of the water-oil interface and does not reflect any variation of the microscopic oil-displacement efficiency of the invading water with the well spacing. In fact the latter was explicitly assumed to be independent of the well spacing in the theoretical treatment of bottom-water-drive reservoirs presented in Chap. 11.

The total economic oil recovery, including that produced after water break-through to abandonment, in bottom-water-drive systems will not increase in direct proportion to the well density, even if $\bar{a} > 3.5$. However, there will probably be some net gain in recovery as the well density is increased until the spacing becomes so small that the local interface elevations below the individual wells develop appreciable overlapping. On the other hand, from an economic standpoint this gain may be of no interest if the recovery per well is not enough to pay for the cost of its drilling and operation. For under conditions where the recovery varies linearly with the well density the sweep efficiency is inherently low, having a maximum of about 13 per cent, at $\bar{a} = 3.5$, even at a minimum well penetration.

If the formation is highly anisotropic and the value of $\bar{a}$ is less than 3.5, for well spacings commonly used in practice, the sweep efficiency will no longer increase linearly with the well density. The gain in recovery per additional well will decrease as the well density is increased, although the sweep efficiency is basically higher for the low $\bar{a}$ and spacings. One may therefore expect that there will be an optimum well density, or spacing, for maximum profit, at which the value of the gain in recovery by drilling an additional well will be just balanced by the operating and investment cost of the well.

On the basis of physical considerations it thus appears that in complete-water-drive systems the local[1] oil-expulsion efficiency due to the water-displacement mechanism should be independent of the well spacing. The latter will affect only the broad areal or volumetric sweep efficiency of the advancing water front, which may have a bearing on the economic ultimate

---

[1] The well spacing could theoretically influence the oil-displacement mechanism by affecting the velocities and pressure gradients at the water-oil interface. While the reality of such effects is still a moot question (cf. Sec. 12.16), the velocity and pressure gradients at the water-oil interface will actually be determined mainly by the total field withdrawals rather than by the well spacing, except when the water is approaching the immediate vicinities of the well bores.

recoveries. In edgewater drives this geometrical-efficiency factor will be more sensitive to the general well pattern and location of the wells than to their absolute spacing. In bottom-water drives the sweep-efficiency factor, as affected by the well spacing, may be of importance in thin producing formations and such as are not highly anisotropic.[1] In general there is no reason to anticipate that either "wide" or "close" well spacing will be inherently advantageous. The number of wells used to exploit a uniform water-drive reservoir should be determined mainly by the economic balance between the cost of drilling and operation and the value of the additional recovery that may be obtained from additional wells through the achievement of greater geometrical sweep efficiency. Except for the development of sufficient withdrawal capacity to bring the operating life of the field into a practical range, the time factor, as related to the possibility of rapid oil recovery through high well densities, should be considered in relation to the danger of accelerated pressure decline resulting from excessive withdrawal rates.

14.5. Physical Aspects of the Well-spacing Problem in Gas-drive Fields; The Physical Ultimate Recovery.—In principle the well-spacing problem for both water-drive and gas-drive reservoirs is essentially comprised of two questions, namely: How does the physical ultimate oil recovery vary with the well spacing? How does the production rate per well vs. time relationship vary with the well spacing? The first pertains to the purely physical relationship between oil expulsion and the well spacing, or distance of drainage. The physical ultimate recovery in strict gas-drive reservoirs refers to the state of complete pressure depletion, where the pressure has declined to atmospheric throughout the reservoir, presumably after an infinite time. The second question, which is also of physical interest, is, however, of primary importance from a practical point of view, since it provides the basis for introducing the economic factors into the problem. Thus, even if the ultimate physical recoveries should be the same for two different well spacings, the economic ultimate recoveries may still be different if the integrated areas under the production-rate vs. time curve to the time of abandonment, as determined by limiting production rates or reservoir pressure, should be different.

In the discussion in the preceding section of water-drive reservoirs it was concluded that the physical ultimate recoveries will be independent of the well spacing, since the efficiency of the water-flushing oil-displacement

---

[1] It should be noted, however, that in most reservoirs bounded by mobile waters the bottom-water-drive mechanism will control directly only the performance of the edge wells immediately overlying the water-oil contact. Unless the formation is strictly uniform, the tendency for a vertical rise of the water table may be masked by the lateral encroachment characterizing the edgewater-drive producing mechanism.

mechanism should not be affected by the distance of travel of the water, and the energy available for displacing the oil may be considered as virtually unlimited.   No attempt was made to construct production-decline curves.   The economic aspects of the problem entered mainly through possible effects of the well spacing on the geometrical sweep efficiency of the invading water, which, in turn, would determine the development of water cuts in the production and the operating costs or abandonment time.

The availability of an unlimited supply of oil-expulsion energy cannot be cited in the case of gas-drive reservoirs as an argument for the lack of dependence of the physical ultimate recovery on the well spacing.   Nor, as previously mentioned, does the finiteness of the energy available in itself imply a variation of recovery with well spacing.   Considered critically the existence of an effect of well spacing on the physical ultimate oil-saturation distribution has been neither proved nor disproved, either theoretically or experimentally.

The only reported experimental data on the saturation distribution in an extended body (4.5 ft) of porous medium after depletion by solution-gas drive[1] indicate a uniform saturation except for end effects, although some decrease in residual-oil saturation on approaching the fluid outlet might be expected, as the result of the increasing total gas flow per unit area as one comes near the outflow boundary.   However, the extent to which this may be compensated by the associated increase in oil flux being moved on approaching the well is quite uncertain.

As previously mentioned the only reported calculations[2] of the saturation distribution in a gas-drive system at complete pressure depletion (atmospheric pressure throughout), using the general equations for heterogeneous-fluid flow, gave lower saturations near the outflow boundary.   But these referred only to a linear system, and the numerical character of the computations did not permit high precision.   On the other hand these showed the distribution to be formally independent of the absolute distance from the low-pressure boundary or of the length of the system and to depend only on the ratio of the distance to the total length.   The ultimate saturation distribution was thus a function only of the general fluid and rock properties.   The total physical gas-drive recovery from a linear column of rock would therefore be independent of the number of fluid-withdrawal centers used for the depletion.[3]   Similar considerations applied to closed radial systems indicate that there, too, the ultimate saturation distribution will depend only on the radial distance expressed as a ratio to the maximum radius and that the saturation at the external closed boundary is inde-

[1] Cf. H. G. Botset, *AIME Trans.*, **140,** 91 (1940).

[2] Muskat and Meres, *loc. cit.*

[3] M. Muskat, *AIME Trans.*, **136,** 37 (1940).

pendent of the radius.[1] This also implies a total recovery per unit area independent of the drainage area per well. Although this conclusion cannot be considered as proved rigorously without further analytical support, there are no fundamental reasons for questioning its validity, nor is there basic theoretical evidence disproving it.[2]

It therefore appears that, subject to the development of evidence to the contrary, the physical ultimate recovery in uniform solution-gas-drive fields may be considered as independent of the well spacing. While it would be a matter of considerable scientific interest to settle this question in a more satisfactory manner, it alone cannot solve the practical well-spacing problem. For, aside from its reference to the state of complete pressure depletion and equalization, it is restricted by the assumption that the producing mechanism is of the pure solution-gas type. If the possibility of fluid segregation and gravity drainage be admitted—as, of course, it must—there is no doubt that such effects will be more pronounced under wide well spacings for fixed well withdrawal rates. To the extent that gravity segregation between the gas and oil may influence the ultimate recovery, variations in well spacing will have a corresponding effect. On the other hand, if gravity be considered as the ultimate controlling recovery agent, the well spacing would again appear to be relegated to a role of minor significance with respect to the physical ultimate recovery.

14.6. Economic Ultimate Recoveries and Well Spacing in Gas-drive Fields.—As previously indicated the economic ultimate recoveries can be determined in principle simply from the production-rate vs. time decline curve. Evidently the integral under such a curve to the time when the production rate has fallen to the abandonment limit will give the economic ultimate recovery. If such curves were constructed and their integrals evaluated for different well spacings, the variation of the economic ultimate recovery with the well spacing would be obtained directly.

In Sec. 10.6 was outlined an approximate procedure for deriving the production-rate vs. time decline curve in gas-drive systems. This, however, was based on the general depletion theory for gas-drive reservoirs, which did not take well spacing into account. Moreover the production rates at any state of depletion were expressed in terms of relative productivity indices, which also did not provide for well-spacing effects. Except for the single numerical analysis of a linear system previously referred to, no serious attempt has been reported of a calculation of the production-rate-decline curves for gas-drive systems, based on Eqs. 7.7(1), when the radius of drainage of the producing well, as created by the presence and interference

---

[1] Muskat and Meres, *loc. cit.*

[2] While these considerations refer only to the solution-gas-drive oil-producing mechanism, capillary pressure effects would themselves lead to complete uniformity of the oil saturation throughout uniform reservoirs when ultimate pressure depletion is reached.

of neighboring wells, was taken into account. One is therefore again forced to resort to approximate procedures of uncertain accuracy.

If, in spite of its obvious limitations in quantitative significance, the production rate per well, Q, be formally expressed in its limiting steady-state form, and the producing pressure differential be taken as a fixed fraction, c, of the reservoir pressure p, Q will be given by

$$Q = \frac{2\pi k_o h c p}{\mu\beta \log r_e/r_w}, \tag{1}$$

where r_e may be considered as the "radius of drainage,"[1] such that πr_e^2 represents the drainage area per well. k_o is the permeability to oil, h the pay thickness, μ the oil viscosity, and β its formation-volume factor. Neglecting, further, any variation in saturation distribution within the drainage area,

$$Q = -\pi f h r_e^2 \frac{\partial}{\partial t}\left(\frac{\rho_o}{\beta}\right), \tag{2}$$

where ρ_o is the oil saturation and f is the porosity. Upon combining Eqs. (1) and (2), t can be formally expressed as

$$\frac{2kct}{fr_e^2 \log r_e/r_w} \equiv \bar{t} = -\int \frac{\mu\beta \, d(\rho_o/\beta)}{p k_o/k}, \tag{3}$$

where k is the homogeneous-fluid permeability and $\bar{t}$ represents a composite time variable.

If the additional assumption be made that ρ_o and p are related as in the general gas-drive depletion history, *i.e.*, by Eqs. 10.3(1), the integral of Eq. (3) can be numerically evaluated to give $\bar{t}$ as a function of p or ρ_o. Such a curve, for the hypothetical gas-drive reservoir producing a 30°API-gravity crude, discussed in Sec. 10.4, is plotted in Fig. 14.2. Since the time variable $\bar{t}$ involves the radius of drainage, r_e, the pressure curve of Fig. 14.2 represents the generalized pressure-decline curve for all well spacings, in the particular hypothetical reservoir under consideration, and under the assumption of the validity of Eqs. (1) and (2). The corresponding production-rate decline is also plotted in Fig. 14.2, where $\overline{Q}$ is defined by

$$\overline{Q} = \frac{Q \log r_e/r_w}{2\pi h k c} = \frac{p k_o/k}{\mu\beta}. \tag{4}$$

---

[1] Equation (1) implies that the initial production rate or productivity index depends on r_e, for which there is no evidence and which is inherently very unlikely. However, as soon as the cumulative withdrawals become sufficient to affect appreciably the pressure midway between the wells, the well interference or the value of r_e will be reflected in the magnitude of Q, although the functional relationship of Eq.(1) will even then be only an approximation at the best. It should be noted, too, that the term "radius of drainage" as used here is a measure only of the linear separation between wells and does not imply a physical limit to fluid movement (cf. Sec. 7.6).

Noting further from Eq. (2) that the average cumulative recovery per unit pore volume is

$$\overline{P} = \frac{\int Q \, dt}{\pi f h r_e^2} = \left(\frac{\rho_o}{\beta}\right)_i - \frac{\rho_o}{\beta}, \tag{5}$$

where the subscript i refers to the initial value, the relation between the recovery and production rate is readily obtained. For the reservoir system

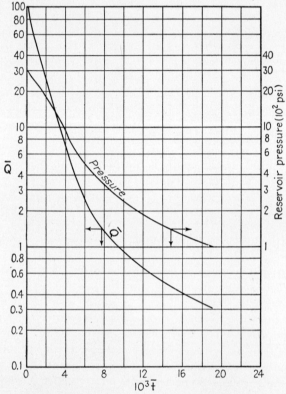

Fig. 14.2. The calculated approximate pressure- and production-rate-decline curves for a hypothetical solution-gas-drive reservoir. $\overline{Q}$ and $\overline{t}$ are the dimensionless production-rate and time parameters defined by Eqs. 14.6(4) and 14.6(3).

used in constructing Fig. 14.2 the variation of $\overline{Q}$ with $\overline{P}$ is plotted in Fig. 14.3. The abscissas of Fig. 14.3 will evidently give the economic ultimate recovery, if in the value of $\overline{Q}$ the production rate for abandonment Q_a be substituted for Q.

It will be clear from the nature of Eq. (4) that, aside from the well-spacing factor, the economic ultimate recovery will be determined by the abandonment rate per millidarcy-foot of the pay. In particular $\overline{P}$ decreases with increasing values of the limiting production rate per millidarcy-foot.

This, of course, is to be expected from general considerations.[1] The variation of $\overline{P}$ with the drainage radius r_e is evidently quite slow, since it enters Eq. (4) only in a logarithmic manner. Thus, on assuming $Q_a = 15$ bbl/day and $kh = 1,250$ md-ft, $\overline{Q}$ becomes $(0.1153/c) \log r_e/r_w$. Hence, for a 10-acre spacing $(r_e = 372.37$ ft), $\overline{Q} = 8.42$ for $c = 0.1$ and $r_w = \frac{1}{4}$, implying an ultimate recovery of 12.4 per cent of the pore space. If the spacing be

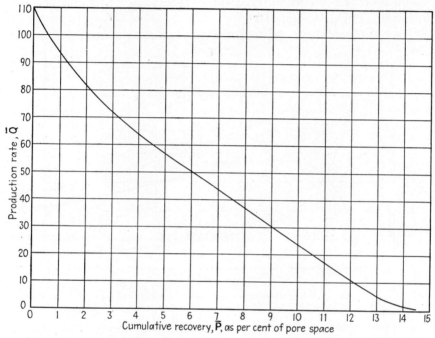

FIG. 14.3. The calculated approximate variation of the production-rate parameter $\overline{Q}$ with the average cumulative recovery $\overline{P}$, for a hypothetical solution-gas-drive reservoir. $\overline{Q}$ and $\overline{P}$ are defined by Eqs. 14.6(4) and 14.6(5).

40 acres per well $(r_e = 744.74$ ft), $\overline{Q}$ will be raised to 9.22, with a recovery of 12.3 per cent of the pore space. This is a reduction of only 1 per cent of the 10 acre per well value.

In carrying through the economic aspects of the well-spacing problem to an evaluation of the ultimate profit, the actual production-rate vs. time curve plays a major role. For the interest factor associated with the operating life may ultimately control the spacing for a given reservoir at which the profit will be a maximum. If the interest rate be i and the

---

[1] The simple dependence of Eqs. (1) to (5) on the product kh does not, of course, provide for a possible variation of the connate-water saturation and permeability-saturation relationship with the permeability. If such effects are taken into account, the variation of the ultimate recovery with k and h may be different (cf. Table 1, p. 832).

production per well during the jth unit time interval be Q_j, the present worth of the future oil production per well will be

$$\sum \frac{Q_j(u-v)}{(1+i)^j},$$

where u is the net price of the oil per barrel, discounting for royalties, and v is the operating cost per barrel, Q_j being expressed as the total production, in barrels, in the jth interest period. If the drilling and investment cost

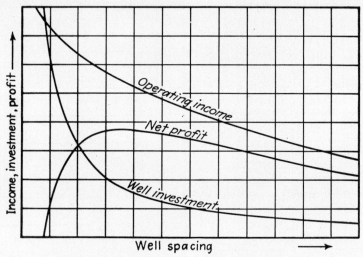

FIG. 14.4. A diagrammatic representation of the well-investment, operating-income, and net profit variations with the well spacing in solution-gas-drive fields.

per well be I, the total present value of the future profits from n wells will be

$$\text{Present worth of profits} = n\left[\sum \frac{Q_j(u-v)}{(1+i)^j} - I\right] = \frac{A}{A_o}\left[\sum \frac{Q_j(u-v)}{(1+i)^j} - I\right], \quad (6)$$

where A is the total productive area, A_o is the drainage area per well, πr_e^2, and it is assumed that all the wells are drilled at the same time.

The general implications of Eq. (6) can be derived without detailed numerical calculations. The summation term in Eq. (6), as a function of the well spacing A_o, will have the general shape of the upper curve in Fig. 14.4. It represents the present worth of the net income from the total future production. It will have a maximum at zero spacing, given by the value of the ultimate recovery at infinite well density, as if it were all recovered at once. The fall in this curve is partly due to the decreasing economic recovery with increasing well spacing (cf. Fig. 14.3) but much more so to the increasing operating life, to abandonment, and to increasing interest discount as the well spacing is increased. As may be inferred from Eq. (3) and Fig. 14.2 the operating life will be approximately proportional to r_e^2, that is, to A_o.

The hyperbolic curve in Fig. 14.4 represents the well-investment term I/A_o. The present worth of the net profit, neglecting taxes, is evidently the difference between the income and investment curves and is also plotted in Fig. 14.4. It has a maximum at the value of the spacing where the slope of the income curve is equal to that of the investment curve. That spacing is the "optimum" spacing.

No scale or numerical values are given in Fig. 14.4 because such numerical evaluation would reflect largely the multitude of assumptions and approximations underlying the calculation of the production-decline curve, as represented by the terms Q_j in Eq. (6). These include not only those referring to the physical description of the depletion process, as expressed by Eqs. (1) and (2), but also the simplified representation of operating conditions, such as the use of the constant c as a measure of the producing pressure differential or well pressure. Moreover the whole basis for Eq. (6) breaks down if the production should be controlled by proration restrictions. Until the producing-well capacities fall to the proration allowables, there will be no natural decline and the operating life will be automatically extended until this limit is reached. The pressure and time when this occurs could be determined from Eq. (4), or its equivalent, on taking Q as the allowable rate. If the proration should impose a limitation on well production rates regardless of the total number of wells, the effect of the time factor in favoring higher well densities will be accentuated. However, under proration programs in which the field withdrawal rate as a whole is given an upper limit, so that the allowable production rates per well are reduced as the well density is increased, the role played by the time factor as related to the well spacing will be greatly minimized. On the other hand, to the extent that withdrawal-rate restrictions may permit more effective natural pressure-maintenance action by water intrusion, the validity of any decline-rate prediction on the basis of pure gas-drive pressure depletion will be nullified.

From the general structure of Eqs. (3) and (4) it is clear that the production decline—unprorated—will become more rapid and the income curve of Fig. 14.4 will become flatter as the permeability of the producing formation is increased. Aside from the associated increase in the absolute value of the economic ultimate recovery, the change in the income curve will be such as to lead a shift in the point of equal slopes of the net income and investment curves to higher well spacings. That is, the economic optimum[1]

---

[1] It should be noted that the economic "optimum" spacing—that of maximum profit—will in general be wider than that giving the maximum profitable economic

spacing will increase with increasing values of kh. Likewise a shift toward lower well densities and higher optimum spacings will result from increased well-investment costs I. Higher limiting abandonment rates Q_a will also require wider spacings to achieve maximum profits.

Perhaps the simplest formal procedure for deriving a closed expression for the profit vs. well-spacing relationship is to assume[1] that the production rate will decline exponentially with time. On assuming, furthermore, the ultimate physical recovery to be independent of the well spacing the production rate can be expressed as

$$Q = Q_i e^{-\alpha kt/r_e^2}, \tag{7}$$

where Q_i represents the initial rate and α is a constant determining the ultimate physical recovery. r_e is the drainage radius per well corresponding to the well spacing. The physical ultimate recovery, per well, P_∞, is

$$P_\infty = \frac{Q_i r_e^2}{k\alpha}, \tag{8}$$

and the cumulative recovery to the time when the production rate has become Q is

$$P = P_\infty \left(1 - \frac{Q}{Q_i}\right). \tag{9}$$

If Q be considered as the abandonment rate Q_a, it follows from Eq. (9) that the economic ultimate recovery will increase linearly with decreasing abandonment rate. Equation (9) also implies that the total economic ultimate recovery is independent of the well spacing.

To evaluate Eq. (6) it is noted that the production during the jth time interval, *i.e.*, between $(j-1)t_o$ and jt_o, is

$$Q_j = P_\infty e^{-(j-1)\alpha kt_o/r_e^2}(1 - e^{-\alpha kt_o/r_e^2}). \tag{10}$$

Hence

$$\sum_1^m \frac{Q_j}{(1+i)^j} = \frac{P_\infty(1 - e^{-g})}{1 + i - e^{-g}}\left[1 - \frac{e^{-mg}}{(1+i)^m}\right], \tag{11}$$

where g is the term $\alpha kt_o/r_e^2$ and the summation limit m represents the total number of interest periods, t_o, in the operating life and is determined by

$$Q_a = Q_i e^{-gm}, \tag{12}$$

where Q_a is the abandonment rate.

---

ultimate oil recovery. It is controlled mainly by the economic importance of the time factor as compared with the well-investment cost, rather than such slow variations in the economic ultimate recovery as may be caused by variations in the well spacing.

[1] Such declines will give constant "loss-ratios" and apparently do describe the observed performance in some gas-drive fields (cf. Sec. 10.11).

On assuming that both the net price of the oil, u, and operating cost per barrel,[1] v, remain constant throughout the producing life, the present worth of the net income will therefore be

$$A \frac{P_o(u - v)(1 - e^{-g})}{1 + i - e^{-g}} \left[1 - \frac{Q_a}{Q_i(1 + i)^m} \right],$$

where P_o is the physical ultimate recovery per unit area (acre). As Q_a/Q_i will in general be very small, the bracket can be approximated[2] by 1, and Eq. (6) will then take the form

$$\text{Present worth of profits} = A \left[\frac{P_o(u - v)(1 - e^{-g})}{1 + i - e^{-g}} - \frac{I}{A_o} \right], \quad (13)$$

where A_o is number of acres per well. The whole effect of the time factor and variation of the economic ultimate recovery with the spacing is thus represented by the term $(1 - e^{-g})/(1 + i - e^{-g})$. This may have a range of 0 to $1/(1 + i)$. Since $g = Q_i t_o/P_\infty = Q_i t_o/P_o A_o$, wide spacings will imply small values of g and of the factor giving the net income. As the well density is increased, g will increase and so will the net operating income. The resulting curve of net operating income vs. well density will therefore be similar to that shown in Fig. 14.4.

The spacing, A_o, for maximum profits implied by Eq. (13) is given by

$$e^{-Q_i t_o/P_o A_o} = 1 + i + \beta - \sqrt{\beta^2 + 2\beta(1 + i)} \quad : \quad \beta = \frac{Q_i i(u - v)t_o}{2I}. \quad (14)*$$

Hence as Q_i is essentially proportional to the permeability, the optimum spacing will here, too, increase with increasing permeability. It will also increase as the well cost I increases. However, it will decrease as the net current profit per barrel, $u - v$, increases, or as the average recovery P_o increases.

While these general conclusions appear inherently reasonable, they are, of course, little more than direct implications of the gross assumptions underlying the analysis, in particular, Eq. (7). Moreover, they also presuppose wide-open producing conditions without proration control. No attempt will therefore be made to "derive" optimum spacing values by

---

[1] The assumption that v remains fixed will obviously not apply in practice, where the artificial lifting cost per barrel in the later life of a well is generally much greater than the cost per barrel of operating a flowing well and its ultimate rise to equality with the net selling price forces abandonment. Since the decrease in $u - v$ could be taken care of by numerical summation of Eq. (6) if its variation were known, the simplifying assumption is made here for convenience in illustrating the general principles involved in applying the economic factors to the well-spacing problem.

[2] With this approximation the values of the profits and optimum spacing become independent of the abandonment rate.

* The validity of Eq. (14) involves the tacit assumption that $2\beta > i^2$.

numerical illustration of the above equations, although such values would probably be of the correct order of magnitude.

Attempts have also been made to derive approximate solutions to the well-spacing problem by calculating the saturation distribution at the time of abandonment, as determined by the fall in production rate to the economic limit. These do not involve directly the calculation of the complete production-decline history. They are based on assumptions regarding the nature of the flow conditions at the time of abandonment. In one[1] it was assumed that each unit area, regardless of its location, contributes equally to the total producing rate from the well, Q_a, so that

$$\frac{2\pi r k_o h}{\mu_o \beta} \frac{dp}{dr} = \frac{(r_e^2 - r^2)Q_a}{r_e^2},\tag{15}$$

where r_e is again the equivalent radius of the drainage area of the well, k_o is the permeability to oil, μ_o, β, the viscosity and formation-volume factor of the oil, and p the pressure at r at the state of abandonment. The interrelation between k_o, or the liquid saturation, and the pressure was fixed by adding the assumption that the gas-oil ratio for the free-gas phase is constant, *i.e.*,

$$\frac{\gamma \beta \mu_o}{\mu_g} \Psi = \text{const},\tag{16}$$

where $\Psi = k_g/k_o$, the ratio of the permeabilities to gas and oil, γ is the gas-phase density, and μ_g is its viscosity. p is assumed known at r_e. To determine the constant in Eq. (16), Eq. (15) was solved for different values of the constant until the residual-oil saturation distribution, when averaged over the whole area to r_e, corresponded to a preassigned average recovery, as determined from the general solution for the gas-drive depletion history by the methods discussed in Chap. 10. The average recoveries for well spacings smaller than that corresponding to r_e were then considered as averages to the various radial distances determined from the basic saturation distribution calculated for the original radius of drainage, r_e. Thus was obtained a variation of the economic ultimate recovery with the well spacing for the same limiting production rate for abandonment, Q_a.

The production-decline rates for various well-spacing developments were derived by calculating the variation of the productivity index with the reservoir pressure or cumulative recovery, as outlined in Chap. 10, and assuming a sequence of declining bottom-hole producing-well pressures. The corresponding well production rates were applied to all wells regardless of the spacing, so that the field production rates were directly proportional

[1] W. H. Barlow and W. B. Berwald, *API Drilling and Production Practice*, 1945, p. 129. The stepwise numerical procedure outlined there is equivalent to the analytical representation of Eqs. (15) and (16).

to the well density at any given reservoir pressure. These were translated as production-rate vs. cumulative-recovery curves by assuming the cumulative recovery at a given spacing and reservoir pressure to be proportional to the ultimate recovery at abandonment for that spacing. On converting these to curves of cumulative recovery vs. time and applying the economic factors, including interest discount, well investment, and operating costs, the variation in profit vs. the well spacing was finally determined.

An application of this procedure to a total reservoir area of 160 acres (r_e = 1,489 ft), with Q_a = 15 bbl/day, kh = 14 darcy-ft, $p(r_e)$ = 75 psi, and an ultimate average recovery for the 160 acres of 20 per cent, showed the residual-oil saturation at abandonment to vary by only 1.6 per cent over the whole distance range. The variation in the recoveries for well spacings less than 160 acres per well approximately paralleled the saturation-distribution curve. On carrying through the economic implications of the recovery curve, using a well investment of $50,000 per well, and an operating cost of $0.20 per barrel, the spacing for maximum profit was found to be 60 acres per well.[1]

A similar treatment has been reported using different assumptions regarding the pressure distribution and the state of flow at the time of abandonment. Here the abandonment production rate Q_a was considered as passing simultaneously through the whole drainage area,[2] as if one had strictly steady-state-flow conditions. Instead of Eq. (15) the pressure distribution was assumed to be given by[3]

$$Q_a = \frac{2\pi k_o hr}{\mu_o \beta} \frac{dp}{dr} = \text{const.} \tag{17}$$

As in the case of the method discussed at the beginning of this section, it is assumed here, too, that the relation between oil saturation and pressure is the same as that determined for the reservoir as a whole, by the methods outlined in Chap. 10 for the solution-gas-drive depletion mechanism. The pressure distribution is then determined by integrating Eq. (17), beginning at the well bore and assuming the well pressure corresponding to Q_a. On applying this procedure for a limiting production rate of 5 bbl/day from a 10-ft 10-md sand with a 0-psi well-bore pressure, the pressure was found to rise to 1,254 psi at 640 ft and the oil saturation to vary from 27.1 to 30.6 per cent between the well bore and 640 ft. The results of similar

---

[1] This numerical value is not to be considered as a "typical" optimum well spacing. Almost any value can be derived by the same procedure by appropriate choices of the permeability and thickness, even if the economic factors be kept fixed (cf. L. F. Elkins, *API Drilling and Production Practice*, 1945, p. 141).

[2] V. Moyer, *AIME Trans.*, **174**, 88 (1948).

[3] Equation (17) is the analytical equivalent of the numerical procedure actually reported.

calculations for other sand conditions and different well spacings are tabulated in Table 1.

TABLE 1.—CALCULATED ECONOMIC ULTIMATE RECOVERIES FOR SOLUTION-GAS-DRIVE RESERVOIRS

Spacing, acres per well	Sand thickness, ft	Permeability, md	Connate-water saturation, %	Oil recovery, % oil in place
0.46	100	10	45	40.6
1.85	100	10	45	40.4
7.38	100	10	45	40.2
29.54	100	10	45	40.0
29.54	10	10	45	32.8
29.54	100	57	30	37.5
29.54	10	57	30	36.8
29.54	100	500	20	32.9
29.54	10	500	20	32.8

These calculations were made assuming for the reservoir fluid data those[1] of the crude oil and natural gas from the Dominguez oil field, California, with an initial formation-volume factor of 1.42, gas solubility of 682 ft^3/bbl, and reservoir pressure and temperature of 3,000 psi and 220°F. The permeability data used were based on those of Leverett and Lewis[2] for unconsolidated sands. While the numerical values listed in Table 1 may be of little absolute significance, their relative magnitudes probably reflect the effect of the different variables to a reasonable approximation. Thus the first four rows show again the very slow decrease in economic[3] ultimate recovery with increasing well spacing. Comparison of the recoveries in Table 1 for cases where only the thickness is varied shows the fractional economic ultimate recoveries to decrease with decreasing pay thickness, for fixed abandonment rates, though this effect is much smaller for the higher permeability formations. This implication may be also inferred from the other treatments of the problem, as expressed by Fig. 14.3 and Eq. (9). The variation of the ultimate recovery with the permeability, as shown in Table 1, is different for the 10- and 100-ft pays and is apparently dominated by the assumed variation in the connate-water saturation.

Neither Table 1 nor the previous graphical and analytical results are to be considered as "solutions" of the well-spacing problem. On the con-

[1] B. H. Sage and W. N. Lacey, *Ind. and Eng. Chemistry*, **28**, 249 (1936).

[2] M. C. Leverett and W. B. Lewis, *AIME Trans.*, **142**, 107 (1941).

[3] Since the relations between the pressure and oil saturation were derived by application of Eq. 10.3(1), they necessarily implied that the *physical* ultimate recoveries were independent of the well spacing.

trary it will be clear from the above discussion that they have all been based on various groups of assumptions for which there is no strict justification. The method leading to Figs. 14.2 and 14.3 implies that the physical ultimate recovery is independent of the well spacing. Equations (7) and (15) also imply this assumption. The constancy of the free-gas-phase gas-oil ratio at abandonment, required by Eq. (16), could not obtain under steady-state heterogeneous-fluid flow and would probably be even a poorer approximation in the case of transient flow. The integration of Eq. (17) makes use of a relationship between the oil saturation and pressure that takes no account of well spacing. Moreover this relationship is applied in a form implying that the local gas-oil ratio at the abandonment state may have a maximum within the drainage area. As there is no rigorous solution of the well-spacing problem available even for any specific simple case,[1] the various assumptions involved in the different treatments outlined above cannot be satisfactorily evaluated to see whether or not they have inherently masked or accentuated any real effect that the well spacing may have on the oil recovery.

The present status of the well-spacing problem in pure solution-gas-drive reservoirs, from the theoretical point of view, thus appears to be one where the physical ultimate recovery is generally considered to be basically independent of the spacing or drainage area per well. This still represents only an assumption, though there is no conclusive evidence to the contrary. The economic ultimate recovery is also beyond the power of rigorous calculation. However, all the approximate methods of analysis thus far proposed lead to essentially the same result, namely, that within the range of physical data pertinent to actual oil reservoirs the economic ultimate recovery in solution-gas-drive reservoirs will increase slowly as the well density is increased. The final economic evaluation of different well-

---

[1] While the approximate treatments reviewed here have been based on assumptions leading directly to pressure and saturation distributions at the state of economic abandonment, which, if the latter corresponded to complete pressure depletion, would also imply saturations which are both uniform in the drainage area and independent of the well spacing, additional recent studies purport to give the complete transient histories. In one (cf. R. G. Loper and J. C. Calhoun, Jr., AIME meetings, Dallas, Tex., October, 1948), the transients are represented by sequences of steady states. As the latter inherently imply limiting states of uniform saturation at ultimate pressure depletion, the reported results that the ultimate recoveries and production histories are substantially independent of the well spacing and rates of withdrawal can hardly be considered as conclusive. In another investigation (cf. C. C. Miller, W. F. Kieschnick, Jr., and E. R. Brownscombe, AIME meetings, Dallas, Tex., October, 1948), the detailed transient history is developed by a considerably more complicated procedure. However, here, too, limited application is made of steady-state approximations, and it is difficult to evaluate their integrated effect on the final results, which again indicate the ultimate recovery to be essentially independent of the well spacing and rate of production.

spacing development programs thus will be primarily controlled by the purely economic factors of well cost, net price of the oil, interest rate, and operating life. These factors alone, even if the economic ultimate recovery be taken as strictly constant, will lead to spacings for maximum profit ranging under different conditions from what would seem to be, in relation to other cases, as excessively "close" to absurdly "wide." If the producing formation is shallow and of low permeability and if the net price of the oil is high, spacings lower than 10 acres per well may yield the maximum profit. For the very deep fields, however, with high drilling costs and good permeability, spacings as wide as 40 acres per well may not suffice to give the maximum profit. In extreme cases, of course, as in condensate fields, where the ultimate recoveries per unit volume of rock are inherently low, the economic factors may require well densities as low as one well to 200 or 400 acres if the operations are to be profitable at all.

Since it is clear that the physical aspects of the well-spacing problem are of minor significance, any reasonable approximation pertaining to the details of the physical recovery processes should suffice for the choice of the well spacing. Of course, as accurate estimates as possible should be made of the ultimate recoveries. And the nature of the producing reservoir with respect to uniformity, oil content, and type of recovery mechanism should be determined before a development program is fixed. However, it is mainly the simple economic balance between drilling cost, net price of the oil, and the probable withdrawal rates, in the light of well producing capacities, proration restrictions, and the general recovery mechanism, that will ultimately fix the optimum spacing.

14.7. Field Observations on the Relation between Well Spacing and Recovery.—As noted in Sec. 14.2, most of the early reported field data pertaining to well spacing have been of questionable significance because of the failure to recognize the influence of factors that may either unduly exaggerate or, conversely, entirely mask whatever real effect, if any, the well spacing may have had on the oil recovery. It is only recently that serious attempts have been made to discriminate between the effect of well spacing on recovery and the influence of other factors. The analysis of recovery data from areas in a single reservoir developed with different well spacings is still beset with difficulties, although in the case of the water-drive Silica Arbuckle pool a critical analysis[1] of the field performance indicates no substantial differences in recovery that can be directly attributed to differences in well density. Previous studies[2] had been interpreted as showing the average recoveries to vary inversely as the average

---

[1] L. F. Elkins, *Oil and Gas Jour.*, **45**, 201 (Nov. 16, 1946).

[2] W. W. Cutler, Jr., *U.S. Bur. Mines Bull.* 228 (1924); H. C. Miller and R. V. Higgins, *U.S. Bur. Mines Rept. Inv.* 3479 (1939).

well separation. While this result has been quoted often as an established theorem, known as "Cutler's rule," it seems most likely in the light of the more recent investigations that the data on which it was based involved extraneous factors which inherently tended to favor greater recoveries from the areas of higher well densities.

Evidence that "Cutler's rule" could be generalized to comparisons of different reservoirs has been cited[1] from the recoveries of the water-drive Mexia-Powell fault-line fields in Texas, all of which produced from the Woodbine Sand. However, a detailed analysis[2] of the reservoir and producing characteristics of these fields showed that within the limits of uncertainty of the basic reservoir data there was no systematic variation of recovery with the well spacing. The final results obtained in this study are listed in Table 2.

TABLE 2.—RECOVERY DATA FOR THE FAULT-LINE FIELDS

Field	Ultimate recovery 10^3 bbl	Sand vol., acre-ft	Recovery, bbl/acre-ft	Spacing, acres per well
Mexia......	93,547	92,600	1,010	4.7
Wortham...	23,500	20,818	1,100	2.1
Currie.....	7,000	7,551	927	5.8
Richland...	6,500	5,933	1,096	2.4
Powell.....	113,500	120,813	939	3.5

By far the most complete study of the recovery from oil reservoirs, with respect to the possible effect of well spacing, is that[3] based on data on 101 fields (2 of the 103 listed are duplicates) submitted to the Special Study Committee on Allocation of Production and Well Spacing, API, Division of Production. As these data are also of inherent interest as a record both of reservoir characteristics and recovery estimates, they are reproduced in full[4] in Tables 3 and 4.

In spite of the extensive character of the tabulations of Tables 3 and 4, they still represent too few fields for a complete quantitative statistical analysis. Thus only 26 of the listed fields produced by dissolved-gas or gas-cap drives alone. The porosities of these varied from 12.5 to 29 per cent and their permeabilities from 7 to 2,000 md. The connate-water saturations ranged from 2 to 40 per cent and the crude gravities from

[1] S. K. Clark, C. W. Tomlinson, and J. S. Royds, *AAPG Bull.*, **28**, 231 (1944).

[2] W. V. Vietti, J. J. Mullane, O. F. Thornton, and A. F. Van Everdingen, *API Drilling and Production Practice*, 1945, p. 160.

[3] R. C. Craze and S. E. Buckley, *API Drilling and Production Practice*, 1945, p. 144.

[4] The last five fields (104–108) have been added from the paper of A. W. Baucum and M. D. Steinle, *Oil and Gas Jour.*, **45**, 199 (July 27, 1946).

836 PHYSICAL PRINCIPLES OF OIL PRODUCTION [CHAP. 14

TABLE 3.—RESERVOIR AND PRODUCTION DATA FOR API STUDY RESERVOIRS*

Study No.	Type of formation	Type of drive	Area of reservoir, acres	Thickness of oil-producing formation, ft	Porosity, %	Interstitial (connate) water, %	Permeability, md	Depth of reservoir, ft	Reservoir temp., °F	Initial reservoir pressure, psi gauge	Current reservoir pressure, psi gauge
1	Sand	Dissolved-gas	13,000	59	20	15	90	2,000	93	750	15
2	Sand	Water	1,000	10	30	35	1,000	6,600	158	2,950	2,880
3	Sand	Water	960	41	18	20	75	3,200	136	1,440	—
4	Lime	Water	4,400	20	Cavernous	—	—	3,540	90	—	—
5	Sand and lime	Gas	10,000	16.5	12.5	20	150	3,800	118	1,300	182
6	Sand	Water	135,076	35	25.22	17	2,000-3,000	3,700	146	1,620	1,015
7	Lime	Water-gas-cap	20,000	100	Cavernous	20	Cavernous	1,000-1,800	82	700	521
8	Dolomite	Water	6,200	30		—	—	3,500		1,375	713
9	Lime	Water-gas-cap	10,040	124	Cavernous	20	Cavernous	4,050-4,174	96	1,500	1,165
10	Sand	Water	4,250	17	32	30	50-5,000	3,500	134	1,980	1,925
11	Sand	Water	550	11.6	27.6	30	2,800	2,200	122	864	774
12	Sand	Dissolved-gas	280	43.5	15.3	40	6.8	2,100	109	—	25
13	Sand	Dissolved-gas-water	1,868	20	28	30	264	4,850	165	1,900	1,245
14	Sand	Water	472	36	22	15	0-1,650	4,800	132	2,048	1,680
15	Sand	Dissolved-gas	24,913	43.3	17	35	125	2,900	120	1,165	45
16	Sand	Water	501	17	34	33	3,000	5,800	165	2,662	2,448
17	Sand	Dissolved-gas	7,598	120	23	2	1,500-2,000	6,500	132	2,630	26
18	Sand	Dissolved-gas	4,000	35.8	20.2	35	395	7,550	198	3,520	1,630
19	Sand	Gas	2,350	36	18	32	80	2,110	86	800	75
20	Sand	Water	1,813	18.3	14	25	118	1,950	87	735	285
21	Sand	Gas	7,550	55	20	20	400	1,100	75†	—	Vacuum
22	Sand	Gas	404	28.8	18.2	25	356	2,525	90	—	
23	Lime	Water	4,500	90	17.6	30	1,010	7,600	207	3,480	2,876
24	Sand	Gas	915	42	13	20	206	7,200	139	2,900	1,280
25	Sand	Gas	1,280	48.3	15	20	—	3,000	133	—	
26	Sand	Water	828	14.6	15	15	328	4,575	130	1,100	750
27	Sand	Water	423	17.1	18.9	20	450	1,700	76	750	250
28	Sand	Gas	608	13.6	17.9	35	100	2,175	85	925	500
29	Sand	Water	2,614	27.9	18.5	39	40	6,600	172	2,180	1,956
30	Sand	Water	1,153	33.9	20.7	36	72	6,750	173	2,220	2,215
31	Sand	Water	712	162	13.4	14.7	365	3,500	100	1,728	1,442
32	Sand	Water	1,350	19.8	17	31.4	90	3,000	100	1,135	200
33	Sand	Water	680	4.88	24	35	—	5,320	155	2,300	2,000†
34	Sand	Water	1,081	10.84	33	30	—	4,750	140	2,100	1,800
35	Sand	Water-gas-cap	3,660	13	25	30	—	5,300	150	2,410	2,200

36	Sand	Dissolved-gas	8,833	57.45	16.2	26.5	25–228	1,850	82	700	20
37	Sand	Water	176	113	28	10	5,000	3,300	140	—	—
38	Sand	Water	105	131.7	28	10	5,000	3,500	140	—	—
39	Sand	Water	17,879	48.21	27	28		5,200	170	2,275	2,014†
40	Sand	Water	4,145	21	27	20		2,300	125	1,250†	250†
41	Sand	Water	8,154	24.4	25	20		2,300	128	1,300†	125†
42	Sand	Gas	946	19.4	27	35		3,900	147	1,680	300†
43	Sand	Water	1,640	66	26.9	25.9	1,335	9,990	210	4,615	4,450
44	Sand	Water	572	16.8	17.9	40	104	1,210	78	440	170
45	Sand	Water	1,913	10.8	17	36.7	152	2,500	97	725	
46	Sand and lime	Gas	1,553	12.7	18	38	100	3,086	100	885	
47	Sand and lime	Gas	640	18.9	18	40	90	3,300	99	1,225	
48	Sand	Gas	3,065	18.9	17.2	35.5	75	3,200	98	1,085	
49	Sand	Gas	1,280	14	23	25		2,100	110	960†	250†
50	Sand	Gas	7,330	17.47	29	20		2,700	125	1,050	100†
51	Sand	Water	1,166	16.4	31	25	1,000–5,000	6,765	164	3,150	3,045
52	Sand	Water	5,475	65.5	27.3	35	1,100	7,045	178	3,265	3,060
53	Sand	Water	340	44.2	27.3	35	1,100	7,045	178	3,265	3,112
54	Sand	Water	515	27.5	27.3	35	1,100	7,045	178	3,265	3,114
55	Sand	Water	1,145	12.3	24	42.5	300	7,045	178	3,265	2,833
56	Lime	Water	1,005	20	16	26	Cavernous	3,230	129	1,470	1,270
57	Sand	Dissolved-gas	4,345	6.9	28	25	600	2,875	145	1,125	768
58	Sand	Water	14,570	70.3	27	35	700	4,980	170	2,250	1,988
59	Sand	Water	3,410	11.9	25	35	400	4,980	170	2,250	1,145
60	Sand	Water	133,400	39.6	26	17	2,000–3,000	3,700	146	1,620	1,006
61	Sand	Water	1,616	28	26.3	35	200	6,630	185	3,060	2,639
62	Sand	Water	758	10.5	26.3	35	200	6,630	185	3,060	2,498
63	Sand	Water	1,865	17.6	28	40	700	8,730	207	4,225	2,482
64	Sand	Water	5,190	20.1	24	35	400	4,370	143	1,980	1,851
65	Sand	Water	225	18	30	34	3,000	5,870	177	2,725	2,586
66	Sand	Water	4,325	12.7	23.3	30	400	5,470	164	2,528	2,117
67	Sand	Dissolved-gas	3,850	15.6	25.5	30	300	5,320	180	2,400	425
68	Sand	Water	250	8.4	20	30	100	3,020	121	1,400	
69	Sand	Dissolved-gas	70	17.5	21.8	25	400–500	1,100	90	400	0
70	Sand	Dissolved-gas	1,400	14.6	22	30	500	2,010	125	820	
71	Sand	Water	1,326	45.2	26.4	35	200	3,980	156	1,800	1,567
72	Sand	Water	201	14.6	25	35	200	3,980	156	1,800	800
73	Sand	Water–dissolved-gas	4,120	13.8	30.7	25	500	2,450	132	1,000	500
74	Sand	Water–gas	1,175	85.6	31.8	30	500	3,470	156	1,550	1,302
75	Sand	Water	560	25.1	25	30	1,300	3,965	133	1,804	1,723

TABLE 3.—Reservoir and Production Data for API Study Reservoirs* (Continued)

Study No.	Type of formation	Type of drive	Area of reservoir, acres	Thickness of oil-producing formation, ft	Porosity, %	Interstitial (connate) water, %	Permeability, md	Depth of reservoir, ft	Reservoir temp, °F	Initial reservoir pressure, psi gauge	Current reservoir pressure, psi gauge
76	Sand	Water	325	9	25	40	600	4,360	140	1,985	1,892
77	Sand	Water	476	16.8	25	40	600	4,920	158	2,253	2,069
78	Sand	Water	7,800	47.3	24.8	25	1,000	4,230	147	1,920	1,132
79	Sand	Water	4,000	80.7	25	35	1,300	5,220	162	2,430	2,292
80	Sand	Water	505	14	24	40	1,000	3,120	146	1,450	1,226
81	Sand	Water	1,200	29.8	25	35	1,300	5,390	163	2,490	2,377
82	Sand	Water	340	50	24	40	1,000	4,350	151	2,000	1,932
83	Sand	Water	11,471	14.4	28.5	32	1,000	5,860	177	2,725	2,448
84	Sand	Water	6,246	56.6	31.5	30	3,400	5,860	177	2,725	2,574
85	Sand	Water	4,500	114	25	20	2,000	2,880	136	1,250	892
86	Sand	Dissolved-gas	303	16.6	23	35	500	1,200	96	330	25
87	Sand	Water	4,262	162.2	30.5	35	1,000	5,950	164	2,705	2,560
88	Sand	Water	335	10.7	32.7	20	1,865	9,500	182	4,400	4,260
89	Sand	Water	410	19	32.7	20	1,865	9,500	182	4,400	4,230
90	Sand	Water	340	23	33	10	2,200	6,590	183	3,040	2,540
91	Sand	Water	673	37.5	33	15	2,200	6,900	183	3,200	2,850
92	Sand	Water	125	37.5	27.7	24	420	5,500	175	2,500	750
93	Sand	Water	300	42	26	25.5	300	6,000	185	2,800	500
94	Sand	Dissolved-gas	286	91	23.5	29.2	130	6,700	196	3,080	700
95	Sand	Dissolved-gas	750	63.7	22.2	31	85	7,400	205	3,300	250
96	Sand	Water	730	115	32	21	945	3,600	136	1,600	150
97	Sand	Water-dissolved-gas	1,765	219	32.3	21.5	720	4,500	150	1,700	20
98	Sand	Water	363	230	31.3	22	640	5,000	160	2,160	200
99	Sand	Water	630	164	30.7	22.5	435	5,900	177	2,530	700
100	Sand	Water-dissolved-gas	839	332	29	23	330	6,500	188	2,830	100
101	Sand	Water	1,900	86	35	20	3,000	2,000	115	450	200
102	Sand	Water	78	242	26.5	24	200	4,700	165	1,900	300
103	Sand	Water	74	61	25.8	24	205	7,300	205	2,710	1,300
104	Sand	Water	570	16.8	17.9	40	104	—	—	425	—
105	Sand	Water	570	9.6	17.0	34	303	—	—	710	—
106	Sand	Water	1,240	12.5	15.3	26.5	95	—	—	1,000	—
107	Sand	Water	1,350	8.2	19.2	37	100	—	—	1,050	—
108	Sand	Water	1,800	13.4	19.6	36	186	—	—	500	—

* Studies 104–108 are taken from A. W. Baucum and W. D. Steinle, *Oil and Gas Jour.*, **45**, 199 (July 27, 1946).
† Estimated

22.5 to 46.4°API.   The lowest reservoir oil viscosity was 0.45 cp and the highest 9.5 cp., and the initial formation-volume factors had a minimum of 1.03 and a maximum of 1.67.   The average well spacings covered a range of 2.8 to 46.8 acres per well.   Since all these factors may be expected a priori to have some effect on the economic recovery, a unique differentiation between their individual effects in such a small group is evidently

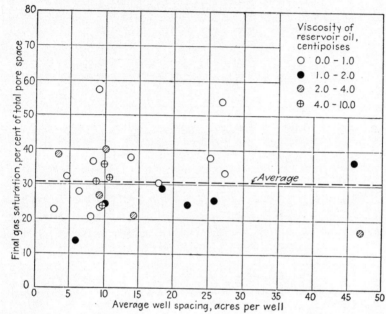

FIG. 14.5.   The relation between the ultimate free-gas saturation at depletion vs. the well spacing, as observed or estimated in gas-drive fields.   (*After Craze and Buckley, API Drilling and Production Practice, 1945.*)

impossible.   Nevertheless it is of interest to note that if the well spacing be considered as the primary variable affecting the recovery, no significant trend is found.   This will be apparent from Fig. 14.5, in which the recoveries listed in Table 4, ranging from 126 to 590 bbl/acre-ft, have been converted to equivalent ultimate free-gas saturations.   From the general theory of gas-drive reservoir performance, as discussed in Chap. 10, it will be clear that in comparative studies of reservoirs with different physical characteristics the ultimate free-gas saturation should represent a more fundamental index of the total recovery than the absolute recovery expressed as barrels per acre-foot or as a fraction of the pore space.

The scatter-diagram character of Fig. 14.5, with the free-gas saturations ranging from about 14 to 57 per cent, obviously implies either that many of the data are inherently subject to large errors or that other factors have

Table 4.—Additional Reservoir and Production Data for API Study Reservoirs

Study No.	Residual-oil gravity, °API at 60°F	Dissolved gas liberated at field trap pressure, ft³/bbl	Saturation pressure at reservoir temp., psi gauge	Formation-vol. factor, initial reservoir bbl/stock-tank bbl	Reservoir oil viscosity at initial conditions, cp	Depletion of ultimate production to date, %	Average well spacing, acres per well	Uniformity of spacing	Estimated ultimate recovery, bbl/acre-ft
1	38	450	500	1.25	3.19*	79	10.2	Uniform	368
2	31	450	2,789	1.195	1.27*	31	26.3	Uniform	600
3	31	136	960	1.07	1.51*	53	37.0	Irregular	410
4	30			1.15*	2.70*	79	40.0	Uniform	200
5	41	404	1,300	1.24	0.66*	42.3	25.3	Nonuniform	232
6	38–39.8	365	740	1.20	1.0	42.2	4.9	Irregular	1,058
7	35	165	720	1.09	1.46*	68.8	36.1	Irregular	200
8	35	670		1.19*	1.36*	48.5	35.5	Nonuniform	267
9	34.5	250	1,500	1.24	1.20	52	38.0	Uniform	161
10	23		1,181	1.20	3.60	69	17.2	Uniform	629
11†	34.8	140		1.092	1.09*	35.8	34.35	Nonuniform	735
12	41			1.28*	1.60*	98	10	Uniform	138
13	40	140	778	1.318	1.30	25	46	Uniform	472
14	40	640	2,048	1.33	0.66*	48	11.2	Uniform	765
15‡	40			1.139	0.70*	97	8.0	Uniform	187
16§	35	574	2,570	1.29	0.61*	44.3	15.7	Uniform	958
17	38	735	2,630	1.356	0.82*	63.5	9.2	Irregular	590
18‖	32–34	762	3,520	1.45	0.88*	49	27.4	Uniform	237
19	42	83		1.100	3.24*	67	9.3	Uniform	297
20	39	115	333	1.065	2.46	39.9	21.6	Uniform	306
21	31			1.060	2.47*	90.2	3.4	Uniform	530
22	37.2			1.100	3.11*	79.7	14.4	Uniform	213
23	39	960	3,355	1.490	0.67	21.9	39.1	Uniform	407
24	41	1,055	2,885	1.667	0.59*	96.7	27.0	Irregular	233
25	38.5			1.250	4.72	86.5	10	Uniform	247

26	33.2	143	531	1.100	1.22*	74.2	45.9	Nonuniform	437
27	26.3	114	758	1.070	16.8	47.2	26.4	Nonuniform	261
28	32	—	—	1.080	2.12*	34.7	46.8	Uniform	180
29	49	447	1,078	1.299	0.40	12.7	65.4	Uniform	280
30	49	110	374	1.105	0.54	2.6	57.7	Uniform	466
31	39.5	540	1,703	1.269	1.00	14.1	25.4	Nonuniform	390
32	38	300	685	1.176	0.94*	63.5	9.6	Uniform	396
33	25	310*	2,300*	1.15*	3.22*	49	37.8	Uniform	518
34	26	400	2,100	1.25	3.17*	59.5	21.2	Uniform	644
35	26	310	2,410	1.17	2.7	28.4	26.5	Uniform	629
36	39	230	520	1.14	0.97*	52.5	8.4	Irregular	354
37	20	—	—	1.09*	11.4*	100	4.8	Irregular	1,111
38	20	—	—	1.09*	11.4*	100	17.5	Irregular	1,121
39	38	750	2,275*	1.25*	0.62*	25.3	18.2	Uniform	742
40	22	160*	1,250*	1.08*	8.7*	59.3	9.0	Uniform	427
41	21	370*	1,300*	1.17*	10.7*	69	9.8	Uniform	398
42	44	482	1,506	1.25	0.46*	67	9.4	Uniform	547
43	39	916	4,615	1.51	0.45	20.4	65.6	Uniform	617
44	36.6	8	214	1.036	5.2	32.8	23.8	Uniform	429
45	38.2	8	117	1.016	3.63	37.5	15.6	Nonuniform	242
46	38.6	184	383	1.149	2.0	49.5	25.9	Uniform	256
47	38	435	985	1.280	1.7	54.5	22.1	Uniform	165
48	37	420	971	1.2	1.04*	56	18.4	Nonuniform	248
49	22.5	100*	960*	1.05*	9.5*	55.2	8.9	Uniform	500
50	24.5	140*	1,050*	1.08*	5.0*	52.5	9.7	Uniform	443
51	29.5	480	3,150	1.27	1.3	67.3	10.1	Uniform	811
52	35	600	3,265	1.30	0.70	15.8	20.0	Uniform	660
53	35	600	3,265	1.30	0.70	20.8	16.2	Uniform	660
54	35	600	3,265	1.30	0.70	18.8	21.4	Uniform	680
55	35	600	3,265	1.30	0.70	70.8	17.0	Uniform	382
56	43	200	715	1.15	0.95	38.1	18.3	Irregular	411
57	46.4	270	1,125	1.18	0.47	26.7	17.9	Nonuniform	460

Table 4.—Additional Reservoir and Production Data for API Study Reservoirs (Continued)

Study No.	Residual-oil gravity, °API at 60°F	Dissolved gas liberated at field trap pressure, ft³/bbl	Saturation pressure at reservoir temp., psi gauge	Formation-vol. factor, initial reservoir bbl/stock-tank bbl	Reservoir oil viscosity at initial conditions, cp	Depletion of ultimate production to date, %	Average well spacing, acres per well	Uniformity of spacing	Estimated ultimate recovery, bbl/acre-ft
58	38.2	565	2,250	1.30	0.58	25.1	18.1	Uniform	661
59	38.2	540	2,250	1.28	0.58	63.1	21.2	Uniform	600
60	39.3	365	740	1.27	0.96	42.1	4.9	Irregular	949
61	42.6	750	3,060	1.37	0.41*	37.6	20.4	Uniform	387
62	42.6	735	3,060	1.39	0.41*	73.3	26.1	Uniform	387
63	38.8	855	4,225	1.45	0.48*	57.5	18.3	Uniform	530
64	23.5	245	1,980	1.09	4.77*	65.4	23.6	Uniform	472
65	38.6	550	2,725	1.35	0.57*	61.4	20.4	Uniform	690
66	32.4	520	2,528	1.28	0.51	68.1	21.6	Uniform	500
67	44.2	800	2,400	1.47	0.45	68.2	13.9	Uniform	343
68	21.5	175	1,125	1.04	10.1*	82.3	31.2	Uniform	357
69	40	100	400	1.10	0.84*	100	2.8	Uniform	285
70	22.5	100	820	1.10	7.8*	52.9	10.8	Uniform	453
71	33.1	365	1,800	1.19	1.03*	28.5	17.4	Irregular	674
72	33.1	365	1,800	1.19	1.03*	58.5	14.4	Irregular	592
73	22	125	1,000	1.12	8.0*	76.4	9.8	Uniform	505
74	28.5	260	1,550	1.16	1.37	39	17.0	Irregular	920
75	22	210	1,804	1.10	4.2	50.5	13.6	Irregular	552
76	19.5	225	1,985	1.11	13.2*	50.4	32.5	Irregular	540
77	23	260	2,253	1.12	4.67*	53.7	16.4	Irregular	540
78	22	20	140	1.05	6.6*	48	8.8	Irregular	430
79	25.5	330	2,430	1.16	1.4	26.6	18.3	Uniform	682
80	24.4	160	1,450	1.08	4.0*	63.8	18.7	Uniform	560
81	25.4	340	2,490	1.16	2.78*	25.2	19.7	Uniform	682
82	25.4	230	2,000	1.10	3.11*	25	14.8	Uniform	500

83	35.7	550	2,725	1.35	0.72*	29.8	20.4	Uniform	625
84	35.7	550	2,725	1.35	0.72*	12.6	20.3	Uniform	771
85	34.8	300	1,250	1.25	0.99*	42.7	7.7	Irregular	683
86	33.1	40	—	1.03	1.62*	51.9	5.9	Irregular	225
87	29.5	440	2,705	1.21	1.5	10.9	19.9	Uniform	595
88	37.9	800	4,400	1.43	0.45	77	37	Irregular	815
89	37	900	4,400	1.43	0.45	86	34	Irregular	1,030
90	43.5	800	3,040	1.33	0.52*	94	23	Irregular	963
91	40.2	800	3,200	1.33	0.54*	99	14	Irregular	1,165
92	36	—	—	1.37	0.67	74	8.3	Irregular	563
93	36	—	—	1.44	0.60	75	10.5	Irregular	410
94	36.5	—	—	1.50	0.56	80	6.4	Irregular	126
95	36	—	—	1.54	0.50	89	4.6	Irregular	180
96	30.5	—	—	1.17	2.0	85	3.5	Irregular	743
97	34	—	—	1.21	1.1	87	2.6	Irregular	688
98	33.5	—	—	1.25	0.95	86	7.0	Irregular	445
99	33	—	—	1.30	0.90	86	5.0	Irregular	443
100	33	—	—	1.38	0.79	89	6.0	Irregular	307
101	16.5	—	—	1.05	158	60	6.5	Irregular	540
102	31	—	—	1.24	1.42	83	7.0	Uniform	402
103	32	—	—	1.35	1.08	80	5.0	Irregular	412
104	—	8	199	1.04	5.2	—	23.8	Uniform	429
105	—	8	100	1.02	3.6	—	13.1	Uniform	430
106	—	280	660	1.18	2.0	—	9.6	Uniform	428
107	—	320	740	1.18	2.0	—	9.6	Uniform	400
108	—	8	100	1.02	4.4	—	10.0	Uniform	482

* Estimated.

† Produced gas and water return begun mid-1945.

‡ Estimated normal recovery (field being repressured).

§ Small gas-return project operating.

|| Estimated normal recovery (field being repressured).

affected the recovery more than the well spacing. In fact, if the data be segregated and plotted separately vs. the well spacing for the different viscosity groups, the data for the two lowest viscosity classes show[1] trends of increasing ultimate gas saturations with increasing well spacing, which is hardly compatible with general physical considerations. Moreover the inclusion in Fig. 14.5 of data for fields which have undoubtedly enjoyed substantial recoveries due to gravity drainage or gas-cap expansion with

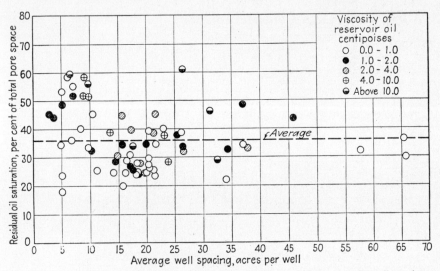

Fig. 14.6.   The relation between the residual-oil saturation at depletion vs. the well spacing, as observed or estimated in water-drive sandstone fields.   (*After Craze and Buckley, API Drilling and Production Practice, 1945.*)

those which have been largely depleted by pure solution-gas drive imposes an additional difficulty in detecting an effect of well spacing, if this should exist. In any case it is clear that the significance of Fig. 14.5 is basically negative and that it serves only to show that the data on gas-drive fields now available do not suffice to exhibit any positive effect of the well spacing on the economic ultimate recovery.

The 74 fields listed in Tables 3 and 4 that have produced largely by water-drive action would appear to provide a better basis for statistical interpretation than the gas-drive reservoirs. Yet it is to be noted that they, too, form a very heterogeneous ensemble. The porosities of the producing formations ranged from 13.4 to 35 per cent, and the permeabilities from 40 to 5,000 md, with connate-water saturation limits of 10 and 42.5 per cent. The produced crude gravities varied from 16.5 to 49°API, and their initial reservoir viscosities from 0.40 to 158 cp. The closest well spacing was 3.5 acres per well, and the widest was 65.6 acres per well.

[1] Cf. R. W. French, *API Drilling and Production Practice*, 1945, p. 155.

Analogous to the use of the ultimate free-gas saturation in the case of gas-drive fields, the recovery index for the water-drive mechanism is most appropriately expressed by the average residual-oil saturation, as a fraction of the pore space. Whereas the estimated recoveries range from 200 to 1,165 bbl/acre-ft, the residual-oil saturations vary from 60.9 to 17.9 per cent. The plot of all the API data on the residual-oil saturation vs. the well spacing for sandstone fields, grouped according to their viscosities, is given in Fig. 14.6.

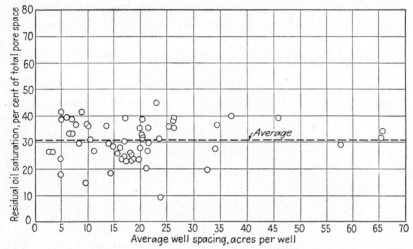

FIG. 14.7. The relation between the well spacing and the observed or estimated residual-oil saturations in depleted water-drive sandstone reservoirs, corrected to equivalent values of the reservoir oil viscosity (1.05 cp), fractional pressure decline at depletion (0.15), and permeability (700 md). (*After Craze and Buckley, API Drilling and Production Practice, 1945.*)

The large spread in the residual-oil saturations, plotted in Fig. 14.6, also here must reflect either major errors in the data or the effect of factors other than the well spacing, or the resultant of both. By replotting the data vs. the reservoir viscosities, and then in sequence the deviations from the average trends against the fractional pressure decline during the producing history and against the formation permeability, it is possible to apply corrections for these factors. The resulting corrected residual-oil saturations, when plotted again against the well spacing, give Fig. 14.7. The intermediate deviation plots appear to indicate that the residual-oil saturation increases with increasing reservoir oil viscosity, with increasing fractional pressure decline, and with decreasing permeability. While these trends are a priori reasonable, it is questionable that they can be considered as "proved" from a strict statistical point of view. The apparent lack of dependence of the corrected residual-oil saturations on the well spacing, as indicated by Fig. 14.7, must therefore here, too, be interpreted as having mainly a negative significance.

Although these data do not suffice uniquely to segregate the effects of the various factors that could influence the ultimate recovery, the very negative character of their implications is in itself of importance. Regardless of the nature of the statistical treatment to which they may be subjected, it is clear that they do not give any evidence for a real variation of the ultimate recovery with the well spacing. They do not prove that such a variation is actually nonexistent. But if any accuracy whatever is attributed to the data as a whole, they do show that the resultant effect of other factors, such as the oil viscosity, formation permeability, and completeness of the water-drive action, may be considerably greater than that of the well spacing.

It may be recalled that the theoretical considerations discussed in the previous sections, although admittedly incomplete and approximate, indicate that the economic ultimate recoveries in gas-drive fields may decrease somewhat with increasing well spacing and that no significant variation is to be expected in complete-water-drive fields, except for bottom-water drives in substantially isotropic formations. At the same time, however, this implied that the physical properties of the reservoir and its fluids and the economic factors would in general play a greater role than the well density alone. This apparently is also the implication of the recovery data as actually observed.

14.8. Interference Phenomena—Transient Effects.—As noted in Sec. 14.3, interference effects between producing wells and withdrawal areas within single reservoirs reflect fluid intercommunication and fluid migration and hence have a bearing on the well-spacing problem from a qualitative point of view. There are many aspects to the general subject of interference phenomena. These include gross regional reservoir interactions, fluid migration between parts of single reservoirs, localized observations of pressure or fluid depletion (or lack of such depletion) associated with new drilling, and field experiments deliberately performed to detect and study well interference. Finally, there is the type of interference that is concerned mainly with the geometrical properties of fluid streamlines as they may be affected by the distance between the localized foci of withdrawals. None of these phases of the subject can be quantitatively interpreted in relation to specific implications regarding a possible variation of oil recovery with the well spacing. However, they do throw light on the gross behavior of fluids in porous media, which underlies the whole subject of oil-reservoir performance.

Mention was also made in Sec. 14.3 of the pressure decline observed at East Hawkins, in Texas, due to fluid withdrawals from the Woodbine Sand through the East Texas field, and that at Village, in Arkansas, resulting from the Smackover Lime production at the near-by Magnolia field. These do not imply in themselves that a gross liquid depletion has occurred at these large distances from the primary foci of fluid withdrawals. But to the extent that density declines or gas evolution must have followed the fall in the pressure, the corresponding excess volumetric fluid expansion must necessarily have been removed from the area of pressure decline. Its magnitude could be calculated by mutliplying the effective compressibility of the fluids by their gross volume and by the amount of pressure reduction.

Gross regional migration within an oil reservoir due to delayed development has often been observed. The Oklahoma City Wilcox reservoir,[1] Oklahoma County, Okla., provides an outstanding example of such a situation. The southern part of the field, below line AA in Fig. 14.8, was developed in 1930–1933. The northward extension of the field was discovered in 1935, and its development was completed in 1936. Whereas the original pressure in the southern area was 2,686 psi, at − 5,260 ft, one of the first completions (Nov. 10, 1935) in the north extension had an initial bottom-hole pressure of 575 psi. It has been calculated that at least 60 million barrels of oil have migrated across AA to the rapidly depleted area to the south, as the result of the regional pressure gradient from north to south. Most of this oil has been apparently

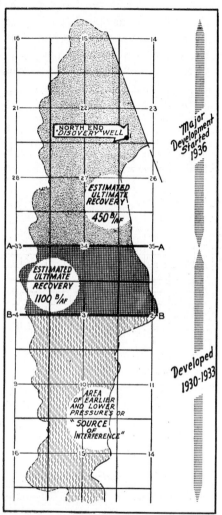

Fig. 14.8.   A diagrammatic sketch of the Oklahoma City field, Oklahoma, showing the development periods and estimated ultimate recoveries. (*After Barnes, Oil and Gas Jour., 1946.*)

withdrawn from the shaded area in Fig. 14.8 between AA and BB. In fact the indicated average recovery in this area is more than 90 per cent of the

[1] Figure 14.8 and some of the above data are taken from K. B. Barnes, *Oil and Gas Jour.*, **45**, 120 (May 18, 1946); cf. also Vietti, Mullane, Thornton, and Van Everdingen, *loc. cit.*

estimated original oil in place, and the recoveries from several quarter sections south of AA have even exceeded the original oil in place by more than 50 per cent. There can be no doubt, that, in this reservoir, depletion of pressure in the northern area has been accompanied by oil depletion and movement. Evidently efficient oil drainage would have been possible in this field with rather wide spacing.

The early development of the Glenn pool, Creek County, Okla., illus-

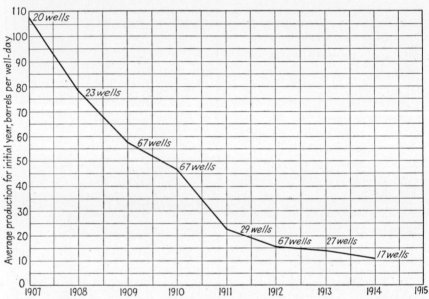

FIG. 14.9.   The history of the first year's average production rates of wells in the Glenn pool, Oklahoma, in successive years of development of the field.   (*After Lewis and Beal, AIME Trans., 1918.*)

trates the systematic depletion in undrilled areas due to previous fluid withdrawals.   In Fig. 14.9[1] are plotted the first year's average producing rates of wells drilled in successive years between 1907 and 1914.   As this field was not subject to proration restrictions, the plotted production rates represent the maximum producing capacities.   Whether the new drilling represented largely "infill"[2] or extension drilling, it is clear that the producing formation was being substantially depleted in the undrilled areas before the latter were completely developed.   As the ultimate recoveries in many of the early-developed shallow Oklahoma fields showed a general increasing trend with the first year's average production rates,[3] Fig. 14.9

[1] This figure is taken from J. O. Lewis and C. H. Beal, *AIME Trans.*, **59**, 492 (1918).

[2] The term "infill" refers to a well location surrounded by wells that are already in production.

[3] Cf. R. Arnold and J. L. Darnell, "Manual for the Oil and Gas Industry," p. 88, John Wiley & Sons, Inc., 1920.

implies that well recoveries at the Glenn pool decreased as the wells were drilled later in its life. It would, therefore, appear that the closer spacing resulting from much of the later drilling was not necessary to achieve effective oil depletion and drainage throughout the reservoir.

The ability of a single well to drain a large area in a permeable formation is demonstrated by the performance of the discovery well in the water-

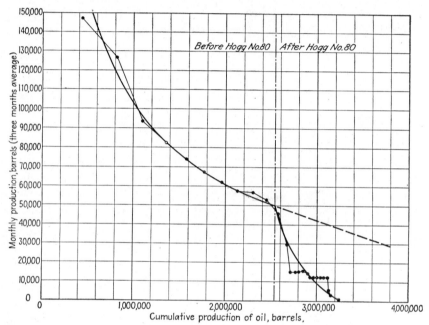

Fig. 14.10. The production history of W. C. Hogg No. 68 well, producing from the "B" Sand in the north side of the West Columbia field, Texas, and its reaction to the completion of Hogg No. 80. (*After Miller, AAPG Bull., 1942.*)

drive Arbuckle Dolomite reservoir of the Ploog pool,[1] Rice County, Kans. Further development of the field did not begin until 30 months after completion of the first well. As the result of this 30-month period of production without interference from other producers it ultimately developed a total recovery of 450,000 bbl, or 31 per cent of the total for the field. The average recovery of each of the remaining nine wells was only 111,000 bbl.*

An outstanding example of the direct interference in the performance of a well by an offset is shown in Fig. 14.10,[2] in which is plotted the production rate, averaged over 3-month intervals, vs. the cumulative production of

[1] Cf. Barnes, *loc. cit.*, from which the data below, on the Dill pool, are also taken.

* Similar and even more striking observations have been reported in the Sinclair-Moren pool, Young County, Tex. [cf. M. G. Cheney, *Petroleum Technology,* **3,** 1 (November, 1940)].

[2] Cf. J. C. Miller, *AAPG Bull.,* **26,** 1441 (1942).

the W. C. Hogg No. 68 well, completed on Jan. 21, 1922, in the "B" Sand in the north side of the West Columbia field, Brazoria County, Tex. This well had produced 2,540,000 bbl with a regular decline in production rate by Sept. 1, 1924, when the offset well, Hogg No. 80, was completed 440 ft updip. As will be seen from Fig. 14.10 the production rate of Hogg No. 68 immediately thereafter dropped sharply. The water-oil ratio of Hogg No. 68 also rose sharply after the completion of No. 80, although the water-oil ratio of Hogg No. 68 and 80 combined continued to follow the previous trend for No. 68. Number 68 was finally abandoned in July, 1933, after a cumulative recovery of 3,243,000 bbl. By extrapolation of the earlier trend, as indicated by the dashed curve, it seems highly probable that if it were not for the interference from Hogg No. 80, the ultimate recovery of Hogg No. 68 would have been at least 2,000,000 bbl greater.

By its very nature, infill drilling represents an attempt either to recover oil that presumably would not be produced by the offset wells or to prevent oil migration to other wells or other leases and operators. If the infill well is drilled considerably later than the surrounding wells and is found to have as high a producing capacity and pressure as did the offsets originally, the lack of drainage by the latter is evidently to be inferred. If, conversely, however, the infill characteristics are comparable with those of the offsets at the time of completion of the infill well, it is clear that drainage to the former has been taking place and that the infill was unnecessary with respect to the over-all recovery from the reservoir. Both types of results have often been observed, ranging from the extreme cases of no interference whatever to those where the formation at the location of the infill well was apparently being depleted just as completely as the areas immediately around the previously drilled offsets. Thus not infrequently a new well in a tight formation, drilled after the field as a whole has been largely depleted and the remaining producers have been put on artificial lift, may be completed as a flowing well with a production rate as great as all the rest of the field. Nor is it uncommon to find an infill well come in merely at the same level of production capacity and pressure as its offsetting producers at the time of its completion, as was observed in the Hunton Lime reservoir of the Dill pool, Okfuskee County, Okla. When a previously undrilled 40-acre location in this reservoir was drilled in 1941, its initial production rate of 101 bbl/day and pressure of 113 psi, as well as its gas-oil ratio, were substantially the same as its offsets, which had been producing since 1936. The initial average pressure and production rates in the field had been 1,725 psi and 1,500 bbl/day. Moreover, whereas the earlier drilled nearest offsets had an estimated average ultimate recovery of 166,000 bbl, that of the infill well appeared to be no greater than 44,000 bbl.

There can be little doubt that the failure of infill or field extension wells in gas-drive reservoirs to give production rates and recoveries as high as those for wells completed early in the life of a field actually represents a depletion of the oil content by drainage to the neighboring producers. Unless the oil should remain supersaturated, a decline in pressure must of necessity lead to gas evolution and oil displacement. There is no reason to expect the oil-expulsion mechanism associated with the gas evolution to be basically different at an undrilled location from that in the area about a producing well. While the time required for the pressure reaction to develop at distant points from a producing well cannot be predicted quantitatively, the physical processes associated with the pressure decline, when and if this occurs, will be substantially the same regardless of the location of the rock with respect to the ultimate point of fluid escape from the reservoir. It is true that the low recoveries from wells completed with low pressures and initial production capacities are due to these very conditions. However, these initial states are to be considered, not as arbitrarily chosen independent parameters, but as representing intermediate stages of oil and gas depletion developed by the same oil-expulsion processes determining the performance of the reservoir as a whole.

In contrast to the above-discussed types of interference phenomena, which are generally observed during the course of the normal development and operation of a field, are the tests deliberately made to detect and measure the interaction between wells through the medium of their common reservoir. In experiments of this kind the time scale is greatly reduced as compared with that involved in observing general depletion interference effects. It is to be expected, therefore, that, while positive interference will imply rapid well-to-well intercommunication and good permeability, the failure to observe interference in a short-period test will not necessarily indicate the lack of long-term communication.

Many variations in procedure can be followed in making interference tests. Among those most commonly used is that in which pressure observations are made on a group of shut-in wells surrounding a central well whose production rate is varied, and the converse procedure in which the observations are made at the central well, which is kept shut in while the surrounding wells are produced.

The type of result observed when the composite effect of a ring of producers on a central shut-in well is measured is illustrated by tests[1] made at Holly Ridge, Tensas Parish, La., producing from the Basal Tuscaloosa Sand at 8,400 ft. The formation is 25 ft thick and has a permeability of

---

[1] Cf. Barnes, *op. cit.*, pp. 120, 157. In this and previous papers by the author in the Apr. 13, Apr. 20, and May 4, 1946, issues the results of the Carmi as well as many other interference experiments are reported.

35 md and porosity of 20 per cent. The oil, undersaturated at the time of the tests, had a viscosity of 0.9 cp and formation-volume factor of 1.54. After shutting in for 30 hr the eight wells, on a 40-acre spacing (well separation of 1,320 ft) surrounding the central well, they were put on production at 250 bbl/day each. The bottom-hole pressure of the central shut-in well dropped from an initial value of 3,700 psi to 3,690 after 20 hr. After 69 hr it had declined to 3,670 psi. Extrapolation of these data indicated that the pressure drop would have been more than 300 psi if the test had been continued for 30 days.

A similar experiment at the Carmi pool, Pratt County, Kans., producing from the Arbuckle Dolomite, at 4,275 ft, gave a pressure drop of 15 psi at the central well of a nine-spot, when the 8 40-acre offsets were pumped at a rate of 5 bbl/hr. The pressure fell 10 psi more when the pumping rate of the surrounding wells was increased to 8 bbl/hr. A further increase to 9 bbl/hr resulted in an additional pressure decline at the central well of 3 psi. While tests of this kind will give greater interference effects than the reverse type, they will represent the integrated resultant of the various producers and their separate contributions cannot be determined without additional special testing.

The results of interference experiments in which measurements were made of the pressure decline in shut-in offsets resulting from the production of a central well are plotted in Figs. 14.11 and 14.12.[1] These were performed in the Silica pool, Rice and Barton Counties, Kans., producing a highly undersaturated oil from the Arbuckle Dolomite. It will be seen from Fig. 14.11 that, whereas the fluid heads in six of the shut-in wells began to decline within a few hours after the central well began pumping, no decline was observed in four of the test wells.[2] The map in Fig. 14.11 shows that the shut-in wells giving no pressure reaction were all northeast of the producer. Evidently there is a much lower degree of fluid intercommunication in the direction northeast of the producer than in other directions. This type of observation is a particularly valuable feature of interference tests in which the central well rather than its offsets is used as the producer. And it may be noted that independent evidence indicated that there is a restriction to fluid movement in the Silica pool between the area to the right of the line through wells 5 and 6 and that to the left of it.

The test illustrated by Fig. 14.12 is of interest in showing that not all short-period interference tests give positive effects even if the reservoir oil is undersaturated. While the average distance of the offsets from the

---

[1] These are reproduced from L. F. Elkins, *Oil and Gas Jour.*, **45**, 201 (Nov. 16, 1946).

[2] The increases in the rates of the rise of these wells after 36 hr were apparently due to shutting in a producer, at 32 hr, east of the test group.

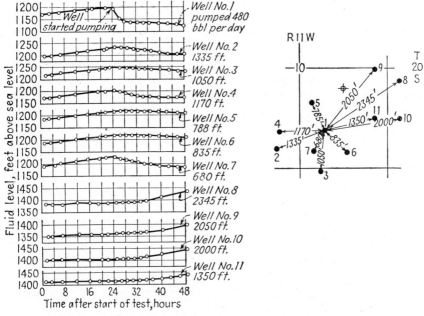

Fig. 14.11.   The plan and results of a well interference test in the Silica pool, Kansas, showing directional interference effects.   (*After Elkins, Oil and Gas Jour., 1946.*)

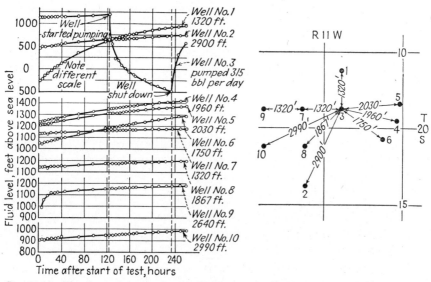

Fig. 14.12.   The plan and results of a well interference test in the Silica pool, Kansas, showing no interference effects.   (*After Elkins, Oil and Gas Jour., 1946.*)

central producers was in this experiment larger than in that of Fig. 14.11, the duration of the test was considerably longer.   Yet in not a single offset shut-in well was any change observed in the trend of pressure rise measured before the central well began producing.

The results of an interesting interference experiment involving several changes in flowing conditions of offset wells are plotted in Fig. 14.13.[1] These were made on water-supply wells of the Scott Street pumping plant of

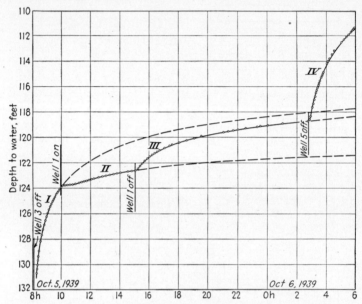

FIG. 14.13.   The results of interference tests on wells of the Scott Street pumping plant in Houston, Tex. Water-level data were taken in Well No. 3.   (*After Jacob, AGU Trans., 1941.*)

Houston, Tex.   Well #3, in which the fluid-head measurements were made, was producing from a group of sands between 553 to 919 ft.   At the beginning of the test, Well #3, which had been producing at a rate of 2,100 gal/min, was shut down.   Its subsequent build-up is given by segment I of Fig. 14.13.   Then, 1 hr 55 min later Well #1, 790 ft southeast of Well #3, was put on the pump at a rate of 250 gal/min.   The retardation in the build-up is plotted as segment II.   After pumping for 5 hr 5 min Well #1 was shut down, resulting in the resumed build-up given by III.   After another period of 11 hr 45 min Well #5, 605 ft southeast of #3 and which had been producing 1,770 gal/min, was shut down.   Again, almost immediately the fluid level in Well #3 began a sharp rise, as plotted in segment IV.   While no attempt will be made here to analyze these data in

[1] Figure 14.13 is taken from C. E. Jacob, *AGU Trans.*, pt. III, p. 744 (1941).

detail,[1] it is clear that the tests prove the existence of most effective inter-communication between the wells involved in the experiments.

If the fluid movement associated with the interference phenomena is of the homogeneous-fluid type, as will obtain in the case of undersaturated liquids, and if the formation is uniform, the compressible-liquid-theory analysis of Chap. 11 may be used in their interpretation. In particular, in the case of direct well-interference tests the effects due to changes in fluid withdrawal at individual wells could be computed by applying Eq. 11.7(3) in the form

$$\Delta p = -\frac{\mu Q\beta}{4\pi k} Ei\left(-\frac{r^2}{4at}\right) = -\frac{70.60\mu Q\beta}{k} Ei\left(-\frac{r^2}{4at}\right); \qquad a = \frac{k}{f\kappa\mu}, \qquad (1)$$

where Δp is the pressure decline (drawdown) at the time t and distance r from a well that has been producing since $t = 0$ at a constant surface rate Q, per unit formation thickness. μ is the liquid viscosity, β its formation-volume factor, and κ its compressibility. k and f are the permeability and porosity of the stratum. The second expression in Eq. (1) gives Δp in psi if k is expressed in millidarcys and Q in barrels per day per foot of pay, and the dimensionless argument of the Ei function is expressed in any consistent units. For large values of t or small values of r the asymptotic expansion of the Ei function leads to a limiting form for the pressure drop given by

$$\Delta p(\text{psi}) = \frac{70.60\mu Q\beta}{k}\left(\log\frac{4at}{r^2} - 0.5772\right), \qquad (2)$$

thus implying a logarithmic increase with time. On the other hand, at large distances from the well or at small values of t, Eq. (1) will approach asymptotically the form

$$\Delta p(\text{psi}) = 1.787\frac{Q\beta te^{-r^2/4at}}{r^2 f\kappa}\left[1 - 0\left(\frac{at}{r^2}\right)\right], \qquad (3)$$

where Q is again expressed as barrels per day per foot, t in days, κ in psi^{-1}, and r in feet.[2] The pressure reaction will therefore decrease very rapidly with increasing distance from the withdrawal well.

To apply Eq. (1) in the interpretation of interference data, assumptions are made for the effective average values of k/μ and a in the region between the producing and observation well. The quantity $4\pi k\,\Delta p/\mu Q\beta$ is then plotted vs. $r^2/4at$, or more conveniently vs. $4at/r^2$. If the curve so obtained follows the functional variation of the Ei function, as required by Eq. (1), the assumed values of $k/\mu\beta$ and a may be considered as correct.

[1] Such an analysis of the data, similar to that discussed below, is given by Jacob (*ibid.*).

[2] These refer to the coefficients of the exponential. In the latter the units must be consistent with the dimensionless character of the argument.

If not, other values are assumed until the curve plotted from the basic Δp vs. t data does fall on the Ei-function curve.

If the reservoir and fluid properties are strictly uniform throughout the area of the tests, the same values of $k/\mu\beta$ and a will give a fit with the Ei curve for the different wells in the test group, correction being made, of course, for the appropriate values of r. Moreover, the drawdown data

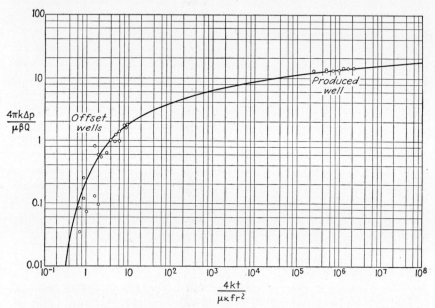

Fig. 14.14. A plot of well interference-test data obtained in the East Texas field. k = permeability; μ = oil viscosity; β = formation-volume factor of oil; κ = oil compressibility; f = porosity; Q = production rate per unit thickness; Δp = pressure drawdown; t = time. Solid curve represents Ei-function variation. (*After Elkins, Oil and Gas Jour., 1946.*)

at the producing well, using for r the well radius, should also fall on the same curve as those for the distant wells. If the data for the different wells are not adjustable to the Ei curve with the same values of $k/\mu\beta$ and a, a nonuniformity in the reservoir is to be inferred.

An example of an almost exact correlation of interference-test data with Eq. (1) is shown in Fig. 14.14.[1] These data were obtained in the East Texas field, where the undersaturated character of the oil and local uniformity of the producing formation were especially favorable for the application of Eq. (1). The curve through the data represents the Ei function, the constants assumed in plotting the ordinate and abscissa values being

[1] Figure 14.14, as well as Fig. 14.15, is taken from L. F. Elkins, *Oil and Gas Jour.*, **45**, 201 (Nov. 16, 1946).

$kh/\mu = 150$ and $k/\mu\kappa f = 2.48 \times 10^6$, where k is expressed in darcys, h in feet, and κ in psi^{-1} and the time in days. Of particular interest is that the data to the 'right of the graph representing the pressure drawdown in the produced well fall on the same curve as those for the observation shut-in offset wells. Moreover the latter were mutually consistent, within experimental errors, and did not require arbitrary adjustments of the basic physical constants.

A quite different result is obtained by plotting the interference-test data

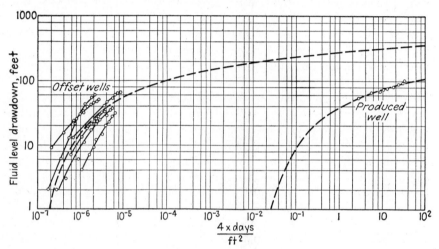

Fig. 14.15. A plot of the well interference-test data of Fig. 14.11 taken in the Silica pool. Dashed curves show Ei-function variations. (*After Elkins, Oil and Gas Jour., 1946.*)

of Fig. 14.11 in the Silica pool, as shown in Fig. 114.5. Here the observed fluid-level drawdowns are plotted directly as the ordinates, and the abscissas represent the variable factor in the argument of the Ei function. It will be seen that the data for the shut-in offset and producing wells fall on widely separated Ei curves. Although it is possible to calculate apparent values of the physical constants from any pair of drawdown observations, assuming the functional relationship of Eq. (1), the validity of the latter presupposes that the physical properties of the reservoir and its fluid are uniform throughout. Any synthesis of resultant effects due to composite systems in "parallel" or "series" will give much more complex relationships, except in special extreme cases. The apparent grouping of the data about separate Ei functions does not in itself imply a simple structure of the reservoir rock or of the mechanism of fluid intercommunication. On the other hand the nature of the spread in the data, as shown by Fig. 14.15, definitely indicates the existence of a much more effective means of fluid intercommunication between the wells than corresponds to the rock and fluid properties immediately about the producing well. The interference reactions at the offset wells are certainly higher or more rapidly developed than would be expected from the behavior of the producing well. This is suggestive of the presence of thin high-permeability streaks in the productive section, although such an interpretation undoubtedly represents a great oversimplification from a quantitative point of view. In any case, the comparative pressure interferences observed at different wells in a group must qualitatively represent measures of the relative fluid-transmitting capacities of the formation lying directly between the corresponding well pairs.

While interference tests, under favorable circumstances, can give valuable qualitative information regarding the continuity and uniformity of reservoir formations, it is important not to overemphasize negative results. If there has been appreciable gas evolution within the pay in the region between the test wells, the effective fluid compressibility may exceed that for undersaturated oils by a factor of 10 or greater. The fluid permeabilities will also be considerably lower. Although Eq. (1) will be inherently inapplicable for quantitatively describing the pressure reactions, the factors determining the scale of the time transients in multiphase-flow systems will undoubtedly be the same as indicated by Eq. (1). Accordingly it may be anticipated that for comparable homogeneous-fluid permeabilities the time required for the development of observable pressure reactions in gas-drive systems may exceed that required for undersaturated oil reservoirs by fiftyfold or even more. The apparent failure to observe interference reactions in tests of only a few days' duration therefore is not to be construed as evidence of the complete absence of fluid intercommunication. On the contrary, positive interference reactions in gas-drive fields with moderate spacings will generally indicate the presence of channels of extremely high permeability, such as an interconnected fracture system in a limestone or dolomite formation, rather than representing the expected result in a uniform-permeability reservoir. In any case, when the existence of the rapid intercommunication is established, its implications with respect to the well-spacing problem must be evaluated in the light of the economic factors discussed previously.

14.9. Interference Phenomena—Steady-state Geometrical Effects; Small Well Groups.—In contrast to the transient types of interference phenomena discussed qualitatively in the preceding section with respect to the general problem of reservoir drainage and recovery the purely geometrical interference effects can be treated quantitatively. Of course, such treatment will involve the restrictive assumptions of steady-state homogeneous-fluid flow. However, since only the geometrical aspects of the subject will be considered, these assumptions should not seriously affect the validity of the results.

The specific problem to be discussed in this section is that of the mutual effect of the various wells in a small group on their steady-state production (or injection) capacity. Evidently, if there were no interference, the combined capacity of a group of wells would be simply the total number times the capacity of a single well if it alone were producing from the reservoir. Lesser values will thus reflect the action of mutual geometrical interference. If the driving force on the group of wells be considered as applied over a circular boundary of radius R, large as compared with the mutual well separations r_{ij}, and with a center lying in the general vicinity of the wells (cf. Fig. 14.16), it may be shown that the

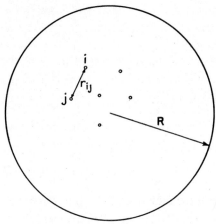

FIG. 14.16. A diagrammatic representation of a small group of wells in a circular area producing by steady-state flow.

average pressure over the surface of the jth well, of radius r_j, may be expressed as

$$p_j = p_e + \frac{\mu\beta}{2\pi k h}\left(Q_j \log \frac{r_j}{R} + \Sigma' Q_i \log \frac{r_{ij}}{R}\right), \qquad (1)$$

where p_e is the average driving pressure, Q_j the surface withdrawal rate of the jth well, k, h, μ, β have their usual significance, and the prime denotes the omission of the term $i = j$.

For the trivial case of a single well, Eq. (1) evidently reduces to the basic radial-flow formula [cf. Eq. 5.1(6)],

$$Q_o = \frac{2\pi k h(p_e - p_w)}{\mu\beta \log R/r_w}, \qquad (2)$$

where r_w is the well radius r_1.

Upon applying Eq. (1) in sequence to the two wells of a two-well pair, of separation d_2 and common well radius r_w and well pressure p_w (cf. Fig. 14.17), it is readily found that

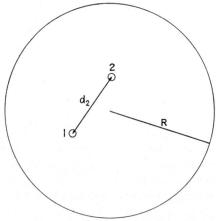

FIG. 14.17. A well pair producing from a large circular area.

$$Q_1 = Q_2 = \frac{2\pi k h(p_e - p_w)}{\mu\beta \log R^2/d_2 r_w}. \qquad (3)$$

The interference effect is expressed by the observations that the total flux capacity from the two wells has a ratio to that of two noninterfering wells of

$$\frac{Q_1 + Q_2}{2Q_o} = \frac{\log R/r_w}{\log R^2/r_w d_2} = 1 - \frac{\log R/d_2}{\log R^2/r_w d_2}, \tag{4}$$

and that each well of the pair has an effective radius $r_w d_2/R$. For example, if R be taken as 5,000 ft and $d_2 = 200$ ft, $r_w = \frac{1}{4}$ ft, each well of the pair will have a flux capacity only 75.5 per cent as great as the equivalent isolated wells. The apparent radius is only 0.01 ft, as compared with its assumed value of $\frac{1}{4}$ ft.

Three wells in an equilateral triangular pattern, of mutual separation d_3, may be similarly shown to have individual well capacities given by

$$Q_1 = Q_2 = Q_3 = \frac{2\pi kh(p_e - p_w)}{\mu\beta \log R^3/r_w d_3^2}, \tag{5}$$

and for four wells in a square pattern of side d_4

$$Q_1 = Q_2 = Q_3 = Q_4 = \frac{2\pi kh(p_e - p_w)}{\mu\beta \log R^4/\sqrt{2} r_w d_4^3}. \tag{6}$$

In a five-spot arrangement (cf. Fig. 14.18), the four outside wells not only show a mutual interference but in addition shield the central well. An application of Eq. (1) leads to the results

$$\left. \begin{aligned} Q_1 = Q_2 = Q_3 = Q_4 &= \frac{2\pi kh(p_e - p_w)}{\mu\beta\Delta} \log \frac{d}{\sqrt{2} r_w^*}, \\ Q_5 &= \frac{2\pi kh(p_e - p_w) \log d/4\sqrt{2} r_w}{\mu\beta\Delta}, \end{aligned} \right\} \tag{7}$$

where

$$\Delta = 4 \log \frac{\sqrt{2} R}{d} \log \frac{d}{\sqrt{2} r_w} + \log \frac{R}{r_w} \log \frac{d}{4\sqrt{2} r_w},$$

so that

$$\frac{Q_5}{Q_1} = 1 - \frac{\log 4}{\log d/\sqrt{2} r_w}, \tag{8}$$

which has the value 0.78 for $d/r_w = 800$. The total flux capacity of the five wells is only 43 per cent of that of five equivalent isolated wells.

Carrying through similar applications to groups of 9 wells in a nine-spot and 16 wells in a square network, and comparing the average well capacities $\overline{Q}_n$ for the various groups containing n wells, one obtains the ratios

$$\overline{Q}_1 \ : \ \overline{Q}_2 \ : \ \overline{Q}_3 \ : \ \overline{Q}_4 \ : \ \overline{Q}_5 \ : \ \overline{Q}_9 \ : \ \overline{Q}_{16}$$
$$= 1 \ : \ 0.755 \ : \ 0.606 \ : \ 0.515 \ : \ 0.430 \ : \ 0.309 \ : \ 0.208;$$

assuming in all cases $R = 5,000$ ft, $r_w = \frac{1}{4}$ ft, and a basic well separation of 200 ft. Thus the mutual interference of the wells in the group of 16 will

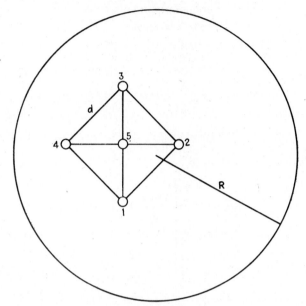

Fig. 14.18. A single five-spot group of wells producing from a large circular area.

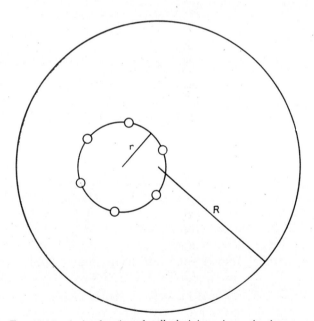

Fig. 14.19. A circular ring of wells draining a large circular area.

be such as to reduce the average individual well capacities to only 20.8 per cent of their capacities if producing alone.

If the wells are distributed uniformly over a circular ring of radius r (cf. Fig. 14.19), Eq. (1) may be shown to give for the total resultant production capacity

$$Q_n = \frac{2\pi k h (p_e - p_w)}{\mu\beta[\log R/r + (1/n) \log r/nr_w]}. \tag{9}$$

This evidently has a maximum corresponding to a single well with a radius r. If the latter be 80 ft, even the infinitely dense distribution of wells on

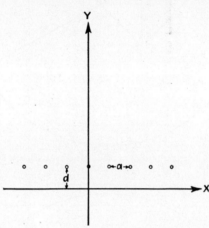

Fig. 14.20. A diagrammatic representation of a linear well array.

the ring will give a production capacity only 2.4 times that of a single well of radius $\frac{1}{4}$ ft.

The above results show the limitations in the production (or injection) capacity of groups of wells due to their interference. Effects of this type may be of importance in the development of complete-water-drive fields and in the planning of gas- or water-injection programs.[1] The achievement of efficient sweep patterns of invading or injected fluids should be, of course, the primary aim in choosing the well locations and density. However, it should be recognized that as the number of wells is increased there will in general be less than proportionate increases in the production or injection capacity of the system as a whole.

14.10. Geometrical Interference Effects in Infinite Well Arrays.—As noted in Sec. 12.2, if a group of wells is distributed on a line with uniform

[1] Such interference effects may also be of practical significance in the performance of water-supply wells producing by gravity. Sand model experiments on such systems have confirmed Eqs. (3) to (7) within the experimental accuracy [cf. H. E. Babbitt and D. H. Caldwell, *Univ. Ill. Eng. Exper. Sta. Bull.*, Series 374 (Jan. 7, 1948)].

spacing so as to form an effectively[1] infinite linear array, the steady-state pressure distribution for homogeneous-fluid flow will be given by

$$p(x,y) = \frac{\mu Q \beta}{4\pi k h} \log \left[\cosh \frac{2\pi(y-d)}{a} - \cos \frac{2\pi x}{a} \right],\qquad (1)$$

where it is assumed that the wells, of production rate Q, lie on a line at a distance d from and parallel to the x axis and with a mutual well separation

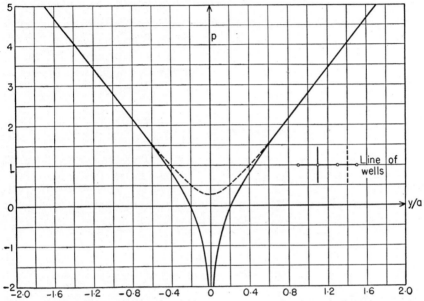

FIG. 14.21.   The calculated steady-state homogeneous-fluid pressure distribution about an infinite array of wells of spacing a, lying on the x axis, with $\mu Q\beta/4\pi kh$ in Eq. 14.10(1) taken as $\frac{1}{2}$. ——, pressure distribution along a normal to the array that passes through a well. ----, pressure distribution along a normal that passes midway between two wells.

a (cf. Fig. 14.20).   The equipressure contours (cf. Fig. 12.4) are periodic along parallels to the array, and the contours in the immediate neighborhood of the wells are circular.   The variation of the pressure normal to the array (cf. Fig. 14.21) becomes almost exactly linear within a distance equal to half the well spacing, as if the array were a continuous line sink or source of constant density $\mu Q\beta/2kha$.

If the linear well array be considered as lying immediately ahead of a linear edgewater boundary forming a line-drive system, the pressure distribution will be given by

$$p = p_e + \frac{\mu Q \beta}{4\pi k h} \log \frac{\cosh 2\pi(y-d)/a - \cos 2\pi x/a}{\cosh 2\pi(y+d)/a - \cos 2\pi x/a},\qquad (2)$$

[1] While actual arrays will never be infinite, they may be treated as such with respect to the wells appreciably removed from the ends of the array.

where p_e is the pressure at the line drive (edgewater), which is considered as lying on the x axis, and the array lies on $y = d$ (cf. Fig. 14.22). Q in Eq. (2) represents the production rate per well and is given, in terms of the pressure differential between the drive and the wells, Δp, by

$$Q = \frac{2\pi kh\,\Delta p}{\mu\beta\,\log\,(\sinh 2\pi d/a)/(\sinh \pi r_w/a)},\tag{3}$$

where r_w is the well radius.

The mutual interference effect of the wells within the linear array is obtained by comparing Eq. (3) with that for the production capacity of a single well subject to the line drive [cf. Eq. 5.2(11)], *i.e.*,

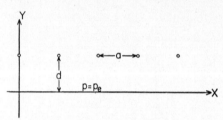

FIG. 14.22. A line array of wells parallel to a line drive at $y = 0$.

$$Q_o = \frac{2\pi kh\,\Delta p}{\mu\beta\,\log\,2d/r_w}.\tag{4}$$

For $d/a = 1$ and $d/r_w = 400$, $Q/Q_o = 0.64$, which implies that the presence of the other wells surrounding any particular one in the infinite array will cut down its production capacity to 64 per cent of its value when producing alone.

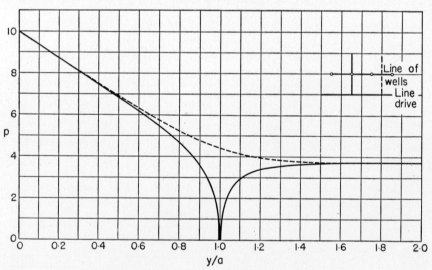

FIG. 14.23. The calculated steady-state homogeneous-fluid pressure distribution normal to an infinite array of wells, of spacing a, at $y = a$, supplied by a line drive at $y = 0$, ———, pressure distribution along a normal that passes through a well. - - - -, pressure distribution along a normal that passes midway between two wells. $\mu Q\beta/4\pi kh$ is taken as $\frac{1}{2}$ and line-drive pressure as 10.

The pressure distribution defined by Eq. (2) has a variation normal to the array as shown in Fig. 14.23. The line drive is assumed to lie on the x axis, and the distance of the well array from the drive is taken as equal to the well spacing within the array, a. As is to be expected, the pressures here are no longer symmetrical about the well array. The line drive induces a regional pressure gradient providing the flux into the wells. Beyond the line array, however, the pressure rapidly approaches constancy within a distance equal to half the well spacing.

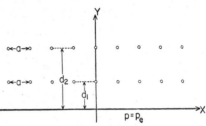

FIG. 14.24. A diagrammatic representation of two line arrays lying parallel to a line drive.

In addition to the interference between the wells within a linear array, parallel arrays show mutual shielding effects. These may be determined as follows: The pressure distribution about two line arrays, as shown in Fig. 14.24, is readily verified to be

$$p = p_e + \frac{\mu\beta}{4\pi kh}\left[Q_1 \log \frac{\cosh 2\pi(y - d_1)/a - \cos 2\pi x/a}{\cosh 2\pi(y + d_1)/a - \cos 2\pi x/a}\right.$$
$$\left. + Q_2 \log \frac{\cosh 2\pi(y - d_2)/a - \cos 2\pi x/a}{\cosh 2\pi(y + d_2)/a - \cos 2\pi x/a}\right], \quad (5)$$

where Q_1 and Q_2 are the production rates, per well, in the two arrays. To determine the geometrical interaction between the arrays the pressures at all the wells may be taken as the same. By imposing this requirement on Eq. (5) the production rates Q_1, Q_2 may be shown to have the ratio

$$\frac{Q_1}{Q_2} = \frac{\log \dfrac{\sinh \pi r_w/a \, \sinh \pi(d_2 + d_1)/a}{\sinh 2\pi d_2/a \, \sinh \pi(d_2 - d_1)/a}}{\log \dfrac{\sinh \pi r_w/a \, \sinh \pi(d_2 + d_1)/a}{\sinh 2\pi d_1/a \, \sinh \pi(d_2 - d_1)/a}}. \quad (6)$$

Since $d_2 > d_1$, $Q_1/Q_2 > 1$. That is, as is to be expected, the first line "shields" the second so that the production capacity of the latter is reduced. This shielding effect may be expressed by the quantity S, defined as

$$S = \frac{Q_1}{Q_1 + Q_2} = \frac{Q_1/Q_2}{1 + Q_1/Q_2}, \quad (7)$$

which is the fraction of the total flux into the two lines of wells produced by the first array. $1 - S$ thus represents the fraction of the total flux that leaks past the first line.

On assuming that the common well radius of all the wells is $\frac{1}{4}$ ft, in all cases, and that, in particular, the well spacing within the lines is 660 ft and the distance between the lines is also 660 ft, the variation of S with d_1, the separation between the first array and the line drive, is plotted as curve I in Fig. 14.25. It will be seen that the shielding decreases as the arrays recede from the line drive but reaches its asymptotic value of 0.671 by the time the first line is removed by a distance equal to half the common

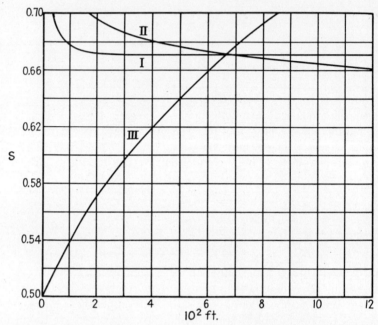

Fig. 14.25. The calculated steady-state homogeneous-fluid shielding effects between two lines of wells producing by a line drive. S = (production capacity per well in first line)/(production capacity per well pair in both lines). Curve I, well spacing and distance between lines = 660 ft; abscissa = distance of first array from line drive. Curve II, all basic distances within and between lines are equal; abscissa = common spacing value. Curve III, well spacing within lines and distance of first array from line drive = 660 ft; abscissa = separation between the two lines.

spacing. In any case the maximum leakage through the first line array is only about one-third of the total flux.

If the spacings within and between the lines are equal to each other and to the distance of the first array from the line drive, that is, $d_1 = a = d_2/2$, the effect of this common spacing on S is plotted as curve II in Fig. 14.25. Here the shielding decreases monotonically with increasing spacing, approaching the limit of $\frac{1}{2}$. The latter would represent no real interference since it corresponds to $Q_1/Q_2 = 1$. This is, of course, to be expected, as the well separations are made arbitrarily large.

If again the well spacing is kept fixed at 660 ft and the separation of the first line array from the line drive at the same value, the variation of S

with the separation between the arrays, $d_2 - d_1$, is that plotted as curve III in Fig. 14.25. The continuous rise in S as the separation $d_2 - d_1$ is increased simply represents the increasing shielding of the first line of wells as the other is moved back.

The pressure distributions normal to the arrays when the well pressures are kept at half that at the line drive ($p = 1$) are plotted for a uniformly spaced system, that is, $d_1 = a = d_2/2$, in Fig. 14.26. The lower average

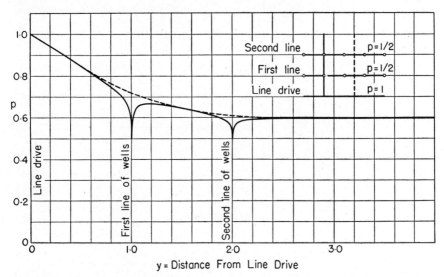

Fig. 14.26. The calculated steady-state homogeneous-fluid pressure distributions normal to two parallel well arrays supplied by a line drive. Distances between the lines and between the first array and the line drive equal the well spacing within the lines, which is taken as unity. Pressure at line drive = 1; pressure at well arrays = $\frac{1}{2}$. ——, pressure distribution along a normal passing through the wells. ----, pressure distribution along a normal that passes midway between the wells.

gradient between the two lines, as compared with that between the line drive and the first line, reflects the differences in flux flowing in these regions. The development of coincidences between the solid and dashed curves at short distances from the well arrays, in Figs. 14.21, 14.23, and 14.26, shows the high degree of localization of the effects of the individual wells in the arrays. At distances from the arrays about half the well spacing the pressure distributions correspond to those due to strictly continuous and uniform line distributions of fluid withdrawal.

When there are three line arrays of wells parallel to and producing by a line drive, an analysis exactly similar to that outlined above for single and double line arrays leads[1] to the shielding effects plotted in Fig. 14.27.

[1] Cf. M. Muskat, "The Flow of Homogeneous Fluids Through Porous Media," Sec. 9.11, McGraw-Hill Book Company, Inc., 1937.

Here the well separation in all the arrays is taken as 660 ft, and the first two lines are also located at 660-ft intervals from the line drive. The abscissas in Fig. 14.27 are the separations between the second and third lines. S_1 is the fraction of the total flux to all three arrays that is carried away by the first array. S_2 is the fraction of the sum of second and third line fluxes that is produced by the second. It will be noted that the curve for S_2 is very similar to curve III in Fig. 14.25, indicating that the first line of wells does not appreciably affect the shielding which the second

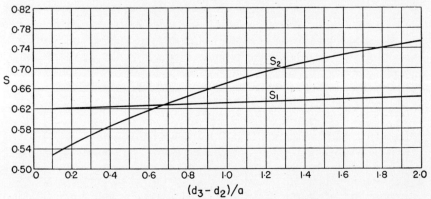

FIG. 14.27. The calculated variation of the steady-state homogeneous-fluid shielding effect in a system of three parallel line arrays with the distance, $d_3 - d_2$, between the second and third lines. $a = 660$ ft = well spacing within the lines, the distance between the first two arrays, and between the first array and the line drive. S_1 = fraction of total fluid produced by the three lines that enters the first line; S_2 = fraction of fluid leaking past the first array that enters the second line. Pressures in all three lines are assumed equal.

exerts on the third line. It will be observed, too, that the leakage past the first line $(1 - S_1)$ to the other two is only slightly greater than the corresponding values shown in Fig. 14.25 for two line arrays. And it may be shown that the total leakage past an initial array to an infinite number of equally spaced arrays will not exceed 37.8 per cent.

The effect of staggering the various line arrays can also easily be treated by the same method as that used for the above-discussed unstaggered systems. Thus for two well arrays the flux ratio for the two lines will be exactly the same as given by Eq. (6), except that hyperbolic functions having arguments that are the sums and differences of the line separations will be changed to hyperbolic cosines. From this it follows that, as long as the array separations are comparable with the well spacings within the lines, the staggering of the lines will have only a negligible influence on the shielding and leakage effects.

While these considerations may not appear to be of practical interest, they can find application in the development of complete-water-drive reservoirs. In monoclinal or unidirectional types of drive, such as approximately obtain in the East Texas field, the shielding effects discussed here will indicate the reduction in flux capacities to be expected from additional drilling behind the wells closest to the edgewater. Evidently essentially the same magnitudes will apply in the case of anticlinal reservoirs in which circular rings of wells may be used instead of line arrays.

14.11. Recovery Factors; Recoverable Reserves—The Problem.—The discussion throughout this work has emphasized mainly the general performance histories of various types of reservoir. The oil recoveries have been referred to largely in an incidental manner as an expression of the end result of the production history. In the case of gas-drive reservoirs the theoretical treatment presented in Chap. 10 did serve to provide automatically predictions of the ultimate recoveries. However, these were considered essentially from a comparative point of view with respect to different types of rock and fluid properties or methods of operation. And in the gas-cap-expansion and water-drive-performance analyses the recovery factors were implicitly introduced as residual-oil parameters, which could affect only the quantitative aspects of the production histories.

From a practical standpoint it is the recovery factor—the fraction of the oil in place that is recovered—or an equivalent measure of the ultimate oil recovery that is of primary importance. If this factor is not sufficiently high to pay for the cost of the drilling and operation of the producing wells, hypothetical considerations or predictions of the performance of the reservoir, if it were to be exploited, will be only of academic significance. It is true, of course, that the ultimate oil recovery is merely the abandonment value of the cumulative recovery. As such it will depend on the nature of the production mechanism and the actual operating history. A prediction of its magnitude before a reservoir is fully developed and its exploitation program established would thus appear to be pure conjecture from a scientific point of view. Yet unless some estimate of the recovery can be and is made early in the development stage, it would be little more than an economic gamble to continue drilling.

The only solution to this dilemma evidently lies in experience and theoretical considerations. From the very first wells, estimates must be made of the reservoir content and the *probable* producing mechanism. A recovery factor is then applied such as experience has shown to obtain in other reservoirs that have similar reservoir and fluid properties and that have produced by the same mechanism. This may be modified by supplementary considerations based on theoretical calculations, when feasible, of the probable reservoir performance and recovery. This is not an ideal solution. It is evidently subject to much uncertainty and fraught with the danger of erroneous assumptions regarding the actual recovery mechanism. Yet it is a necessary procedure if oil-field development is not to degenerate into blind guesswork.

It will be clear from the previous chapters that the identification of the actual production mechanism and the prediction of the future trend of the performance are not easy problems. Such identification becomes increasingly difficult as the available producing history is shortened. And its prediction immediately following the completion of the first well may seem utterly worthless. Yet it must be recognized that literally thousands of oil reservoirs have already been discovered and exploited. These provide an enormous fund of information and experience regarding the general characteristics of many geologic horizons and the types of oil reservoir usually found in them. For example, it is undoubtedly true, in a statistical sense, that the finding of water-drive reservoirs in many of the strata thus far explored most extensively in California is improbable. Conversely, future discoveries of Arbuckle Limestone reservoirs in Kansas may be expected, in a statistical sense, to be highly undersaturated and controlled by complete water drives. Of course, it is to be anticipated that geological and geophysical exploration will lead to the discovery of many new productive horizons and formations. And deeper drilling will also undoubtedly reveal previously unexploited oil-bearing zones. Nevertheless a correlation and accumulation of previous experience will in many cases serve as guides of statistical significance for the "guessing" of the probable producing mechanism, when supplemented by observations of the bottom-hole pressure, the bubble point of the oil, the presence of gas caps, the occurrence of faulting, and other structural characteristics of the reservoir.

It is in the light of these considerations that some of the available data and theoretical predictions relating to recovery factors or their equivalents will be reviewed in the next three sections. The figures given are not to be interpreted as directly applicable quantitatively to specific reservoirs. They will be of significance only as reported observations or theoretical estimates and may serve to indicate the range of expected recoveries rather than exact predictions for particular reservoirs not yet fully developed.

As noted in Sec. 9.2, while the gross classification of reservoirs into gas-drive and water-drive systems is convenient and suffices for the purpose of treating their general production performance, it is desirable to use a different classification in considering the oil recoveries. In particular, gas-cap-expansion or gravity-drainage reservoirs must then be differentiated from the solution-gas-drive systems. And the partial-water-drive reservoirs will be more appropriately grouped with those producing by the complete-water-drive mechanism rather than by gas drive. The following discussion of recovery factors will be divided accordingly.

14.12. Recovery Factors in Gas-drive Reservoirs.—From the standpoint of actual field observations, the most complete systematic record of recoveries from gas-drive reservoirs is that given in Tables 3 and 4. Of the 103 field studies 25 are indicated as simple gas-drive systems. The recovery factors for these are listed in Table 5. The recovery values in

TABLE 5.—OBSERVED AND ESTIMATED RECOVERY FACTORS OF GAS-DRIVE RESERVOIRS

Field study No.	Bbl/ acre-ft	% initial oil	% pore space	Ultimate free-gas saturation, %	Residual-oil saturation, %
1	368	35	24	39	47
5	232	37	24	35	45
12	138	25	12	23	38
15	187	25	14	19	46
18	237	34	15	33	32
19	297	34	21	25	44
21	530	45	34	38	43
22	213	22	15	18	57
24	233	48	23	53	28
25	247	33	21	34	46
28	180	22	13	16	49
36	354	33	21	28	46
42	547	50	26	37	28
46	256	34	18	24	38
47	165	25	12	23	37
48	248	35	19	27	37
49	500	39	28	30	57
50	443	27	20	23	57
57	460	33	21	30	45
67	343	36	17	36	34
69	285	25	17	23	52
70	453	42	27	31	39
86	225	20	13	14	53
94	126	15	7	27	44
95	180	23	10	31	38

barrels per acre-foot are taken from Table 4. The others are calculated values, rounded off to two significant figures, using the other pertinent data of Tables 3 and 4. It will be noted that the recoveries in barrels per acre-foot vary from 126 to 547, the per cent initial oil from 15 to 50, the per cent pore-space recovery from 7 to 34, and the ultimate free-gas saturation from 14 to 53 per cent.[1]

[1] The discrepancies between the free-gas saturations listed in Table 5 and those plotted in Fig. 14.5 are probably due to differences between the assumed values of the formation-volume factors at depletion. Similar discrepancies in the residual-oil saturations of Fig. 14.6 and Table 6 presumably arise for the same reason.

While with such spreads in the recovery factors a number as small as 25 can hardly be of statistical significance, it is nevertheless of interest to note the frequency distribution among the various types of recovery factors. These are plotted in Fig. 14.28. No detailed statistical analyses of these distributions are warranted, although it seems likely that the extreme maximal values may represent fields in which the solution-gas-drive recoveries have been supplemented by the partial-water-drive or gravity-drainage mechanisms. Including all the 25 recovery values, the median values, indicated by the letter M in Fig. 14.28, are 248 bbl/acre-ft, 33 per cent of the initial oil in place, 20 per cent of the pore space, and an ultimate free-gas saturation of 28 per cent.

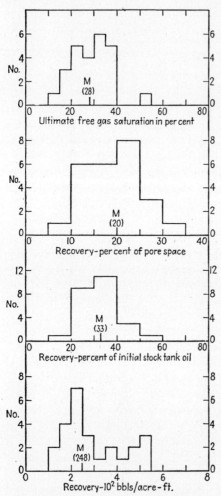

Qualitatively the observed or estimated field recoveries, as listed in Table 5 and plotted in Fig. 14.28, agree with the corresponding values computed by the theory of solution-gas-drive reservoirs in Chap. 10. Thus, for the comparative calculations on crudes of different gravities (cf. Fig. 10.9), the recovery factor ranges between 20 and 40°API crudes were approximately 20 to 31 per cent of the initial oil, 12 to 15 per cent of the pore space, and 22 to 37 per cent of free-gas saturation. These all lie within the frequency plots of Fig. 14.28.

Fig. 14.28. The frequency distributions of the recovery factors of 25 gas-drive reservoirs, as plotted from Table 5. M indicates median value.

Several discrepancies, however, are to be noted. As a whole, the ratio of the per cent of initial oil to per cent of pore-space recovery is lower for the field data than for the calculated results. The same is true with respect to the ratio of the ultimate free-gas saturation to the per cent of pore-space recovery. The reason for this lies in the low average initial formation-volume factors listed in Table 4 for the actual reservoirs as compared with those used in the comparative theoretical calculations (cf. Chap. 10, Table 3). If R_p, R_i denote the recoveries as fractions of the pore space and initial oil, ρ_g the ultimate free-gas saturation, ρ_w the connate-water saturation, and β_i, β_f the initial and final formation-volume factors, it is readily shown that

$$\left.\begin{aligned} \rho_g &= (1 - \rho_w)\left(1 - \frac{\beta_f}{\beta_i}\right) + R_p\beta_f, \\ R_i &= \frac{\beta_i R_p}{1 - \rho_w}. \end{aligned}\right\} \tag{1}$$

The use of small values of β_i will evidently lead to relatively low values of ρ_g/R_p and R_i/R_p.

More serious, and perhaps more significant, however, appears to be the fact that the field-estimated R_p's are definitely greater than those calculated. Whereas the maximal calculated values of the recovery as a per cent of the pore space for strict solution-gas-drive reservoirs without appreciable gas caps are of the order of 16 to 17 per cent, the *median* of the field data is 20 per cent. Since the theoretical calculations referred to were all based on a permeability-saturation curve having a 10 per cent equilibrium gas saturation, whereas field observations usually do not show such equilibrium saturations, and the gas-to-oil permeability ratios determined from field data lie generally higher than the laboratory-determined data (cf. Sec. 10.12), just the reverse situation would have been expected. Part of the reason may lie in the average low formation-volume factors of the oils (cf. Table 4) which would materially reduce the shrinkage effects in limiting the solution-gas-drive recoveries (cf. Sec. 10.4). It is doubtful, however, that this is the primary cause of the discrepancy, since a number of listed values of the initial formation-volume factors actually appear to be too low in relation to the other related data of Tables 3 and 4. While other factors may be underestimations of the gross productive reservoir volumes and the average porosities, it seems likely that the contributions of other recovery mechanisms—gravity drainage and water flushing—have also been responsible, at least in part, for the apparently high recoveries. It will be noted from Table 4 that a number of the solution-gas-drive fields were highly depleted at the time their ultimate-recovery estimates were made. Even if during the period of flush production the reservoirs were produced wide open without material aid from gravity drainage or water entry, the latter may lead to substantially increased recoveries during the period of "settled" production if the operations are continued to very low rates. Unfortunately, too little is known about these older fields to determine definitely whether these discrepancies are real or only apparent.

On the other hand, from a practical standpoint, it appears that most reservoirs which apparently are controlled predominantly by the solution-gas-drive mechanism nevertheless may be expected actually to yield over their whole producing lives ultimate recoveries appreciably greater than those indicated by purely theoretical calculations, if based on permeability-saturation data similar to those used in Chap. 10.

Neither the theoretical considerations of Chap. 10 nor the observed field recovery data discussed here make explicit reference to the rates of production. As pointed out in Sec. 10.2 the analytical formulation of the solution-gas-drive depletion problem inherently ignores the time factor and any effects that production rates may have on the performance or recovery. Field observations are also basically limited in this regard, simply because a field can be produced only once, and the results of different operating conditions or rates of withdrawal can only be conjectured. It has nevertheless been tacitly assumed in this work that ultimate recoveries of gas-drive reservoirs are essentially independent of the producing rates. While fundamentally this has been nothing more than an assumption, it has been accepted here as valid in uniform reservoirs because there is no conclusive evidence to the contrary, and the physical mechanism of the process of oil expulsion by solution-gas drives does not indicate that a direct effect of production rate should be expected. Of course, this proposition refers only to the solution-gas-drive process itself, and this limitation has been suggested frequently by the terms "strict" and "pure" solution-gas drive. Although such conditions may well represent extreme idealizations seldom occurring in practice, they constitute a well-defined mechanism subject to physical analysis.

On the other hand, it should be emphasized that to the extent that actual reservoirs are not controlled exclusively by the "strict" solution-gas-drive mechanism, the production rates may have a significant effect on the recovery. For such variations from the solution-gas-drive behavior, which will be due to gas segregation and gravity drainage or water intrusion, are basically rate-sensitive. The latter will both[1] generally tend to lead to higher recoveries. And their contributions to the reservoir performance and recovery will be accentuated the more slowly the reservoir is depleted. Accordingly, where either or both of these supplementary recovery agents are potentially available, the recoveries may be expected to increase with decreasing total fluid-withdrawal rates. Of course, this rate sensitivity

---

[1] It is assumed here that such gas segregation as may develop will form gas caps above the exposed sections in the well bore rather than continuous channels for gas by-passing. The latter would, of course, lead to accelerated gas depletion and reduced recoveries. Likewise, the water intrusion may also result in lower recoveries if it simply channels through high-permeability strata and forces premature well abandonment.

will not be a fixed characteristic for all reservoirs. In some it may be totally negligible, whereas in others it may be an important factor in determining the conditions for achieving maximum operating efficiency. The role that it may play can be ascertained only by a detailed study of the particular reservoir of interest.

It should also be noted that all the theoretical considerations developed in Chap. 10 regarding gas-drive reservoirs implied that the latter are substantially uniform throughout. It is clear that, if the reservoir is stratified or nonuniform with a wide range of permeability variation, the highest permeability zones will be depleted first. If the field is produced "wide open," the higher permeability strata will be depleted rapidly and the production capacity of the total section may fall to the limit for profitable operation before substantial depletion has occurred in the tight parts of the pay.[1] If the withdrawal rates are reduced, the increased time of operation will permit a longer period of cross flow from the low- to the high-permeability zones. It is uncertain, however, that the increased time factor for cross flow will more than counterbalance the decreased pressure differentials inducing the drainage between the parts of varying permeability. If so, this will be the result of the nonlinearity of the system, since it may be shown that if it is basically linear, in a mathematical sense, the withdrawal rates will not affect the recoveries.

In view of the wide range of variation of all the factors that may affect the ultimate recovery, it is not to be expected that the small number of recovery data listed in Table 5 would suffice to give any significant indication of the individual effects of these parameters. While there is little evidence that the theory of solution-gas-drive performance, as developed in Chap. 10, is quantitatively applicable to actual reservoirs with respect to absolute magnitudes, there is no reason to doubt the correctness of the theoretical predictions regarding the *relative* recoveries when the physical characteristics are varied. Thus the theoretically predicted increase of the ultimate free-gas saturation and recovery as a fraction of the initial-oil content with increasing crude API gravity (cf. Fig. 10.9) may be considered at least as highly probable. It is very likely, too, that the absolute recovery, as a fraction of the pore space, will show a maximum at an intermediate crude gravity, as compared with very low values for heavy and highly viscous oils, and with the values for the very light crudes whose high gas solubilities will also result in low recoveries due to excessive shrinkage effects. The effect of the connate-water content, which in itself may lead to decreased recoveries with decreasing saturations (cf. Sec.

---

[1] For example, in a four-zone formation, depleted as a unit and comprised of a 90-ft 100-md, a 50-ft 10-md, a 10-ft 1-md, and a 50-ft 0.1-md stratum, the percentages of their ultimate solution-gas-drive recoveries, at the time their composite producing rate has fallen to 8 bbl/day, have been calculated to be 100, 92, 53, and 12, respectively (cf. Keller, Tracy, and Roe, *loc. cit.*).

10.4), may well be masked by effects due to related changes in the permeability-saturation characteristics. For the lower connate-water saturations will generally be associated with relatively higher permeabilities in geologically similar formations (cf. Fig. 3.12). Even for the same permeability-saturation relationship it is to be expected that the economic ultimate recovery will increase with increasing permeability. And for unconsolidated sands there are some laboratory data indicating[1] that the equilibrium relative permeabilities decrease with increasing permeability, which would imply lower equilibrium liquid saturations and higher solution-gas-drive physical recoveries. On the other hand there is no basis available at present for attempting even semiquantitative predictions[2] of general variations in the recovery factor with the permeability. And there is no evidence suggesting that the porosity itself plays any role whatever in determining the solution-gas-drive recoveries, when expressed in any form in which the pore space is the basic unit of volume, except as it may be associated indirectly with changes in the permeability.

Finally it should be observed that the numerical data discussed above refer only to the simple solution-gas-drive depletion processes, in which only the gas normally associated with the oil as solution gas is used for oil expulsion. It will be recalled from Chap. 10 (cf. Sec. 10.5) that such recoveries may be appreciably increased by the supplementary sweeping action of gas-cap gas if produced through the oil zone, even without the aid of gravity drainage. And if part or all of the produced gas be returned to the reservoir, still greater recoveries can be achieved (cf. Secs. 10.7 and 10.8) under favorable circumstances. No attempt will be made here to give rule-of-thumb estimates of the increases to be expected, as they would be misleading and inherently subject to great error. Since the simple depletion recoveries evidently vary widely, as is clear from Fig. 14.28, with the detailed reservoir conditions and characteristics, so will the results from gas-injection operations. As discussed in Sec. 10.13, both clear-cut failures and outstanding successes have been experienced in gas-return operations. Generalizations based on either extreme are totally unwarranted. Only by a thorough analysis of the rock, fluid, and structural conditions in the specific reservoir under consideration can even semiquantitative estimates of the expected recoveries from gas injection as well as normal depletion be made for the particular reservoir of interest. If oil reservoirs were the

[1] Cf. M. Muskat, R. D. Wyckoff, H. G. Botset, and D. W. Reed, *AIME Trans.*, **123**, 69 (1937).

[2] These remarks refer only to the physical processes determining the recovery factor. The effect of the permeability on the economic ultimate recovery from solution-gas-drive reservoirs may be estimated from the approximate theoretical considerations of Sec. 14.6 (cf., for example, Fig. 14.3 and Table 1).

product of a manufacturing process of fixed and rigidly controlled specifications, it is extremely doubtful that their recoveries for the same basic producing mechanism would show such variations as indicated by Fig. 14.28, even if subjected to different operating histories.

14.13. Recovery Factors in Water-drive Reservoirs.—Table 6 lists the recovery factors from 69 sand reservoirs, which presumably have been producing only —or largely—by the water-drive mechanism, as derived from Tables 3 and 4. The values of recovery as per cent of the initial oil, per cent of pore space, and per cent of residual-oil saturation were calculated from the barrels per acrefoot recoveries, by using the other related data in Tables 3 and 4. The residual-oil saturations, in reservoir measure, are probably low, since it was arbitrarily assumed that the formation-volume factors at depletion were those corresponding to 100 psi.

It will be seen that the barrels per acre-foot recoveries range from 242 to 1,165, the per cent initial oil from 24 to 78, the per cent pore space from 18 to 54, and residual-oil saturation from 16 to 59 per cent. The frequency distributions of these recovery factors are plotted in Fig. 14.29. Although it would be possible to apply standard statistical methods

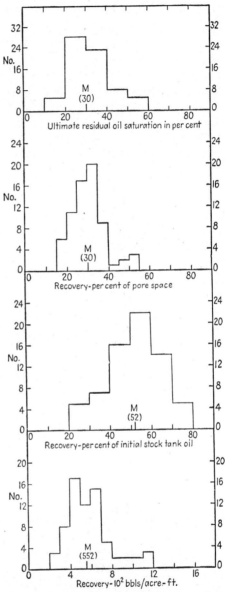

Fig. 14.29. The frequency distributions of the recovery factors of 69 water-drive reservoirs, as plotted from Table 6. M indicates median value.

to convert these curves into equivalent Gaussian distributions with suitably adjusted scales and asymmetry factors, so as to indicate "most probable" re-

Table 6.—Observed and Estimated Recovery Factors of Water-drive Reservoirs

Field study No.	Bbl/ acre-ft	% initial oil	% pore space	Residual- oil saturation, %
2	600	47	26	31
3	410	55	29	25
6	1,058	78	54	16
10	629	43	25	35
11	735	54	34	32
14	765	70	45	21
16	958	70	36	17
20	306	40	28	45
26	437	49	38	42
27	261	24	18	59
29	280	42	20	31
30	466	50	29	32
31	390	56	38	32
32	396	51	30	30
33	518	49	28	30
34	644	45	25	33
37	1,111	62	51	33
38	1,121	62	52	32
39	742	61	35	24
40	427	28	20	56
41	398	30	21	50
43	617	60	30	22
44	429	53	31	28
45	242	29	18	47
51	811	57	34	28
52	660	62	31	20
53	660	62	31	20
54	680	64	32	19
55	382	46	21	26
58	661	63	32	20
59	600	61	31	22
61	387	40	19	32
62	387	41	19	31
63	530	59	24	19
64	472	43	25	36
65	690	61	30	21
66	500	51	28	29
68	357	34	23	46
71	674	60	33	23
72	592	56	31	26

TABLE 6.—OBSERVED AND ESTIMATED RECOVERY FACTORS OF WATER-DRIVE RESERVOIRS (*Continued*)

Field study No.	Bbl/ acre-ft	% initial oil	% pore space	Residual-oil saturation, %
75	552	45	28	37
76	540	52	28	27
77	540	52	28	27
78	430	31	22	52
79	682	63	35	22
80	560	54	30	27
81	682	63	35	22
82	500	49	27	29
83	625	56	28	24
84	771	61	32	22
85	683	55	35	31
87	595	47	25	31
88	815	57	32	26
89	1,030	73	41	17
90	963	56	38	33
91	1,165	71	46	20
92	563	47	26	32
93	410	39	20	34
96	743	44	30	40
98	445	29	18	48
99	443	31	19	45
101	540	26	20	58
102	402	32	20	45
103	412	37	21	39
104	429	53	31	27
105	430	50	33	32
106	428	58	36	26
107	400	50	27	27
108	482	50	32	31

coveries and probabilities of deviations from them, it is doubtful that such representations would be of physical significance.[1] As noted in Sec. 14.7 the water-drive recoveries (residual-oil saturations) show definite trends of

[1] It should be noted also that the individual recovery factors listed in Tables 5 and 6 are probably in many cases of questionable accuracy, not only because of meager data underlying the original reservoir content evaluations, but even more so because of the limited states of depletion at which the recovery estimates were made and the gross extrapolations involved in predicting the "ultimate" recoveries (cf. Table 4).

variation with the reservoir oil viscosity, formation permeability, and pressure decline, so that the observed differences in recovery are not merely chance fluctuations. It may be observed, however, that the median values, indicated by the letter M, are 552 bbl/acre-ft, 52 per cent initial oil, 30 per cent pore space, and 30 per cent residual oil.

In contrast to the case of gas-drive reservoirs, no simple comparison can be made between these estimated field recoveries and those anticipated theoretically. In fact, in the discussions of complete-water-drive reservoir performance in Chap. 11 only incidental reference was made to recovery factors. The latter did not appear directly in the theory of the pressure and production performance. Rather it played the role of a parameter, which could be introduced independently to translate the volume of water intrusion into equivalent reservoir area invaded. And it is only in the theoretical treatment of partial-water-drive reservoirs that this procedure was actually applied explicitly (cf. Sec. 10.18).

In principle it would appear possible to calculate theoretically ultimate water-drive recoveries by applying the requirement that abandonment is determined by a limiting value of the water-oil ratio. This ratio, R_w, can be formally expressed as [cf. Eq. 10.2(4)]

$$R_w = \frac{\mu_o \beta_o}{\mu_w \beta_w} \frac{k_w}{k_o}, \tag{1}$$

where the subscripts w, o refer to water and oil, and μ, β, k are the viscosity, formation-volume factor, and permeability, respectively. Under flow conditions in which both oil and water are supplied to the system, Eq. (1) would indeed determine the saturation distribution and, in particular, the reservoir oil saturation, if R_w and the pressure were fixed and the permeability-saturation characteristics were known (cf. Sec. 8.3). Such calculations have been reported[1] to show the variation of the residual-oil saturation with the oil viscosity as implied by Eq. (1) to be similar to that observed.

It is doubtful, however, that the phase distribution criterion represented by Eq. (1) is basically applicable to the ultimate recovery problem. The mechanism of oil displacement by water flushing is inherently different from that of simultaneous water and oil flow. Laboratory studies indicate that water invasion in water-wet sands proceeds by the advance of an oil-water "front" in which the first passage of the water through an individual pore gives a virtually complete flushing of the oil removable from that pore.[2]

[1] Craze and Buckley, *loc. cit.*

[2] Cf., for example, J. N. Breston and R. V. Hughes, *Journal of Petroleum Technology*, **1**, 100 (1949), who found in a series of water-flooding experiments on two consolidated sand cores that before the first water break-through the average oil recoveries were 94.6 and 88.0 per cent of the ultimate recoveries obtained at the time of development of 100:1 water-oil ratios. In some similar flooding experiments Holmgren

Behind the front the oil saturation appears to be immediately reduced to its ultimate residual-oil value, without substantial decrease or stripping by the continued subsequent flow of water. The residual oil may be visualized as being broken up into a dispersed and discontinuous globular distribution, or isolated local masses of oil enveloping a small number of pores, with substantially zero permeability. The magnitude of the residual-oil saturation is thus locally determined by the detailed microscopic pore structure and capillary properties, rather than a balance in relative permeabilities of both mobile components of a two-phase system.

The almost universal gradual development of water cuts following first water break-through in producing wells may appear to completely contradict the above-described picture.[1] Such observations, however, can be readily explained otherwise than by a continued stripping *locally* of the residual oil left after the first throughflow of the invading water. The imperfection of the purely geometrical sweep pattern would necessarily lead to a gradual build-up in the water cut even in a strictly uniform formation with sharp local water-oil fronts. The continued spreading of the conelike vertical or areal cusps after first water entry will automatically result in the simultaneous production of water and oil regardless of the detailed microscopic oil-displacement efficiency. More important, undoubtedly, is the inherent local variability and permeability stratification characterizing virtually all reservoir rocks. The superposition of the sequence of water break-through in the individual localized regions of different permeability[2]

[*Petroleum Technology*, **11**, 1 (July, 1948)] found the water saturation at first break-through to increase by only 3.3 per cent by the time the efflux was 100 per cent water. While in practice the water-flooding recoveries after first water break-through represent a substantial part of the total recoveries (cf. Sec. 12.15), this phase of water flooding does not appear to be of major significance from the point of view of the microscopic physical processes involved.

[1] The transition-zone theory of water invasion [cf. S. E. Buckley and M. C. Leverett, *AIME Trans.*, **146**, 107–116 (1942)] also is not in basic contradiction with the views expressed here, since that is tacitly based on the *assumption* of incomplete initial-oil displacement. While there are other uncertainties associated with the detailed development of that theory, it would nevertheless lead to sharp water-oil boundaries if the oil saturation behind the water front were considered as immobile. Rather sharp fronts are directly indicated by X-ray tracer studies of water-oil displacement processes [cf. R. L. Boyer, F. Morgan, and M. Muskat, *AIME Trans.*, **170**, 15 (1947)]. On the other hand the above considerations are limited to conditions where the water saturation ahead of the invading water is the "irreducible" and immobile interstitial-water content. Within water-oil transition zones, there may be, of course, simultaneous flow of oil and water even without extraneous water entry.

[2] Even if the apparent stripping or subordinate phase of oil flow after initial water break-through be visualized as the result of the spreading of the water displacement into tighter pore channels on a semimicroscopic scale, the resultant flow behavior still would not correspond to the steady-state conditions of multiphase flow implied by

will evidently lead to a continuous growth in water cut even though in each pore the water flow may change sharply from 0 to 100 per cent. When only simple stratification is involved, the development of water production could be computed in the same way as that of dry gas in cycling operations (cf. Sec. 13.8), corrected for the difference in effective mobility between the water and oil.[1]

On the basis of these considerations the actual ultimate recoveries will represent the integrated resultants of those in the localized parts of the formation in which there has been complete water invasion and in which the oil saturations have been reduced to their ultimate residual values. Although the effect of the relative viscosities will not be simply controlled by Eq. (1), it is still to be expected that the average residual-oil saturations when the gross water-oil ratio reaches the abandonment value will increase with increasing viscosity. The resistance to invasion and flushing of the tighter zones will evidently be greater if the oil viscosity is high, and their flooding will be less complete by the time of abandonment.[2] On the other hand it may well be that in the case of very viscous reservoir oils the water may break through in a manner similar to that of a nonwetting phase, without immediately reducing the oil saturation to its ultimate residual value. Additional research will be needed definitely to "prove" either displacement mechanism. On the other hand, from a practical point of view, since there are very few permeability-saturation data available for the application of Eq. (1) and any application that might be attempted would involve the unsolved problems of averaging different permeability strata, the assumptions of a gross average "irreducible" residual-oil saturation or applications of the stratification theory neglecting the oil-stripping

---

Eq. (1). In fact the failure to expel all the recoverable oil by the time of first water break-through even in laboratory experiments is probably due to the different rates of advance of the invading water in the pores of different effective radii, rather than a continued reduction in oil saturation in individual pore channels after they have suffered an initial flooding action, unless the oil displacement results from a capillary imbibition of the water rather than a flooding action.

[1] Such calculations have been carried out by H. Dykstra and R. L. Parsons, API meetings, Los Angeles, Calif., May, 1948. While the main reason may have been analytical simplicity, the stripping phase was neglected in these computations, although the flooding experiments of these authors did show appreciable oil recoveries after the initial water break-through.

[2] The stratification theory of water invasion into oil-saturated reservoirs actually shows this type of sensitivity to the oil viscosity; cf. H. Dykstra and R. L. Parsons, API meetings, Los Angeles, Calif., May, 1948. In fact, both the latter investigation, based on a probability-permeability distribution, and independent studies of the author, based on exponential and linear-permeability distributions, show that a greater range of economic recoveries due to oil-viscosity differences can be explained in this way than by Eq. (1).

phase, will probably be as accurate for estimating the ultimate recovery as a calculation by Eq. (1), even if the latter should be confirmed as a physically valid criterion.

The rate of withdrawal will also affect the ultimate recoveries from water-drive fields in an indirect manner. As seen in Chap. 11 the *pressure* performance of water-drive reservoirs is basically rate-sensitive. This in itself, however, does not imply an effect of the rate of fluid movement on the actual oil-displacement mechanism. On the other hand, from a practical and economic standpoint, rapid pressure declines, caused by excessive withdrawal rates, will shorten the flowing life and may lead to earlier abandonment, with lower recoveries, than relatively low production rates. There is little evidence that the fall in pressure below the bubble point and associated gas release will fundamentally lower the microscopic displacement efficiency.[1] But here, too, there will be a gain in preventing a pressure decline far below the bubble point related to the shrinkage properties of the reservoir oil. If the latter is trapped behind the water front above the bubble-point pressure, only part of it will represent unrecovered stock-tank oil. The same reservoir volume of oil trapped at a low or near-atmospheric pressure will be equivalent to an almost equal volume of stock-tank oil. Thus, if the average residual oil in a formation with 25 per cent connate water is 30 per cent, the ultimate recovery if the residual oil is trapped at the bubble-point pressure, with a formation volume factor of 1.25, will be 36 per cent of the pore space and 60 per cent of the initial oil. But if the flushing has taken place at a pressure for which the formation volume factor is 1.05, the 30 per cent residual-oil saturation will correspond to recoveries of only 31.4 per cent of the pore space and 52.4 per cent of the oil initially in place.

Except for these qualitative considerations, indicating that partial water drives resulting from sustained withdrawal rates which are greater than the maximum steady-state supply capacity of the aquifer may have lower recoveries than complete water drives, there are no data providing definite comparisons between partial- and complete-water-drive recoveries. Undoubtedly many of the fields listed in Tables 3, 4, and 6 as water-drive reservoirs actually were produced by partial water drives. Quite possibly the lower ranges in the recovery factors indicated in Fig. 14.29 reflect the action of partial rather than complete water drives. This is suggested by the fact that the residual-oil saturations corresponding to the API data of Tables 3 and 4 showed an increasing trend with increasing pressure de-

---

[1] In fact recent laboratory experiments indicate that water invasion into a sand containing a free-gas phase may leave somewhat lower oil saturations than in flushing a sand filled with oil and water; cf. H. Dykstra and R. L. Parsons, API meetings, Los Angeles, Calif., May. 1948.

clines during the producing life. On the other hand the fault-line fields, which were produced without proration or other restrictions and most certainly were not controlled by the complete-water-drive mechanism, have given recoveries that are comparable with the very highest observed for any water-drive fields (cf. Table 2).

Unfortunately the theoretical treatment of partial-water-drive reservoirs cannot give a conclusive answer to the question of comparative recoveries by partial and complete water drives (cf. Sec. 10.18) since the ultimate recovery factor is in effect arbitrarily introduced into the theory in the form of a residual-oil assumption. However, even if the residual oil after water flushing be assumed the same under partial- as well as complete-water-drive operation, the theory does indicate that the economic ultimate recoveries should, in general, decrease as the withdrawal rates are made large compared with the supply capacity of the aquifer. The much slower pressure declines and tendency for pressure stabilization at high pressures when the withdrawal rates are restricted (cf. Fig. 10.45) should lead to substantially higher economic recoveries, even though the microscopic efficiency of the oil-displacement mechanism may be no greater than under wide-open production. Although the support from field observations for these theoretical implications is rather meager, it is generally assumed that withdrawal-rate control in reservoirs which are subject to water-drive action will be conducive to the achievement of greater economic ultimate recoveries, provided that the reservoirs are inherently susceptible to both microscopic and macroscopic water-drive recoveries of high efficiency.

To the extent that Tables 5 and 6 and Figs. 14.28 and 14.29 have any statistical significance it appears highly probable that the "average" or "typical" water-drive reservoir will yield higher absolute recoveries, for equal reservoir volumes, than the average gas-drive reservoir. It should be noted, however, that this is not a necessary or universal rule. These data themselves show that there is considerable overlapping in the ranges of observed and estimated recoveries from the two types of producing mechanism. Although the data do not indicate that the region of overlapping, in which the gas-drive recoveries even exceed those for the water-drive reservoirs, is associated with any particular reservoir characteristic, it may be anticipated that an equivalence of recovery by the two mechanisms may occur when the formation has a high connate-water saturation. The range of water saturations involved may be estimated by considering the residual-oil saturation as the recovery index. The residual-oil saturation in a gas-drive reservoir is evidently 1 minus the sum of the water and free-gas saturations. On taking as a "probable" ultimate gas saturation 28 per cent—the median of the data in Table 5—, it follows that, for water saturations of 42 per cent or greater, the residual-oil saturation by *gas drive* will not exceed 30 per cent, the observed median value for the water-drive reservoirs. Because of the higher average pressure of residual-oil trapping in water-drive reservoirs, equal residual-oil saturations will still imply higher recoveries under the water-drive mechanism. On the other hand in many gas-drive reservoirs the contribution of gravity drainage may lead to appreciably higher average free-gas saturations than the assumed value of 28 per cent. While no precise dividing point can be fixed, it appears likely that if the connate-water saturations are 45 per cent or greater, the inherent recovery efficiency of gas-drive mechanisms will be comparable with that of the water-drive. The optimum producing method will then be determined largely by economic factors other than the magnitude of the absolute recovery.[1] Such situations may well arise in the exploitation of formations having appreciable clay content, in which the connate-water saturations are often quite high. The inherently low water permeabilities frequently associated with "dirty sands" (cf. Sec. 3.7) may also imply such low potential water-intrusion rates from adjoining aquifers that a water-drive operation may be both impractical and uneconomic.

14.14. Recovery Factors under Gravity Drainage.—As noted in Sec. 9.2, from a recovery standpoint it is appropriate to consider gravity-drainage or gas-cap-expansion reservoirs separately from those producing by the solution-gas-drive mechanism. On the other hand there are literally no data of statistical significance giving observed recoveries in fields that have been primarily controlled by the gravity-drainage mechanism. Of the three field studies of Tables 3 and 4 in which gas-cap expansion was recognized as a factor, all apparently were also subject to water-drive action. Moreover in two of these cases the producing formations were cavernous limestones with low per acre-foot recoveries, and in the other, producing from a sandstone, the recovery was relatively high.

The physical basis for the concept of gravity drainage as a means of oil recovery is the simple observation that as long as the oil phase has a non-vanishing permeability it will of necessity move in the direction of the net force acting on it. Since the oil will always be subject to the downward force of gravity, it will therefore tend to "drain" downflank in a reservoir if the other potentially opposing forces, due to pressure differentials and capillary pressures, do not exceed the gravity force. In order to give full

---

[1] If the formation is highly variable in character, the by-passing and oil trapping under water drive may lead to lower resultant recoveries than by solution-gas drive, even if the microscopic oil-displacement efficiencies for the two mechanisms are comparable. Such reservoirs, and especially highly fractured limestones, may also show a sensitivity of recovery and differential water invasion of the different permeability components to the rate of fluid withdrawal if the primary oil-displacement mechanism is the solution-gas drive. Such effects, however, are not to be confused with those related to capillary cross flow discussed in Sec. 7.10.

play to the gravity forces the pressure gradients must be vanishing, so that the gravity drainage is "free." Only capillary forces will then limit the downward oil flow induced by gravity. As discussed in Sec. 7.9, capillary forces will determine the initial equilibrium fluid distribution in the interphase transition zones. Below the gas-oil transition zone, however, the capillary forces also will affect the fluid permeabilities. Except for the time factor the ultimate recovery which can be obtained by the gravity-drainage mechanism will thus be simply determined by the residual-oil saturation at which the oil permeability becomes vanishing. It is the magnitude of this residual-oil saturation that is the crux of the evaluation of the process of gravity drainage as an oil-recovery mechanism.

Although it is becoming rather common practice to inject gas for pressure maintenance into reservoirs with gas caps but limited water drives, field data on gravity-drainage reservoirs for statistical evaluation of their recovery factors will probably not be available for some time. Laboratory evidence is also very meager. There have been reports[1] of experiments demonstrating the existence of gravity drainage of wetting phases in porous materials. Desaturation capillary-pressure vs. saturation tests of the type made for connate-water determinations (cf. Sec. 3.11) simulate, in some respects, experimentation on gravity-drainage processes. However, these relate to the phase wetting the porous medium, whereas it is the behavior of the nonwetting phase—oil—in a three-phase system that determines the role of gravity drainage as an oil-recovery mechanism. Only one study of such systems has thus far been reported.[2] In this it was found that not only will the residual-oil saturation left by capillary-displacement processes corresponding to gravity drainage generally be comparable with those following water displacement but it may even be lower in some cases. This is in accord with the physical criterion of mobility as the limiting factor in all[3] fluid-displacement processes. The limit of mobility, in turn, may be expected to correspond to a breakup of the oil into a dispersed and discontinuous phase.

The upper saturation limit for the globular discontinuous distribution of the oil phase should be determined largely by the microscopic pore structure and geometry of the porous medium and hence should be approximately the same whether it is created by water displacement or gravity drainage. It is conceivable, however, that the values of the interfacial

[1] D. L. Katz, *AIME Trans.*, **146**, 28 (1942); R. F. Stahl, W. A. Martin, and R. L. Huntington, *AIME Trans.*, **151**, 138 (1943).

[2] H. J. Welge, *Petroleum Technology*, **11**, 1 (September, 1948).

[3] It is only because of the loss of the *driving* force as the reservoir becomes depleted of its gas content and pressure that this limit is not reached in solution-gas-drive reservoirs.

tensions and the microscopic-flow processes will also affect the absolute value of the saturation at which the local continuity of the oil phase may be destroyed. The construction of a detailed physical theory of these phenomena merely to rationalize the few data available at present is hardly warranted. On the other hand it appears reasonable to assume that, as indicated by the experiments referred to above, the residual-oil saturations resulting from the gravity-drainage oil-recovery mechanism should generally be comparable with that associated with water displacement, i.e., of the order of 20 to 35 per cent.[1] Of course in both cases the microscopic recovery efficiency will be reduced in practice owing to effects of permeability stratification and other inhomogeneities. The time factor involved in reaching the limiting oil saturation under gravity drainage will also limit the actual recoveries. As a whole, however, it appears that the oil potentially recoverable by gravity drainage and gas-cap expansion should not differ greatly from that to be expected from water drives. Such few field observations as have been reported seem to be in agreement with this general conclusion.

As discussed in Sec. 10.16 the role played by gravity drainage is inherently rate-sensitive. If the fluid-withdrawal rates are large compared with the downflank oil drainage, the performance will be controlled by the solution-gas-drive mechanism. While gravity drainage may continue throughout the producing life, its rate will be reduced by the lower oil permeabilities in the oil zone associated with the gas-drive depletion process. The total volume of pay desaturated by gravity drainage by the time the production rates become too low for continued operation may then be a small fraction of the whole reservoir volume. The net gravity-drainage contribution to the resultant recovery will be correspondingly low. Moreover, even if production could be continued after pressure depletion at such rates as could be supplied only by the gravity drainage, the stock-tank equivalent of the residual oil left in the formation will still be higher than if the residual oil were trapped in the pores in an "inflated" state at high pressures. The achievement of the maximum potential recoveries by the gravity-drainage mechanism will therefore require careful control of the withdrawal rates, location of the producing wells below the gas-oil contact, and all such operating practices as are economically practicable for minimizing the reduction in oil permeability. It is the economic feasibility of exercising these controls that will determine the importance of the gravity-drainage mechanism under actual operating conditions.

---

[1] And for the same reasons as those discussed in the preceding section with respect to the recoveries by water drive, even gravity-drainage recoveries should not significantly exceed those under solution-gas drive when the connate-water saturation is high (of the order of 45 to 50 per cent).

The effect of the various parameters on the general performance of gravity-drainage systems has been discussed in Sec. 10.16. However, these will control largely the time scale and economic factors, rather than the inherent potentialities of the gravity-drainage recovery mechanism. The construction of even an approximate relation between the actual recovery factors and the reservoir parameters is at present beyond the scope of either field experience or laboratory-developed information, except as such factors may be computed by the methods of Chap. 10 for individual reservoirs.

14.15. Recoverable Reserves.[1]—In the previous discussion the recovery factors have been expressed in various alternative forms, including the fraction of the pore space and the fraction of the initial-oil content represented by the total ultimate recovery, the ultimate free-gas saturation, the barrels of recovery per acre-foot of oil pay, and the residual-oil saturation. Even if these were known with certainty, they evidently would not suffice alone to evaluate specific reservoirs with respect to their actual cumulative oil recovery. Additional information will be needed.

In the simplest cases, where a recovery factor expressed in barrels per acre-foot may be considered as known or subject to reasonably accurate prediction, the total recovery from the particular reservoir of interest will then be, obviously, the acre-foot of reservoir pay times the recovery factor, i.e., the value given by the formula

$$\text{Cumulative recovery (bbl)} = FAh, \qquad (1)$$

where A is the productive area in acres, h the net effective pay thickness in feet, and F the recovery factor in barrels per acre-foot. However, the two components comprising the gross reservoir volume, i.e., the productive area A and net formation thickness h, are also often uncertain. The extent of the productive area evidently cannot be accurately ascertained until the field has been sufficiently developed to delimit both the external closure of the oil-bearing zone by bounding waters, faulting, permeability, or porosity pinch-out, or their equivalents, and the internal boundaries, such as may be created by gas caps or truncation of the pay. While the establishment of these limits may involve considerable delay after the completion of the discovery well, it is usually not beset with great inherent difficulty if the reservoir appears to contain enough recoverable reserves to warrant full development.

The determination of the net average pay thickness is often a more difficult problem. The gross thickness of the oil-bearing section can usually be established by a combination of electrical logging, geological-sample in-

---

[1] The discussion in this section refers only to crude-oil reservoirs. The problem of reserves estimation in condensate reservoirs has already been reviewed in Chap. 13.

spection, and core analysis. Shale and barren zones within the productive formation are readily identified and eliminated. However, if there is a wide range of variation of permeability along the well bore, the lower limit of permeability to be included within the "effective" thickness becomes a matter of arbitrary choice. This arbitrariness arises from economic factors. Evidently, if an oil-bearing formation producing by a solution-gas drive is largely comprised of strata exceeding 100 md in permeability, it is unlikely that oil-saturated sections having permeabilities lower than 1 md will be substantially depleted by the time the producing rate has fallen to the limit for profitable operation. Likewise, if the reservoir were producing by a complete water drive, the break-through of water in the high-permeability strata may force abandonment long before appreciable water invasion and flushing have developed in the tight parts of the section, unless the former can be effectively plugged off. On the other hand, in massive limestone or dolomitic reservoirs where the average permeability of the intergranular matrix may not exceed 5 md, the exclusion of such oil-saturated pay as may have only 1 to 2 md permeability would lead to gross underestimation of the recoverable reserves.

No simple fixed formulas will solve this problem. A statistical approach,[1] though also basically arbitrary, may at least tend to minimize the "personal equation." A "five percentile" or an equivalent statistical measure might conceivably provide a basis for limiting the "effective" productive pay thickness. Since the tight zones will generally provide some recovery through cross flow into the more permeable strata and inherently suffer some depletion or water invasion, even though much slower than the highly permeable sections, the lower limit for inclusion as "effective" pay should be made quite low so as to compensate for the complete exclusion of the still tighter parts. Moreover, in establishing lower limits, consideration should be given to the type of producing mechanism and the possibility that extraneous fluid injection may be undertaken during the producing life.

The potential recovery from apparently nonproductive parts of the reservoir should also be taken into account in fixing the productive area. The latter should not be limited merely to that encompassed within the producing units assigned to the wells. At the reservoir limits there will generally lie a band of acreage with an average effective pay thickness too small to warrant complete development. Since this area can drain into

---

[1] Evidence that permeability distributions may be amenable to statistical analysis has been given by J. Law, *AIME Trans.*, **155**, 202 (1944), who showed that in the Dominguez field, California, the sample permeabilities appear to follow Gaussian distributions with the logarithm of the permeability as a base; cf. also A. C. Bulnes, *AIME Trans.*, **165**, 223 (1946).

the actual boundary wells, it should be included as part of the productive area, though it may ultimately be given a low weighting due to its limited thickness and greater distance from the producing wells.

To translate recovery factors expressed as fractions of the pore space into equivalent total recoveries the average porosity must be determined as well as the gross productive volume. For the cumulative recovery is then

$$\text{Cumulative recovery (bbl)} = 7,758.4 F A h f, \qquad (2)$$

where f is the fractional porosity and F is now the recovery factor as a fraction of the pore space. Since f represents the average porosity of the net effective pay, the determination of the latter combined with the corresponding core-analysis data will provide the required values to be used in Eq. (2).

When the recovery factor F is expressed as an ultimate free-gas saturation, as is appropriate in the case of gas-drive reservoirs, the cumulative recovery may be calculated by the formula

$$\text{Cumulative recovery (bbl)} = 7,758.4 \left[\frac{F}{\beta_f} - (1 - \rho_w) \left(\frac{1}{\beta_f} - \frac{1}{\beta_i} \right) \right] A h f, \qquad (3)$$

where ρ_w is the connate-water saturation, β_i the initial formation-volume factor of the reservoir oil, and β_f its value at abandonment, and it is assumed that none of the connate water is produced and that there has been no water entry during the producing life. A, h, and f have the same meaning as in Eqs. (1) and (2). The new factors involved here are ρ_w, β_i, and β_f. While, as discussed in Chap. 3, the determination of the true connate-water saturation long has been beset with serious difficulties, it now appears possible to establish the value of ρ_w, at least for the individual core samples, rather satisfactorily by such methods as oil-base coring or capillary-pressure experiments. The initial formation-volume factor β_i can be readily measured by conventional p-V-T experiments on bottom-hole or recombined-separator samples of the reservoir fluids. The value of β_f depends on the pressure assumed for the ultimate economic-depletion state. This will not be known in advance with certainty. However, it will often be possible to estimate the abandonment pressure within 100 psi, taking into account the depth of the formation and the initial well potentials or permeability and thickness of the reservoir stratum. The corresponding uncertainty in β_f will then be no greater than that associated with the composite value of the other factors in the equation.

The use of the residual-oil saturation as a measure of the recovery factor F, as is convenient in evaluating complete-water-drive or gravity-drainage reservoirs, requires the same basic data as in Eq. (3). The formula giving the ultimate recovery is then

$$\text{Cumulative recovery (bbl)} = 7{,}758.4 \left(\frac{1 - \rho_w}{\beta_i} - \frac{F}{\beta_f} \right) A h f. \qquad (4)$$

While there may be somewhat greater uncertainty regarding the value of β_f to be used in water-drive reservoirs than in gas-drive systems, the errors caused by those made in β_f will be proportional to the residual-oil saturation in both cases and hence[1] may be no more serious in Eq. (4) than in Eq. (3).

A recovery factor F as a fraction of the initial stock-tank oil in place implies a cumulative recovery given by

$$\text{Cumulative recovery (bbl)} = \frac{7{,}758.4 F (1 - \rho_w) A h f}{\beta_i}. \qquad (5)$$

This requires no data beyond those needed in the application of Eqs. (3) and (4). In the use of this type of recovery factor, however, the initial-oil content of the reservoir, under favorable conditions, may be determined independently by application of the material-balance equation. As pointed out in Chap. 8 the material-balance method is often subject to serious uncertainty of interpretation. However, in solution-gas-drive reservoirs in which the size of the gas cap, if this exists, can be determined at least in comparison with the volume of the oil zone, the material-balance method should provide reasonably accurate oil-volume estimates. Such application may be especially valuable in limestone or dolomite reservoirs, where the accurate determination of the individual factors in Eq. (5) pertaining to the rock may be particularly difficult.

It should be emphasized that the use of single values for the various physical parameters and recovery factors in the above equations does not imply that such will be inherently appropriate for all parts of a common reservoir. While it is convenient in gross reservoir evaluations to use average properties and recovery factors which can be directly inserted in Eqs. (1) to (5), the establishment of these averages is one of the most difficult aspects of reservoir analysis. Perhaps the simplest of these averaging problems are the determinations of the effective reservoir bulk volume, represented by the composite factor $A h$, and the initial formation-volume factor of the oil phase. The averaging of the porosity is usually not too difficult if the formation has been effectively cored. But the connate-water saturation, which is used in Eqs. (2) to (5) to fix the initial oil saturation,[2] requires an independent averaging, which generally must be

---

[1] The residual-oil saturations for gas-drive recovery (cf. Table 5) have a median value of 45 per cent, as compared with 30 per cent for water-drive reservoirs (cf. Fig. 14.29).

[2] The use of $1 - \rho_w$ for the initial oil saturation in Eqs. (2) to (5) implies, of course, the assumption that the reservoir is in its virgin condition. For the estimation of recoveries after partial depletion has taken place, the terms $1 - \rho_w$ should be replaced by ρ_{oi}, and the latter must be evaluated independently.

based on data which are far less complete and satisfactory than those for the porosity. Finally, the recovery factors themselves are so difficult of precise prediction even under ideal conditions that attempts to differentiate between different parts of a reservoir often seem hardly warranted. However, such situations, when they arise, should be considered as reflecting only an adaptation to a realistic limitation of the present state of the science of reservoir engineering, and as a challenge for future progress, rather than as conformance to an established principle that the local recoveries from nonuniform reservoirs will actually be the same throughout.

14.16. Some Outstanding Problems in the Physics of Oil Production.— From the strictly scientific point of view the material presented in this work may well appear to be hardly sufficiently crystallized to be considered as the subject matter of a science. Much of the discussion has been clothed in multitudes of qualifying remarks, and many of the conclusions have been expressed merely as possibilities or at best as probabilities of occurrence. Virtually all the specific analytical and numerical considerations have been restricted to "ideal" systems, with the foreknowledge that these have never been observed to occur in practice.

Perhaps in no other science presuming to deal quantitatively with its subject matter is the latter so ill defined as in the physics of oil production or reservoir engineering. Not only is each actual reservoir "specimen" in itself of almost infinite complexity, but the ensemble of all those already known and those which one may expect to be discovered contains no strictly "duplicate" samples. Virtually all experiments that may be performed on actual reservoirs not only are irreversible in the thermodynamic sense but are moreover essentially destructive of the specimen, with respect to the basic parameters defining its state prior to the experiment. Individual experimental observations involving extended periods of reservoir performance are not subject to repetition, either to test reproducibility or systematic cause-and-effect relationships. In its operational sense the "principle of uncertainty," which is usually considered as limited to the realm of microscopic physics, constitutes the very essence of applied reservoir engineering as a science.

It is because of the inherent sample variability within the general class of oil-producing reservoirs that their study does not lead to generalized and universally applicable quantitative conclusions. However frustrating the situation may appear to be, no two actual reservoirs will perform in exactly the same manner, nor will they react identically to operating control. Nor is this to be expected. Even if all the implications of a set of assumed reservoir and operating conditions could be rigorously derived, it would be sheer coincidence if the predicted behavior has or ever will be observed *quantitatively* among actual oil reservoirs, except as the numerical assumptions were deliberately chosen from those of a specific developed reservoir. As the detailed physical characteristics of the members within the individual major reservoir groups, whether classified according to recovery mechanism, performance, or structure, vary over wide ranges, so will the quantitative aspects of their performance and recoveries. Hence, except in referring to a particular reservoir whose properties are completely specified, general predictions of performance and recovery must of necessity be limited to *probabilities* of occurrence, and with wide ranges of *possible* deviations from the general expected trends. Any pretense that actual reservoir behavior will quantitatively follow universal functional histories, independently of their unique and individual characteristics, would simply deny the basic deterministic foundation of macroscopic physics.

The only alternative to hopeless attempts to treat quantitatively each specific known reservoir is the discussion of idealized prototypes. To give such treatments some degree of practical significance the numerical values of the parameters defining the particular systems illustrating the analytical developments have been chosen to lie within the range known to obtain in practice. Accordingly in *order of magnitude* the quantitative aspects of the behavior of the illustrative examples should agree with those to be observed in such actual reservoirs as satisfy the broad physical assumptions underlying the analysis. However, it would still be purely accidental if there were exact agreement between these prototype idealized predictions and the corresponding observations in any specific reservoir.

Perhaps the outstanding unsolved problem in the physics of oil production is that of treating nonuniform flow systems. This is not merely the problem of averaging varying parameters, such as the permeability, connate-water saturation, thickness, or ultimate recovery factors, as discussed in the last section. Rather it is the *dynamical* statistics of nonuniform reservoirs that would give an expression to the differential behavior of and interaction between localized regions of a common reservoir of different properties. Even the relatively simple case of multiphase transient flow in a stratified horizon has been given no satisfactory treatment in which the cross flow between the zones of different permeability is taken into account, although it is possible to construct the analytical formulation of this problem and solve the equations by laborious numerical procedures. It is because of the tremendous complexity of problems of this type that thus far all reported analytical studies of heterogeneous-fluid systems have been restricted to the idealized "uniform reservoir." Among the many questions intimately related to the dynamics of nonuniform systems are (1) their equivalence to uniform reservoirs with the average properties of the inhomogeneous system; (2) the effect of intermittent operations, including shifting of withdrawal wells, on the recoveries and performance; and

(3) the statistics of the sweep geometry for invasion of extraneous fluids. While it will probably never be feasible to gather sufficient data completely to define the nature of the inhomogeneities in any specific reservoir, the basic reasons for such discrepancies as occur between observed and calculated uniform reservoir performance will remain uncertain until some estimate can be made at least of the order of magnitude of the effects of the reservoir nonuniformity.

Even for strictly uniform reservoirs the complete dynamical treatment for multiphase flow still remains an unsolved problem. As was noted in Chap. 10 the effects of wells as foci of withdrawals have been neglected in most theories thus far developed for the prediction of complete gas-drive reservoir histories. From a strictly physical point of view, such effects, if any, as well spacing and withdrawal rate may have on solution-gas-drive recoveries can be determined only by a solution of the fundamental equations 7.7(1) in which both the pressure distributions within the reservoir and time factor are explicitly taken into account. Until such analyses are carried through, the physical foundation of the well-spacing problem in gas-drive fields and their withdrawal-rate sensitivity will rest mainly on plausible hypotheses.

A rather disturbing situation that has been crystallizing in the last several years, as detailed reservoir data have been accumulated, is the increasing evidence that strict thermodynamic equilibrium apparently does not always obtain in all reservoirs at the time of exploitation.[1] Bubble-point pressures at common datum levels within an interconnected reservoir presumably are not always the same, within experimental errors, even in undersaturated reservoirs. And in reservoirs overlain by gas caps, under-saturation has apparently been observed below the gas-oil contact. While many of these observations may be associated with changes in character of the oil, that in itself is a manifestation of the apparent lack of equilibrium. In fact, not infrequently, decreases in API gravity with increasing depths are observed[2] that are far larger than would be expected[3] merely under gravitational-equilibrium separation. The heavy and tarry oils often found near water-oil contacts undoubtedly are in neither diffusion nor gravitational equilibrium with the overlying lighter oils in the same formation. Presumably such agents as may be causing local changes in the character

[1] It is also far from established that complete phase equilibrium obtains in all reservoirs throughout their producing lives. While no "proved" cases of nonequilibrium performance have been reported, there is laboratory evidence that supersaturation in fluid-carrying porous media can occur.

[2] Cf., for example, H. D. Hedberg, L. C. Sass, and H. J. Funkhouser, AAPG Bull., 31, 2089 (1947).

[3] M. Muskat, Phys. Rev., 35, 1384 (1930); B. H. Sage and W. N. Lacey, AIME Trans., 132, 120 (1939).

of the oil have been exerting their influence at a more rapid rate than the thermodynamic potential gradients which would tend to establish substantial uniformity. Even if the heavy oils found near the water-oil contacts entered the reservoirs as such, their continued localization in the reservoir still implies that the diffusion forces have been quite slow even on a geologic-time scale. Virtually as a matter of necessity, all these nonequilibrium phenomena have been almost completely ignored in the quantitative study of reservoir behavior. This problem, however, must be faced, and it merits serious study.

The clarification of the whole complex of details of fluid-displacement processes still must be considered as an outstanding problem. As has been previously discussed, not only is the satisfactory treatment of the gravity-drainage recovery mechanism beset with formidable analytical difficulties, but the estimation of the residual-oil saturations and ultimate recoveries is almost in the state of speculation by analogy. The capillary phenomena in three-phase systems, which determine the residual-oil saturation left by gravity drainage, as well as the detailed structure of interphase transition zones, are in some respects uncertain even with respect to their qualitative features. The literature is completely lacking in three-phase permeability-saturation characteristics of consolidated porous media, although literally all gas-drive-flow processes involve three phases. The commonly made assumption that the gas-to-oil permeability ratios are functions only of the total liquid content, and hence can be determined from two-phase measurements, certainly can have no universal validity. Much basic heterogeneous-fluid experimentation is urgently needed definitely to establish the quantitative aspects of the effects involved.

The detailed fluid dynamics and performance of nonoölitic or "intermediate" limestones and dolomites are an almost totally unexplored phase of reservoir engineering. Such reservoirs may be considered as a special type of the nonuniform reservoir referred to earlier. However, they deserve particular study, not only because limestones as oil-producing formations are of great practical importance, but also because their inhomogeneities are of an essentially local character. The basic concept of the multiphase permeability-saturation relationship is of questionable significance in fractured limestones, unless limited only to the intergranular component of the composite rock mass. While the gross reservoir performance of limestone reservoirs is in many respects substantially similar to that of sand formations, very little is understood of the internal averaging processes that apparently lead to such similarities.

Clay-containing sands also constitute a subject worthy of much study and effort in developing a complete understanding of the physics of oil production. The specification of their properties, as porosity, permeability, and connate-water content, does not in itself give a full description of their dynamical characteristics. Intergranular hydratable clays will undoubtedly affect materially the permeability-saturation relationships of the porous medium. Since it has become recognized that clays may markedly influence the permeability to water, inspection of reservoir rocks has shown that "dirty" sands are far more prevalent than previously anticipated. They can no longer be treated merely as anomalous and exceptional occurrences. On the contrary their unique properties will require thorough investigation if the interpretation and prediction of the performance of dirty-sand reservoirs are to be placed on a sound foundation.

The above suggestions for further study do not, of course, exhaust the list that could be made of the outstanding "unknowns" of reservoir engineering. Many more will be apparent to the readers of this work. Those briefly discussed here do not indicate an order of importance or priority but are to be considered mainly as illustrative of the types of problem to which future research must be devoted. It is to be anticipated that as these are studied further, still others will be revealed which will demand additional investigation. And it may also be expected that the solution of these problems will lead to even stronger evidence than now available that each specific reservoir must be thoroughly studied individually in order completely to understand or predict its behavior. Although generalizations of physical principles will certainly be developed, their quantitative application in practice will undoubtedly always remain a matter of fitting the broad methods of analysis to the unique properties and characteristics of the particular reservoir of interest.

14.17. Summary.—There is no completely satisfactory theory for predicting the variation of either physical or economic ultimate recoveries of oil reservoirs with the well spacing. Physical and theoretical considerations do not provide a mechanism such as would indicate a substantial variation of the physical ultimate recovery with the well spacing. Nor is such a variation suggested by direct laboratory experimentation. On the other hand there is a great variety of evidence that oil- and water-reservoir fluids can move through intercommunicating and continuous porous media over distances comparable with gross reservoir dimensions. There are likewise many observations, such as infill drilling records, showing that reservoir strata may be depleted of both fluid content and pressure at points distant from or in between producing wells, as a result of continued fluid withdrawals from the latter. There are no significant observations indicating that the local efficiency of oil displacement by water or gas is appreciably affected by the distance which these fluids have moved from their original location.

If the economic ultimate recovery be limited by minimal well production rates, such recoveries in solution-gas-drive fields would be expected, from physical considerations, to increase with decreasing well spacing. The rate of such variation may be either negligibly slow or of practical importance, depending on the reservoir properties and limiting abandonment production rates. In edgewater-drive reservoirs even the economic ultimate recoveries should be substantially independent of the well spacing. Theory indicates that in bottom-water-drive reservoirs the economic ultimate recoveries will increase with increasing well density. This increase should be approximately linear in isotropic formations. But when the effective average vertical permeability is 1 per cent or less than the average horizontal permeability, the economic ultimate recoveries should increase but slowly, if at all, with decreasing well spacing, within the range of well spacings commonly used in practice. Such effects, when observable, result from variations in the geometrical sweep efficiency of the rising water-oil interface with the well spacing, rather than a sensitivity of the local oil-displacement efficiency of the water-flushing process to well separations.

Observed and estimated actual ultimate recoveries in 27 gas-drive reservoirs do not disclose an effect of the well spacing on the recovery (cf. Fig. 14.5). Similar data on 74 water-drive fields likewise do not indicate any effect of well spacing on the actual recoveries such as may be segregated from other factors that probably exert a greater influence on the recovery than the well spacing (cf. Figs. 14.6 and 14.7). While field observations do not prove that variations of ultimate recoveries with the well spacing are strictly nonexistent, the inherent scattering of the data indicates that reservoir and operating conditions and well location in general are of greater importance than the well spacing in determining the ultimate recoveries.

Short-period pressure-interference tests, if positive, establish only the existence of rapid fluid intercommunication between the test wells. If negative, they only set upper limits to combinations of the reservoir and fluid properties that determine the transmissibility of pressure reactions. In fact observable reactions should not be expected in tests of only a few days' duration if the reservoir contains an appreciable gas saturation in the oil zone, unless there are exceptionally high permeability channels extending between the test wells. In any case negative transient interference-test results do not alone imply the insufficiency of the well spacing for long-term drainage and recovery.

In practical operations the well spacing should be determined primarily on the basis of reservoir continuity and the economic factors pertinent to the particular reservoir of interest. The initial well-spacing plan should be made as wide as possible, and such as will still permit a definition of the reservoir limits, the continuity of the formation, and general structure and reservoir conditions. The wells should be initially located in the productive area on such a pattern as will permit infill development, if necessary, without destroying the possibilities of uniform drainage from the productive area. Extension of the original pattern should be undertaken only after the information developed from the primary plan indicates the need for closer spacing due to lack of reservoir continuity and after the producing mechanism has been sufficiently well established to show that the recoverable oil in the reservoir will support additional drilling.

The minimum number of wells is evidently one in each separate productive unit having sufficient recoverable reserves to provide a pay-out for at least one well. The maximum number of wells to be drilled is given simply by the value of the economic ultimate recovery from the productive acreage, or any separate unit in it, taking into account royalty discount, market price of the oil, and estimated operating cost, divided by the drilling and completion cost per well. The optimum well spacing, between the minimum and maximum, from the point of view of profit, is to be determined largely on the basis of the economic importance of the time factor. Except for this factor the optimum number of wells would be the minimum number. While in no case can this optimum spacing be determined rigorously, even approximate calculations will depend markedly upon the producing mechanism and the nature of the production limitations that may be imposed by regulatory bodies or arranged through voluntary cooperative agreement.

In complete-water-drive fields the well density should be only so great as will provide the allowed field withdrawals. The latter, if feasible, should be limited to the capacity of the aquifer to replace the withdrawals without continued and excessive pressure decline. In partial-water-drive fields the well density should again be controlled by such total field withdrawals as will be conducive to maximum effectiveness of the water drive and an approximation to complete-water-drive performance.

In solution-gas-drive reservoirs without proration limitations the well density should be such as to permit the maximum gain from the shortened operating life and decreased interest discounts, as compared with the cost of the additional wells beyond the absolute minimum required for drainage purposes only. Uniform solution-gas-drive fields whose withdrawal rates are restricted by proration or equivalent regulations should be developed only to such a density as will provide the field allowable, except as the latter may be fixed in relation to the number of producing wells. The well density in gas-cap-expansion reservoirs should be only so high as will provide either the field allowables or such total withdrawals as will not greatly exceed the rate of downstructure oil gravity drainage, provided that such withdrawal rates are not too low for profitable operation.

With respect to the general efficiency of oil recovery the proper location of the wells will be of greater effect than the number of wells drilled. Except in solution-gas-drive reservoirs the total field withdrawal rates will be the controlling factor in determining the gross reservoir performance and recovery efficiency, as compared with the number of wells used. When fluid injection is applied to a reservoir as a direct oil-displacing medium, the location and density of the wells should be chosen so as to achieve the optimum geometrical sweeping pattern. The well density itself will be of little importance except as it may affect the sweep efficiency.

While not pertinent to the well-spacing problem from the recovery standpoint, the purely geometrical interference between wells is of interest in itself and may be of importance in general field-development programs. Using a steady-state homogeneous-fluid treatment it can be shown that the mere presence of a neighboring well within 200 ft of another, in a circular area of 5,000 ft radius, will reduce the well capacity of each to only 75 per cent of its capacity if each alone were draining the area [cf. Eq. 14.9(4)]. The central well in a square of 200-ft sides, draining a similar area, will have a capacity only 78 per cent of that of each of the 4 surrounding wells. Similarly the total production capacity of 16 wells in a square network, with a 200-ft well separation, will be no more than 3.3 times that of a single isolated well. Wells located on a ring will suffer similar mutual interference effects. Thus the maximum steady-state production (or injection) capacity of any number of wells placed on an 80-ft ring lying near the center of a 5,000-ft-radius circular area is only 2.4 times that of a single well.

These geometrical-interference reactions also affect the production capacities of groups of wells distributed in linear arrays parallel to a line drive. Except for end effects, wells in such a linear array lying 100 ft from the line drive and mutually spaced by 100 ft will have production capacities only 64 per cent as great as would a single well 100 ft from the line drive. Within distances equal to half the well spacing the pressure distributions about such linear arrays become almost identical with that due to strictly continuous and uniform linear distributions of fluid withdrawal or injection (cf. Figs. 14.21 and 14.23).

Linear well arrays also will mutually interfere with each other if supplied by a common steady-state fluid source, as a line drive. Thus if two parallel line arrays of wells are supplied by the same line drive, the array closer to the line drive will shield the more distant array so that the latter will produce at a lower rate even for the same well pressure. As is to be expected the shielding effect of the first line of wells will decrease as the well spacing within the line increases, as both arrays recede from the line drive, and as the separation between the lines decreases. At spacings within and between the lines of the order of 660 ft it is found that only about one-third of the total flux from the line drive leaks past the first line array, so that the production capacity of the latter is approximately twice that of the second line. Similar homogeneous-fluid treatments of three line arrays show that for uniform well spacings of 660 ft throughout the first line will capture about 63 per cent of the total flux and that the flux into the second will be about 67 per cent of that leaking past the first line. These shielding effects will be unaffected by the staggering of the line arrays as long as their separations are comparable with the well spacings within the lines. These geometrical-interference phenomena should be taken into account in the development of water-drive fields.

Although the ultimate recovery of an oil reservoir is essentially only the integrated effect of its whole performance history, it represents a basic criterion for evaluation of the reservoir from an economic standpoint. As the details of the performance during the producing life will vary with the values of the rock and fluid physical parameters, as well as the operating conditions, so will the resultant value of the ultimate recovery. It is to be expected, therefore, that observed recoveries will be spread over wide ranges even for the same basic recovery mechanism. Thus among 25 gas-drive fields on which recovery data have been reported, the recoveries varied from 126 to 547, with a median of 248, in barrels per acre-foot; from 15 to 50, with a median of 33, as per cent of the initial-oil content; from 7 to 34, with a median of 20, as per cent of the pore space; and from 14 to 53, with a median of 28 per cent, as ultimate free-gas saturation. These ranges include those indicated by the purely theoretical calculations of Chap. 10. However, the observed and estimated recoveries expressed as a fraction of the pore space definitely appear to be greater, on the average, than those which would be anticipated from the strict solution-gas-drive theory. While the reason for the discrepancy is not certain, it seems likely that a major factor has been the contributions to the solution-gas-drive recoveries by gravity drainage or water invasion.

Field data are more numerous on water-drive-reservoir recovery. Among 69 fields for which such data have been tabulated (cf. Table 6) the barrels per acre-foot recoveries range from 242 to 1,165, with a median of 552, the per cent initial oil from 24 to 78, with a median of 52 per cent, the per cent of pore space from 18 to 54, with a median of 30 per cent, and the per cent residual-oil saturation from 16 to 59, with a median of 30 per cent. These large spreads evidently indicate that the water-drive recovery mechanism also cannot be associated with sharply defined and unique recovery factors independently of the reservoir characteristics. In fact a study of the data for the water-drive reservoirs indicates trends of decreasing resid-ual-oil saturations with increasing permeability, decreasing oil viscosity, and decreasing pressure decline during the producing life.

In contrast to solution-gas-drive reservoirs there is no self-contained theory available for predicting water-drive recoveries that automatically gives the terminal fluid saturations in the reservoir at abandonment. The recovery factor must be introduced independently in terms of the residual-oil saturation assumed for the water-invaded area. While an application of the permeability-saturation characteristics for the producing formation [cf. Eq. 14.13(1)] will presumably give the residual-oil saturation at the time when the water-oil ratios will reach any preassigned abandonment limit, this procedure is of doubtful significance. It refers to a simultaneous flow of oil and water in which both have mobility. It seems likely, however, that the microscopic process of oil expulsion by water flushing in water-wet sands above water-oil transition zones proceeds by the advance of an oil-water "front," behind which the oil saturations are, for practical purposes, immediately reduced to a state of vanishing permeability and a discontinuous distribution. It is this residual-oil saturation, determined by the microscopic geometry and capillary properties of the rock and fluids, that measures the local oil recovery. The gross reservoir recovery will be this value reduced by the incompleteness of flushing of the whole reservoir rock at the time of abandonment, due to the imperfection of the geometrical sweep pattern and the differential degrees of water invasion in different parts of the formation caused by permeability stratification and inhomogeneities.

For gravity-drainage reservoirs there are so few interpretable field data available that one must appeal directly to theoretical considerations. Since gravity drainage is simply the process of downstructure oil movement under the force of gravity, the drainage will continue, if unopposed by pressure gradients, until the oil permeability falls to zero. This may be expected to occur when its saturation is reduced so that it becomes broken up into a discontinuous distribution. This saturation should be of the same order of magnitude as that limiting oil expulsion by water flushing. Accordingly the recoveries by gravity drainage and gas-cap expansion should, under favorable circumstances, be comparable with those which can be obtained by complete water drives.

Whether the recovery mechanism is that of solution-gas drive, water flushing, or gravity drainage, it has not been established that these processes are locally and inherently rate-sensitive. On the other hand in all cases there are indirect effects that, together with economic factors, will often lead to reduced recoveries at excessive production rates and higher recoveries when the withdrawal rates are restricted. In gas-drive reservoirs, in which there is a potential source of increased recovery by gravity drainage or water intrusion, the contributions of the latter will be increased at reduced withdrawal rates. High production rates in water-drive reservoirs will lead to more rapid declines in pressure, even for equal cumulative productions, shorten the flowing life, increase operating costs, and tend to force abandonment at lower total recoveries. Moreover the residual-oil saturations left in the water-invaded strata at low pressures will represent a greater equivalent as stock-tank oil than if trapped at pressures for which the formation-volume factor is high.[1] If the withdrawal rates are uncontrolled in gravity-drainage reservoirs, the oil saturation and permeability in the oil zone may be so reduced as greatly to lower the downflank gravity drainage. Prevention of gas coning and break-through and bleeding of the gas-cap gas will also be more difficult. In addition the residual-oil saturation, if this oil is left in the expanded gas cap at low pressures, will be equivalent to a greater stock-tank volume of oil than if it were trapped at higher pressures resulting from lower production rates with lower gas-oil ratios.

Even if the recovery "factor" of a reservoir has been estimated, it must be translated into its equivalent total oil recovery before a reservoir as a whole can be evaluated economically. When the recovery factor is known as barrels per acre-foot, the total ultimate oil recovery will be simply the recovery factor times the gross reservoir volume [cf. Eq. 14.15(1)]. The latter requires a determination of the productive area and net effective pay thickness. The productive area is ultimately defined by the field development and the nonproductive wells outside the field limits. The gross pay thickness is usually established easily by logging, coring, and geological-sample inspection. The discrimination between "effective" and "non-effective" pay in a common oil-bearing horizon is, however, often subject to much uncertainty. The setting of a lower permeability limit for the parts of the formation that presumably will not contribute appreciably to the recovery during the producing life is basically arbitrary. No formula is available for this purpose, and the choice must be made on the admittedly unscientific basis of "good judgment."

If the recovery factor is expressed as a per cent of the pore space, the average reservoir porosity, in addition to its gross volume, must be known for a conversion to the equivalent cumulative oil recovery [cf. Eq. 14.15(2)]. When it is given as an ultimate free-gas saturation or as a residual-oil saturation, the interstitial-water saturation and initial and final formation-volume factors of the oil are also required for calculating the ultimate oil recovery. The same data, excluding the final formation-volume factor of the oil, enter the formula giving the expected cumulative recovery in terms of a recovery factor expressed as a fraction of the initial oil in place. In this latter case, however, the total initial oil in place can also be determined directly, under favorable circumstances, by the application of the material-balance method to the reservoir-performance observations.

---

[1] These generalized conclusions should not, however, be considered as universally valid (cf. Sec. 14.13).

The physics of oil production is in no sense a "completed" science. The broad physical principles underlying the subject are reasonably well established. It has had sufficient application to justify faith in the soundness of its foundations. But its scope is far wider than its present development encompasses.

The elements comprising the subject matter of this science are so variable among themselves and the individual members—the reservoirs—are so extremely complex that it is very doubtful that it will ever achieve the state of complete description such as characterizes the classical mechanics or electromagnetic theory. Accordingly it is virtually impossible to crystallize the developments and discussions of its manifold aspects in the form of universal and definitive conclusions. Beyond the physical principles themselves, there is no universal validity to any quantitative assertion presuming to encompass even a restricted class of actual reservoirs. If treated statistically the physics of oil production would be governed by a statistics of variables—such as the leaves of a tree—rather than identical particles, as gas molecules. Such conclusions as may be derived pertaining to reservoirs and their performance must be carefully qualified to emphasize that any generalization represents only possibilities or probabilities rather than well-defined descriptions of specific reservoirs which actually occur in practice. The probability is very low indeed that any hypothetical reservoir, constructed purely for illustrative purposes, will be duplicated by any reservoir in existence.

Aside from its inherent basic complexity the physics of oil production is beset with many specific unsolved problems. One of the most serious of these is the treatment of nonuniform reservoirs. The gross statistical averaging into equivalent uniform systems undoubtedly suffices in many respects. Transient phenomena, however, which lead to differential depletion in parts of a common reservoir having different properties, cannot be represented so simply. Certainly the averaging processes will be different from those for the steady-state dynamics. In any case, until at least a few simple nonuniform systems are rigorously treated the quantitative significance of approximation and simplifying procedures cannot be evaluated. Unfortunately, however, even the strictly uniform reservoir has not yet been subjected to a complete analysis in which effects of pressure gradients and withdrawal rates have been rigorously taken into account.

Indications that at the time of discovery reservoirs may not necessarily be in a state of complete thermodynamic equilibrium warrant thorough study. The complexity of quantitative reservoir analysis evidently will be increased manifold if nonequilibrium phenomena associated either with original reservoir conditions or the producing life should have to be introduced in performance theory. The possibility of occurrence of supersaturation, suggested in recent experimental studies, must be given serious consideration.   The details of fluid-displacement processes in porous media are also a subject too little explored as yet.   Many more three-phase capillary and permeability-saturation data must be determined to clarify many of the conjectural aspects of recovery mechanisms, and especially that of gravity drainage.

Clay-containing sands and intermediate limestones comprising reservoir formations have been almost completely neglected as subjects of special quantitative study.   Yet the assumption that their performance will be quantitatively similar to that of clean sandstones has no supporting evidence.   Although such reservoirs can no longer be considered as exceptional, they have been given very little intensive study and investigation from the point of view of fluid dynamics.

Compared with other major sciences the research effort thus far devoted to the physics of oil production has been almost infinitesimal.   Nevertheless it has been possible to develop the basic foundations for this complex subject, which already provide at least a semiquantitative description and correlation of many of the major features of oil-reservoir performance. Further effort on an expanded scale is certain to lead to an understanding of the physical phenomena of oil production which will compare favorably with that achieved in other applied physics sciences.

APPENDIX I

CONVERSION FACTORS AND VARIOUS CONSTANTS[1]

Length

1 cm = 0.39370 in.
   = 0.032808 ft
   = 6.2137 × 10^{-6} mile

1 in. = 2.5400 cm

1 ft = 30.480 cm

1 mile = 1.6093 × 10^5 cm

Area

1 cm² = 0.15500 in.²

1 m² = 10.764 ft²
   = 3.8610 × 10^{-7} sq mile
   = 2.4710 × 10^{-4} acre

1 in.² = 6.4516 cm²

1 ft² = 929.03 cm²
   = 2.2957 × 10^{-5} acre

1 mile² = 2.5900 × 10^6 m²
   = 640 acres

1 acre = 4.0469 × 10^3 m²
   = 43,560 ft²

Volume

1 cc = 0.99997 ml
   = 3.5314 × 10^{-5} ft³
   = 2.6417 × 10^{-4} gal
   = 6.2898 × 10^{-6} bbl
   = 8.1071 × 10^{-10} acre-ft

1 ft³ = 28.317 × 10^3 cc
   = 7.4805 gal
   = 0.17811 bbl
   = 2.2957 × 10^{-5} acre-ft

1 gal = 3.7854 × 10^3 cc
   = 0.13368 ft³
   = 2.3810 × 10^{-2} bbl
   = 3.0689 × 10^{-6} acre-ft

1 bbl = 158.99 × 10^3 cc
   = 5.6146 ft³
   = 42 gal
   = 1.2889 × 10^{-4} acre-ft

1 acre-ft = 1.2335 × 10^9 cc
   = 43,560 ft³
   = 3.2585 × 10^5 gal
   = 7,758.4 bbl

Pressure

1 atm = 76 cm Hg at 0°C (of density 13.5951 gm/cc under normal gravity, *i.e.*, 980.665 cm/sec²)

1 atm = 1.01325 × 10^6 dynes/cm²
   = 14.696 psi
   = 33.899 ft H_2O at 4°C under normal gravity
   = 1,033.2 cm H_2O
   = 1,033.2 gm/cm²

1 psi = 6.8046 × 10^{-2} atm
   = 6.8948 × 10^4 dynes/cm²
   = 2.3067 ft H_2O
   = 70.308 cm H_2O
   = 70.307 gm/cm²

1 dyne/cm² = 9.8692 × 10^{-7} atm
   = 1.4504 × 10^{-5} psi
   = 3.3456 × 10^{-5} ft H_2O
   = 1.0197 × 10^{-3} cm H_2O
   = 1.0197 × 10^{-3} gm/cm²

1 cm H_2O (4°C) = 9.6782 × 10^{-4} atm
   = 0.014223 psi
   = 980.64 dynes/cm²
   = 0.99998 gm/cm²
   = 0.073556 cm Hg (0°C)

[1] Listed to five significant figures.

$$1 \text{ ft } H_2O \ (4°C) = 0.029499 \text{ atm}$$
$$= 29.890 \times 10^3 \text{ dynes/cm}^2$$
$$= 2.2420 \text{ cm Hg}$$
$$= 0.43352 \text{ psi}$$
$$= 30.479 \text{ gm/cm}^2$$

Volumetric Rate

$$1 \text{ cc/sec} = 2.1186 \times 10^{-2} \text{ ft}^3/\text{min}$$
$$= 1.5850 \times 10^{-2} \text{ gal/min}$$
$$= 0.54344 \text{ bbl/day}$$

$$1 \text{ gal/min} = 0.13368 \text{ ft}^3/\text{min}$$
$$= 34.286 \text{ bbl/day}$$
$$= 63.091 \text{ cc/sec}$$

$$1 \text{ bbl/day} = 3.8990 \times 10^{-3} \text{ ft}^3/\text{min}$$
$$= 2.9167 \times 10^{-2} \text{ gal/min}$$
$$= 1.8401 \text{ cc/sec}$$

$$1 \text{ (cc/sec)/atm} = 3.6979 \times 10^{-2} \text{ (bbl/day)/psi}$$
$$= 4.6758 \times 10^{-4} \text{ (gal/min)/ft } H_2O$$

$$1 \text{ (bbl/day)/psi} = 1.2644 \times 10^{-2} \text{ (gal/min)/ft } H_2O$$
$$= 27.042 \text{ (cc/sec)/atm}$$

$$1 \text{ (gal/min)/ft } H_2O = 2.1387 \times 10^3 \text{ (cc/sec)/atm}$$
$$= 79.087 \text{ (bbl/day)/psi}$$

Density

$$1 \text{ gm/cc} = 62.428 \text{ lb/ft}^3$$
$$= 8.3454 \text{ lb/gal}$$
$$= 350.51 \text{ lb/bbl}$$

$$1 \text{ lb/ft}^3 = 0.13368 \text{ lb/gal}$$
$$= 5.6146 \text{ lb/bbl}$$
$$= 0.016018 \text{ gm/cc}$$

$$1 \text{ lb/gal} = 42 \text{ lb/bbl}$$
$$= 0.11983 \text{ gm/cc}$$
$$= 7.4805 \text{ lb/ft}^3$$

$$1 \text{ lb/bbl} = 2.8530 \times 10^{-3} \text{ gm/cc}$$
$$= 0.17811 \text{ lb/ft}^3$$
$$= 0.023810 \text{ lb/gal}$$

Density of water at 60°F $= 0.99901$ gm/cc
$$= 62.366 \text{ lb/ft}^3$$
$$= 8.3372 \text{ lb/gal}$$

Density of air at 1 atm, 60°F $= 0.0012232$ gm/cm^3
$$= 0.076362 \text{ lb/ft}^3$$
$$= 0.010208 \text{ lb/gal}$$

Molar Volume

Volume of 1 gm mole at 1 atm, 0°C $= 22.414 \times 10^3$ cc
Volume of 1 lb mole at 1 atm, 0°C $\ = 359.04$ ft^3
Volume of 1 lb mole at 1 atm, 60°F $= 379.48$ ft^3

Viscosity Conversion Formula

$$\mu/d = 0.226t - \frac{195}{t} \text{ (for oils of 100 sec Saybolt or less)}$$

$$= 0.220t - \frac{135}{t} \text{ (for oils of 100 sec Saybolt or greater)}$$

where μ = absolute viscosity in centipoises
d = density of oil at same temperature
t = "Saybolt seconds" (universal)

APPENDIX II

LAPLACE'S EQUATION IN CURVILINEAR COORDINATES

Because of the fundamental role played by Laplace's equation in the flow of homogeneous fluids through porous media, and by the intermediate equation of continuity in all hydrodynamic systems, it is desirable to have available generalized coordinate expressions for the vector operators involved in these equations. These may be developed as follows:

Suppose a set of orthogonal curvilinear coordinates

$$\alpha(x,y,z) = \text{const};$$
$$\beta(x,y,z) = \text{const}; \qquad (1)$$
$$\gamma(x,y,z) = \text{const}$$

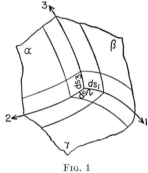

Fig. 1

to be defined for every point of the region of interest. Their intersections may be considered to define three mutually orthogonal directions as indicated in Fig. 1. The net flux corresponding to a vector velocity $\bar{v}$ out of a differential volume element of sides ds_1, ds_2, ds_3, or the divergence of the vector $\bar{v}$, is then evidently

$$ds_1\,ds_2\,ds_3\,\text{div}\,\bar{v} = \frac{\partial}{\partial s_1}\,(v_1\,ds_2\,ds_3)ds_1 + \frac{\partial}{\partial s_2}\,(v_2\,ds_3\,ds_1)ds_2 + \frac{\partial}{\partial s_3}\,(v_3\,ds_1\,ds_2)ds_3, \quad (2)$$

where v_1, v_2, v_3 are the components of $\bar{v}$ parallel to the directions 1, 2, 3. Denoting the ratios of the differential changes in the coordinates α, β, γ to the actual differential elements of length ds_1, ds_2, ds_3 by h_1, h_2, h_3, so that

$$d\alpha = h_1\,ds_1; \qquad d\beta = h_2\,ds_2; \qquad d\gamma = h_3\,ds_3, \qquad (3)$$

and dividing by the volume element $ds_1\,ds_2\,ds_3$, the above expression for the divergence becomes

$$\text{div}\,\bar{v} = h_1h_2h_3\left[\frac{\partial}{\partial\alpha}\left(\frac{v_1}{h_2h_3}\right) + \frac{\partial}{\partial\beta}\left(\frac{v_2}{h_3h_1}\right) + \frac{\partial}{\partial\gamma}\left(\frac{v_3}{h_1h_2}\right)\right]. \qquad (4)$$

If now $\bar{v}$ be considered as the gradient of a potential function Φ, so that $\bar{v} = \nabla\Phi$, Eq. (4) becomes

$$\text{div}\,\bar{v} = \nabla^2\Phi = h_1h_2h_3\left[\frac{\partial}{\partial\alpha}\left(\frac{h_1}{h_2h_3}\frac{\partial\Phi}{\partial\alpha}\right) + \frac{\partial}{\partial\beta}\left(\frac{h_2}{h_3h_1}\frac{\partial\Phi}{\partial\beta}\right) + \frac{\partial}{\partial\gamma}\left(\frac{h_3}{h_1h_2}\frac{\partial\Phi}{\partial\gamma}\right)\right]. \qquad (5)$$

This expression set equal to zero represents the curvilinear-coordinate transformation of the Laplace equation $\nabla^2 \Phi = 0$.

Thus for the cartesian system (x,y,z) it is clear that $h_1 = h_2 = h_3 = 1$, so that the normal Laplacian form of Eq. 4.5(1) is obtained. For the cylindrical-coordinate system (r,θ,z) (cf. Fig. 4.1), $(h_1,h_2,h_3) = (1,1/r,1)$, leading to Eq. 4.5(4). And for the spherical-coordinate system (r,θ,χ) (cf. Fig. 4.2), $(h_1,h_2,h_3) = (1,1/r,1/r \sin \theta)$, from which Eq. 4.5(7) follows at once. In a similar manner the corresponding equation for any other coordinate system may be derived, once the appropriate (h_1,h_2,h_3) have been found.

AUTHOR INDEX

SUBJECT INDEX

A

Acid treatment, 242, 250*ff*.
Allegany field, 694, 695, 705
Anahuac field, 103, 166, 167
Anisotropic formations, 262*ff*.
  bottom-water drives in, 620*ff*.
  casing perforations in, 218, 219
  partially penetrating wells in, 264*ff*.
  stream functions in, 617
Apparent gas densities, 69
  of methane and ethane, 71, 72
Atlanta field, 585, 586

B

Bay City field, 106
Beattie-Bridgeman equation, 38
Bell Run field, 106
Bellvue field, 457
Benton field, 740, 790
Bessel functions, 208, 224, 257, 542, 609
Big Creek field, 585
Big Lake field, 102, 783
Bilateral cycling pattern, 760*ff*.
Binary hydrocarbon systems, 40*ff*.
  behavior of, in critical region, 42*ff*.
  effect of composition on, 49*ff*.
Bottom-water drives, 234, 368, 604*ff*., 818, 819
  analytical treatment of, 608*ff*.
  anisotropy in, 620*ff*.
  physical representation of, 604*ff*.
  sweep efficiency in, 612*ff*.
  well capacities in, 608*ff*.
  well spacing effects in, 620*ff*.
Boundary conditions, 183, 190, 406
Boyle point, 39
Boyle's law, 34
Bradford field, 170, 694, 695, 700, 701
Breckenridge field, 102
Bubble point, 31*ff*.

Bubble-point curve, 33*ff*., 41
Buckner field, 373, 585*ff*., 603
Bulk-volume determinations, 115*ff*.
Burbank field, 106

C

Caddo field, 171
Canal field, 464, 478, 479
Capillary phenomena, 304*ff*.
  dynamical effects of, 321*ff*.
Capillary pressures, 162, 295, 304*ff*.
  connate water determination by, 162, 163
  drainage curves of, 310, 316
  imbibition curves of, 310, 316, 317
Carmi pool, 851, 852
Carterville field, 457
Casing perforations, 215*ff*.
  in anisotropic strata, 218, 219
  effect of perforation density on, 218
Casmelia field, 10
Charles's law, 34
Chip coring, 151, 708
Circular cycling pattern, 758, 759
Classical hydrodynamics, 181*ff*.
Coalinga field, 601
Coefficient of storage, 541
Coefficient of thermal expansion, 38
Compressibility coefficient, 39
Compressibility factor, 34, 58
Compressible liquid flow, 188, 528*ff*.
Condensate, 2, 48
Condensate reservoirs, 2, 48, 738*ff*.
  condensation about well bores in, 793, 794
  cycling of, 754*ff*.
  depletion history of, 743*ff*.
  field observations on, 782*ff*.
  fluid composition in, 740
  practical aspects of development of, 787*ff*.
  revaporization effects in, 794*ff*.

915